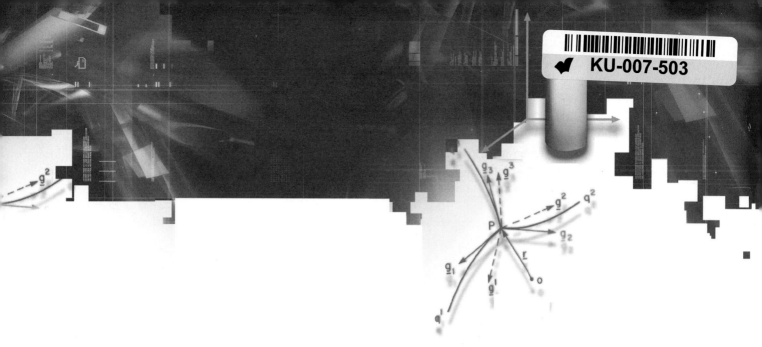

# Advanced Mathematics for
# Engineering
## and
# Science

by

## C F Chan Man Fong
## D De Kee

Tulane University, USA

## P N Kaloni

University of Windsor, Canada

 **World Scientific**
New Jersey • London • Singapore • Hong Kong

*Published by*

World Scientific Publishing Co. Pte. Ltd.

5 Toh Tuck Link, Singapore 596224

*USA office:* Suite 202, 1060 Main Street, River Edge, NJ 07661

*UK office:* 57 Shelton Street, Covent Garden, London WC2H 9HE

**British Library Cataloguing-in-Publication Data**
A catalogue record for this book is available from the British Library.

ISBN    981-238-291-7
ISBN    981-238-292-5 (pbk)

Printed in Singapore.

# PREFACE

The objective of this book is to provide a mathematical text at the third year level and beyond, appropriate for students of engineering and sciences. It is a book of applicable mathematics. We have avoided the approach of listing only the techniques followed by a few examples, without explaining why the techniques work. Thus we have provided not only the know how but also the know why. Equally it is not written as a book of pure mathematics with a list of theorems followed by their proofs. Our emphasis is to help students develop an understanding of mathematics and its applications. We have refrained from using clichés like "it is obvious" and "it can be shown", which might be true only to a mature mathematician. In general, we have been generous in writing down all the steps in solving the example problems. Contrary to the opinion of the publisher of S. Hawking's book, *A Short History of Time*, we believe that, for students, every additional equation in the worked examples will double the readership.

Many engineering schools offer little mathematics beyond the second year level. This is not a desirable situation as junior and senior year courses have to be watered-down accordingly. For graduate work, many students are handicapped by a lack of preparation in mathematics. Practicing engineers reading the technical literature, are more likely to get stuck because of a lack of mathematical skills. Language is seldom a problem. Further self-study of mathematics is easier said than done. It demands not only a good book but also an enormous amount of self-discipline. The present book is an appropriate one for self-study. We hope to have provided enough motivation, however we cannot provide the discipline!

The advent of computers does not imply that engineers need less mathematics. On the contrary, it requires more maturity in mathematics. Mathematical modelling can be more sophisticated and the degree of realism can be improved by using computers. That is to say, engineers benefit greatly from more advanced mathematical training. As Von Karman said: "There is nothing more practical than a good theory". The black box approach to numerical simulation, in our opinion, should be avoided. Manipulating sophisticated software, written by others, may give the illusion of doing advanced work, but does not necessarily develop one's creativity in solving real problems. A careful analysis of the problem should precede any numerical simulation and this demands mathematical dexterity.

The book contains ten chapters. In Chapter one, we review freshman and sophomore calculus and ordinary differential equations. Chapter two deals with series solutions of differential equations. The concept of orthogonal sets of functions, Bessel functions, Legendre polynomials, and the Sturm Liouville problem are introduced in this chapter. Chapter three covers complex variables: analytic functions, conformal mapping, and integration by the method of residues. Chapter four is devoted to vector and tensor calculus. Topics covered include the divergence and Stokes' theorem, covariant and contravariant components, covariant differentiation, isotropic and objective tensors. Chapters five and six consider partial differential equations, namely Laplace, wave, diffusion and Schrödinger equations. Various analytical methods, such as separation of variables, integral transforms, Green's functions, and similarity solutions are discussed. The next two chapters are devoted to numerical methods. Chapter seven describes methods of solving algebraic and ordinary differential equations. Numerical integration and interpolation are also included in this chapter. Chapter eight deals with numerical solutions of partial differential equations: both finite difference and finite element techniques are introduced. Chapter nine considers calculus of variations. The Euler-Lagrange equations are derived and the transversality and subsidiary conditions are discussed. Finally, Chapter ten, which is entitled Special Topics, briefly discusses phase space, Hamiltonian mechanics, probability theory, statistical thermodynamics and Brownian motion.

Each chapter contains several solved problems clarifying the introduced concepts. Some of the examples are taken from the recent literature and serve to illustrate the applications in various fields of engineering and science. At the end of each chapter, there are assignment problems labeled a or b. The ones labeled b are the more difficult ones.

There is more material in this book than can be covered in a one semester course. An example of a typical undergraduate course could cover Chapter two, parts of Chapters four, five and six, and Chapter seven.

A list of references is provided at the end of the book. The book is a product of close collaboration between two mathematicians and an engineer. The engineer has been helpful in pinpointing the problems engineering students encounter in books written by mathematicians.

We are indebted to many of our former professors, colleagues, and students who indirectly contributed to this work. Drs. K. Morrison and D. Rodrigue helped with the programming associated with Chapters seven and eight. Ms. S. Boily deserves our warmest thanks for expertly typing the bulk of the manuscript several times. We very much appreciate the help and contribution of  Drs. D. Cartin , Q. Ye and their staff at World Scientific.

*New Orleans*                                                              C.F. Chan Man Fong

*December 2002*                                                              D. De Kee

P.N. Kaloni

# CONTENTS

**References**                                                                                                                                                                                 **861**

**Appendices**                                                                                                                                                                                 **867**

**Author Index**                                                                                                                                                                               **875**

**Subject Index**                                                                                                                                                                              **877**

# CHAPTER 1

# REVIEW OF CALCULUS AND ORDINARY DIFFERENTIAL EQUATIONS

## 1.1   FUNCTIONS OF ONE REAL VARIABLE

The search for functional relationships between variables is one of the aims of science. For simplicity, we shall start by considering two real variables. These variables can be quantified, that is to say, to each of the two variables we can associate a set of real numbers. The rule which assigns to each real number of one set a number of the other set is called a **function**. It is customary to denote a function by f. Thus if the rule is to square, we write

$$y = f(x) = x^2 \qquad\qquad (1.1\text{-}1a,b)$$

The variable y is known as the **dependent variable** and x is the **independent variable**. It is important not to confuse the function (rule) f with the value $f(x)$ of that function at a point x. A function does not always need to be expressed as an algebraic expression as in Equation (1.1-1). For example, the price of a litre of gas is a function of the geographical location of the gas station. It is not obvious that we can express this function as an algebraic expression. But we can draw up a table listing the geographical positions of all gas stations and the price charged at each gas station. Each gas station can be numbered and thus to each number of this set there exists another number in the set of prices charged at the corresponding gas station. Thus, the definition of a function as given above is general enough to include most of the functional relationships between two variables encountered in science and engineering.

The function f might not be applicable (defined) over all real numbers. The set of numbers for which f is applicable is called the **domain** of f. Thus if f is extracting the square root of a real number, f is not applicable to negative numbers. The domain of f in this case is the set of non-negative numbers. The **range** of f is the set of values that f can acquire over its domain. Figure 1.1-1 illustrates the concept of domain and range. The function f is said to be **even** if $f(-x) = f(x)$ and **odd** if $f(-x) = -f(x)$. Thus, $f(x) = x^2$ is even since

$$f(-x) = (-x)^2 = x^2 = f(x) \qquad\qquad (1.1\text{-}2a,b,c)$$

while the function $f(x) = x^3$ is odd because

$$f(-x) = (-x)^3 = -x^3 = -f(x) \qquad\qquad (1.1\text{-}3a,b,c)$$

A function is **periodic** and of period $T$ if

$$f(x + T) = f(x) \qquad\qquad (1.1\text{-}4)$$

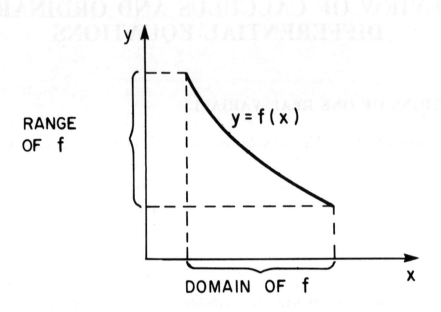

**FIGURE 1.1-1    Domain and range of a function**

An example of a periodic function is $\sin x$ and its period is $2\pi$.

A function $f$ is **continuous** at the point $x_0$ if $f(x)$ tends to the same limit as $x$ tends to $x_0$ from both sides of $x_0$ and the limit is $f(x_0)$. This is expressed as

$$\lim_{x \to x_{0+}} f(x) = f(x_0) = \lim_{x \to x_{0-}} f(x) \qquad\qquad (1.1\text{-}5a,b)$$

The notation $\lim\limits_{x \to x_{0+}}$ means approaching $x_0$ from the right side of $x_0$ (or from above) and $\lim\limits_{x \to x_{0-}}$ the limit as $x_0$ is approached from the left (or from below). An alternative equivalent definition of continuity of $f(x)$ at $x = x_0$ is, given $\varepsilon > 0$, there exists a number $\delta$ (which can be a function of $\varepsilon$) such that whenever $|x - x_0| < \delta$, then

$$\left| f(x) - f(x_0) \right| < \varepsilon \qquad\qquad (1.1\text{-}6)$$

This is illustrated in Figure 1.1-2.

If the function $f(x)$ is continuous in a closed interval $[a, b]$, it is continuous at every point $x$ in the interval $a \le x \le b$.

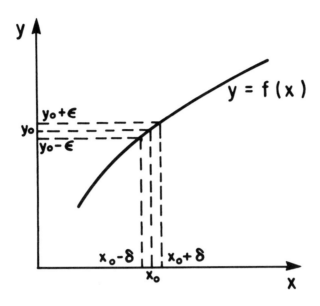

**FIGURE 1.1-2    Continuity of a function**

## 1.2    DERIVATIVES

We might be interested not only in the values of a function at various points $x$ but also at its rate of change.  For example, waiting at the corner of "walk and don't walk", one might want to know, not only the position of a car but also its speed before crossing the road.  The average rate of change of $f(x)$ in an interval $[x_0 + \Delta x, x_0]$ is defined as

$$\frac{\Delta f}{\Delta x} = \frac{f(x_0 + \Delta x) - f(x_0)}{\Delta x} \tag{1.2-1}$$

The rate of change of $f$ at $x_0$, which is the **derivative** of $f$ at $x_0$, is defined as

$$f'(x_0) = \lim_{\Delta x \to 0} \frac{\Delta f}{\Delta x} = \lim_{\Delta x \to 0} \frac{f(x_0 + \Delta x) - f(x_0)}{\Delta x} \tag{1.2-2a,b}$$

We have assumed that the limit in Equations (1.2-2a,b) exists and $f$ is thus differentiable at $x = x_0$. The derivative of $f$ with respect to $x$ is also denoted as $\dfrac{df}{dx}$.

The second derivative of $f$ is the derivative of $f'$ and is denoted by $f''$ or $\dfrac{d^2 f}{dx^2}$.  Likewise higher derivatives can be defined and the $n^{th}$ derivative is written either as $f^{(n)}$ or $\dfrac{d^n f}{dx^n}$.

Geometrically, $f'(x_0)$ is the tangent to the curve $f(x)$ at the point $x = x_0$.

## Rules for Differentiation

(i)      If $y$ is a function of $z$ and $z$ is a function of $x$

$$\frac{dy}{dx} = \frac{dy}{dz}\frac{dz}{dx} \quad \textbf{(chain rule)}$$
(1.2-3)

(ii)     If $u$ and $v$ are differentiable functions of $x$

$$\frac{d}{dx}\left(\frac{u}{v}\right) = \frac{v\frac{du}{dx} - u\frac{dv}{dx}}{v^2}$$
(1.2-4)

(iii)    $\frac{d}{dx}(uv) = u\frac{dv}{dx} + v\frac{du}{dx}$
(1.2-5)

(iv)     $\dfrac{d^n(uv)}{dx^n} = u\dfrac{d^n v}{dx^n} + n\dfrac{du}{dx}\dfrac{d^{n-1} v}{dx^{n-1}} + ... + \binom{n}{r}\dfrac{d^r u}{dx^r}\dfrac{d^{n-r} v}{dx^{n-r}} + ... + \dfrac{d^n u}{dx^n} v$
(1.2-6)

where $\binom{n}{r} = \dfrac{n!}{(n-r)!\, r!}$, $\quad n!\, [= n\,(n-1)\,...\,1]$ is the **factorial** of $n$.

Rule (iv) is known as **Leibnitz rule** (one of them!).

## Mean Value Theorem

If $f(x)$ is continuous in the closed interval $a \le x \le b$, and $f(x)$ is differentiable in the open interval $a < x < b$, there exists a point $c$ in $(a, b)$, such that

$$f'(c) = \frac{f(b) - f(a)}{b - a}$$
(1.2-7)

From Equation (1.2-7), we deduce that if $f'(c) = 0$ for every $c$ in $(a, b)$, then $f$ is a constant. If $f'(c) > 0$ for every $c$ in $(a, b)$, then $f(x)$ is an increasing function, that is to say, as $x$ increases $f(x)$ increases. Conversely, if $f'(c) < 0$ for every $c$ in $(a, b)$, $f(x)$ is a decreasing function of $x$.

## Cauchy Mean Value Theorem

If $f$ and $g$ are continuous in $[a, b]$ and differentiable in $(a, b)$, there exist a number $c$ in $(a, b)$ such that

$$\frac{f(b) - f(a)}{g(b) - g(a)} = \frac{f'(c)}{g'(c)}$$
(1.2-8)

If $g(x) = x$, then Equation (1.2-8) reduces to Equation (1.2-7).

## L'Hôpital's Rule

$\lim\limits_{x \to x_0} f(x) = 0$ and $\lim\limits_{x \to x_0} g(x) = 0$, the $\lim\limits_{x \to x_0} \dfrac{f(x)}{g(x)}$ is indeterminate. But if the $\lim\limits_{x \to x_0} \dfrac{f'(x)}{g'(x)}$ exists

$$\lim_{x \to x_0} \frac{f(x)}{g(x)} = \lim_{x \to x_0} \frac{f'(x)}{g'(x)} \tag{1.2-9}$$

If $\lim\limits_{x \to x_0} \dfrac{f'(x)}{g'(x)}$ does not exist but $\lim\limits_{x \to x_0} \dfrac{f^{(n)}(x)}{g^{(n)}(x)}$ exists for some value of n, Equation (1.2-9) can be replaced by

$$\lim_{x \to x_0} \frac{f(x)}{g(x)} = \lim_{x \to x_0} \frac{f^{(n)}(x)}{g^{(n)}(x)} \tag{1.2-10}$$

The same rule applies if the $\lim\limits_{x \to x_0} f(x) = \infty$ and $\lim\limits_{x \to x_0} g(x) = \infty$, the $\lim\limits_{x \to x_0} \dfrac{f(x)}{g(x)}$ is indeterminate. The rule holds for $x \longrightarrow \infty$ or $x \longrightarrow -\infty$. Other indeterminate forms, such as the difference of two quantities tending to infinity, must first be reduced to one of the indeterminate forms discussed here before applying the rule.

## Taylor's Theorem

If $f(x)$ is continuous and differentiable

$$f(x) = f(x_0) + (x - x_0) f'(x_0) + \frac{(x - x_0)^2}{2} f''(x_0) + \dots + \frac{(x - x_0)^n}{n!} f^{(n)}(x_0) + R_n \tag{1.2-11}$$

where $R_n$ is the remainder term.

There are various ways of expressing the remainder term $R_n$. The simplest one is probably Lagrange's expression which may be written as

$$R_n = \frac{(x - x_0)^{n+1}}{(n + 1)!} f^{(n+1)} [x_0 + \theta (x - x_0)] \tag{1.2-12}$$

where $0 < \theta < 1$.

$R_n$ is the result of a summation of the remaining terms, and represents the error made by truncating the series at the $n^{th}$ term. Note that we are expanding about a point $x_0$ which belongs to an interval $(x_0, x)$. Therefore, the remainder term for each point $x$ in the interval will generally be different. The maximum truncation error, associated with the evaluation of a function, at different values of x within the considered interval, determines the value of $\theta$ in Equation (1.2-12).

In Equation (1.2-11), we have expanded $f(x)$ about the point $x_0$, and if $x_0$ is the origin, we are dealing with **Maclaurin series**. The Taylor series expansion is widely used as a method of approximating a function by a polynomial.

## Maximum and Minimum

We might need to know the **extreme** (maximum or minimum) values of a function and this can be obtained by finding the derivatives of the function. Thus, if the function f has an **extremum** at $x_0$

$$\Delta f = f(x_0 + h) - f(x_0) \qquad (1.2\text{-}13)$$

must have the same sign irrespective of the sign of h.

If $\Delta f$ is positive, f has a **minimum** at $x_0$ and if $\Delta f$ is negative, f has a **maximum** at $x_0$. Figure 1.2-1 defines such extrema. From Equation (1.2-11), we see that Equation (1.2-13) can be written as

$$\Delta f = h f'(x_0) + \frac{h^2}{2} f''(x_0) + \dots \qquad (1.2\text{-}14)$$

where $h = x - x_0$.

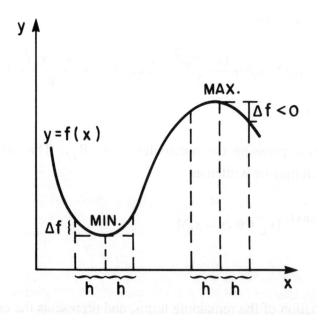

**FIGURE 1.2-1    Extremum of a function  f**

From Equation (1.2-14) we deduce that the condition for f to have an extremum at $x_0$ is

$$f'(x_0) = 0 \qquad\qquad (1.2\text{-}15)$$

The conditions for f to have a maximum or minimum at $x = x_0$ are

$$f''(x_0) > 0, \quad f \text{ has a minimum} \qquad\qquad (1.2\text{-}16)$$

$$f''(x_0) < 0, \quad f \text{ has a maximum} \qquad\qquad (1.2\text{-}17)$$

But if $f''(x_0) = 0$, we cannot deduce that f has an extreme value at $x_0$. We need to consider higher derivatives until we obtain a $f^{(n)}(x_0)$ which is non-zero. Thus, the general criteria for extreme values are

if $f(x)$ is defined in [a, b] and $x_0$ is an interior point of (a, b), and if $f^{(n)}(x_0)$ exists and is non-zero, but $f'(x_0) = f''(x_0) = ... = f^{(n-1)}(x_0) = 0$, $f(x)$ has an extreme value at $x_0$ if n is even. If $f^{(n)}(x_0) < 0$, f has a maximum at $x = x_0$ and if $f^{(n)}(x_0) > 0$, f has a minimum at $x = x_0$. If n is odd, $f(x)$ does not have an extreme value at $x = x_0$.

***Example 1.2-1***. Find the extreme values of $f(x) = x^3$, if they exist.

On differentiating, we have

$$f'(x) = 3x^2, \qquad f''(x) = 6x, \qquad f'''(x) = 6 \qquad\qquad (1.2\text{-}18a,b,c)$$

From Equation (1.2-18a, b), we see that

$$f'(0) = f''(0) = 0 \qquad\qquad (1.2\text{-}19a,b)$$

Thus we need to consider higher derivatives and the next one $f'''(0)$ happens to be non-zero. From the criteria given earlier we deduce that f does not have an extreme value at the origin. The origin is neither a maximum nor a minimum, it is a point of inflection, as can be seen by drawing the curve given by $f(x) = x^3$.

## 1.3   INTEGRALS

An integral can be considered to be an **antiderivative**. Thus, if we know that the derivative of $F(x)$ is $f(x)$ $[= F'(x)]$, an integral of $f(x)$ is $F(x)$. For example, the derivative of $\frac{1}{3}x^3$ is $x^2$, and an integral of $x^2$ is $\frac{1}{3}x^3$. Note that we have used the article an. Since the derivative of a constant is zero, $F(x)$ is arbitrary to the extent of an arbitrary constant. The integral we have defined is known as an **indefinite integral** which is usually denoted by the symbol $\int$ . Thus, we write

$$F(x) = \int f(x)\, dx = \int_a^x f(t)\, dt \qquad\qquad (1.3\text{-}1a,b)$$

where $a$ is an arbitrary constant of integration. Equations (1.3-1a,b) define a function of $x$ in terms of a dummy variable $t$.

The integral may be interpreted as the area enclosed by the curve $y = f(x)$ and the x-axis. For the area to be definite, we need to fix the ordinates, such as, $x = a$ and $x = b$. Thus, if $A$ is the area bounded by the curve $y = f(x)$, the x-axis and the ordinates $x = a$, $x = b$

$$A = \int_a^b f(x)\, dx \qquad\qquad (1.3\text{-}2)$$

Equation (1.3-2) defines a **definite integral**; the limits $x = a$ and $x = b$ are given. We can convert the indefinite integral in Equation (1.3-1) to a definite integral if $x = b$. In this case, we usually write

$$F(b) - F(a) = \int_a^b f(x)\, dx = \Big[\, F(x) \,\Big]_a^b \qquad\qquad (1.3\text{-}3a,b)$$

Thus to evaluate a definite integral analytically, we first need to find an indefinite integral. There are tables of integrals, where the indefinite integrals of standard functions are given. Below we list some of the general methods of integration.

**Integration by Parts**

If $f$ and $g$ are functions of $x$

$$\frac{d}{dx}\,(fg) = f'g + fg' \qquad\qquad (1.3\text{-}4)$$

It follows from Equation (1.3-4) that

$$\int f\frac{dg}{dx}\, dx = [fg] - \int \frac{df}{dx}\, g\, dx \qquad\qquad (1.3\text{-}5)$$

***Example 1.3-1.*** Integrate $\int e^{ax} \sin bx\, dx$.

We integrate by parts, identifying from Equation (1.3-5) $e^{ax}$ as $f(x)$ and $\sin bx$ as $\dfrac{dg}{dx}$. Carrying out the integration, we have

$$\int e^{ax} \sin bx \, dx = \left[ \frac{-e^{ax} \cos bx}{b} \right] + \frac{a}{b} \int e^{ax} \cos bx \, dx \qquad (1.3\text{-}6)$$

On integrating $\int e^{ax} \cos bx \, dx$ by parts again, we obtain

$$\int e^{ax} \cos bx \, dx = \left[ \frac{e^{ax}}{b} \sin bx \right] - \frac{a}{b} \int e^{ax} \sin bx \, dx \qquad (1.3\text{-}7)$$

Combining Equations (1.3-6, 7) yields

$$\left(1 + \frac{a^2}{b^2}\right) \int e^{ax} \sin bx \, dx = -\frac{1}{b} \left[ e^{ax} \cos bx \right] + \frac{a}{b^2} \left[ e^{ax} \sin bx \right] \qquad (1.3\text{-}8)$$

Hence

$$\int e^{ax} \sin bx \, dx = \frac{e^{ax}}{a^2 + b^2} \left[ a \sin bx - b \cos bx \right] \qquad (1.3\text{-}9)$$

**Integration by Substitution**

Certain integrals $\int f(x) \, dx$ can be easily evaluated if we substitute $x$ by a function $\phi(z)$ say. Since

$$x = \phi(z), \qquad dx = \phi'(z) \, dz \qquad (1.3\text{-}10a,b)$$

It follows from Equations (1.3-10a,b) that

$$\int f(x) \, dx = \int f[\phi(z)] \, \phi'(z) \, dz \qquad (1.3\text{-}11)$$

***Example 1.3-2.*** Integrate $\int \sqrt{a^2 - x^2} \, dx$, where $a$ is a constant.

Substitute $x$ by $a \sin z$, so that

$$dx = a \cos z \, dz \qquad (1.3\text{-}12)$$

$$\int \sqrt{a^2 - x^2} \, dx = \int \left( \sqrt{a^2 - a^2 \sin^2 z} \right) a \cos z \, dz \qquad (1.3\text{-}13a)$$

$$= a^2 \int \cos^2 z \, dz \; = \; \frac{a^2}{2} \int (\cos 2z + 1) \, dz \qquad\qquad (1.3\text{-}13\text{b,c})$$

$$= \frac{a^2}{2} \left[ \frac{1}{2} \sin 2z + z \right] \qquad\qquad (1.3\text{-}13\text{d})$$

Returning to the original variable x, we have

$$z = \text{arc} \sin \left( \frac{x}{a} \right) \qquad\qquad (1.3\text{-}14\text{a})$$

$$\sin 2z \; = \; 2 \sin z \cos z \; = \; \frac{2x}{a} \sqrt{1 - \left( \frac{x}{a} \right)^2} \qquad\qquad (1.3\text{-}14\text{b,c})$$

Thus

$$\int \sqrt{a^2 - x^2} \, dx \; = \; \frac{a^2}{2} \left[ \frac{x}{a^2} \sqrt{a^2 - x^2} + \text{arc} \sin \left( \frac{x}{a} \right) \right] \qquad\qquad (1.3\text{-}15)$$

●

In evaluating finite integrals, it is often simpler to express the limits of integration in terms of the new variable z.

In the method of substitution, the key is to find a substitution such that the integral is reduced to a standard form.

## Integration of Rational Functions

A rational function of x is a function of the form $f(x) / g(x)$, where $f(x)$ and $g(x)$ are polynomials in x. The rational function can be expressed as a sum of partial fractions and can thus be integrated.

***Example 1.3-3.*** Integrate $\displaystyle\int \frac{5x + 2}{x^3 - 8} \, dx$.

The function $\dfrac{5x + 2}{x^3 - 8}$ can be expressed as a sum of partial fractions as follows.

$$\frac{5x + 2}{x^3 - 8} = \frac{5x + 2}{(x - 2)(x^2 + 2x + 4)} = \frac{A}{x - 2} + \frac{Bx + C}{x^2 + 2x + 4} \qquad (1.3\text{-}16\text{a,b})$$

where A, B and C are constants.

By comparing powers of x, we obtain

$$A = 1, \qquad B = -1, \qquad C = 1 \tag{1.3-17a,b,c}$$

Thus

$$\int \frac{5x + 2}{x^3 - 8} \, dx = \int \frac{dx}{x - 2} + \int \frac{(1 - x) \, dx}{x^2 + 2x + 4} \tag{1.3-18}$$

The integral $\int \dfrac{dx}{x - 2}$ is standard and

$$\int \frac{dx}{x - 2} = \ell n \, (x - 2) \tag{1.3-19}$$

The second integral on the right side of Equation (1.3-18) can be evaluated as follows

$$\int \frac{(1 - x) \, dx}{x^2 + 2x + 4} = -\int \frac{(x + 1 - 2) \, dx}{x^2 + 2x + 4} = -\int \frac{(x + 1) \, dx}{x^2 + 2x + 4} + 2 \int \frac{dx}{(x + 1)^2 + 3} = -I_1 + 2I_2 \tag{1.3-20a,b,c}$$

To evaluate $I_1$, we make the following substitution

$$z = x^2 + 2x + 4, \qquad dz = (2x + 2) \, dx \tag{1.3-21a,b}$$

$$\int \frac{(x + 1) dx}{x^2 + 2x + 4} = \frac{1}{2} \int \frac{dz}{z} = \frac{1}{2} \ell n \, (x^2 + 2x + 4) \tag{1.3-22a,b}$$

To evaluate $I_2$, we let

$$(x + 1) = \sqrt{3} \tan \theta, \qquad dx = \sqrt{3} \sec^2 \theta \, d\theta \tag{1.3-23a,b}$$

$$\int \frac{dx}{(x + 1)^2 + 3} = \int \frac{\sqrt{3} \sec^2\theta d\theta}{3 \, (\tan^2\theta + 1)} = \int \frac{\sqrt{3} \sec^2\theta d\theta}{3 \sec^2\theta} = \frac{1}{\sqrt{3}} \theta = \frac{\text{arc tan}}{\sqrt{3}} \left( \frac{x + 1}{\sqrt{3}} \right) \tag{1.3-24a,b,c,d}$$

Combining Equations (1.3-19 to 24d), we obtain

$$\int \frac{5x + 2}{x^3 - 8} \, dx = \ell n \, (x - 2) - \frac{1}{2} \ell n \, (x^2 + 2x + 4) + \frac{2}{\sqrt{3}} \text{ arc tan} \left( \frac{x + 1}{\sqrt{3}} \right) \tag{1.3-25}$$

•

In the past, considerable efforts were devoted to finding methods to express integrals in closed form and in terms of elementary functions. Contour integration, in the theory of complex analysis (Chapter 3) can be used to evaluate real integrals. Currently, a popular approach is to resort to numerical methods (Chapter 7).

**Some Theorems**

(i)    $$\int_a^b f(x)\,dx = -\int_b^a f(x)\,dx$$    (1.3-26)

(ii)    $$\int_a^c f(x)\,dx = \int_a^b f(x)\,dx + \int_b^c f(x)\,dx$$    (1.3-27)

(iii)    $$\int_{-a}^a f(x)\,dx = \begin{cases} 2\int_0^a f(x)\,dx, & \text{if } f(x) \text{ is even} \\[2ex] 0, & \text{if } f(x) \text{ is odd} \end{cases}$$    (1.3-28a,b)

(iv)    $$\int_0^a f(x)\,dx = \int_0^a f(a-x)\,dx$$    (1.3-29)

(v)    **First mean value theorem**

If M and m are the upper and lower bounds respectively of $f(x)$ in $(a, b)$

$$m(b-a) \le \int_a^b f(x)\,dx \le M(b-a)$$    (1.3-30a,b)

(vi)    **Generalized first mean value theorem**

Under the conditions on $f(x)$ given in (v), and for $g(x) > 0$ everywhere in (a, b)

$$m\int_a^b g(x)\,dx \le \int_a^b f(x)\,g(x)\,dx \le M\int_a^b g(x)\,dx$$    (1.3-31a,b)

The above mean value theorems provide bounds on integrals and can be useful in error analysis.

## 1.4   FUNCTIONS OF SEVERAL VARIABLES

So far, we have considered functions of one variable only. In science and engineering, we often encounter one variable which depends on several other independent variables. For example, the volume of gas depends on the temperature and the pressure. For simplicity, we shall consider functions of two independent variables $x$ and $y$. In most cases, the extension to $n$ variables $x_1$, $x_2$, ..., $x_n$ is obvious.

The dependent variable $u$ is said to be a function of the independent variables $x$ and $y$ if to every pair of values of $(x, y)$ one can assign a value of $u$. In this case we write

$$u = f(x, y) \tag{1.4-1}$$

The domain of $f$ is the set of values of $(x, y)$ over which $f$ is applicable. The range of $f$ is the set of values that $u$ may have over the domain of $f$.

The function $f(x, y)$ is continuous at $(x_0, y_0)$ if given $\varepsilon > 0$, there exists a $\delta$, such that whenever

$$\sqrt{(x - x_0)^2 + (y - y_0)^2} < \delta \tag{1.4-2}$$

$$\left| f(x, y) - f(x_0, y_0) \right| < \varepsilon \tag{1.4-3}$$

## 1.5   DERIVATIVES

Since $u$ is a function of two variables $x$ and $y$, we may calculate the rate of change of $u$ with respect to $x$, holding $y$ fixed. This is the **partial derivative** of $u$ with respect to $x$ and is denoted by $\dfrac{\partial u}{\partial x}$. Other notations are: $\dfrac{\partial f}{\partial x}$, $f_x$ or $u_x$. Thus

$$\frac{\partial f}{\partial x} = \lim_{\Delta x \to 0} \frac{f(x + \Delta x, y) - f(x, y)}{\Delta x} \tag{1.5-1}$$

Similarly, $\dfrac{\partial f}{\partial y}$ is defined as

$$\frac{\partial f}{\partial y} = \lim_{\Delta y \to 0} \frac{f(x, y + \Delta y) - f(x, y)}{\Delta y} \tag{1.5-2}$$

The computation of $\dfrac{\partial f}{\partial x}$ is the same as in the case of one independent variable. Here, $y$ is treated as a constant. Similarly to compute $\dfrac{\partial f}{\partial y}$, we consider $x$ to be a constant. Since $f_x$ and $f_y$ are functions of $x$ and $y$, their partial derivatives with respect to $x$ and $y$ may exist. They are defined as

$$\frac{\partial}{\partial x}\left(\frac{\partial f}{\partial x}\right) = \lim_{\Delta x \to 0} \frac{f_x(x + \Delta x, y) - f_x(x, y)}{\Delta x} \qquad (1.5\text{-}3)$$

$$\frac{\partial}{\partial y}\left(\frac{\partial f}{\partial x}\right) = \lim_{\Delta y \to 0} \frac{f_x(x, y + \Delta y) - f_x(x, y)}{\Delta y} \qquad (1.5\text{-}4)$$

The **second-order partial derivatives** are denoted as

$$\frac{\partial}{\partial x}\left(\frac{\partial f}{\partial x}\right) = \frac{\partial^2 f}{\partial x^2} = f_{xx} \qquad (1.5\text{-}5a,b)$$

$$\frac{\partial}{\partial y}\left(\frac{\partial f}{\partial x}\right) = \frac{\partial^2 f}{\partial y \partial x} = f_{yx} \qquad (1.5\text{-}6a,b)$$

$$\frac{\partial}{\partial x}\left(\frac{\partial f}{\partial y}\right) = \frac{\partial^2 f}{\partial x \partial y} = f_{xy} \qquad (1.5\text{-}7a,b)$$

$$\frac{\partial}{\partial y}\left(\frac{\partial f}{\partial y}\right) = \frac{\partial^2 f}{\partial y^2} = f_{yy} \qquad (1.5\text{-}8a,b)$$

We note that $f_{yx}$ means taking the partial derivative of $f$ with respect to $x$ first and then with respect to $y$, whereas for $f_{xy}$ the order of differentiation is reversed. One may wonder if the order of differentiation is important. If $f_{xy}$ is continuous then the order is not important. In practice, this is generally the case and

$$f_{xy} = f_{yx} \qquad (1.5\text{-}9)$$

Likewise higher partial derivatives can be defined and computed. If the partial derivatives are continuous, then the order of differentiation is not important.

In an xy-coordinate system, the first order partial derivatives $f_x$ and $f_y$ may be regarded as the rate of change of $f$ along the $x$ and y-axis respectively. We can also define and compute the rate of change of $f$ along any arbitrary line in the xy-plane. Such a rate of change is known as a **directional derivative** and is denoted by $\frac{\partial f}{\partial n}$, where the vector $\underline{n}$ is parallel to the line along which we wish to determine the rate of change. Thus $\frac{\partial f}{\partial n}$ at a point $(x_0, y_0)$ along a line that makes an angle $\theta$ with the x-axis is defined as

$$\frac{\partial f}{\partial n} = \lim_{\rho \to 0} \left[\frac{f(x_0 + \rho \cos \theta, y_0 + \rho \sin \theta) - f(x_0, y_0)}{\rho}\right] \qquad (1\text{-}5\text{-}10)$$

where $\rho$ is the distance of any point on the line from $(x_0, y_0)$. This situation is illustrated in Figure 1.5-1.

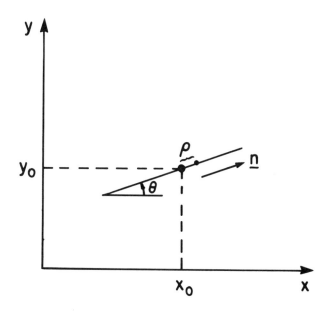

**FIGURE 1.5-1    Directional derivative along a line parallel to $\underline{n}$**

We can rewrite Equation (1.5-10) as

$$\frac{\partial f}{\partial n} = \lim_{\rho \to 0} \left[ \frac{f(x_0 + \rho \cos\theta, y_0 + \rho \sin\theta) - \overbrace{f(x_0, y_0 + \rho \sin\theta) + f(x_0, y_0 + \rho \sin\theta)}^{\substack{0 \\ \|}} - f(x_0, y_0)}{\rho} \right]$$

(1.5-11a)

Note that, in the first two terms on the right side of Equation (1.5-11a), we have kept $y_0 + \rho \sin\theta = y_1$ constant.

$$\frac{\partial f}{\partial n} = \lim_{\rho \to 0} \left[ \frac{f(x_0 + \rho \cos\theta, y_1) - f(x_0, y_1) + f(x_0, y_1) - f(x_0, y_0)}{\rho} \right]$$

(1.5-11b)

This amounts to considering $f$ to be a function of $x$ only in the first two terms. Expanding this function of $x$ in a Taylor series yields: $f(x_0, y_1) + \frac{\partial f}{\partial x} \rho \cos \theta$. Similarly, one can consider $f$ to be a function of $y$ only in the last two terms, resulting in: $f(x_0, y_0) + \frac{\partial f}{\partial y} \rho \sin \theta$. We deduce that

$$\lim_{\rho \to 0} \left[ \frac{f(x_0, y_1) + \frac{\partial f}{\partial x} \rho \cos \theta - f(x_0, y_1)}{\rho} \right] = \frac{\partial f}{\partial x} \cos \theta \tag{1.5-12a}$$

and similarly

$$\lim_{\rho \to 0} \left[ \frac{f(x_0, y_0) + \frac{\partial f}{\partial y} \rho \sin \theta - f(x_0, y_0)}{\rho} \right] = \frac{\partial f}{\partial y} \sin \theta \tag{1.5-12b}$$

Therefore

$$\frac{\partial f}{\partial n} = \frac{\partial f}{\partial x} \cos \theta + \frac{\partial f}{\partial y} \sin \theta \tag{1.5-13}$$

Thus if $f_x$ and $f_y$ are known, we can compute $\frac{\partial f}{\partial n}$.

## Total Derivatives

We now determine the change in u, $\Delta u$, when both $x$ and $y$ change simultaneously to $x + \Delta x$ and to $y + \Delta y$ respectively. Then

$$\Delta u = f(x + \Delta x, y + \Delta y) - f(x, y) \tag{1.5-14a}$$

$$= f(x + \Delta x, y + \Delta y) - f(x, y + \Delta y) + f(x, y + \Delta y) - f(x, y) \tag{1.5-14b}$$

The observations made following Equation (1.5-11) are applicable to Equation (1.5-14b), and on taking the limits $\Delta x \longrightarrow 0$, $\Delta y \longrightarrow 0$, we obtain

$$du = df = \frac{\partial f}{\partial x} dx + \frac{\partial f}{\partial y} dy \tag{1.5-15a,b}$$

The existence of df guarantees the existence of $f_x$ and $f_y$, but the converse is not true. For df to exist we require not only the existence of $f_x$ and $f_y$, but we also require f to be continuous.

The **differential** df may be regarded as a function of 4 independent variables x, y, dx and dy. **Higher differentials** $d^2f$, $d^3f$, ... , $d^n f$ can also be defined. Thus

$$d^2f = d(df) = d\left(\frac{\partial f}{\partial x}\,dx\right) + d\left(\frac{\partial f}{\partial y}\,dy\right) \tag{1.5-16a,b}$$

Substituting $d$ by $\dfrac{\partial}{\partial x}\,dx + \dfrac{\partial}{\partial y}\,dy$ yields

$$d^2f = \frac{\partial}{\partial x}\left(\frac{\partial f}{\partial x}\,dx\right)dx + \frac{\partial}{\partial y}\left(\frac{\partial f}{\partial x}\,dx\right)dy + \frac{\partial}{\partial x}\left(\frac{\partial f}{\partial y}\,dy\right)dx + \frac{\partial}{\partial y}\left(\frac{\partial f}{\partial y}\,dy\right)dy \tag{1.5-16c}$$

$$= \frac{\partial^2 f}{\partial x^2}\,(dx)^2 + 2\,\frac{\partial^2 f}{\partial x \partial y}\,dx\,dy + \frac{\partial^2 f}{\partial y^2}\,(dy)^2 \tag{1.5-16d}$$

It can be shown by induction that

$$d^n f = \frac{\partial^n f}{\partial x^n}\,(dx)^n + \binom{n}{1}\frac{\partial^n f}{\partial x^{n-1}\partial y}\,(dx)^{n-1}\,dy + \ldots + \binom{n}{r}\frac{\partial^n f}{\partial x^{n-r}\partial y^r}\,(dx)^{n-r}\,(dy)^r + \ldots + \frac{\partial^n f}{\partial y^n}\,(dy)^n \tag{1.5-17}$$

In order to remember Equation (1.5-17), one can rewrite it as

$$d^n f = \left(\frac{\partial}{\partial x}\,dx + \frac{\partial}{\partial y}\,dy\right)^n f \tag{1.5-18}$$

In Equation (1.5-18), the right side can be expanded formally as a binomial expansion.

If both $x$ and $y$ are functions of another variable $t$, then from Equation (1.5-15) we have

$$\frac{du}{dt} = \frac{\partial f}{\partial x}\,\frac{dx}{dt} + \frac{\partial f}{\partial y}\,\frac{dy}{dt} \tag{1.5-19}$$

If $x$ and $y$ are functions of another set of independent variables $r$ and $s$, then

$$dx = \frac{\partial x}{\partial r}\,dr + \frac{\partial x}{\partial s}\,ds \tag{1.5-20a}$$

$$dy = \frac{\partial y}{\partial r}\,dr + \frac{\partial y}{\partial s}\,ds \tag{1.5-20b}$$

Substituting $dx$ and $dy$ in Equation (1.5-15), we obtain

$$du = \left(\frac{\partial f}{\partial x}\,\frac{\partial x}{\partial r} + \frac{\partial f}{\partial y}\,\frac{\partial y}{\partial r}\right)dr + \left(\frac{\partial f}{\partial x}\,\frac{\partial x}{\partial s} + \frac{\partial f}{\partial y}\,\frac{\partial y}{\partial s}\right)ds \tag{1.5-21}$$

It follows from Equation (1.5-21) that

$$\frac{\partial u}{\partial r} = \frac{\partial f}{\partial r} = \frac{\partial f}{\partial x}\frac{\partial x}{\partial r} + \frac{\partial f}{\partial y}\frac{\partial y}{\partial r} \qquad (1.5\text{-}22a,b)$$

$$\frac{\partial u}{\partial s} = \frac{\partial f}{\partial s} = \frac{\partial f}{\partial x}\frac{\partial x}{\partial s} + \frac{\partial f}{\partial y}\frac{\partial y}{\partial s} \qquad (1.5\text{-}22c,d)$$

Equations (1.5-22a, b, c, d) again express the chain rule.

***Example 1.5-1.*** In rectangular Cartesian coordinates system, f is given by

$$f = x^2 + y^2 \qquad (1.5\text{-}23)$$

We change to polar coordinates (r, θ). The transformation equations are

$$x = r\cos\theta, \qquad y = r\sin\theta \qquad (1.5\text{-}24a,b)$$

Calculate $\dfrac{\partial f}{\partial r}$ and $\dfrac{\partial f}{\partial \theta}$ .

From Equations (1.5-22a, b, c, d), we have

$$\frac{\partial f}{\partial r} = \frac{\partial f}{\partial x}\frac{\partial x}{\partial r} + \frac{\partial f}{\partial y}\frac{\partial y}{\partial r} \qquad (1.5\text{-}25a)$$

$$\frac{\partial f}{\partial \theta} = \frac{\partial f}{\partial x}\frac{\partial x}{\partial \theta} + \frac{\partial f}{\partial y}\frac{\partial y}{\partial \theta} \qquad (1.5\text{-}25b)$$

Computing the partial derivatives yields

$$\frac{\partial f}{\partial x} = 2x , \qquad\qquad \frac{\partial f}{\partial y} = 2y \qquad (1.5\text{-}26a,b)$$

$$\frac{\partial x}{\partial r} = \cos\theta , \qquad\qquad \frac{\partial y}{\partial r} = \sin\theta \qquad (1.5\text{-}26c,d)$$

$$\frac{\partial x}{\partial \theta} = -r\sin\theta , \qquad\qquad \frac{\partial y}{\partial \theta} = r\cos\theta \qquad (1.5\text{-}26e,f)$$

Substituting Equation (1.5-26a to f) into Equations (1.5-25a, b), we obtain

$$\frac{\partial f}{\partial r} = 2x\cos\theta + 2y\sin\theta = 2r \qquad (1.5\text{-}27a,b)$$

$$\frac{\partial f}{\partial \theta} = 2x\,(-r\sin\theta) + 2y\,(r\cos\theta) = 0 \qquad (1.5\text{-}27c,d)$$

Equations (1.5-27a, b, c, d) can be obtained by substituting Equations (1.5-24) into Equation (1.5-23) and thus $f$ is expressed explicitly as a function of $r$ and $\theta$ and the partial differentiation can be carried out. In many cases, the substitution can be very complicated.

## 1.6   IMPLICIT FUNCTIONS

So far we have considered $u$ as an **explicit** function of $x$ and $y$ [$u = f(x, y)$]. There are examples where it is more convenient to express $u$ **implicitly** as a function of $x$ and $y$. For example, in thermodynamics an equation of state which could be given as $T = f(P, V)$ is usually written, implicitly as

$$f(P, V\ T) = 0 \tag{1.6-1}$$

where $P$ is the pressure, $V$ is the volume and $T$ is the temperature.

The two-parameter Redlich and Kwong (1949) equation is expressed as

$$\left[P + \frac{n^2 a}{T^{1/2} V(V + nb)}\right][V - nb] - nRT = 0 \tag{1.6-2}$$

where $a$ and $b$ are two parameters, $R$ is the gas constant and $n$ the number of moles.

In theory, we can solve for $T$ in Equation (1.6-2) and express $T$ as a function of $P$ and $V$. Then by partial differentiation, we can obtain $\frac{\partial T}{\partial P}$ and other partial derivatives. But as can be seen from Equation (1.6-2) it is not easy to solve for $T$, it implies solving a cubic equation. Even if we solve for $T$, the resulting function will be even more complicated than Equation (1.6-2), and finding $\frac{\partial T}{\partial P}$ (say) will be time consuming. It is simpler to differentiate the expression in Equation (1.6-2) partially with respect to $P$ and then deduce $\frac{\partial T}{\partial P}$. We shall show how this is done by reverting to the variables $x$, $y$ and $u$.

We now consider an implicit function written as

$$f(x, y, u) = 0 \tag{1.6-3}$$

In terms of the thermodynamic example, we can think of

$$x = P, \qquad y = V, \qquad u = T \tag{1.6-4a,b,c}$$

The variables $x$, $y$ and $u$ are not all independent. We are free to choose any one of them as the dependent variable. Since $f = 0$, $df = 0$ and from Equation (1.5-15), we obtain

$$df = \frac{\partial f}{\partial x}\, dx + \frac{\partial f}{\partial y}\, dy + \frac{\partial f}{\partial u}\, du = 0 \qquad\qquad (1.6\text{-}5a,b)$$

Equation (1.6-5) is still true for $f = $ constant, since $df = 0$ in this case also.

The partial derivative $\dfrac{\partial u}{\partial x}$ is obtained from Equation (1.6-5) by putting $dy = 0$ (since $y$ is kept as a constant). So

$$\frac{\partial u}{\partial x} = -\left(\frac{\partial f}{\partial x}\right) \bigg/ \left(\frac{\partial f}{\partial u}\right) \qquad\qquad (1.6\text{-}6)$$

Similarly

$$\frac{\partial u}{\partial y} = -\left(\frac{\partial f}{\partial y}\right) \bigg/ \left(\frac{\partial f}{\partial u}\right) \qquad\qquad (1.6\text{-}7)$$

We could equally obtain $\dfrac{\partial u}{\partial x}$, $\dfrac{\partial u}{\partial y}$ by differentiating $f$ from Equation (1.6-3) and using the chain rule (Equations 1.5-22a, b). Thus taking $u$ as the dependent variable and differentiating partially with respect to $x$, we have

$$\frac{\partial f}{\partial x} + \frac{\partial f}{\partial u}\,\frac{\partial u}{\partial x} = 0 \qquad\qquad (1.6\text{-}8)$$

It then follows that

$$\frac{\partial u}{\partial x} = -\left(\frac{\partial f}{\partial x}\right) \bigg/ \left(\frac{\partial f}{\partial u}\right) \qquad\qquad (1.6\text{-}9)$$

Similarly $\dfrac{\partial u}{\partial y}$ and higher derivatives such as $\dfrac{\partial^2 u}{\partial x^2}$ can be computed.

If, in addition to Equation (1.6-3), the variables $x, y$ and $u$ are related by another equation written as

$$g(x, y, u) = 0 \qquad\qquad (1.6\text{-}10)$$

we essentially have two equations involving three variables. We choose the only independent variable to be $x$. Differentiating $f$ and $g$ with respect to $x$, yields

$$\frac{\partial f}{\partial x} + \frac{\partial f}{\partial y}\,\frac{\partial y}{\partial x} + \frac{\partial f}{\partial u}\,\frac{\partial u}{\partial x} = 0 \qquad\qquad (1.6\text{-}11a)$$

$$\frac{\partial g}{\partial x} + \frac{\partial g}{\partial y}\,\frac{\partial y}{\partial x} + \frac{\partial g}{\partial u}\,\frac{\partial u}{\partial x} = 0 \qquad\qquad (1.6\text{-}11b)$$

Equations (1.6-11a, b) form a system of two algebraic equations involving two unknowns $\dfrac{\partial y}{\partial x}$ and $\dfrac{\partial u}{\partial x}$. The solutions are

$$\frac{\partial y}{\partial x} = \frac{-\begin{vmatrix} f_x & f_u \\ g_x & g_u \end{vmatrix}}{J} \tag{1.6-12a}$$

$$\frac{\partial u}{\partial x} = \frac{\begin{vmatrix} f_x & f_y \\ g_x & g_y \end{vmatrix}}{J} \tag{1.6-12b}$$

$$J = \begin{vmatrix} f_y & f_u \\ g_y & g_u \end{vmatrix} \tag{1.6-12c}$$

Thus for $\dfrac{\partial y}{\partial x}$ and $\dfrac{\partial u}{\partial x}$ to exist, the Jacobian $J$ must not vanish.

## 1.7   SOME THEOREMS

### Euler's Theorem

A function $f(x, y, u)$ is a **homogenous** function of degree $n$ if

$$f(\alpha x, \alpha y, \alpha u) = \alpha^n f(x, y, u) \tag{1.7-1}$$

Defining new variables, which we indicate by a star (*), we write

$$x^* = \alpha x, \qquad y^* = \alpha y, \qquad u^* = \alpha u \tag{1.7-2a,b,c}$$

Equation (1.7-1) becomes

$$f(x^*, y^*, u^*) = \alpha^n f(x,y,u) \tag{1.7-3}$$

Differentiating Equation (1.7-3) with respect to the parameter $\alpha$ yields

$$\frac{\partial f}{\partial x^*} \frac{\partial x^*}{\partial \alpha} + \frac{\partial f}{\partial y^*} \frac{\partial y^*}{\partial \alpha} + \frac{\partial f}{\partial u^*} \frac{\partial u^*}{\partial \alpha} = n\alpha^{n-1} f(x, y, u) \tag{1.7-4}$$

Choosing $\alpha = 1$, Equation (1.7-4) becomes

$$x \frac{\partial f}{\partial x} + y \frac{\partial f}{\partial y} + u \frac{\partial f}{\partial u} = n f(x, y, u) \tag{1.7-5}$$

Note that we substitute $\dfrac{\partial x^*}{\partial \alpha}$ $(= x)$ etc. into Equation (1.7-4) before setting $\alpha = 1$, otherwise we would be trying to differentiate by a constant.

Equation (1.7-5) is known as **Euler's theorem**.

The generalization of Equation (1.7-5) is

$$\left( x \frac{\partial}{\partial x} + y \frac{\partial}{\partial y} + u \frac{\partial}{\partial u} \right)^r f(x, y, u) = n(n-1) \dots (n-r+1) f(x, y, u) \tag{1.7-6}$$

## Taylor's Theorem

**Taylor's theorem** for functions of one variable can be extended to functions of several variables. For simplicity we give the formula for two independent variables.

$$f(x_0+h, y_0+k) = f(x_0, y_0) + \left\{ h \frac{\partial f}{\partial x} + k \frac{\partial f}{\partial y} \right\} + \frac{1}{2} \left\{ h^2 \frac{\partial^2 f}{\partial x^2} + 2hk \frac{\partial^2 f}{\partial x \partial y} + k^2 \frac{\partial^2 f}{\partial y^2} \right\} + \dots$$

$$\dots + \frac{1}{n!} \left\{ h^n \frac{\partial^n f}{\partial x^n} + \binom{n}{1} h^{n-1} k \frac{\partial^n f}{\partial x^{n-1} \partial y} + \dots + \binom{n}{r} h^{n-r} k^r \frac{\partial^n f}{\partial x^{n-r} \partial y^r} + \dots + k^n \frac{\partial^n f}{\partial y^n} \right\} + R_n \tag{1.7-7}$$

The remainder term $R_n$ is given by

$$R_n = \frac{1}{(n+1)!} \left\{ h^{n+1} \frac{\partial^{n+1} f}{\partial x^{n+1}} + h^n k \frac{\partial^{n+1} f}{\partial x^n \partial y} + \dots + k^{n+1} \frac{\partial^{n+1} f}{\partial y^{n+1}} \right\} \tag{1.7-8}$$

The derivatives in Equation (1.7-7) are to be evaluated at the point $(x_0, y_0)$ and those in Equation (1.7-8) at the point $(x_0 + \theta h, y_0 + \theta k)$ and $0 \le \theta \le 1$.

## 1.8   INTEGRAL OF A FUNCTION DEPENDING ON A PARAMETER

The function $f(x, y)$ is a function of two variables $x$ and $y$ and we may integrate the function with respect to $y$ holding $x$ fixed. We then obtain an integral which is a function of $x$ and we may consider $x$ as a parameter. Thus if we integrate $f(x, y)$ between two fixed points $y = a$ and $y = b$, we have

$$I(x) = \int_a^b f(x, y) \, dy \tag{1.8-1}$$

Differentiating I with respect to x results in

$$\frac{dI}{dx} = \int_a^b \frac{\partial f}{\partial x} \, dy \tag{1.8-2}$$

If the limits of integration are not fixed but are functions of x, we integrate with respect to y from a point on a curve given by $y = u(x)$ to a point on another curve given by $y = v(x)$

$$I(x) = \int_{u(x)}^{v(x)} f(x, y) \, dy \tag{1.8-3}$$

Note that if we were integrating at another point on the same curve, the limits of integration would read $u(x_1)$ and $u(x_2)$ which is equivalent to integrating from a to b.

I may be treated as a function of three variables x, $v(x)$ and $u(x)$. Using the results in Section 1.5, $\frac{dI}{dx}$ is given by

$$\frac{dI}{dx} = \frac{\partial I}{\partial x} + \frac{\partial I}{\partial v} \frac{dv}{dx} + \frac{\partial I}{\partial u} \frac{du}{dx} \tag{1.8-4}$$

From the definitions of the partial derivatives $\frac{\partial I}{\partial x}$, $\frac{\partial I}{\partial v}$ and $\frac{\partial I}{\partial u}$, we have, via Equation (1.8-2)

$$\frac{\partial I}{\partial x} = \int_{u(x)}^{v(x)} \frac{\partial f}{\partial x} \, dy \tag{1.8-5}$$

Note that $u(x)$ and $v(x)$ are not fixed!

To evaluate the partial derivative $\frac{\partial I}{\partial v}$, we fix the variables x and $u(x)$. Equation (1.3-1) yields

$$\frac{\partial F(x)}{\partial x} = \frac{\partial}{\partial x} \int_a^x f(t) \, dt = \frac{\partial}{\partial x} [F(x) - F(a)] = F'(x) = f(x) \tag{1.8-6a,b,c,d}$$

Therefore, identifying y with t and $v(x)$ with x, we have

$$\frac{\partial I}{\partial v} = \frac{\partial}{\partial v} \int_{u(x)}^{v(x)} f(x, y) \, dy = f[x, v(x)] \tag{1.8-7a,b}$$

Similarly

$$\frac{\partial I}{\partial u} = -f[x, u(x)] \tag{1.8-8}$$

After appropriate substitution, Equation (1.8-4) becomes

$$\frac{dI}{dx} = \int_{u(x)}^{v(x)} \frac{\partial f}{\partial x} \, dy + f(x, v) \frac{dv}{dx} - f(x, u) \frac{du}{dx} \tag{1.8-9}$$

Equation (1.8-9) is known as **Leibnitz rule**.

***Example 1.8-1.*** Let I be given by

$$I = \int_{y=x}^{y=x^2} xy \, dy \tag{1.8-10}$$

Calculate $\frac{dI}{dx}$ by

(a)     using Equation (1.8-9),

(b)     integrate and obtain I explicitly as a function of x and then differentiate. Figure 1.8-1 shows a projection in the xy-plane of the integration path.

From Equation (1.8-9)

$$\frac{dI}{dx} = \int_{x}^{x^2} y \, dy + (x)(x^2)(2x) - (x)(x) \tag{1.8-11a}$$

$$= \left[ \frac{y^2}{2} \right]_{x}^{x^2} + 2x^4 - x^2 \tag{1.8-11b}$$

$$= \frac{5x^4}{2} - \frac{3x^2}{2} \tag{1.8-11c}$$

On integrating directly

$$I = \left[ \frac{xy^2}{2} \right]_{x}^{x^2} = \frac{x^5}{2} - \frac{x^3}{2} \tag{1.8-12a,b}$$

It follows from Equation (1.8-12b) that

$$\frac{dI}{dx} = \frac{5x^4}{2} - \frac{3x^2}{2} \tag{1.8-12c}$$

Both methods result in the same expression for $\frac{dI}{dx}$, as they should.  There are instances when it is not possible to evaluate $I$ explicitly and one has to use Leibnitz's rule.

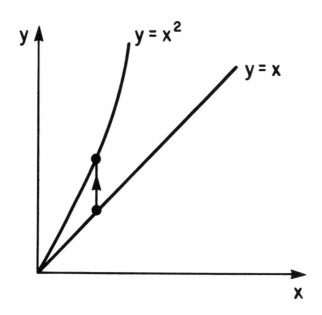

**FIGURE 1.8-1     Integration described by Equation (1.8-10)**

## 1.9    ORDINARY DIFFERENTIAL EQUATIONS (O.D.E.) – DEFINITIONS

A differential equation is an equation involving one dependent variable and its derivatives with respect to one or more independent variables.  If only one independent variable is involved, it is an **ordinary differential equation** (O.D.E.) and if more than one independent variable is involved, it is a **partial differential equation** (P.D.E.).  Many laws and relations in science, engineering, economics, and other fields of applied science are expressed as differential equations.

The highest order derivative occurring in the differential equation determines the **order** of the differential equation.  The **degree** of a differential equation is determined by the power to which the highest derivative is raised.  A differential equation is **linear** only if the dependent variable and its derivatives occur to the first degree.  Otherwise it is **non-linear**.

For example, $y' = 5y$ is an O.D.E. ($y$ is a function of $x$ only) of order one ($y'$ is the highest order derivative) and of degree one ($y'$ is raised to the power one) and is linear. $(y'')^3 + y = x$ is of order two, of degree three and is non-linear.

A function $f(x)$ is a solution of a given differential equation on some interval, if $f(x)$ is defined and differentiable on that interval and if the equation becomes an identity when $y$ and $y^{(n)}$ are replaced by $f(x)$ and $f^{(n)}(x)$ respectively.

For example, we can easily verify that $f(x) = e^{ax}$ is a solution of the equation $y' = ay$. Indeed, $f'(x) = a e^{ax} = y'$ and the right side $(ay)$ is of course $af(x) = ae^{ax}$. There are several types of ordinary differential equations. Examples of first-order differential equations are separable equations, exact differential equations, linear differential equations, homogeneous linear equations, etc. For each of these types, there exists a known, standardized, procedure to arrive at a solution. Starting at Section 1.10, we summarize the approach, leading to the solution of several of the types of ordinary differential equations encountered in practice. In Section 1.19, we look at the modeling problem.

## 1.10  FIRST-ORDER DIFFERENTIAL EQUATIONS

The standard form of a first-order differential equation is as follows

$$M(x, y)\, dx + N(x, y)\, dy = 0 \qquad\qquad (1.10\text{-}1)$$

or $\quad y' = -\dfrac{M(x, y)}{N(x, y)} \qquad\qquad (1.10\text{-}2)$

Equations of this type occur in problems dealing with orthogonal trajectories, growth, decay, and chemical reactions.

## 1.11  SEPARABLE FIRST-ORDER DIFFERENTIAL EQUATIONS

For the case where $M$ is a function of $x$ only and $N$ is a function of $y$ only, a straightforward integration will yield a result, as follows

$$\int M(x)\, dx = -\int N(y)\, dy \qquad\qquad (1.11\text{-}1)$$

***Example 1.11-1.*** Solve $y\, dx - x^2\, dy = 0$. $\qquad\qquad (1.11\text{-}2)$

Dividing both sides of the equation by $x^2 y$ results in the appropriate form

$$\frac{dx}{x^2} - \frac{dy}{y} = 0 \qquad\qquad (1.11\text{-}3)$$

Integration yields

$$\ell n \ y = \ell n \ c - \frac{1}{x} \qquad (1.11\text{-}4)$$

where c is the constant of integration.

This can be written as

$$\ell n \ y - \ell n \ c = -\frac{1}{x} \qquad (1.11\text{-}5)$$

that is

$$\ell n \ \frac{y}{c} = -\frac{1}{x} \qquad (1.11\text{-}6)$$

or $\qquad y = c \, e^{-1/x} \qquad (1.11\text{-}7)$

***Example 1.11-2***. In a constant volume batch reactor, the rate of disappearance of reactant A can be given by

$$\frac{dc_A}{dt} = -k \ f(c_A) \qquad (1.11\text{-}8)$$

Solve Equation (1.11-8) for the case where $f(c_A) = c_A$.

$$\frac{dc_A}{c_A} = -k \ dt \qquad (1.11\text{-}9)$$

$$\ell n \ c_A = -kt + \ell n \ c \qquad (1.11\text{-}10)$$

$$\ell n \ \frac{c_A}{c} = -kt \qquad (1.11\text{-}11)$$

$$c_A = ce^{-kt} \qquad (1.11\text{-}12)$$

## 1.12 HOMOGENEOUS FIRST-ORDER DIFFERENTIAL EQUATIONS

If $M(x, y)$ and $N(x, y)$ in Equation (1.10-1) are homogeneous polynomials of the same degree, then the substitution $y = ux$ or $x = vy$ will generate a separable first-order differential equation.

$x^2 - 3xy + y^2$ is an example of a homogeneous polynomial of degree two. $x + y - 1$ is not a homogeneous polynomial.

***Example 1.12-1.*** Solve

$$(x + y)\, dy - (x - y)\, dx = 0 \qquad (1.12\text{-}1)$$

$$\frac{dy}{dx} = \frac{(x - y)}{(x + y)} \qquad (1.12\text{-}2)$$

The substitution $y = ux$ defines $y'$ as

$$\frac{dy}{dx} = \left(\frac{du}{dx}\right) x + (1)\, u \qquad (1.12\text{-}3)$$

We now have

$$x\left(\frac{du}{dx}\right) + u = \frac{x - ux}{x + ux} = \frac{x\,(1 - u)}{x\,(1 + u)} \qquad (1.12\text{-}4,\ 5)$$

$$x\left(\frac{du}{dx}\right) + u = \frac{(1 - u)}{(1 + u)} \qquad (1.12\text{-}6)$$

$$x\left(\frac{du}{dx}\right) = \frac{(1 - u)}{(1 + u)} - u = \frac{1 - 2u - u^2}{1 + u} \qquad (1.12\text{-}7,\ 8)$$

$$\frac{dx}{x} = \frac{(1 + u)}{1 - 2u - u^2}\, du \qquad (1.12\text{-}9)$$

Integration yields

$$\ln x + \ln c = \int \frac{dP}{-2P} \qquad (1.12\text{-}10)$$

where $P = 1 - 2u - u^2$.

Therefore

$$\ln cx = -\frac{1}{2} \ln\left(1 - 2u - u^2\right) \qquad (1.12\text{-}11)$$

$$\ln cx^{-2} = \ln\left(1 - 2u - u^2\right) \qquad (1.12\text{-}12)$$

$$\frac{c}{x^2} = 1 - 2u - u^2 \qquad (1.12\text{-}13)$$

Replacing $u$ by $\frac{y}{x}$, we finally obtain

$$\frac{c}{x^2} = 1 - \frac{2y}{x} - \frac{y^2}{x^2} \tag{1.12-14}$$

or $\quad x^2 - 2xy - y^2 = c \tag{1.12-15}$

●

NOTE: An equation such as

$$\frac{dy}{dx} = \frac{a_1 x + b_1 y + c_1}{a_2 x + b_2 y + c_2} \tag{1.12-16}$$

where $a_1$, $b_1$, $c_1$, $a_2$, $b_2$ and $c_2$ are constants can be reduced to a homogeneous equation if through a change of variables, we manage to do away with the constants $c_1$ and $c_2$.

Think of the numerator and the denominator as representing two intersecting straight lines. If we translate the origin of the coordinate system to their point of intersection, that is to the solution $(\alpha, \beta)$ of the system

$$\begin{cases} a_1 x + b_1 y + c_1 = 0 \\ a_2 x + b_2 y + c_2 = 0 \end{cases} \tag{1.12-17, 18}$$

we then obtain a situation where the directions of the lines are preserved ($a_i$ and $b_i$ remain unchanged) and the coefficients $c_i$ vanish. Since the coordinates of this point of intersection are $(\alpha, \beta)$, we perform the following change of variables

$$\begin{cases} X = x - \alpha \\ Y = y - \beta \end{cases} \tag{1.12-19, 20}$$

Substitutions in Equation (1.12-16) yields

$$\frac{d(Y + \beta)}{d(X + \alpha)} = \frac{dY}{dX} = \frac{a_1(X + \alpha) + b_1(Y + \beta) + c_1}{a_2(X + \alpha) + b_2(Y + \beta) + c_2} \tag{1.12-21a,b}$$

$$\frac{dY}{dX} = \frac{a_1 X + b_1 Y + \underline{a_1 \alpha + b_1 \beta + c_1}}{a_2 X + b_2 Y + \underline{a_2 \alpha + b_2 \beta + c_2}} \tag{1.12-22}$$

where the underlined terms add up to zero by virtue of system (1.12-17,18) with solutions $x = \alpha$ and $y = \beta$. We are left with the homogeneous equation

$$\frac{dY}{dX} = \frac{a_1 X + b_1 Y}{a_2 X + b_2 Y} \qquad (1.12\text{-}23)$$

The solution of such a reducible equation is thus obtained by

i)        solving equation (1.12-23) [a first-order homogeneous equation],

ii)       determining the solution ($\alpha$ and $\beta$) of system (1.12-17,18). This requires the determinant

$$\begin{vmatrix} a_1 & b_1 \\ a_2 & b_2 \end{vmatrix} \neq 0 \quad \text{and}$$

iii)      replacing $X$ and $Y$ in the solution of (1.12-23) by $x - \alpha$ and $y - \beta$ respectively.

***Example 1.12-2.***  Solve

$$\frac{dy}{dx} = \frac{x - y - 3}{x + y - 1} \qquad (1.12\text{-}24)$$

i)        We first solve $\dfrac{dY}{dX} = \dfrac{X - Y}{X + Y}$ using the substitution $Y = uX$. This leads to Equation (1.12-15).

ii)       Next we determine the values $\alpha$ and $\beta$ by solving the system

$$\begin{cases} x - y = 3 \\ x + y = 1 \end{cases} \qquad (1.12\text{-}25, 26)$$

We obtain $\alpha = 2$ and $\beta = -1$.

iii)      We now substitute $X$ and $Y$ in the solution of part (i) by $(x - 2)$ and $(y + 1)$. The final solution is thus given by

$$x^2 - 2xy - y^2 - 6x + 2y = c - 7 = \text{constant} \qquad (1.12\text{-}27)$$

●

A problem arises if the determinant $(a_1 b_2 - b_1 a_2) = 0$; that is, the coefficients of $x$ and $y$ in the linear equations are multiples of one another. That is to say, the two lines are parallel and do not intersect.

Introducing a variable $z = a_1 x + b_1 y$ yields a relation of the form $\dfrac{dz}{dx} = f$ and $f$ is a function of $z$ only, a separable first-order equation.

***Example 1.12-3.*** Solve

$$\frac{dy}{dx} = \frac{2x - 7y + 1}{6x - 21y - 1} \tag{1.12-28}$$

Note that

$$\frac{2}{6} = \frac{-7}{-21} = \frac{1}{3} \tag{1.12-29a,b}$$

Let $z = 2x - 7y$ then

$$\frac{dz}{dx} = 2 - 7\frac{dy}{dx} \tag{1.12-30}$$

$$= 2 - 7\left(\frac{2x - 7y + 1}{6x - 21y - 1}\right) \tag{1.12-31}$$

$$= 2 - 7\left(\frac{z + 1}{3z - 1}\right) \tag{1.12-32}$$

or $\quad \dfrac{dz}{dx} = -\dfrac{z + 9}{3z - 1}$ \hfill (1.12-33)

This equation can be solved by separating the variables to yield

$$3z - 28 \ln(z+9) = -x + c' \tag{1.12-34}$$

where $c'$ is a constant.

Replacing $z$ by $2x - 7y$ yields the solution to the problem

$$6x - 21y - 28 \ln(2x - 7y + 9) = -x + c' \tag{1.12-35}$$

or $\quad 7x - 21y - 28 \ln(2x - 7y + 9) = c'$ \hfill (1.12-36)

or $\quad x - 3y - 4 \ln(2x - 7y + 9) = c$ \hfill (1.12-37)

## 1.13  TOTAL OR EXACT FIRST-ORDER DIFFERENTIAL EQUATIONS

A given differential equation, could have been obtained by differentiating an implicit function. For example, one can determine by inspection that the equation

$$x \, dy + y \, dx = 0 \tag{1.13-1}$$

results from expanding

$$d(x\,y) = 0 \tag{1.13-2}$$

Integration yields

$$x\,y = \text{constant} \tag{1.13-3}$$

or     $y = \dfrac{c}{x}$ (1.13-4)

An equation such as

$$x^2\,dy + 2\,x\,dx = 0 \tag{1.13-5}$$

can be solved by inspection, after multiplication by an appropriate "integrating" factor.  In this particular case, multiplication by $e^y$ results in the following **total differential equation**

$$d(x^2\,e^y) = 0 \tag{1.13-6}$$

thus     $x^2\,e^y = c$ (1.13-7)

and     $y = \ell n\,\dfrac{c}{x^2} = -2\,\ell n\,cx$ (1.13-8, 9)

The following test allows one to determine if the equation

$$M\,dx + N\,dy = 0 \tag{1.10-1}$$

is a total (or exact) differential equation.  We suspect that the equation is exact.  That is, Equation (1.10-1) could be represented by $dF(x, y) = 0$.  This can be written as

$$\frac{\partial F}{\partial x}\,dx + \frac{\partial F}{\partial y}\,dy = 0 \tag{1.13-10}$$

Therefore

$$M = \frac{\partial F}{\partial x} \quad \text{and} \quad N = \frac{\partial F}{\partial y} \tag{1.13-11, 12}$$

So M and N are partial derivatives of the same function F. Furthermore, assuming that F and its partial derivatives, of at least order two, are continuous in the region of interest, we note that

$$\frac{\partial^2 F}{\partial x \partial y} = \frac{\partial^2 F}{\partial y \partial x} \tag{1.13-13}$$

That is to say

$$\frac{\partial M}{\partial y} = \frac{\partial N}{\partial x} \tag{1.13-14}$$

If our starting equation satisfies relation (1.13-14), we know that we are dealing with the total derivative of a function of two variables. Equating that function to a constant yields the solution $F(x, y) = \text{constant}$. One way of determining $F$ is as follows: since $\frac{\partial F}{\partial x} = M$, a "partial" integration with respect to $x$ (keep $y$ constant) yields $F$. The "constant" of integration will in general be an arbitrary function of $y$, which will disappear on differentiating with respect to $x$. That is to say

$$F = \int M \, dx + f(y) \tag{1.13-15}$$

$f(y)$ can then be determined from $\frac{\partial F}{\partial y} = N$, as illustrated next.

**Example 1.13-1.** Solve

$$(3x^2 - 6xy + 4\cos y) \, dx + (2y - 3x^2 - 4x\sin y - \frac{1}{y}) \, dy = 0 \tag{1.13-16}$$

$$\frac{\partial M}{\partial y} = -6x - 4\sin y \tag{1.13-17}$$

$$\frac{\partial N}{\partial x} = -6x - 4\sin y \tag{1.13-18}$$

Therefore

$$\frac{\partial F}{\partial x} = 3x^2 - 6xy + 4\cos y \tag{1.13-19}$$

and    $$F = \int (3x^2 - 6xy + 4\cos y) \, dx = x^3 - 3x^2y + 4x\cos y + f(y) \tag{1.13-20}$$

$f(y)$ is now determined by substitution of the previous equation for $F$ in $\frac{\partial F}{\partial y} = N$.

That is

$$\frac{\partial}{\partial y}\left[x^3 - 3x^2y + 4x\cos y + f(y)\right] = 2y - 3x^2 - 4x\sin y - \frac{1}{y} \tag{1.13-21}$$

or    $$-3x^2 - 4x\sin y + f'(y) = 2y - 3x^2 - 4x\sin y - \frac{1}{y} \tag{1.13-22}$$

$$f'(y) = 2y - \frac{1}{y} \tag{1.13-23}$$

$$f(y) = y^2 - \ell n\, y + c \tag{1.13-24}$$

Hence, the solution is

$$x^3 - 3\,x^2 y + 4\,x \cos y + y^2 - \ell n\, y = \text{constant} \tag{1.13-25}$$

●

If Equation (1.10-1) is not exact, we can try to make it exact by multiplying with an **integrating factor** I. Equation (1.10-1) becomes

$$IMdx + INdy = 0 \tag{1.13-26}$$

which is exact if

$$\frac{\partial}{\partial y}\,(IM) = \frac{\partial}{\partial x}\,(IN) \tag{1.13-27}$$

That is

$$I\frac{\partial M}{\partial y} + M\frac{\partial I}{\partial y} = I\frac{\partial N}{\partial x} + N\frac{\partial I}{\partial x} \tag{1.13-28}$$

or

$$\frac{1}{I}\left(N\frac{\partial I}{\partial x} - M\frac{\partial I}{\partial y}\right) = \frac{\partial M}{\partial y} - \frac{\partial N}{\partial x} \tag{1.13-29}$$

In general it is not easy to find I but there are special cases when there is a standard procedure to obtain I.

(i)     I is a function of x only. Then $\dfrac{\partial I}{\partial y} = 0$, and Equation (1.13-29) becomes

$$\frac{1}{I}\frac{dI}{dx} = \frac{\partial M/\partial y - \partial N/\partial x}{N} \tag{1.13-30}$$

The assumption that I is a function of x only implies that the right side of Equation (1.13-30) is also a function of x only. Therefore, the integrating factor I can be determined and Equation (1.13-26) is exact and can be solved.

(ii)    I is a function of $y$ only. Then as in (i), we have

$$\frac{1}{I}\frac{dI}{dy} = \frac{\partial N/\partial x - \partial M/\partial y}{M} \tag{1.13-31}$$

The right side is a function of $y$ only and allows $I$ to be determined, leading to a solution.

***Example 1.13-2.***  Solve

$$(3x^2 - y^2)\,dy - 2xy\,dx = 0 \tag{1.13-32}$$

by finding an appropriate integrating factor.

$$\frac{\partial M}{\partial y} = -2x \tag{1.13-33}$$

$$\frac{\partial N}{\partial x} = 6x \tag{1.13-34}$$

In this case the equation is not exact, we note that

$$\left(\frac{\partial N}{\partial x} - \frac{\partial M}{\partial y}\right) / M = \frac{8x}{-2xy} = -\frac{4}{y} \tag{1.13-35a,b}$$

is a function of $y$ only.

From Equation (1.13-31), we obtain

$$\frac{1}{I}\frac{dI}{dy} = -\frac{4}{y} \quad \Rightarrow \quad \frac{dI}{I} = -\frac{4}{y}\,dy \tag{1.13-36a,b}$$

$$I = y^{-4} \tag{1.13-37}$$

Multiplying Equation (1.13-32) by the integrating factor $y^{-4}$, we have

$$(3x^2y^{-4} - y^{-2})\,dy - 2xy^{-3}\,dx = 0 \tag{1.13-38}$$

Equation (1.13-38) is exact, we can proceed as in Example 1.13-1.

$$\frac{\partial F}{\partial x} = -2xy^{-3} \tag{1.13-39}$$

$$F = -x^2y^{-3} + f(y) \tag{1.13-40}$$

$$\frac{\partial F}{\partial y} = 3x^2 y^{-4} + \frac{df}{dy} = 3x^2 y^{-4} - y^{-2} \qquad (1.13\text{-}41a,b)$$

Therefore

$$\frac{df}{dy} = -y^{-2} \qquad (1.13\text{-}42)$$

$$f = y^{-1} + c \qquad (1.13\text{-}43)$$

The solution is

$$F = \text{constant} \quad \Rightarrow \quad -x^2 y^{-3} + y^{-1} = c \qquad (1.13\text{-}44a,b)$$

## 1.14  LINEAR FIRST-ORDER DIFFERENTIAL EQUATIONS

The standard form is given by

$$\frac{dy}{dx} + P(x)y = Q(x) \qquad (1.14\text{-}1)$$

Note that $P$ and $Q$ are not functions of $y$. The integrating factor $I(x)$ is given by

$$I(x) = \exp \int P(x)\, dx \qquad (1.14\text{-}2)$$

Multiplying both sides of Equation (1.14-1) by $I(x)$ yields a left side which is equal to $\dfrac{d(Iy)}{dx}$.
Direct integration produces the solution as follows

$$\frac{d}{dx}[yI] = Q(x)I(x) \qquad (1.14\text{-}3)$$

and

$$y = I^{-1}\left[\int^x Q(\xi)\, I(\xi)\, d\xi + c\right] \qquad (1.14\text{-}4)$$

***Example 1.14-1***.  Solve

$$y' - 2xy = x \qquad (1.14\text{-}5)$$

$$I(x) = \exp \int -2x\, dx = \exp -\frac{2x^2}{2} \qquad (1.14\text{-}6, 7)$$

Multiplying both sides of the equation by $e^{-x^2}$ yields

$$e^{-x^2}y' - 2xe^{-x^2}y = xe^{-x^2} \tag{1.14-8}$$

or $\qquad \dfrac{d}{dx}(ye^{-x^2}) = xe^{-x^2} \tag{1.14-9}$

Integrating yields

$$\int \frac{d}{dx}(ye^{-x^2})\,dx = \int xe^{-x^2}\,dx \tag{1.14-10}$$

$$ye^{-x^2} = -\frac{1}{2}e^{-x^2} + c \tag{1.14-11}$$

and finally

$$y = ce^{x^2} - \frac{1}{2}$$

●

Another procedure which will generate a solution for Equation (1.14-1) is as follows: first one solves the homogeneous equation

$$y' + P(x)\,y = 0 \tag{1.14-13}$$

to yield the homogeneous solution $y_h$. We then propose the solution of Equation (1.14-1) to be of the form of $y_h$, where we replace the constant of integration by a function of $x$, as illustrated in Example 1.14-3. This method is known as the **method of variation of parameters**.

***Example 1.14-2.*** The rate equations for components A, B and C involved in the following first-order reactions

$$A \xrightarrow{k_1} B \xrightarrow{k_2} C$$

are written as

$$\frac{dc_A}{dt} = -k_1 c_A \tag{1.14-14}$$

$$\frac{dc_B}{dt} = k_1 c_A - k_2 c_B \tag{1.14-15}$$

$$\frac{dc_C}{dt} = k_2 c_B \qquad (1.14\text{-}16)$$

We wish to solve for the time evolution of the concentration $c_B$, given that at $t = 0$, $c_A = c_{A_0}$ and $c_B = c_C = 0$.

Equation (1.14-14) has been treated in Example 1.11-2 and yields

$$c_A = c_{A_0} e^{-k_1 t} \qquad (1.14\text{-}17)$$

Combining Equations (1.14-15, 17) leads to

$$\frac{dc_B}{dt} + k_2 c_B = k_1 c_{A_0} e^{-k_1 t} \qquad (1.14\text{-}18)$$

which is of the form of Equation (1.14-1) with $P(t) = k_2$ and $Q(t) = k_1 c_{A_0} e^{-k_1 t}$.

Note that the independent variable is now $t$. The integrating factor $I(t) = e^{k_2 t}$.

Multiplying Equation (1.14-18) by $I(t)$ leads to

$$\frac{d}{dt}\left(c_B e^{k_2 t}\right) = k_1 c_{A_0} e^{(k_2 - k_1) t} \qquad (1.14\text{-}19)$$

Integration yields

$$c_B e^{k_2 t} = \frac{k_1 c_{A_0} e^{(k_2 - k_1) t}}{k_2 - k_1} + c \qquad (1.14\text{-}20)$$

The constant $c$ is evaluated from the condition $c_B = 0$ at $t = 0$. That is

$$0 = \frac{k_1 c_{A_0}}{k_2 - k_1} + c \qquad (1.14\text{-}21)$$

Finally we obtain

$$c_B = k_1 c_{A_0} \left(\frac{e^{-k_1 t} - e^{-k_2 t}}{k_2 - k_1}\right) \qquad (1.14\text{-}22)$$

**Example 1.14-3.** Solve

$$y' - \frac{2y}{x + 1} = (x + 1)^3 \qquad (1.14\text{-}23)$$

The homogeneous equation

$$y' - \frac{2y}{x+1} = 0 \tag{1.14-24}$$

or $\quad \dfrac{y'}{y} = \dfrac{2}{x+1}$           (1.14-25)

results in the solution

$$y_h = c(x+1)^2 \tag{1.14-26}$$

We now propose the solution of the given problem to be of the form $y = u(x+1)^2$ where $u$ is a function of $x$ which is to be determined. Substitution of this solution, in the given problem results in the following expression

$$u'(x+1)^2 + 2u(x+1) - \frac{2u(x+1)^2}{(x+1)} = (x+1)^3 \tag{1.14-27}$$

Note that since $(x+1)^2$ is the homogeneous solution, the terms in $u$ have to cancel.

We then obtain

$$u' = x + 1 \tag{1.14-28}$$

and on integrating, we can write $u$ as

$$u = \frac{1}{2}(x+1)^2 + c \tag{1.14-29}$$

The solution to the problem is thus given by

$$y = \left[ \frac{1}{2}(x+1)^2 + c \right](x+1)^2 \tag{1.14-30}$$

or $\quad y = c(x+1)^2 + \dfrac{1}{2}(x+1)^4$       (1.14-31)

The second term on the right side (that is choosing $c = 0$) is a **particular solution** $y_p$. The general solution $y = y_h + y_p$.

## 1.15 BERNOULLI'S EQUATION

The standard form of this equation is given by

$$y' + P(x)\, y = Q(x)\, y^n \tag{1.15-1}$$

This equation can be reduced to the form of a linear first-order equation through the following substitution

$$y^{1-n} = v \qquad (1.15-2)$$

First one divides both sides of the equation by $y^n$. This yields the following equation

$$y^{-n} \frac{dy}{dx} + P(x) y^{1-n} = Q(x) \qquad (1.15-3)$$

Then let $v = y^{1-n}$. This means that

$$\frac{dv}{dx} = \frac{d}{dx} (y^{1-n}) = (1-n) y^{-n} \frac{dy}{dx} \qquad (1.15-4, 5)$$

Equation (1.15-3) reduces now to the following linear first-order equation

$$\left(\frac{1}{1-n}\right) \frac{dv}{dx} + P(x)v = Q(x) \qquad (1.15-6)$$

or

$$\frac{dv}{dx} + (1-n) P(x) v = (1-n) Q(x) \qquad (1.15-7)$$

which is linear in $v$.

***Example 1.15-1.*** Solve

$$\frac{dy}{dx} - y = x y^5 \qquad (1.15-8)$$

Dividing by $y^5$ yields

$$y^{-5} \frac{dy}{dx} - y^{-4} = x \qquad (1.15-9)$$

Let

$$v = y^{-4} \quad \Rightarrow \quad \frac{dv}{dx} = -4y^{-5} \frac{dy}{dx} \qquad (1.15-10, 11)$$

Combining Equations (1.15-9 to 11), we obtain

$$-\frac{1}{4} \frac{dv}{dx} - v = x \quad \text{or} \quad \frac{dv}{dx} + 4v = -4x \qquad (1.15-12, 13)$$

This equation can now be solved with the following integrating factor

$$I(x) = e^{4 \int dx} = e^{4x} \qquad (1.15-14, 15)$$

to yield

$$v e^{4x} = -xe^{4x} + \frac{1}{4} e^{4x} + c \qquad (1.15\text{-}16)$$

Substituting $y^{-4}$ for $v$ produces

$$y^{-4} e^{4x} = -x e^{4x} + \frac{1}{4} e^{4x} + c \qquad (1.15\text{-}17)$$

and finally

$$\frac{1}{y^4} = -x + \frac{1}{4} + c e^{-4x} \qquad (1.15\text{-}18)$$

## 1.16 SECOND-ORDER LINEAR DIFFERENTIAL EQUATIONS WITH CONSTANT COEFFICIENTS

The standard form is given by

$$y'' + A y' + B y = Q(x) \qquad (1.16\text{-}1)$$

where $A$ and $B$ are constants.

An alternative form of Equation (1.16-1) is

$$L(y) = Q(x) \qquad (1.16\text{-}2)$$

where the **linear differential operator** $L(y)$ is defined by

$$L(y) = y'' + A y' + B y \qquad (1.16\text{-}3)$$

The **homogeneous** differential equation is

$$L(y) = 0 \qquad (1.16\text{-}4)$$

If $y_1$ and $y_2$ are two linearly independent solutions of the homogeneous equation $L(y) = 0$, then by the principle of superposition, $c_1 y_1 + c_2 y_2$ is also a solution where $c_1$ and $c_2$ are constants. That is

$$L(c_1 y_1 + c_2 y_2) = c_1 L(y_1) + c_2 L(y_2) = 0 \qquad (1.16\text{-}5a,b)$$

As mentioned earlier, the general solution to $L(y) = Q(x)$ is given by

$$y = y_h + y_p \qquad (1.16\text{-}6)$$

Solving the homogeneous equation

$$y'' + Ay' + By = 0 \qquad\qquad (1.16\text{-}7)$$

will generate $y_h$.

We propose $y_h$ to be of the form $e^{\alpha x}$. Substitution of this function and its derivatives into the homogeneous equation yields

$$e^{\alpha x}(\alpha^2 + A\alpha + B) = 0 \qquad\qquad (1.16\text{-}8)$$

Since $e^{\alpha x} \neq 0$, we have that

$$\alpha^2 + A\alpha + B = 0 \qquad\qquad (1.16\text{-}9)$$

This is known as the **characteristic or auxiliary equation**. This equation has two solutions, $\alpha_1$ and $\alpha_2$. Since we are dealing with a second order equation, we also have two constants of integration and the solution $y_h$ is given by

$$y_h = c_1 e^{\alpha_1 x} + c_2 e^{\alpha_2 x} \qquad\qquad (1.16\text{-}10)$$

The second order characteristic equation could have

i)        two real and distinct roots (the case where $D^2 = A^2 - 4B > 0$).

          In this case, $y_h$ is given by equation (1.16-10).

ii)       two equal real roots (the case where $D = 0$).

          In this case, $y_h$ is  given by

$$y_h = c_1 e^{\alpha_1 x} + c_2 x e^{\alpha_1 x} \qquad\qquad (1.16\text{-}11)$$

iii)      two complex conjugate roots (the case where $D^2 < 0$)

               $y_h$ is now given by

$$y_h = c_1 e^{(a+ib)x} + c_2 e^{(a-ib)x} \qquad\qquad (1.16\text{-}12)$$

where $\alpha_{1,2} = -\dfrac{A}{2} \pm i\,\dfrac{\sqrt{4B - A^2}}{2} = a \pm ib$

This complex solution can be transformed into a real one, using the Euler formula to yield

$$y_h = e^{ax} (c_3 \cos bx + c_4 \sin bx) \tag{1.16-13}$$

where $c_3$ and $c_4$ are constants and real. $\tag{1.16-14, 15}$

Next we have to determine the particular solution $y_p$. This can be done by proposing a solution, based on the form of $Q(x)$, the right side of the equation. Substitution of the proposed form and of its derivatives into the equation to be solved followed by equating the coefficients of the like terms allows one to determine the values of the introduced constants as illustrated next. This solution technique is referred to as the **method of undetermined coefficients**.

***Example 1.16-1.*** Solve

$$y'' - 3y' + 2y = x^2 + 1 \tag{1.16-16}$$

The characteristic equation

$$\alpha^2 - 3\alpha + 2 = 0 \tag{1.16-17}$$

has the following roots: $\alpha_1 = 1$ and $\alpha_2 = 2$ so that

$$y_h = c_1 e^x + c_2 e^{2x} \tag{1.16-18}$$

Since the right side of the equation is of order two, we propose for $y_p$ the quadratic polynomial $ax^2 + bx + c$, and proceed via substitution in the given equation to determine the values of the constants a, b and c.

$$y_p = ax^2 + bx + c \tag{1.16-19}$$

$$y_p' = 2ax + b \tag{1.16-20}$$

$$y_p'' = 2a \tag{1.16-21}$$

Substitution yields

$$2a - 3 (2ax + b) + 2 (ax^2 + bx + c) = x^2 + 1 \tag{1.16-22}$$

$$2a - 6ax - 3b + 2ax^2 + 2bx + 2c = x^2 + 1 \tag{1.16-23}$$

$$2ax^2 + x (2b - 6a) + 2a - 3b + 2c = x^2 + 1 \tag{1.16-24}$$

Equating the coefficients of $x^2$ yields

$$2a = 1 \quad \text{and} \quad a = \frac{1}{2} \tag{1.16-25a,b}$$

Equating the coefficients of $x$ yields

$$2b - 6a = 0 \qquad\qquad (1.16\text{-}26)$$

$$2b = 6a = 3 \quad \text{and} \quad b = \frac{3}{2} \qquad (1.16\text{-}27\text{a,b, } 28)$$

Equating the coefficients of $x^0$ yields

$$2a - 3b + 2c = 1 \qquad\qquad (1.16\text{-}29)$$

$$1 - \frac{9}{2} + 2c = 1 \quad \text{or} \quad c = \frac{9}{4} \qquad (1.16\text{-}30\text{a,b})$$

$y_p$ is thus given by

$$y_p = \frac{x^2}{2} + \frac{3}{2} x + \frac{9}{4} \qquad\qquad (1.16\text{-}31)$$

and the solution to the problem is

$$y = y_h + y_p = c_1 e^x + c_2 e^{2x} + \frac{1}{4} (2x^2 + 6x + 9) \qquad (1.16\text{-}32)$$


●

The form of the $y_p$ to be chosen depends on $Q(x)$ and on the homogeneous solution $y_h$, which in turn depends on $L(y)$. Examples are

(a)     $Q(x)$ is a polynomial.

$$Q(x) = q_0 + q_1 x + q_2 x^2 + .... + q_n x^n \qquad (1.16\text{-}33)$$

Try

$$y_p = a_0 + a_1 x + .... + a_n x^n \qquad\qquad (1.16\text{-}34)$$

if $A \neq 0$, $B \neq 0$.

If $B = 0$, $A \neq 0$, try

$$y_p = x (a_0 + a_1 x + .... + a_n x^n) \qquad (1.16\text{-}35)$$

If both $B$ and $A$ are equal to zero, then the solution can be obtained by direct integration.

(b)     $Q(x) = \sin qx$ or $\cos qx$ or a combination of both. $\qquad (1.16\text{-}36)$

In this case an appropriate $y_p$ is

$$y_p = a_0 \cos qx + b_0 \sin qx \qquad (1.16\text{-}37)$$

If $\cos qx$ (or/and $\sin qx$) is not a solution of the homogeneous equation. If $\cos qx$ (or/and $\sin qx$) is present in $y_h$ then we need to try

$$y_p = x (a_0 \cos qx + b_0 \sin qx) \qquad (1.16\text{-}38)$$

(c)    $Q(x) = e^{qx}.$ \qquad (1.16\text{-}39)

The particular integral will be of the form

$$y_p = a_0 e^{qx} \qquad (1.16\text{-}40)$$

if $e^{qx}$ is not a term in $y_h$.

If $e^{qx}$ occurs in $y_h$, then $y_p$ is

$$y_p = x (a_0 e^{qx}) \qquad (1.16\text{-}41)$$

If $xe^{qx}$ is also part of the homogeneous solution, then

$$y_p = x^2 (a_0 e^{qx}) \qquad (1.16\text{-}42)$$

We can also determine the particular solution for a combination of the cases examined in (a to c) by intelligent guess work. We need to find the number of undetermined coefficients that can be determined to fit the differential equation. The particular solution can also be obtained by the method of variation of parameters (see Examples 1.14-3, 18-1).

*Example 1.16-2.* Levenspiel (1972) describes a flow problem with diffusion, involving a first-order chemical reaction. The reaction is characterized by the rate expression $\dfrac{dc_A}{dt} = -kc_A$.

The average velocity in the tubular reactor is $\langle v_z \rangle$. The concentration of A is given at the inlet $(z = 0)$ of the reactor by $c_{A_0}$ and at the outlet $(z = L)$ by $c_{AL}$.

A mass balance on A leads to the following differential equation which we wish to solve

$$-\mathcal{D}_{AB} \frac{d^2 c_A}{dz^2} + \langle v_z \rangle \frac{dc_A}{dz} + kc_A = 0 \qquad (1.16\text{-}43)$$

We can write Equation (1.16-43) in standard form as follows

$$\frac{d^2 c_A}{dz^2} - \frac{\langle v_z \rangle}{\mathcal{D}_{AB}} \frac{dc_A}{dz} - \frac{k}{\mathcal{D}_{AB}} c_A = 0 \tag{1.16-44}$$

We now propose a solution of the form $c_A = e^{\alpha z}$. As before, substitution leads to the characteristic (or auxiliary) equation

$$\alpha^2 - \frac{\langle v_z \rangle}{\mathcal{D}_{AB}} \alpha - \frac{k}{\mathcal{D}_{AB}} = 0 \tag{1.16-45}$$

The result in terms of the concentration $c_A$ is given by

$$c_A = c_1 e^{\alpha_1 z} + c_2 e^{\alpha_2 z} \tag{1.16-46}$$

where $\alpha_{1,2} = \dfrac{\langle v_z \rangle}{2 \mathcal{D}_{AB}} \left[ 1 \pm \sqrt{1 + \dfrac{4k \mathcal{D}_{AB}}{\langle v_z \rangle^2}} \right]$

The constants $c_1$ and $c_2$ are evaluated from the conditions

$$c_A = c_{A_0} \qquad \text{at } z = 0 \tag{1.16-47}$$

$$c_A = c_{AL} \qquad \text{at } z = L \tag{1.16-48}$$

This leads to the following final result

$$c_A = \left[ c_{A_0} - \left( \frac{c_{AL} - c_{A_0} e^{\alpha_1 L}}{e^{\alpha_2 L} - e^{\alpha_1 L}} \right) \right] e^{\alpha_1 z} + \frac{c_{AL} - c_{A_0} e^{\alpha_1 L}}{e^{\alpha_2 L} - e^{\alpha_1 L}} e^{\alpha_2 z} \tag{1.16-49}$$

## 1.17  SOLUTIONS BY LAPLACE TRANSFORM

Linear differential equations can sometimes be reduced to algebraic equations, which are easier to solve. A way of achieving this is by performing a so called Laplace transform.

The **Laplace transform** $L[f(t)]$ of a function $f(t)$ is defined as

$$L[f(t)] = F(s) = \int_0^\infty f(t) e^{-st} dt \tag{1.17-1}$$

where the integral is assumed to converge.

The Laplace transform is linear. That is to say

$$L[c_1 f_1(t) + c_2 f_2(t) + ....] = c_1 L[f_1(t)] + c_2 L[f_2(t)] + ....$$                    (1.17-2)

where $c_1$, $c_2$ are constants.

For a given differential equation, the procedure to follow is

i)    take the Laplace transform of the equation,

ii)   solve the resulting algebraic equation; that is to say, obtain $F(s)$,

iii)  invert the transform, to obtain the solution $f(t)$. This is usually done by consulting tables of Laplace transforms.

The Laplace transform of a variety of functions are tabulated in most mathematical tables. Table 1.17-1 gives some useful transforms. Next, we state some theorems without proof.

i)    **Initial-value theorem**

$$\lim_{t \to 0} f(t) = \lim_{s \to \infty} s F(s)$$                    (1.17-3)

ii)   **Final-value theorem**

$$\lim_{t \to \infty} f(t) = \lim_{s \to 0} s F(s)$$                    (1.17-4)

iii)  **Translation of a function**

$$L[e^{-at} f(t)] = F(s + a)$$                    (1.17-5)

iv)   **Derivatives of transforms**

$$L[t^n f(t)] = (-1)^n \frac{d^n F(s)}{ds^n}$$                    (1.17-6)

v)    **Convolution**

$$E(s) = F(s) G(s)$$

$$= L\left[ \int_0^t f(t-u) g(u) du \right]$$                    (1.17-7)

Convolution is used when $E(s)$ does not represent the Laplace transform of a known function but is the product of the Laplace transform of two functions whose transforms are known.

## TABLE  1.17-1

## Laplace Transforms

| Function | Transform |
|---|---|
| 1 | $\dfrac{1}{s}$ |
| $t^n$ | $\dfrac{n!}{s^{n+1}}, \quad n = 1, 2, \ldots$ |
| $e^{at}$ | $\dfrac{1}{s-a}$ |
| $\sin at$ | $\dfrac{a}{s^2 + a^2}$ |
| $\cos at$ | $\dfrac{s}{s^2 + a^2}$ |
| $\dfrac{d^n f(t)}{dt^n}$ | $s^n F(s) - \displaystyle\sum_{k=1}^{n} s^{n-k} \dfrac{d^{k-1} f(0)}{dt^{k-1}}$ |
| $\displaystyle\int_0^t f(t)\, dt$ | $\dfrac{F(s)}{s}$ |
| $\dfrac{1}{\sqrt{t}} \exp(-x^2/4t)$ | $\sqrt{\left(\dfrac{\pi}{s}\right)} \exp(-x\sqrt{s})$ |
| $\dfrac{x}{2\, t^{3/2}} \exp(-x^2/4t)$ | $\sqrt{\pi} \exp(-x\sqrt{s})$ |
| $\operatorname{erfc}(x/2\sqrt{t})$ | $\dfrac{1}{s} \exp(-x\sqrt{s})$ |
| $\left(t + \dfrac{x^2}{2}\right) \operatorname{erfc}(x/2\sqrt{t}) - x\sqrt{\dfrac{t}{\pi}} \exp(-x^2/4t)$ | $\dfrac{e^{-x\sqrt{s}}}{s^2}$ |
| $\dfrac{1}{2} e^{at} \left\{ e^{-x\sqrt{a}} \operatorname{erfc}\left[\dfrac{x}{2\sqrt{t}} - \sqrt{(at)}\right] + e^{x\sqrt{a}} \operatorname{erfc}\left[\dfrac{x}{2\sqrt{t}} + \sqrt{(at)}\right] \right\}$ | $\dfrac{e^{-x\sqrt{s}}}{s - \alpha}$ |
| $\begin{cases} 0, & 0 < t < k \\ (t-k)^{\mu-1}/\Gamma(\mu), & t > k \end{cases}$ | $e^{-ks}/s^{\mu}, \quad \mu > 0$ |
| $a^n J_n(at)$ | $\dfrac{(\sqrt{s^2 + a^2} - s)^n}{\sqrt{s^2 + a^2}}, \quad n > -1$ |

## Heaviside Step Function and Dirac Delta Function

The Heaviside step function is a function which is equal to zero for $-\infty \le t < 0$ and is equal to 1 for all positive $t$. This can be expressed as

$$H(t) = \begin{cases} 0, & -\infty \le t < 0 \\ \\ 1, & t > 0 \end{cases} \qquad (1.17\text{-}8a,b)$$

$H(t)$ is discontinuous at $t = 0$.

The Laplace transform of $H(t)$ is given by

$$L[H(t)] = \int_0^\infty e^{-st} H(t)\, dt \qquad (1.17\text{-}9a)$$

$$= \left[ -\frac{1}{s} e^{-st} \right]_0^\infty \qquad (1.17\text{-}9b)$$

$$= \frac{1}{s} \qquad (1.17\text{-}9c)$$

We can generalize the Heaviside step function for the case where the discontinuity occurs at $t = a$. In this case, we have

$$H(t-a) = \begin{cases} 0, & t < a \\ \\ 1, & t > a \end{cases} \qquad (1.17\text{-}10a,b)$$

$$L[H(t-a)] = \int_0^\infty e^{-st} H(t-a)\, dt \qquad (1.17\text{-}11a)$$

$$= \int_a^\infty e^{-st}\, dt \qquad (1.17\text{-}11b)$$

$$= \left[ -\frac{1}{s} e^{-st} \right]_a^\infty \qquad (1.17\text{-}11c)$$

$$= \frac{1}{s} e^{-sa} \qquad (1.17\text{-}11d)$$

An arbitrary function $f(t)$ can be shifted over a distance $a$, by multiplying $f(t)$ by $H(t-a)$. This will result in the quantity $f(t-a)\, H(t-a)$. Figure 1.17-1 illustrates this translation graphically.

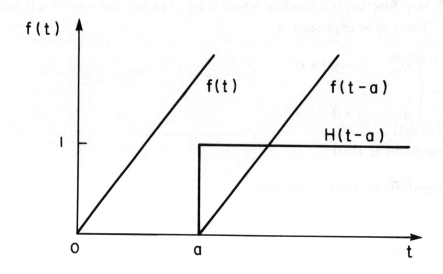

**FIGURE 1.17-1    Translation of a function   f(t)   over a distance  a**

Taking the Laplace transform, we obtain

$$L\left[f(t-a)\,H(t-a)\right] = \int_0^\infty e^{-st}\,f(t-a)\,H(t-a)\,dt \qquad (1.17\text{-}12a)$$

$$= \int_a^\infty e^{-st}\,f(t-a)\,dt \qquad (1.17\text{-}12b)$$

$$= \int_0^\infty e^{-s(t'+a)}\,f(t')\,dt' \qquad (1.17\text{-}12c)$$

$$= e^{-sa}\int_0^\infty e^{-st'}\,f(t')\,dt' \qquad (1.17\text{-}12d)$$

$$= e^{-sa}\,F(s) \qquad (1.17\text{-}12e)$$

where  F  is the Laplace transform of  f.

In Equation (1.17-12c) we have used the transformation

$$t' = t-a \qquad (1.17\text{-}13)$$

The Dirac delta function is defined as

$$\delta(t) = 0, \quad \text{everywhere except at } t = 0 \tag{1.17-14a}$$

$$\int_{-\infty}^{\infty} \delta(t)\, dt = 1 \tag{1.17-14b}$$

$$\int_{-\infty}^{\infty} f(t)\, \delta(t)\, dt = f(0) \tag{1.17-14c}$$

The Laplace transform of $\delta(t)$ can be obtained using Equation (1.17-14c).

$$L\big[\delta(t)\big] = \int_{0}^{\infty} e^{-st}\, \delta(t)\, dt \tag{1.17-15a}$$

$$= \int_{-\infty}^{\infty} e^{-st}\, \delta(t)\, dt \tag{1.17-15b}$$

$$= 1 \tag{1.17-15c}$$

The limits in Equation (1.17-14b, c) need not to be $(-\infty, \infty)$, they could be $(-\varepsilon_1, \varepsilon_2)$ for any positive $\varepsilon_1$, $\varepsilon_2$.

Thus

$$\int_{-\infty}^{x} \delta(t)\, dt = 0, \quad \text{if } x < 0 \tag{1.17-16a}$$

$$= 1, \quad \text{if } x > 0 \tag{1.17-16b}$$

$$= H(x) \tag{1.17-16c}$$

where $H(x)$ is the Heaviside step function.

By formally differentiating both sides of Equation (1.17-16c) we obtain, via Equation (1.3-1)

$$\delta(x) = \frac{dH}{dx} \tag{1.17-17}$$

Equation (1.17-17) indicates that $\delta(x)$ is not an ordinary function, because $H(x)$ is not continuous at $x = 0$ and therefore does not have a derivative at $x = 0$. The derivative $\frac{dH}{dx}$ is zero in any interval that does not include the origin. For formal computational purposes, we may regard the derivative of the Heaviside step function to be the Dirac delta function.

If the discontinuity is at $t = a$, then

$$\delta\,(t - a) = 0\,, \quad \text{everywhere except at } t = a \tag{1.17-18a}$$

$$\int_{-\infty}^{\infty} \delta(t - a)\,dt = 1 \tag{1.17-18b}$$

$$\int_{-\infty}^{\infty} f(t)\,\delta(t - a)\,dt = \int_{a-\varepsilon_1}^{a+\varepsilon_2} f(t)\,\delta(t - a)\,dt = f(a) \tag{1.17-18c,d}$$

$$L\left[\delta(t - a)\right] = \int_{0}^{\infty} e^{-st}\,\delta(t - a)\,dt = e^{-sa} \tag{1.17-18e,f}$$

$$\delta(t - a) = \frac{dH}{dt}\,(t - a) \tag{1.17-18g}$$

We may interpret $\delta\,(t - a)$ to be an impulse at $t = a$ and this will be considered in the next section. We will illustrate the usefulness of the Heaviside and Dirac functions in Example 1.17-4 and in Problems 35a and 36b.

*Example 1.17-1*. Compute the Laplace transform of sin (at).

Applying the definition, we write

$$L\left[\sin\,(at)\right] = \int_{0}^{\infty} e^{-st}\,\sin\,at\;dt \tag{1.17-19}$$

Referring to Equation (1.3-9) we observe that the integral on the right side is given by

$$\frac{e^{-st}}{s^2 + a^2}\,\left[-s\,\sin\,at - a\,\cos\,at\right]\Bigg|_{0}^{\infty}$$

Evaluation of this term for $t = \infty$ and $t = 0$ yields the solution $\dfrac{a}{s^2 + a^2}$ as given in Table 1.17-1.

*Example 1.17-2*. Solve the following second order differential equation

$$\frac{d^2h}{dt^2} + \frac{a}{\lambda}\frac{dh}{dt} + \frac{1}{\lambda^2}h = f(t) \tag{1.17-20}$$

subject to the conditions $h = 0$ at $t = 0$ and $\frac{dh}{dt} = 0$ at $t = 0$.

Using Table 1.17-1, we determine the following Laplace transform

$$\frac{d^2h}{dt^2} \rightarrow s^2H(s) - sh(0) - h'(0) = s^2H(s) \tag{1.17-21a,b}$$

$$\frac{dh}{dt} \rightarrow sH(s) - h(0) = sH(s) \tag{1.17-21c,d}$$

$$h \rightarrow H(s) \tag{1.17-22a}$$

$$f(t) \rightarrow F(s) \tag{1.17-22b}$$

Substituting into Equation (1.17-20) yields the following algebraic equation

$$s^2H(s) + \frac{as}{\lambda}H(s) + \frac{H(s)}{\lambda^2} = F(s) \tag{1.17-23}$$

Solving for $H(s)$, we obtain

$$H(s) = \frac{\lambda^2 F(s)}{1 + a\lambda s + \lambda^2 s^2} \tag{1.17-24}$$

To invert $H(s)$ requires knowledge of $F(s)$ and therefore of $f(t)$. For the simplified case where $f(t)$ is assumed to be a constant, say $c_0$, $F(s)$ equals $\frac{c_0}{s}$.

Equation (1.17-24) can then be written as

$$H(s) = \frac{\lambda^2 c_0}{s(1 + a\lambda s + \lambda^2 s^2)} = \frac{\lambda^2 c_0}{s(s - s_1)(s - s_2)} \tag{1.17-25a,b}$$

where $s_{1,2} = \left(-a\lambda \pm \lambda\sqrt{(a^2 - 4)}\right) / 2\lambda^2$ \qquad (1.17-26)

This can be further expressed in terms of partial fractions, which are easier to invert. That is

$$H(s) = \frac{A}{s} + \frac{B}{s - s_1} + \frac{C}{s - s_2} \tag{1.17-27}$$

We determine the constants A, B and C by multiplying Equation (1.17-27) respectively by $(s - s_2)$, $(s - s_1)$ and $s$ and setting in turn $s = s_2$, $s = s_1$ and $s = 0$. This leads to

$$C = (s - s_2) H(s)\big|_{s=s_2} = \frac{\lambda^2 c_0}{s_2 (s_2 - s_1)} \tag{1.17-28a,b}$$

$$B = (s - s_1) H(s)\big|_{s=s_1} = \frac{\lambda^2 c_0}{s_1 (s_1 - s_2)} \tag{1.17-29a,b}$$

$$A = s H(s)\big|_{s=0} = \frac{\lambda^2 c_0}{s_1 s_2} \tag{1.17-30a,b}$$

Equation (1.17-27) becomes

$$H(s) = \frac{\lambda^2 c_0}{s\, s_1 s_2} + \frac{\lambda^2 c_0}{s_1 (s_1 - s_2) (s - s_1)} + \frac{\lambda^2 c_0}{s_2 (s_2 - s_1) (s - s_2)} \tag{1.17-31}$$

Inversion results in $h(t)$ which is given by

$$h(t) = \frac{\lambda^2 c_0}{s_1 s_2} + \frac{\lambda^2 c_0}{s_1 (s_1 - s_2)} e^{s_1 t} + \frac{\lambda^2 c_0}{s_2 (s_2 - s_1)} e^{s_2 t} \tag{1.17-32}$$

The behavior of $h(t)$ depends on the values of $s_1$ and $s_2$ and therefore on the values of $a$ and $\lambda$ [see Equation (1.17-26)].

In particular, if $a > 2$, $s_1$ and $s_2$ are real and $h$ changes with $t$ in a non-oscillatory fashion. If $a < 2$, $s_1$ and $s_2$ are complex and one has an oscillatory response (see Section 1.16 iii). For $a = 2$, a critically damped response is obtained.

***Example 1.17-3***. Solve the kinetic expression given by Equation (1.14-18), subject to the condition $c_B = 0$ at $t = 0$, by Laplace transform. Invert using convolution.

The Laplace transform of the equation is

$$s\, C_B(s) + k_2 C_B(s) = k_1 c_{A_0} \left( \frac{1}{s + k_1} \right) \tag{1.17-33}$$

$$C_B(s) = \frac{k_1 c_{A_0}}{(s + k_1)(s + k_2)} \tag{1.17-34}$$

Using convolution (Equation 1.17-7), we write

$$c_B(t) = k_1 c_{A_0} \int_0^t e^{-k_1(t-u)} e^{-k_2 u} \, du \tag{1.17-35a}$$

$$= k_1 c_{A_0} \int_0^t e^{-k_1 t - u(k_2 - k_1)} \, du \tag{1.17-35b}$$

$$= k_1 c_{A_0} e^{-k_1 t} \int_0^t e^{-u(k_2 - k_1)} \, du \tag{1.17-35c}$$

$$= \frac{-k_1 c_{A_0} e^{-k_1 t}}{(k_2 - k_1)} e^{-u(k_2 - k_1)} \Big|_0^t \tag{1.17-35d}$$

$$c_B(t) = \frac{-k_1 c_{A_0} e^{-k_1 t}}{(k_2 - k_1)} \left[ e^{-t(k_2 - k_1)} - 1 \right] \tag{1.14-20}$$

***Example 1.17-4.*** A slab of a viscoelastic material was at rest and at time $t = 0$, it is suddenly sheared by an amount $\gamma_0$, as illustrated in Figure 1.17-2. Determine the shear stress in the material as a function of time.

We assume the relation between the shear stress $(\tau_{yx})$ and the shear rate $(\dot{\gamma}_{yx})$ to be given by

$$\tau_{yx} + \lambda_0 \dot{\tau}_{yx} = -\mu \dot{\gamma}_{yx} \tag{1.17-36}$$

$\mu$, $\lambda_0$ are the viscosity and relaxation time respectively, and the dot over the quantities denote differentiation with respect to time.

The shear is given by

$$\gamma_{yx} = \gamma_0 H(t) \tag{1.17-37}$$

Using Equation (1.17-17), we have

$$\dot{\gamma}_{yx} = \gamma_0 \delta(t) \tag{1.17-38}$$

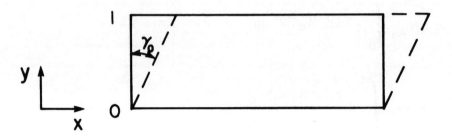

**FIGURE 1.17-2    Shear deformation of a slab**

Substituting Equation (1,17-38) into Equation (1.17-36), we obtain

$$\tau_{yx} + \lambda_0 \dot{\tau}_{yx} = -\mu\, \gamma_0\, \delta(t) \tag{1.17-39}$$

Taking the Laplace transform of Equation (1.17-39) and assuming that $\tau_{yx} = 0$ at $t = 0$, we get

$$T_{yx}(s) + \lambda_0\, s\, T_{yx}(s) = -\mu\, \gamma_0 \tag{1.17-40}$$

$T_{yx}(s)$, the Laplace transform of $\tau_{yx}$, is given from Equation (1.17-40) by

$$T_{yx}(s) = \frac{-\mu\, \gamma_0}{\lambda_0\, (s + 1/\lambda_0)} \tag{1.17-41}$$

The inverse of $T_{yx}(s)$ can be seen from Table 1.17-1 to be

$$\tau_{yx}(t) = \frac{-\mu\, \gamma_0}{\lambda_0}\, e^{-t/\lambda_o} \tag{1.17-42}$$

Thus the shear stress decays to zero exponentially.

## 1.18  SOLUTIONS USING GREEN'S FUNCTIONS

The Green's function method is a powerful method to solve boundary value problems and can be used not only for ordinary differential equations but also for partial differential equations and integral equations. Our purpose in introducing the Green's function now is to show the generality of the method. Once a Green's function has been obtained for a given operator and a set of boundary conditions, then the solution to the boundary value problem can be written as an integral. Thus the Green's function is equivalent to the integrating factor in the first-order equation.

The solution of a non-homogeneous boundary value problem can be obtained if the Green's function $G$ is known for the homogeneous equation.

The solution of Equation (1.16-1) with $Q(x)$ replaced by $f(x)$, subject to the conditions $y = 0$ at $x = x_0$ and $x = x_1$ can be written as

$$y(x) = \int_{x_0}^{x_1} G(x,t) \, f(t) \, dt \tag{1.18-1}$$

where $G(x, t)$ is the **Green's function** for this problem.

The Green's function is a function of two variables, $x$ and $t$. In Equation (1.18-1), we integrate with respect to $t$ so as to obtain a solution $y$ at point $x$. This requires that during the process of constructing $G$, we regard the differential equation which $G$ has to satisfy, as a differential equation in $t$. The boundary conditions will be applied at $t = x_0$ and $t = x_1$.

The Green's function for any $x$ has to satisfy conditions such as

i)      $G(x, t)$ is a solution of the homogeneous Equation (1.16-4), except at $t = x$. That is to say

$$\frac{d^2 G}{dt^2} + A \frac{dG}{dt} + BG = 0 \tag{1.18-2}$$

ii)      $G$ satisfies the boundary conditions. That is $G = 0$ at $t = x_0$ and $t = x_1$.

iii)      $G$ is continuous at $t = x$. That is

$$G\Big|_{t=x_-} = G\Big|_{t=x_+} \tag{1.18-3}$$

iv)       The first derivative is discontinuous at $t = x$.

$$\left. \frac{dG}{dt} \right|_{t = x_+} - \left. \frac{dG}{dt} \right|_{t = x_-} = 1 \qquad\qquad (1.18\text{-}4)$$

More generally, this difference equals the reciprocal of the coefficient of the highest derivative in Equation (1.18-2).

v)        G is symmetric.

$$G(x, t) = G(t, x) \qquad\qquad (1.18\text{-}5)$$

*Example 1.18-1.*  Solve the equation

$$\frac{d^2y}{dx^2} + y = f(x) \qquad\qquad (1.18\text{-}6)$$

subject to $y(0) = y(1) = 0$ and identify the Green's function.

The solution to the differential equation is given, in general, by Equation (1.16-6).  Having obtained the solution to Equation (1.18-6), it will be possible to deduce the Green's function for the operator $L = \dfrac{d^2}{dx^2} + 1$.  We can then show that the Green's function satisfies conditions (i to v).  Having obtained the Green's function allows us to write down the solution to Equation (1.18-6) for any $f(x)$.

The solution to the homogeneous equation is

$$y_h = c_1 \cos x + c_2 \sin x \qquad\qquad (1.18\text{-}7)$$

The particular solution $y_P$ can be obtained via the method of variation of parameters, as follows.  We try a solution of the form

$$y_P = a(x) \cos x + b(x) \sin x \qquad\qquad (1.18\text{-}8)$$

where we replace the constants $c_1$ and $c_2$ by functions of $x$.

We then calculate

$$y_P' = -a(x) \sin x + b(x) \cos x + a'(x) \cos x + b'(x) \sin x \qquad\qquad (1.18\text{-}9)$$

Note that $y_P$ has to satisfy Equation (1.18-6). However, $y_P$ contains two unknown (arbitrary) functions, $a(x)$ and $b(x)$. One should realize that on substituting $y_P''$ and $y_P$ in Equation (1.18-6) one generates only one equation to determine the functions $a(x)$ and $b(x)$. It becomes therefore necessary to impose one additional condition. Such a condition should be chosen to simplify the problem. The obvious condition is

$$a'(x) \cos x + b'(x) \sin x = 0 \tag{1.18-10}$$

Next, we compute $y_P''$

$$y_P'' = -a'(x) \sin x + b'(x) \cos x - a(x) \cos x - b(x) \sin x \tag{1.18-11}$$

Substitution into Equation (1.18-6) yields

$$-a'(x) \sin x + b'(x) \cos x = f(x) \tag{1.18-12}$$

We can now use Equation (1.18-10) to replace $a'(x)$ or $b'(x)$. Replacing $b'(x)$, Equation (1.18-12) becomes

$$-a'(x) \sin x - a'(x) \frac{\cos^2 x}{\sin x} = f(x) \tag{1.18-13}$$

which reduces to

$$a'(x) = -f(x) \sin x \tag{1.18-14}$$

Integration yields

$$a(x) = -\int_{c_1}^{x} f(t) \sin t \, dt \tag{1.18-15}$$

As in Equation (1.3-1), $t$ is a dummy variable.

Also, from Equation (1.18-10), we have

$$b'(x) = \frac{-a'(x) \cos x}{\sin x} \tag{1.18-16}$$

On integrating, we obtain

$$b(x) = \int_{c_2}^{x} f(t) \cos t \, dt \qquad (1.18\text{-}17)$$

The constants $c_1$ and $c_2$ can be determined from the boundary conditions.

We note that when $f(x) = 0$, $a' = b' = 0$. That is, $a$ and $b$ are constants (say $c_1$ and $c_2$ respectively) and we obtain $y_h$.

In other words, introducing $c_1$ and $c_2$ as limits of integration in Equations (1.18-15, 17) will generate a solution $y_p$ which in fact includes $y_h$. A similar situation was encountered in Example 1.14-3.

Thus the general solution of Equation (1.18-6) may be written as

$$y = -\cos x \int_{c_1}^{x} f(t) \sin t \, dt + \sin x \int_{c_2}^{x} f(t) \cos t \, dt \qquad (1.18\text{-}18)$$

The boundary condition $y = 0$ at $x = 0$ implies

$$0 = \int_{c_1}^{0} f(t) \sin t \, dt \qquad (1.18\text{-}19)$$

which leads to $c_1 = 0$.

Imposing the boundary condition at $x = 1$ yields

$$0 = -\cos(1) \int_{0}^{1} f(t) \sin t \, dt + \sin(1) \int_{c_2}^{1} f(t) \cos t \, dt \qquad (1.18\text{-}20)$$

This may be written as

$$0 = -\cos(1) \int_{0}^{1} f(t) \sin t \, dt + \sin(1) \left[ \int_{c_2}^{0} f(t) \cos t \, dt + \int_{0}^{1} f(t) \cos t \, dt \right] \qquad (1.18\text{-}21a)$$

$$= -\cos(1) \int_{0}^{1} f(t) \sin t \, dt + \sin(1) \int_{0}^{1} f(t) \cos t \, dt + \sin(1) \int_{c_2}^{0} f(t) \cos t \, dt \qquad (1.18\text{-}21b)$$

$$= \int_0^1 f(t) \sin(1-t)\, dt + \sin(1)\int_{c_2}^0 f(t) \cos t\, dt \qquad (1.18\text{-}21c)$$

Thus

$$\int_{c_2}^0 f(t) \cos t\, dt = -\frac{1}{\sin(1)} \int_0^1 f(t) \sin(1-t)\, dt \qquad (1.18\text{-}22)$$

Recall that we are in the process of determining $G$, in order to write the solution as in Equation (1.18-1). To achieve this, we wish to determine $G$ in two regions: $t < x$ and $t > x$, because $G$ has several properties to satisfy at the points $t = x_-$ and $t = x_+$.

Thus we write Equation (1.18-18) as

$$y(x) = -\cos x \int_{c_1}^x f(t) \sin t\, dt + \sin x \left[ \int_{c_2}^0 f(t) \cos t\, dt + \int_0^x f(t) \cos t\, dt \right] \qquad (1.18\text{-}23)$$

Combining Equations (1.18-19, 22), we obtain

$$y(x) = -\cos x \int_0^x f(t) \sin t\, dt + \sin x \int_0^x f(t) \cos t\, dt - \frac{\sin x}{\sin(1)} \int_0^1 f(t) \sin(1-t)\, dt \qquad (1.18\text{-}24a)$$

$$= \int_0^x f(t) \sin(x-t)\, dt - \frac{\sin x}{\sin(1)} \int_0^1 f(t) \sin(1-t)\, dt \qquad (1.18\text{-}24b)$$

$$= \int_0^x f(t) \sin(x-t)\, dt - \frac{\sin x}{\sin(1)} \left[ \int_0^x f(t) \sin(1-t)\, dt + \int_x^1 f(t) \sin(1-t)\, dt \right] \qquad (1.18\text{-}24c)$$

$$= \int_0^x f(t) \sin(x-t)\, dt - \frac{\sin x}{\sin(1)} \int_0^x f(t) \sin(1-t)\, dt - \frac{\sin x}{\sin(1)} \int_x^1 f(t) \sin(1-t)\, dt \qquad (1.18\text{-}24d)$$

$$= \int_0^x f(t) \left[ \frac{\sin(1)\sin(x-t) - \sin x \sin(1-t)}{\sin(1)} \right] dt - \int_x^1 \left[ \frac{f(t) \sin x \sin(1-t)}{\sin(1)} \right] dt \qquad (1.18\text{-}24e)$$

$$= \int_0^x f(t) \left[ \frac{\sin t \ \sin(x - 1) \ dt}{\sin(1)} \right] dt + \int_x^1 f(t) \left[ \frac{\sin x \ \sin(t - 1)}{\sin(1)} \right] dt \tag{1.18-24f}$$

$$y(x) = \int_0^1 f(t) \ G(x, t) \ dt \tag{1.18-1}$$

where $G(x, t)$ is the Green's function and is given by

$$G(x, t) = \begin{cases} \dfrac{\sin t \ \sin (x - 1)}{\sin (1)}, & 0 \le t \le x \\[3mm] \dfrac{\sin x \ \sin (t - 1)}{\sin (1)}, & x \le t \le 1 \end{cases} \tag{1.18-25a, b}$$

From Equations (1.18-25a, b) we can deduce that  $G$  satisfies conditions (iv, v) given earlier.  In particular, we deduce that

$$G' \Big|_{t=x_-} = \frac{\cos x \ \sin (x - 1)}{\sin (1)} \tag{1.18-26a}$$

$$G' \Big|_{t=x_+} = \frac{\sin x \ \cos (x - 1)}{\sin (1)} \tag{1.18-26b}$$

and

$$G' \Big|_{t=x_+} - G' \Big|_{t=x_-} = \frac{\sin x \ \cos (x - 1) - \cos x \ \sin (x - 1)}{\sin (1)} = 1 \tag{1.18-27a,b}$$

The solution to the problem therefore is given by Equation (1.18-1) where  $G(x, t)$,  which is independent of  $f(t)$,  is given by Equation (1.18-25a, b).  As mentioned earlier, the solution of Equation (1.18-6) can now be determined for any  $f(x)$,  via Equations (1.18-25a, b).  The boundary conditions will be satisfied automatically.  The Green's function can of course also be constructed from the condition given by Equations (1.18-2 to 5).  This exercise is addressed in Example 1.18-2 and in Problem 37b.

The Green's functions for several frequently used operators and boundary conditions are given in the following table.  The boundaries in Table 1.18-1 are all chosen to be  $x = 0$  and  $x = 1$.  That is to say, the solution to a given problem is to be written as in Equation (1.18-24f) where the appropriate function  $G(x, t)$  should be chosen from the right side column in Table 1.18-1.

## TABLE 1.18-1

### Green's Functions

| Operator L(y) | Boundary Conditions | G (x, t) |
| --- | --- | --- |
| 1. $y''$ | $y(0) = y(1) = 0$ | $t(x-1), \ x \geq t$<br>$x(t-1), \ x \leq t$ |
| 2. $y''$ | $y(0) = y'(1) = 0$ | $-t, \ x \geq t$<br>$-x, \ x \leq t$ |
| 3. $y'' + \lambda^2 y$ | $y(0) = y(1) = 0$ | $\dfrac{\sin \lambda t \ \sin \lambda (x-1)}{\lambda \sin \lambda}, \ x \geq t$<br>$\dfrac{\sin \lambda x \ \sin \lambda (t-1)}{\lambda \sin \lambda}, \ x \leq t$ |
| 4. $y'' - \lambda^2 y$ | $y(0) = y(1) = 0$ | $\dfrac{\sinh \lambda t \ \sinh \lambda (x-1)}{\lambda \sinh \lambda}, \ x \geq t$<br>$\dfrac{\sinh \lambda x \ \sinh \lambda (t-1)}{\lambda \sinh \lambda}, \ x \leq t$ |

***Example 1.18-2.*** Consider the transverse displacement of a string of unit length fixed at its two ends, $x = 0$ and $x = 1$. If $y$ is the displacement of the string from its equilibrium position, as a result of a force distribution $f(x)$, then $y$ satisfies

$$-T \, \frac{d^2 y}{dx^2} = f(x) \tag{1.18-28}$$

where $T$ is the tension in the string.

The boundary conditions are

$$y(0) = y(1) = 0 \tag{1.18-29a,b}$$

We now construct the Green's function for the operator

$$L(y) = y'' = \frac{d^2}{dx^2} \tag{1.18-30a,b}$$

Since $G$ has to satisfy

$$\frac{d^2 G}{dt^2} = 0 \qquad (1.18\text{-}31)$$

$G(x, t)$ is given by

$$G(x, t) = c_1(x) + c_2(x) t , \qquad t < x \qquad (1.18\text{-}32a)$$

$$= c_3(x) + c_4(x) t , \qquad t > x \qquad (1.18\text{-}32b)$$

Applying the boundary conditions on $t$ yields

$$c_1 = 0 \qquad (1.18\text{-}33a)$$

$$c_3 + c_4 = 0 \qquad (1.18\text{-}33b)$$

The continuity condition implies

$$xc_2 = c_3 + xc_4 \qquad (1.18\text{-}34)$$

The jump discontinuity can be expressed as

$$c_4 - c_2 = 1 \qquad (1.18\text{-}35)$$

From Equations (1.18-33 to 35) we can solve for $c_1$ to $c_4$. The result are

$$c_1 = 0 , \qquad c_2 = (x - 1), \qquad c_3 = -x, \qquad c_4 = x \qquad (1.18\text{-}36a,b,c,d)$$

Thus $G(x, t)$ is given by

$$G(x, t) = t(x - 1) , \qquad t < x \qquad (1.18\text{-}37a)$$

$$= x(t - 1) , \qquad t > x \qquad (1.18\text{-}37b)$$

which is given in Table 1.18-1.

The displacement $y$ at any point $x$ is then given by Equation (1.18-1) as

$$y(x) = -\int_0^1 G(x, t) \frac{f(t)}{T} dt \qquad (1.18\text{-}38)$$

As an example, we can consider $f(x)$ to be an impulsive force. That is to say, the force acts at one point only, say at $x = \xi$. Thus $f(x)$ will be zero everywhere except at the point $x = \xi$, where it is not defined. Mathematically $f(x)$ can be represented by the Dirac delta function $\delta$. We write $f(x)$ as

$$f(x) = \delta(x - \xi) \tag{1.18-39}$$

The delta function is not an ordinary function but it can be regarded as a limit of a sequence of functions. An example of such a sequence is

$$\lim_{\varepsilon \to 0} \frac{1}{\sqrt{\pi \varepsilon}} \exp\left[-\left(\frac{x}{\varepsilon}\right)^2\right] = \delta(x) \tag{1.18-40}$$

An alternative approach, introduced by Schwartz (1957), is to consider the Dirac delta function as a functional or distribution, given by Equations (1.17-14c). Both approaches produce equivalent results. The main properties of the Dirac delta function are given in Section 1.17.

Combining Equations (1.18-38, 39), we obtain

$$y(x) = -\int_0^1 \frac{G(x, t)\,\delta(t - \xi)}{T}\, dt \tag{1.18-41a}$$

$$= -\frac{G(x, \xi)}{T} \tag{1.18-41b}$$

Equation (1.18-41b) provides a physical interpretation of the Green's function. G represents the displacement at x due to a point force applied at $\xi$.

## 1.19  MODELLING OF PHYSICAL SYSTEMS

In order to generate the equation(s) representing the physical situation of interest, it is useful to proceed according to the following steps.

1.    Draw a sketch of the problem. Indicate information such as pressures, temperatures, etc.

2.    Make sure you understand the physical process(es) involved.

3.    Formulate a model in mathematical terms. That is to say, determine the equation(s) as well as the boundary and/or initial condition(s).

4.    Non-dimensionalize the equation(s). This may reduce the number of variables and it allows for the identification of controlling variables such as, for example, a Reynolds number. It also allows one to easily compare situations which are dimensionally quite different. For example, one can compare flow situations in pipes of small to very large diameters in terms of a dimensionless variable $\xi = \frac{r}{R}$ which will always vary from 0 to 1.

5.    Determine the limiting forms (asymptotic solutions).

6.      Solve the equation(s).

7.      Verify your solution(s)!  Do they make sense physically?  Do all terms have the same
        dimensions?

As a first example, we consider the flow of a Newtonian fluid in a pipe.  We can imagine thin
concentric cylindrical sheets of liquid sliding past each other.  Figure 1.19-1 illustrates the problem
which is to be modelled subject to the following assumptions

—  steady-state flow;

—  laminar flow:  Re $= \dfrac{D \langle v_z \rangle \rho}{\mu} < 2100$  and $v_z$ is a function of r only;

—  the fluid is incompressible:  $\rho$ = constant;

—  there are no end effects; that is, the piece of pipe of length  L  which we are considering is located
   far from either end of the pipe which is very long.

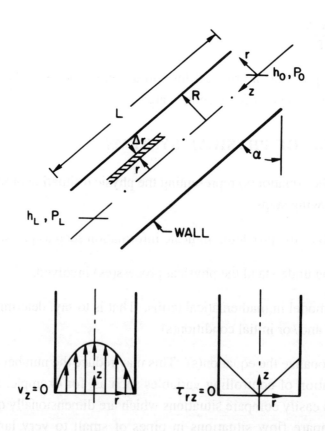

FIGURE 1.19-1      Flow in a pipe with associated velocity profile $v_z$ (r)
                   and shear stress profile $\tau_{rz}$

Cylindrical coordinates are used since the geometry of the problem is cylindrical. This makes it mathematically easier (less calculations) to apply the boundary conditions.

The differential equation representing this physical situation is obtained by applying a momentum (or force) balance on a shell of thickness $\Delta r$. The balance is given as follows

$$\begin{pmatrix} \text{rate of } z \\ \text{momentum in} \end{pmatrix} - \begin{pmatrix} \text{rate of } z \\ \text{momentum out} \end{pmatrix} + \begin{pmatrix} \text{net sum of forces} \\ \text{in } z\text{-direction} \end{pmatrix} = 0 \tag{1.19-1}$$

The right side of Equation (1.19-1) is zero because of the steady-state assumption (no accumulation!). Analyzing the problem, one has to realize that $z$ momentum is transported according to two mechanisms: (i) by convection due to bulk flow, and (ii) by viscous (molecular) transport. The forces acting on the system (cylindrical shell of thickness $\Delta r$ and length L) are: (i) the pressure force in the z-direction, and (ii) the z-component of the gravity force.

The individual contributions making up the force balance are

a)      z-momentum in by convection at $z = 0$.

This contribution is obtained by multiplying the volumetric flowrate associated with the appropriate area ($2\pi r\Delta r$), with $\rho v_z$. That is

z-momentum in

$$\left( v_z\, 2\pi r\Delta r \right) \rho\, v_z \big|_{z=0}$$

by convection at $z = 0$

b)      Similarly we write

z-momentum out

$$\left( v_z\, 2\pi r\Delta r \right) \rho\, v_z \big|_{z=L}$$

by convection at $z = L$

Note that the quantities 2, $\pi$, r, $\Delta r$ and $\rho$ are not changing. Since the flow is laminar, $v_z$ is not going to change with $z$ so that the net contribution to the force balance of the z-momentum associated with convection is zero.

c)      z-momentum in by viscous transport at $r$.

This contribution is obtained by multiplying the shear stress $\tau_{rz}$ (force per area) by the appropriate area ($2\pi rL$). That is

z-momentum in

$$\tau_{rz}\left(2\pi rL\right)\big|_r$$

by viscous transport at $r$

d)      Similarly we write

z-momentum in                                    $\tau_{rz}(2\pi rL)\big|_{r+\Delta r}$

by viscous transport at $r + \Delta r$

Note that since we are dealing with a Newtonian fluid, the shear stress $\tau_{rz}$ is given by

$$\tau_{rz} = -\mu \frac{dv_z}{dr} \qquad\qquad\qquad (1.19\text{-}2)$$

$v_z$ is a function of $r$, otherwise $\frac{dv_z}{dr}$ would be zero. The Newtonian viscosity $\mu$ is constant and therefore the left side of Equation (1.19-2) has to be a function of $r$ and we will have a net contribution from the z-momentum by viscous transport to the force balance.

e)      The contribution from the pressure force is obtained by multiplying the pressure (force per area) at $z = 0$ ($P_0$ in Figure 1.19-1) by the appropriate area ($2\pi r\Delta r$). That is

pressure force at $z = 0$ \qquad $P_0\,(2\pi r\Delta r)$

f)      Similarly we write

pressure force at $z = L$ \qquad $P_L\,(2\pi r\Delta r)$

g)      The z-contribution to the force balance due to the gravity acting on the system (shell) is given by the weight of the fluid in the shell. That is, the volume ($2\pi r\Delta rL$) multiplied by $\rho g$ and by $\cos\alpha$ (z-contribution)

gravity force \qquad\qquad $(2\pi r\Delta rL)\,\rho g \cos\alpha$

Note that $L\cos\alpha = h_0 - h_L$, the difference in height of the positions at $z = 0$ and $z = L$ in Figure 1.19-1. The gravity force can therefore be written as: $2\pi r\,\rho g\,(h_0 - h_L)\,\Delta r$.

Substitution of the terms in c), d), e), f) and g) into the force balance yields

$$\tau_{rz}(2\pi rL)\big|_r - \tau_{rz}(2\pi rL)\big|_{r+\Delta r} + (P_0 - P_L)\,2\pi r\Delta r + 2\pi r\rho\,g(h_0 - h_L)\,\Delta r = 0 \qquad (1.19\text{-}3)$$

We can divide Equation (1.19-3) by $2\pi L\Delta r$ to obtain

$$\frac{r\,\tau_{rz}\big|_r - r\,\tau_{rz}\big|_{r+\Delta r}}{\Delta r} + \left(\frac{P_0 - P_L}{L}\right)r + r\rho g\left(\frac{h_0 - h_L}{L}\right) = 0 \qquad (1.19\text{-}4)$$

Note that we did not divide by $r$. We established earlier that $\tau_{rz}$ is a function of $r$.

We are in the process of setting up a differential equation relating $\tau_{rz}$ to $r$! Recall that the derivative of a function $y = f(x)$ with respect to $x$ at the point $x = x_0$ is defined by

$$\lim_{\Delta x \to 0} \frac{f(x_0 + \Delta x) - f(x_0)}{\Delta x} = \lim_{\Delta x \to 0} \frac{\Delta y}{\Delta x} = \frac{df}{dx} \qquad (1.19\text{-}5a,b)$$

In Equation (1.19-4), we have a term involving the quantity $r\,\tau_{rz}|_r - r\,\tau_{rz}|_{r+\Delta r}$, that is, $r\,\tau_{rz}$ initial (at $r$) – $r\,\tau_{rz}$ final (at $r + \Delta r$), so that we are dealing with $-\Delta(r\,\tau_{rz})$.

Taking the limit for $\Delta r \longrightarrow 0$ of Equation (1.19-4) yields

$$\lim_{\Delta r \to 0} \left( \frac{r\,\tau_{rz}|_r - r\,\tau_{rz}|_{r+\Delta r}}{\Delta r} \right) + \frac{r}{L}\left[(P_0 - P_L) + \rho g(h_0 - h_L)\right] = 0 \qquad (1.19\text{-}6)$$

or via Equation (1.19-5)

$$-\frac{d}{dr}(r\,\tau_{rz}) + r\left(\frac{\mathcal{P}_0 - \mathcal{P}_L}{L}\right) = 0 \qquad (1.19\text{-}7)$$

Equation (1.19-7) is the desired differential equation with the potential $\mathcal{P}$ defined by

$$\mathcal{P} = P + \rho g h \qquad (1.19\text{-}8)$$

So the potential at $z = 0$ is given by: $\mathcal{P}_0 = P_0 + \rho g h_0$ and the potential at $z = L$ is: $\mathcal{P}_L = P_L + \rho g h_L$.

Note that the potential drop per length $\left(\dfrac{\mathcal{P}_0 - \mathcal{P}_L}{L}\right)$ becomes the pressure drop per length when $h_0 = h_L$ (horizontal flow).

The differential Equation (1.19-7) is solved as follows, to yield the linear shear stress profile shown in Figure 1.19-1.

$$\frac{d}{dr}(r\,\tau_{rz}) = r\left(\frac{\mathcal{P}_0 - \mathcal{P}_L}{L}\right) \qquad (1.19\text{-}9)$$

$$\int d(r\,\tau_{rz}) = \int r\left(\frac{\mathcal{P}_0 - \mathcal{P}_L}{L}\right) dr \qquad (1.19\text{-}10)$$

$$r\,\tau_{rz} = \frac{r^2}{2}\left(\frac{\mathcal{P}_0 - \mathcal{P}_L}{L}\right) + C_1 \qquad (1.19\text{-}11)$$

$$\tau_{rz} = \frac{r}{2}\left(\frac{\mathcal{P}_0 - \mathcal{P}_L}{L}\right) + \frac{C_1}{r} \qquad (1.19\text{-}12)$$

To evaluate $C_1$, we look at the physical situation for $r = 0$. We cannot accept an infinitely large force (per area). Therefore $C_1 = 0$ and $\tau_{rz}$ is a linear function of $r$.

$$\tau_{rz} = \frac{r}{2}\left(\frac{\mathcal{P}_0 - \mathcal{P}_L}{L}\right) \tag{1.19-13}$$

The velocity profile is obtained by replacing the left side of Equation (1.19-13) by $-\mu\frac{dv_z}{dr}$. The minus sign is as a result of following the convention by Bird et al. (1960).

We now have a differential equation relating $v_z$ to $r$ given by

$$\frac{dv_z}{dr} = -\left(\frac{\mathcal{P}_0 - \mathcal{P}_L}{2\mu L}\right)r \tag{1.19-14}$$

This equation is solved as follows

$$\int dv_z = -\left(\frac{\mathcal{P}_0 - \mathcal{P}_L}{2\mu L}\right)\int r\, dr \tag{1.19-15}$$

$$v_z = -\left(\frac{\mathcal{P}_0 - \mathcal{P}_L}{2\mu L}\right)\frac{r^2}{2} + C_2 \tag{1.19-16}$$

The physics of the situation allows us to assume that the velocity is zero at the wall. That is, $v_z = 0$ at $r = R$. Applying this to Equation (1.19-16) allows us to evaluate the integration constant $C_2$.

$$0 = -\left(\frac{\mathcal{P}_0 - \mathcal{P}_L}{2\mu L}\right)\frac{R^2}{2} + C_2 \tag{1.19-17}$$

So

$$C_2 = \left(\frac{\mathcal{P}_0 - \mathcal{P}_L}{4\mu L}\right)R^2 \tag{1.19-18}$$

Substitution into Equation (1.19-16) yields

$$v_z = \left(\frac{\mathcal{P}_0 - \mathcal{P}_L}{4\mu L}\right)R^2 - \left(\frac{\mathcal{P}_0 - \mathcal{P}_L}{4\mu L}\right)r^2 \tag{1.19-19}$$

That is, $v_z$ is related to the square of $r$ (a parabola as in Figure 1.19-1).

The relation for $v_z$ is usually written as

$$v_z = \left(\frac{\mathcal{P}_0 - \mathcal{P}_L}{4\mu L}\right)R^2\left[1 - \left(\frac{r}{R}\right)^2\right] \tag{1.19-20}$$

Note that the maximum velocity occurs at the center of the tube. That is, at $r = 0$.

The average velocity $\langle v_z \rangle$ is obtained by dividing the flow rate by the cross-sectional area, as follows

$$\langle v_z \rangle = \frac{\int_0^{2\pi} \int_0^R v_z \, r \, dr \, d\theta}{\int_0^{2\pi} \int_0^R r \, dr \, d\theta} \qquad (1.19\text{-}21)$$

The denominator yields $\pi R^2$, as it should. The numerator is the product of a point velocity $(v_z)$ and an infinitesimal area $(r \, dr \, d\theta)$. Integrating this product yields a flow rate, as it should.

Combining Equations (1,19-20, 21) yields

$$\langle v_z \rangle = \frac{2\pi \left( \frac{\mathcal{P}_0 - \mathcal{P}_L}{4\mu L} \right) R^2 \int_0^R \left[ 1 - \left( \frac{r}{R} \right)^2 \right] r \, dr}{\pi R^2} \qquad (1.19\text{-}22)$$

We introduce a dimensionless variable $\xi = \frac{r}{R}$. The integral can then be written as

$$\int_0^1 \left( 1 - \xi^2 \right) \xi \, d\xi = \frac{1}{4} \qquad (1.19\text{-}23)$$

The constant $R^2$ necessary to non-dimensionalize $r \, dr$ is obtained from the denominator.

Note also that the limits of integration have to be adjusted according to the newly introduced variable $\xi$. Indeed, as the upper limit goes to $R$, the new variable $\xi$, goes to one. The lower limit remains unchanged.

Note that this integral is now independent of $R$ and is valid for pipes of any radius. Equation (1.19-22) reduces to

$$\langle v_z \rangle = 2 \left( \frac{\mathcal{P}_0 - \mathcal{P}_L}{4\mu L} \right) R^2 \left( \frac{1}{4} \right) = \left( \frac{\mathcal{P}_0 - \mathcal{P}_L}{8\mu L} \right) R^2 \qquad (1.19\text{-}24\text{a,b})$$

***Example 1.19-2.*** As a second example, we consider the flow of a non-Newtonian fluid in a pipe.

We will assume the fluid to obey the power-law. That is to say

$$\tau_{rz} = -\mu \left| \frac{dv_z}{dr} \right|^n \qquad (1.19\text{-}25)$$

At this point, one should realize that Equation (1.19-13) is still valid.  In fact, it is valid for any fluid, since no assumption as to the type of fluid has to be made to derive the shear stress distribution.  We now proceed by computing the velocity distribution.  Equation (1.19-13) now becomes

$$\tau_{rz} = \frac{r}{2}\left(\frac{\mathcal{P}_0 - \mathcal{P}_L}{L}\right) = -\mu\left|\frac{dv_z}{dr}\right|^n \qquad\qquad (1.19\text{-}26)$$

and the shear rate $\left(\dfrac{dv_z}{dr}\right)$ is given by

$$\left(\frac{dv_z}{dr}\right) = \left(\frac{\mathcal{P}_0 - \mathcal{P}_L}{2\mu L}\right)^{1/n}\left(-r^{1/n}\right) \qquad\qquad (1.19\text{-}27)$$

Integration yields the velocity profile.

$$v_z = \left(\frac{\mathcal{P}_0 - \mathcal{P}_L}{2\mu L}\right)^{1/n}\int -r^{1/n}\,dr \qquad\qquad (1.19\text{-}28a)$$

$$= \left(\frac{\mathcal{P}_0 - \mathcal{P}_L}{2\mu L}\right)^{1/n}\left(\frac{-r^{1/n+1}}{\frac{1}{n}+1}\right) + C_2 \qquad\qquad (1.19\text{-}28b)$$

The constant $C_2$ is obtained, using the same (no slip) boundary condition as in Example 1.12-1, and the velocity profile is

$$v_z = \left(\frac{\mathcal{P}_0 - \mathcal{P}_L}{2\mu L}\right)^{1/n}\frac{nR^{1/n+1}}{n+1}\left[1 - \left(\frac{r}{R}\right)^{1/n+1}\right] \qquad\qquad (1.19\text{-}29)$$

Note that for $n = 1$, the power-law fluid reduces to the Newtonian fluid.  That is to say, Equation (1.19-29) reduces to Equation (1.19-20).

The average velocity is obtained by substituting Equation (1.19-29) into Equation (1.19-21), following the procedure given in Example 1.19-1.

$$\langle v_z \rangle = \frac{2}{R^2}\left(\frac{\mathcal{P}_0 - \mathcal{P}_L}{2\mu L}\right)^{1/n}\frac{nR^{1/n+1}}{n+1}\int_0^R\left[1 - \left(\frac{r}{R}\right)^{1/n+1}\right]dr \qquad\qquad (1.19\text{-}30a)$$

$$= \frac{R}{\frac{1}{n}+3}\left[\frac{(\mathcal{P}_0 - \mathcal{P}_L)}{2\mu L}R\right]^{1/n} \qquad\qquad (1.19\text{-}30b)$$

We are now in a position to compare the velocity profiles in dimensionless form and to evaluate asymptotic behavior, as n tends to zero and to infinity. To achieve this, we plot $\dfrac{v_z}{\langle v_z \rangle}$ versus $\dfrac{r}{R}$.

Dividing Equation (1.19-29) by Equation (1.19-30c) yields

$$\frac{v_z}{\langle v_z \rangle} = \frac{3n+1}{n+1} \left[ 1 - \left(\frac{r}{R}\right)^{1/n+1} \right]$$

(1.19-31)

Different values of n yield different profiles. The ones of interest are tabulated in Table 1.19-1 and are shown in Figure 1.19-2.

## TABLE 1.19-1

### Velocity profiles

| n | $\dfrac{v_z}{\langle v_z \rangle}$ |
|:---:|:---:|
| 0 | $1\left[1-\left(\dfrac{r}{R}\right)^{\infty}\right] = \begin{cases} 1, & 0 \le \dfrac{r}{R} < 1 \\ 0, & \dfrac{r}{R} = 1 \end{cases}$ |
| 1 | $2\left[1-\left(\dfrac{r}{R}\right)^{2}\right]$ |
| $\infty$ | $3\left[1-\left(\dfrac{r}{R}\right)^{1}\right]$ |

We note that for n = 1, we obtain the Newtonian result (parabolic profile). The asymptotic solutions (linear profiles) are obtained for n = 0 and n = ∞. These profiles are illustrated in Figure 1.19-2. For n = 0, the slope is zero and for n = ∞, we obtain a maximum slope of 3. For intermediate values of n, the slope must lie between 0 and 3. In practice, n is usually between 0 and 1 and the observed velocity profiles are indeed more blunt than the Newtonian one.

In this example, we continued to use parameter $\mu$ as in Example 1.19-1. However, note that the dimensions of $\mu$ depend on n and $\mu$ has the same dimension in both examples when n = 1.

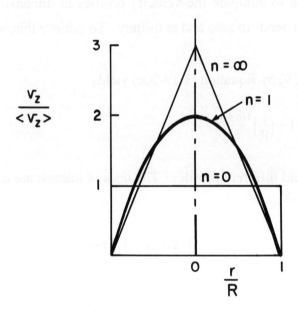

**FIGURE 1.19-2**    **Dimensionless velocity profiles**

## PROBLEMS

1a.    Differentiate the following functions with respect to x

  (i)    arc sin x

  (ii)    arc cos $\sqrt{1-x^2}$

  (iii)    $\ell n \, (x^2 \, \ell n \, x)$

Answer:    (i) and (ii) $1/\sqrt{1-x^2}$

(iii) $(1 + 2 \, \ell n \, x)/x \, \ell n \, x$

2a.    The circumference C and the area A of a circle of radius r are given respectively by

$$C = 2\pi r, \qquad A = \pi r^2$$

Use the chain rule to compute $\dfrac{dA}{dC}$ .                                        Answer: r

3b.    If $y = e^{-\alpha x} / (1 - x)^\alpha$, where $\alpha$ is a constant, show that

$$(1 - x) \frac{dy}{dx} = \alpha x y$$

Use Leibnitz rule to show that

$$(1 - x)\, y^{(n+1)} - (n + \alpha x)\, y^{(n)} - n\alpha y^{(n-1)} = 0$$

where $y^{(n)} = \dfrac{d^n y}{dx^n}$.

4a.    Find the limit as $x \longrightarrow 0$ of the following functions

(i) $\dfrac{\sin x}{x}$        (ii) $\dfrac{x \cos x}{\sin x}$        (iii) $\dfrac{e^x - 1}{\sinh x}$        (iv) $x^x$

(Hint: in (iv), take $\ell n$)                                    Answer (i) to (iv):  1

5a.    Use Taylor's theorem to obtain an expression for $\cos x$ about $x = \pi/2$. Find the remainder term if only the first three terms are retained. What is the maximum error if we use the expansion you have derived to compute the value of $\cos x$ in the interval $\pi/4 \le x \le 3\pi/4$ ?

6b.    The viscosity $\eta$ of a non-Newtonian fluid is an even function of the shear rate $\dot{\gamma}$. If $\dot{\gamma}$ is small, such that $\dot{\gamma}^4$ and higher powers of $\dot{\gamma}$ can be neglected, express $\eta$ as a polynomial in $\dot{\gamma}$, using Maclaurin's expansion. If it is known that $\eta$ is given by the rational function

$$\eta = \frac{a + b\dot{\gamma}^2}{1 + c\dot{\gamma}^2}$$

write down the coefficients of the expansion you have obtained in terms of a, b, and c.

7a.    Integrate the following

(i) $\displaystyle\int x\, e^x\, dx$        (ii) $\displaystyle\int \cos(\ell n\, x)\, dx$        (iii) $\displaystyle\int x\sqrt{x^2 + 1}\; dx$

(iv) $\displaystyle\int e^x \sin e^x\, dx$        (v) $\displaystyle\int \frac{2x^2 + x + 1}{(x - 1)^2 (x + 3)}\, dx$

Answer: (i) $e^x (x - 1)$;   (ii) $\dfrac{x}{2}\Big[\cos(\ell n\, x) + \sin(\ell n\, x)\Big]$;   (iii) $\dfrac{1}{3}\big(1 + x^2\big)^{3/2}$

(iv) $-\cos e^x$ ;   (v) $\ell n\, (x^2 + 2x - 3) - (x - 1)^{-1}$

8a.    Sketch the curve $y = \cos x$ in the range $0 \le x \le \pi$. Find the area of the region enclosed by $y = \cos x$, the x-axis, $x = 0$, and $x = \pi$.

**9b.** The area of the surface of revolution $S$ obtained by rotating the curve $y = f(x)$ about the x-axis from $x = a$ to $x = b$ is given by

$$S = 2\pi \int_a^b y \sqrt{1 + \left(\frac{dy}{dx}\right)^2} \; dx$$

Compute $S$ for $y = e^{-x}$, $0 \le x \le \infty$.    Answer: $\pi \left(\sqrt{2} + \sinh^{-1} 1\right)$

**10a.** Find $f_x$, $f_y$, $f_{xy}$ for the following functions

(i) $f(x, y) = x^3 - x^2 y^2 + y^3$

(ii) $f(x, y) = \ell n \, (x^2 + y^2)$

(iii) $f(x, y) = x \cos y + y \sin x$

**11b.** The pressure $P$, volume $V$, and temperature $T$ for 1 mole of an ideal gas are related by the equation

$$PV = RT$$

where $R$ is a constant.

Find the change in volume if both the temperature and pressure are increased by 1%.

**12b.** The temperature $T$ in a body is given by

$$T = T_0 \, (1 + ax + by) \, e^{cz}$$

where $a$, $b$, $c$, and $T_0$ ($> 0$) are constants. Find the rate of change of $T$ along $x$, $y$, and z-axes at the origin. In addition, find the direction in which the temperature changes most rapidly at the origin.

Answer: $\underline{n} = \dfrac{(a, b, c)}{\sqrt{a^2 + b^2 + c^2}}$

**13a.** Show that if the independent variables $x$ and $t$ are changed to

$$\xi = x + ct, \qquad \eta = x - ct,$$

the wave equation

$$\frac{\partial^2 u}{\partial t^2} = c^2 \frac{\partial^2 u}{\partial x^2}, \quad c \text{ is a constant,}$$

is transformed to

$$\frac{\partial^2 u}{\partial \xi \, \partial \eta} = 0$$

14a.    Compute $\dfrac{dy}{dx}$ if

$$y^2 - \sin xy + x^2 = 4 \,.$$

15a.    The pressure $P$, the volume $V$ and the temperature $T$ are given by

$$\left(P + \frac{\alpha}{V^2}\right)(V - \beta) = RT$$

where $\alpha$, $\beta$, and $R$ are constants. This is the Van der Waals' equation.

Compute $\dfrac{\partial T}{\partial P}$ and $\dfrac{\partial V}{\partial T}$ .

16b.    The volumetric flow rate $Q$ of a non-Newtonian fluid in a circular tube of radius $R$ is given by

$$Q(R, \tau_R) = \frac{\pi R^3}{\tau_R^3} \int_0^{\tau_R} \dot{\gamma}_{rz} \, \tau_{rz}^2 \, d\tau_{rz}$$

where $\tau_R$ and $\tau_{rz}$ are the shear stress at the wall and at any point in the tube respectively. $\dot{\gamma}_{rz}$ is the shear rate, which is a function of the shear stress. By differentiating with respect to $\tau_R$,

show that the shear rate at the wall $\dot{\gamma}_R$ is given by

$$\dot{\gamma}_R = \frac{\tau_R}{\pi R^3}\left[\frac{3Q}{\tau_R} + \frac{dQ}{d\tau_R}\right]$$

Also, show that the viscosity $\eta\,(\dot{\gamma}_R)$, which is defined as the ratio of the shear stress to the shear rate, can be written as

$$\eta\left(\dot{\gamma}_R\right) = \frac{\tau_R}{\left(Q/\pi R^3\right)}\left[3 + \frac{d\left(\ell n\, Q/\pi r^3\right)}{d\left(\ell n\, \tau_R\right)}\right]^{-1}$$

The above result is known as the Weissenberg-Rabinowitsch equation.

17a.    Determine the order and degree of

(i)    $\dfrac{d^2y}{dx^2} + 3\dfrac{dy}{dx} = \cos x$                    (ii)   $(y')^2 - x\,y' = 0$

18a.    Verify that $e^{-2x}(c_1 \cos x + c_2 \sin x)$ is a solution of $y'' + 4y' + 5y = 0$.

19a.    Solve the following separable first-order differential equations

(i)      $x^2\, dx + 3y^3\, dy = 0$                                   Answer: $\dfrac{x^3}{3} = -\dfrac{3}{4}y^4 + c$

(ii)     $xy\, dx + \sqrt{1-x^2}\, dy = 0$                           Answer: $y = c\, e^{\sqrt{1-x^2}}$

(iii)    $y' = \dfrac{\sin x}{\cos y}$ $\left(\text{given that at } x = \pi,\, y = \dfrac{\pi}{2}\right)$            Answer: $\sin y = -\cos x$

20b.    In a constant-volume batch reactor, the rate of disappearance of reactant A can be given by

$$\frac{dC_A}{dt} = -k\,C_A^n$$

Solve for $C_A$ given that $C_A = C_{Ao}$ at $t = 0$. Discuss the obtained result for the cases $n > 1$ and $n < 1$.

21b.    In the reversible chemical reaction $A \underset{k_2}{\overset{k_1}{\underset{\leftarrow}{\rightarrow}}} B + C$, the amount of component A, broken down at time $t$ is represented by $\chi$.

$$\frac{d\chi}{dt} = k_1\left(A_0 - \chi\right) - k_2\,\chi^2$$

with $A_0$ representing the initial concentration of chemical A.

Assuming the rate constants $k_1$ and $k_2$ to be related as follows: $k_1 = \dfrac{A_0}{2}\,k_2$, show that

$$k_1 = \frac{1}{3t} \ln\left(\frac{A_0 + \chi}{A_0 - 2\chi}\right)$$

given the initial condition that $\chi = 0$ at $t = 0$.

22a. Solve the following homogeneous first-order differential equations

(i) $(x + y)\,dx + x\,dy = 0$ $\qquad\qquad$ Answer: $x^2 + 2xy = c$

(ii) $x\,dy - y\,dx = \sqrt{x^2 + y^2}\,dx$ $\qquad$ Answer: $y + \sqrt{x^2 + y^2} = c\,x^2$

(iii) $(2\sqrt{St} - S)\,dt + t\,dS = 0$ $\qquad$ Answer: $t\exp\sqrt{\frac{S}{t}} = c$

23b. A mixture of liquids A and B is boiling in a vessel. The volumes of the components are $V_A(t)$ and $V_B(t)$. At $t = 0$, the initial volumes are $V_A(0)$ and $V_B(0)$. The evaporation of component A is proportional to the volume $V_A(t)$ of A.

That is to say, $\dfrac{dV_A(t)}{dt} = -\alpha V_A(t)$

The evaporation of component B is related to the evaporation of component A as follows

$$\frac{dV_B(t)}{dt} = \frac{dV_A(t)}{dt} - \beta V_B(t)\,.$$

Show that

$$\beta \ln\frac{V_A(t)}{V_A(0)} = \alpha \ln\left[\frac{\alpha V_A(t) + (\beta - \alpha) V_B(t)}{\alpha V_A(0) + (\beta - \alpha) V_B(0)}\right]$$

24a. Solve the following reducible first-order differential equations

(i) $(y - 3x)\,dy + (y - 3x + 2)\,dx = 0$ $\qquad$ Answer: $2(y + x) - \ln(2y - 6x + 1) = c$

(ii) $(2y + x + 1)\,dx = (2x + 4y + 3)\,dy$ $\qquad$ Answer: $4x - 8y - \ln(4x + 8y + 5) = c$

25a. Solve the following exact first-order differential equations

(i) $x\,dx + y\,dy = (x^2 + y^2)\,dy$ $\qquad$ Answer: $\ln(x^2 + y^2) = 2y + c$

(ii) $y\,e^{xy}\,dx + x\,e^{xy}\,dy = 0$ $\qquad$ Answer: $e^{xy} = c$

26a. Solve the following linear first-order differential equations

(i)     $y' + \dfrac{2}{x} y = x$ ;   $y(1) = 0$                               Answer: $y = \dfrac{1}{4}(-x^{-2} + x^2)$

(ii)    $y' - 7y = e^x$                                                       Answer: $y = c\, e^{7x} - \dfrac{1}{6} e^x$

27b.    Viscoelastic behavior can be described via a Maxwell model given by

$$\tau_{yx} + \lambda_0 \dfrac{d}{dt}\, \tau_{yx} = -\mu\, \dot{\gamma}_{yx}$$

where $\tau_{yx}$ is the shear stress, $\dot{\gamma}_{yx}$ is the shear rate, $\mu$ is the viscosity, and $\lambda_0$ is the relaxation time.  Show, using the appropriate integrating factor that the Maxwell model can be written as

$$\tau_{yx} = -\int_{-\infty}^{t} \left[ \dfrac{\mu}{\lambda_0}\, e^{-(t-\,t')/\lambda_0} \right] \dot{\gamma}_{yx}(t')\, dt'$$

where the dummy integration variable $t'$ is interpreted as a time in the past.  That is, $t' < t$.

28a.    Complete Example 1.14-2 by computing the time evolution of $c_C$.  Discuss the particular cases where $k_2 \gg k_1$ and $k_2 \ll k_1$.

29b.    Solve the following Bernouilli equations

(i)     $\dfrac{dy}{dx} + 2xy = -x\, y^4$                                     Answer: $y^{-3} = -\dfrac{1}{2} + c\, e^{3x^2}$

(ii)    $2\,(x^2 + 7x - 8)\dfrac{dy}{dx} + (6x + 21)\, y = 3\,(x + 8)^2\, y^{5/3}$

                                                                             Answer: $y^{-2/3} = (x + 8)\,[1 + c\,(x - 1)]$

30a.    Solve the following second-order differential equations with constant coefficients

(i)     $y'' - 4y' + 7y = e^{2x}$           Answer: $y = e^{2x}\,(c_1 \cos \sqrt{3}\, x + c_2 \sin \sqrt{3}\, x) + \dfrac{e^{2x}}{3}$

(ii)    $y'' + 4y = 4 \cos 2x$                          Answer: $y = (c_1 + x) \sin 2x + c_2 \cos 2x$

31a.    Solve the following differential equations via Laplace transform

(i)     $y'(t) - 5y(t) = 0$  (subject to initial condition $y(0) = 2$)          Answer: $y(t) = 2\, e^{5t}$

(ii)    $y''(t) + y(t) = 2$  ($y(0) = 0$ and $y'(0) = 3$)          Answer: $y(t) = 2 + 3 \sin t - 2 \cos t$

32b.   Solve the following system of differential equations via Laplace transform

$$\begin{cases} y_1'' = y_1 + 3y_2 \\ \\ y_2'' = 4y_1 - 4e^t \end{cases}$$

for the following initial conditions: $y_1(0) = 2$, $y_1'(0) = 3$, $y_2(0) = 1$, $y_2'(0) = 2$.

Answer:  $y_1 = e^t + e^{2t}$
$y_2 = e^{2t}$

33a.   The differential equation, describing the motion of a manometer fluid, subject to a sudden pressure difference $\Delta P = P_a - P_b$ is given by

$$\frac{d^2k}{dt^2} + \frac{6\mu}{R^2\rho}\frac{dk}{dt} + \frac{3gk}{2L} = 0$$

where $k = 2h - \left(\dfrac{\Delta P}{\rho g}\right)$

$\rho$ is the density of the fluid, $\mu$ is the fluid viscosity, $L$ is the length of the manometer fluid, $R$ is the manometer tube radius, and $2h$ is identified in Figure 1.P-33. Determine the relation between $k$ and $t$, for a step change in $\Delta P$, at $t = 0$. The conditions on $h$ at $t = 0$ are: $h = 0$ and $\dfrac{dh}{dt} = 0$.

34a.   The differential equation governing the mixing process, illustrated by Figure 1.P-34, is given by the following unsteady-state macroscopic mass balance

$$\frac{d}{dt}m_{tot} = -\Delta w$$

where $m_{tot} = V\rho_0(t)$ is the total mass, and where $w = \rho Q$. $Q$ is the volumetric flow rate and $\rho$ is the density.

At steady state, the mass balance reduces to

$$w_{is} - w_{0s} = 0$$

At time $t = 0$, a step change $\Delta w_i$ is imposed.

Determine $w_0(t)$, the mass flow rate at the outlet, using the Laplace transform method.

Answer: $w_0(t) = \left(w_{is} + \Delta w_i\right) + \dfrac{V}{Q}e^{-\frac{V}{Q}t}\left(w_{0s} - w_{is} - \Delta w_i\right)$

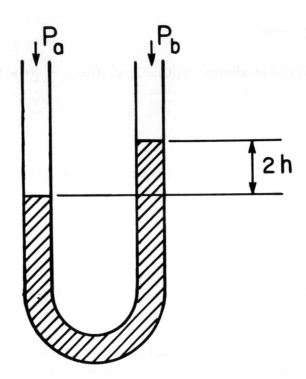

**FIGURE 1.P-33    Motion of a manometer fluid**

35a.    In Problem 27b, it was deduced that the constitutive equation of a Maxwell fluid can be written
as

$$\tau_{yx} = -\frac{\mu}{\lambda_0} \int_{-\infty}^{t} e^{-(t-t')/\lambda_0} \, \dot{\gamma}_{yx}(t') \, dt'$$

A Maxwell fluid has been at rest and, at time $t = a$, a shear of magnitude $\gamma_0$ was suddenly
imposed. Using the properties of the Heaviside and Dirac functions, deduce the shear stress
$\tau_{yx}$ at time $t$. Consider separately the cases when $t > a$ and $t < a$.

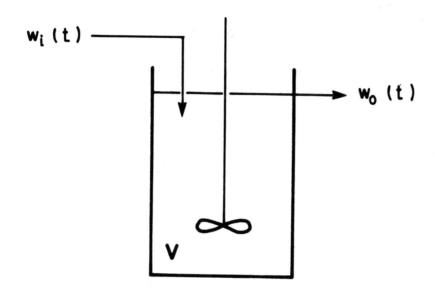

**FIGURE 1.P-34    Mixing process at constant volumetric flow rate**

36b.   Solve the differential equation

$$\frac{dy}{dt} + y = \delta(t - a)$$

with  $y(0) = 1,\ a > 0,$

(i)      by the Laplace transform method,

(ii)     by finding the integrating factor.

Consider the cases  $t > a$  and  $t < a$ separately.                Answer:  $t > a,\ y = e^{-t} + e^{a-t}$

$$t < a,\ y = e^{-t}$$

37b.   Construct the Green's function, needed to solve Equation (1.18-6), by applying the conditions given by Equations (1.18-2 to 5).  That is to say

(i)      determine the solution to the homogeneous equation for the cases  $t < x$  and  $t > x$;

(ii)     use the boundary and continuity conditions to determine the relations which have to be satisfied by the constants;

(iii)    use the jump discontinuity condition of Equation (1.18-4) to determine the remaining equation needed to solve for the constants that were generated in step (i).

Then solve Equation (1.18-6) for $f(x) = \sin x$, using Equation (1.18-1).

Answer: $\dfrac{1}{2} \dfrac{\cos 1}{\sin 1} \sin x - \dfrac{1}{2} x \cos x$

38a.    Solve the boundary-value problem

$$y'' - y = e^x, \qquad 0 \le x \le 1$$

$$y(0) = y(1) = 0$$

by the Green's function method. Choose the appropriate Green's function from Table 1.18-1. Verify that the Green's function you have chosen satisfies conditions (i) to (v).

Answer : $y = \dfrac{x}{2} e^x - \dfrac{e \sinh x}{2 \sinh 1}$

39b.    A uniform string of unit length and of mass $m$ lies along the x-axis. Its two ends $x = 0$ and $x = 1$ are fixed. The tension is $T$. A periodic force $f(x) \cos \omega t$ is applied to the string. $\omega$ is a constant and $t$ is the time. If $y$ is the vertical displacement of the string, then $y$ satisfies the following equations

$$c^2 \frac{\partial^2 y}{\partial x^2} - \frac{\partial^2 y}{\partial t^2} = f(x) \cos \omega t$$

$$c^2 = T / m$$

$$y(0, t) = y(1, t) = 0.$$

Assume that $y$ is of the form

$$y(x, t) = h(x) \cos \omega t$$

and determine the differential equation and the boundary conditions that $h$ has to satisfy. Solve for $h$ using the appropriate Green's function. Hence determine $y$ for the following $f$

(i)      $f(x) = a_0 \delta(x - 1/2)$, $a_0$ is a constant and $\delta$ is the Dirac delta function.

(ii)     $f(x) = x$.

40b.    A first-order reaction $A \longrightarrow$ products takes place in a tubular reactor of radius $R$ and length $L$ in the $z$ direction.

Perform a mass balance over a differential element of thickness $\Delta z$, to generate the appropriate expression for the flux $N_{Az}$ defined by Bird et al. (1960).

Differentiate $N_{Az}$ with respect to $z$ to generate Equation (1.16-44).

41b. In order to accelerate the aeration of a Newtonian fluid, a continuous belt has been introduced in the fluid reservoir, as shown in Figure 1.P-41. The belt has a width W and a velocity V. The belt transports a laminar film of liquid, up to a wall C as illustrated. As the air penetration in the liquid film depends on the film thickness, we wish to establish a relation between the film thickness $\delta$ and the velocity V. Show that this can be done in the following way.

(i) Perform a momentum (or force) balance over a length L to obtain the following shear stress distribution

$$\tau_{xz} = \rho\, g\, x \cos \beta$$

List all assumptions made.

(ii) Combine Newton's law with this equation and show the velocity profile to be

$$v_z = -V + \frac{\rho\, g \cos \beta}{2\mu}\left(\delta^2 - x^2\right)$$

(iii) Calculate the flow rate

$$Q = \int_0^w \int_0^\delta v_z\, dx\, dy$$

and deduce from the physics of the situation that

$$\delta^2 = \frac{3\mu V}{\rho\, g \cos \beta}$$

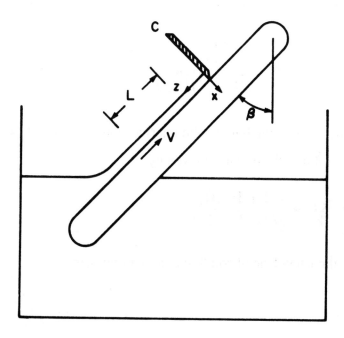

**FIGURE 1.P-41    Aeration of a fluid**

42b.   In the design of spray dryers, one is interested in knowing the time required to solidify the liquid droplets. Estimate the time $t_f$ required to solidify a droplet of radius R if

(i)     the droplet is initially at the melt temperature $T_0$ and the surrounding air is at $T_\infty$ as shown in Figure 1.P-42,

(ii)    the heat transfer coefficient h at the solid gas interface is constant,

(iii)   the sensible heat required to cool the droplet from $T_0$ to $T_\infty$ is negligible compared to the latent heat of fusion.

Perform an energy balance on the solidified spherical portion and show that the temperature profile is given by

$$\frac{T - T_0}{T_s - T_0} = \frac{R_f^{-1} - r^{-1}}{R_f^{-1} - R^{-1}}$$

In performing this calculation, note that the heat flux $q_r$ is linearly related to the temperature gradient $\frac{dT}{dr}$ as follows

$$q_r = -k \frac{dT}{dr}$$

This is Fourier's law. The constant k is the thermal conductivity.

Show that the heat loss to the surrounding air is given by

$$Q = 4\pi R^2 q_r \big|_{r=R} = \frac{4\pi R^2 h (T_0 - T_\infty)}{\left(\dfrac{R}{R_f} - 1\right)\left(\dfrac{R\,h}{k}\right) + 1}$$

Given that the heat loss at the liquid-solid interface ($R_f$) is $-\rho\, \Delta\hat{H}_f\, 4\pi R_f^2 \dfrac{dR_f}{dt}$, show that the time required to solidify the droplet is given by

$$t_f = \left(\frac{1}{6} \frac{R\,h}{k} + \frac{1}{3}\right)\left[\frac{\rho\, R\, \Delta\hat{H}_f}{h\,(T_0 - T_\infty)}\right]$$

where $\Delta\hat{H}_f$ is the latent heat of solidification per unit mass.

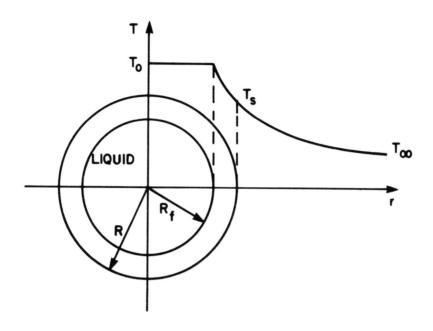

**FIGURE 1.P-42**     **Solidification of a droplet**

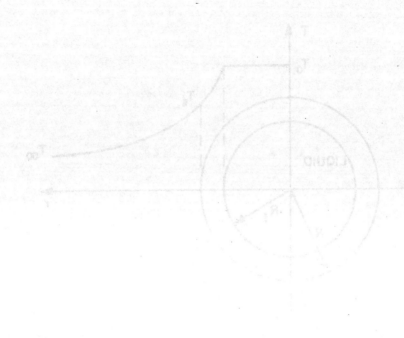

FIGURE 1.P-42   Solidification of a droplet

# CHAPTER 2

# SERIES SOLUTIONS AND SPECIAL FUNCTIONS

## 2.1  DEFINITIONS

In Chapter 1, we reviewed several of the standard techniques used to solve ordinary differential equations (O.D.E.'s). In particular, we have seen that second order linear differential equations with constant coefficients admit a solution in the form of an exponential function. Since the derivatives of exponential functions are also exponential functions, the differential equation reduces to an algebraic equation [see Equations (1.16-8, 9)]. If the coefficients of the differential equation are not constants, then the solution is not of an exponential form. In this chapter, we develop methods of solving O.D.E.'s with variable coefficients. As in Chapter 1, we consider second order O.D.E.'s. Higher order O.D.E.'s can be solved by the same method.

A second order O.D.E. can be written as

$$a_2(x) \, y'' + a_1(x) \, y' + a_0(x) \, y = a(x) \tag{2.1-1}$$

where $a_2(x) \neq 0$ and $'$ denotes differentiation with respect to $x$.

Equation (2.1-1) can be written in standard form as

$$y'' + p_1 \, y' + p_2 \, y = b \tag{2.1-2a}$$

where $\quad p_1 = a_1 / a_2, \qquad p_2 = a_0 / a_2, \qquad b = a / a_2 \tag{2.1-2b,c,d}$

We recall that if $b$ is zero, Equation (2.1-2a) is homogeneous. The solution of Equation (2.1-2a) can be written as

$$y = y_h + y_p \tag{2.1-3}$$

where $y_h$ is the homogeneous solution and $y_p$ is the particular integral. If $b$ is zero, we have only the homogeneous solution. The particular integral (or particular solution) is usually obtained by the method of variation of parameters which is described in Example 1.18-1 and Section 2.5.

The solution of the homogeneous equation in the neighborhood of a point $x_0$ is assumed to be given by a power series in $(x - x_0)$. The form of this series depends on the nature of the point $x_0$. The point $x_0$ is an **ordinary point** if both $p_1$ and $p_2$ are analytic at the point $x_0$. We recall that a function $f(x)$ is analytic at $x_0$ if its Taylor series about $x_0$ exists. That is to say, $f(x)$ can be represented by the series

$$f(x) = \sum_{n=0}^{\infty} a_n (x - x_0)^n \tag{2.1-4a}$$

$$a_n = f^{(n)}(x_0) / n! \tag{2.1-4b}$$

If one or the other (or both) coefficients $(p_1, p_2)$ is not analytic at $x_0$, $x_0$ is a **singular point**. From Equations (2.1-2b, c), it can be seen that if $a_2(x_0)$ is zero and either $a_1(x_0)$ or $a_0(x_0)$ is non-zero, then $x_0$ is a singular point.

*Example 2.1-1*. Analyze the following equations for ordinary and singular points.

i)       $y'' + x\, y' + (x^2 - 4)\, y = 0$                                                               (2.1-5a)

ii)      $(x - 1)\, y'' + x\, y' + \dfrac{1}{x}\, y = 0$                                                            (2.1-5b)

iii)     $x^2 (x - 2)^2\, y'' + 2(x - 2)\, y' + (x + 1)\, y = 0$                          (2.1-5c)

We proceed by evaluating the functions $p_1$, $p_2$, and $b$ and by determining the presence or absence of singular points.

i)       This equation is in standard form and

$$p_1 = x, \qquad p_2 = x^2 - 4, \qquad \text{and } b = 0 \tag{2.1-6a,b,c}$$

These functions are analytic everywhere and all points are ordinary points.

ii)      The standard form of this equation is given by

$$y'' + \frac{x}{x-1}\, y' + \frac{1}{x(x-1)}\, y = 0 \tag{2.1-7a}$$

$$p_1 = \frac{x}{x-1}, \qquad p_2 = \frac{1}{x(x-1)}, \qquad \text{and } b = 0 \tag{2.1-7b,c,d}$$

$p_1$ is not analytic at $x = 1$ and $p_2$ is not analytic at $x = 0$ and at $x = 1$. The singular points are therefore at $x = 0$ and at $x = 1$ and all other points are ordinary points.

iii)     The standard form of this equation is given by

$$y'' + \frac{2(x-2)}{x^2(x-2)^2}\, y' + \frac{x+1}{x^2(x-2)^2}\, y = 0 \tag{2.1-8a}$$

$$p_1 = \frac{2}{x^2(x-2)} , \qquad p_2 = \frac{x+1}{x^2(x-2)^2} , \qquad \text{and } b = 0 \qquad (2.1\text{-}8b,c,d)$$

Here, the singular points are at $x = 0$ and at $x = 2$.

If $x_0$ is a singular point of the differential equation and $(x - x_0) p_1$ as well as $(x - x_0)^2 p_2$ are both analytic at $x_0$, $x_0$ is a **regular singular point** of the differential equation. If not, $x_0$ is an **irregular singular point**.

***Example 2.1-2.*** Determine which of the singular points in Example 2.1-1 are regular singular points, assuming $x_0 = 0$.

i)      All points are ordinary points.

ii)

$$(x - x_0) p_1 = x p_1 = \frac{x^2}{x-1} \qquad (2.1\text{-}9a,b)$$

$$(x - x_0)^2 p_2 = x^2 p_2 = \frac{x}{x-1} \qquad (2.1\text{-}9c,d)$$

Both $(x - x_0) p_1$ and $(x - x_0)^2 p_2$ are analytic at $x = 0$ and therefore $x_0 = 0$ is a regular singular point.

iii)

$$(x - x_0) p_1 = x p_1 = \frac{2}{x(x-2)} \qquad (2.1\text{-}10a,b)$$

$$(x - x_0)^2 p_2 = x^2 p_2 = \frac{x+1}{(x-2)^2} \qquad (2.1\text{-}10c,d)$$

$x p_1$ is not analytic at $x = 0$ and therefore $x_0 = 0$ is an irregular singular point.

Note that had we chosen $x_0 = 2$ in part (iii), we would be dealing with

$$(x - 2) p_1 = \frac{2}{x^2} \qquad \text{and} \qquad (x - 2)^2 p_2 = \frac{x+1}{x^2} \qquad (2.1\text{-}11a,b)$$

which are analytic at $x_0 = 2$ and therefore $x_0 = 2$ would be a regular singular point.

●

In this chapter, we use power series extensively and we next summarize their properties.

## 2.2   POWER SERIES

The series on the right side of Equation (2.1-4a) is a **power series** in $(x - x_0)$. By translating the origin to $x_0$, we can write the series as $\sum_{n=0}^{\infty} a_n x^n$. The series converges at a point $x$ if the $\lim_{m \to \infty} \sum_{n=0}^{m} a_n x^n$ exists and the sum of the series is the value of this limit. Every power series converges at $x = 0$ and its sum is $a_0$. Not all power series converge for all non-zero values of $x$. There are three possibilities: (a) the series converges for all values of $x$, (b) the series converges for some values of $x$ and diverges for other values of $x$, and (c) the series diverges for all non-zero values of $x$. We illustrate this situation by the following examples.

(a)  $\quad \sum_{n=0}^{\infty} x^n/(n!) = 1 + x + x^2/2! + x^3/3! + ...$    (2.2-1a)

(b)  $\quad \sum_{n=0}^{\infty} x^n = 1 + x + x^2 + x^3 + ...$    (2.2-1b)

(c)  $\quad \sum_{n=0}^{\infty} n! \, x^n = 1 + x + 2! \, x^2 + 3! \, x^3 + ...$    (2.2-1c)

The first series represents $\exp(x)$ and is convergent for all values of $x$. The second series is a geometric series and its sum is $1/(1-x)$ and is convergent if $|x| < 1$ and diverges if $|x| \ge 1$. The third series diverges for all values of $x$ ($\ne 0$). If a series converges for $|x| < R$ and diverges for $|x| > R$, $R$ is the **radius of convergence**. It is usual to state that $R$ is zero for series that converge only at $x = 0$ and $R$ is infinite for series that converge for all values of $x$. Thus all power series have a radius of convergence. A series may or may not converge at its radius of convergence.

In many cases, the radius of convergence $R$ can be determined by the **ratio test**. We recall that the series $\sum_{n=0}^{\infty} u_n$ is convergent if the ratio

$$\lim_{n \to \infty} \left| \frac{u_{n+1}}{u_n} \right| = L$$    (2.2-2)

is less than one ($L < 1$). If $L > 1$, the series is divergent. No conclusion can be drawn if $L = 1$.

Applying this test to the series $\sum_{n=0}^{\infty} a_n x^n$, we have ($x \ne 0$)

$$\lim_{n \to \infty} \left| \frac{a_{n+1} \, x^{n+1}}{a_n \, x^n} \right| = \lim_{n \to \infty} \left| \frac{a_{n+1}}{a_n} \right| |x| = L \qquad (2.2\text{-}3a,b)$$

Thus if

$$\lim_{n \to \infty} \left| \frac{a_{n+1}}{a_n} \right| |x| < 1 \quad \Rightarrow \quad |x| < \lim_{n \to \infty} \left| \frac{a_n}{a_{n+1}} \right| \qquad (2.2\text{-}4a,b)$$

the series is convergent. If

$$|x| > \lim_{n \to \infty} \left| \frac{a_n}{a_{n+1}} \right| \qquad (2.2\text{-}5)$$

the series is divergent. It follows that the radius of convergence R is given by

$$R = \lim_{n \to \infty} \left| \frac{a_n}{a_{n+1}} \right| \qquad (2.2\text{-}6)$$

The **comparison test** is another simple method of determining whether a series is convergent or divergent. If $\sum\limits_{n=0}^{\infty} u_n$ is convergent, the series $\sum\limits_{n=0}^{\infty} v_n$ is convergent if $v_n \leq K \, u_n$ ; if $\sum\limits_{n=0}^{\infty} u_n$ is divergent, the series $\sum\limits_{n=0}^{\infty} v_n$ is divergent if $v_n \geq K \, u_n$ for all $n$ and any positive constant K. This test can also be stated as follows. If the ratio $v_n/u_n$ tends to a finite non-zero limit as $n \longrightarrow \infty$, then $\sum\limits_{n=0}^{\infty} v_n$ converges or diverges according as $\sum\limits_{n=0}^{\infty} u_n$ converges or diverges.

The series $\sum\limits_{n=1}^{\infty} 1/n$ is divergent and the series $\sum\limits_{n=1}^{\infty} 1/n^2$ is convergent.

***Example 2.2-1.*** Discuss the convergence of the following power series.

a) $\quad \sum\limits_{n=0}^{\infty} x^n/(1+n)^2$
b) $\quad \sum\limits_{n=0}^{\infty} x^n/(1+n)$
c) $\quad \sum\limits_{n=0}^{\infty} (x-2)^{2n}/(1+2n)$

a) $\qquad R = \lim_{n \to \infty} \left| \frac{(2+n)^2}{(1+n)^2} \right| = \lim_{n \to \infty} \left| \frac{(1+2/n)^2}{(1+1/n)^2} \right| = 1 \qquad (2.2\text{-}7a,b,c)$

The series is convergent if $|x| < 1$ and is divergent if $|x| > 1$. For $x = 1$, the series becomes $\sum_{n=0}^{\infty} 1/(1+n)^2$ and by comparison with the convergent series $\sum_{n=1}^{\infty} 1/n^2$, we deduce that the series is also convergent if $|x| = 1$.

b)  $\qquad\qquad R = \lim_{n \to \infty} \left| \frac{2+n}{1+n} \right| = 1$  $\qquad\qquad\qquad\qquad\qquad$ (2.2-8)

As in a), the series is convergent if $|x| < 1$ and divergent if $|x| > 1$. For $x = 1$, the series becomes $\sum_{n=0}^{\infty} 1/(1+n)$ and since $\sum_{n=1}^{\infty} 1/n$ is divergent, it follows that $\sum_{n=0}^{\infty} 1/(1+n)$ is also divergent. If $x = -1$, the series $\sum_{n=0}^{\infty} (-1)^n/(1+n)$ is an alternating series. We apply **Leibnitz's test**, which states that for an alternating series $\sum_{n=0}^{\infty} (-1)^n u_n$, if $u_n > u_{n+1} > 0$ and $\lim_{n \to \infty} u_n$ is zero, the series is convergent. In the present example, the series is convergent if $x = -1$.

c)  $\qquad$ In this case, we translate the origin and write

$\qquad\qquad x^* = x - 2$  $\qquad\qquad\qquad\qquad\qquad\qquad\qquad\qquad\qquad\qquad$ (2.2-9)

The series becomes $\sum_{n=0}^{\infty} (x^*)^{2n}/(1+2n)$.

$\qquad\qquad R = \lim_{n \to \infty} \left| \frac{3+2n}{1+2n} \right| = 1$  $\qquad\qquad\qquad\qquad\qquad$ (2.2-10a,b)

The series is convergent if $|x^*| < 1$ or $|x - 2| < 1$ and is divergent if $|x^*| > 1$ or $|x - 2| > 1$. In this example $x^*$ is raised to an even power and there is no need to separately consider the cases $x^* = 1$ and $x^* = -1$. To $|x^*| = 1$ corresponds $x = 1$ or $x = 3$. At both of these values of $x$, the series behaves as $\sum_{n=0}^{\infty} 1/n$ and is divergent.

●

We make use of the following properties of power series.

If the series $\sum_{n=0}^{\infty} a_n x^n$ converges for $|x| < R$ $(R \neq 0)$ and its sum is denoted by $f(x)$, that is to say

$$f(x) = \sum_{n=0}^{\infty} a_n x^n \qquad (2.2\text{-}11)$$

the series can be differentiated term by term as many times as is required. The differentiated series have the same radius of convergence $R$ and converge to the corresponding derivatives of $f(x)$. The series can also be integrated term by term and the resulting series represents the integral of $f(x)$.

If $\sum_{n=0}^{\infty} b_n x^n$ is another power series with radius of convergence $R$, the two series $\sum_{n=0}^{\infty} a_n x^n$ and $\sum_{n=0}^{\infty} b_n x^n$ can be added and multiplied in the same way as polynomials.

If two power series converge to the same sum throughout their interval of convergence, their coefficients are equal. Further discussions on power series are given in Chapter 3.

## 2.3   ORDINARY POINTS

Consider the standard form of the second order O.D.E. as given by Equation (2.1-2a). We seek a solution in the neighborhood of $x_0$. Without loss of generality, we can set $x_0$ to be zero. If $x_0$ is non-zero, we can translate the origin and write

$$x^* = x - x_0 \qquad (2.3\text{-}1a)$$

In the new variable $x^*$, $x = x_0$ corresponds to $x^* = 0$.

If we are required to find the solution at points near infinity, we change the independent variable from $x$ to $x_1$ and write

$$x_1 = 1/x \qquad (2.3\text{-}1b)$$

To points near $x$ at infinity correspond to points near $x_1$ at the origin.

The functions $p_1$ and $p_2$ are analytic at $x_0 \ (= 0)$ and their Taylor series are

$$p_1(x) = \sum_{n=0}^{\infty} p_{1n} x^n \qquad (2.3\text{-}2a)$$

$$p_2(x) = \sum_{n=0}^{\infty} p_{2n} x^n \qquad (2.3\text{-}2b)$$

where $p_{1n} = p_1^{(n)}(0)$ and $p_{2n} = p_2^{(n)}(0)$      $(2.3\text{-}2c,d)$

The homogeneous form of Equation (2.1-2a) can be written as

$$y" + \left( \sum_{n=0}^{\infty} p_{1n} x^n \right) y' + \left( \sum_{n=0}^{\infty} p_{2n} x^n \right) y = 0 \qquad (2.3-3)$$

The solution $y$ also has a Taylor series near the origin and $y$ can be expressed as

$$y = \sum_{n=0}^{\infty} c_n x^n \qquad (2.3-4a)$$

where $c_n$ $(n = 0, 1, ... )$ are constants.

Differentiating term by term yields

$$y' = \sum_{n=1}^{\infty} n c_n x^{n-1} \qquad (2.3-4b)$$

$$y" = \sum_{n=2}^{\infty} n(n-1) c_n x^{n-2} \qquad (2.3-4c)$$

Substituting Equations (2.3-4a to c) into Equation (2.3-3) results in an equation of the form

$$k_0 + k_1 x + k_2 x^2 + ... + k_n x^n + ... = 0 \qquad (2.3-5)$$

where $k_i$ are known expressions in terms of $p_{1j}$, $p_{2k}$, and $c_\ell$ .

Equation (2.3-5) is true for all $x$ and this implies that each of the $k_i$ $(i = 0, 1, ... )$ is zero. The **recurrence equation** $(k_i = 0)$ allows us to determine $c_n$ in terms of $p_{1i}$ and $p_{2j}$. The next example illustrates the method of obtaining a **series solution**.

***Example 2.3-1***.  Obtain the power series solution of

$$(1 + x^2) y" + 2xy' - 2y = 0 \qquad (2.3-6)$$

in the neighborhood of the origin.

In standard form, Equation (2.3-6) is written as

$$y" + \frac{2x}{1 + x^2} y' - \frac{2}{1 + x^2} y = 0 \qquad (2.3-7)$$

The functions $p_1$ $[= 2x/(1 + x^2)]$ and $p_2$ $[= -2/(1 + x^2)]$ are analytic at the origin. We seek a solution given by Equation (2.3-4a) with $y'$ and $y''$ given by Equations (2.3-4b, c) respectively. For ease of computation, we write Equations (2.3-4a to c) such that the summation index starts from zero. To achieve this, we write $(n = r, \quad n = r + 1, \quad n = r + 2)$ in Equations (2.3-4a, b, c) respectively. These equations are now written as

$$y = \sum_{r=0}^{\infty} c_r x^r \tag{2.3-8a}$$

$$y' = \sum_{r=0}^{\infty} (r + 1) c_{r+1} x^r \tag{2.3-8b}$$

$$y'' = \sum_{r=0}^{\infty} (r + 2)(r + 1) c_{r+2} x^r \tag{2.3-8c}$$

To avoid having to expand $(1 + x^2)^{-1}$, we work with Equation (2.3-6) rather than with Equation (2.3-7). Substituting Equations (2.3-8a to c) into Equation (2.3-6), we obtain

$$\sum_{r=0}^{\infty} [(r + 2)(r + 1) c_{r+2} x^r + (r + 2)(r + 1) c_{r+2} x^{r+2} + 2 (r + 1) c_{r+1} x^{r+1} - 2c_r x^r] = 0$$

$$\tag{2.3-9}$$

Comparing powers of $x$, we have

$x^0$: $\quad 2c_2 - 2c_0 = 0 \quad \Rightarrow \quad c_2 = c_0$ $\hfill$ (2.3-10a,b)

$x^1$: $\quad 6c_3 + 2c_1 - 2c_1 = 0 \quad \Rightarrow \quad c_3 = 0$ $\hfill$ (2.3-10c,d)

$x^2$: $\quad 12c_4 + 2c_2 + 4c_2 - 2c_2 = 0 \quad \Rightarrow \quad c_4 = -c_2/3$ $\hfill$ (2.3-10e,f)

.
.
.

$x^s$: $\quad (s + 2)(s + 1) c_{s+2} + s(s - 1) c_s + 2s c_s - 2c_s = 0 \quad \Rightarrow \quad c_{s+2} = -(s - 1) c_s/(s + 1)$ (2.3-10g,h)

We note from Equation (2.3-10h) that we have a formula relating $c_{s+2}$ and $c_s$ and this implies that we can separate the solution into an even and an odd solution. That is to say, we obtain $c_2$, $c_4$, $c_6$, ... in terms of $c_0$ and $c_3$, $c_5$, $c_7$, ... in terms of $c_1$. In the present example, $c_3$ is zero and consequently all the coefficients with an odd index greater than or equal to three are zero. It follows that the solution can be written as

$$y = c_0 [1 + x^2 - x^4/3 + x^6/5 - ... ] + c_1 x \tag{2.3-11}$$

where $c_0$ and $c_1$ are arbitrary constants.

Note that Equation (2.3-6) is a second order O.D.E and has two linearly independent solutions $y_1$ and $y_2$. The general solution is given by the linear combination of $y_1$ and $y_2$. There is no loss of generality in setting $c_0$ and $c_1$ to be equal to 1. The **fundamental solutions** $y_1$ and $y_2$ can be written as

$$y_1 = 1 + x^2 - x^4/3 + x^6/5 - \dots, \qquad y_2 = x \qquad (2.3\text{-}12a,b)$$

The **general solution** $y$ is expressed as

$$y = Ay_1 + By_2 \qquad (2.3\text{-}13)$$

where A and B are arbitrary constants.

The general solution of a second order equation involves two arbitrary constants (A and B) and they are determined by the initial conditions [$y(0)$ and $y'(0)$] or by the boundary conditions.

In Example 2.3-1, $y_2$ ($= x$) has only one term and is a valid solution for all values of $x$. The solution $y_1$ is in the form of an infinite series and is valid as long as the series is convergent. From Equation (2.3-11), it can be seen that $y_1$ can be written as

$$y_1 = 1 + x (x - x^3/3 + x^5/5 - \dots) \qquad (2.3\text{-}14)$$

The infinite series is convergent for $|x| < 1$. At $|x| = 1$, the series is an alternating series and by Leibnitz's test, it is convergent. The solution $y_1$ is valid for $|x| \le 1$.

On expanding $p_1(x)$ and $p_2(x)$ in powers of $x$, we obtain

$$p_1(x) = 2x (1 - x^2 + x^4 - x^6 + \dots) \qquad (2.3\text{-}15a)$$

$$p_2(x) = -2 (1 - x^2 + x^4 - x^6 + \dots) \qquad (2.3\text{-}15b)$$

We note that the series in Equations (2.3-15a, b) are convergent for $|x| < 1$ and $p_1(x)$ and $p_2(x)$ are analytic in the interval $|x| < 1$. The point ($x = 1$) is a singular point (see Example 3.6-5). We observe that the series solution about an ordinary point is valid in an interval that extends at least up to the nearest singular point of the differential equation. This observation is not restricted to Example 2.3-1 and can be generalized to all power series solutions about an ordinary point. It must be pointed out that it does not follow that there is no analytic solution that goes beyond the critical point of the differential equation. In Example 2.3-1, the solution $y_2$ ($= x$) is valid for all values of $x$. The presence of singular points in a differential equation does not imply that all the solutions are singular at the singular points. Further examples will follow to illustrate this statement.

## 2.4 REGULAR SINGULAR POINTS AND THE METHOD OF FROBENIUS

We recall that $x_0$ $(=0)$ is a regular point if $x\,p_1(x)$ and $x^2 p_2(x)$ are analytic. That is to say, $x\,p_1(x)$ and $x^2 p_2(x)$ can be expanded as

$$x\,p_1(x) = a_0 + a_1 x + a_2 x^2 + \dots \tag{2.4-1a}$$

$$x^2 p_2(x) = b_0 + b_1 x + b_2 x^2 + \dots \tag{2.4-1b}$$

Substituting Equations (2.4-1a, b) into Equation (2.1-2a) and considering the homogeneous case, we obtain

$$y'' + (a_0/x + a_1 + a_2 x + \dots)\,y' + (b_0/x^2 + b_1/x + b_2 + \dots)\,y = 0 \tag{2.4-2}$$

We seek a solution in the neighborhood of the origin and the leading terms of Equation (2.4-2) are

$$y'' + (a_0/x)\,y' + (b_0/x^2)\,y = 0 \tag{2.4-3a}$$

or $\quad x^2 y'' + a_0 x y' + b_0 y = 0 \tag{2.4-3b}$

Equation (2.4-3b) is **Euler's (Cauchy's) equidimensional equation**. It admits a solution of the form

$$y = x^r \tag{2.4-4}$$

On differentiating and substituting into Equation (2.4-3b), we obtain

$$x^r [r(r-1) + a_0 r + b_0] = 0 \tag{2.4-5}$$

To obtain a non-trivial solution $(x^r \neq 0)$, we require

$$r(r-1) + a_0 r + b_0 = 0 \tag{2.4-6}$$

Equation (2.4-6) is a quadratic in $r$ and has two roots $r_1$ and $r_2$. The two linearly independent solutions are $x^{r_1}$ and $x^{r_2}$.

Alternatively Equation (2.4-3b) can be transformed to an equation with constant coefficients by writing

$$x = e^t \tag{2.4-7}$$

Using the chain rule, we obtain

$$\frac{dy}{dx} = e^{-t}\frac{dy}{dt} \tag{2.4-8a}$$

$$\frac{d^2y}{dx^2} = e^{-t}\left(-e^{-t}\frac{dy}{dt} + e^{-t}\frac{d^2y}{dt^2}\right) \tag{2.4-8b}$$

Combining Equations (2.4-3b, 8a, b) yields

$$\frac{d^2y}{dt^2} + (a_0 - 1)\frac{dy}{dt} + b_0 y = 0 \tag{2.4-9}$$

The coefficients in Equation (2.4-9) are constants and can be solved by the methods described in Chapter 1.

***Example 2.4-1.*** Solve the following Euler equations.

a)      $4x^2 y'' + 4xy' - y = 0$                                                (2.4-10a)

b)      $x^2 y'' - xy' + y = 0$                                                  (2.4-10b)

a)      In this example, Equation (2.4-6) is

$$4\,r\,(r-1) + 4\,r - 1 = 0 \tag{2.4-11}$$

The two roots are

$$r_1 = 1/2 \quad \text{and} \quad r_2 = -1/2 \tag{2.4-12a,b}$$

The two linearly independent solutions are

$$y_1 = x^{1/2}, \qquad y_2 = x^{-1/2} \tag{2.4-13a,b}$$

The general solution is a linear combination given by

$$y = c_1 x^{1/2} + c_2 x^{-1/2} \tag{2.4-13c}$$

b)      In this case, Equation (2.4-6) becomes

$$r(r-1) - r + 1 = 0 \tag{2.4-14a}$$

or      $(r-1)^2 = 0$                                                    (2.4-14b)

We have a double root $(r = 1)$ and we have only one solution

$$y_1 = x \tag{2.4-15}$$

To obtain the other solution, we note that if $f(r)$ has a double root at $r_0$, then

$$f(r_0) = 0, \qquad f'(r_0) = 0 \tag{2.4-16a, b}$$

This means that if Equation (2.4-6) has a double root, $x^r$ and $\dfrac{\partial}{\partial r}(x^r)$ are solutions of the differential equation. To differentiate with respect to $r$, we write $x^r$ as $\exp(r \ln x)$. That is to say

$$\frac{\partial}{\partial r}(x^r) = \frac{\partial}{\partial r}[\exp(r \ln x)] \tag{2.4-17a}$$

$$= (\ln x)\exp(r \ln x) \tag{2.4-17b}$$

$$= x^r \ln x \tag{2.4-17c}$$

The other linearly independent solution is

$$y_2 = x \ln x \tag{2.4-18}$$

Alternatively, by changing the independent variable $x$ to $t$ [Equation (2.4-7)], Equation (2.4-10b) can be written as

$$\frac{d^2 y}{dt^2} - 2\frac{dy}{dt} + y = 0 \tag{2.4-19}$$

The solutions are [see Equation (1.6-11)]

$$y_1 = e^t = x \tag{2.4-20a,b}$$

$$y_2 = t\,e^t = x \ln x \tag{2.4-20c,d}$$

●

Note that the exponent $r$ is not necessarily an integer. This suggests that if we retain all the terms on the right side of Equation (2.4-1a, b), we should try a solution of the form

$$y = x^r \sum_{n=0}^{\infty} c_n x^n \tag{2.4-21}$$

where $c_0$ is not zero and $r$ is any real or complex number.

The series is differentiated term by term and substituted in the differential equation. On comparing powers of x, we obtain a set of algebraic equations. The equation associated with the lowest power of x is a quadratic equation in r [see Equation (2.4-6)] and is the **indicial equation**. The other equations are the **recurrence formulae** and are used to determine the coefficients $c_n$ [see Equations (2.3-9a to h)]. In general, the indicial equation yields two distinct values of r which are denoted $r_1$ and $r_2$. The two linearly independent solutions are

$$y_1 = \sum_{n=0}^{\infty} c_n x^{n+r_1} \tag{2.4-22a}$$

$$y_2 = \sum_{n=0}^{\infty} c_n x^{n+r_2} \tag{2.4-22b}$$

If the two roots coincide (see Example 2.4-1b) or the two roots differ by an integer, $y_1$ and $y_2$ are not linearly independent and we have to modify our method. In the examples that follow, we consider the three possible cases: the roots of the indicial equation are distinct and do not differ by an integer; the two roots are coincident; and the two roots differ by an integer. This method of solving a differential equation is called the **method of Frobenius**.

*Example 2.4-2*. Obtain a power series solution to the following equation

$$2xy'' + (x+1) y' + 3y = 0 \tag{2.4-23}$$

in the neighborhood of the origin.

In this example

$$p_1(x) = (x+1)/2x, \qquad p_2(x) = 3/2x \tag{2.4-24a,b}$$

$$x p_1(x) = (x+1)/2, \qquad x^2 p_2(x) = 3x/2 \tag{2.4-24c,d}$$

From Equations (2.4-24a to d), we deduce that x = 0 is a regular singular point. We seek a solution of the form

$$y = \sum_{n=0}^{\infty} c_n x^{n+r} \tag{2.4-25}$$

On differentiating, we obtain

$$y' = \sum_{n=0}^{\infty} (n+r) c_n x^{n+r-1} \tag{2.4-26a}$$

$$y'' = \sum_{n=0}^{\infty} (n+r)(n+r-1) c_n x^{n+r-2} \tag{2.4-26b}$$

Substituting Equations (2.4-25, 26a, b) in Equation (2.4-23) yields

$$\sum_{n=0}^{\infty} [2(n+r)(n+r-1) c_n x^{n+r-1} + (n+r) c_n x^{n+r} + (n+r) c_n x^{n+r-1} + 3 c_n x^{n+r}] = 0$$

$$\tag{2.4-27}$$

We compare powers of x. The lowest power of x is $x^{r-1}$

$$x^{r-1}: \quad 2r(r-1) c_0 + r c_0 = 0 \implies c_0 [r(2r-1)] = 0 \implies r_1 = 1/2 \text{ and } r_2 = 0 \tag{2.4-28a-d}$$

$$x^r: \quad 2(1+r)(r) c_1 + r c_0 + (1+r) c_1 + 3 c_0 = 0 \implies c_1 = -[(3+r) c_0]/[(r+1)(2r+1)]$$

$$\tag{2.4-28e,f}$$

.

.

.

$$x^{r+s}: \quad 2(r+s+1)(r+s) c_{s+1} + (s+r) c_s + (s+r+1) c_{s+1} + 3 c_s = 0 \implies$$

$$c_{s+1} = -[(s+r+3) c_s]/[(s+r+1)(2s+2r+1)] \tag{2.4-28g,h}$$

Substituting the value of $r = 1/2$ into Equation (2.4-28h) leads to

$$c_{s+1} = -[(2s+7) c_s]/[2(2s+3)(s+1)] \tag{2.4-29}$$

We can compute $c_1$, $c_2$, $c_3$, ... and they are

$$c_1 = -7 c_0/[2 \cdot 3] \tag{2.4-30a}$$

$$c_2 = -9 c_1/[2 \cdot 5 \cdot 2] = 7 \cdot 9 c_0/[2^2 \cdot 5 \cdot 3 \cdot 2] \tag{2.4-30b,c}$$

$$c_3 = -11 c_2/[2 \cdot 7 \cdot 3] = -7 \cdot 9 \cdot 11 c_0/[2^3 \cdot 7 \cdot 5 \cdot 3 \cdot 2 \cdot 3] \tag{2.4-30d,e}$$

We denote the solution corresponding to $r = 1/2$ by $y_1$ and it can be written as

$$y_1 = c_0 \sqrt{x} \left[ 1 - \frac{7}{6} x + \frac{21}{40} x^2 - \frac{77}{560} x^3 + ... \right] \tag{2.4-31}$$

From Equations (2.4-29, 30a to e), we deduce that

$$c_{s+1} = \frac{(-1)^{s+1} (2s+7)(2s+5) ... 11 \cdot 9 \cdot 7 c_0}{2^{s+1} (2s+3)(2s+1) ... 7 \cdot 5 \cdot 3 \cdot (s+1)(s) ... 3 \cdot 2} \tag{2.4-32}$$

The right side can be simplified by the following identities

$$(2s + 7) \ldots 7 = \frac{(2s + 7)! \; 3!}{6! \; 2^s \; (s + 3)!} \tag{2.4-33a}$$

$$(2s + 3) \ldots 3 = \frac{(2s + 3)!}{2^{s+1} \; (s + 1)!} \tag{2.4-33b}$$

Combining Equations (2.4-32, 33a, b) yields

$$c_{s+1} = \frac{(-1)^{s+1} \; (2s + 7)! \; c_0}{2^s \; 5! \; (s + 3)! \; (2s + 3)!} \tag{2.4-34}$$

Setting  $n = s + 1$,  Equation (2.4-34) becomes

$$c_n = \frac{(-1)^n \; (2n + 5)! \; c_0}{2^{n-1} \; 5! \; (n + 2)! \; (2n + 1)!} \tag{2.4-35}$$

The solution  $y_1$  can be written as

$$y_1 = c_0 \sqrt{x} \; \sum_{n=0}^{\infty} \frac{(-1)^n \; (2n + 5)! \; x^n}{2^{n-1} \; 5! \; (n + 2)! \; (2n + 1)!} \tag{2.4-36}$$

To obtain the other solution  $y_2$,  we consider the case  $r_2 = 0$.  We substitute this value of  $r_2$  in Equation (2.4-28h) and to avoid confusion we denote  $c_s$  by  $b_s$  for the case  $r_2 = 0$.  Equation (2.4-28h) becomes

$$b_{s+1} = -(s + 3) \; b_s / [(s + 1) \; (2 \; s + 1)] \tag{2.4-37}$$

The coefficients  $b_1$, $b_2$, $b_3$, ... are

$$b_1 = -3b_0 \tag{2.4-38a}$$

$$b_2 = -4b_1 / [2 \cdot 3] = 4 \cdot 3 \, b_0 / [2 \cdot 3] \tag{2.4-38b,c}$$

$$b_3 = -5b_2 / [3 \cdot 5] = -5 \cdot 4 \cdot 3 \, b_0 / [2 \cdot 3 \cdot 3 \cdot 5] \tag{2.4-38d,e}$$

From Equations (2.4-37, 38a to e), we deduce that

$$b_{s+1} = \frac{(-1)^{s+1} \; (s + 3) \; (s + 2) \ldots 5 \cdot 4 \cdot 3 \, b_0}{(s + 1) \; s \ldots 3 \cdot 2 \cdot (2s + 1) \; (2s - 1) \ldots 5 \cdot 3} \tag{2.4-39a}$$

$$= \frac{(-1)^{s+1}\,(s+3)!\,2^{s+1}\,(s+1)!\,b_0}{2\,(s+1)!\,(2s+2)!} \tag{2.4-39b}$$

$$= \frac{(-1)^{s+1}\,2^{s}\,(s+3)!\,b_0}{(2s+2)!} \tag{2.4-39c}$$

Again setting $s+1=n$, we find that the solution $y_2$ can be written as

$$y_2 = b_0 \sum_{n=0}^{\infty} \frac{(-1)^n\,2^{n-1}\,(n+2)!\,x^n}{(2n)!} \tag{2.4-40}$$

The general solution $y$ is a linear combination of $y_1$ and $y_2$ and can be written as

$$y = c_0 \sqrt{x} \sum_{n=0}^{\infty} \frac{(-1)^n\,(2n+5)!\,x^n}{2^{n-1}\,5!\,(n+2)!\,(2n+1)!} + b_0 \sum_{n=0}^{\infty} \frac{(-1)^n\,2^{n-1}\,(n+2)!\,x^n}{(2n)!} \tag{2.4-41}$$

The arbitrary constants $c_0$ and $b_0$ are determined from the initial conditions or the boundary conditions.

From Equations (2.2-6, 4-29, 37), we deduce that the radii of convergence of the two series are

$$R = \lim_{s \to \infty} \left| \frac{c_s}{c_{s+1}} \right| = \lim_{s \to \infty} \left| \frac{2\,(2s+3)\,(s+1)}{(2s+7)} \right| = \infty \tag{2.4-42a,b,c}$$

$$R = \lim_{s \to \infty} \left| \frac{b_s}{b_{s+1}} \right| = \lim_{s \to \infty} \left| \frac{(s+1)\,(2s+1)}{(s+3)} \right| = \infty \tag{2.4-43a,b,c}$$

The solutions $y_1$ and $y_2$ are valid for all values of $x$.

***Example 2.4-3***. Obtain a series solution to the equation

$$x^2\,y'' + x\,y' + x^2\,y = 0 \tag{2.4-44}$$

in the neighborhood of the origin.

In this example, we have

$$p_1 = 1/x, \qquad p_2 = 1 \tag{2.4-45a,b}$$

$$x\,p_1 = 1, \qquad x^2\,p_2 = x^2 \tag{2.4-45c,d}$$

The origin is a regular singular point and the series solution is of the form

$$y = \sum_{n=0}^{\infty} c_n x^{n+r} \tag{2.4-46}$$

Differentiating and substituting in Equation (2.4-44) yields

$$\sum_{n=0}^{\infty} \left[ (n+r)(n+r-1) c_n x^{n+r} + (n+r) c_n x^{n+r} + c_n x^{n+r+2} \right] = 0 \tag{2.4-47}$$

Comparing the powers of $x$, starting with the lowest power ($n = 0$), we have

$x^r$:    $c_0 [r(r-1)+r] = 0 \implies r = 0$ (double root)                                 (2.4-48a,b)

$x^{r+1}$:    $c_1 [(1+r)r + (1+r)] = 0 \implies c_1 = 0$ $(r=0)$                         (2.4-48c,d)

$x^{r+2}$:    $c_2 [(2+r)(1+r) + (2+r)] + c_0 = 0 \implies c_2 = -c_0/(2+r)^2$          (2.4-48e,f)
.
.
.

$x^{r+s}$:    $c_s [(s+r)(s+r-1) + (s+r)] + c_{s-2} = 0 \implies c_s = -c_{s-2}/(s+r)^2$   (2.4-48g,h)

Note that with $r = 0$, we deduce that $c_1$ is zero. From Equation (2.4-48h), we deduce that if $s$ is odd, $c_s$ is zero. We consider the case where $s$ is even and we write $s = 2m$ and Equation (2.4-48h) becomes

$$c_{2m} = -c_{2m-2}/(2m)^2 \tag{2.4-49}$$

The coefficients $c_2$, $c_4$, $c_6$ ... are given by

$$c_2 = -c_0/2^2 \tag{2.4-50a}$$

$$c_4 = -c_2/4^2 = c_0/[2^2 \cdot 4^2] \tag{2.4-50b,c}$$

$$c_6 = -c_4/6^2 = -c_0/[2^2 \cdot 4^2 \cdot 6^2] \tag{2.4-50d,e}$$

From Equations (2.4-49, 50 a to e), we deduce that

$$c_{2m} = (-1)^m c_0/[2^2 \cdot 4^2 \cdot 6^2 ... (2m)^2] \tag{2.4-51a}$$

$$= (-1)^m c_0/[(2^{2m})(m!)^2] \tag{2.4-51b}$$

One series solution can be written as

$$y_1(x) = c_0 \sum_{m=0}^{\infty} \frac{(-1)^m x^{2m}}{2^{2m} (m!)^2} \qquad (2.4\text{-}52)$$

Equation (2.4-44) is **Bessel's equation** of order zero and the equation of arbitrary order will be considered later. It is customary to set $c_0 = 1$ and to denote $y_1(x)$ by $J_0(x)$. Equation (2.4-52) becomes

$$J_0(x) = \sum_{m=0}^{\infty} \frac{(-1)^m x^{2m}}{2^{2m} (m!)^2} \qquad (2.4\text{-}53)$$

In this example, the indicial equation [Equation (2.4-48a)] has a double root and the second linearly independent solution is not readily available. Example 2.4-1b suggests that we look for a solution of the form

$$y = \ell n \, x \, J_0(x) + \sum_{n=1}^{\infty} b_n x^n \qquad (2.4\text{-}54)$$

Note that if the double root of the indicial equation is $r_0$ ($\neq 0$), the form of the series solution is

$$y = (\ell n \, x) \, y_1(x) + \sum_{n=0}^{\infty} b_n x^{n+r_0} \qquad (2.4\text{-}55)$$

Differentiating $y$, we obtain

$$y' = \ell n \, x \, J_0'(x) + J_0(x)/x + \sum_{n=1}^{\infty} n \, b_n x^{n-1} \qquad (2.4\text{-}56a)$$

$$y'' = \ell n \, x \, J_0''(x) + 2J_0'(x)/x - J_0(x)/x^2 + \sum_{n=2}^{\infty} n \, (n-1) \, b_n x^{n-2} \qquad (2.4\text{-}56b)$$

Substituting Equations (2.4-54, 56a, b) in Equation (2.4-44) yields

$$\ell n \, x \left[ x^2 J_0'' + x \, J_0' + x^2 J_0 \right] + 2x J_0' + \sum_{n=2}^{\infty} n \, (n-1) \, b_n x^n + \sum_{n=1}^{\infty} n \, b_n x^n + \sum_{n=1}^{\infty} b_n x^{n+2} = 0 \qquad (2.4\text{-}57)$$

The function $J_0$ is a solution of Equation (2.4-44) and the terms inside the square bracket equal zero. From Equation (2.4-53), we obtain

$$J_0'(x) = \sum_{m=1}^{\infty} \frac{(-1)^m (2m) x^{2m-1}}{2^{2m} (m!)^2} \tag{2.4-58}$$

Substituting Equation (2.4-58) in Equation (2.4-57) and changing the indices appropriately such that the summation index s starts from one (see Example 2.3-1), we obtain

$$\sum_{s=1}^{\infty} \left[ \frac{(-1)^s x^{2s}}{2^{2s-2} s! (s-1)!} + s (s+1) b_{s+1} x^{s+1} + s b_s x^s + b_s x^{s+2} \right] = 0 \tag{2.4-59}$$

To obtain the coefficients $b_s$, we compare powers of x.

$x^1$:    $b_1 = 0$ (2.4-60a)

$x^2$:    $-1 + 2b_2 + 2b_2 = 0 \quad \Rightarrow \quad b_2 = 1/4$ (2.4-60b,c)

$x^3$:    $6b_3 + 3b_3 + b_1 = 0 \quad \Rightarrow \quad b_3 = -b_1/9 = 0$ (2.4-60d,e,f)

One observes that $b_s$ is zero if s is odd and we need to consider the even powers of x only.

$x^{2m}$:    $\dfrac{(-1)^m}{2^{2m-2} m! (m-1)!} + 2m (2m-1) b_{2m} + 2m b_{2m} + b_{2m-2} = 0$ (2.4-61)

The recurrence formula is

$$b_{2m} = -\left[ \frac{(-1)^m}{2^{2m-2} m! (m-1)!} + b_{2m-2} \right] \Big/ (2m)^2 \tag{2.4-62}$$

The coefficient $b_2$ is known [Equation (2.4-60c)] and from Equation (2.4-62), we can obtain $b_4$, $b_6$, ... as follows

$$b_4 = -\left[ \frac{1}{4 \cdot 2} + b_2 \right] \Big/ 16 = -\frac{3}{128} \tag{2.4-63a,b}$$

$$b_6 = -\left[ \frac{-1}{2^4 3! 2!} + b_4 \right] \Big/ 36 = \frac{11}{13824} \tag{2.4-63c,d}$$

The second solution, which is linearly independent of $J_0$, is denoted by $y_2$ and is given by

$$y_2 = \ell n \, x \, J_0(x) + \frac{x^2}{4} - \frac{3x^4}{128} + \frac{11x^6}{13824} \cdots \tag{2.4-64}$$

From Equations (2.4-62, 63a to d), we can verify that the general term $b_{2m}$ can be written as

$$b_{2m} = \frac{(-1)^{m-1}}{2^{2m}(m!)^2}\left(1 + \frac{1}{2} + \frac{1}{3} + \dots + \frac{1}{m}\right) \tag{2.4-65}$$

The solution $y_2$ is

$$y_2 = \ell n \, x \, J_0(x) + \sum_{m=1}^{\infty} b_{2m} x^{2m} \tag{2.4-66}$$

The general solution of Equation (2.4-44) is the linear combination of $J_0$ and $y_2$ and is

$$y = A J_0 + B y_2 \tag{2.4-67}$$

where A and B are constants.

Note the presence of $\ell n \, x$ in $y_2$. This implies that $y_2$ is singular at the origin. If the physics of the problem require y to be finite at the origin, we require B to be zero.

**Example 2.4-4.** Solve the equation

$$x^2(x^2 - 1)\, y'' - (x^2 + 1)\, x\, y' + (x^2 + 1)\, y = 0 \tag{2.4-68}$$

in the neighborhood of the origin.

In this example, $p_1$ and $p_2$ are given respectively by

$$p_1 = -\frac{(x^2 + 1)}{x(x^2 - 1)}, \qquad p_2 = \frac{(x^2 + 1)}{x^2(x^2 - 1)} \tag{2.4-69a,b}$$

The singular points are at $x = 0$, $x = 1$, and $x = -1$. We further note that $x = 0$ is a regular singular point and we assume a solution of the form

$$y = \sum_{n=0}^{\infty} c_n x^{r+n} \tag{2.4-70}$$

Differentiating and substituting the resulting expressions in Equation (2.4-68) yields

$$\sum_{n=0}^{\infty} \Big[ (n+r)(n+r-1)\, c_n \, x^{r+n+2} - (n+r)(n+r-1)\, c_n \, x^{r+n} - (n+r)\, c_n \, x^{r+n+2} - (n+r)\, c_n \, x^{r+n}$$
$$+ c_n \, x^{r+n+2} + c_n \, x^{r+n} \Big] = 0 \tag{2.4-71}$$

Comparing powers of x, we obtain

$$x^r: \qquad c_0 [-r(r-1)-r+1] = 0 \quad \Rightarrow \quad r^2 - 1 = 0 \quad \Rightarrow \quad r = \pm 1 \qquad \text{(2.4-72a,b,c)}$$

$$x^{r+1}: \quad c_1 [-(1+r)r-r+1] = 0 \quad \Rightarrow \quad c_1 (r^2 + 2r - 1) = 0 \quad \Rightarrow \quad c_1 = 0 \ (r = \pm 1) \quad \text{(2.4-72d,e,f)}$$

$$x^{r+2}: \quad c_0 [r(r-1)-r+1] + c_2 [-(2+r)(1+r)-(2+r)+1] = 0 \Rightarrow c_2 = c_0 (r-1)^2 / [(2+r)^2 - 1]$$

$$\text{(2.4-72g,h)}$$

.
.
.

$$x^{r+s}: \quad c_{s-2} [(s+r-2)(s+r-3)-(s+r-2)+1] + c_s [-(s+r)(s+r-1)-(s+r)+1] = 0$$

$$\Rightarrow \quad c_s = c_{s-2} [(s+r-2)(s+r-4)] / [(s+r)^2 - 1] \qquad \text{(2.4-72i,j)}$$

We deduce from Equations (2.4-72f, j) that $c_1 = c_3 = c_5 = \ldots = 0$. Substituting the value of $r = 1$ in Equation (2.4-72h) leads to $c_2 = 0$ and Equation (2.4-72j) implies that $c_4 = c_6 = \ldots = 0$. The solution $y_1$ corresponding to $r = 1$ is a polynomial (one term only) and is

$$y_1 = c_0 x \qquad \text{(2.4-73)}$$

To determine $y_2$, the solution corresponding to $r = -1$, we substitute this value of $r$ in Equation (2.4-72h) and we observe that the numerator $(r-1)^2$ is non-zero and the denominator $[(2+r)^2 - 1]$ is zero. This implies that $c_2$ is infinity and we must seek a solution of the form

$$y = y_1 \ln x + \sum_{n=0}^{\infty} b_n x^{n-1} \qquad (b_0 \neq 0) \qquad \text{(2.4-74a)}$$

$$= c_0 x \ln x + \sum_{n=0}^{\infty} b_n x^{n-1} \qquad \text{(2.4-74b)}$$

On differentiating term by term, we obtain

$$y' = c_0 (\ln x + 1) + \sum_{n=0}^{\infty} (n-1) b_n x^{n-2} \qquad \text{(2.4-75a)}$$

$$y'' = c_0/x + \sum_{n=0}^{\infty} (n-1)(n-2) b_n x^{n-3} \qquad \text{(2.4-75b)}$$

Substituting $y$, $y'$, and $y''$ in Equation (2.4-68) yields

$$\sum_{n=0}^{\infty} \left[ (n-1)(n-2) b_n x^{n+1} - (n-1)(n-2) b_n x^{n-1} - (n-1) b_n x^{n+1} - (n-1) b_n x^{n-1} \right.$$

$$\left. + b_n x^{n+1} + b_n x^{n-1} \right] - 2c_0 x = 0 \qquad \text{(2.4-76)}$$

We now compare powers of x.

$$x^{-1}: \quad b_0(-2+1+1) = 0 \tag{2.4-77a}$$

$$x^0: \quad b_1 = 0 \tag{2.4-77b}$$

$$x^1: \quad 2b_0 + b_0 - b_2 + b_0 + b_2 - 2c_0 = 0 \quad \Rightarrow \quad b_0 = c_0/2 \tag{2.4-77c,d}$$

$$x^3: \quad -(3)(2)b_4 - b_2 - 3b_4 + b_2 + b_4 = 0 \quad \Rightarrow \quad b_4 = 0 \tag{2.4-77e,f}$$

From Equations (2.4-76, 77a to f), we deduce that

$$b_1 = b_3 = \ldots = 0, \qquad b_0 \neq 0, \qquad b_2 = b_4 = \ldots = 0 \tag{2.4-78a,b,c}$$

The solution $y_2$, corresponding to $r = -1$, is given by

$$y_2 = c_0 \, x \, \ell n \, x + b_0/(2x) \tag{2.4-79}$$

Without loss of generality, we can set $c_0 = b_0 = 1$ and the two linearly independent solutions $y_1$ and $y_2$ can be written as

$$y_1 = x, \qquad y_2 = x \, \ell n \, x + 1/(2x) \tag{2.4-80a,b}$$

The general solution of Equation (2.4-68) is

$$y = A \, y_1 + B \, y_2 \tag{2.4-81}$$

where A and B are arbitrary constants.

Note that $y_2$ is singular at the origin and if the general solution y is finite at the origin, B has to be zero.

***Example 2.4-5.*** Find a series solution to the equation

$$x^2 \, y'' + x \, y' + (x^2 - 1/4) \, y = 0 \tag{2.4-82}$$

valid near the origin.

The origin is a regular singular point $[x \, p_1(x) = 1, \; x^2 \, p_2(x) = (x^2 - 1/4)]$ and we seek a solution in the form of Equation (2.4-70). Proceeding as in the previous example, we obtain

$$\sum_{n=0}^{\infty} \left[ (n+r)(n+r-1) \, c_n \, x^{n+r} + (n+r) \, c_n \, x^{n+r} + c_n \, x^{n+r+2} - (1/4) \, c_n \, x^{n+r} \right] = 0 \tag{2.4-83}$$

Comparing powers of x yields

$x^r$:    $c_0 [r (r - 1) + r - 1/4] = 0 \Rightarrow r = \pm 1/2$                     (2.4-84a,b)

$x^{r+1}$:  $c_1 [r (1 + r) + (1 + r) - 1/4] = 0 \Rightarrow c_1 = 0$ if $r = 1/2$          2.4.84c)

                                    $\Rightarrow c_1$ is arbitrary if $r = -1/2$          (2.4-84d)

$\cdot$
$\cdot$
$\cdot$

$x^{r+s}$:  $c_s [(s+r) (s+r-1) + (s+r) - 1/4] + c_{s-2} = 0 \Rightarrow c_s = -c_{s-2}/[(s+r)^2 - 1/4]$   (2.4-84e,f)

We now substitute the value of $r = 1/2$ in Equation (2.4-84f) and we obtain

$$c_s = -c_{s-2}\big/[s (s + 1)]$$                                     (2.4-85)

From Equations (2.4-84c, 85), we note that all the coefficients with an odd index are zero. Writing $s = 2m$, we obtain

$$c_{2m} = -c_{2(m-1)}\big/[(2m) (2m + 1)]$$                             (2.4-86)

From Equation (2.4-86), we compute the first few coefficients and they are

$$c_2 = -c_0/[2 \bullet 3]$$                                          (2.4-87a)

$$c_4 = -c_2/[4 \bullet 5] = c_0/5!$$                                 (2.4-87b,c)

$$c_6 = -c_4/[6 \bullet 7] = c_0/7!$$                                 (2.4-87d,e)

From Equations (2.4-86, 87a to e), we deduce that

$$c_{2m} = (-1)^m c_0/(2m + 1)!$$                                     (2.4-88)

The solution $y_1$ is given by

$$y_1 = c_0 \sum_{m=0}^{\infty} \frac{(-1)^m x^{2m+1/2}}{(2m+1)!} = c_0 x^{-1/2} \sum_{m=0}^{\infty} \frac{(-1)^m x^{2m+1}}{(2m+1)!}$$     (2.4-89a,b)

In Equation (2.4-89b), the sum is $\sin x$ and $y_1$ can be written in **closed form** as

$$y_1 = c_0 x^{-1/2} \sin x$$                                        (2.4-90)

To obtain $y_2$, we substitute the value of $r = -1/2$ in Equation (2.4-84f) and we obtain

$$c_s = -c_{s-2} / [s (s-1)] \tag{2.4-91}$$

The denominator is zero if $s = 1$ and it is not possible to determine $c_1$. From Equation (2.4-84c), we deduce that $c_1$ is arbitrary and we can compute $c_3$, $c_5$, ... All these coefficients are non-zero. The coefficients with even indices can be written as multiples of $c_0$ and those with odd indices are multiples of $c_1$. We compute the first few coefficients as follows

$$c_2 = -c_0/(2) , \qquad c_4 = -c_2/(4 \cdot 3) = c_0/4! , \qquad c_6 = -c_4/(6 \cdot 5) = -c_0/6!, \quad ... \tag{2.4-92a}$$

$$c_3 = -c_1/(3 \cdot 2) , \qquad c_5 = -c_3/(5 \cdot 4) = c_1/5! , \qquad c_7 = -c_5/(7 \cdot 6) = -c_1/7!, \quad ... \tag{2.4-92b}$$

From Equations (2.4-91, 92a, b), we verify that

$$c_{2m} = (-1)^m c_0/(2m)! , \qquad c_{2m+1} = (-1)^m c_1/(2m+1)! \tag{2.4-93a,b}$$

The solution $y_2$ can be written as

$$y_2 = x^{-1/2} \sum_{m=0}^{\infty} \left[ \frac{(-1)^m c_0 x^{2m}}{(2m)!} + \frac{(-1)^m c_1 x^{2m+1}}{(2m+1)!} \right] \tag{2.4-94}$$

Note that the second term on the right side of Equation (2.4-94) is $y_1$ (with $c_1$ replacing $c_0$). The summation $\sum_{m=0}^{\infty} (-1)^m x^{2m}/(2m)!$ is $\cos x$ and the two linearly independent solutions $y_1$ and $y_2$ are, to the extent of a multiplicative constant,

$$y_1 = x^{-1/2} \sin x , \qquad y_2 = x^{-1/2} \cos x \tag{2.4-95a,b}$$

The general solution of Equation (2.4-82) is

$$y = A y_1 + B y_2 \tag{2.4-96}$$

where $A$ and $B$ are arbitrary constants.

Observe that $y_2$ is singular at the origin and $y_1$ is finite at the origin. For solutions which remain finite at the origin, we require $B$ to be zero.

*Example 2.4-6*. Gupta and Douglas (1967) considered a steady state diffusion problem, associated with a first order irreversible reaction involving isobutylene and spherical cation exchange resin particles.

Equation (1) of their paper can be written as (see Equation A.IV-3)

$$0 = \mathcal{D}_{AB} \left[ \frac{1}{r^2} \frac{\partial}{\partial r} \left( r^2 \frac{\partial c_A}{\partial r} \right) \right] + R_A \tag{2.4-97a}$$

where $R_A = -kc_A$ \hfill (2.4-97b)

Assuming $c_A$ to be a function of $r$ only (symmetry!), Equation (2.4-97a) reduces to an O.D.E. given by

$$\mathcal{D}_{AB} \left( r^2 \frac{d^2 c_A}{dr^2} + 2r \frac{dc_A}{dr} \right) - kr^2 c_A = 0 \tag{2.4-98a}$$

The boundary conditions are

$$c_A = c_{A_i} \qquad \text{at } r = R \tag{2.4-98b}$$

$$\frac{dc_A}{dr} = 0 \qquad \text{at } r = 0 \tag{2.4-98c}$$

We now solve Equations (2.4-98a–c) for $c_A$ via the method of Frobenius.

Note that the origin $(r = 0)$ is a regular singular point. We seek a solution of the form

$$c_A = \sum_{n=0}^{\infty} c_n r^{n+\rho} \tag{2.4-99}$$

In Equation (2.4-99), the exponent involves $\rho$, so as to avoid confusion with the radial variable $r$.

Differentiating Equation (2.4-99) and substituting the resulting expressions in Equation (2.4-98a) yields

$$\sum_{n=0}^{\infty} [c_n (n+\rho)(n+\rho-1) r^{n+\rho} + 2(n+\rho) c_n r^{n+\rho} - K c_n r^{n+\rho+2}] = 0 \tag{2.4-100a}$$

where $K = k / \mathcal{D}_{AB}$ \hfill (2.4-100b)

Comparing powers of $r$, we have

$$r^\rho: \quad c_0(\rho)(\rho-1) + 2\rho c_0 = 0 \;\Rightarrow\; \rho(\rho+1) = 0 \;\Rightarrow\; \rho = 0 \text{ and } \rho = -1 \tag{2.4-101a,b,c,d}$$

$$r^{1+\rho}: \quad c_1(1+\rho)(\rho) + 2(1+\rho) c_1 = 0 \;\Rightarrow\; c_1(1+\rho)(2+\rho) = 0$$

$$\Rightarrow \begin{cases} c_1 = 0, & \text{if } \rho = 0 \\[2mm] c_1 \text{ is arbitrary}, & \text{if } \rho = -1 \end{cases} \tag{2.4-101e,f,g,h}$$

.
.
.

$r^{s+\rho}$:  $c_s(s+\rho)(s+\rho-1) + 2(s+\rho)c_s - Kc_{s-2} = 0 \Rightarrow c_s = \dfrac{Kc_{s-2}}{(s+\rho)(s+\rho+1)}$ $\qquad$ (2.4-101i,j)

To determine $c_s$, we substitute the values of $\rho$ in Equation (2.4-101j)

$$c_s = \frac{Kc_{s-2}}{s(s+1)}, \qquad \text{for } \rho = 0 \qquad\qquad (2.4\text{-}102)$$

Equation (2.4-101h) states that $c_1 = 0$ if $\rho = 0$.

Therefore, $c_s = 0$ if $s$ is odd. We proceed by replacing $s$ by $2m$ and Equation (2.4-102) becomes

$$c_{2m} = \frac{Kc_{2m-2}}{2m(2m+1)} \qquad\qquad (2.4\text{-}103)$$

We compute the first few coefficients as follows

$$c_2 = \frac{Kc_0}{2\cdot 3}, \qquad c_4 = \frac{Kc_2}{4\cdot 5} = \frac{K^2c_0}{5!} \qquad\qquad (2.4\text{-}104a,b)$$

and deduce that

$$c_{2m} = \frac{K^m c_0}{(2m+1)!} \qquad\qquad (2.4\text{-}105)$$

One solution is

$$c_{A_1} = c_0 \sum_{m=0}^{\infty} \frac{K^m r^{2m}}{(2m+1)!} \qquad\qquad (2.4\text{-}106)$$

Substituting the other value of $\rho$ in Equation (2.4-101i), we obtain

$$c_s = \frac{Kc_{s-2}}{(s-1)(s)} \qquad\qquad (2.4\text{-}107)$$

In this case, we have two arbitrary constants, $c_0$ and $c_1$. This implies an even as well as an odd solution. To compute the even solution, we replace $s$ by $2m$ and Equation (2.4-107) becomes

$$c_{2m} = \frac{Kc_{2m-2}}{(2m-1)(2m)} \qquad\qquad (2.4\text{-}108)$$

As before, we obtain

$$c_{2m} = \frac{K^m c_0}{2m!} \qquad\qquad (2.4\text{-}109)$$

For the case where $s$ is odd, we write

$$c_{2m+1} = \frac{K c_{2m-1}}{(2m)(2m+1)} \qquad\qquad (2.4\text{-}110)$$

In terms of $c_1$, we obtain

$$c_{2m+1} = \frac{K^m c_1}{(2m+1)!} \qquad\qquad (2.4\text{-}111)$$

Another solution to Equation (2.4-98a) is

$$c_{A_2} = c_0 \sum_{m=0}^{\infty} \frac{K^m r^{2m-1}}{(2m)!} + c_1 \sum_{m=0}^{\infty} \frac{K^m r^{2m}}{(2m+1)!} \qquad\qquad (2.4\text{-}112)$$

Note that the second term on the right side of Equation (2.4-112) is equal to the right side of Equation (2.4-106).

Therefore, Equation (2.4-112) is the general solution of Equation (2.4-98a).

The solution can be written in a closed form by observing that

$$\cosh x = \sum_{m=0}^{\infty} \frac{x^{2m}}{(2m)!} \quad , \qquad \sinh x = \sum_{m=0}^{\infty} \frac{x^{2m+1}}{(2m+1)!} \qquad\qquad (2.4\text{-}113\text{a,b})$$

Equation (2.4-112) can be written as

$$c_A = \frac{c_0}{r} \cosh r \sqrt{K} + \frac{c_1}{r} \sinh r \sqrt{K} \qquad\qquad (2.4\text{-}114)$$

Differentiation yields

$$\frac{dc_A}{dr} = -\frac{c_0}{r^2} \cosh r \sqrt{K} + c_0 \frac{\sqrt{K}}{r} \sinh r \sqrt{K} - \frac{c_1}{r^2} \sinh r \sqrt{K} + c_1 \frac{\sqrt{K}}{r} \cosh r \sqrt{K}$$

$$(2.4\text{-}115)$$

and applying the boundary conditions requires

$$c_{A_i} = \frac{c_0}{R} \cosh R \sqrt{K} + \frac{c_1}{R} \sinh R \sqrt{K} \qquad\qquad (2.4\text{-}116a)$$

$$0 = [-c_0 \cosh r \sqrt{K} + c_0 r \sqrt{K} \sinh r \sqrt{K} - c_1 \sinh r \sqrt{K} + c_1 r \sqrt{K} \cosh r \sqrt{K}] / r^2$$

(2.4-116b)

On evaluating Equation (2.4-116b), as $r \longrightarrow 0$, we obtain $c_0 = 0$.

We now have

$$c_{A_i} = \frac{c_1}{R} \sinh R \sqrt{K}$$

(2.5-117)

and substituting for $c_1$ in Equation (2.4-114) yields the final result

$$c_A = \frac{c_{A_i} R \sinh r \sqrt{K}}{r \sinh R \sqrt{K}}$$

(2.4-118)

which is Equation (2) in Gupta and Douglas (1967).

Before summarizing the Frobenius method, we note that Equation (2.4-98a) can be transformed to a much simpler equation involving constant coefficients.

Replacing $c_A$ by $\frac{u(r)}{r}$ transforms the equation to

$$\frac{d^2 u}{dr^2} - Ku = 0$$

(2.4-119)

with solution

$$u = c_0 \cosh r \sqrt{K} + c_1 \sinh r \sqrt{K}$$

(2.4-120)

and with $c_A$ correctly given by Equation (2.4-114).

●

We summarize Frobenius's method of finding a solution near a regular singular point as follows. We assume that the solution is of the form

$$y = \sum_{n=0}^{\infty} c_n x^{n+r}, \quad c_0 \neq 0$$

(2.4-121)

We differentiate the series term by term and substitute the resulting expressions in the differential equation. On setting the coefficient of the lowest power of $x$ to zero, we obtain a quadratic (for a second order equation) equation in $r$. We denote the two solutions by $r_1$ and $r_2$. We now consider the following three cases.

(a)     $r_1$ and $r_2$ are distinct and do not differ by an integer. By comparing powers of $x$, we obtain a recurrence formula that allows us to obtain $c_1$, $c_2$, $c_3$, ... in terms of $c_0$ and $r_1$ (or $r_2$).

The two linearly independent solutions are

$$y_1 = \sum_{n=0}^{\infty} c_n (r_1) x^{n+r_1}$$                              (2.4-122a)

$$y_2 = \sum_{n=0}^{\infty} c_n (r_2) x^{n+r_2}$$                              (2.4-122b)

(b)     $r_1 = r_2$. In this case, one solution $y_1$ is given by Equation (2.4-122a). To obtain the other solution, we assume a solution of the form

$$y = y_1 \ln x + \sum_{n=0}^{\infty} b_n x^{n+r_1}$$                         (2.4-123)

We proceed as in case (a) to obtain $b_n$.

(c)     $r_1$ and $r_2$ differ by an integer. Let us assume that $r_1 > r_2$. The solution $y_1$ can be obtained as in (a). We try to compute the coefficients $c_s$ with the value of $r = r_2$. If all the coefficients can be computed as in Examples 2.4-5 and 6, the second linearly independent solution $y_2$ is obtained. If, in the computation of $c_s$, we have to divide by zero as in Example 2.4-4, we assume a solution in the form of Equation (2.4-123) and proceed to calculate $b_n$. Note that in case (b) one of the solutions always has a $\ln x$ term whereas in case (c) this is not the case (see Examples 2.4-5 and 6).

If one solution is known, we can obtain the second solution by the **method of variation of parameters** and this is explained in the next section.

## 2.5    METHOD OF VARIATION OF PARAMETERS

We use the method of variation of parameters to find a second linearly independent solution of Equation (2.1-2a) if one solution is known. Let $y_1$ be a solution of Equation (2.1-2a) and we assume a second solution to be given by

$$y(x) = u(x) y_1(x)$$                                                        (2.5-1)

On differentiating and substituting $y$, $y'$, and $y''$ in Equation (2.1-2a), we obtain

$$u \left( y_1'' + p_1 y_1' + p_2 y_1 \right) + u' \left( 2y_1' + p_1 y_1 \right) + u'' y_1 = 0 \qquad (2.5\text{-}2)$$

Since $y_1$ is a solution of Equation (2.1-2a), Equation (2.5-2) simplifies to

$$u' \left( 2y_1' + p_1 y_1 \right) + u'' y_1 = 0 \qquad (2.5\text{-}3)$$

We substitute $u'$ by $v$ and Equation (2.5-3) can be written as

$$\int \frac{dv}{v} = - \int \left( \frac{2y_1' + p_1 y_1}{y_1} \right) dx \qquad (2.5\text{-}4)$$

On integrating, we obtain

$$\ell n \, v = -2 \ell n \, y_1 - \int p_1 \, dx \qquad (2.5\text{-}5a)$$

or $\qquad v = \left( 1/y_1^2 \right) \exp \left( - \int p_1 \, dx \right) \qquad (2.5\text{-}5b)$

One further integration yields

$$u = \int \left[ \left( 1/y_1^2 \right) \exp \left( - \int p_1 \, dx \right) \right] dx \qquad (2.5\text{-}6)$$

The function $u$ is generally not a constant and the second linearly independent solution $y_2$ is

$$y_2 = u \, (x) \, y_1 (x) \qquad (2.5\text{-}7)$$

***Example 2.5-1.*** Obtain a second linearly independent solution of Equation (2.4-68) given that one solution $y_1$ is $x$.

We assume that the second solution is

$$y = x \, u \, (x) \qquad (2.5\text{-}8)$$

We substitute $y$, $y'$, and $y''$ from Equation (2.5-8) in Equation (2.4-68) and we obtain

$$u'' \left[ x^3 \left( x^2 - 1 \right) \right] + u' \left[ 2x^2 \left( x^2 - 1 \right) - x^2 \left( x^2 + 1 \right) \right] = 0 \qquad (2.5\text{-}9)$$

We denote $u'$ by $v$ and Equation (2.5-9) becomes

$$x \left( x^2 - 1 \right) \frac{dv}{dx} + \left( x^2 - 3 \right) v = 0 \qquad (2.5\text{-}10)$$

Equation (2.5-10) can be written as

$$\frac{dv}{v} = \frac{-(x^2 - 3)}{x(x^2 - 1)} \, dx \tag{2.5-11a}$$

$$= \left( \frac{1}{x - 1} + \frac{1}{x + 1} - \frac{3}{x} \right) dx \tag{2.5-11b}$$

On integrating, we obtain

$$v = c_1 (x^2 - 1)/x^3 \tag{2.5-12}$$

where $c_1$ is a constant.

The function $u$ is obtained by integrating $v$ and is found to be

$$u = c_1 [\ell n \, x + 1/(2x^2)] + K \tag{2.5-13}$$

where $K$ is a constant.

Combining Equations (2.5-8, 13) yields

$$y = c_1 x [\ell n \, x + 1/(2x^2)] + K x \tag{2.5-14}$$

The second linearly independent solution can now be identified to be

$$y_2 = x [\ell n \, x + 1/(2x^2)] \tag{2.5-15}$$

There is no loss of generality in setting $c_1 = 1$ and $K x$ is $y_1$. Equation (2.5-15) is exactly Equation (2.4-80b).

## 2.6   STURM LIOUVILLE PROBLEM

In the previous section, it is stated that the general solution of a second order O.D.E. is a linear combination of two linearly independent solutions. To determine the two constants, we need to impose two conditions. In many physical problems, the conditions are imposed at the boundaries and these problems are boundary value problems. Many of the second order boundary value problems can be stated as follows

$$(r \, y')' + (q + \lambda p) \, y = 0, \qquad a \le x \le b \tag{2.6-1a}$$

subject to

$$a_1 \, y \, (a) + a_2 \, y' \, (a) = 0 \tag{2.6-1b}$$

$$b_1 \, y \, (b) + b_2 \, y' \, (b) = 0 \tag{2.6-1c}$$

where r, p, and q are continuous real functions of x, $\lambda$ is a constant parameter (possibly complex), $a_1$, $a_2$, $b_1$, and $b_2$ are constants.

The system defined by Equations (2.6-1a to c) is the Sturm Liouville problem. Equation (2.6-1a) can be written as

$$r\,y'' + r'\,y' + (q + \lambda p)\,y = 0 \tag{2.6-1d}$$

Many of the equations in mathematical physics are special cases of Equations (2.6-1a or d) and some of them are given next.

(a)     Simple harmonic equation

$$y'' + \lambda y = 0 \tag{2.6-2a}$$

(In this case, $r = p = 1$, $q = 0$.)

(b)     Legendre equation

$$(1 - x^2)\,y'' - 2x\,y' + \ell\,(\ell + 1)\,y = 0 \tag{2.6-2b}$$

or     $$[(1 - x^2)\,y']' + \ell\,(\ell + 1)\,y = 0 \tag{2.6-2c}$$

[In this case, $r = (1 - x^2)$, $q = 0$, $p = 1$, $\lambda = \ell\,(\ell + 1)$.]

(c)     Bessel equation

We denote the independent variable by $\bar{x}$ and write Bessel's equation as

$$\bar{x}^2 \frac{d^2y}{d\bar{x}^2} + \bar{x}\,\frac{dy}{d\bar{x}} + (\bar{x}^2 - v^2)\,y = 0 \tag{2.6-2d}$$

On setting $\bar{x} = x\sqrt{\lambda}$, Equation (2.6-2d) becomes

$$x^2 y'' + x\,y' + (\lambda x^2 - v^2)\,y = 0 \tag{2.6-2e}$$

or     $$(x\,y')' + (-v^2/x + \lambda x)\,y = 0 \tag{2.6-2f}$$

(In this case, $r = x$, $q = -v^2/x$, $p = x$.)

(d)      Hermite equation

$$y'' - 2xy' + \mu y = 0 \tag{2.6-2g}$$

or      $(e^{-x^2} y')' + \mu e^{-x^2} y = 0 \tag{2.6-2h}$

(In this case,  $r = p = e^{-x^2}$,  $q = 0$,  $\lambda = \mu$.)

In general, the non-trivial solution ($y \neq 0$) of Equation (2.6-1a) depends on $\lambda$ and it is only for some values of $\lambda$ that the boundary conditions [Equations (2.6-1b, c)] can be satisfied.

These values of $\lambda$ are the eigenvalues (characteristic values) and the corresponding functions $y(x, \lambda)$ are the eigenfunctions (characteristic functions).

***Example 2.6-1.***  Solve the equation

$$y'' + \lambda y = 0 \tag{2.6-3a}$$

subject to the conditions

$$y(0) = 0, \qquad y'(\pi) = 0 \tag{2.6-3b,c}$$

We assume $\lambda$ to be real and it can be positive, zero, or negative. We consider these three cases separately.

(a)      $\lambda < 0$. For convenience, we set $\lambda = -m^2$ and the solution of Equation (2.6-3a) is

$$y = c_1 e^{mx} + c_2 e^{-mx} \tag{2.6-4}$$

where $c_1$ and $c_2$ are constants.

To satisfy Equations (2.6-3b, c), we require

$$0 = c_1 + c_2 \tag{2.6-5a}$$

$$0 = m(c_1 e^{m\pi} - c_2 e^{-m\pi}) \tag{2.6-5b}$$

The only solution of Equations (2.6-5a, b) is

$$c_1 = c_2 = 0 \tag{2.6-6a,b}$$

This leads to the trivial solution ($y = 0$).

(b)     $\lambda = 0$. The solution is now given by

$$y = c_3 + c_4 x \tag{2.6-7}$$

where $c_3$ and $c_4$ are constants.

The boundary conditions imply

$$c_3 = c_4 = 0 \tag{2.6-8a,b}$$

Again the only possible solution is the trivial solution.

(c)     $\lambda > 0$. We set $\lambda = n^2$. The solution is

$$y = c_5 \sin nx + c_6 \cos nx \tag{2.6-9}$$

where $c_5$ and $c_6$ are constants.

Applying Equations (2.6-3b, c) yields

$$0 = c_6 \tag{2.6-10a}$$

$$0 = c_5 \, n \cos n\pi \tag{2.6-10b}$$

Equation (2.6-10b) implies that either $c_5$ is zero which leads to a trivial solution or that $\cos n\pi$ is zero which provides the non-trivial solution. The cos function has multiple zeros and $\cos n\pi$ is zero if

$$n = (2s + 1) / 2, \quad s = 0, 1, 2, ... \tag{2.6-11}$$

The system defined by Equations (2.6-3a to c) has an infinite number of eigenvalues and they are given by

$$\lambda_s = \left(\frac{2s + 1}{2}\right)^2, \quad s = 0, 1, 2, ... \tag{2.6-12}$$

The corresponding eigenfunctions are

$$y_s = \sin\left(\frac{2s + 1}{2}\right) x \tag{2.6-13}$$

Note that the eigenvalues are real, positive, and discrete. Such properties are associated for example with discrete energy levels in quantum mechanics.

●

Next we discuss the general properties of the Sturm Liouville problem.

## The Eigenvalues Are Real

Suppose $\lambda$ is complex and this implies that $y$ is also complex. Taking the complex conjugate of Equations (2.6-1a to c) and noting that $r$, $p$, $q$, $a_1$, $a_2$, $b_1$ and $b_2$ are real, we obtain

$$(r\,\bar{y}')' + (q + \bar{\lambda}\,p)\,\bar{y} = 0 \tag{2.6-14a}$$

$$a_1\,\bar{y}\,(a) + a_2\,\bar{y}'\,(a) = 0 \tag{2.6-14b}$$

$$b_1\,\bar{y}\,(b) + b_2\,\bar{y}'\,(b) = 0 \tag{2.6-14c}$$

where $\bar{y}$ is the complex conjugate of $y$.

We multiply Equation (2.6-1a) by $\bar{y}$ and Equation (2.6-14a) by $y$ and subtract one from the other to yield

$$(\lambda - \bar{\lambda})\,p\,\bar{y}\,y = y\,(r\,\bar{y}')' - \bar{y}\,(r\,y')' \tag{2.6-15}$$

On integrating, we obtain

$$(\lambda - \bar{\lambda}) \int_a^b p\,\bar{y}\,y\,dx = \int_a^b [y\,(r\,\bar{y}')' - \bar{y}\,(r\,y')']\,dx \tag{2.6-16a}$$

$$= \left[y\,r\,\bar{y}' - \bar{y}\,r\,y'\right]_a^b - \int_a^b (y'\,r\,\bar{y}' - \bar{y}'\,r\,y')\,dx \tag{2.6-16b}$$

$$= 0 \tag{2.6-16c}$$

To obtain Equation (2.6-16c), we have used boundary conditions [Equations (2.6-1b, c, 14b, c)]. Since the eigenfunctions are non-trivial, Equation (2.6-16c) implies that $\lambda = \bar{\lambda}$, that is to say, $\lambda$ is real.

## The Eigenfunctions Are Orthogonal

Let $\lambda_n$ and $\lambda_m$ ($\lambda_n \neq \lambda_m$) be two eigenvalues and their corresponding functions are $y_n$ and $y_m$. The functions $y_n$ and $y_m$ satisfy

$$(r\,y_n')' + (q + \lambda_n\,p)\,y_n = 0 \tag{2.6-17a}$$

$$a_1\,y_n(a) + a_2\,y_n'(a) = 0, \qquad b_1\,y_n(b) + b_2\,y_n'(b) = 0 \tag{2.6-17b,c}$$

$$(r \, y_m')' + (q + \lambda_m \, p) \, y_m = 0 \tag{2.6-18a}$$

$$a_1 \, y_m(a) + a_2 \, y_m'(a) = 0 , \qquad b_1 \, y_m(b) + b_2 \, y_m'(b) = 0 \tag{2.6-18b,c}$$

We proceed as in (i), that is to say, we multiply Equations (2.6-17a, 18a) by $y_m$ and $y_n$ respectively, subtract one from the other and integrate the resulting expression to yield

$$(\lambda_n - \lambda_m) \int_a^b p \, y_n \, y_m \, dx = 0 \tag{2.6-19}$$

We have assumed that $\lambda_n \neq \lambda_m$ and it follows that

$$\int_a^b p \, y_n \, y_m \, dx = 0 \tag{2.6-20}$$

The functions $y_n$ and $y_m$ are **orthogonal** with respect to the weight $p(x)$. If $\lambda_n = \lambda_m$, Equation (2.6-20) is no longer true and

$$\int_a^b p \, y_n^2 \, dx = I_n^2 \quad (\neq 0) \tag{2.6-21}$$

The eigenfunctions can be **normalized** and we define the **normalized eigenfunction** $y_n^*$ to be

$$y_n^* = y_n / I_n \tag{2.6-22}$$

Equations (2.6-20, 21) can be expressed as

$$\int_a^b p \, y_n^* \, y_m^* \, dx = \int_a^b \frac{p \, y_n \, y_m}{I_n \, I_m} \, dx = \delta_{nm} \tag{2.6-23a,b}$$

where $\delta_{nm}$ is the Kronecker delta and is defined by

$$\delta_{nm} = \begin{cases} 0 , & \text{if } n \neq m \\ \\ 1 , & \text{if } n = m \end{cases} \tag{2.6-24a,b}$$

## 2.7   SPECIAL FUNCTIONS

The solutions of Equation (2.6-3a) are hyperbolic functions ($\lambda < 0$), polynomial ($\lambda = 0$), and trigonometric functions ($\lambda > 0$). The properties of these functions are well known. The solutions of Equations (2.6-2b, e, g) are referred to as special functions and their properties have been investigated and recorded. Since the beginning of the eighteenth century, many such functions have been considered and their properties are listed in Erdélyi et al. (1953, 1955) and in Abramowitz and Stegun (1970). In this section, we consider Legendre polynomials and Bessel functions.

### Legendre's Functions

Legendre's equation can be written as

$$(1 - x^2)\, y'' - 2\, xy' + ky = 0 \tag{2.7-1}$$

where $k$ is a constant.

In Chapter 5, it is shown that if Laplace's equation in spherical coordinates is solved by the method of separation of variables and if axial symmetry is assumed, the equation in the $\theta$-direction is Legendre's equation with $x = \cos\theta$. Thus the poles ($\theta = 0, \pi$) correspond to $x = \pm 1$.

We seek a solution near the origin, which is an ordinary point, and propose $y$ to be of the form

$$y = \sum_{n=0}^{\infty} c_n x^n \tag{2.7-2}$$

On differentiating term by term and substituting $y$, $y'$, and $y''$ in Equation (2.7-1), one obtains

$$\sum_{n=2}^{\infty} n\,(n-1)\,c_n\,x^{n-2} - \sum_{n=2}^{\infty} n\,(n-1)\,c_n\,x^n - 2\sum_{n=1}^{\infty} n\,c_n\,x^n + k\sum_{n=0}^{\infty} c_n\,x^n = 0 \tag{2.7-3}$$

Following the procedure described in Example 2.3-1, we make a change in the indices so that in all cases the summation starts from $r = 0$ to $\infty$. Equation (2.7-3) becomes

$$\sum_{r=0}^{\infty} \left[ (r+2)\,(r+1)\,c_{r+2}\,x^r - (r+2)\,(r+1)\,c_{r+2}\,x^{r+2} - 2\,(r+1)\,c_{r+1}\,x^{r+1} + k\,c_r\,x^r \right] = 0 \tag{2.7-4}$$

We now compare powers of $x$

$$x^0: \quad 2c_2 + kc_0 = 0 \quad \Rightarrow \quad c_2 = -(k/2)\,c_0 \tag{2.7-5a,b}$$

$x^1$:     $3 \cdot 2\, c_3 - 2 c_1 + k c_1 = 0 \;\Rightarrow\; c_3 = c_1\,(2-k)/6$                    (2.7-5c,d)

$x^2$:     $4 \cdot 3\, c_4 - 2 c_2 - 2 \cdot 2\, c_2 + k c_2 = 0 \;\Rightarrow\; c_4 = c_2\,(6-k)/12$          (2.7-5e,f)

.
.
.

$x^s$:     $(s+2)(s+1)\, c_{s+2} - s(s-1)\, c_s - 2s\, c_s + k c_s = 0 \;\Rightarrow\; c_{s+2} = c_s\,(s^2+s-k)/(s+2)(s+1)$

(2.7-5g,h)

The solution can be separated into an even function and an odd function. The even function involves $c_0$, $c_2$, $c_4$, ... and we denote this function by $y_1$. The odd function $y_2$ is in terms of $c_1$, $c_3$, $c_5$, ... From the recurrence formula [Equation (2.7-5h)], $c_2$, $c_4$, ... can be expressed in terms of $c_0$ and $c_3$, $c_5$, ... in terms of $c_1$. The general solution can be written as

$$y = c_0\, y_1 + c_1\, y_2 \qquad\qquad (2.7\text{-}6)$$

where $c_0$ and $c_1$ are arbitrary constants.

From Equation (2.7-5h), we deduce that the radius of convergence $R$ is

$$R = \lim_{n \to \infty} \left| \frac{c_{n+2}}{c_n} \right| = \lim_{n \to \infty} \left| \frac{n^2+n-k}{n^2+3n+1} \right| = 1 \qquad\qquad (2.7\text{-}7a,b,c)$$

To examine the validity of the solution at $x = 1$, we consider the special case of $k = 0$.

Equation (2.7-5h) now becomes

$$c_{s+2} = s\, c_s /(s+2) \qquad\qquad (2.7\text{-}8)$$

We note from Equation (2.7-5b) that in this case $c_2$ is zero and this implies $c_4 = c_6 = ... = 0$ and $y_1$ is given by

$$y_1 = c_0 \qquad\qquad (2.7\text{-}9)$$

In this case, the even solution is a constant and is valid for all values of $x$ including $x = \pm 1$. From Equation (2.7-8), we deduce that $y_2$ is given by

$$y_2 = c_1\, x\,(1 + x^2/3 + x^4/5 + x^6/7 + ... ) \qquad\qquad (2.7\text{-}10)$$

For values of $x = \pm 1$, we determine that the series diverges by comparing it with $\sum 1/n$. We note that $x = \pm 1$ are singular points and, for this special value of $k$, one solution is valid at the singular points and the other is not. This result can be extended to general values of $k$. For convenience, we set $k = \ell\,(\ell + 1)$ and Equation (2.7-5h) becomes

$$c_{s+2} = c_s \, (s^2 + s - \ell^2 - \ell) / (s+2)(s+1) = c_s \, (s - \ell)(s + \ell + 1) / (s+2)(s+1) \qquad (2.7\text{-}11)$$

From Equation (2.7-11), we observe that if $\ell$ is a non-negative integer, $c_{\ell+2}$ is zero and so are $c_{\ell+4}$, $c_{\ell+6}$, ... Thus the infinite series becomes a polynomial and the solution is valid for all values of x. In particular, if $\ell$ is even, the even solution (a polynomial) is valid for all values of x and, if $\ell$ is odd, the odd solution is a polynomial. The case we considered earlier is $k = 0$ ($\ell = 0$) and the even solution is a constant. In general, for any integer $\ell$, the degree of the polynomial is $\ell$. These polynomials are the **Legendre polynomials** and are denoted by $P_\ell(x)$. The constants $c_0$ (or $c_1$) are chosen such that $P_\ell(1)$ is unity. The first few Legendre polynomials are shown in Figure 2.7-1 and are

$$P_0(x) = 1, \qquad\qquad P_1(x) = x \qquad\qquad (2.7\text{-}12a,b)$$

$$P_2(x) = \frac{1}{2}(3x^2 - 1), \qquad P_3(x) = \frac{1}{2}(5x^2 - 3x) \qquad (2.7\text{-}12c,d)$$

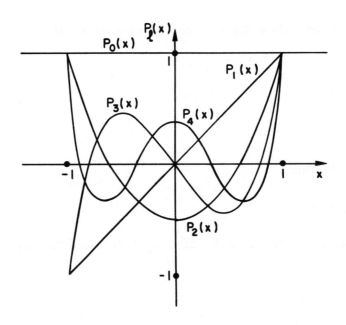

**FIGURE 2.7-1    Legendre polynomials**

A relatively easy method of computing $P_\ell(x)$ is to use **Rodrigues' formula** which can be written as

$$P_\ell(x) = \frac{1}{2^\ell \, \ell!} \frac{d^\ell}{dx^\ell} (x^2 - 1)^\ell \qquad (2.7\text{-}13)$$

The Legendre polynomials can also be obtained by using the **generating function** $(1-2xt+t^2)^{-1/2}$. Expanding this function in powers of t yields

$$\frac{1}{\sqrt{1-2xt+t^2}} = \sum_{\ell=0}^{\infty} t^\ell P_\ell(x) \tag{2.7-14}$$

Another method of determining $P_\ell(x)$ is to use the recurrence formula which can be written as

$$(\ell+1)P_{\ell+1}(x) = (2\ell+1)xP_\ell(x) - \ell P_{\ell-1}(x) \tag{2.7-15}$$

Knowing $P_0$ and $P_1$, we can calculate $P_2$, $P_3$, ...

An important property of the Legendre polynomials is the orthogonal property. From Equations (2.6-2c, 20, 21), we deduce that the Legendre polynomials are orthogonal with respect to weight one. The orthogonal property can be written as

$$\int_{-1}^{1} P_\ell(x) P_m(x)\, dx = \frac{2}{2\ell+1}\delta_{\ell m} \tag{2.7-16}$$

where $\delta_{\ell m}$ is the Kronecker delta.

The second linearly independent solution of Equation (2.7-1) can be obtained by the method of variation of parameters. We denote the solution by $Q_\ell(x)$ and assume that

$$Q_\ell(x) = u(x) P_\ell(x) \tag{2.7-17}$$

On differentiating $Q_\ell(x)$ twice and substituting the resulting expressions in Equation (2.7-1) with $k = \ell(\ell+1)$, we obtain

$$u[(1-x^2)P_\ell'' - 2xP_\ell' + \ell(\ell+1)P_\ell] + (1-x^2)(u''P_\ell + 2u'P_\ell') - 2xu'P_\ell = 0 \tag{2.7-18}$$

Since $P_\ell$ is a solution of Legendre's equation, Equation (2.7-18) simplifies to

$$(1-x^2)P_\ell u'' + u'[2(1-x^2)P_\ell' - 2xP_\ell] = 0 \tag{2.7-19}$$

On writing $u' = v$, Equation (2.7-19) can be written as

$$\frac{dv}{dx} + 2v[(P_\ell'/P_\ell) - x/(1-x^2)] = 0 \tag{2.7-20}$$

Equation (2.7-20) is a first order O.D.E. and the integrating factor I.F. is

$$\text{I.F.} = \exp \int \left[ (2P_\ell'/P_\ell) - 2x/(1-x^2) \right] dx \qquad (2.7\text{-}21a)$$

$$= \exp \left[ 2\ln P_\ell + \ln (1-x^2) \right] \qquad (2.7\text{-}21b)$$

$$= (1-x^2) \, P_\ell^2 \qquad (2.7\text{-}21c)$$

Equation (2.7-20) can be written as

$$\frac{d}{dx} \left[ v (1-x^2) P_\ell^2 \right] = 0 \qquad (2.7\text{-}22)$$

The solution is

$$v = C / \left[ (1-x^2) P_\ell^2 \right] \qquad (2.7\text{-}23)$$

where $C$ is a constant.

It follows that $Q_\ell (x)$ is given by

$$Q_\ell (x) = C P_\ell (x) \int^x \frac{d\xi}{(1-\xi^2) P_\ell^2 (\xi)} \qquad (2.7\text{-}24)$$

The functions $P_\ell (x)$ and $Q_\ell (x)$ are **Legendre's polynomials of the first and second kind**.

***Example 2.7-1***. Calculate the Legendre polynomials of the second kind, $Q_0 (x)$ and $Q_1 (x)$.

From Equations (2.7-12a, 24), we have

$$Q_0 (x) = C \int^x \frac{d\xi}{(1-\xi^2)} \qquad (2.7\text{-}25a)$$

$$= \frac{C}{2} \int^x \left[ \frac{1}{1-\xi} + \frac{1}{1+\xi} \right] d\xi \qquad (2.7\text{-}25b)$$

$$= \frac{C}{2} \ln \left[ \frac{1+x}{1-x} \right] \qquad (2.7\text{-}25c)$$

It is usual to choose $C$ to be one and $Q_0 (x)$ is given by

$$Q_0(x) = \frac{1}{2} \ln\left[\frac{1+x}{1-x}\right] \tag{2.7-26}$$

Expanding $Q_0(x)$ about the origin yields

$$Q_0(x) = \frac{1}{2}[x - x^2/2 + x^3/3 - x^4/4 + ... - (-x - x^2/2 - x^3/3 - x^4/4 - ...)] \tag{2.7-27a}$$

$$= x + x^3/3 + ... \tag{2.7-27b}$$

From Equations (2.7-12b, 24), we obtain

$$Q_1(x) = Cx \int^x \frac{d\xi}{\xi^2(1-\xi^2)} \tag{2.7-28a}$$

$$= Cx \int^x \left[\frac{1}{\xi^2} + \frac{1}{2}\left(\frac{1}{1-\xi} + \frac{1}{1+\xi}\right)\right] d\xi \tag{2.7-28b}$$

$$= Cx \left[-\frac{1}{x} + \frac{1}{2} \ln\left(\frac{1+x}{1-x}\right)\right] \tag{2.7-28c}$$

As in the case of $Q_0(x)$, we choose $C$ to be one and $Q_1(x)$ is given by

$$Q_1(x) = \frac{x}{2} \ln\left(\frac{1+x}{1-x}\right) - 1 \tag{2.7-29}$$

Expanding the $\ln$ function in powers of $x$, we obtain

$$Q_1(x) = -1 + \frac{1}{2}(x^2 + x^4/3 + ... ) \tag{2.7-30}$$

From Equations (2.7-26, 29), we note that $Q_0$ and $Q_1$ have singularities at $x = \pm 1$. The infinite series given by Equations (2.7-27b, 30) are the infinite series solutions of Legendre's equation and are valid for $|x| < 1$. For $\ell = 0$ ($\ell$ is even), the even solution $P_0(x)$ (= 1) is valid at $|x| = 1$ and the odd solution $Q_0(x)$ is not valid at $|x| = 1$. Likewise for $\ell = 1$ ($\ell$ is odd), the odd solution $P_1(x)$ (= x) is valid at $|x| = 1$ and the even solution $Q_1(x)$ is not valid at $|x| = 1$.

●

The other Legendre functions of the second kind can be computed from the recurrence formula

$$(\ell+1)Q_{\ell+1} = x(1+2\ell)Q_\ell - \ell Q_{\ell-1} \tag{2.7-31}$$

Note that the recurrence formulae for both $P_\ell$ [Equation (2.7-15)] and $Q_\ell$ [Equation (2.7-31)] are identical.

The general solution of Legendre's equation can be written as

$$y = A P_\ell(x) + B Q_\ell(x) \qquad\qquad (2.7\text{-}32)$$

where A and B are constants and $\ell$ is a non-negative integer.

The function $Q_\ell(x)$ is singular at $|x| = 1$ and if we require the solution $y$ to be finite at $|x| = 1$, B has to be zero.

The **associated Legendre equation** can be written as

$$(1 - x^2) \frac{d^2 y}{dx^2} - 2x \frac{dy}{dx} + [\ell(\ell+1) - m^2/(1 - x^2)] \, y = 0 \qquad\qquad (2.7\text{-}33)$$

where $\ell$ and m are integers.

If m is zero, Equation (2.7-33) reduces to the standard Legendre equation [Equation (2.7-1)]. Equation (2.7-33) is derived from Laplace's equation in spherical coordinates [see Equation (5.5-37b)]. The additional term $m^2 y / (1 - x^2)$ represents the non-symmetric contribution.

We start by considering the simplest case (m = 1) and Equation (2.7-33) becomes

$$(1 - x^2) \frac{d^2 y}{dx^2} - 2x \frac{dy}{dx} + [\ell(\ell+1) - (1 - x^2)^{-1}] \, y = 0 \qquad\qquad (2.7\text{-}34)$$

One would be tempted to introduce a series solution; however, the following procedure provides us with an ingenious way of solving the problem.

The Legendre polynomials $P_\ell(x)$ satisfy the equation

$$(1 - x^2) P_\ell'' - 2x P_\ell' + \ell(\ell+1) P_\ell = 0 \qquad\qquad (2.7\text{-}35)$$

Differentiating with respect to x yields

$$(1 - x^2) P_\ell''' - 4x P_\ell'' + \ell(\ell+1) P_\ell' = 0 \qquad\qquad (2.7\text{-}36)$$

A new function $w(x)$ is defined by

$$w(x) = (1 - x^2)^{1/2} P_\ell' \qquad\qquad (2.7\text{-}37a)$$

or $\quad P_\ell' = (1 - x^2)^{-1/2} \, w(x)$ $\hfill$ (2.7-37b)

On differentiating, we obtain

$$P_\ell'' = (1 - x^2)^{-1/2} \, w' + x \, (1 - x^2)^{-3/2} \, w \qquad (2.7\text{-}38a)$$

$$P_\ell''' = (1 - x^2)^{-1/2} \, w'' + 2x \, (1 - x^2)^{-3/2} \, w' + w \, (1 + 2x^2) \, (1 - x^2)^{-5/2} \qquad (2.7\text{-}38b)$$

Substituting Equations (2.7-37b, 38a, b) in Equation (2.7-36) yields

$$(1 - x^2) \, w'' - 2xw' + w \, [\ell \, (\ell + 1) - (1 - x^2)^{-1}] = 0 \qquad (2.7\text{-}34)$$

This is Equation (2.7-34) with $y = w$. By convention, the solution of the associated Legendre equation is denoted by $P_\ell^m(x)$. From Equation (2.7-37a), we deduce that

$$P_\ell^1(x) = (1 - x^2)^{1/2} \, P_\ell'(x) \qquad (2.7\text{-}39)$$

In the general case, $P_\ell^m(x)$ is given by

$$P_\ell^m(x) = (1 - x^2)^{m/2} \, \frac{d^m P_\ell}{dx^m} \qquad (2.7\text{-}40)$$

Similarly the **associated Legendre function of the second kind** can be computed from the formula

$$Q_\ell^m(x) = (1 - x^2)^{m/2} \, \frac{d^m Q_\ell}{dx^m} \qquad (2.7\text{-}41)$$

The functions $Q_\ell^m(x)$ are singular at $|x| = 1$.

The general solution of Equation (2.7-33b) is

$$y = A \, P_\ell^m(x) + B \, Q_\ell^m(x) \qquad (2.7\text{-}42)$$

where $A$ and $B$ are constants. If $y$ is finite at $|x| = 1$, $B$ has to be zero.

The polynomial $P_\ell$ is of degree $\ell$ and we note from Equation (2.7-40) that if $m > \ell$, $P_\ell^m(x)$ is zero. The function $P_\ell^m$ is defined for $\ell \geq m$. Note that due to the term $(1 - x^2)^{m/2}$ in Equation (2.7-40), $P_\ell^m$ is a polynomial iff $m$ is even.

Properties of the Legendre and the associated Legendre functions of both kinds are listed in the references cited earlier.

## Bessel Functions

Bessel functions were introduced by Bessel in 1824, in the discussion of a problem in astronomy. Bessel's equation occurs in the solution of Laplace's equation in cylindrical coordinates [see Equation (5.5-8a, 1b)]. It can be written as

$$x^2 y'' + xy' + (x^2 - v^2)\, y = 0 \tag{2.7-43}$$

where $v$ is a constant.

The origin is a regular singular point and we seek a solution of the form

$$y = \sum_{n=0}^{\infty} c_n x^{n+r} \tag{2.7-44}$$

Differentiating term by term and substituting the resulting expressions in Equation (2.7-43) yields

$$\sum_{n=0}^{\infty} [(n+r)(n+r-1) c_n x^{n+r} + (n+r) c_n x^{n+r} + c_n x^{n+r+2} - v^2 c_n x^{n+r}] = 0 \tag{2.7-45}$$

Comparing powers of $x$, we obtain

$$x^r: \quad c_0 [r(r-1) + r - v^2] = 0 \quad \Rightarrow \quad r = \pm v \ (c_0 \neq 0) \tag{2.7-46a,b}$$

$$x^{r+1}: \quad c_1 [r(r+1) + (r+1) - v^2] = 0 \quad \Rightarrow \quad c_1 = 0 \tag{2.7-46c,d}$$

.
.
.

$$x^{r+s}: \quad c_s [(s+r)(s+r-1) + (s+r) - v^2] + c_{s-2} = 0 \quad \Rightarrow \quad c_s = -c_{s-2} / [(s+r)^2 - v^2] \tag{2.7-46e,f}$$

For $r = v$, Equation (2.7-46f) becomes

$$c_s = -c_{s-2} / [s(s+2v)] \tag{2.7-47}$$

From Equation (2.7-47), we can compute the even terms $c_2$, $c_4$, $c_6$, ... Writing $s = 2p$, Equation (2.7-47) can now be written as

$$c_{2p} = -c_{2p-2} / [2^2 p(p+v)] \tag{2.7-48}$$

The first few coefficients are

$$c_2 = -c_0 / [2^2 (1 + v)] \tag{2.7-49a}$$

$$c_4 = -c_2 / [2^2 (2) (2 + v)] = c_0 / [2^4 (2) (1 + v) (2 + v)] \tag{2.7-49b,c}$$

$$c_6 = -c_4 / [2^2 (3) (3 + v)] = -c_0 / [2^6 (2) (3) (1 + v) (2 + v) (3 + v)] \tag{2.7-49d,e}$$

Equations (2.7-49a to e) suggest that

$$c_{2p} = (-1)^p c_0 / [2^{2p} (p!) (1 + v) (2 + v) \dots (p + v)] \tag{2.7-50}$$

It can be verified that Equation (2.7-50) satisfies Equation (2.7-48). One solution of Equation (2.7-43) can be written as

$$y = c_0 \sum_{p=0}^{\infty} (-1)^p x^{2p+v} / [2^{2p} (p!) (1 + v) (2 + v) \dots (p + v)] \tag{2.7-51}$$

If $v$ is a positive integer, the product $(1 + v) (2 + v) \dots (p + v)$ can be written as $(p + v)!/v!$. To give a meaning to $v!$ when $v$ is not an integer, we define the **gamma function** $\Gamma(v)$ by

$$\Gamma(v) = \int_0^{\infty} t^{v-1} e^{-t} dt , \quad v > 0 \tag{2.7-52}$$

The condition that $v$ is positive is necessary so as to ensure the convergence of the integral.

$$\Gamma(v+1) = \int_0^{\infty} t^v e^{-t} dt \tag{2.7-53}$$

On integrating by parts, we obtain

$$\Gamma(v+1) = \left[ -t^v e^{-t} \right]_0^{\infty} + v \int_0^{\infty} t^{v-1} e^{-t} dt \tag{2.7-54a}$$

$$= v \, \Gamma(v) \tag{2.7-54b}$$

From Equation (2.7-53), we deduce that

$$\Gamma(1) = \int_0^{\infty} e^{-t} dt = 1 \tag{2.7-55a,b}$$

Combining Equations (2.7-54b, 55b) yields

$$\Gamma(2) = 1, \quad \Gamma(3) = 2\,\Gamma(2) = 2 \bullet 1, \quad \Gamma(4) = 3\,\Gamma(3) = 3 \bullet 2 \bullet 1 \qquad \text{(2.7-56a-e)}$$

Generalizing Equations (2.7-56a to e), we obtain

$$\Gamma(v+1) = v! \quad \text{and} \quad 0! = 1 \qquad \text{(2.7-57a,b)}$$

for $v \geq 0$.

Equation (2.7-54b) can be used to define $\Gamma(v)$ for all values of $v$ ($\neq 0, -1, -2, \ldots$). That is to say, we define $\Gamma(v)$ as

$$\Gamma(v) = [\Gamma(v+1)]/v \qquad \text{(2.7-58)}$$

If $-1 < v < 0$, then $0 < v+1 < 1$ and $\Gamma(v+1)$ is defined. It follows from Equation (2.7-58) that $\Gamma(v)$ is defined. Similarly if $-2 < v < -1$, then $v+1$ lies between $-1$ and $0$, and $\Gamma(v+1)$ has just been defined. Similarly, the function $\Gamma(v)$ is defined for all negative non-integers. From Equation (2.7-58), we deduce that $\Gamma(0)$ can be defined as

$$\Gamma(0) = \lim_{v \to 0} \Gamma(v) = \lim_{v \to 0} \frac{\Gamma(v+1)}{v} = \pm\infty \qquad \text{(2.7-59a,b,c)}$$

It follows from Equation (2.7-58), that $\Gamma(v)$ is $\pm\infty$ for all negative integers. The graph of $\Gamma(v)$ is shown in Figure 2.7-2. We note that (see Chapter 4, Problem 9b)

$$\Gamma(1/2) = \sqrt{\pi} \qquad \text{(2.7-60)}$$

Equation (2.7-51) can be written, in terms of gamma functions, for all positive $v$, as

$$y = c_0 \sum_{p=0}^{\infty} (-1)^p \, x^{2p+v} \, \Gamma(v+1) / [2^{2p} \, (p!) \, \Gamma(p+v+1)] \qquad \text{(2.7-61)}$$

By choosing $c_0$ to be $1/[2^v \, \Gamma(v+1)]$, we obtain the Bessel function of the first kind of order $v$ and it is denoted by $J_v(x)$. That is to say

$$J_v(x) = \sum_{p=0}^{\infty} (-1)^p \, x^{2p+v} / [2^{2p+v} \, (p!) \, \Gamma(p+v+1)] \qquad \text{(2.7-62a)}$$

$$= \sum_{p=0}^{\infty} (-1)^p \, (x/2)^{2p+v} / [(p!) \, \Gamma(p+v+1)] \qquad \text{(2.7-62b)}$$

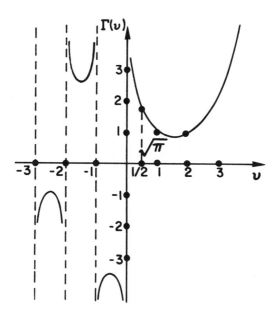

**FIGURE 2.7-2   Gamma function**

Thus one solution of Equation (2.7-43) is $J_v(x)$. If $v$ is not an integer or zero, the other solution is obtained by considering the other root of the indicial equation [Equation (2.7-46b)]. That is to say, the other solution is $J_{-v}(x)$ and is written as

$$J_{-v}(x) = \sum_{p=0}^{\infty} (-1)^p (x/2)^{2p-v} / [(p!) \, \Gamma(p - v + 1)] \tag{2.7-63}$$

Note that whereas $J_v(x)$ has no singularity at the origin, $J_{-v}(x)$ is singular at the origin.

For $v$ ($\neq 0$ or an integer), the general solution of Equation (2.7-43) is

$$y = A \, J_v(x) + B \, J_{-v}(x) \tag{2.7-64}$$

where A and B are constants.

If the solution is finite at the origin, B must be zero.

We recall that if the two roots of the indicial equation are coincident or differ by an integer, the two linearly independent solutions are not obtained in a straight forward manner as described earlier for non-integral values of $v$. If $v$ is zero, the two roots are coincident and this case has been considered in Example 2.4-3. If the two roots differ by an integer, this implies that $2v$ ($r_1 = v$, $r_2 = -v$) is an integer. We consider the two cases where $2v$ is an odd or an even integer separately. In Example

2.4-5, we solved the case where $2v$ is one. We recall that if $2v$ is one, $c_1$ is not necessarily zero [see Equation (2.4-84d)] and we have one solution starting with $c_0$ and the other starting with $c_1$. By assigning to $c_0$ and $c_1$ the values given earlier, we obtain $J_{1/2}(x)$ and $J_{-1/2}(x)$. Similarly, $J_v(x)$ and $J_{-v}(x)$ are obtained in the case where $2v$ is an odd integer. If $2v$ is an even integer, $v$ is an integer and from Equation (2.7-63) we note that the series starts from $p = v$ since $\Gamma(p - v + 1)$ is $\pm\infty$ for $p \leq v$. Writing $q = p - v$, Equation (2.7-63) can be written as

$$J_{-v}(x) = \sum_{q=0}^{\infty} (-1)^{q+v} (x/2)^{2q+v} / [(q + v)! \, (q!)] \tag{2.7-65}$$

Comparing Equations (2.7-62b, 65), we deduce that

$$J_{-v}(x) = (-1)^v J_v(x) \tag{2.7-66}$$

If $v$ is an integer, $J_v(x)$ and $J_{-v}(x)$ are not linearly independent.

Since $J_v(x)$ is known, the other linearly independent solution can be obtained by the method of variation of parameters. This method yields

$$y_2 = J_v(x) \int \frac{dx}{x \, [J_v(x)]^2} \tag{2.7-67}$$

The solution $y_2$ is usually not considered. Instead, the **Bessel function of the second kind** $Y_v(x)$ is defined as

$$Y_v(x) = [J_v(x) \cos v\pi - J_{-v}(x)] / \sin v\pi \tag{2.7-68}$$

From Equation (2.7-66), we conclude that both the numerator and the denominator on the right side of Equation (2.7-68) are zero. By applying l'Hôpital's rule, we deduce that $Y_v(x)$ exists in the limit as $v$ tends to an integer. Thus, the general solution is

$$y = A J_v(x) + B Y_v(x) \tag{2.7-69}$$

where $A$ and $B$ are constants.

The function $Y_v(x)$ has a $\ln(x/2)$ term (see Problem 19b) and is singular at the origin. If $y$ is finite at the origin, $B$ is zero.

If $v$ is not an integer, the solution of Equation (2.7-43) is given by Equation (2.7-64 or 69), but if $v$ is an integer only Equation (2.7-69) is valid.

We recall that there are circumstances when it is preferable to work with $\exp(\pm ix)$ rather than with $\sin x$ and $\cos x$. Equally, there are circumstances when it is preferable to choose **Hankel functions** (Bessel functions of the third kind) of order $\nu$ as solutions of Equation (2.7-43) instead of $J_\nu(x)$ and $Y_\nu(x)$. These Hankel functions $H_\nu^{(1)}$ and $H_\nu^{(2)}$ are defined by

$$H_\nu^{(1)} = J_\nu(x) + i\, Y_\nu(x) \tag{2.7-70a}$$

$$H_\nu^{(2)} = J_\nu(x) - i\, Y_\nu(x) \tag{2.7-70b}$$

The functions $H_\nu^{(1)}$ and $H_\nu^{(2)}$ are linearly independent and the general solution of Equation (2.7-43) is a linear combination of $H_\nu^{(1)}$ and $H_\nu^{(2)}$. From Equations (2.7-70a, b), we obtain

$$J_\nu(x) = \frac{1}{2}\,(H_\nu^{(1)} + H_\nu^{(2)}) \tag{2.7-71a}$$

$$Y_\nu(x) = \frac{i}{2}\,(H_\nu^{(2)} - H_\nu^{(1)}) \tag{2.7-71b}$$

The functions $J_0$ and $J_1$ are shown in Figure 2.7-3 and $Y_0$ and $Y_1$ in Figure 2.7-4. Table 2.7-1 lists some properties of Bessel functions.

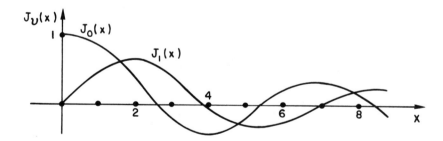

**FIGURE 2.7-3**    **Bessel functions of the first kind**

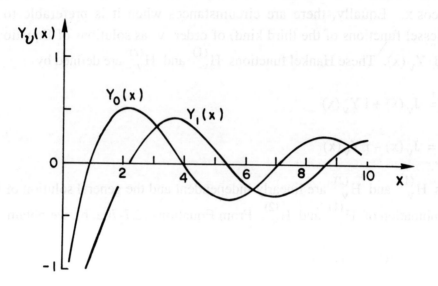

**FIGURE 2.7-4    Bessel functions of the second kind**

The functions $J_{-\nu}(x)$, $Y_\nu(x)$, $H_\nu^{(1)}(x)$ and $H_\nu^{(2)}(x)$ have the same recurrence relations as the function $J_\nu(x)$.

## TABLE 2.7-1

### Properties of Bessel functions

(i)     $[x^\nu J_\nu(x)]' = x^\nu J_{\nu-1}(x)$

(ii)    $[x^{-\nu} J_\nu(x)]' = -x^{-\nu} J_{\nu+1}(x)$

(iii)   $J_{\nu-1}(x) + J_{\nu+1}(x) = \dfrac{2\nu}{x} J_\nu(x)$

(iv)    $J_{\nu-1}(x) - J_{\nu+1}(x) = 2J_\nu'(x)$

(v)     $\displaystyle\int x^\nu J_{\nu-1}(x)\, dx = x^\nu J_\nu(x) + \text{constant}$

(vi)    $\displaystyle\int x^{-\nu} J_{\nu+1}(x)\, dx = x^{-\nu} J_\nu(x) + \text{constant}$

The function $J_v(x)$ has an infinite number of zeros and we denote the zeros by $\lambda_n$, $n = 1, 2, ...$
That is to say

$$J_v(\lambda_n) = 0, \qquad n = 1, 2, ... \tag{2.7-72}$$

The orthogonal property of $J_v(x)$ can be written as

$$\int_0^1 x\, J_v(\lambda_m x)\, J_v(\lambda_n x)\, dx = \begin{cases} 0 & \text{if } m \neq n \\[2mm] \frac{1}{2}\,[J_{v+1}(\lambda_n)]^2 & \text{if } m = n \end{cases} \tag{2.7-73a,b}$$

***Example 2.7-2.*** Compute $J_{3/2}(x)$, using the results of Example 2.4-5.

In Example 2.4-5, we have solved the Bessel equation for the case $v = \pm 1/2$ and the two linearly independent solutions were $x^{-1/2} \sin x$ and $x^{-1/2} \cos x$ [Equations (2.4-95a, b)]. We recall that in the definition of $J_v(x)$, the multiplicative constant $(c_0)$ is $1/[2^v\, \Gamma(v+1)]$.

Using the properties of the gamma function, we deduce that $J_{1/2}(x)$ and $J_{-1/2}(x)$ are given by

$$J_{1/2}(x) = \sqrt{\frac{2}{\pi x}}\, \sin x \tag{2.7-74}$$

$$J_{-1/2}(x) = \sqrt{\frac{2}{\pi x}}\, \cos x \tag{2.7-75}$$

From Table 2.7-1, we obtain

$$J_{3/2}(x) = \frac{1}{x}\, J_{1/2}(x) - J_{-1/2}(x) \tag{2.7-76a}$$

$$= \sqrt{\frac{2}{\pi x}} \left( \frac{\sin x}{x} - \cos x \right) \tag{2.7-76b}$$

## Modified Bessel's Equation

The **modified Bessel's equation** is of frequent occurrence in applied mathematics [see Equation (5.5-8a)] and can be written as

$$x^2\, y'' + x\, y' - (x^2 + v^2)\, y = 0 \tag{2.7-77}$$

Comparing Equations (2.7-43, 77), we note that they differ only in the coefficient of $y$. By writing $z = ix$, Equation (2.7-77) becomes

$$z^2 \frac{d^2y}{dz^2} + z \frac{dy}{dz} + (z^2 - v^2) \, y = 0 \qquad\qquad (2.7\text{-}78)$$

Equation (2.7-78) is exactly Equation (2.7-43) with $z$ being replaced by $x$ and the linearly independent solutions of Equation (2.7-77) are $J_v(ix)$ and $J_{-v}(ix)$. That is to say, they are Bessel functions with purely imaginary argument. Usually, the solution needs to be given in the form of real variables and $J_v(ix)$ and $J_{-v}(ix)$ are not in a suitable form.

We seek a series solution as described previously for Bessel's equation. The solution we obtain can be denoted by $I_v(x)$ as follows

$$I_v(x) = \sum_{p=0}^{\infty} (x/2)^{2p+v} / [(p!)\, \Gamma(p + v + 1)] \qquad\qquad (2.7\text{-}79)$$

Comparing Equations (2.7-62b, 79), we deduce that

$$I_v(x) = i^{-v} J_v(ix) = e^{-i\pi v/2} J_v(ix) \qquad\qquad (2.7\text{-}80a,b)$$

Note that we have already established that the solution of Equation (2.7-77) is $J_v(ix)$ and to obtain the real part of the solution we multiply $J_v(ix)$ by a complex constant $(i^{-v})$. The function $I_v(x)$, defined by Equation (2.7-79), is real. If $v$ is not an integer, the two linearly independent solutions are $I_v(x)$ and $I_{-v}(x)$. If $v$ is an integer, we define a new function $K_v$ as

$$K_v(x) = \frac{\pi}{2} \left[ \frac{I_{-v} - I_v}{\sin v\pi} \right] \qquad\qquad (2.7\text{-}81)$$

The function $K_v(x)$ is linearly independent of $I_v(x)$ and the limit as $v$ tends to an integer is defined. The two linearly independent solutions of Equation (2.7-77) for all values of $v$ (including integral values) are $I_v(x)$ and $K_v(x)$. The properties of all Bessel functions are given in Watson (1966).

Many second order linear differential equations can be transformed to Bessel's equation (or to another standard equation) by a suitable substitution. Kamke (1959) and Murphy (1960) list several such possibilities.

**Example 2.7-3.** A simplified form of the equation governing the linear stability of a Newtonian fluid flowing between two parallel walls can be written as (Rosenhead, 1963, p. 524)

$$\frac{d^4 \chi_0}{d\eta^4} - i\eta \frac{d^2 \chi_0}{d\eta^2} = 0 \qquad\qquad (2.7\text{-}82)$$

where $\chi_0$ is related to the stream function and $\eta$ is related to the distance from the centre of the channel. Both quantities are complex and we wish to solve Equation (2.7-82).

Equation (2.7-82) can be reduced to a second order equation by writing

$$u = \frac{d^2 \chi_0}{d\eta^2} \tag{2.7-83}$$

Equation (2.7-82) becomes

$$\frac{d^2 u}{d\eta^2} - i\eta u = 0 \tag{2.7-84}$$

We now write

$$u = \eta^{1/2} v \tag{2.7-85}$$

Differentiating and substituting into Equation (2.7-84), we obtain

$$\eta^{1/2} \frac{d^2 v}{d\eta^2} + \eta^{-1/2} \frac{dv}{d\eta} - \frac{1}{4} \eta^{-3/2} v - i\eta^{3/2} v = 0 \tag{2.7-86}$$

We further transform the independent variable $\eta$ by introducing $z$ as

$$z = \frac{2}{3} (i\eta)^{3/2} \tag{2.7-87}$$

The chain rule yields

$$\frac{dv}{d\eta} = i (i\eta)^{1/2} \frac{dv}{dz} = i \left(\frac{3}{2} z\right)^{1/3} \frac{dv}{dz} \tag{2.7-88a,b}$$

$$\frac{d^2 v}{d\eta^2} = - \left(\frac{3}{2} z\right)^{2/3} \frac{d^2 v}{dz^2} - \frac{1}{2} \left(\frac{3}{2} z\right)^{-1/3} \frac{dv}{dz} \tag{2.7-88c}$$

Substituting Equations (2.7-88b, c) in Equation (2.7-86) yields

$$z^2 \frac{d^2 v}{dz^2} + z \frac{dv}{dz} + (z^2 - 1/9) v = 0 \tag{2.7-89}$$

Equation (2.7-89) is Bessel's equation of order 1/3. Since we are dealing with complex functions, we write the solution in terms of Hankel functions. The two linearly independent solutions of Equation

(2.7-89) are $H_{1/3}^{(1)}(z)$ and $H_{1/3}^{(2)}(z)$. It follows that the fundamental (linearly independent) solutions of Equation (2.7-84) are

$$u_1 = \eta^{1/2} H_{1/3}^{(1)}\left[\frac{2}{3}(i\eta)^{3/2}\right] \tag{2.7-90a}$$

$$u_2 = \eta^{1/2} H_{1/3}^{(2)}\left[\frac{2}{3}(i\eta)^{3/2}\right] \tag{2.7-90b}$$

On integrating twice, we obtain two solutions for $\chi_0$.

## 2.8    FOURIER SERIES

In Section 2.6, we have seen that the Sturm-Liouville system generates orthogonal eigenfunctions. Trigonometric functions $\sin nx$, $\cos nx$, Legendre polynomials $P_\ell(x)$ and Bessel functions $J_v(\lambda_m x)$ are among the eigenfunctions we have encountered. We recall that the eigenfunctions $y_n(x)$ and $y_m(x)$ are orthogonal with respect to weight $p(x)$ if

$$\int_a^b p(x)\, y_n(x)\, y_m(x)\, dx = \begin{cases} 0 & \text{if } m \neq n \\ \\ I_n^2 & \text{if } m = n \end{cases} \tag{2.8-1a,b}$$

We assume that the integral exists and this means that the functions are **square integrable**. These square integrable functions generate a space ($L_2$ space) and since $n$ can be infinity, the dimension of the space is infinite.

The orthogonal property of $y_n$ reminds us of the property of orthogonal vectors and we state some of the properties of vectors, which are presented in more detail in Chapter 4. Usually, the dimension of the space is finite and if the dimension is $n$, we can choose $n$ vectors $\underline{g}_1, \underline{g}_2, \dots, \underline{g}_n$ which are linearly independent as bases. These bases are orthogonal (not orthonormal) if

$$\underline{g}_i \bullet \underline{g}_j = \langle \underline{g}_i, \underline{g}_j \rangle = \begin{cases} 0 & \text{if } i \neq j \\ \\ I_i^2\ (\neq 0) & \text{if } i = j \end{cases} \tag{2.8-2a,b}$$

The product $\underline{g}_i \bullet \underline{g}_j$ (or $\langle \underline{g}_i, \underline{g}_j \rangle$) is the **scalar (dot or inner) product** and $I_i$ is the **magnitude** of the vector $\underline{g}_i$. If $\underline{u}$ is any vector in the space, $\underline{u}$ can be expressed as

$$\underline{u} = \sum_{i=1}^{n} c_i\, \underline{g}_i \tag{2.8-3}$$

where $c_i$ are constants and are obtained by forming the dot product of $\underline{u}$ with $\underline{g}_j$. That is to say

$$\langle \underline{u}, \underline{g}_j \rangle = \sum_{i=1}^{n} c_i \langle \underline{g}_i, \underline{g}_j \rangle = c_j I_j^2 \qquad \text{(2.8-4a,b)}$$

We deduce that

$$c_j = \langle \underline{u}, \underline{g}_j \rangle / I_j^2 = \langle \underline{u}, \underline{g}_j \rangle / \langle \underline{g}_j, \underline{g}_j \rangle \qquad \text{(2.8-5a,b)}$$

Having been inspired by these properties of vector spaces, we now continue with function spaces and we regard the functions $y_n$ as the basis of the infinite space, and the integral in Equations (2.8-1a, b) as the inner product. The magnitude $y_n$, which is denoted by $\| y_n \|$ is the **norm** of $y_n$. That is to say

$$\| y_n \|^2 = \langle y_n, y_n \rangle = \int_a^b p(x) \, y_n^2(x) \, dx = I_n^2 \qquad \text{(2.8-6a,b,c)}$$

We now represent a function $f(x)$ by a linear combination of $y_n$ and write

$$f(x) \approx \sum_{n=1}^{N} c_n y_n(x) \qquad \text{(2.8-7)}$$

To determine the constants $c_n$, we make use of the orthogonal property of $y_n$. Forming the inner product, we find that $c_n$ is given by

$$c_n = \langle f, y_n \rangle / \| y_n \|^2 \qquad \text{(2.8-8a)}$$

$$\langle f, y_n \rangle = \int_a^b p(x) \, f(x) \, y_n(x) \, dx \qquad \text{(2.8-8b)}$$

$$\| y_n \|^2 = \int_a^b p(x) \, [y_n(x)]^2 \, dx \qquad \text{(2.8-8c)}$$

We need to examine in what sense the sum (letting $N \longrightarrow \infty$) approximates $f(x)$ in the interval [a, b], with $c_n$ defined by Equation (2.8-8a). Usually, we consider **pointwise approximation** and the difference between the sum and the function $f(x)$ is small for all values of $x$ in the interval. The sum converges to $f(x)$ if

$$\left| f(x) - \sum_{n=1}^{N} c_n y_n \right| < \varepsilon \qquad \text{(2.8-9)}$$

for each $\varepsilon > 0$ whenever $N > N_\varepsilon$ and for all $x$ in the interval.

Another approximation which is widely used in the treatment of experimental data is the **least square approximation**. Using the least square criterion, we require that the integral $\int_a^b p(x) [f(x) - \sum c_n y_n]^2 \, dx$ be a minimum. We demonstrate that this is the present case. Let

$$D = \int_a^b p(x) [f - \sum_{n=1}^N c_n y_n]^2 \, dx = \int_a^b p(x) [f^2 - 2f \sum_{n=1}^N c_n y_n + (\sum_{n=1}^N c_n y_n)^2] \, dx$$

$$(2.8\text{-}10a,b)$$

$D$ is a function of $c_n$ and $D$ is a minimum if

$$\frac{\partial D}{\partial c_s} = 0, \quad s = 1, 2, \ldots \tag{2.8-11a}$$

On differentiating, we obtain

$$\frac{\partial D}{\partial c_s} = \int_a^b p(x) [-2f y_s + 2y_s \sum_{n=1}^N c_n y_n] \, dx \tag{2.8-11b}$$

and using the orthogonal property of $y_n$, we write

$$\frac{\partial D}{\partial c_s} = \int_a^b p(x) [-2f y_s + 2c_s y_s^2] \, dx \tag{2.8-11c}$$

Minimizing $D$ implies that $c_s$ is given by Equation (2.8-8a). The approximation in this case is in the least square sense. If $D \longrightarrow 0$ as $N \longrightarrow \infty$, the sum $\sum_{n=1}^N c_n y_n$ **converges in the mean** to $f(x)$. The series is the **Fourier series** and the coefficient $c_n$ is the **Fourier coefficient**. If for every $f(x)$ for which $\int_a^b p(x) f^2 \, dx$ is finite, $\sum_{n=1}^N c_n y_n$ converges in the mean to $f(x)$, the set of functions $(y_1, y_2, \ldots)$ is **complete** and the space they span is a **Hilbert space**.

Note that in this case the requirement is that the integrals exist and the function can have a jump discontinuity in the interval. The function $y_n$ can be continuous though $f(x)$ can be discontinuous. If $f(x)$ has a **jump discontinuity** at $x_0$, the sum $\sum_{n=1}^N c_n y_n$ converges to $\frac{1}{2}[f(x_{0+}) + f(x_{0-})]$, that is to say, to the mean value of $f(x)$ as $x$ approaches $x_0$ from the left and from the right.

## Trigonometric Fourier Series

The trigonometric functions $\sin x$ and $\cos x$ are periodic with period $2\pi$. A periodic function $f(x)$ of period $T$ is defined by Equation (1.1-14).

For simplicity, we assume $f(x)$ to be defined in the interval $[-\pi, \pi]$. If $f(x)$ is defined in the interval $[a, b]$, then the transformation

$$x^* = 2\pi [x - (a + b)/2] / (b - a) \tag{2.8-12}$$

transforms the interval $[a, b]$ to $[-\pi, \pi]$.

The orthogonal property of $\cos x$ and $\sin x$ can be written as

$$\int_{-\pi}^{\pi} \sin nx \cos mx \, dx = 0, \quad \text{for all } m \text{ and } n \tag{2.8-13a}$$

$$\int_{-\pi}^{\pi} \sin nx \sin mx \, dx = \int_{-\pi}^{\pi} \cos nx \cos mx \, dx = \pi \, \delta_{nm} \tag{2.8-13b,c,d}$$

where $\delta_{nm}$ is the Kronecker delta.

The functions $\sin x$ and $\cos x$ form an orthogonal basis in the interval $[-\pi, \pi]$ with unit weight $[p(x) = 1]$. If $f(x)$ is periodic and of period $2\pi$ and $\int_{-\pi}^{\pi} f^2 \, dx$ is bounded, the Fourier series converges in the mean to $f(x)$. That is to say

$$f(x) \approx \frac{1}{2} a_0 + \sum_{n=1}^{\infty} [a_n \cos nx + b_n \sin nx] \tag{2.8-14a}$$

$$a_n = \frac{1}{\pi} \int_{-\pi}^{\pi} f(x) \cos nx \, dx \tag{2.8-14b}$$

$$b_n = \frac{1}{\pi} \int_{-\pi}^{\pi} f(x) \sin nx \, dx \tag{2.8-14c}$$

The coefficients $a_n$ and $b_n$ given by Equations (2.8-14b, c) are special cases of Equation (2.8-8a). Note that we have written $\frac{1}{2} a_0$ and not $a_0$ so that the formula for $a_n$ [Equation (2.8-14b)] is applicable to $a_0$. Recall that if $f(x)$ has a jump discontinuity at $x_0$, the series converges to $\frac{1}{2}[f(x_{0+}) + f(x_{0-})]$.

***Example 2.8-1.*** Determine the Fourier series for the function $f(x)$ defined in the interval $(-\pi, \pi)$ by

$$f(x) = x, \quad -\pi < x < \pi \tag{2.8-15}$$

and periodic with period $2\pi$.

To what value does the series converge at $x = \pi/2$ and at $x = \pi$ ?

From Equations (2.8-14b, c), we obtain

$$a_0 = \frac{1}{\pi} \int_{-\pi}^{\pi} x \, dx = \frac{1}{\pi} \left[ \frac{x^2}{2} \right]_{-\pi}^{\pi} = 0 \tag{2.8-16a,b,c}$$

$$a_n = \frac{1}{\pi} \int_{-\pi}^{\pi} x \cos nx \, dx = \frac{1}{\pi} \left[ \frac{x \sin nx}{n} \right]_{-\pi}^{\pi} - \frac{1}{\pi} \int_{-\pi}^{\pi} \frac{\sin nx}{n} \, dx = 0 \tag{2.8-16d,e,f}$$

$$b_n = \frac{1}{\pi} \int_{-\pi}^{\pi} x \sin nx \, dx = \frac{1}{\pi} \left[ -\frac{x \cos nx}{n} \right]_{-\pi}^{\pi} + \frac{1}{\pi} \int_{-\pi}^{\pi} \frac{\cos nx}{n} \, dx = -\frac{2(-1)^n}{n} \tag{2.8-16g,h,i}$$

Substituting $a_n$ and $b_n$ in Equation (2.8-14a), we obtain

$$x \approx -2 \sum_{n=1}^{\infty} \frac{(-1)^n}{n} \sin nx \tag{2.8-17}$$

The periodic function $f(x)$ is sketched in Figure 2.8-1. The function is continuous at $x = \pi/2$ and the series converges to the function. That is to say

$$\pi/2 = -2 \sum_{n=1}^{\infty} \frac{(-1)^n}{n} \sin(n\pi/2) \tag{2.8-18a}$$

$$= -2 \sum_{s=0}^{\infty} \frac{(-1)^{2s+1}}{(2s+1)} \sin[(2s+1)\pi/2] \tag{2.8-18b}$$

$$= 2 \sum_{s=0}^{\infty} \frac{(-1)^s}{(2s+1)} \tag{2.8-18c}$$

We note that $\sin(n\pi/2)$ is zero when $n$ is even and if $n$ is odd, we write $n = (2s+1)$ with $s = 0, 1, 2, \ldots$ and $\sin[(2s+1)\pi/2] = (-1)^s$.

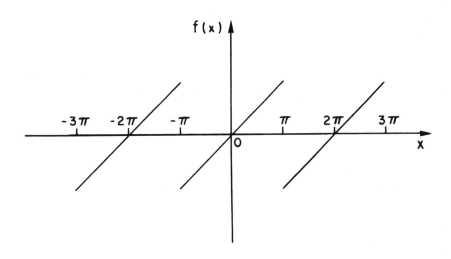

**FIGURE 2.8-1    Periodic function of period $2\pi$**

From Equation (2.8-18c), we deduce that

$$\sum_{s=0}^{\infty} \frac{(-1)^s}{2s+1} = \pi/4 \tag{2.8-19}$$

At $x = \pi$, the function is discontinuous and the sum converges to $\frac{1}{2}[f(\pi-0) + f(\pi+0)]$. That is to say

$$\frac{1}{2}[f(\pi-0) + f(\pi+0)] = -2\sum_{n=1}^{\infty} \frac{(-1)^n}{n} \sin n\pi \tag{2.8-20}$$

From Figure 2.8-1, $f(\pi-0)$ is $\pi$, $f(\pi+0)$ is $-\pi$, and $\sin n\pi$ is zero. Both sides of Equation (2.8-20) are zero.

●

Even and odd functions are defined in Chapter 1.

The functions $\cos x$, $x^{2n}$ ($n$ is an integer) are even functions and the functions $\sin x$, $x^{(2n+1)}$ are odd functions. If $f(x)$ is an even function

$$f'(0) = 0 \qquad\qquad (2.8\text{-}21)$$

and if $f(x)$ is odd

$$f(0) = 0 \qquad\qquad (2.8\text{-}22)$$

The product of two even (or odd) functions is an even function and the product of an even and an odd function is an odd function. If $f(x)$ is an even function, all coefficients $b_n$ are zero and Equations (2.8-14a to c) reduce to

$$f(x) \approx \frac{1}{2} a_0 + \sum_{n=1}^{\infty} a_n \cos nx \qquad\qquad (2.8\text{-}23a)$$

$$a_n = \frac{2}{\pi} \int_0^{\pi} f(x) \cos nx \, dx \qquad\qquad (2.8\text{-}23b)$$

$$b_n = 0 \qquad\qquad (2.8\text{-}23c)$$

Similarly, if $f(x)$ is an odd function, we obtain

$$f(x) \approx \sum_{n=1}^{\infty} b_n \sin nx \qquad\qquad (2.8\text{-}24a)$$

$$b_n = \frac{2}{\pi} \int_0^{\pi} f(x) \sin nx \, dx \qquad\qquad (2.8\text{-}24b)$$

Note that we have made use of the fact that

$$\int_{-\pi}^{\pi} f(x) \, dx = \begin{cases} 0, & \text{if } f(x) \text{ is odd} \\[2ex] 2\int_0^{\pi} f(x) \, dx, & \text{if } f(x) \text{ is even} \end{cases} \qquad (2.8\text{-}25a,b)$$

In many situations, it is known that $f(x)$ has a period of $2\pi$ but is defined only in the interval $0 < x < \pi$. Mathematically, it is possible to define $f(x)$ to be even or odd, but usually the physics of the problem dictates whether $f(x)$ is even or odd (see Chapter 5).

Consider a rod of length $\pi$ in the interval $0 < x < \pi$ with temperature distribution $T(x)$. If the end $x = 0$ is insulated, it implies that at $x = 0$, $dT/dx$ is zero. From Equation (2.8-21), we deduce that $T$ is even. If the temperature at $x = 0$ is kept at a constant temperature $T_0$, by writing

$$f(x) = T - T_0 \qquad (2.8\text{-}26)$$

we obtain $f(0) = 0$. From Equation (2.8-22), we deduce that $T$ is an odd function.

***Example 2.8-2.*** The function $f(x)$ is periodic and of period $2\pi$. It is defined by

$$f(x) = x, \qquad 0 < x < \pi \qquad (2.8\text{-}27a)$$

Obtain the Fourier series of $f(x)$ if: (a) $f$ is odd; and (b) $f$ is even.

Case (a) is considered in Example 2.8-1. The function $f$ in case (b) is sketched in Figure 2.8-2.

From Equations (2.8-16e, f, 23b), we obtain

$$a_n = \frac{2}{\pi} \int_0^\pi x \cos nx \, dx \qquad (2.8\text{-}27b)$$

$$= \frac{2}{\pi} \left[ \frac{\cos nx}{n^2} \right]_0^\pi \qquad (2.8\text{-}27c)$$

$$= -\frac{4}{\pi (2s + 1)^2}, \qquad s = 0, 1, 2, \ldots \qquad (2.8\text{-}27d)$$

$$a_0 = \frac{2}{\pi} \int_0^\pi x \, dx = \pi \qquad (2.8\text{-}27e)$$

Substituting $a_n$ in Equation (2.8-23a) yields

$$x \approx \frac{\pi}{2} - \frac{4}{\pi} \sum_{s=0}^\infty \frac{\cos (2s + 1) x}{(2s + 1)^2} \qquad (2.8\text{-}28)$$

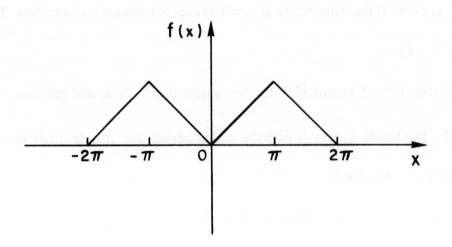

**FIGURE 2.8-2    Even function**

Note that both series on the right side of Equations (2.8-17, 28) represent the same function in the interval $0 < x < \pi$, but not in the interval $-\pi < x < 0$.

At $x = 0$, the function is continuous and from Equation (2.8-28), we deduce that

$$\sum_{s=0}^{\infty} \frac{1}{(2s+1)^2} = \frac{\pi^2}{8} \tag{2.8-29}$$

●

If $f(x)$ has a Fourier series, that is to say, $f(x)$ can be represented by Equations (2.8-14a to c), the series can be integrated term by term. This means that

$$\int_{x_0}^{x} f(x)\, dx = \frac{1}{2} \int_{x_0}^{x} a_0\, dx + \sum_{n=1}^{\infty} \int_{x_0}^{x} (a_n \cos nx + b_n \sin nx)\, dx \tag{2.8-30}$$

where $x_0$ is arbitrary.

If $f(x)$ has a Fourier series, the series can be differentiated term by term, if $f'(x)$ is piecewise continuous and $f(-\pi) = f(\pi)$.

Under the conditions stated earlier, we can write

$$\frac{1}{2} [f'(x-0) + f'(x+0)] = \sum_{n=1}^{\infty} [n(-a_n \sin nx + b_n \cos nx)] \qquad (2.8\text{-}31)$$

where $a_n$ and $b_n$ are defined by Equations (2.8-14b, c).

Equations (2.8-14a to c) can be written in an alternative form. Recall that

$$\cos nx = \frac{1}{2} [e^{inx} + e^{-inx}] \qquad (2.8\text{-}32a)$$

$$\sin nx = \frac{1}{2i} [e^{inx} - e^{-inx}] \qquad (2.8\text{-}32b)$$

Substituting Equations (2.8-32a, b) in Equation (2.8-14a) yields

$$f(x) \approx \frac{1}{2} a_0 + \frac{1}{2} \sum_{n=1}^{\infty} [a_n (e^{inx} + e^{-inx}) - ib_n (e^{inx} - e^{-inx})] \qquad (2.8\text{-}33a)$$

$$\approx \frac{1}{2} a_0 + \frac{1}{2} \sum_{n=1}^{\infty} [e^{inx} (a_n - ib_n) + e^{-inx} (a_n + ib_n)] \qquad (2.8\text{-}33b)$$

$$\approx \sum_{n=-\infty}^{\infty} c_n e^{inx} \qquad (2.8\text{-}33c)$$

where

$$c_n = \frac{1}{2} (a_n - ib_n) = \frac{1}{2\pi} \int_{-\pi}^{\pi} f(x) [\cos nx - i \sin nx] \, dx = \frac{1}{2\pi} \int_{-\pi}^{\pi} f(x) e^{-inx} \, dx$$

$$(2.8\text{-}33d,e,f)$$

Note that Equation (2.8-33f) is valid for both positive and negative values of $n$.

***Example 2.8-3.*** Determine the Fourier series for $f(x)$ defined by

$$f(x) = e^x, \quad -\pi < x < \pi \qquad (2.8\text{-}34)$$

and with period $2\pi$. This function is shown in Figure 2.8-3. Calculate the sum of the series at $x = \pi$.

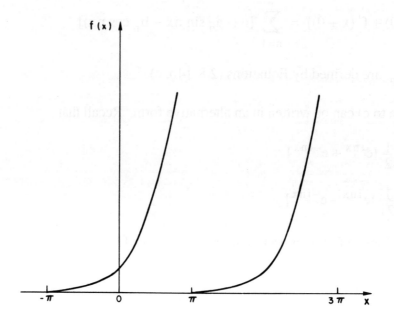

**FIGURE 2.8-3    Periodic function $e^x$**

The coefficients $c_n$ are given by [Equation (2.8-33f)]

$$c_n = \frac{1}{2\pi} \int_{-\pi}^{\pi} e^x \, e^{-inx} \, dx \qquad\qquad (2.8\text{-}35a)$$

$$= \frac{1}{2\pi} \left[ \frac{e^{(1-in)x}}{1-in} \right]_{-\pi}^{\pi} \qquad\qquad (2.8\text{-}35b)$$

$$= \frac{(1+in)}{2\pi\,(1+n^2)} \left[ e^{\pi}\,(\cos n\pi - i \sin n\pi) - e^{-\pi}\,(\cos n\pi + i \sin n\pi) \right] \qquad (2.8\text{-}35c)$$

$$= \frac{(1+in)}{\pi\,(1+n^2)} \left[ (-1)^n \sinh \pi \right] \qquad\qquad (2.8\text{-}35d)$$

Combining Equations (2.8-33c, 35d) yields

$$e^x \approx \frac{\sinh \pi}{\pi} \sum_{n=-\infty}^{\infty} \frac{(-1)^n\,(1+in)\,e^{inx}}{(1+n^2)} \qquad\qquad (2.8\text{-}36a)$$

$$\approx \frac{\sinh \pi}{\pi} \sum_{n=-\infty}^{\infty} (-1)^n \frac{(\cos nx - n \sin nx) + i (n \cos nx + \sin nx)}{(1 + n^2)} \tag{2.8-36b}$$

Taking the real part, we obtain

$$e^x \approx \frac{\sinh \pi}{\pi} \sum_{n=-\infty}^{\infty} \frac{(-1)^n (\cos nx - n \sin nx)}{(1 + n^2)} \tag{2.8-37a}$$

$$\approx \frac{\sinh \pi}{\pi} \left[ 1 + 2 \sum_{n=1}^{\infty} \frac{(-1)^n (\cos nx - n \sin nx)}{(1 + n^2)} \right] \tag{2.8-37b}$$

The imaginary part can be shown to be zero in the following way.

$$\sum_{n=-\infty}^{\infty} (n \cos nx + \sin nx) = \sum_{n=-\infty}^{-1} (n \cos nx + \sin nx) + \sum_{n=1}^{\infty} (n \cos nx + \sin nx) \tag{2.8-38a}$$

$$= \sum_{m=\infty}^{1} [(-m) \cos (-mx) + \sin (-mx)] + \sum_{n=1}^{\infty} (n \cos nx + \sin nx) \tag{2.8-38b}$$

$$= \sum_{m=1}^{\infty} (-m \cos mx - \sin mx) + \sum_{n=1}^{\infty} (n \cos nx + \sin nx) \tag{2.8-38c}$$

$$= 0 \tag{2.8-38d}$$

We replaced $n$ by $-m$ in the first sum on the right side of Equation (2.8-38b) and we made use of the fact that cos is an even function and sin is an odd function.

At $x = \pi$, the function is discontinuous and the series converges in the mean. That is to say

$$\frac{1}{2} [f(\pi - 0) + f(\pi + 0)] = \frac{\sinh \pi}{\pi} \left[ 1 + 2 \sum_{n=1}^{\infty} \frac{(-1)^n \cos n\pi}{1 + n^2} \right] \tag{2.8-39}$$

From Figure 2.8-3, we deduce

$$\frac{1}{2} [e^\pi + e^{-\pi}] = \frac{\sinh \pi}{\pi} \left[ 1 + 2 \sum_{n=1}^{\infty} \frac{1}{1 + n^2} \right] \tag{2.8-40a}$$

$$\cosh \pi = \frac{\sinh \pi}{\pi} \left[ 1 + 2 \sum_{n=1}^{\infty} \frac{1}{1 + n^2} \right] \tag{2.8-40b}$$

**Example 2.8-4.** The deflection  w  of a uniform beam of length  $\ell$  with an elastic support under a given external load  p  is given by

$$E I \frac{d^4 w}{dx^4} + k w = p(x) \tag{2.8-41}$$

where  E I  is the flexural rigidity and  k  is the modulus of the elastic support.

If the beam is supported at the ends, the boundary conditions are

$$w = \frac{d^2 w}{dx^2} = 0 \tag{2.8-42a,b}$$

at  $x = 0$  and  $x = \ell$.

Assume that the load is constant and is applied on the interval  $\ell/3 < x < 2\ell/3,$  as shown in Figure 2.8-4.  Obtain the deflection  w (x).  The derivation of Equation (2.8-41) is given in von Karman and Biot (1940).

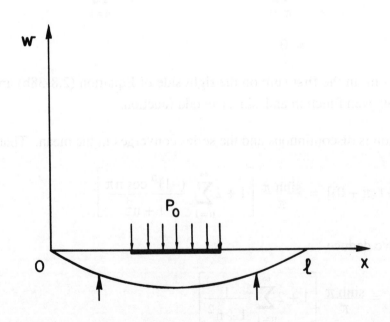

**FIGURE 2.8-4    Beam on elastic foundation under an external load**

The function $p(x)$ can be written as

$$p(x) = \begin{cases} 0, & 0 < x < \ell/3 \\ p_0, & \ell/3 < x < 2\ell/3 \\ 0, & 2\ell/3 < x < \ell \end{cases} \qquad (2.8\text{-}43a,b,c)$$

where $p_0$ is the constant external load.

The function $p(x)$ can be considered to be an odd function of period $2\ell$. Its Fourier series is given by

$$p(x) \approx \sum_{n=1}^{\infty} b_n \sin \frac{n\pi x}{\ell} \qquad (2.8\text{-}44a)$$

$$b_n = \frac{2}{\ell} \int_0^\ell p(x) \sin \frac{n\pi x}{\ell} \, dx = (2p_0/\ell) \int_{\ell/3}^{2\ell/3} \sin \frac{n\pi x}{\ell} \, dx \qquad (2.8\text{-}44b,c)$$

$$= -\frac{2p_0}{n\pi} \left[ \cos \frac{2n\pi}{3} - \cos \frac{n\pi}{3} \right] = \frac{4p_0}{n\pi} \sin \frac{n\pi}{2} \sin \frac{n\pi}{6} \qquad (2.8\text{-}44d,e)$$

We seek a solution of the form

$$w = \sum_{n=1}^{\infty} w_n \sin \frac{n\pi x}{\ell} \qquad (2.8\text{-}45)$$

Note that $w$ automatically satisfies the boundary conditions. Differentiating $w$ four times and substituting the resulting expression in Equation (2.8-41) yields

$$EI \left[ \sum_{n=1}^{\infty} \frac{n^4 \pi^4}{\ell^4} w_n \sin \frac{n\pi x}{\ell} \right] + k \sum_{n=1}^{\infty} w_n \sin \frac{n\pi x}{\ell} = \sum_{n=1}^{\infty} b_n \sin \frac{n\pi x}{\ell} \qquad (2.8\text{-}46a)$$

We deduce that

$$w_n \left[ EI \frac{n^4 \pi^4}{\ell^4} + k \right] = b_n \qquad (2.8\text{-}46b)$$

or $\quad w_n = b_n \ell^4 / (EIn^4\pi^4 + k\ell^4) \qquad (2.8\text{-}46c)$

The coefficients $b_n$ are given by Equation (2.8-44e), $w_n$ can be calculated from Equation (2.8-46c). On substituting $w_n$ in Equation (2.8-45), we obtain $w$.

## Fourier Integral

The transformation from a function $f(x)$ having a period $2\pi$ to a function $f(x^*)$ having a period $2L$ is obtained by writing

$$x = x^* \pi / L \qquad (2.8\text{-}47)$$

Combining Equations (2.8-33c, f, 47) yields

$$f(x^*) \approx \sum_{n=-\infty}^{\infty} \left[ \left( \frac{1}{2L} \int_{-L}^{L} f(x^*) \exp(-i\alpha_n x^*) \, dx^* \right) \exp(i\alpha_n x^*) \right] \qquad (2.8\text{-}48a)$$

where $\alpha_n = n\pi / L$ $\qquad (2.8\text{-}48b)$

Let $\quad \Delta \alpha_n = (n+1)\pi / L - n\pi / L = \pi / L$ $\qquad (2.8\text{-}49a,b)$

Substituting Equation (2.8-49b) in Equation (2.8-48a), we obtain

$$f(x^*) \approx \sum_{n=-\infty}^{\infty} \left[ \frac{\Delta \alpha_n}{2\pi} \int_{-L}^{L} f(\xi) \exp(-i\alpha_n \xi) \, d\xi \, \exp(i\alpha_n x^*) \right] \qquad (2.8\text{-}50a)$$

$$\approx \sum_{n=-\infty}^{\infty} \frac{\Delta \alpha_n}{2\pi} \int_{-L}^{L} f(\xi) \exp[i\alpha_n (x^* - \xi)] \, d\xi \qquad (2.8\text{-}50b)$$

On letting $L \longrightarrow \infty$, Equation (2.8-50b) becomes

$$f(x^*) \approx \frac{1}{2\pi} \int_{-\infty}^{\infty} \left[ \int_{-\infty}^{\infty} f(\xi) \, e^{i\alpha(x^* - \xi)} \, d\xi \right] d\alpha \qquad (2.8\text{-}51a)$$

$$\approx \frac{1}{2\pi} \int_{-\infty}^{\infty} \int_{-\infty}^{\infty} f(\xi) \, [\cos \alpha (x^* - \xi) + i \sin \alpha (x^* - \xi)] \, d\xi \, d\alpha \qquad (2.8\text{-}51b)$$

$$\approx \frac{1}{\pi} \int_{0}^{\infty} \left[ \int_{-\infty}^{\infty} f(\xi) \cos \alpha (x^* - \xi) \, d\xi \right] d\alpha \qquad (2.8\text{-}51c)$$

$$\approx \frac{1}{\pi} \int_0^\infty \left[ \int_{-\infty}^\infty f(\xi) (\cos \alpha x^* \cos \alpha \xi + \sin \alpha x^* \sin \alpha \xi) \, d\xi \right] d\alpha \qquad (2.8\text{-}51d)$$

$$\approx \frac{1}{\pi} \int_0^\infty \left[ A(\alpha) \cos \alpha x^* + B(\alpha) \sin \alpha x^* \right] d\alpha \qquad (2.8\text{-}51e)$$

where

$$A(\alpha) = \int_{-\infty}^\infty f(\xi) \cos \alpha \xi \, d\xi , \qquad B(\alpha) = \int_{-\infty}^\infty f(\xi) \sin \alpha \xi \, d\xi \qquad (2.8\text{-}51f,g)$$

Note that $\sin \alpha$ is an odd function so $\int_{-\infty}^\infty \sin \alpha \, d\alpha$ is zero. The quantities $A(\alpha)$ and $B(\alpha)$ exist if $\int_{-\infty}^\infty |f(\xi)| \, d\xi$ converges. The right side of Equations (2.8-51a to e) is the **Fourier integral** of $f(x^*)$. If $f(x^*)$ has a jump discontinuity at a point $x_0^*$, the integral converges to $\frac{1}{2} [f(x_{0+}^*) + f(x_{0-}^*)]$. If $f(\xi)$ is even, $B(\alpha)$ is zero and if $f(\xi)$ is odd, $A(\alpha)$ is zero.

***Example 2.8-5.*** Determine the Fourier integral of the following functions

(a) $\qquad f(x) = \begin{cases} 1, & |x| < 1 \\ \\ 0, & |x| > 1 \end{cases}$ $\qquad\qquad\qquad\qquad\qquad\qquad\qquad$ (2.8-52a,b)

(b) $\qquad f(x) = \begin{cases} 1, & 0 < x < 1 \\ -1, & -1 < x < 0 \\ \\ 0, & |x| > 1 \end{cases}$ $\qquad\qquad\qquad\qquad\qquad$ (2.8-53a,b,c)

Figure 2.8-5 illustrates these two functions.

(a) $\qquad f(x)$ is even and $B(\alpha)$ is zero.

$$A(\alpha) = 2 \int_0^\infty f(\xi) \cos \alpha \xi \, d\xi = 2 \int_0^1 \cos \alpha \xi \, d\xi = \frac{2 \sin \alpha}{\alpha} \qquad (2.8\text{-}54a,b,c)$$

$$f(x) \approx \frac{1}{\pi} \int_0^\infty \frac{2 \sin \alpha}{\alpha} \cos \alpha x \, d\alpha \qquad (2.8\text{-}55)$$

The function is continuous everywhere except at $x = \pm 1$. We deduce that

$$\int_0^\infty \frac{\sin \alpha \cos \alpha x}{\alpha} \, d\alpha = \begin{cases} \pi/2, & 0 \le |x| < 1 \\[2mm] \pi/4, & x = \pm 1 \\[2mm] 0, & |x| > 1 \end{cases} \qquad (2.8\text{-}56\text{a,b,c})$$

On setting $x = 0$, we obtain

$$\int_0^\infty \frac{\sin \alpha}{\alpha} \, d\alpha = \pi/2 \qquad (2.8\text{-}57)$$

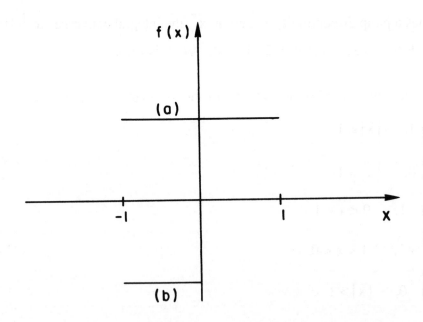

**FIGURE 2.8-5   (a) Even function; (b) Odd function**

(b)      $f(x)$ is odd and $A(\alpha)$ is zero.

$$B(\alpha) = 2 \int_0^1 \sin \alpha \xi \, d\xi = \frac{2}{\alpha} (1 - \cos \alpha) \qquad (2.8\text{-}58\text{a,b})$$

$$f(x) \approx \frac{2}{\pi} \int_0^\infty \frac{(1 - \cos \alpha)}{\alpha} \sin \alpha x \, d\alpha \tag{2.8-59}$$

The function is continuous everywhere except at $x = 0$ and at $x = \pm 1$. We deduce that

$$\int_0^\infty \frac{(1 - \cos \alpha) \sin \alpha x}{\alpha} \, d\alpha = \begin{cases} 0, & x = 0 \\ \pi/2, & 0 < x < 1 \\ \pi/4, & x = 1 \\ 0, & x > 1 \end{cases} \tag{2.8-60a-d}$$

***Example 2.8-6.*** Extend the problem considered in Example 2.8-4 to an infinite beam extending from $-\infty$ to $\infty$. The appropriate boundary condition is that $w$ vanishes at infinity. Assume that $p(x)$ is given by

$$p(x) = \begin{cases} p_0, & |x| < a \\ 0, & |x| > a \end{cases} \tag{2.8-61a,b}$$

where $p_0$ and $a$ are constants.

We represent $p(x)$ and $w(x)$ by their Fourier integrals and write

$$p(x) \approx \frac{1}{\pi} \int_0^\infty [A(\alpha) \cos \alpha x + B(\alpha) \sin \alpha x] \, d\alpha \tag{2.8-62a}$$

$$w(x) \approx \frac{1}{\pi} \int_0^\infty [C(\alpha) \cos \alpha x + D(\alpha) \sin \alpha x] \, d\alpha \tag{2.8-62b}$$

Differentiating $w(x)$ four times and substituting the resulting expression together with the expression for $p$ in Equation (2.8-41) yields

$$\frac{(\alpha^4 EI + k)}{\pi} \int_0^\infty [C(\alpha) \cos \alpha x + D(\alpha) \sin \alpha x] \, d\alpha = \frac{1}{\pi} \int_0^\infty [A(\alpha) \cos \alpha x + B(\alpha) \sin \alpha x] \, d\alpha \tag{2.8-63}$$

We deduce that

$$(\alpha^4 EI + k) \, C(\alpha) \; = \; A(\alpha) \tag{2.8-64a}$$

$$(\alpha^4 EI + k) \, D(\alpha) \; = \; B(\alpha) \tag{2.8-64b}$$

From Equations (2.8-51f, g, 61a, b), we obtain

$$A(\alpha) \; = \; \int_{-\infty}^{\infty} p(x) \cos \alpha x \; dx \; = \; 2p_0 \int_{0}^{a} \cos \alpha x \; dx \; = \; (2p_0 \sin \alpha a)/\alpha \tag{2.8-65a,b,c}$$

$$B(\alpha) \; = \; \int_{-\infty}^{\infty} p(x) \sin \alpha x \; dx \; = \; p_0 \int_{-a}^{a} \sin \alpha x \; dx \; = \; 0 \tag{2.8-65d,e,f}$$

Combining Equations (2.8-64a to 65f) yields

$$C(\alpha) \; = \; 2p_0 \sin \alpha a \, / \, [\alpha \, (\alpha^4 EI + k)] \tag{2.8-66a}$$

$$D(\alpha) \; = \; 0 \tag{2.8-66b}$$

Substituting Equations (2.8-66a, b) in Equation (2.8-62b) yields

$$w(x) \; = \; \frac{2p_0}{\pi} \int_{0}^{\infty} \frac{\sin \alpha a \cos \alpha x}{\alpha \, (\alpha^4 EI + k)} \; d\alpha \tag{2.8-67}$$

Examples of Legendre-Fourier and Bessel-Fourier series are given in Chapter 5.

## 2.9   ASYMPTOTIC SOLUTIONS

We have obtained convergent series solution in the neighborhood of $x_0 \, (= 0)$ if $x_0$ is an ordinary or regular singular point. If $x_0$ is an irregular singular point, there is no method of generating a convergent series solution. In some cases, we can obtain a formal series solution which is a good approximation for small values of $x$. This is illustrated in the next example.

***Example 2.9-1***. Find a power series that satisfies the equation

$$x^3 \, y'' + (x^2 + x) \, y' - y \; = \; 0 \tag{2.9-1}$$

The origin $(x = 0)$ is an irregular singular point, that is to say, $x \, (x^2 + x)/x^3$ and $x^2 \, (-1)/x^3$ do not have a Taylor series about $x = 0$. We seek a formal series solution as given by Equation (2.3-8a). Substituting Equations (2.3-8a to c) in Equation (2.9-1) yields

$$\sum_{r=0}^{\infty} \left[ (r+2)(r+1)c_{r+2} x^{r+3} + (r+1)c_{r+1} x^{r+2} + (r+1)c_{r+1} x^{r+1} - c_r x^r \right] = 0 \qquad (2.9\text{-}2)$$

Comparing powers of x, we obtain

$x^0$: $\qquad c_0 = 0$ $\hspace{7cm}$ (2.9-3a)

$x^1$: $\qquad c_1 - c_1 = 0 \;\Rightarrow\; c_1$ is arbitrary $\hspace{4cm}$ (2.9-3b)

$x^2$: $\qquad c_1 + 2c_2 - c_2 = 0 \;\Rightarrow\; c_2 = -c_1$ $\hspace{4cm}$ (2.9-3c,d)

$x^3$: $\qquad 2c_2 + 2c_2 + 3c_3 - c_3 = 0 \;\Rightarrow\; c_3 = -2c_2 = 2c_1$ $\hspace{2.5cm}$ (2.9-3e,f)

.

.

.

$x^s$: $\qquad (s-1)(s-2)c_{s-1} + (s-1)c_{s-1} + s\,c_s - c_s = 0 \;\Rightarrow\; c_s = -(s-1)c_{s-1}$ $\hspace{0.5cm}$ (2.9-3g,h)

The formal series that satisfies the differential equation is

$$c_1 \left[ x - x^2 + 2x^3 + (-1)^{s+1}\, 2{\cdot}3{\cdot}4 \ldots (s-1)\, x^s + \ldots \right] = c_1 \sum_{s=1}^{\infty} (-1)^{s+1}(s-1)!\, x^s \qquad (2.9\text{-}4)$$

Can this divergent series [Equation (2.9-4)] be used to calculate the values of y for small values of x ? The answer is yes. Using the divergent series, we find that $y(0.1)$ can be written as

$$y(0.1) \approx c_1 \left[ 10^{-1} - 10^{-2} + 2 \times 10^{-3} - 6 \times 10^{-4} + 2{\cdot}4 \times 10^{-4} + \ldots \right] \qquad (2.9\text{-}5)$$

We note that the magnitude of the terms decreases as the order increases and if we require $y(0.1)$ to be accurate to two decimal places, the first two terms are sufficient and $y(0.1) = 0.99\, c_1$.

The divergent series $s_n(x)$ $[= \sum_{r=0}^{n} c_r x^r]$ is an **asymptotic series expansion** of a function $f(x)$. The power series $s_n(x)$ represents $f(x)$ asymptotically as $x \longrightarrow 0$ if

$$x^{-n} [f(x) - s_n] \longrightarrow 0, \text{ for all } n \geq 0 \text{ as } x \longrightarrow 0 \qquad (2.9\text{-}6)$$

The error in approximating a function by a convergent series decreases as the number of terms increases. We are certain of attaining the required accuracy by taking a sufficient number of terms. From Equation (2.9-6), it is seen that the approximation improves as the magnitude of x decreases for a fixed value of n. Since the asymptotic series is divergent, the error may increase as n increases. Usually, for asymptotic series, only a few terms are required. We can increase the value of n as long as the value of the $n^{th}$ term is less than the value of the $(n-1)^{th}$ term. We stop at the $(n-1)^{th}$ term if the value of the $n^{th}$ term is greater than that of the $(n-1)^{th}$ term.

Asymptotic series are widely used in the solution of differential equations and for evaluating integrals. In applied mathematics, one often neither knows nor cares if the series converges or not [Van Dyke (1975)]. Even if a convergent series is available, it is sometimes profitable to consider an asymptotic series as shown in the next example.

***Example 2.9-2***. Solve Bessel's equation of order zero [Equation (2.4-44)] for large values of x.

We first transform Equation (2.4-44) to its **normal form**. A second order differential equation is in its normal form if the first derivative $(dy/dx)$ is not present. To achieve this, we write

$$y = u(x) v(x) \tag{2.9-7}$$

Differentiating and substituting the resulting expressions in Equation (2.4-44) yields

$$x^2 (u''v + 2u'v' + uv'') + x (u'v + uv') + x^2 uv = 0 \tag{2.9-8a}$$

or     $$x^2 v u'' + u' (2x^2 v' + xv) + u (x^2 v'' + xv' + x^2 v) = 0 \tag{2.9-8b}$$

We now impose the condition

$$2x^2 v' + xv = 0 \tag{2.9-9}$$

The solution is

$$v = x^{-1/2} \tag{2.9-10}$$

Equation (2.9-8b) becomes

$$u'' + u (1 + 1/4x^2) = 0 \tag{2.9-11}$$

For large values of x, Equation (2.9-11) is approximately the simple harmonic equation [Equation (2.6-3a)] and the solution is $e^{ix}$. This suggests that we seek a solution of the form

$$u = e^{ix} \sum_{n=0}^{\infty} c_n x^{-n} \tag{2.9-12}$$

Note that, in this case, we look for a solution as $x \longrightarrow \infty$, therefore we expand in reciprocal powers of x.

On differentiating, we obtain

$$u' = i e^{ix} \sum_{n=0}^{\infty} c_n x^{-n} - e^{ix} \sum_{n=1}^{\infty} n c_n x^{-n-1} \tag{2.9-13a}$$

$$= i e^{ix} \sum_{r=0}^{\infty} c_r x^{-r} - e^{ix} \sum_{r=0}^{\infty} (r+1) c_{r+1} x^{-r-2} \qquad (2.9\text{-}13b)$$

$$= e^{ix} \sum_{r=0}^{\infty} [i c_r x^{-r} - (r+1) c_{r+1} x^{-r-2}] \qquad (2.9\text{-}13c)$$

The second derivative $u''$ is given by

$$u'' = e^{ix} \sum_{r=0}^{\infty} [-c_r x^{-r} - 2i(r+1) c_{r+1} x^{-r-2} + (r+1)(r+2) c_{r+1} x^{-r-3}] \qquad (2.9\text{-}13d)$$

Substituting Equations (2.9-13a, d) in Equation (2.9-11) and multiplying by $e^{-ix}$, we obtain

$$\sum_{r=0}^{\infty} [-c_r x^{-r} - 2i(r+1) c_{r+1} x^{-r-2} + (r+1)(r+2) c_{r+1} x^{-r-3} + c_r x^{-r} + (1/4) c_r x^{-r-2}] = 0$$

$$(2.9\text{-}14)$$

Comparing powers of $x$ yields

$$x^0: \quad -c_0 + c_0 = 0 \quad \Rightarrow \quad c_0 \text{ is arbitrary} \qquad (2.9\text{-}15a)$$

$$x^{-1}: \quad -c_1 + c_1 = 0 \quad \Rightarrow \quad c_1 \text{ is arbitrary} \qquad (2.9\text{-}15b)$$

$$x^{-2}: \quad -c_2 - 2ic_1 + c_2 + \frac{1}{4} c_0 = 0 \quad \Rightarrow \quad c_1 = -ic_0/4 \cdot 2 \qquad (2.9\text{-}15c,d)$$

$$x^{-3}: \quad -c_3 - 4ic_2 + 2c_1 + c_3 + \frac{1}{4} c_1 = 0 \quad \Rightarrow \quad c_2 = -i(9c_1/4)/4 \qquad (2.9\text{-}15e,f)$$

$$\cdot$$
$$\cdot$$
$$\cdot$$

$$x^{-s}: \quad -c_s - 2i(s-1) c_{s-1} + (s-2)(s-1) c_{s-2} + c_s + 1/4 \, c_{s-2} = 0$$

$$\Rightarrow \quad 2i(s-1) c_{s-1} = c_{s-2} (s^2 - 3s + 9/4) \quad \Rightarrow \quad c_{s-1} = -i c_{s-2} (s - 3/2)^2 / 2(s-1)$$

$$(2.9\text{-}15g,h,i)$$

We note that the coefficients $c_s$ are alternately real and imaginary.

As usual, to allow for the tabulation of Bessel functions, we set $c_0$ to be equal to one and $u$ can be written as

$$u = e^{ix} \left[ 1 - \frac{i}{4 \cdot 2 \, x} - \frac{3^2}{4^2 \cdot 2^2 \cdot 2 x^2} + \frac{i \cdot 3^2 \cdot 5^2}{4^3 \cdot 2^3 \cdot 3! \, x^3} + \frac{3^2 \cdot 5^2 \cdot 7^2}{4^4 \cdot 2^4 \cdot 4! \, x^4} + \ldots \right] \qquad (2.9\text{-}16a)$$

$$= e^{ix} (w_r - i\, w_i) \qquad\qquad (2.9\text{-}16b)$$

$$\text{where}\quad w_r = 1 - 3^2/[4^2 \cdot 2^2 \cdot 2x^2] + 3^2 \cdot 5^2 \cdot 7^2/[4^4 \cdot 2^4 \cdot 4!\, x^4] + \dots \qquad (2.9\text{-}16c)$$

$$w_i = 1/[4 \cdot 2\, x] - 3^2 \cdot 5^2/[4^3 \cdot 2^3 \cdot 3!\, x^3] + \dots \qquad\qquad (2.9\text{-}16d)$$

Combining Equations (2.9-7, 10, 16b) yields

$$y = x^{-1/2} e^{ix} (w_r - i\, w_i) \qquad\qquad (2.9\text{-}17a)$$

$$= x^{-1/2} (\cos x + i \sin x)\, (w_r - i\, w_i) \qquad\qquad (2.9\text{-}17b)$$

$$= x^{-1/2} [w_r \cos x + w_i \sin x + i\, (w_r \sin x - w_i \cos x)] \qquad (2.9\text{-}17c)$$

The two real fundamental solutions $y_1$ and $y_2$ are given by

$$y_1 = (y + \bar{y})/2 = x^{-1/2} (w_r \cos x + w_i \sin x) \qquad\qquad (2.9\text{-}18a,b)$$

$$y_2 = (y - \bar{y})/2i = x^{-1/2} (w_r \sin x - w_i \cos x) \qquad\qquad (2.9\text{-}18c,d)$$

where $\bar{y}$ is the complex conjugate of $y$.

The general solution of Equation (2.4-44), as $x \longrightarrow \infty$, is given by the linear combination of $y_1$ and $y_2$. The solution of Equation (2.4-44) is the Bessel function of order zero. That is to say, as $x \longrightarrow \infty$

$$J_0(x) \approx A\, y_1 + B\, y_2 \qquad\qquad (2.9\text{-}19)$$

where $A$ and $B$ are arbitrary constants.

From Equations (2.9-16c, d), we deduce that

$$\lim_{x \to \infty} w_r = 1, \qquad \lim_{x \to \infty} w_i = 0 \qquad\qquad (2.9\text{-}20a,b)$$

Differentiating both sides of Equation (2.9-19) yields

$$J_0'(x) \approx A\, y_1' + B\, y_2' \qquad\qquad (2.9\text{-}21)$$

From Equations (2.9-16c, d, 18b, d, 20a, b), we deduce that the leading terms of $y_1'$ and $y_2'$, as $x \longrightarrow \infty$, are

$$y_1' \approx x^{-1/2} (-w_r \sin x) \approx -x^{-1/2} \sin x \qquad (2.9\text{-}22a,b)$$

$$y_2' \approx x^{-1/2} \cos x \qquad (2.9\text{-}22c)$$

Combining Equations (2.9-19 to 22c) yields

$$\lim_{x \to \infty} x^{1/2} J_0(x) = A \cos x + B \sin x \qquad (2.9\text{-}23a)$$

$$\lim_{x \to \infty} x^{1/2} J_0'(x) = -A \sin x + B \cos x \qquad (2.9\text{-}23b)$$

The solution is

$$A = \lim_{x \to \infty} x^{1/2} [ J_0(x) \cos x - J_0'(x) \sin x] \qquad (2.9\text{-}24a)$$

$$B = \lim_{x \to \infty} x^{1/2} [ J_0(x) \sin x + J_0'(x) \cos x] \qquad (2.9\text{-}24b)$$

The integral representation of $J_0(x)$ is (see Problem 22b)

$$J_0(x) = \frac{1}{\pi} \int_0^\pi \cos (x \cos \theta) \, d\theta \qquad (2.9\text{-}25)$$

It follows that

$$J_0'(x) = -\frac{1}{\pi} \int_0^\pi [\sin (x \cos \theta)] \cos \theta \, d\theta \qquad (2.9\text{-}26)$$

Substituting Equations (2.9-25, 26) in Equation (2.9-24a) yields

$$A = \lim_{x \to \infty} \frac{x^{1/2}}{\pi} \int_0^\pi [\cos x \cos (x \cos \theta) + \sin x \cos \theta \sin (x \cos \theta)] \, d\theta \qquad (2.9\text{-}27a)$$

$$= \lim_{x \to \infty} \frac{x^{1/2}}{2\pi} \int_0^\pi [\cos (x + x \cos \theta) + \cos (x - x \cos \theta)$$
$$+ \cos \theta \{\cos (x - x \cos \theta) - \cos (x + x \cos \theta)\}] \, d\theta \qquad (2.9\text{-}27b)$$

$$= \lim_{x \to \infty} \frac{x^{1/2}}{2\pi} \int_0^\pi [(1 - \cos \theta) \cos (x + x \cos \theta) + (1 + \cos \theta) \cos (x - x \cos \theta)] \, d\theta \qquad (2.9\text{-}27c)$$

$$= \lim_{x \to \infty} \frac{x^{1/2}}{\pi} \int_0^{\pi} [(\sin^2 \theta/2) \cos (2x \cos^2 \theta/2) + (\cos^2 \theta/2) \cos (2x \sin^2 \theta/2)] \, d\theta \quad (2.9\text{-}27\text{d})$$

To evaluate the first integral on the right side of Equation (2.9-27d), we write

$$\phi = (\sqrt{2x}) \cos \theta/2 \qquad (2.9\text{-}28)$$

On differentiating, we obtain

$$d\phi = -\frac{1}{2} (\sqrt{2x}) \sin \theta/2 \, d\theta \qquad (2.9\text{-}29)$$

It follows from Equations (2.9-28, 29) that

$$\int_0^{\pi} \sin^2 \theta/2 \, \cos (2x \cos^2 \theta/2) \, d\theta = \int_{\sqrt{2x}}^0 (\cos \phi^2) \left(-\sqrt{\frac{2}{x}}\right) \left(1 - \frac{\phi^2}{2x}\right)^{1/2} d\phi \qquad (2.9\text{-}30\text{a})$$

$$= \sqrt{\frac{2}{x}} \int_0^{\sqrt{2x}} \left(1 - \frac{\phi^2}{2x}\right)^{1/2} (\cos \phi^2) \, d\phi \qquad (2.9\text{-}30\text{b})$$

Therefore

$$\lim_{x \to \infty} \frac{x^{1/2}}{\pi} \int_0^{\pi} \sin^2 \theta/2 \, \cos (2x \cos^2 \theta/2) \, d\theta$$

$$= \lim_{x \to \infty} \frac{x^{1/2}}{\pi} \left(\frac{2}{x}\right)^{1/2} \int_0^{\sqrt{2x}} \left(1 - \frac{\phi^2}{2x}\right)^{1/2} (\cos \phi^2) \, d\phi \qquad (2.9\text{-}31\text{a})$$

$$= \frac{\sqrt{2}}{\pi} \int_0^{\infty} \cos \phi^2 \, d\phi \qquad (2.9\text{-}31\text{b})$$

$$= \frac{1}{2\sqrt{\pi}} \qquad (2.9\text{-}31\text{c})$$

Similarly, B and the associated integrals can be evaluated. It is found that

$$A = B = \frac{1}{\sqrt{\pi}} \qquad (2.9\text{-}32\text{a,b})$$

Combining Equations (2.9-18a, b, 19, 32a, b) yields

$$J_0 \approx \frac{x^{-1/2}}{\sqrt{\pi}} [w_r (\cos x + \sin x) + w_i (\sin x - \cos x)] \tag{2.9-33a}$$

$$\approx \left(\frac{2}{\pi x}\right)^{1/2} [w_r \cos (x - \pi/4) + w_i \sin (x - \pi/4)] \tag{2.9-33b}$$

From Equation (2.9-15i), we deduce, using the ratio test, that the series representations of $w_r$ and $w_i$ [Equations (2.9-16c, d)] are divergent. However, to determine $J_0(6)$ accurate to five places of decimals, we need to consider only the first seven terms of the series (four for $w_r$ and three for $w_i$) and we obtain

$$J_0(6) = 0.15064 \tag{2.9-34}$$

The function $J_0(x)$ can also be represented by a convergent series [Equation (2.4-53)] and, after adding the first twenty one terms, we obtain

$$J_0(6) = 0.15067 \tag{2.9-35}$$

and this value is accurate to four decimal places only.

●

An asymptotic series is often more useful than a convergent series. Since Poincaré's pioneering work, considerable progress has been made in the understanding of the asymptotic series. From the two examples we have considered, we can postulate that the asymptotic solution of a differential equation can be of the form

$$y \approx \{\exp [\lambda (x)]\} x^r \sum_{n=0}^{\infty} c_n x^{-n} \tag{2.9-36}$$

as $x \longrightarrow \infty$.

In Example 2.9-1, $\lambda = r = 0$ and it is not common for an asymptotic solution to be simply a power series. Usually, exponential functions are involved as in Example 2.9-2, where

$$\lambda (x) = i x, \qquad r = -1/2 \tag{2.9-37a,b}$$

Further details on asymptotic expansions can be found in Cesari (1963), Nayfeh (1973), Van Dyke (1964), and Wasow (1965).

## Parameter Expansion

Many physical problems involve a parameter which can be small or large. In fluid dynamics, we have the Reynolds number which can be small (Stokes flow) or large (boundary-layer flow). We consider the case when the parameter is small and we denote it by $\varepsilon$. If the parameter is large, we consider its reciprocal. We seek a solution in powers of $\varepsilon$ and often the series solution is an asymptotic series. The next example illustrates the method of parameter expansion.

***Example 2.9-3.*** A particle of unit mass is thrown vertically upwards (y-direction) with an initial velocity $v_0$. If the air resistance at speed $v$ is assumed to be $\varepsilon v^2$, where $\varepsilon$ is a small positive constant, determine the maximum height $y_m$ reached by the particle at time $t_m$.

We take the origin at the surface of the earth. The equation of motion (Newton's second law of motion) is

$$\ddot{y} = -[g + \varepsilon \dot{y}^2] \tag{2.9-38}$$

where $g$ is the gravitational acceleration and the dot denotes differentiation with respect to time.

The initial conditions are

$$y(0) = 0, \qquad \dot{y}(0) = v_0 \tag{2.9-39a,b}$$

We have identified $\varepsilon$ to be a small parameter and we start by expanding all functions of interest in powers of $\varepsilon$. That is

$$y(t) = y_0(t) + \varepsilon y_1(t) + \dots \tag{2.9-40a}$$

$$t_m = t_0 + \varepsilon t_1 + \dots \tag{2.9-40b}$$

Substituting Equation (2.9-40a) in Equations (2.9-38, 39a, b), we obtain respectively

$$\ddot{y}_0 + \varepsilon \ddot{y}_1 + \dots = -[g + \varepsilon(\dot{y}_0^2 + 2\varepsilon \dot{y}_0 \dot{y}_1 + \dots)] \tag{2.9-41a}$$

$$y_0(0) + \varepsilon y_1(0) + \dots = 0 \tag{2.9-41b}$$

$$\dot{y}_0(0) + \varepsilon \dot{y}_1(0) + \dots = v_0 \tag{2.9-41c}$$

At time $t_m$, the particle is at rest and this is expressed, using Taylor's expansion, as

$$\dot{y}(t_m) = \dot{y}_0(t_0 + \varepsilon t_1 + \dots) + \varepsilon \dot{y}_1(t_0 + \varepsilon t_1 + \dots) + \dots \tag{2.9-42a}$$

$$= \dot{y}_0(t_0) + \varepsilon t_1 \ddot{y}_0(t_0) + \dots + \varepsilon \dot{y}_1(t_0) + \varepsilon^2 t_1 \dot{y}_1(t_0) + \dots \qquad (2.9\text{-}42\text{b})$$

$$= \dot{y}_0(t_0) + \varepsilon [t_1 \ddot{y}_0(t_0) + \dot{y}_1(t_0)] + \dots = 0 \qquad (2.9\text{-}42\text{c,d})$$

Comparing powers of $\varepsilon$, we obtain from Equations (2.9-41a to c)

$$\varepsilon^0: \quad \ddot{y}_0 = -g, \qquad y_0(0) = 0, \qquad \dot{y}_0(0) = v_0 \qquad (2.9\text{-}43\text{a,b,c})$$

$$\varepsilon^1: \quad \ddot{y}_1 = -\dot{y}_0^2, \qquad y_1(0) = 0, \qquad \dot{y}_1(0) = 0 \qquad (2.9\text{-}43\text{d,e,f})$$

From Equations (2.9-43a to c), we deduce

$$\dot{y}_0 = -gt + v_0 \qquad (2.9\text{-}44\text{a})$$

$$y_0 = -\frac{1}{2} gt^2 + v_0 t \qquad (2.9\text{-}44\text{b})$$

Substituting Equation (2.9-44a) in Equation (2.9-43d) yields

$$\ddot{y}_1 = -[g^2 t^2 - 2 v_0 g t + v_0^2] \qquad (2.9\text{-}45)$$

The solution subject to Equations (2.9-43e, f) can be written as

$$\dot{y}_1 = -[\frac{1}{3} g^2 t^3 - v_0 g t^2 + v_0^2 t] \qquad (2.9\text{-}46\text{a})$$

$$y_1 = -[\frac{1}{12} g^2 t^4 - \frac{1}{3} v_0 g t^3 + \frac{1}{2} v_0^2 t^2] \qquad (2.9\text{-}46\text{b})$$

From Equations (2.9-42c, 44a), we obtain

$$t_0 = v_0 / g \qquad (2.9\text{-}47)$$

Combining Equations (2.9-42d, 43a, 46a, 47) yields

$$t_1 = -\dot{y}_1(t_0) / \ddot{y}_0(t_0) = -\frac{1}{3}(v_0^3 / g^2) \qquad (2.9\text{-}48\text{a,b})$$

Substituting Equations (2.9-47, 48b) in Equation (2.9-40b) yields

$$t_m = v_0 / g - \varepsilon [\frac{1}{3}(v_0^3 / g^2)] + \dots \qquad (2.9\text{-}49)$$

The effect of the air resistance is to reduce $t_m$.

The maximum height $y_m$ is obtained from Equations (2.9-44b, 46b, 49) and can be written as

$$y_m = -\frac{1}{2} g \left[ v_0/g - (\varepsilon/3) (v_0^3/g^2) + ... \right]^2 + v_0 \left[ v_0/g - (\varepsilon/3) (v_0^3/g^2) + ... \right]$$

$$- \varepsilon \left[ \frac{1}{12} g^2 (v_0/g + ...)^4 - \frac{1}{3} v_0 g (v_0/g + ...)^3 + \frac{1}{2} v_0^2 (v_0/g + ...)^2 \right] \qquad (2.9\text{-}50a)$$

$$= (v_0/2g) - \varepsilon (v_0^4/4g^2) \qquad (2.9\text{-}50b)$$

As expected, the air resistance lowers the maximum height.

The present problem can be solved exactly. Replacing $\ddot{y}$ by $(v\, dv/dy)$, Equation (2.9-38) becomes

$$v \frac{dv}{dy} = -[g + \varepsilon v^2] \qquad (2.9\text{-}51a)$$

or $\qquad \dfrac{v\, dv}{g + \varepsilon v^2} = -dy \qquad (2.9\text{-}51b)$

The solution of Equation (2.9-51b) subject to Equations (2.9-39a, b) is

$$y = \frac{1}{2\varepsilon} \ln \left[ \frac{g + \varepsilon v_0^2}{g + \varepsilon v^2} \right] \qquad (2.9\text{-}52)$$

The maximum height $y_m$ is given by

$$y_m = \frac{1}{2\varepsilon} \ln \left( 1 + \varepsilon v_0^2/g \right) \qquad (2.9\text{-}53a)$$

$$= \frac{1}{2\varepsilon} \left[ \varepsilon v_0^2/g - \varepsilon^2 v_0^4/2g^2 + ... \right] \qquad (2.9\text{-}53b)$$

$$= v_0^2/2g - \varepsilon v_0^4/4g^2 + ... \qquad (2.9\text{-}53c)$$

Equation (2.9-50b) is exactly Equation (2.9-53c).

●

It is not uncommon for the small parameter to be associated with non-linear terms, as in Example 2.9-3, and unlike the present example, it is often impossible to obtain an exact solution. By the perturbation method, the non-linear system reduces to a linear system. The non-linear terms become the non-homogeneous terms [Equation (2.9-43d)]. The equations can then be solved and an approximate solution is obtained.

The **regular perturbation** (straight forward expansion) method as described in the preceding example may generate a solution which is not valid throughout the region of interest (usually at infinity or at the origin). In this case, we have a **singular perturbation**. We next introduce several examples involving a singular perturbation.

i)      The function $\exp(-\varepsilon x)$ can be approximated as

$$\exp(-\varepsilon x) = \sum_{n=0}^{\infty} (-1)^n (\varepsilon x)^n / n! = 1 - \varepsilon x + \varepsilon^2 x^2 + \ldots \qquad (2.9\text{-}54\text{a,b})$$

The infinite series is convergent and, as an asymptotic expansion, we consider the first three terms as shown in Equation (2.9-54b). If $x < (1/\varepsilon)$, the approximation is valid, but if $x \gg (1/\varepsilon)$, the approximation is no longer valid and the magnitude of the first three terms can exceed one while $\exp(-\varepsilon x) \leq 1$ for all $x \geq 0$. Thus the approximation of $\exp(-\varepsilon x)$, $(\varepsilon \ll 1)$ by the first three terms, is valid for $x < (1/\varepsilon)$ but not for $x \gg (1/\varepsilon)$. In the case of a convergent series, we need to include more terms as the value of $x$ increases. There is a limit to the number of terms that can be included. If we are required to evaluate $\exp(-\varepsilon x)$ for large values of $x$, the convergent series [Equation (2.9-54a)] is not useful (see Example 2.9-2).

ii)     The function $\sqrt{x + \varepsilon}$ can be approximated as

$$\sqrt{x + \varepsilon} = \sqrt{x}\,(1 + \varepsilon/x)^{1/2} = \sqrt{x}\,(1 + \varepsilon/2x - \varepsilon^2/8x^2 + \ldots) \qquad (2.9\text{-}55\text{a,b})$$

Except for the first term, $(\varepsilon = 0)$, all the other terms are singular at the origin.

iii)    If the coefficient of the highest derivative in a differential equation is $\varepsilon$, the solution in powers of $\varepsilon$ is often singular (see Section 2.4 and Example 2.9-4).

Several methods have been developed to extend the validity of the solution. The basic idea is to use more than one scale. This is discussed in the next example.

*Example 2.9-4*. In Example 1.14-2, the rate equations for components A, B, and C involved in first order reactions are solved. In more complicated cases, it is not possible to obtain the analytic solution and the **quasi-steady state approximation** is introduced. This implies that the left side in Equation (1.14-15) is set to zero. Discuss the validity of this approximation. Follow Bowen et al. (1963) and write Equations (1.14-14 to 16) in dimensionless form. Show that the solution by the regular perturbation is singular.

Setting $dB/dt = 0$, we deduce from Equation (1.14-15) that

$$c_B = (k_1/k_2)\,c_A \qquad (2.9\text{-}56)$$

From the exact solutions [Equations (1.14-17, 22)], we obtain

$$c_B = \frac{k_1 c_A [1 - \exp(k_1 - k_2)t]}{k_2 (1 - k_1/k_2)}$$

(2.9-57)

Comparing Equations (2.9-56, 57), we find that the approximate solution [Equation (2.9-56)] is valid if

$$k_1/k_2 \approx 0, \qquad \exp(k_1 - k_2)t \approx 0$$

(2.9-58a,b)

The approximation is valid if $k_2 \gg k_1$ and $t \longrightarrow \infty$. Even if $k_2 \gg k_1$, the approximation is not valid for small values of $t$ and this restriction will be discussed later.

We introduce the following dimensionless quantities

$$c_A^* = c_A/c_{A_0}, \qquad c_B^* = c_B/c_{A_0}, \qquad c_C^* = c_C/c_{A_0}$$

(2.9-59a,b,c)

$$\varepsilon = k_1/k_2, \qquad t^* = k_1 t$$

(2.9-59d,e)

Using Equations (2.9-59a to e), Equations (1.14-15 to 16) becomes

$$\frac{dc_A^*}{dt^*} = -c_A^*$$

(2.9-60a)

$$\varepsilon \frac{dc_B^*}{dt^*} = -c_B^* + \varepsilon c_A^*$$

(2.9-60b)

$$\varepsilon \frac{dc_C^*}{dt^*} = c_B^*$$

(2.9-60c)

The initial conditions are

$$c_A^*(0) = 1, \qquad c_B^*(0) = c_C^*(0) = 0$$

(2.9-61a,b,c)

On solving Equations (2.9-60a to c) subject to Equations (2.9-61a, b, c), we obtain

$$c_A^* = e^{-t^*}$$

(2.9-62a)

$$c_B^* = \varepsilon (1 - \varepsilon)^{-1} [e^{-t^*} - e^{-t^*/\varepsilon}]$$

(2.9-62b)

$$c_C^* = (1 - \varepsilon)^{-1} [e^{-t^*} - \varepsilon e^{-t^*/\varepsilon}] - 1$$

(2.9-62c)

For simplicity, we consider only Equations (2.9-60a, b) and solve them by the regular perturbation method. We express $c_A^*$ and $c_B^*$ as

$$c_A^* = c_{A_0}^* + \varepsilon c_{A_1}^* + \varepsilon^2 c_{A_2}^* + \ldots \tag{2.9-63a}$$

$$c_B^* = c_{B_0}^* + \varepsilon c_{B_1}^* + \varepsilon^2 c_{B_2}^* + \ldots \tag{2.9-63b}$$

Substituting Equations (2.9-63a, b) in Equations (2.9-60a, b) and comparing powers of $\varepsilon$, we obtain

$$\varepsilon^0: \quad \frac{d c_{A_0}^*}{d t^*} = -c_{A_0}^* \tag{2.9-64a}$$

$$0 = c_{B_0}^* \tag{2.9-64b}$$

$$\varepsilon^1: \quad \frac{d c_{A_1}^*}{d t^*} = -c_{A_1}^* \tag{2.9-64c}$$

$$\frac{d c_{B_0}^*}{d t^*} = -c_{B_1}^* + c_{A_0}^* \tag{2.9-64d}$$

$$\varepsilon^2: \quad \frac{d c_{A_2}^*}{d t^*} = -c_{A_2}^* \tag{2.9-64e}$$

$$\frac{d c_{B_1}^*}{d t^*} = -c_{B_2}^* + c_{A_1}^* \tag{2.9-64f}$$

The initial conditions are

$$c_{A_0}^*(0) = 1, \quad c_{A_1}^*(0) = c_{A_2}^*(0) = \ldots = 0 \tag{2.9-65a,b,c}$$

$$c_{B_0}^*(0) = c_{B_1}^*(0) = c_{B_2}^*(0) = \ldots = 0 \tag{2.9-65d,e,f}$$

Solving Equations (2.9-64a to 65d), we obtain

$$c_{A_0}^* = e^{-t^*}, \quad c_{A_1}^* = c_{A_2}^* = 0 \tag{2.9-66a,b,c}$$

$$c_{B_0}^* = 0, \quad c_{B_1}^* = c_{A_0}^*, \quad c_{B_2}^* = 0 \tag{2.9-66d,e,f}$$

The approximate solutions for $c_A^*$ and $c_B^*$ are

$$c_A^* \approx e^{-t^*}, \qquad c_B^* \approx \varepsilon e^{-t^*} \qquad\qquad\qquad\qquad (2.9\text{-}67a,b)$$

Equation (2.9-62b) can be written as

$$c_B^* \approx \varepsilon (1 + \varepsilon + ...) [e^{-t^*} - e^{-t^*/\varepsilon}] \qquad\qquad\qquad (2.9\text{-}68)$$

Comparing Equations (2.9-67b, 68), we note that Equation (2.9-67b) is a valid approximation in the case $\varepsilon \longrightarrow 0$, $t \neq 0$. At time $t \approx 0$, the approximation is not valid. The solutions given by Equations (2.9-56, 67b) are not uniformly valid and do not satisfy the initial condition $[c_B^*(0) = 0]$. The coefficient of $dc_B^*/dt^*$ is $\varepsilon$. Expanding $c_B^*$ in powers of $\varepsilon$, the differential equation [Equation (2.9-60b)] reduces to a system of algebraic equations [Equations (2.9-64b, d, f)] and no arbitrary constant can be introduced so as to satisfy the initial conditions. Note that in Equations (2.9-64d, f), $c_{B_0}^*$ and $c_{B_1}^*$ are known quantities and are defined in Equations (2.9-64b, d) respectively. We further note that the zeroth approximation

$$c_{A_0}^* = e^{-t^*}, \qquad c_{B_0}^* = 0 \qquad\qquad\qquad\qquad (2.9\text{-}69a,b)$$

is uniformly valid. This zeroth order solution cannot be improved because $c_{B_1}^*$ is not uniformly valid. This implies that the quasi-steady state method is valid for the zeroth approximation and not for higher approximations.

A similar situation exists in fluid mechanics. The Stokes solution of the flow past a sphere is uniformly valid but the higher approximation is not valid and is known as the **Whitehead paradox**. A thorough discussion on the Whitehead paradox is given in Van Dyke (1975).

It is the presence of the term $e^{-t^*/\varepsilon}$ in Equation (2.9-62b) that contributes to the singularity. On taking the limit as $\varepsilon \longrightarrow 0$ first and then as $t^* \longrightarrow 0$, the limit of $e^{-t^*/\varepsilon}$ is zero. On reversing the limiting process, the limit of $e^{-t^*/\varepsilon}$ is one. This implies that

$$\lim_{\substack{\varepsilon \to 0 \\ t^* \to 0}} e^{-t^*/\varepsilon} \neq \lim_{\substack{t^* \to 0 \\ \varepsilon \to 0}} e^{-t^*/\varepsilon} \qquad\qquad\qquad (2.9\text{-}70)$$

We further note that $e^{-t^*/\varepsilon}$ changes rapidly near $t^* = 0$. This suggests that to obtain a uniformly valid solution we need to seek the solution in two separate regions. We obtain a solution near $t^* = 0$, which is the **inner region or boundary layer** where $\varepsilon$ is important, and a solution for $t^* \gg 0$, which is the **outer region**. We use two different time scales, one for the inner solution and one for the outer solution. This perturbation method is the **matched asymptotic expansion** method (boundary layer method) and was introduced by Prandtl to resolve **d'Alembert's paradox** in fluid mechanics (see Chapter 3).

In the present example, $c_A^*$ [Equation (2.9-67a)] is uniformly valid and we need to solve only Equation (2.9-60b) by the method of matched asymptotic expansion.

The time scale for the inner region is

$$T = t^*/\varepsilon \qquad (2.9\text{-}71)$$

Using Equation (2.9-71), Equation (2.9-60b) becomes

$$\frac{dc_B^*}{dT} = -c_B^* + \varepsilon c_A^* \qquad (2.9\text{-}72)$$

Note that, by our choice of $T$, the coefficient of $dc_B^*/dT$ is one and not $\varepsilon$. The independent variable for the inner region is chosen such that the coefficient of the highest derivative is not $\varepsilon$. We denote the inner solution by $c_B^{(i)*}$ and it is expanded as

$$c_B^{(i)*} = c_{B_0}^{(i)*} + \varepsilon c_{B_1}^{(i)*} + \dots \qquad (2.9\text{-}73)$$

Substituting Equation (2.9-73) in Equation (2.9-72) and comparing powers of $\varepsilon$, we obtain

$$\varepsilon^0: \qquad \frac{dc_{B_0}^{(i)*}}{dT} = -c_{B_0}^{(i)*} \qquad (2.9\text{-}74\text{a})$$

$$\varepsilon^1: \qquad \frac{dc_{B_1}^{(i)*}}{dT} = -c_{B_1}^{(i)*} + c_{A_0}^* \qquad (2.9\text{-}74\text{b})$$

The initial conditions can be applied in the inner region and they are

$$c_{B_0}^{(i)*}(0) = c_{B_1}^{(i)*}(0) = 0 \qquad (2.9\text{-}74\text{c,d})$$

The solution of Equation (2.9-74a) that satisfies Equation (2.9-74c) is

$$c_{B_0}^{(i)*} = 0 \qquad (2.9\text{-}75)$$

Combining Equations (2.9-66a, 71, 74b) yields

$$\frac{d\,c_{B_1}^{(i)*}}{dT} + c_{B_1}^{(i)*} = e^{-\varepsilon T} = 1 - \varepsilon T + ... \approx 1 \tag{2.9-76a,b,c}$$

On solving Equation (2.9-76c) subject to Equation (2.9-74d), we obtain

$$c_{B_1}^{(i)*} = 1 - e^{-T} \tag{2.9-77}$$

The inner solution $c_B^{(i)*}$ can be written as

$$c_B^{(i)*} = \varepsilon(1 - e^{-T}) + ... \tag{2.9-78}$$

For the outer region, $t^*$ is the time scale and the solution given by Equation (2.9-67b) is valid for $t^* \gg 0$. The outer solution $c_B^{(o)*}$ is given by Equation (2.9-67b). Note that since all conditions are at time $t^* = 0$, no condition can be imposed on the outer solution. As mentioned earlier, the outer solution is determined by an algebraic equation and there is no arbitrary constant. If the outer solution is obtained by solving a differential equation, the arbitrary constant is determined by using a **matching principle**. The simplest one was proposed by Prandtl and can be expressed as

$$\lim_{T \to \infty} c_B^{(i)*} = \lim_{t^* \to 0} c_B^{(o)*} \tag{2.9-79}$$

In the present example, Equation (2.9-79) is identically satisfied. That is to say

$$\lim_{T \to \infty} \varepsilon(1 - e^{-T}) = \lim_{t^* \to 0} \varepsilon e^{-t^*} = \varepsilon \tag{2.9-80a,b,c}$$

The matching principle can be interpreted as follows. There exists a region where both the outer and the inner solutions are valid and are equal. The solution which is uniformly valid is the **composite solution** and is denoted by $c_B^{(c)*}$. This solution is given by

$$c_B^{(c)*} = c_B^{(i)*} + c_B^{(o)*} + \left[c_B^{(i)*}\right]^o = c_B^{(i)*} + c_B^{(o)*} + \left[c_B^{(o)*}\right]^i \tag{2.9-81a,b}$$

where $\left[c_B^{(i)*}\right]^o$ and $\left[c_B^{(o)*}\right]^i$ denote the outer limit ($T \to \infty$) of the inner solution and the inner limit ($t^* \to 0$) of the outer solution respectively. Equation (2.9-79) implies

$$\left[c_B^{(i)*}\right]^o = \left[c_B^{(o)*}\right]^i \tag{2.9-82}$$

Combining Equations (2.9-67b, 78, 80c, 81a) yields

$$c_B^{(c)*} = \varepsilon(1 - e^{-T}) + \varepsilon e^{-t^*} - \varepsilon + O(\varepsilon^2) \tag{2.9-83a}$$

$$\approx \varepsilon [e^{-t^*} - e^{-t^*/\varepsilon}] \tag{2.9-83b}$$

Note that $c_B^{(c)*}$ satisfies condition $c_B^* = 0$.

For further details on singular perturbation, one can consult Nayfeh (1973) or Van Dyke (1975).

## PROBLEMS

1a. Determine the singular points of the following equations. Are the singular points regular or irregular?

(i) $\quad (1 - x^3) \dfrac{d^2y}{dx^2} + x^2 \dfrac{dy}{dx} + x^4 y = 0$

(ii) $\quad x \dfrac{dy}{dx} + \sin x \dfrac{dy}{dx} + y \cos x = 0$

(iii) $\quad x^2 \dfrac{d^2y}{dx^2} + \cos x \dfrac{dy}{dx} + y \sin x = 0$

(iv) $\quad x \dfrac{d^2y}{dx^2} + e^x \dfrac{dy}{dx} + x^3 y = 0$

2a. Determine the radius of convergence of the following series. Do they converge on the circle of convergence?

(i) $\quad \displaystyle\sum_{n=0}^{\infty} \dfrac{x^n}{2^n}$   (iii) $\quad \displaystyle\sum_{n=0}^{\infty} \dfrac{x^n}{(2n+1)^2}$

(ii) $\quad \displaystyle\sum_{n=1}^{\infty} \dfrac{x^n}{n}$   (iv) $\quad \displaystyle\sum_{n=0}^{\infty} \dfrac{(-1)^n x^n}{n}$

3a. Compute a series solution of the equation

$$(1 + x) y' - (x + 2) y = 0$$

Compare the series solution with the exact solution.

Answer: $(1 + x) e^x$

4b. Hermite's equation can be written as

$$\frac{d^2y}{dx^2} - 2x \frac{dy}{dx} + 2ny = 0$$

Show that its general solution can be written as the sum of two infinite series and are of the form

$$y = a_0 \left[ 1 - \frac{2n}{2!} x^2 + \frac{2^2 n (n-2)}{4!} x^4 - ... \right] + a_1 \left[ x - \frac{2(n-1)}{3!} x^3 + ... \right]$$

Determine the radius of convergence of the series.

If  n  is a non-negative integer, one of the series terminates and the polynomial is denoted by $H_n(x)$.  Calculate the first four polynomials, that is, $H_0, H_1, H_2$, and $H_3$.  To specify $H_n$ completely, the constant $a_0$ (or $a_1$) is chosen such that the coefficient of the highest power of  x  ($x^n$) is $2^n$.  Verify that the polynomials $H_0$ to $H_3$ are given by Rodrigues's formula

$$H_n(x) = (-1)^n e^{x^2} \frac{d^n e^{-x^2}}{dx^n}$$

Use Rodrigues's formula to deduce that

$$\int_{-\infty}^{\infty} e^{-x^2} [H_n(x)]^2 dx = (2^n n! \sqrt{\pi})$$

[Hint: Integrate by parts, note that the coefficient of $x^n$ in $H_n$ is $2^n$ and deduce the $n^{th}$ derivative of $H_n$.  $\int_0^{\infty} e^{-x^2} dx = \sqrt{\pi}/2$ ]

5a.    Find a series solution of the following equations

(i)    $\dfrac{d^2y}{dx^2} - xy = 0$                                      (ii)    $\dfrac{d^2y}{dx^2} + xy + y = 0$

6a.    Find the general solutions of the following equations

(i)    $x^2 \dfrac{d^2y}{dx^2} + 5x \dfrac{dy}{dx} - 5y = 0$                      Answer:  $Ax^{-5} + Bx$

(ii)    $x^2 \dfrac{d^2y}{dx^2} - 3x \dfrac{dy}{dx} + 4y = 0$                      Answer:  $Ax^2 + Bx^2 \ln x$

(iii)     $(x-1)^2 \dfrac{d^2y}{dx^2} + (x-1)\dfrac{dy}{dx} - y = 0$              Answer:  $A(x-1) + B(x-1)^{-1}$

7b.     Solve the equation

$$x^2 y'' + x y' - y = x$$

subject to

$$y(1) = y'(1) = 0$$

Answer:  $\dfrac{(1-x^2)}{4x} + \dfrac{x}{2}\ell n\,x$

8a.     Solve the following equations

(i)     $4x\dfrac{d^2y}{dx^2} + 3\dfrac{dy}{dx} + 3y = 0$

(ii)     $x^2\dfrac{d^2y}{dx^2} - 3x\dfrac{dy}{dx} + 4(x+1)y = 0$

(iii)     $x\dfrac{d^2y}{dx^2} + 2\dfrac{dy}{dx} + xy = 0$

9b.     Deduce a series solution for the equation

$$x^2\dfrac{d^2y}{dx^2} + (x^2+x)\dfrac{dy}{dx} - y = 0$$

Express the series in a closed form.  Determine the solution that satisfies the conditions $y(1) = 1$,  $y'(1) = -1$.

Answer:  $\dfrac{(x-1)}{x} + \dfrac{e^{(1-x)}}{x}$

10a.     Show that  $r = 0$  is a double root of the indicial equation of

$$x\dfrac{d^2y}{dx^2} + (1-x^2)\dfrac{dy}{dx} + 4xy = 0$$

Find the solution that is finite at the origin.

Answer:  $1 - x^2 + x^4/8$

11b.   Sheppard and Eisenklam (1983) have considered the dispersion of an ideal gas through porous masses of solid particles.  The gas is assumed to be compressible and after some simplifications (details are given in the paper), the equation to be solved is

$$N_{DO} (1+Bx) \frac{d^2C}{dx^2} - (1+Bx+\frac{B}{2} N_{DO}) \frac{dC}{dx} + \left[\frac{B}{2} - (1+Bx)^{3/2}(\tau_0) (p)\right] C = 0$$

where  C  is the Laplace transform of the concentration,  $N_{DO}$  is the dispersion number,  B  is related to the permeability of the porous mass,  $\tau_0$  is a residence time,  and  p  is the Laplace transform variable.

Show that the origin is an ordinary point.  On expanding all quantities in powers of  x  up to  $x^5$, obtain  C  in a power series of  x  up to  $x^5$.

12a.   Deduce that the indicial equation of

$$x^2 \frac{d^2y}{dx^2} + (x - x^2) \frac{dy}{dx} - xy = 0$$

is  $r^2 = 0$.

Obtain one series solution which is non-singular at the origin and verify that it is  $e^x$.  Use the method of variation of parameters to show that the second solution is  $\int (e^{-x}/x)\, dx$.  By expanding  $e^{-x}$  in powers of  x, find the second series solution.

13b.   Jenson and Jeffreys (1963) have considered the problem of the temperature distribution in a transverse fin of triangular cross-section.  If  T  denotes the fin temperature and  $T_A$  denotes the air temperature, show from an energy balance that  y  $(= T - T_A)$  satisfies the equation

$$x (b - x) \frac{d^2y}{dx^2} + (b - 2x) \frac{dy}{dx} - \beta (b - x) y = 0$$

where  x  is the distance from the rim of the fin, as shown in Figure 2.P-13b.  The outer radius of the fin is   b,   $\beta = h/k \sin \alpha$,   h   is the heat transfer coefficient,   k   is the thermal conductivity of the fin, and   $\alpha$   is the half angle of the vertex of the triangular fin.

The appropriate boundary conditions are

at  x  =  0,          y remains finite

at  x  =  b – a,     y  =  $T_B - T_A$

$T_B$ is the temperature of the pipe and a is the radius of the pipe.

Show that the origin (x = 0) is a regular singular point.

Obtain a series solution for y if a = 5 cm, b = 15 cm, $\beta$ = 380 W/mK, $T_A$ = 288 K, and $T_B$ = 373 K.

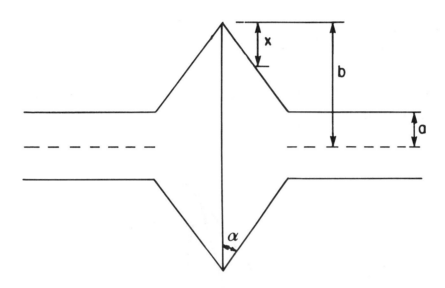

**FIGURE 2.P-13b**    **Transverse cooling fin**

14a.    Show that the eigenvalues $\lambda$ of the Sturm-Liouville problem

$$\frac{d^2 y}{dx^2} + \lambda y = 0$$

$$\alpha_1 y(0) + \alpha_2 y'(0) = 0, \quad \beta_1 y(\pi) + \beta_2 y'(\pi) = 0$$

are given by

$$(\alpha_1 \beta_1 + \lambda \alpha_2 \beta_2) \tan \pi \sqrt{\lambda} = \sqrt{\lambda} \, (\beta_1 \alpha_2 - \alpha_1 \beta_2)$$

15b.    Consider the following electrostatic problem, where a positive charge e is placed at point A, a distance a from the origin, as shown in Figure 2.P-15b. A negative charge $-e$ is placed at B, a distance $-a$ from the origin. The potential u at any point P is given by

$$u = e/r_1 - e/r_2$$

where $r_1 = AP$, $r_2 = BP$. Using the cosine rule, we obtain

$$r_1^{-1} = r^{-1} \left[ 1 - (2a/r) \cos \theta + a^2/r^2 \right]^{-1/2}$$

$$r_2^{-1} = r^{-1} \left[ 1 + (2a/r) \cos \theta + a^2/r^2 \right]^{-1/2}$$

Use Equation (2.7-14) to express $r_1^{-1}$ and $r_2^{-1}$ in terms of the Legendre polynomials. Hence, show that

$$u = \frac{2e}{r} \sum_{s=0}^{\infty} \left( \frac{a}{r} \right)^{2s+1} P_{2s+1} (\cos \theta)$$

Find the limiting value of $u$ in the case $a \longrightarrow 0$, $ea \longrightarrow \mu$ $(\neq 0)$. This limiting case corresponds to a dipole.

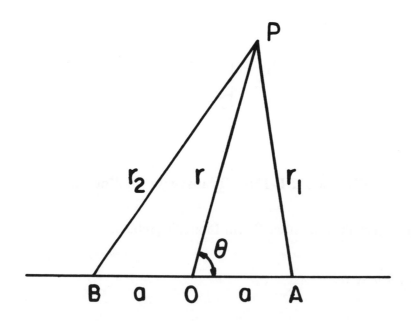

**FIGURE 2.P-15b      Charges $(\pm e)$ at A and B**

16a.    Use Rodrigues's formula [Equation (2.7-13)] to compute $P_0$, $P_1$, and $P_2$.

17b.    Show that

$$\int_{-1}^{1} P_2^2 \, dx = 2/5$$

[Hint: Use Rodrigues's formula, integrate by parts, and note that $P_2^{''} = 3$.]

18a.  Calculate the associated Legendre functions $P_2^1$ and $P_2^2$ from Equation (2.7-40). Verify that they satisfy the associated Legendre equation.

$$\text{Answer: } 3x\sqrt{1-x^2} \ ; \ \ 3(1-x^2)$$

19b.  Apply l'Hôpital's rule to Equation (2.7-68) and deduce that

$$Y_n(x) = \frac{1}{\pi} \lim_{v \to n} \left[ \frac{\partial J_v}{\partial v} - (-1)^n \frac{\partial J_{-v}}{\partial v} \right]$$

where $n$ is an integer.

Assuming that $J_v(x)$ and $J_{-v}(x)$ are given by Equations (2.7-62b, 63), show that $Y_n(x)$ has a $\ell n(x/2)$ term.

20a.  From Example 2.7-2 and Table 2.7-1, deduce that

(i) $\quad J_{-3/2}(x) = -\sqrt{\dfrac{2}{\pi x}} \left( \dfrac{\cos x}{x} + \sin x \right)$

(ii) $\quad J_{5/2}(x) = \sqrt{\dfrac{2}{\pi x}} \left( \dfrac{3\sin x}{x^2} - \dfrac{3\cos x}{x} - \sin x \right)$

(iii) $\quad J_0'(x) = \sqrt{\dfrac{1}{2\pi x}} \left( \cos x - \sin x \right) = -J_1(x)$

21b.  Choudhury and Jaluria (1994) obtained an analytical solution for the transient temperature distribution in a moving rod. They assumed the existence of a steady temperature and, to obtain the steady temperature, they had to solve the equation [their equation (12)]

$$\frac{1}{R} \frac{d}{dR} \left( R \frac{dR_1}{dR} \right) + \lambda^2 R_1 = 0$$

where $R$ is the dimensionless radial distance, $R_1$ is the dependent variable, and $\lambda$ is a constant. The boundary conditions [their equation (14)] are

$$\left. \frac{dR_1}{dR} \right|_{R=0} = 0 , \qquad \left. \frac{dR_1}{dR} \right|_{R=1} = - \text{Bi } R_1(1)$$

where Bi is the surface Biot number.

Obtain $R_1$ in terms of Bessel functions and show that the eigenvalues $\lambda_n$ are given by

$$\lambda_n J_1(\lambda_n) = \text{Bi } J_0(\lambda_n)$$

22b.   The generating function for Bessel's functions of integral order is $\exp[(x/2)(t-1/t)]$. Expanding the generating function yields

$$\exp[(x/2)(t-1/t)] = \sum_{\substack{r=0 \\ s=0}}^{\infty} \frac{(-1)^s}{r!\,s!} \left(\frac{x}{2}\right)^{r+s} t^{r-s} = \sum_{n=-\infty}^{\infty} t^n J_n(x)$$

Show that on substituting $t = e^{i\theta}$ and equating the real and imaginary parts, we obtain

$$\cos(x \sin \theta) = J_0(x) + 2 \sum_{n=1}^{\infty} J_{2n}(x) \cos 2n\theta$$

$$\sin(x \sin \theta) = 2 \sum_{n=0}^{\infty} J_{2n+1}(x) \sin[(2n+1)\theta]$$

Multiply the first equation by $\cos m\theta$, the second equation by $\sin m\theta$, where $m$ is an integer, and use Equations (2.8-13a to d) to deduce that

$$J_n(x) = \frac{1}{\pi} \int_0^\pi \cos(n\theta - x \sin\theta)\, d\theta$$

This is the integral formula for Bessel's function.

Verify that

$$J_0(x) = \frac{1}{\pi} \int_0^\pi \cos(x \sin\theta)\, d\theta$$

satisfies Equation (2.4-44).

23a.   Represent the function $f(x) = 3x^3 - 2x + 1$ defined in the interval $-1 \le x \le 1$ by a linear combination of Legendre polynomials.

Answer: $P_0 - 0.2\, P_1 + 1.2\, P_3$

24b.   The function $f(x)$ is defined by

$$f(x) = \begin{cases} 1 & 0 \le x < 1/2 \\ \dfrac{1}{2} & x = 1/2 \\ 0 & 1/2 < x \le 1 \end{cases}$$

and is represented by $\displaystyle\sum_{n=1}^{\infty} c_n J_0(\lambda_n x)$, where the $\lambda_n$ are the zeros of $J_0(x)$. Show that the coefficients $c_n$ are given by

$$c_n = J_1(\lambda_n/2) / [\lambda_n J_1(\lambda_n)]^2$$

25a. The function $f(x)$ is periodic and of period $2\pi$. It is defined by

$$f(x) = \begin{cases} -1 & -\pi < x < 0 \\ \\ 1 & 0 < x < \pi \end{cases}$$

Sketch $f(x)$ over the interval $-3\pi$ to $3\pi$. Show that its Fourier series is

$$f(x) \approx \frac{4}{\pi} \sum_{s=0}^{\infty} \frac{\sin(2s+1)x}{(2s+1)}$$

Deduce that

$$\sum_{s=0}^{\infty} \frac{(-1)^s}{(2s+1)} = \frac{\pi}{4}$$

26a. Represent the function

$$f(x) = \begin{cases} 0 & -\pi < x < 0 \\ \\ 1 & 0 < x < \pi \end{cases}$$

by a complex Fourier series [Equation (2.8-33c)].

$$\text{Answer: } \frac{1}{2} - \frac{i}{\pi} \sum_{s=-\infty}^{\infty} \frac{e^{i(2s+1)x}}{(2s+1)}$$

27b. The equation governing the motion of a damped harmonic oscillator under the influence of an external periodic force $f(t)$ is

$$\frac{d^2 y}{dt^2} + k \frac{dy}{dt} + n^2 y = f(t)$$

where $k$ and $n$ are constants.

Find the Fourier series of $f(t)$, if $f(t)$ is periodic of period $2T$ and is equal to $ct/2T$ in the interval $(0, 2T)$.

Assume that $y(t)$ is also periodic of period $2T$ and write its Fourier series expansion. Differentiate the series term by term and substitute the resulting expression in the differential equation. Determine the Fourier coefficients by comparing the coefficients of the constant term, the cos terms, and the sin terms.

$$\text{Answer: } a_0 = c/n^2$$
$$a_s = ks\pi c^2 T^3 / [\alpha_s^2 + k^2 s^2 \pi^2 T^2]$$
$$b_s = cT^2 \alpha_s / [\alpha_s^2 + k^2 s^2 \pi^2 T^2]$$
$$\alpha_s^2 = s^2 \pi^2 - n^2 T^2$$

28a.    Determine the Fourier integral representation of

$$f(x) = \begin{cases} e^{-x} & x > 0 \\ \\ 0 & x < 0 \end{cases}$$

Deduce that

(i)    $$\int_0^\infty \frac{d\alpha}{1 + \alpha^2} = \frac{\pi}{2}$$            Answer: $A(\alpha) = 1/(1 + \alpha^2)$

(ii)    $$\int_0^\infty \frac{\cos \alpha \, d\alpha}{1 + \alpha^2} = \frac{\pi}{2e}$$            Answer: $B(\alpha) = \alpha/(1 + \alpha^2)$

29a.    Show that the equation

$$x^2 \frac{d^2y}{dx^2} + (3x - 1)\frac{dy}{dx} + y = 0$$

has an irregular singular point at the origin.

Find a power series $\left( \sum\limits_{r=0}^{\infty} c_r x^r \right)$ that satisfies the equation. Calculate the radius of convergence of the series and comment on the validity of the series as a solution of the equation.

30a.  Solve the equation

$$\frac{d^2y}{dx^2} + \frac{dy}{dx} - \varepsilon y^2 = 0$$

subject to initial conditions

$$y(0) = 1, \qquad \frac{dy}{dx}\bigg|_{x=0} = 0$$

by the method of perturbation, assuming $\varepsilon$ to be small.  Calculate the first two terms of the expansion.

Answer: $1 + \varepsilon(-1 + x + e^{-x})$

31b.  Find a two term expansion valid for small $\varepsilon$ for the solution of

$$\frac{d^2y}{dx^2} + \varepsilon y^2 = 0$$

$$y(0) = 0, \qquad y(1 + \varepsilon) = 1$$

[Hint: Expand all quantities in powers of $\varepsilon$.]

Answer: $x - \frac{\varepsilon x}{12}(11 + x^3)$

32b.  The equation of a simple pendulum is

$$\frac{d^2y}{dt^2} + \omega^2 \sin y = 0$$

where $y$ is the angle of inclination and $\omega$ is the frequency.

Substituting the expansion of $\sin y$ in the equation of motion yields

$$\frac{d^2y}{dt^2} + \omega^2 (y - y^3/6 + \dots) = 0$$

Assuming $\omega^2/6$ to be small, the equation governing the motion of a pendulum can be written as

$$\frac{d^2y}{dt^2} + \omega^2 y - \varepsilon y^3 = 0$$

Suppose that the initial conditions are

$$y\,(0)\,=\,1\,,\qquad \frac{dy}{dt}\Bigg|_{t=0}\,=\,0$$

Show that the solution to order $\varepsilon$ is

$$y\,=\,\cos\,\omega t + \varepsilon\,[(\cos\,\omega t - \cos\,3\omega t)\,/\,32\omega^2 + (3t\,\sin\,\omega t)\,/\,8\omega]$$

The presence of the term $3t\,\sin\,\omega t$ (secular term) renders the solution invalid for large values of $t$. To obtain a uniformly valid solution, we can use the method of multiple scales. We introduce two time scales $T_0$ and $T_1$ and write

$$t\,=\,T_0 + \varepsilon\,T_1$$

Substitute $t$ in the solution and determine $T_1$ such that the secular term is eliminated. Determine the uniformly valid solution to order $\varepsilon$.

Answer: $T_1\,=\,3T_0/8\omega^2$

# CHAPTER 3

# COMPLEX VARIABLES

## 3.1 INTRODUCTION

The inadequacy of the real number system (rational and irrational numbers) in solving algebraic equations was known to mathematicians in the past. It therefore became necessary to extend the real number system, so as to obtain meaningful solutions to simple equations such as

$$x^2 + 1 = 0 \qquad\qquad (3.1\text{-}1)$$

For quite sometime, it appears that equations, which could not be solved in the domain of real numbers, were solved by accepting $\sqrt{-1}$ as a possible number. This notation, however, has had its own shortcomings. Euler was the first to introduce the symbol $i$ for $\sqrt{-1}$ with the basic property

$$i^2 = -1 \qquad\qquad (3.1\text{-}2)$$

(Electrical engineers use $j$ to denote $\sqrt{-1}$.)

He also established the relationships between complex numbers and trigonometric functions. However, in those times, no actual meaning could be assigned to the expression $\sqrt{-1}$. It was, therefore, called an "**imaginary**" (as opposed to real) number. This usage still prevails in the literature.

It was not until around 1800 that sound footing was given to the complex number system by Gauss, Wessel, and Argand. Gauss proved that every algebraic equation with real coefficients has **complex roots** of the form $c + i d$. Real roots are special cases, when $d$ is zero. Argand proposed a graphical representation of complex numbers. The concept of a function was subsequently extended to **complex functions** of the type

$$w = f(z) \qquad\qquad (3.1\text{-}3)$$

where $z \ (= x + i y)$ is the independent variable.

The concept of **complex variables** is a powerful and a widely used tool in mathematical analysis. The theory of differential equations has been extended within the domain of complex variables. Complex integral calculus has found a wide variety of applications in evaluating integrals, inverting power series, forming infinite products, and asymptotic expansions. Applied mathematicians, physicists, and engineers make extensive use of complex variables. It is indispensable for students in mathematical, physical, and engineering sciences to have some knowledge of the theory of complex analysis.

## 3.2    BASIC PROPERTIES OF COMPLEX NUMBERS

We can write a **complex number** $z$ as

$$z = x + iy \tag{3.2-1}$$

where $x$ and $y$ are real numbers.

The numbers $x$ and $y$ are the real and imaginary parts of $z$ respectively and are denoted by $\mathrm{Re}\,(z)$ and $\mathrm{Im}\,(z)$.

We can also regard $z$ as an **ordered pair** of real numbers. As in vector algebra, we write $z$ as $(x, y)$.

Just as a real number can be represented by a point on a line, a complex number can be represented by a point in a plane. This representation is the Argand diagram and is shown in Figure 3.2-1.

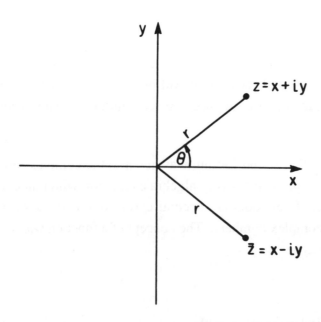

**FIGURE 3.2-1    Argand diagram**

Two complex numbers are equal if and only if (iff) their real and imaginary parts are equal. If $z_1$ $(= x_1 + iy_1)$ and $z_2$ $(= x_2 + iy_2)$ are equal, it implies

$$x_1 = x_2, \qquad y_1 = y_2 \tag{3.2-2a,b}$$

Thus a complex equation is equivalent to two real equations. The addition and multiplication can be handled in the same way as for real numbers and, whenever $i^2$ appears, it is replaced by $-1$. The **commutative, associative, and distributive laws** hold. We list some of the results

$$z_1 + z_2 = (x_1 + x_2, \, y_1 + y_2) \tag{3.2-3a}$$

$$z_1 - z_2 = (x_1 - x_2, \, y_1 - y_2) \tag{3.2-3b}$$

$$z_1 \bullet z_2 = (x_1 x_2 - y_1 y_2, \, x_1 y_2 + x_2 y_1) = z_2 \bullet z_1 \tag{3.2-3c,d}$$

$$\frac{z_1}{z_2} = \frac{x_1 + iy_1}{x_2 + iy_2} = \frac{(x_1 + iy_1)(x_2 - iy_2)}{(x_2 + iy_2)(x_2 - iy_2)} = \frac{[x_1 x_2 + y_1 y_2 + i(x_2 y_1 - x_1 y_2)]}{x^2 + y^2} \tag{3.2-3e,f,g}$$

The **complex conjugate** $\bar{z}$ of the complex number $z \; (= x + iy)$ is defined as

$$\bar{z} = x - iy \tag{3.2-4}$$

In the Argand diagram, it is the reflection of $z$ about the x-axis and it is shown in Figure 3.2-1.

So far we used only the rectangular Cartesian system. We can also use the **polar coordinate system** $(r, \theta)$. From Figure 3.2-1, we find

$$x = r \cos \theta, \qquad y = r \sin \theta \tag{3.2-5a,b}$$

Inverting Equations (3.2-5a, b), we obtain

$$r = \sqrt{x^2 + y^2} \tag{3.2-6a}$$

$$\tan \theta = y \, / \, x \tag{3.2-6b}$$

The number $r$ is called the **modulus or the absolute value of z** and is denoted by $|z|$. It can be regarded as the length of the vector represented by $z$. The absolute value of $z$ is also given by

$$|z| = \sqrt{z \bar{z}} \tag{3.2-7}$$

$\theta$ is the **argument or amplitude** of $z$. It is written as $\arg z = \theta$. Hence we can write

$$z = x + iy = r \, (\cos \theta + i \sin \theta) \tag{3.2-8a,b}$$

Since any multiple of $2\pi$ radians may be added to $\theta$ without changing the value of $z$, we specify $-\pi < \theta \le \pi$ as the **principal value** of $\arg z$, and denote it by $\text{Arg } z$. The polar representation is useful for computational purposes, as shown next.

First we calculate the product of two complex numbers. Let

$$z_1 = r_1 (\cos \theta_1 + i \sin \theta_1), \qquad z_2 = r_2 (\cos \theta_2 + i \sin \theta_2) \qquad (3.2\text{-}9a,b)$$

Then

$$z_1 z_2 = r_1 r_2 [\cos \theta_1 \cos \theta_2 - \sin \theta_1 \sin \theta_2 + i (\cos \theta_1 \sin \theta_2 + \cos \theta_2 \sin \theta_1)] \qquad (3.2\text{-}10a)$$

$$= r_1 r_2 [\cos (\theta_1 + \theta_2) + i \sin (\theta_1 + \theta_2)] \qquad (3.2\text{-}10b)$$

From this equation, we note that

$$|z_1 z_2| = r_1 r_2 \qquad (3.2\text{-}11a)$$

$$\text{Arg} (z_1 z_2) = \text{Arg } z_1 + \text{Arg } z_2 + 2 n \pi, \qquad n = 0, \pm 1, \pm 2, \dots \qquad (3.2\text{-}11b)$$

Generalizing Equation (3.2-10b), we write

$$z_1 z_2 z_3 \dots z_n = r_1 r_2 r_3 \dots r_n [\cos (\theta_1 + \theta_2 + \dots + \theta_n) + i \sin (\theta_1 + \theta_2 + \dots + \theta_n)] \qquad (3.2\text{-}12)$$

Setting

$$z_1 = z_2 = \dots = z_n = z = r (\cos \theta + i \sin \theta) \qquad (3.2\text{-}13)$$

in Equation (3.2-12), we obtain

$$z^n = r^n (\cos n\theta + i \sin n\theta) \qquad (3.2\text{-}14)$$

This is known as **De Moivre's theorem** (formula). We have deduced it for positive integral exponents, but it is true for all rational values of n.

To perform a division, in polar form, we write

$$\frac{z_1}{z_2} = \frac{r_1 (\cos \theta_1 + i \sin \theta_1)}{r_2 (\cos \theta_2 + i \sin \theta_2)} \qquad (3.2\text{-}15a)$$

$$= \frac{r_1}{r_2} \frac{(\cos \theta_1 + i \sin \theta_1)(\cos \theta_2 - i \sin \theta_2)}{(\cos \theta_2 + i \sin \theta_2)(\cos \theta_2 - i \sin \theta_2)} \qquad (3.2\text{-}15b)$$

$$= \frac{r_1}{r_2} \left[ \frac{(\cos \theta_1 \cos \theta_2 + \sin \theta_1 \sin \theta_2) + i (\cos \theta_2 \sin \theta_1 - \sin \theta_2 \cos \theta_1)}{\cos^2 \theta_2 + \sin^2 \theta_2} \right] \qquad (3.2\text{-}15c)$$

$$= \frac{r_1}{r_2} \left[ \cos\left(\theta_1 - \theta_2\right) + i \sin\left(\theta_1 - \theta_2\right) \right] \tag{3.2-15d}$$

Thus we note that

$$\left| \frac{z_1}{z_2} \right| = \frac{r_1}{r_2} \tag{3.2-16a}$$

$$\text{Arg}\left( \frac{z_1}{z_2} \right) = \text{Arg } z_1 - \text{Arg } z_2 + 2n\pi, \qquad n = 0, \pm 1, \pm 2, \dots \tag{3.2-16b}$$

If we set $z_1$ to be one in Equation (3.2-15d), we obtain

$$\frac{1}{z_2} = \frac{1}{r_2} \left[ \cos\left(-\theta_2\right) + i \sin\left(-\theta_2\right) \right] \tag{3.2-17a}$$

$$= \frac{1}{r_2} \left[ \cos \theta_2 - i \sin \theta_2 \right] \tag{3.2-17b}$$

***Example 3.2-1.*** Determine $(1 + i)^{10,000}$.

We use the polar form and write

$$(1 + i) = \sqrt{2} \left( \cos \frac{\pi}{4} + i \sin \frac{\pi}{4} \right) \tag{3.2-18}$$

Therefore

$$(1 + i)^{10,000} = 2^{\frac{10,000}{2}} \left( \cos \frac{10,000}{4} \pi + i \sin \frac{10,000}{4} \pi \right) \tag{3.2-19a}$$

$$= 2^{5,000} \left( \cos 2,500 \, \pi + i \sin 2,500 \, \pi \right) \tag{3.2-19b}$$

$$= 2^{5,000} \tag{3.2-19c}$$

●

To find the $n^{\text{th}}$ root of a complex number, we use the polar representation. Suppose we want to find the $n^{\text{th}}$ root of $z_1$. Let $z_0$ be the $n^{\text{th}}$ root. By definition

$$z_0^n = z_1 \tag{3.2-20}$$

Writing $z_0$ and $z_1$ as

$$z_0 = r_0 \left( \cos \theta_0 + i \sin \theta_0 \right) \tag{3.2-21a}$$

$$z_1 = r_1 (\cos \theta_1 + i \sin \theta_1) \qquad (3.2\text{-}21b)$$

Equation (3.2-20) becomes

$$r_0^n (\cos n\theta_0 + i \sin n\theta_0) = r_1 (\cos \theta_1 + i \sin \theta_1) \qquad (3.2\text{-}22)$$

Equating the real and imaginary parts, we obtain

$$r_1 \cos \theta_1 = r_0^n \cos n\theta_0 \qquad (3.2\text{-}23a)$$

$$r_1 \sin \theta_1 = r_0^n \sin n\theta_0 \qquad (3.2\text{-}23b)$$

The solution is

$$r_0 = r_1^{1/n}, \qquad \theta_0 = \frac{\theta_1}{n} + \frac{2\pi k}{n} \qquad (3.2\text{-}24a,b)$$

where $k$ is an integer.

Since, for any integer $k$ given by

$$k = pn + m, \qquad 0 \le m \le n \qquad (3.2\text{-}25)$$

where $m$ is a remainder, we obtain all the values of $\theta_0$ which produce the $n$ distinct values of $z_0$ by choosing $k = 0, 1, 2, 3, \ldots , n - 1$. Any other value of $k$ would yield a value of $\theta_0$ which differed from the one obtained earlier by a multiple of $2\pi$ radians.

Geometrically these roots lie on a circle of radius $r_1^{1/n}$ whose arguments differ by $\frac{2\pi}{n}$, with the first argument being $\frac{\theta_1}{n}$. Thus the number of $n^{th}$ roots of a complex number is $n$.

*Example 3.2-2*. Determine all the $n^{th}$ roots of unity.

Since

$$1 = \cos 0 + i \sin 0 \qquad (3.2\text{-}26)$$

we have

$$1^{1/n} = \cos \left(0 + \frac{2\pi k}{n}\right) + i \sin \left(0 + \frac{2\pi k}{n}\right) \qquad (3.2\text{-}27a)$$

$$= \cos \left(\frac{2\pi}{n} k\right) + i \sin \left(\frac{2\pi}{n} k\right) \qquad (3.2\text{-}27b)$$

Denoting the roots by $\rho_0, \rho_1, \ldots , \rho_{n-1}$, we obtain from Equation (3.2-27b)

$$\rho_0 = \cos 0 + i \sin 0 = 1 \tag{3.2-28a}$$

$$\rho_1 = \cos\left(\frac{2\pi}{n}\right) + i \sin\left(\frac{2\pi}{n}\right) \tag{3.2-28b}$$

$$\rho_2 = \cos\left(\frac{4\pi}{n}\right) + i \sin\left(\frac{4\pi}{n}\right) \tag{3.2-28c}$$

.
.
.

$$\rho_{n-1} = \cos\left[\frac{(n-1)\pi}{n}\right] + i \sin\left[\frac{(n-1)\pi}{n}\right] \tag{3.2-28d}$$

We list some of the properties of the conjugate of $z$

1.  $\overline{z_1 \pm z_2} = \bar{z}_1 \pm \bar{z}_2$  (3.2-29a)

2.  $\overline{z_1 \cdot z_2} = \bar{z}_1 \cdot \bar{z}_2$  (3.2-29b)

3.  $\left(\overline{\dfrac{1}{z_1}}\right) = \dfrac{1}{\bar{z}_1} \qquad (z_1 \neq 0)$  (3.2-29c)

4.  $\left(\overline{\overline{z}_1}\right) = z_1$  (3.2-29d)

and some of the properties of the absolute value of $z$

1.  $|z| = |\bar{z}|$  (3.2-30a)

2.  $z \cdot \bar{z} = |z|^2 = (\text{Re } z)^2 + (\text{Im } z)^2$  (3.2-30b)

3.  $\text{Re } z \leq |z|, \qquad |\text{Im } z| \leq |z|$  (3.2-30c)

4.  $|z_1 z_2| = |z_1| \, |z_2|$  (3.2-30d)

5.  $\left|\dfrac{1}{z_1}\right| = \dfrac{1}{|z_1|} \qquad (z_1 \neq 0)$  (3.2-30e)

6.  $|z_1 + z_2| \leq |z_1| + |z_2|$  (3.2-30f)

***Example 3.2-3.*** Show that $|z_1 + z_2| \leq |z_1| + |z_2|$.

This is the triangle inequality.

We have

$$|z_1 + z_2|^2 = (z_1 + z_2)(\overline{z_1 + z_2}) = (z_1 + z_2)(\bar{z}_1 + \bar{z}_2) \tag{3.2-31a,b}$$

$$= z_1 \bar{z}_1 + z_1 \bar{z}_2 + z_2 \bar{z}_1 + z_2 \bar{z}_2 \qquad (3.2\text{-}31\text{c})$$

Observe that

$$z_1 + \bar{z}_1 = 2\,\mathrm{Re}\,(z_1) \qquad (3.2\text{-}32)$$

Hence

$$(z_1 \bar{z}_2 + \bar{z}_1 z_2) = 2\,\mathrm{Re}\,(z_1 \bar{z}_2) \qquad (3.2\text{-}33)$$

It follows that

$$|z_1 + z_2|^2 = |z_1|^2 + 2\,\mathrm{Re}\,(z_1 \bar{z}_2) + |z_2|^2 \qquad (3.2\text{-}34\text{a})$$

$$\leq |z_1|^2 + 2\,|z_1 \bar{z}_2| + |z_2|^2 \qquad (3.2\text{-}34\text{b})$$

$$\leq |z_1|^2 + 2\,|z_1|\,|z_2| + |z_2|^2 \qquad (3.2\text{-}34\text{c})$$

$$\leq (|z_1| + |z_2|)^2 \qquad (3.2\text{-}34\text{d})$$

This implies

$$\big[\,|z_1 + z_2| - (|z_1| + |z_2|)\,\big]\,\big[\,|z_1 + z_2| + (|z_1| + |z_2|)\,\big] \leq 0 \qquad (3.2\text{-}35\text{a})$$

Hence

$$|z_1 + z_2| \leq (|z_1| + |z_2|) \qquad (3.2\text{-}35\text{b})$$

●

<u>Warning</u>: We emphasize that complex numbers are not ordered. Expressions such as $z_1 > z_2$ or $z_1 < z_3$ have no meaning unless $z_1$, $z_2$ and $z_3$ are all real.

## 3.3    COMPLEX FUNCTIONS

The concept of functions, limits, continuity and derivatives discussed in the calculus of real variables can be extended to complex variable calculus. Before considering these concepts, we first define **curves**, **regions** and **domains** in the complex plane.

Let $x(t)$ and $y(t)$ be two continuous functions of a real parameter t, defined for $a \leq t \leq b$, so that the equation

$$z = x(t) + i\,y(t) \qquad (3.3\text{-}1)$$

defines a curve C that joins the point

$$z(a) = x(a) + i\, y(a) \qquad\qquad (3.3\text{-}2a)$$

to the point

$$z(b) = x(b) + i\, y(b) \qquad\qquad (3.3\text{-}2b)$$

If $z(a)$ and $z(b)$ are equal, the curve C is a **closed curve**. The curve C is **simple** if it does not cross itself. For example

$$z = \cos t + i \sin t \qquad (0 \le t \le 2\pi) \qquad\qquad (3.3\text{-}3a)$$

is a simple closed curve. It is in fact the unit circle. It can also be expressed as

$$|z| = 1 \qquad\qquad (3.3\text{-}3b)$$

The set of all points z which satisfy the inequality

$$|z - z_0| < \delta \qquad\qquad (3.3\text{-}4)$$

is called the $\delta$ **neighborhood** of the point $z_0$. It consists of all points z lying inside but not on the circle of radius $\delta$ with the center at $z_0$. This is illustrated in Figure 3.3-1.

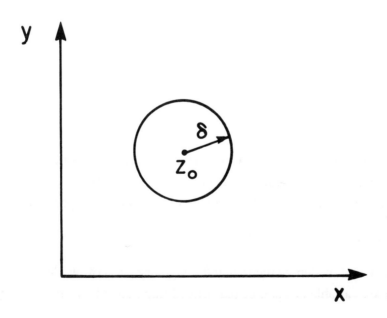

**FIGURE 3.3-1**    $\delta$ **neighborhood of** $z_0$

Similarly, the inequality $|z - z_0| > \delta$ represents the exterior of the circle.

A point $z_0$ is said to be an **interior point** of a set S whenever there is some neighborhood of $z_0$ that contains only points of S ; it is called an **exterior point** of S if there exists a neighborhood of $z_0$ which does not contain points of S. If $z_0$ is neither an interior point nor an exterior point of S, it is a **boundary point** of S.

If every point of a set S is an interior point of S, S is an **open set**. Clearly an open set contains none of its boundary points. If a set S contains all its boundary points , S is **closed**. The **closure** $\overline{S}$ of a set S is the closed set consisting of all points in S as well as the boundary of S.

An open set S is a **connected set** if every pair of points $z_1, z_2$ in S can be joined by a polygonal line that lies entirely in S. The open set $|z| < 1$ is connected and so is the annulus $1 < |z| < 2$. The set of all points in the plane that do not lie on $|z| = 1$ is an open set which is not connected. This is because we cannot join a point inside the unit circle to a point outside the unit circle without crossing the unit circle. The unit circle does not belong to the set.

An open set that is connected is a **domain**. A domain together with some, none or all of its boundary points is a **region**. A set that is formed by taking the union of a domain and its boundary is a **closed region**.

Finally, the set of all points z $(= x + i y)$ such that $y > 0$ is the **upper half plane**. Similarly, for $y < 0$, we have the **lower half plane**. The conditions $x > 0$ and $x < 0$ define the **right half plane** and the **left half plane** respectively.

We now define **functions of a complex variable**.

Let S be a set of complex numbers and let z, which varies in S, be a complex variable. A function f defined on a set S of complex numbers assigns to each z in S a unique complex number w. We write

$$w = f(z) \tag{3.3-5}$$

The number w is the **image of z under f**. The set S is the **domain** of definition of $f(z)$ and the set of all images $f(z)$ is the **range** of $f(z)$. Just as the variable z is decomposed into real and imaginary parts, w can be decomposed into a real and an imaginary part. We write

$$w = f(z) = u(x, y) + i v(x, y) \tag{3.3-6a,b}$$

The real functions u and v denote the real and imaginary parts of w. We note that a complex valued function of a complex variable is a pair of real valued functions of two real variables.

***Example 3.3-1***.  Express the function

$$w = f(z) = z^2 + 2z - 3\overline{z} \tag{3.3-7a,b}$$

in the form of Equation (3.3-6b). Then find the value of $f(1 + i)$.

$$w = (x + i y)^2 + 2 (x + i y) - 3 (x - i y) \tag{3.3-8a}$$

$$= [(x^2 - y^2) + 2x - 3x] + i [2xy + 2y + 3y] \tag{3.3-8b}$$

$$= (x^2 - x - y^2) + i (2xy + 5y) \tag{3.3-8c}$$

Comparing Equations (3.3-6b, 8c), we have

$$u = x^2 - x - y^2, \qquad v = 2xy + 5y \tag{3.3-9a,b}$$

$$f (1 + i) = u (1, 1) + i v (1, 1) = -1 + 7i \tag{3.3-10a,b}$$

●

Let $f(z)$ be defined in some neighborhood of $z_0$ with the possible exception of the point $z_0$ itself. We define the **limit of $f(z)$** as $z$ approaches $z_0$ to be a number $\ell$, if for any $\varepsilon > 0$, there is a positive number $\delta$ such that

$$\left| f(z) - \ell \right| < \varepsilon \quad \text{whenever} \quad 0 < \left| z - z_0 \right| < \delta \tag{3.3-11}$$

We adopt the same notation as in the case of a real variable and write

$$\lim_{z \to z_0} f(z) = \ell \tag{3.3-12a}$$

or $\qquad f(z) \longrightarrow \ell \quad \text{as} \quad z \longrightarrow z_0$ \hfill (3.3-12b)

We note that in the present case, $z$ may approach $z_0$ from any direction in the complex plane and the limit is independent of the direction.

An equivalent definition in terms of $u (x, y)$ and $v (x, y)$ is

$$\lim_{z \to z_0} f(z) = \ell = \ell_1 + i \, \ell_2 \tag{3.3-13}$$

Equating the real and imaginary parts, we have

$$\lim_{(x, y) \to (x_0, y_0)} u (x, y) = \ell_1 \tag{3.3-14a}$$

$$\lim_{(x, y) \to (x_0, y_0)} v (x, y) = \ell_2 \tag{3.3-14b}$$

If a limit exists, it is unique. The concept of limit is illustrated in Figure 3.3-2.

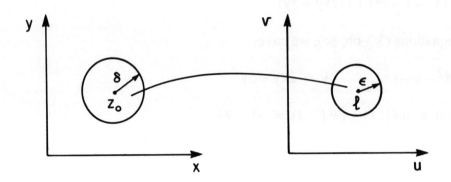

**FIGURE 3.3-2    Limit of f(z) defined by Equations (3.3-14a, b)**

Let $f(z)$ be a function of the complex variable $z$ defined for all values of $z$ in some neighborhood of $z_0$. The function $f(z)$ is **continuous** at $z_0$, if the following three conditions are satisfied

i)       $\lim\limits_{z \to z_0} f(z)$ exists                                                                      (3.3-15a)

ii)      $f(z_0)$ exists                                                                                           (3.3-15b)

iii)     $\lim\limits_{z \to z_0} f(z) = f(z_0)$                                                                (3.3-15c)

The **derivative** of $f(z)$ at $z_0$ written as $f'(z_0)$ is defined as

$$f'(z_0) = \lim_{z \to z_0} \frac{f(z) - f(z_0)}{z - z_0} \qquad (3.3\text{-}16)$$

provided the limit exists.

An alternative definition is to let

$$w = f(z) \qquad (3.3\text{-}17a)$$

$$\Delta w = f(z) - f(z_0), \qquad \Delta z = z - z_0 \qquad (3.3\text{-}17b,c)$$

$$f'(z) = \lim_{\Delta z \to 0} \frac{\Delta w}{\Delta z} = \lim_{\Delta z \to 0} \left[ \frac{\Delta u\,(x,\,y)}{\Delta z} + i\,\frac{\Delta v\,(x,\,y)}{\Delta z} \right] \qquad (3.3\text{-}17\text{d,e})$$

We also write $f'(z)$ as $\dfrac{dw}{dz}$.

If $f(z)$ is **differentiable** at $z_0$, it is continuous at $z_0$. As in the case of real variables, the converse is not true.

***Example 3.3-2.*** Show that the continuous function

$$f(z) = |z|^2 \qquad (3.3\text{-}18)$$

is not differentiable everywhere.

Let $z_0$ be any point in the complex plane, then using Equation (3.3-17d) we have

$$f'(z_0) = \lim_{\Delta z \to 0} \frac{|z|^2 - |z_0|^2}{\Delta z} \qquad (3.3\text{-}19\text{a})$$

$$= \lim_{\Delta z \to 0} \frac{z\,\bar{z} - z_0\,\bar{z}_0}{\Delta z} \qquad (3.3\text{-}19\text{b})$$

$$= \lim_{\Delta z \to 0} \frac{\left(z_0 + \Delta z\right)\left(\bar{z}_0 + \Delta \bar{z}\right) - z_0\,\bar{z}_0}{\Delta z} \qquad (3.3\text{-}19\text{c})$$

$$= \lim_{\Delta z \to 0} \left[ \bar{z}_0 + z_0\,\frac{\Delta \bar{z}}{\Delta z} \right] \qquad (3.3\text{-}19\text{d})$$

Equation (3.3-19c) is obtained by using Equations (3.3-17c, 2-29a).

If $z_0$ is the origin, $\bar{z}_0$ is also zero and $f'(0)$ exists and is zero. If $z_0$ is not the origin, we express $\Delta z$ in polar form and write

$$\Delta z = \Delta r\,(\cos \theta + i \sin \theta) \qquad (3.3\text{-}20\text{a})$$

$$\Delta \bar{z} = \Delta r\,(\cos \theta - i \sin \theta) \qquad (3.3\text{-}20\text{b})$$

Substituting Equations (3.3-20a, b) into Equation (3.3-19d), we obtain

$$f'(z_0) = \lim_{\Delta r \to 0} \left[ \bar{z}_0 + z_0\,\frac{(\cos \theta - i \sin \theta)}{(\cos \theta + i \sin \theta)} \right] \qquad (3.3\text{-}21\text{a})$$

$$= \left[ \bar{z}_0 + z_0 \frac{(\cos\theta - i\sin\theta)^2}{\cos^2\theta + \sin^2\theta} \right] \qquad\qquad (3.3\text{-}21b)$$

$$= \bar{z}_0 + z_0 (\cos 2\theta - i\sin 2\theta) \qquad\qquad (3.3\text{-}21c)$$

Thus $f'(z_0)$ depends on $\theta$ and is not unique. Also the limit does not exist. For example, if $z$ approaches $z_0$ along the real axis ($\theta = 0$), and along the imaginary axis ($\theta = \pi/2$), we obtain respectively from Equation (3.3-21c)

$$f'(z_0) = \bar{z}_0 + z_0 \qquad\qquad (3.3\text{-}22a)$$

$$f'(z_0) = \bar{z}_0 - z_0 \qquad\qquad (3.3\text{-}22b)$$

The function $f$ is not differentiable everywhere except at the origin where $f'(0)$ is zero irrespective of $\theta$.

●

All the familiar rules, such as the rules for differentiating a constant, integer power of $z$, sum, difference, product and quotient of differentiable functions as well as the chain rule of differential calculus of real variables hold in the case of complex variables.

Functions which are differentiable at a single point are not of great interest. We thus define a broad class of functions. A function $f(z)$ is analytic in a domain D, if $f(z)$ is defined and has a derivative at every point in D. The function $f(z)$ is **analytic** at $z_0$ if its derivative exists at each point $z$ in some neighborhood of $z_0$. Synonyms for analytic are **regular and holomorphic**.

If $f(z)$ is analytic for all finite values of $z$, $f(z)$ is an **entire function**. Points where $f(z)$ ceases to be analytic are singular points.

The basic criterion for analyticity of a complex function $f(z)$ is given by the **Cauchy-Riemann** conditions.

**Theorem 1**

A necessary condition for the function $f(z)$ to be analytic in a domain D is that the four partial derivatives $\frac{\partial u}{\partial x}$, $\frac{\partial u}{\partial y}$, $\frac{\partial v}{\partial x}$ and $\frac{\partial v}{\partial y}$ exist and satisfy the equations

$$\frac{\partial u}{\partial x} = \frac{\partial v}{\partial y}, \qquad \frac{\partial u}{\partial y} = -\frac{\partial v}{\partial x} \qquad\qquad (3.3\text{-}23a,b)$$

Proof: Since $f(z)$ is differentiable at any point $z_0$ in D, then $f'(z_0)$ must exist.

This implies that the ratio $[f(z) - f(z_0)] / (z - z_0)$ must tend to a definite limit as $z \longrightarrow z_0$, irrespective of the path taken. We choose to approach $z_0$ along a line parallel to the real axis (y = constant), shown as path 1 in Figure 3.3-3. The increment $\Delta z$ is simplified to

$$\Delta z = z - z_0 = x - x_0 = \Delta x \qquad\qquad (3.3\text{-}24a,b,c)$$

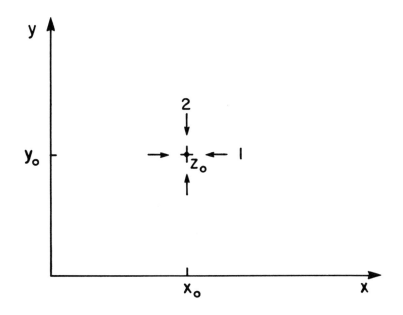

**FIGURE 3.3-3    Two paths approaching $z_0$**

Using Equation (3.3-17e), $f'(z_0)$ is given by

$$f'(z_0) = \lim_{\Delta x \to 0} \frac{u(x_0 + \Delta x, y_0) - u(x_0, y_0)}{\Delta x} + i \lim_{\Delta x \to 0} \frac{v(x_0 + \Delta x, y_0) - v(x_0, y_0)}{\Delta x}$$

$$(3.3\text{-}25)$$

Since $f'(z_0)$ exists, the two limits on the right side of Equation (3.3-25) exist. They are in fact the partial derivatives of u and v with respect to x. Equation (3.3-25) can be written as

$$f'(z_0) = \frac{\partial u}{\partial x} + i \frac{\partial v}{\partial x} \qquad\qquad (3.3\text{-}26)$$

Next we approach $z_0$ along a line parallel to the imaginary axis (x = constant), shown as path 2 in Figure 3.3-3. In this case

$$\Delta z = i(y - y_0) = i \Delta y \qquad\qquad (3.3\text{-}27a,b)$$

The analogues of Equations (3.3-25, 26) are

$$f'(z_0) = \lim_{\Delta y \to 0} \frac{u(x_0, y_0 + \Delta y) - u(x_0, y_0)}{i\,\Delta y} + i \lim_{\Delta y \to 0} \frac{v(x_0, y_0 + \Delta y) - v(x_0, y_0)}{i\,\Delta y}$$

(3.3-28a)

$$= -i\frac{\partial u}{\partial y} + \frac{\partial v}{\partial y}$$

(3.3-28b)

Comparing the real and imaginary parts of Equations (3.3-26, 28b), we obtain

$$\frac{\partial u}{\partial x} = \frac{\partial v}{\partial y}$$

(3.3-29a)

$$\frac{\partial u}{\partial y} = -\frac{\partial v}{\partial x}$$

(3.3-29b)

These are the **Cauchy-Riemann conditions**. They are the necessary but not sufficient conditions for a function to be analytic. This is shown in Example 3.3-3.

Note that $f'(z_0)$ is given by Equations (3.3-26 or 28b).

***Example 3.3-3.*** Show that the function

$$f(z) = \begin{cases} \dfrac{x^3 - y^3}{x^2 + y^2} + i\left(\dfrac{x^3 + y^3}{x^2 + y^2}\right), & z \neq 0 \\[3mm] 0, & z = 0 \end{cases}$$

(3.3-30a,b)

satisfies the Cauchy-Riemann conditions at the origin, but $f'(0)$ does not exist.

From the definition of partial derivatives, we have

$$\frac{\partial u}{\partial x}(0, 0) = \lim_{x \to 0} \frac{u(x, 0) - u(0, 0)}{x} = \lim_{x \to 0} \frac{(x^3/x^2) - 0}{x} = 1$$

(3.3-31a,b,c)

$$\frac{\partial u}{\partial y}(0, 0) = \lim_{y \to 0} \frac{u(0, y) - u(0, 0)}{y} = \lim_{y \to 0} \frac{(-y^3/y^2) - 0}{y} = -1$$

(3.3-31d,e,f)

Similarly

$$\frac{\partial v}{\partial x}(0, 0) = 1, \qquad \frac{\partial v}{\partial y}(0, 0) = 1$$

(3.3-31g,h)

Hence, the Cauchy-Riemann conditions are satisfied at (0, 0). We now show that $f'(z)$ does not exist at (0, 0).

Let z vary along the line

$$y = x \qquad\qquad (3.3\text{-}32)$$

$$f(z) = u + i\,v = 0 + i\,x \qquad\qquad (3.3\text{-}33a,b)$$

$$f'(0) = \lim_{z \to 0} \frac{f(z) - f(0)}{z - 0} = \lim_{x \to 0} \frac{i\,x}{x + i\,x} = \frac{i}{1+i} = \frac{i(1-i)}{(1+i)(1-i)} = \frac{1}{2} + \frac{1}{2}\,i \qquad (3.3\text{-}34a\text{-}e)$$

However, if the origin is approached along the x-axis (y = 0), we can write using Equation (3.3-26)

$$f'(0) = \frac{\partial u}{\partial x}(0, 0) + i\,\frac{\partial v}{\partial x}(0, 0) = 1 + i \qquad\qquad (3.3\text{-}35a,b)$$

Since the two limits are different, $f'(0)$ does not exist.

## Theorem 2

A function $f(z)$ is analytic in a domain D, provided the four first partial derivatives of $u(x, y)$ and $v(x, y)$ exist, satisfy the Cauchy-Riemann equations at each point of D and are continuous.

Note that, in theorem 2, we require the partial derivatives to be continuous.

The Cauchy-Riemann conditions can also be written in polar form. From Equations (3.2-6a, b) we obtain

$$\frac{\partial r}{\partial x} = \cos\theta, \qquad\qquad \frac{\partial r}{\partial y} = \sin\theta \qquad\qquad (3.3\text{-}36a,b)$$

$$\frac{\partial \theta}{\partial x} = -\frac{\sin\theta}{r}, \qquad\qquad \frac{\partial \theta}{\partial y} = \frac{\cos\theta}{r} \qquad\qquad (3.3\text{-}36c,d)$$

Using the chain rule, we have

$$\frac{\partial u}{\partial x} = \frac{\partial u}{\partial r}\frac{\partial r}{\partial x} + \frac{\partial u}{\partial \theta}\frac{\partial \theta}{\partial x} = \cos\theta\,\frac{\partial u}{\partial r} - \frac{\sin\theta}{r}\frac{\partial u}{\partial \theta} \qquad\qquad (3.3\text{-}37a,b)$$

$$\frac{\partial u}{\partial y} = \sin\theta\,\frac{\partial u}{\partial r} + \frac{\cos\theta}{r}\frac{\partial u}{\partial \theta} \qquad\qquad (3.3\text{-}37c)$$

$$\frac{\partial v}{\partial x} = \cos \theta \frac{\partial v}{\partial r} - \frac{\sin \theta}{r} \frac{\partial v}{\partial \theta} \qquad (3.3\text{-}37d)$$

$$\frac{\partial v}{\partial y} = \sin \theta \frac{\partial v}{\partial r} + \frac{\cos \theta}{r} \frac{\partial v}{\partial \theta} \qquad (3.3\text{-}37e)$$

Combining Equations (3.3-23a, b, 37b to e) yields

$$\cos \theta \frac{\partial u}{\partial r} - \frac{\sin \theta}{r} \frac{\partial u}{\partial \theta} = \sin \theta \frac{\partial v}{\partial r} + \frac{\cos \theta}{r} \frac{\partial v}{\partial \theta} \qquad (3.3\text{-}38a)$$

$$\sin \theta \frac{\partial u}{\partial r} + \frac{\cos \theta}{r} \frac{\partial u}{\partial \theta} = - \cos \theta \frac{\partial v}{\partial r} + \frac{\sin \theta}{r} \frac{\partial v}{\partial \theta} \qquad (3.3\text{-}38b)$$

Multiplying Equations (3.3-38a, b) by $\cos \theta$ and $\sin \theta$ respectively, and adding the resulting expressions yields

$$\frac{\partial u}{\partial r} = \frac{1}{r} \frac{\partial v}{\partial \theta} \qquad (3.3\text{-}39a)$$

Similarly, eliminating $\dfrac{\partial u}{\partial r}$ and $\dfrac{\partial v}{\partial \theta}$ from Equations (3.3-38a, b), we obtain

$$\frac{\partial v}{\partial r} = - \frac{1}{r} \frac{\partial u}{\partial \theta} \qquad (3.3\text{-}39b)$$

Equations (3.3-39a, b) are the **Cauchy-Riemann conditions in polar form**.

Combining Equations (3.3-26, 37b, d, 39a, b) yields

$$f'(z) = \frac{dw}{dz} = \left( \cos \theta - i \sin \theta \right) \frac{\partial w}{\partial r} \qquad (3.3\text{-}40a,b)$$

***Example 3.3-4***. Show that the function $e^x (\cos y + i \sin y)$ is analytic and determine its derivative.

We have

$$f(z) = u + i v = e^x \cos y + i e^x \sin y \qquad (3.3\text{-}41a,b)$$

$$\frac{\partial u}{\partial x} = e^x \cos y, \qquad \frac{\partial u}{\partial y} = - e^x \sin y \qquad (3.3\text{-}41c,d)$$

$$\frac{\partial v}{\partial x} = e^x \sin y, \qquad \frac{\partial v}{\partial y} = e^x \cos y \qquad (3.3\text{-}41e,f)$$

The functions u, v, and their partial derivatives satisfy the Cauchy-Riemann equations and are continuous functions of x and y. Hence f(z) is analytic.

The derivative

$$f'(z) = \frac{\partial u}{\partial x} + i \frac{\partial v}{\partial x} = e^x \cos y + i e^x \sin y \qquad (3.3\text{-}42a,b)$$

$$= e^x (\cos y + i \sin y) \qquad (3.3\text{-}42c)$$

which is identical to the given function. We will see in Section 3.4 that this function is $e^z$.

●

Let f(z) be an analytic function. Assume that the mixed second derivatives of u and v exist and are equal. From Equations (3.3-29a, b), we have

$$\frac{\partial^2 v}{\partial x \, \partial y} = \frac{\partial^2 u}{\partial x^2} = -\frac{\partial^2 u}{\partial y^2} \qquad (3.3\text{-}43a,b)$$

$$\frac{\partial^2 u}{\partial x \, \partial y} = \frac{\partial^2 v}{\partial y^2} = -\frac{\partial^2 v}{\partial x^2} \qquad (3.3\text{-}43c,d)$$

From Equations (3.3-43b, d), we deduce that both u and v satisfy the equations

$$\frac{\partial^2 u}{\partial x^2} + \frac{\partial^2 u}{\partial y^2} = 0 \qquad (3.3\text{-}44a)$$

$$\frac{\partial^2 v}{\partial x^2} + \frac{\partial^2 v}{\partial y^2} = 0 \qquad (3.3\text{-}44b)$$

Equations (3.3-44a, b) are **Laplace's equations** which will be solved in Chapter 5. The solution of Laplace's equation is a **harmonic function**. Both the real and imaginary parts of an analytic function are harmonic functions. The functions u and v of the analytic function are also **conjugate functions** (harmonic conjugates).

If a harmonic function u (x, y) is given in some domain D, we can determine the harmonic conjugate v (x, y) through the Cauchy-Riemann relations. The analytic function f(z) can then be determined.

*Example 3.3-5*. Show that

$$u (x, y) = x^3 - 3 x y^2 \qquad (3.3\text{-}45)$$

is a harmonic function.  Find the harmonic conjugate function  $v\,(x, y)$  and the analytic function  $f(z)$.

Taking the partial derivatives of  u,  we obtain

$$\frac{\partial u}{\partial x} \; = \; 3x^2 - 3y^2, \qquad\qquad \frac{\partial u}{\partial y} \; = \; -6xy \qquad\qquad\qquad (3.3\text{-}46a,b)$$

$$\frac{\partial^2 u}{\partial x^2} \; = \; 6x\,, \qquad\qquad \frac{\partial^2 u}{\partial y^2} \; = \; -6x \qquad\qquad\qquad (3.3\text{-}46c,d)$$

From Equations (3.3-46c, d), we deduce that  u  satisfies Laplace's equation, and  u  is harmonic.

From Equation (3.3-29a), we obtain

$$v \; = \; \int \left(3x^2 - 3y^2\right) dy \; + \; g(x) \qquad\qquad\qquad (3.3\text{-}47a)$$

$$= \; 3x^2 y - y^3 + g(x) \qquad\qquad\qquad (3.3\text{-}47b)$$

where  $g(x)$  is an arbitrary function.

The function  v  also has to satisfy Equation (3.3-29b) and this implies

$$-6xy \; = \; -\left[6xy + g'(x)\right] \qquad\qquad\qquad (3.3\text{-}48)$$

From Equation (3.3-48), we deduce that  $g'(x)$  is zero and  g  is a constant  c.  The function  v  can then be written as

$$v \; = \; 3x^2 y - y^3 + c \qquad\qquad\qquad (3.3\text{-}49)$$

The function  f  is given by

$$f \; = \; \left(x^3 - 3xy^2\right) + i\left(3x^2 y - y^3\right) + c \qquad\qquad\qquad (3.3\text{-}50a)$$

$$= \; (x + i\,y)^3 + c \; = \; z^3 + c \qquad\qquad\qquad (3.3\text{-}50b,c)$$

●

Laplace's equation can be written in terms of  z  and  $\bar{z}$.  From Equation (3.2-1), we deduce

$$x \; = \; \left(\frac{z + \bar{z}}{2}\right), \qquad\qquad y \; = \; \left(\frac{z - \bar{z}}{2i}\right) \qquad\qquad\qquad (3.3\text{-}51a,b)$$

Therefore

$$\frac{\partial}{\partial z} = \frac{\partial x}{\partial z}\frac{\partial}{\partial x} + \frac{\partial y}{\partial z}\frac{\partial}{\partial y} = \frac{1}{2}\left(\frac{\partial}{\partial x} - i\frac{\partial}{\partial y}\right) \qquad (3.3\text{-}52a,b)$$

$$\frac{\partial}{\partial \bar{z}} = \frac{\partial x}{\partial \bar{z}}\frac{\partial}{\partial x} + \frac{\partial y}{\partial \bar{z}}\frac{\partial}{\partial y} = \frac{1}{2}\left(\frac{\partial}{\partial x} + i\frac{\partial}{\partial y}\right) \qquad (3.3\text{-}53a,b)$$

$$\frac{\partial}{\partial z}\bullet\frac{\partial}{\partial \bar{z}} = \frac{1}{4}\left(\frac{\partial}{\partial x} - i\frac{\partial}{\partial y}\right)\left(\frac{\partial}{\partial x} + i\frac{\partial}{\partial y}\right) \qquad (3.3\text{-}54)$$

$$4\frac{\partial}{\partial z}\bullet\frac{\partial}{\partial \bar{z}} = \left(\frac{\partial^2}{\partial x^2} + \frac{\partial^2}{\partial y^2}\right) \qquad (3.3\text{-}55)$$

That is

$$\frac{\partial^2 \varphi}{\partial x^2} + \frac{\partial^2 \varphi}{\partial y^2} = 4\frac{\partial^2 \varphi}{\partial z\,\partial \bar{z}} \qquad (3.3\text{-}56)$$

If $\varphi(x, y)$ satisfies Laplace's equation, it follows that

$$\frac{\partial^2 \varphi}{\partial z\,\partial \bar{z}} = 0 \qquad (3.3\text{-}57)$$

and the general solution is

$$\varphi = f(z) + g(\bar{z}) \qquad (3.3\text{-}58)$$

This solution is often used in two dimensional physical problems.

***Example 3.3-6.*** Obtain the **complex potential** $\Phi(z)$ for a two dimensional irrotational flow of an incompressible fluid.

We choose the rectangular Cartesian coordinate system and let $(v_x, v_y)$ be the velocity components. The flow is **irrotational** and this implies that the **vorticity** (curl $\underline{v}$) is zero. This can be expressed as

$$\frac{\partial v_x}{\partial y} - \frac{\partial v_y}{\partial x} = 0 \qquad (3.3\text{-}59)$$

From Appendix I, we obtain the equation of continuity for an incompressible flow as follows

$$\frac{\partial v_x}{\partial x} + \frac{\partial v_y}{\partial y} = 0 \qquad (3.3\text{-}60)$$

From Equations (3.3-59, 60), we deduce

$$\frac{\partial v_x}{\partial y} = \frac{\partial v_y}{\partial x} \qquad\qquad\qquad\qquad (3.3\text{-}61a)$$

$$\frac{\partial v_x}{\partial x} = -\frac{\partial v_y}{\partial y} \qquad\qquad\qquad\qquad (3.3\text{-}61b)$$

Substituting $v_x$ for u and $v_y$ for $(-v)$, Equations (3.3-61a, b) are identical to Equations (3.3-29a, b). That is, the components $v_x$ and $v_y$ satisfy the Cauchy-Riemann relations.

We introduce two functions $\phi(x, y)$ and $\psi(x, y)$ such that

$$v_x = \frac{\partial \phi}{\partial x} = \frac{\partial \psi}{\partial y} \qquad\qquad\qquad\qquad (3.3\text{-}62a,b)$$

$$v_y = \frac{\partial \phi}{\partial y} = -\frac{\partial \psi}{\partial x} \qquad\qquad\qquad\qquad (3.3\text{-}62c,d)$$

From Equations (3.3-62a to d), we note that $\phi$ and $\psi$ satisfy the Cauchy-Riemann relations and are therefore harmonic functions. This can also be verified by combining Equations (3.3-59, 60, 62a to d).

The function $\phi$ is the **potential** and $\psi$ is the **stream function**. They are conjugate functions. The combination $\phi + i\psi$ is an analytic function and the **complex potential** is given by

$$\Phi(z) = \phi + i\psi \qquad\qquad\qquad\qquad (3.3\text{-}63)$$

Differentiating $\Phi(z)$ with respect to z and using Equations (3.3-26, 62a to d), we have

$$\frac{d\Phi}{dz} = \frac{\partial \phi}{\partial x} + i\frac{\partial \psi}{\partial x} = v_x - iv_y \qquad\qquad\qquad (3.3\text{-}64a,b)$$

Similarly using Equation (3.3-28b, 62a to d), we have

$$\frac{d\Phi}{dz} = -i\frac{\partial \phi}{\partial y} + \frac{\partial \psi}{\partial y} = v_x - iv_y \qquad\qquad\qquad (3.3\text{-}65a,b)$$

Using the complex function $\Phi$, we can obtain both the potential $(\mathrm{Re}\,\Phi)$ and the stream function $(\mathrm{Im}\,\Phi)$. The derivative of $\Phi$ yields both velocity components.

●

The concept of a complex potential is widely used in hydrodynamics, and will be discussed further in this chapter. Applications of complex potential in electrostatics are given in Ferraro (1956).

## 3.4  ELEMENTARY FUNCTIONS

The definition of many elementary functions, such as polynomial, exponential and logarithmic, can be extended to complex variables.  It is usually defined such that for real values of the independent variable z, the functions become identical to the functions considered in the calculus of real variables.  However, the complex functions may have properties which the real functions do not possess.  One such property is the possibility of multiple values mentioned in connection with Equation (3.2-8a, b).

A **polynomial** function $P_n(z)$ of degree  n  is defined as

$$P_n(z) = a_n z^n + a_{n-1} z^{n-1} + \ldots + a_1 z + a_0 \tag{3.4-1}$$

where  $a_n \neq 0, a_{n-1}, \ldots, a_0$  are all complex constants.  Similarly, a function

$$w(z) = P(z) / Q(z) \tag{3.4-2}$$

in which  $P(z)$  and  $Q(z)$  are polynomials, defines a **rational algebraic function**.

The **exponential** function is denoted by  $e^z$  [or $\exp(z)$]  and is defined as

$$e^z = 1 + z + \frac{z^2}{2!} + \ldots + \frac{z^n}{n!} + \ldots \tag{3.3-3a}$$

$$= \sum_{n=0}^{\infty} \frac{z^n}{n!} \tag{3.4-3b}$$

Setting the real part of  z  to zero, we have

$$e^{iy} = \sum_{n=0}^{\infty} \frac{(iy)^n}{n!} = \sum_{m=0}^{\infty} (-1)^m \frac{y^{2m}}{(2m)!} + i \sum_{m=0}^{\infty} (-1)^m \frac{y^{2m+1}}{(2m+1)!} \tag{3.4-4a,b}$$

$$= \cos y + i \sin y \tag{3.4-4c}$$

Equation (3.4-4c) is **Euler's formula**.

The exponential function can be written as

$$e^z = e^{x+iy} = e^x (\cos y + i \sin y) \tag{3.4-5a,b}$$

For real  z  (y = 0),  $e^z$  reduces to $e^x$.

The moduli

$$\left| e^{iy} \right| = \left| \cos y + i \sin y \right| = 1 \tag{3.4-6a,b}$$

$$\left| e^z \right| = \left| e^x \right| \left| e^{iy} \right| = e^x \tag{3.4-6c,d}$$

The modulus of $e^z$ is $e^x$ and the argument of $e^z$ is $y$.

If the value of $y$ is increased by $2k\pi$ (k is an integer), the values of $\sin y$ and $\cos y$ remain unchanged and $e^z$ is periodic with period $2\pi$. That is to say,

$$e^z = e^{z + 2i\,k\pi} \tag{3.4-7}$$

Because of the periodicity, we need to consider only the strip

$$-\pi < y \le \pi \tag{3.4-8}$$

Some of the properties of $e^z$ are:

(a)     $e^z$ is analytic.

(b)     $e^z = 1$   implies   $z = 2n\pi i$ (n is an integer) $\tag{3.4-9a,b}$

(c)     $e^{-z} = \dfrac{1}{e^z}$     $(e^z \ne 0)$ $\tag{3.4-9c}$

(d)     $\dfrac{d}{dz}\left(e^z\right) = e^z$ $\tag{3.4-9d}$

(e)     If $z_1\,(= x_1 + i\,y_1)$ and $z_2\,(= x_2 + i\,y_2)$ are two complex numbers

$$e^{z_1} \bullet e^{z_2} = e^{z_1 + z_2} \tag{3.4-9e}$$

(f)     $e^{z_1} = e^{z_2}$   implies   $z_1 - z_2 = 2n\pi i$ (n is an integer) $\tag{3.4-9f,g}$

(g)     If $w$ is an analytic function of $z$

$$\frac{d}{dz}\left(e^w\right) = e^w \frac{dw}{dz} \tag{3.4-9h}$$

***Example 3.4-1.*** Show that $e^{\bar{z}}$ is not an analytic function of $z$ in any domain.

We have

$$e^{\bar{z}} = e^{x - iy} = e^x \bullet e^{-iy} = e^x (\cos y - i \sin y) = u + i\,v \tag{3.4-10a,b,c,d}$$

$$u = e^x \cos y, \qquad\qquad v = -e^x \sin y \tag{3.4-10e,f}$$

$$\frac{\partial u}{\partial x} = e^x \cos y, \qquad\qquad \frac{\partial u}{\partial y} = -e^x \sin y \tag{3.4-10g,h}$$

$$\frac{\partial v}{\partial x} = -e^x \sin y, \qquad\qquad \frac{\partial v}{\partial y} = -e^x \cos y \qquad\qquad\qquad (3.4\text{-}10\text{i,j})$$

It can be seen that Equations (3.4-10g to j) do not satisfy Equations (3.3-29a, b) for any finite values of $x$. Thus the Cauchy-Riemann relations are not satisfied and $e^{\bar{z}}$ is not analytic.

●

The trigonometric functions are defined as

$$\sin z = \frac{e^{iz} - e^{-iz}}{2i}, \qquad\qquad \cos z = \frac{e^{iz} + e^{-iz}}{2} \qquad\qquad (3.4\text{-}11\text{a,b})$$

$$\tan z = \frac{\sin z}{\cos z}, \qquad\qquad \cot z = \frac{\cos z}{\sin z} \qquad\qquad\qquad (3.4\text{-}11\text{c,d})$$

$$\sec z = \frac{1}{\cos z}, \qquad\qquad \operatorname{cosec} z = \frac{1}{\sin z} \qquad\qquad\qquad (3.4\text{-}11\text{e,f})$$

Since $e^z$ is analytic for all $z$, $\sin z$ and $\cos z$ are also analytic for all $z$. The complex trigonometric functions have the same properties as the real functions. We list some of them:

(a)   $\sin z$ and $\cos z$ are periodic with period $2\pi$

(b)   $\dfrac{d}{dz} (\sin z) = \cos z$ $\qquad\qquad\qquad\qquad\qquad\qquad\qquad\qquad$ (3.4-12a)

(c)   $\dfrac{d}{dz} (\cos z) = -\sin z$ $\qquad\qquad\qquad\qquad\qquad\qquad\qquad$ (3.4-12b)

(d)   $\cos z$ is an even function $[\cos (-z) = \cos z]$ and $\sin z$ is an odd function $[\sin (-z) = -\sin z]$

(e)   $e^{iz} = \cos z + i \sin z$ $\qquad\qquad\qquad\qquad\qquad\qquad\qquad\qquad$ (3.4-12c)

   Euler's formula holds for complex variables

(f)   $\sin^2 z + \cos^2 z = 1$ $\qquad\qquad\qquad\qquad\qquad\qquad\qquad\qquad$ (3.4-12d)

(g)   $\sin (z_1 \pm z_2) = \sin z_1 \cos z_2 \pm \cos z_1 \sin z_2$ $\qquad\qquad\qquad$ (3.4-12e)

(h)   $\cos (z_1 \pm z_2) = \cos z_1 \cos z_2 \mp \sin z_1 \sin z_2$ $\qquad\qquad\qquad$ (3.4-12f)

(i)   $\sin \left[ (2n+1) \dfrac{\pi}{2} - z \right] = \cos z$ ($n$ is an integer) $\qquad\qquad$ (3.4-12g)

(j)   $\sin (x + i y) = \sin x \cosh y + i \cos x \sinh y$ $\qquad\qquad\qquad$ (3.4-12h)

(k)   $\cos (x + i y) = \cos x \cosh y - i \sin x \sinh y$ $\qquad\qquad\qquad$ (3.4-12i)

(l)     The zeros of $\sin z$ and $\cos z$ are respectively

$$z = n\pi, \quad z = (2n+1)\frac{\pi}{2} \quad \text{(n is an integer)} \tag{3.4-12 j,k}$$

**Hyperbolic sine (sinh z) and cosine (cosh z)** of $z$ are defined as

$$\sinh z = \frac{e^z - e^{-z}}{2}, \qquad \cosh z = \frac{e^z + e^{-z}}{2} \tag{3.4-13a,b}$$

The other hyperbolic functions are

$$\tanh z = \frac{\sinh z}{\cosh z}, \qquad \coth z = \frac{\cosh z}{\sinh z} \tag{3.4-13c,d}$$

$$\text{sech } z = \frac{1}{\cosh z}, \qquad \text{cosech } z = \frac{1}{\sinh z} \tag{3.4-13e,f}$$

The complex hyperbolic functions have the same properties as the real functions. Some of these are:

(a)     They are analytic

(b)     $\dfrac{d}{dz}(\sinh z) = \cosh z$        (3.4-14a)

(c)     $\dfrac{d}{dz}(\cosh z) = \sinh z$        (3.4-14b)

(d)     $\cosh z$ is an even function, $\sinh z$ is an odd function

(e)     $\cosh^2 z - \sinh^2 z = 1$        (3.4-14c)

(f)     $\sinh(x + iy) = \cos y \sinh x + i \sin y \cosh x$        (3.4-14d)

(g)     $\cosh(x + iy) = \cos y \cosh x + i \sin y \sinh x$        (3.4-14e)

(h)     $\sinh(iz) = i \sin z, \quad i \sinh z = \sin(iz)$        (3.4-14f,g)

(i)     $\cosh(iz) = \cos z, \quad \cosh z = \cos(iz)$        (3.4-14h,i)

(j)     $\sinh\left(\frac{i\pi}{2} - z\right) = i \cosh z$        (3.4-14j)

(k)     The zeros of $\sinh z$ and $\cosh z$ are respectively

$$z = n\pi i, \quad z = (2n+1)\frac{\pi i}{2} \quad \text{(n is an integer)} \tag{3.4-14k,l}$$

We recall that for real variables, if $x$ is any positive real number and

$$x = e^u \qquad (3.4\text{-}15)$$

then

$$u = \ell n \, x \qquad (3.4\text{-}16)$$

We now extend the definition of $\ell n$ to the complex variables and write

$$z = e^w \qquad (3.4\text{-}17)$$

It follows that we can define $\ell n \, z$ as

$$w = \ell n \, z \qquad (3.4\text{-}18)$$

Separating $w$ into its real and imaginary parts $(w = u + i \, v)$, Equation (3.4-17) becomes

$$z = e^u \, (\cos v + i \sin v) \qquad (3.4\text{-}19)$$

From Equation (3.4-19), we obtain

$$|z| = e^u \qquad (3.4\text{-}20a)$$

$$\arg z = v \qquad (3.4\text{-}20b)$$

Combining Equations (3.4-18, 20a, b) yields

$$\ell n \, z = \ell n \, |z| + i \arg z \qquad (3.4\text{-}21)$$

Since $\arg z$ can differ by multiples of $2\pi$, we restrict the definition to the **principal values** of $\arg z$ (Arg z). We then have the principal value of $\ell n \, z$ and we denote it by Ln z. That is to say, Ln z is defined as

$$Ln \, z = \ell n \, |z| + i \arg z \qquad (3.4\text{-}22a)$$

$$-\pi < \arg z \le \pi \qquad (3.4\text{-}22b)$$

The function $\ell n \, z$ as defined by Equation (3.4-21) is a multiple-valued function and since the $\arg z$ can differ by multiples of $2\pi$, we can write

$$\ell n \, z = Ln \, z \pm 2n\pi i \qquad (n \text{ is an integer}) \qquad (3.4\text{-}23)$$

If $z$ is a real positive number, then Arg z is zero and the definition of Ln z is identical to the $\ell n$ in the theory of real variables.

Some of the properties of $\ell n \, z$ (Ln z) are:

(a)     $\ell n\ 1\ =\ 2n\pi i$                                                                                 (3.4-24a)

        $\ell n\ (-1)\ =\ (2n+1)\ \pi i$      (n  is an integer)                                    (3.4-24b)

(b)     $Ln\ i\ =\ i\ \dfrac{\pi}{2}$                                                                   (3.4-24c)

(c)     $Ln\ (-1-i)\ =\ \dfrac{1}{2}\ Ln\ 2 - i\ \dfrac{3\ \pi}{4}$                          (3.4-24d)

(d)     $\ell n\ (z_1\ z_2)\ =\ \ell n\ z_1 + \ell n\ z_2$                                        (3.4-24e)

(e)     $\ell n\ (e^z)\ =\ z + 2n\pi i\ ,\quad Ln\ (e^z)\ =\ z$                              (3.4-24f,g)

(f)     $e^{\ell n\ z}\ =\ z$                                                                                (3.4-24h)

(g)     $\ell n\left(\dfrac{z_1}{z_2}\right)\ =\ \ell n\ z_1 - \ell n\ z_2$                    (3.4-24i)

Note that  Ln z  is not defined at the origin  (|z| = 0).  The negative real axis is a line of discontinuity since the imaginary part of  Ln z  has a jump discontinuity of  $2\pi$  on crossing that line.  We make a cut, as shown in Figure 3.4-1, in the complex plane to remove the origin and the negative real axis.  In the resulting domain  D,  Ln z  is analytic.  The derivative of  Ln z  is given by

$$\dfrac{d}{dz}\ (Ln\ z)\ =\ \dfrac{1}{z}$$                                                          (3.4-25a)

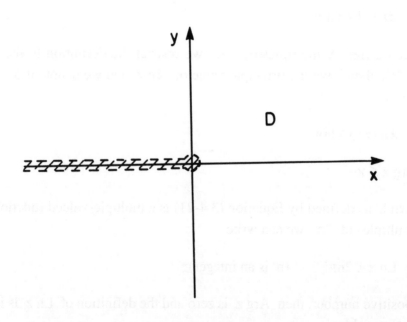

**FIGURE 3.4-1    Domain D in which Ln z is analytic**

Since $\ell n\, z$ and $Ln\, z$ differ only by an arbitrary multiple by $2\pi i$,

$$\frac{d}{dz}\,(\ell n\, z) = \frac{1}{z} \tag{3.5-25b}$$

If $\alpha$ is any complex number, we define $z^\alpha$ to be

$$z^\alpha = \exp\,(\alpha\,\ell n\, z) \tag{3.4-26}$$

Similarly the function $\alpha^z$ is defined as

$$\alpha^z = \exp\,(z\,\ell n\,\alpha) \tag{3.4-27}$$

In general $z^\alpha$ and $\alpha^z$ are multiple-valued functions since $\ell n\, z$ is a multiple-valued function. The principal values of $z^\alpha$ and $\alpha^z$ are obtained by substituting $\ell n\, z$ by $Ln\, z$ in Equations (3.4-26, 27).

**Example 3.4-2.** Find the values of    (i) $\ell n\, e$,    (ii) $Ln\,(1-i)$,    (iii) $(-i)^i$.

(i)     $\ell n\, e = \ell n\,|e| + i\,\arg e$                                      (3.4-28a)

$$= 1 + i\,(0 + 2n\pi) = 1 + 2n\pi i \tag{3.4-28b}$$

(ii)     $Ln\,(1-i) = \ell n\,|1-i| + i\,Arg\,(1-i)$           (3.4-29a)

$$= \frac{1}{2}\,\ell n\, 2 - \frac{\pi}{4}\,i \tag{3.4-29b}$$

(iii)     Taking the principal value, we have

$$(-i)^i = e^{i\,Ln\,(-i)} \tag{3.4-30a}$$

$$= e^{i\,(-i\pi/2)} = e^{\pi/2} \tag{3.4-30b,c}$$

**Example 3.4-3.** If $\alpha$ is a complex number, is $1^\alpha$ always equal to $1$?

From Equation (3.4-27) and noting that $1$ can be written as $e^{2\pi n i}$, we have

$$1^\alpha = \exp\,[\alpha\,\ell n\, e^{2\pi n i}] \tag{3.4-31a}$$

$$= \exp\,[\alpha\,(2\pi n i)] \tag{3.4-31b}$$

$$= \exp\,[-2\pi n\beta + 2\pi n i\,\gamma] \tag{3.4-31c}$$

$$= \exp\,[-2\pi n\beta]\,\{\cos 2\pi n\,\gamma + i\,\sin 2\pi n\,\gamma\} \tag{3.4-31d}$$

To obtain Equation (3.4-31b), we have made use of Equation (3.4-24f) and we have expressed the complex number $\alpha$ as $\gamma + i\beta$.

Thus in general $1^\alpha$ is not equal to $1$. Note that if $n$ is zero (consider only the principal value) $1^\alpha$ is equal to $1$.

## 3.5   COMPLEX INTEGRATION

In Section 3.2, we stated that a complex plane is required to display complex numbers. This two dimensional aspect has an effect on complex integration. A definite integral of a complex function of a complex variable is defined on the curve joining the two end-points of the integral in the complex plane.

To begin with, we consider the definite integral of a complex valued function of a real variable $t$ over a given interval $a \le t \le b$. Let

$$f(t) = u(t) + i \, v(t), \quad a \le t \le b \tag{3.5-1}$$

and we assume that $u(t)$ and $v(t)$ are continuous functions of $t$. We define

$$\int_a^b f(t) \, dt = \int_a^b u(t) \, dt + i \int_a^b v(t) \, dt \tag{3.5-2}$$

Note that both $\int_a^b u(t) \, dt$ and $\int_a^b v(t) \, dt$ are real.

Several properties of real integrals are carried over to complex integrals. For example, if $f(t)$ and $g(t)$ are complex functions, then

$$\int_a^b [f(t) + g(t)] \, dt = \int_a^b f(t) \, dt + \int_a^b g(t) \, dt \tag{3.5-3a}$$

$$\int_a^b f(t) \, dt = \int_a^c f(t) \, dt + \int_c^b f(t) \, dt \tag{3.5-3b}$$

$$\int_a^b \alpha \, f(t) \, dt = \alpha \int_a^b f(t) \, dt \tag{3.5-3c}$$

where $\alpha \, (= \beta + i \, \gamma)$ is a complex constant, $\beta$ and $\gamma$ are real.

$$\int_a^b f(t) \, dt = -\int_b^a f(t) \, dt \tag{3.5-3d}$$

**_Example 3.5-1_.** Evaluate $\displaystyle\int_0^{\pi/4} t \, e^{it} \, dt$.

Since the integrand is a complex valued function of a real variable t, we use Equation (3.5-2) and write

$$\int_0^{\pi/4} t \, e^{it} \, dt = \int_0^{\pi/4} t \cos t \, dt + i \int_0^{\pi/4} t \sin t \, dt \tag{3.5-4a}$$

$$= (t \sin t + \cos t) \Big|_0^{\pi/4} + i \, (-t \cos t + \sin t) \Big|_0^{\pi/4} \tag{3.5-4b}$$

$$= \left( \frac{\pi \sqrt{2}}{8} + \frac{\sqrt{2}}{2} - 1 \right) + i \left( \frac{\sqrt{2}}{2} - \frac{\pi \sqrt{2}}{8} \right) \tag{3.5-4c}$$

●

In Section 3.3, a curve C joining the points $z(a)$ to $z(b)$ is given by

$$z(t) = x(t) + i \, y(t), \qquad a \le t \le b \tag{3.5-5}$$

If $z(a) = z(b)$ is the only point of intersection, C is a **simple closed curve**. The **orientation** is defined by moving from $z(a)$ to $z(b)$ as t increases. This is illustrated in Figure 3.5-1. The unit circle

$$z = e^{it}, \qquad 0 \le t \le 2\pi \tag{3.5-6}$$

is a simple closed curve oriented in the counterclockwise direction. But

$$z = -e^{it}, \qquad 0 \le t \le 2\pi \tag{3.5-7}$$

is a unit circle oriented in the clockwise direction.

The complex function $z(t)$ in Equation (3.5-5) is differentiable if both $x(t)$ and $y(t)$ are differentiable for $a \le t \le b$. The derivative $z'(t)$ is given by

$$z'(t) = x'(t) + i \, y'(t), \qquad a \le t \le b \tag{3.5-8}$$

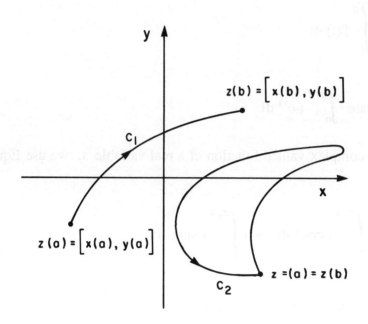

**FIGURE 3.5-1**    Simple $(C_1)$ and simple closed $(C_2)$ curves in the complex plane. Arrows indicate orientation

Curve $C$ is **smooth** if $z'(t)$ is continuous and non-zero in the interval. If $C$ is smooth, the **differential arc length** is given by

$$ds = \sqrt{[x'(t)]^2 + [y'(t)]^2}\ dt = |z'(t)|\ dt \qquad\qquad (3.5\text{-}9a,b)$$

and the length $L$ of the curve $C$ is given by

$$L = \int_a^b |z'(t)|\ dt \qquad\qquad (3.5\text{-}10)$$

If $C$ is given by Equation (3.5-5) and if we let $(-C)$ be the curve that traces the same set of points in the reverse order, curve $(-C)$ is given by

$$z(t) = x(-t) + i\,y(-t), \qquad -b \le t \le -a \qquad\qquad (3.5\text{-}11a)$$

A curve $C$ that is constructed by joining a finite number of smooth curves end to end is called a **contour** (or path). A formula for representing the line segment joining two points $z_1$ and $z_2$ in a complex plane is

$$z = z_1 + t\,(z_2 - z_1), \qquad 0 \le t \le 1 \qquad\qquad (3.5\text{-}11b)$$

The integral of $f(z)$ along a curve $C$ joining the points $z(a)$ to $z(b)$ is a **contour (line) integral** and is written as

$$\int_C f(z)\,dz = \int_a^b f(z)\,z'(t)\,dt \qquad (3.5\text{-}12)$$

Differentiating $z(t)$ from Equation (3.5-5), substituting the resulting expression in Equation (3.5-12), and decomposing $f$ into its real and imaginary parts, yields

$$\int_C f(z)\,dz = \int_C (u + i\,v)\,(dx + i\,dy) \qquad (3.5\text{-}13a)$$

$$= \int_C (u\,dx - v\,dy) + i \int_C (v\,dx + u\,dy) \qquad (3.5\text{-}13b)$$

The contour integral has similar properties to those of integrals of a complex function of a real variable. That is to say, we can replace the real variable $t$ by the complex variable $z$ in Equations (3.5-3a to d). Contour integration in the real two-dimensional plane will be considered in Chapter 4.

For complex integrals, we have the following inequalities:

(i)
$$\left| \int_a^b f(t)\,dt \right| \leq \int_a^b |f(t)|\,dt \qquad (3.5\text{-}14a)$$

(ii)
$$\left| \int_C f(z)\,dz \right| \leq \int_C |f(z)|\,dz \leq ML \qquad (3.5\text{-}14b)$$

where $M$ is the upper bound of $|f(z)|$ and $L$ is the length of the contour $C$. Note that in Equations (3.5-14a, b), we have taken the moduli of all complex quantities and we are dealing with real quantities. It was stressed in Section 3.2 that complex numbers are not ordered and inequalities have meaning only when associated with real numbers. In the theory of definite integrals of real functions, the length $L$ in Equation (3.5-14b) is replaced by the length of the interval of integration ($= b - a$, where $a$ and $b$ are the limits of integration).

***Example 3.5-2.*** Evaluate the integral $\displaystyle \int_C (z - z_0)^n\,dz$

where  n  is an integer and  C  is a circle of radius  r  with centre at  $z_0$  and is described in the counterclockwise direction.

The equation of a circle of radius  r  with centre at  $z_0$  is given by

$$z = z_0 + r\,e^{it}, \qquad 0 \le t \le 2\pi \tag{3.5-15}$$

On differentiating, we obtain

$$dz = i\,r\,e^{it}\,dt \tag{3.5-16}$$

Using Equations (3.5-15, 16), the integral becomes

$$\int_C (z - z_0)^n \, dz = \int_0^{2\pi} r^n \, e^{int} \, i\,r\,e^{it} \, dt \tag{3.5-17a}$$

$$= i\,r^{n+1} \int_0^{2\pi} e^{i(n+1)t} \, dt \tag{3.5-17b}$$

$$= i\,r^{n+1} \int_0^{2\pi} \left[\cos(n+1)t + i\sin(n+1)t\right] dt \tag{3.5-17c}$$

$$= \frac{i\,r^{n+1}}{(n+1)} \left[\sin(n+1)t - i\cos(n+1)t\right]_0^{2\pi} \tag{3.5-17d}$$

Since  cos  and  sin  are periodic, the integral is zero except when  n  is  −1.  In this case, Equation (3.5-17c) reduces to

$$\int_C (z - z_0)^{-1} \, dz = i \int_0^{2\pi} dt \tag{3.5-18a}$$

$$= 2\pi i \tag{3.5-18b}$$

**Example 3.5-3**.  Using Equation (3.5-14b), show that

$$\left| \int_C \frac{dz}{z^4} \right| \le 4\sqrt{2} \tag{3.5-19}$$

where  C  is the segment joining the point (0, i) to (1, 0) as shown in Figure 3.5-2.

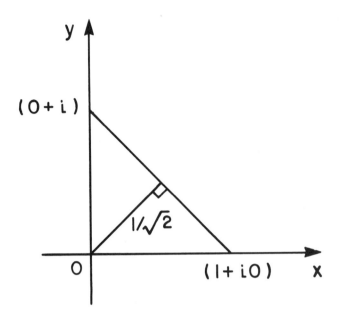

**FIGURE 3.5-2    Path of integration**

The closest point on the segment joining the points $(0, i)$ and $(1,0)$ to the origin is the mid-point of the segment, as shown in Figure 3.5-2. The distance from the origin to that point is $1/\sqrt{2}$. As $z$ varies along the segment, its distance from the origin $|z|$ varies and its minimum value is $1/\sqrt{2}$. The maximum value of $\left|\dfrac{1}{z}\right|$ is $1/(1/\sqrt{2})$ and is $\sqrt{2}$.

The upper bound $M$ of $\left|\dfrac{1}{z^4}\right|$ is given by

$$M = \left(\sqrt{2}\right)^4 = 4 \qquad (3.5\text{-}20a,b)$$

The length $L$ of the segment is $\sqrt{2}$.

It follows from Equation (3.5-14b) that

$$\left|\int_C \frac{dz}{z^4}\right| \le 4\sqrt{2} \qquad (3.5\text{-}19)$$

•

We note that a simple closed contour $C$ divides the complex plane into two domains. One domain is bounded and is referred to as the interior of $C$, and the other domain is unbounded and is the exterior of $C$, as shown in Figure 3.5-3.

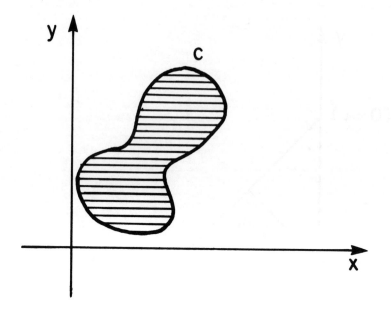

**FIGURE 3.5-3    Interior (shaded) and exterior domains**

Domain D is **simply connected** if every simple closed curve C in D encloses only points of D. In other words, there are no holes in a simply connected domain. A domain that is not simply connected is **multiply connected**. Figure 3.5-4 shows examples of simply connected and multiply connected domains.

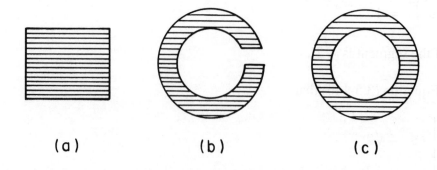

**FIGURE 3.5-4    Simply connected (a, b) and multiply connected domains**

## Cauchy's Theorem

If $f(z)$ is analytic in a simply connected domain $D$ and if $C$ is a simple closed contour that lies in $D$, then

$$\int_C f(z)\,dz = 0 \tag{3.5-21}$$

The integral round a closed contour is also denoted by $\oint f(z)\,dz$.

Proof: The integral can be written as in Equation (3.5-13b). That is to say

$$\int_C f(z)\,dz = \int_C (u\,dx - v\,dy) + i\int_C (v\,dx + u\,dy) \tag{3.5-22}$$

We transform each of the integrals on the right side of Equation (3.5-22) to a double integral using the two-dimensional Stokes (Green's) theorem (see Section 4.4). We have

$$\int_C (u\,dx - v\,dy) = \iint_S \left(-\frac{\partial v}{\partial x} - \frac{\partial u}{\partial y}\right) dx\,dy \tag{3.5-23a}$$

$$\int_C (v\,dx + u\,dy) = \iint_S \left(\frac{\partial u}{\partial x} - \frac{\partial v}{\partial y}\right) dx\,dy \tag{3.5-23b}$$

where $S$ is the domain enclosed by $C$.

Since $f(z)$ is analytic, $u$ and $v$ satisfy the Cauchy-Riemann relations (Equations 3.3-29a, b) and the right side of Equations (3.5-23a, b) are zero. It follows that Equation (3.5-21) holds.

A consequence of Cauchy's theorem is the concept of **path independence**. Consider the integral round a closed contour $C$. Let $z_1$ and $z_2$ be two arbitrary points on $C$ and let them divide $C$ into two arcs $C_1$ and $C_2$ as shown in Figure 3.5-5. We have

$$\int_C f(z)\,dz = \int_{C_1} f(z)\,dz + \int_{C_2} f(z)\,dz = 0 \tag{3.5-24a,b}$$

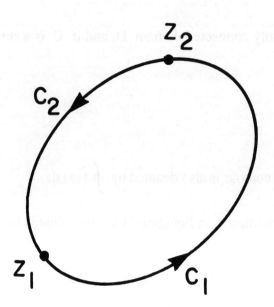

**FIGURE 3.5-5    Path independence**

We deduce

$$\int_{C_1} f(z)\, dz = -\int_{C_2} f(z)\, dz = \int_{-C_2} f(z)\, dz \qquad\qquad (3.5\text{-}25a,b)$$

Note that $-C_2$ is the arc obtained by changing the orientation of $C_2$ and is the curve joining $z_1$ and $z_2$. Equation (3.5-25b) implies that the integral of $f(z)$ from $z_1$ to $z_2$ is the same whether we integrate along $C_1$ or $-C_2$. Thus the value of the integral is independent of the path and depends only on the end points.

***Example 3.5-4***.  Evaluate $\int z^2\, dz$ along each of the straight lines OA, OB and AB as illustrated in Figure 3.5-6.

The parametric equations of the lines (Equation 3.5-11b) are

along OA:      $x(t) = t,$        $y(t) = 0,$        $0 \le t \le 1$                                   (3.5-26a,b)

along OB:      $x(t) = t,$        $y(t) = t,$        $0 \le t \le 1$                                   (3.5-26c,d)

along AB:      $x(t) = 1,$        $y(t) = t,$        $0 \le t \le 1$                                   (3.5-26e,f)

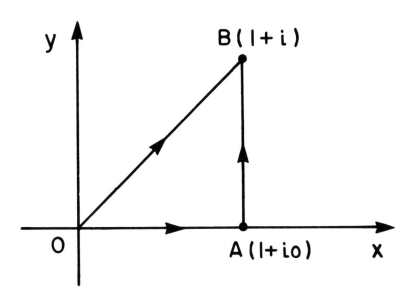

**FIGURE 3.5-6    Contour integration**

The function $z^2$ can be written as

$$z^2 = (x + i y)^2 = (x^2 - y^2) + 2 i x y = u + i v \qquad (3.5\text{-}27\text{a, b,c})$$

with    $dz = dx + i\, dy$ $\qquad\qquad\qquad\qquad\qquad\qquad\qquad (3.5\text{-}27\text{d})$

Using Equation (3.5-13b), we have

$$\int_{OA} z^2\, dz = \int_0^1 t^2\, dt = \frac{1}{3} \qquad\qquad\qquad (3.5\text{-}28\text{a,b})$$

$$\int_{OB} z^2\, dz = \int_0^1 \left(-2t^2\right) dt + i \int_0^1 \left(2t^2\right) dt = -\frac{2}{3} + \frac{2\,i}{3} \qquad (3.5\text{-}28\text{c,d})$$

$$\int_{AB} z^2\, dz = \int_0^1 (-2t)\, dt + i \int_0^1 \left(1 - t^2\right) dt = -1 + \frac{2\,i}{3} \qquad (3.5\text{-}28\text{e,f})$$

If $C$ is the closed contour $OABO$, then

$$\int_C z^2\,dz = \int_{OA} z^2\,dz + \int_{AB} z^2\,dz + \int_{BO} z^2\,dz \qquad (3.5\text{-}29\text{a})$$

$$= \frac{1}{3} + \left(-1 + \frac{2\,i}{3}\right) - \left(-\frac{2}{3} + \frac{2\,i}{3}\right) = 0 \qquad (3.5\text{-}29\text{b,c})$$

verifying Cauchy's theorem.

•

A multiply connected domain can be transformed to a simply connected domain by making suitable cuts. In Figure 3.5-7, the doubly connected domain has been cut by $L_1$ and $L_2$ and the resulting domain is simply connected. For a doubly connected region, one pair of cuts is sufficient and for a triply connected region two pairs of cuts are needed so as to obtain a simply connected region. Note the orientation of the curves in Figure 3.5-7. As we move along the curves, the area enclosed by them are on the left. The curve enclosing the simply connected domain D is $C\,L_1\,C_1\,L_2\,C$. By Cauchy's theorem

$$\int_{CL_1C_1L_2C} f(z)\,dz = \int_C f(z)\,dz + \int_{L_1} f(z)\,dz + \int_{C_1} f(z)\,dz + \int_{L_2} f(z)\,dz = 0 \qquad (3.5\text{-}30\text{a,b})$$

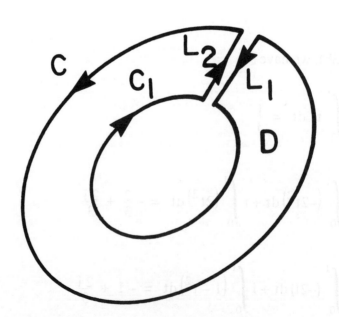

**FIGURE 3.5-7    Transformation of a doubly connected domain
to a simply connected one**

We note that $L_1$ and $L_2$ are in opposite direction (that is, their limits of integration are interchanged) and so their contributions will cancel. Equation (3.5-30b) becomes

$$\oint_C f(z)\,dz = -\oint_{C_1} f(z)\,dz \qquad (3.5\text{-}31)$$

We note that $C$ is oriented anticlockwise and $C_1$ is oriented clockwise. We reverse the orientation of $C_1$ and denote $(-C_1)$ by $C_1'$, Equation (3.5-31) then becomes

$$\oint_C f(z)\,dz = \oint_{C_1'} f(z)\,dz \qquad (3.5\text{-}32)$$

Equation (3.5-32) can be generalized to the case where there is more than one curve inside $C$. Suppose we have $n$ simple closed contours which we denote as $C_j$ ($j = 1, 2, ... , n$) inside another simple closed contour $C$. The regions interior to each contour $C_j$ have no common points. That is to say, the contours $C_j$ do not intersect each other. Let the function $f(z)$ be analytic in domain $D$ which contains all contours and the region between them. We then write

$$\oint_C f(z)\,dz = \sum_{j=1}^{n} \oint_{C_j} f(z)\,dz \qquad (3.5\text{-}33)$$

All the contours $C$ and $C_j$ are oriented in the same direction, usually in the anticlockwise direction which, by convention, is the positive direction.

***Example 3.5-5.*** Evaluate $\displaystyle\oint_C \frac{dz}{z-a}$ where $C$ is a simple closed curve. Consider the following two cases:

(i)     the point $z = a$ is outside $C$,

(ii)    the point $z = a$ is inside $C$.

(i)     Since the point $z = a$ is outside $C$, the function $\dfrac{1}{z-a}$ is analytic everywhere. Via Cauchy's theorem

$$\oint_C \frac{dz}{z-a} = 0 \qquad (3.5\text{-}34)$$

(ii)    In this case, the function is singular at $z = a$ and we enclose it by a circle $C_1$ of radius $\varepsilon$
with centre at the point of singularity as shown in Figure 3.5-8. On $C$ and $C_1$ and the region
enclosed by the two curves, $f(z)$ is analytic and using Equation (3.5-33), we have

$$\int_C \frac{dz}{z-a} = -\int_{C_1} \frac{dz}{z-a} \qquad\qquad (3.5\text{-}35)$$

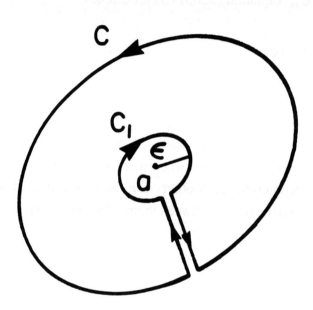

**FIGURE 3.5-8    Contour integral with point $z = a$ inside C**

To evaluate the integral along $C_1$, we write

$$z = a + \varepsilon\, e^{i\theta}, \qquad 0 \le \theta \le 2\pi \qquad\qquad (3.5\text{-}36)$$

The integral then becomes

$$-\int_{C_1} \frac{dz}{z-a} = \int_0^{2\pi} \frac{i\,\varepsilon e^{i\theta}}{\varepsilon\, e^{i\theta}}\, d\theta = i \int_0^{2\pi} d\theta = 2\pi i \qquad\qquad (3.5\text{-}37a,\,b,c)$$

From Equations (3.5-35, 37c), we deduce

$$\int_C \frac{dz}{z-a} = 2\pi i \tag{3.5-38}$$

## Cauchy's Integral Formula

Let $f(z)$ be analytic in a simply connected domain $D$ and let $C$ be a simple closed positively oriented contour that lies in $D$. For any point $z_0$ which lies interior to $C$, we have

$$\int_C \frac{f(z)}{z-z_0} \, dz = 2\pi i \, f(z_0) \tag{3.5-39}$$

In Example 3.5-5, we have verified Equation (3.5-39) for the case $f(z)$ equals to one.

## Integral Formulae for Derivatives

If $f(z)$ is analytic in $D$, it has derivatives of all orders in $D$ which are also analytic functions in $D$. The values of these derivatives at a point $z_0$ in $D$ are given by the formulae

$$f'(z_0) = \frac{1}{2\pi i} \int_C \frac{f(z)}{(z-z_0)^2} \, dz \tag{3.5-40a}$$

$$f''(z_0) = \frac{2!}{2\pi i} \int_C \frac{f(z)}{(z-z_0)^3} \, dz \tag{3.5-40b}$$

$$\vdots$$

$$f^{(n)}(z_0) = \frac{n!}{2\pi i} \int_C \frac{f(z)}{(z-z_0)^{n+1}} \, dz \tag{3.5-40c}$$

In Equations (3.5-40a to c), $C$ is any simple closed path in $D$ which encloses $z_0$.

We omit the proofs of these results but consider their applications. We also note that if one derivative exists, all derivatives exist. This is only true for complex variables.

***Example 3.5-6***. Find the value of the integral $\displaystyle\int_C \frac{e^z \cosh z}{(z-\pi)^3} \, dz$

where contour C is a square whose sides are: $x = \pm 4$ and $y = \pm 4$, described in the positive direction.

The region D, the contour C and the point $z_0$ are shown in Figure 3.5-9. From Equation (3.5-40b), we identify

$$f(z) = e^z \cosh z \qquad\qquad\qquad (3.5\text{-}41\text{a})$$

$$z_0 = \pi \qquad\qquad\qquad (3.5\text{-}41\text{b})$$

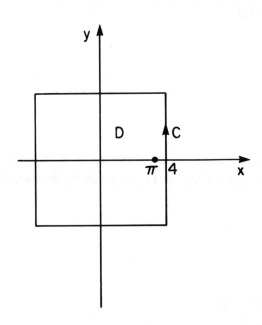

**FIGURE 3.5-9    Integration around a square**

The conditions for Equation (3.5-40b) are satisfied and it follows that

$$\int_C \frac{e^z \cosh z}{(z - \pi)^3} \, dz = i\pi \left. \frac{d^2}{dz^2} \left( e^z \cosh z \right) \right|_{z = \pi} \qquad\qquad (3.5\text{-}42)$$

Carrying out the differentiation, we obtain

$$\frac{d^2}{dz^2} \left( e^z \cosh z \right) = 2 \, e^z \left( \sinh z + \cosh z \right) \qquad\qquad (3.5\text{-}43\text{a})$$

$$= 2e^{2\pi} \quad \text{at } z = \pi \qquad\qquad (3.5\text{-}43\text{b,c})$$

Combining Equations (3.5-42, 43b) yields

$$\oint_C \frac{e^z \cosh z}{(z-\pi)^3} \, dz = 2i\pi e^{2\pi} \tag{3.5-44}$$

***Example 3.5-7.*** Let $f(z)$ be analytic inside and on the circle of radius $R$ with its centre at the origin. Let $z_0 \left(= r_0 \, e^{i\theta_0}\right)$ be any point inside $C$ as shown in Figure 3.5-10. Show that

$$f(z_0) = \frac{1}{2\pi} \int_0^{2\pi} \frac{\left(R^2 - r_0^2\right) f\left(R \, e^{i\psi}\right)}{R^2 - 2r_0 \cos\left(\theta_0 - \psi\right) + r_0^2} \, d\psi \tag{3.5-45}$$

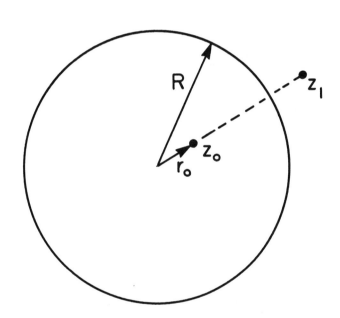

**FIGURE 3.5-10    Point $z_0$ and its inverse point $z_1$**

Obtain the real and imaginary parts of $f(z_0)$.

Since $z_0$ is inside $C$, using the Cauchy's integral formula (Equation 3.5-39), we have

$$f(z_0) = \frac{1}{2\pi i} \int_C \frac{f(z) \, dz}{z - z_0} \tag{3.5-46}$$

We define the inverse point $z_1$ of $z_0$ with respect to $C$ to be

$$z_1 = \frac{R^2}{\bar{z}_0} = \frac{R^2}{r_0}\, e^{i\theta_0} \tag{3.4-47a,b}$$

The point $z_1$ is outside $C$ and from Cauchy's theorem (Equation 3.5-21), we have

$$\frac{1}{2\pi i} \int_C \frac{f(z)}{z - \left(R^2/\bar{z}_0\right)}\, dz = 0 \tag{3.5-48}$$

Subtracting Equation (3.5-48) from Equation (3.5-46), we obtain

$$f(z_0) = \frac{1}{2\pi i} \int_C \left(\frac{1}{z - z_0} - \frac{1}{z - R^2/\bar{z}_0}\right) f(z)\, dz \tag{3.5-49a}$$

$$= \frac{1}{2\pi i} \int_C \frac{\left(z_0 - R^2/\bar{z}_0\right)}{\left(z - z_0\right)\left(z - R^2/\bar{z}_0\right)}\, f(z)\, dz \tag{3.5-49b}$$

We change to the polar form and write

$$z = R\, e^{i\psi} \tag{3.5-50a}$$

$$dz = i\, R\, e^{i\psi}\, d\psi \tag{3.5-50b}$$

$$z_0 = r_0\, e^{i\theta_0} \tag{3.5-50c}$$

Equation (3.5-49b) becomes

$$f(z_0) = \frac{1}{2\pi i} \int_0^{2\pi} \left\{ \frac{e^{i\theta_0}\left(r_0 - R^2/r_0\right) f\!\left(R\, e^{i\psi}\right)}{\left[R\, e^{i\psi} - r_0\, e^{i\theta_0}\right]\left[R\, e^{i\psi} - \left(R^2/r_0\right) e^{i\theta_0}\right]} \right\} i\, R\, e^{i\psi}\, d\psi \tag{3.5-51a}$$

$$= \frac{1}{2\pi} \int_0^{2\pi} \left[ \frac{e^{i(\theta_0+\psi)}\left(r_0^2 - R^2\right) f\!\left(R\, e^{i\psi}\right)}{\left(R\, e^{i\psi} - r_0\, e^{i\theta_0}\right)\left(r_0\, e^{i\psi} - R\, e^{i\theta_0}\right)} \right] d\psi \tag{3.5-51b}$$

$$= \frac{1}{2\pi} \int_0^{2\pi} \frac{\left(R^2 - r_0^2\right) f\!\left(R\, e^{i\psi}\right)}{\left(R\, e^{i\psi} - r_0\, e^{i\theta_0}\right)\left(R\, e^{-i\psi} - r_0\, e^{-i\theta_0}\right)}\, d\psi \tag{3.5-51c}$$

$$= \frac{1}{2\pi} \int_0^{2\pi} \frac{\left(R^2 - r_0^2\right) f\!\left(R\, e^{i\psi}\right)}{R^2 - 2\, R\, r_0 \cos\left(\theta_0 - \psi\right) + r_0^2}\, d\psi \tag{3.5-51d}$$

To obtain the real and imaginary parts of $f(z_0)$, we write

$$f(z_0) = u(r_0, \theta_0) + i\,v(r_0, \theta_0) \tag{3.5-52a}$$

$$f(R\,e^{i\psi}) = u(R, \psi) + i\,v(R, \psi) \tag{3.5-52b}$$

Substituting Equations (3.5-52a, b) in Equation (3.5-51d) and equating the real and imaginary parts, we obtain

$$u(r_0, \theta_0) = \frac{1}{2\pi} \int_0^{2\pi} \frac{(R^2 - r_0^2)\,u(R, \psi)}{R^2 - 2\,R\,r_0\cos(\theta_0 - \psi) + r_0^2}\,d\psi \tag{3.5-53a}$$

$$v(r_0, \theta_0) = \frac{1}{2\pi} \int_0^{2\pi} \frac{(R^2 - r_0^2)\,v(R, \psi)}{R^2 - 2\,R\,r_0\cos(\theta_0 - \psi) + r_0^2}\,d\psi \tag{3.5-53b}$$

Equations (3.5-51d, 53a, b) are **Poisson's integral formulae** and are important in potential theory. We note that the values of $f$ at any point inside $C$ can be determined from the values of $f$ on $C$. The functions $u$ and $v$ are harmonic functions, that is to say, are solutions of Laplace's equation, and we deduce that the solution of Laplace's equation is determined by the values of the function at the boundary only.

## Morera's Theorem

If $f$ is a continuous function in a simply connected domain $D$ and if

$$\int_C f(z)\,dz = 0 \tag{3.5-54}$$

for every closed contour $C$ in $D$, $f(z)$ is analytic in $D$.

Goursat proved Cauchy's theorem (Equation 3.5-21) requiring only the existence and not the continuity of $f'(z)$. Morera's theorem implies that $f(z)$ is analytic as a consequence of Cauchy's theorem.

## Maximum Modulus Principle

If $f$ is a continuous analytic function and is not a constant in a closed bounded domain $D$, then $|f(z)|$ has its maximum value on the boundary $C$ and not at an interior point. If $M$ is the maximum value of $|f(z)|$ on $C$,

$$|f(z)| < M \quad \text{for all z in D} \tag{3.5-55a}$$

If $f(z)$ is constant,

$$|f(z)| \le M \quad \text{for all z in D} \tag{3.5-55b}$$

The maximum modulus principle is true for harmonic functions but not for any smooth real valued functions of two real variables.

## 3.6   SERIES REPRESENTATIONS OF ANALYTIC FUNCTIONS

In this section, the equivalence between analytic functions and power series is explored. The concept of sequences, series and power series of complex numbers are in many cases similar to those of real variables.

### Sequences and Series

Let $z_1, z_2, \ldots, z_n$ be a **sequence** of complex numbers. A sequence $\{z_n\}$ **converges** to $z_0$ if

$$\lim_{n \to \infty} z_n = z_0 \tag{3.6-1}$$

Alternatively Equation (3.6-1) is stated as $z_n \longrightarrow z_0$ as $n \longrightarrow \infty$. Equation (3.6-1) implies that for every $\varepsilon > 0$, there corresponds a positive integer $N$, such that

$$|z_n - z_0| < \varepsilon \quad \text{for all n} > N \tag{3.6-2}$$

$N$ depends on $\varepsilon$.

If the limit does not exist, then the sequence $\{z_n\}$ **diverges**. Separating $z_n$ and $z_0$ into their real and imaginary parts, Equation (3.6-1) becomes

$$\lim_{n \to \infty} x_n = x_0 \tag{3.6-3a}$$

$$\lim_{n \to \infty} y_n = y_0 \tag{3.6-3b}$$

If the sequence $\{z_n\}$ converges, then the sequences $\{x_n\}$ and $\{y_n\}$ also converge.

***Example 3.6-1.*** Discuss the convergence of $\{(1 + i)^n\}$.

We write $z_n [= (1 + i)^n]$ in its polar form

$$z_n = \left[ \sqrt{2} \left( \cos \frac{\pi}{4} + i \sin \frac{\pi}{4} \right) \right]^n \tag{3.6-4a}$$

$$= 2^{n/2} \left[ \cos \frac{n\pi}{4} + i \sin \frac{n\pi}{4} \right] \tag{3.6-4b}$$

The sequence $2^{n/2} \cos \frac{n\pi}{4}$ oscillates and does not converge. Since the real part of the sequence $\{z_n\}$ does not converge, the sequence $\{z_n\}$ diverges. The same thing can be said for the imaginary part.

●

Let $\{z_n\}$ be a sequence and let $s_1, s_2, s_3, \ldots, s_n$ be the partial sum defined as follows

$$s_1 = z_1, \quad s_2 = z_1 + z_2, \quad s_3 = z_1 + z_2 + z_3, \quad s_n = z_1 + z_2 + \ldots + z_n \tag{3.6-5a-d}$$

If $n \longrightarrow \infty$, we have an **infinite series**.

If the sequence of the partial sums $\{s_n\}$ is **convergent**, that is to say

$$\lim_{n \to \infty} s_n = s \tag{3.6-6}$$

exists, the series $\sum_n z_n$ is convergent and the complex number $s$ is the sum of the series. If $\{s_n\}$ diverges, the series **diverges**.

A necessary (but insufficient) condition for a complex series $\sum_{n=1}^{\infty} z_n$ to be convergent is that the $\lim_{n \to \infty} z_n$ vanishes. That is to say, $s_n - s_{n-1}$ in Equation (3.6-5d) tends to zero for large $n$.

The sum of a convergent series of complex numbers can be found by computing the sum of its real and imaginary parts. A series is **absolutely convergent** if $\sum_{n=1}^{\infty} |z_n|$ is convergent.

*Example 3.6-2.* Discuss the convergence of the geometric series

$$\sum_{n=0}^{\infty} z_n = 1 + z + z^2 + \ldots + z^n + \ldots \tag{3.6-7}$$

The partial sum $s_n$ is given by

$$s_n = \frac{1 - z^{n+1}}{1 - z}$$ (3.6-8)

If $|z| < 1$, we deduce

$$\lim_{n \to \infty} s_n = \frac{1}{1 - z}$$ (3.6-9)

The series $\sum z^n$ converges to $\frac{1}{1 - z}$.

The function $\frac{1}{1 - z}$ is analytic inside the circle $|z| < 1$ and has the representation

$$\frac{1}{1 - z} = \sum_{n = 0}^{\infty} z_n$$ (3.6-10)

If $|z| > 1$, the series diverges.

**Comparison Test**

Let $\sum_{n = 0}^{\infty} M_n$ be a convergent series with real non-negative terms. If, for all $n$ greater than $N$,

$$|z_n| \le M_n$$ (3.6-11a)

the series $\sum z_n$ also converges absolutely.

**Ratio Test**

Let $\sum_{n = 0}^{\infty} z_n$ be a complex series and

$$\lim_{n \to \infty} \frac{|z_{n+1}|}{|z_n|} = L$$ (3.6-12)

If $L < 1$, the series converges absolutely and if $L > 1$, the series diverges. No conclusion can be drawn if $L$ is one.

Note that in both tests we use the absolute value of $z_n$ since complex numbers are not ordered.

*Example 3.6-3*. Determine the convergence of the series $\displaystyle\sum_{n=1}^{\infty} \frac{(n+i)^2}{2^n}$.

Using the ratio test, we have

$$\frac{|z_{n+1}|}{|z_n|} \; = \; \frac{|n+1+i|^2}{2^{n+1}} \; \frac{2^n}{|n+i|^2} \; = \; \frac{1}{2} \frac{\left[(n+1)^2+1\right]}{n^2+1} \tag{3.6-13a,b}$$

Taking the limit as $n$ tends to infinity yields $\frac{1}{2}$.

Since the limit is less than one, the series converges.

●

A series of the form

$$\sum_{n=0}^{\infty} c_n (z-z_0)^n \; = \; c_0 + c_1 (z-z_0) + \; ... \; + c_n (z-z_0)^n + \; ... \tag{3.6-14}$$

where $z$ is a complex variable, $z_0$, $c_0$, $c_1$, ... are complex numbers is a **power series**. By a change of origin, we can set $z_0$ to be zero. In Chapter 2, we have shown that every power series of a real variable has a radius of convergence $R$. This result can be extended to complex variables. Every series has a radius of convergence $R$ ($0 \le R < \infty$) and the series converges absolutely if $|z-z_0| < R$ and diverges if $|z-z_0| > R$. On the circle of convergence ($|z-z_0| = R$), the series may converge at some points and may diverge at other points. When $R$ is zero, the series converges only at $z_0$ and when $R$ is infinity, the series converges for all values of $z$.

The radius of convergence depends on the absolute values of $|c_n|$. If the sequence $\sqrt[n]{|c_n|}$ converges to the limit $L$, the radius of convergence $R$ is given by

$$R \; = \; \frac{1}{L} \tag{3.6-15}$$

An alternative equation for $R$ is

$$\frac{1}{R} \; = \; \lim_{n \to \infty} \left| \frac{c_{n+1}}{c_n} \right| \tag{3.6-16}$$

if the limit exists.

A power series represents a continuous function at every point inside its circle of convergence.

***Example 3.6-4.*** Determine the radius of convergence of the series $\displaystyle\sum_{n=0}^{\infty} \frac{(-1)^n z^{2n}}{2^{2n} (n!)^2}$ and determine the function it represents.

From Equation (3.6-14), we identify

$$
c_n = \begin{cases} 0, & \text{if } n \text{ is odd} \\[2em] \dfrac{(-1)^{n/2}}{2^n \left[(n/2)!\right]^2}, & \text{if } n \text{ is even} \end{cases} \qquad\qquad (3.6\text{-}17a,b)
$$

For odd values of n

$$
\sqrt[n]{|c_n|} = 0 \qquad\qquad (3.6\text{-}18a)
$$

For even values of n

$$
\lim_{n \to \infty} \sqrt[n]{|c_n|} = \lim_{n \to \infty} \frac{1}{2 \left[(n/2)!\right]^{2/n}} = 0 \qquad\qquad (3.6\text{-}18b,c)
$$

Thus R is infinity and the series converges absolutely for all values of z. In Chapter 2, we have defined Bessel functions of a real variable. Replacing x by z, we find that the series represents the complex Bessel function of order zero.

## Taylor Series

We now consider the expansion of an analytic function $f(z)$ as a power series.

Let $f(z)$ be analytic everywhere inside the circle C with centre at $z_0$ and radius R. At each point z inside C,

$$
f(z) = \sum_{n=0}^{\infty} \frac{f^{(n)}(z_0)}{n!} (z - z_0)^n \qquad\qquad (3.6\text{-}19)
$$

That is, the power series converges to $f(z)$ when $|z - z_0| < R$.

We first prove the theorem when $z_0$ is the origin. Let z be any point inside the circle C of radius R, as shown in Figure 3.6-1. Let $C_1$ be a circle with radius $R_1 < R$, and let $\xi$ denote a point lying on $C_1$. Since z is interior to $C_1$ and $f(z)$ is analytic, we have, using the Cauchy's integral formula

$$f(z) = \frac{1}{2\pi i} \int\limits_{C_1} \frac{f(\xi)\, d\xi}{\xi - z} \tag{3.6-20}$$

Note that $C_1$ has to be positively oriented.

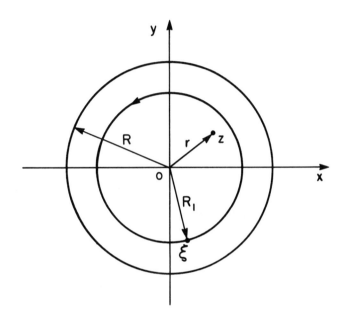

**FIGURE 3.6-1    Illustration for the proof of Taylor series**

We expand $\dfrac{1}{\xi - z}$ in powers of $\dfrac{z}{\xi}$ ($< 1$), as

$$\frac{1}{\xi - z} = \frac{1}{\xi}\left[1 - \frac{z}{\xi}\right]^{-1} \tag{3.6-21a}$$

$$= \frac{1}{\xi}\left[1 + \frac{z}{\xi} + \left(\frac{z}{\xi}\right)^2 + \ldots + \left(\frac{z}{\xi}\right)^{n-1} + \frac{(z/\xi)^n}{1 - (z/\xi)}\right] \tag{3.6-21b}$$

$$= \left[\frac{1}{\xi} + \frac{1}{\xi^2} z + \frac{1}{\xi^3} z^2 + \ldots + \frac{1}{\xi^n} z^{n-1} + z^n \frac{1}{(\xi - z)\xi^n}\right] \tag{3.6-21c}$$

Multiplying each term by $\dfrac{f(\xi)}{2\pi i}$ and integrating around the circle, we have

$$\frac{1}{2\pi i}\int_{C_1}\frac{f(\xi)}{\xi-z}\,d\xi = \frac{1}{2\pi i}\left[\int_{C_1}\frac{f(\xi)}{\xi}\,d\xi + z\int_{C_1}\frac{f(\xi)}{\xi^2}\,d\xi + \ldots + z^n\int_{C_1}\frac{f(\xi)}{(\xi-z)\xi^n}\,d\xi\right]$$

$$(3.6\text{-}22)$$

Using Equations (3.5-39, 40a to c), we obtain

$$f(z) = f(0) + \frac{f'(0)}{1!}z + \frac{f''(0)}{2!}z^2 + \ldots + \frac{f^{n-1}(0)}{(n-1)!}z^{n-1} + R_n(z) \qquad (3.6\text{-}23a)$$

where

$$R_n(z) = \frac{z^n}{2\pi i}\int_{C_1}\frac{f(\xi)\,d\xi}{(\xi-z)\xi^n} \qquad (3.6\text{-}23b)$$

To evaluate the remainder term $R_n$, we note that

$$|\xi-z| \ge \left||\xi|-|z|\right| = R_1 - r \qquad (3.6\text{-}24)$$

If $M$ denotes the maximum value of $|f(\xi)|$ on $C_1$, we write the absolute value of the remainder as follows

$$|R_n(z)| \le \frac{r^n}{2\pi}\frac{M\,2\pi R_1}{(R_1-r)R_1^n} = \frac{M R_1}{(R_1-r)}\left(\frac{r}{R_1}\right)^n \qquad (3.6\text{-}25)$$

Since $(r/R_1)$ is less than one, it follows from Equation (3.6-25) that

$$\lim_{n\to\infty} R_n = 0 \qquad (3.6\text{-}26)$$

We proved that $f(z)$ has a power series representation given by Equation (3.6-23a) with $R_n$ tending to zero as $n$ tends to infinity. That is to say, $f(z)$ has an infinite series representation. The proof was restricted to the case where $z_0$ is the origin. This infinite series is the **Maclaurin series** which is a special case of the **Taylor series**. To extend the proof to the case where $z_0$ is not the origin, we need to shift the origin to $z_0$. We define a function $g(z)$ to be

$$g(z) = f(z+z_0) \qquad (3.6\text{-}27)$$

Since $f(z)$ is analytic in $|z-z_0| < R$, $g(z)$ is analytic in $|(z+z_0)-z_0| < R$, which is $|z| < R$. Thus $g(z)$ has a Maclaurin series expansion which is written as

$$g(z) = \sum_{n=0}^{\infty} \frac{g^{(n)}(0)}{n!} z^n, \quad |z| < R \tag{3.6-28}$$

Replacing $z$ by $(z - z_0)$ in Equations (3.6-27, 28) yields

$$g(z - z_0) = f(z) \tag{3.6-29}$$

$$g(z - z_0) = \sum_{n=0}^{\infty} \frac{g^{(n)}(0)}{n!} (z - z_0)^n, \quad |z - z_0| < R \tag{3.6-30}$$

Combining Equations (3.6-29, 30), we obtain

$$f(z) = \sum_{n=0}^{\infty} \frac{f^{(n)}(z_0)}{n!} (z - z_0)^n, \quad |z - z_0| < R \tag{3.6-31}$$

Note that in the case of complex variables, if $f(z)$ is analytic in $|z - z_0| < R$, the Taylor series represents the function, the remainder term $R_n$ tends to zero as $n$ tends to infinity. In the calculus of real variables, the remainder term $R_n$ (Equation 1.2-12) does not have this property. In the theory of complex variables, an analytic function has a power series representation and the power series is analytic. If $f(z)$ is analytic in a domain containing $z_0$, and $z_1$ is the nearest singular point to $z_0$, the Taylor series (Equation 3.6-31) is convergent in the domain $|z - z_0| < |z_1 - z_0|$.

Since the Taylor series is convergent for an analytic function, term by term differentiation and integration are permissible. The radii of convergence of the differentiated and integrated series are the same as that of the original series.

***Example 3.6-5.*** Expand $1/(1 + z^2)$ in a Taylor (Maclaurin) series about the origin. Determine its radius of convergence.

The singularities of $\dfrac{1}{1 + z^2}$ are

$$z = i, \qquad z = -i \tag{3.6-32}$$

The function $\dfrac{1}{1 + z^2}$ is analytic in the domain

$$|z| < 1 \tag{3.6-33}$$

Using Equation (3.6-23a), we obtain

$$\frac{1}{1 + z^2} = 1 - z^2 + z^4 - z^6 + \dots \tag{3.6-34}$$

Since $\dfrac{1}{1+z^2}$ is analytic in the domain $|z| < 1$, the radius of convergence of the series is one.

The radius of convergence can also be deduced from Equation (3.6-15) and is found to be one, as expected. Note that, in the case of real variables, the radius of convergence can be deduced from

the series expansion only. The function $1/(1+x^2)$ has no singularity along the real line.

## Laurent Series

If the function $f(z)$ is not analytic at a point $z_0$, it does not have a Taylor series about $z_0$. Instead it can be represented by a power series with both positive and negative powers of $(z - z_0)$. This series is a **Laurent series**. Laurent's theorem can be stated as follows. Let $D$ be the annular region bounded by two concentric circles $C_0$ and $C_1$ with centre $z_0$ and radii $R_1$ and $R_2$ respectively, as shown in Figure 3.6-2. Let $f(z)$ be analytic within $D$ and on $C_0$ and $C_1$. At every point $z$ inside $D$, $f(z)$ can be represented by a Laurent series which can be written as

$$f(z) = \sum_{n=0}^{\infty} a_n (z - z_0)^n + \sum_{n=1}^{\infty} \frac{b_n}{(z - z_0)^n} \qquad (3.6\text{-}35a)$$

where

$$a_n = \frac{1}{2\pi i} \int_{C_1} \frac{f(\xi)\ d\xi}{(\xi - z_0)^{n+1}}, \qquad n = 0, 1, 2, \dots \qquad (3.6\text{-}35b)$$

$$b_n = \frac{1}{2\pi i} \int_{C_0} \frac{f(\xi)\ d\xi}{(\xi - z_0)^{-n+1}}, \qquad n = 1, 2, \dots \qquad (3.6\text{-}35c)$$

The integrals around $C_0$ and $C_1$ are to be taken in the anticlockwise direction.

The proof of this theorem is similar to that of Taylor's theorem. We surround the point $z$ by a circle $\gamma$ as shown in Figure 3.6-2. Let $\xi$ be any point on the curves $C_0$, $C_1$ or $\gamma$. From Equation (3.5-33), we have

$$\int_{C_1} \frac{f(\xi)}{\xi - z}\ d\xi - \int_{C_0} \frac{f(\xi)}{\xi - z}\ d\xi - \int_{\gamma} \frac{f(\xi)}{\xi - z}\ d\xi = 0 \qquad (3.6\text{-}36)$$

where $C_1$, $C_0$ and $\gamma$ are considered to be in the anticlockwise direction.

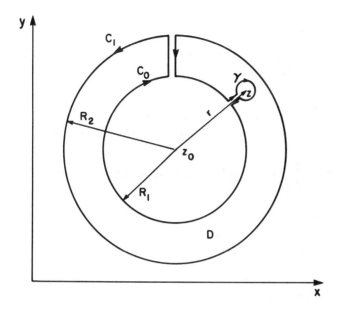

**FIGURE 3.6-2**    **Illustration for the proof of convergence of Laurent series**

Using Equation (3.5-39), we write the last integral as

$$2\pi i \, f(z) \; = \; \int_{\gamma} \frac{f(\xi) \, d\xi}{\xi - z} \tag{3.6-37}$$

Combining Equations (3.6-36, 37) yields

$$f(z) \; = \; \frac{1}{2\pi i} \int_{C_1} \frac{f(\xi) \, d\xi}{\xi - z} \; - \; \frac{1}{2\pi i} \int_{C_0} \frac{f(\xi) \, d\xi}{\xi - z} \tag{3.6-38}$$

On $C_1$, $|\xi| > |z|$ and we have the same situation as in the case of the Taylor series. Equations (3.6-21 to 26) can be carried over and we have

$$\frac{1}{2\pi i} \int_{C_1} \frac{f(\xi) \, d\xi}{\xi - z} \; = \; \sum_{n=0}^{\infty} a_n (z - z_0)^n \tag{3.6-39}$$

where the $a_n$ are given by Equation (3.6-35b).

On $C_0$, $|z| > |\xi|$ and, in this case, we write

$$\frac{-1}{\xi - z} = \frac{1}{(z - z_0) - (\xi - z_0)} = \frac{1}{(z - z_0)} \left[ 1 - \frac{(\xi - z_0)}{(z - z_0)} \right]^{-1} \tag{3.6-40a,b}$$

$$= \frac{1}{(z - z_0)} + \frac{(\xi - z_0)}{(z - z_0)^2} + \ldots + \frac{(\xi - z_0)^{n-1}}{(z - z_0)^n} + \frac{(\xi - z_0)^n}{(z - z_0)^n (z - \xi)} \tag{3.6-40c}$$

Multiplying Equation (3.6-40c) by $\frac{1}{2\pi i} f(\xi)$ and integrating around $C_0$, we obtain

$$-\frac{1}{2\pi i} \int_{C_0} \frac{f(\xi) \, d\xi}{\xi - z} = \frac{b_1}{(z - z_0)} + \frac{b_2}{(z - z_0)^2} + \ldots + \frac{b_n}{(z - z_0)^n} + T_n \tag{3.6-41}$$

where $b_n$ are given by Equation (3.6-35c).

The remainder $T_n$ is given by

$$T_n = \frac{1}{2\pi i} \frac{1}{(z - z_0)^n} \int_{C_0} \frac{(\xi - z_0)^n f(\xi) \, d\xi}{(z - \xi)} \tag{3.6-42}$$

Let $K$ be the maximum value of $|f(\xi)|$ on $C_0$. For any $\xi$ on $C_0$, we have

$$\left| z - \xi \right| = \left| (z - z_0) - (\xi - z_0) \right| \geq \left| z - z_0 \right| - \left| \xi - z_0 \right| = r - R_1 \tag{3.6-43a,b,c}$$

It follows that

$$|T_n| \leq \frac{1}{2\pi r^n} \frac{K R_1^n 2\pi R_1}{(r - R_1)} = \frac{K}{r/R_1 - 1} \left( \frac{R_1}{r} \right)^n \tag{3.6-44a,b}$$

Since $R_1 < r$, $|T_n| \longrightarrow 0$ as $n \longrightarrow \infty$. Hence

$$-\frac{1}{2\pi i} \int_{C_0} \frac{f(\xi) \, d\xi}{(\xi - z)} = \sum_{n=1}^{\infty} b_n (z - z_0)^{-n} \tag{3.6-45}$$

Combining Equations (3.6-38, 39, 45), we obtain Equations (3.6-35a to c). The series in Equations (3.6-39, 45) converges in the domain $|z - z_0| < R_2$ and $|z - z_0| > R_1$ respectively. Consequently the series in Equation (3.6-35a) converges in the annulus $R_1 < |z - z_0| < R_2$.

## Comments

1.      If $f(z)$ is analytic in a region $|z - z_0| < R$, the coefficients $b_n$ in the Laurent series are zero. The Laurent series reduces to the Taylor series about the point $z_0$.

2.      The Laurent series in a specified annulus $R_1 < |z - z_0| < R_2$ is unique. That is to say, if, by any method, we have obtained a series representation for an analytic function in the given annulus, that series is the Laurent series. In the examples that follow, we shall derive a representation of $f(z)$ by methods other than via Equations (3.6-35a to c).

3.      Since $f(z)$ is analytic in the annulus, the contours $C_1$ and $C_0$ in Equations (3.6-39, 45) can be replaced by any circle $C$ lying in the annulus. That is to say, the contour integrals can be taken around the same curve $C$, as long as $C$ lies in the annulus $R_1 < |z - z_0| < R_2$.

*Example 3.6-6.* Expand

$$f(z) = \frac{z + 3}{\left(z^2 - z - 2\right) z} \tag{3.6-46}$$

in powers of $z$ in the following regions

(i)     within the unit circle about the origin,

(ii)    within the annular region between concentric circles about the origin having radii 1 and 2 respectively,

(iii)   exterior to the circle of radius 2.

The function $f(z)$ can be decomposed in partial fractions as

$$f(z) = \frac{z + 3}{z(z - 2)(z + 1)} \tag{3.6-47a}$$

$$= -\frac{3}{2z} + \frac{5}{6(z - 2)} + \frac{2}{3(z + 1)} \tag{3.6-47b}$$

(i)     When $0 < |z| < 1$, we write $f(z)$ as

$$f(z) = -\frac{3}{2z} - \frac{5}{12}\left(1 - \frac{z}{2}\right)^{-1} + \frac{2}{3}(1 + z)^{-1} \tag{3.6-48a}$$

$$= -\frac{3}{2z} - \frac{5}{12}\sum_{n=0}^{\infty}\left(\frac{z}{2}\right)^n + \frac{2}{3}\sum_{n=0}^{\infty}(-1)^n z^n \tag{3.6-48b}$$

(ii)    When $1 < |z| < 2$, we write

$$f(z) = -\frac{3}{2z} - \frac{5}{12\left(1 - \frac{z}{2}\right)} + \frac{2}{3z}\frac{1}{\left(1 + \frac{1}{z}\right)} \tag{3.6-49a}$$

$$= -\frac{3}{2z} - \frac{5}{12}\sum_{n=0}^{\infty}\left(\frac{z}{2}\right)^n + \frac{2}{3z}\sum_{n=0}^{\infty}(-1)^n (z)^{-n} \tag{3.6-49b}$$

(iii)   When $|z| > 2$, we have

$$f(z) = -\frac{3}{2z} + \frac{5}{6z\left(1 - \frac{2}{z}\right)} + \frac{2}{3z}\frac{1}{\left(1 + \frac{1}{z}\right)} \tag{3.6-50a}$$

$$= -\frac{3}{2z} + \frac{5}{6z}\sum_{n=0}^{\infty}\left(\frac{2}{z}\right)^n + \frac{2}{3z}\sum_{n=0}^{\infty}(-1)^n \frac{1}{z^n} \tag{3.6-50b}$$

Note that the series in Equations (3.6-48b, 49b, 50b) converge in the region indicated. In (i), (ii) and (iii), we have written $f(z)$ in a form such that when we expand the appropriate expression as a binomial series, the expansion is valid. The series we have obtained are the Laurent series expanded about the origin which is a singular point of $f(z)$. The other two singular points of $f(z)$ are at $z = -1$ and $z = 2$. The domain is divided into regions such that in each region the function is analytic.

**Example 3.6-7.** Prove that

$$\cosh\left(z + \frac{1}{z}\right) = a_0 + \sum_{n=1}^{\infty} a_n\left(z^n + \frac{1}{z^n}\right), \qquad |z| > 0 \tag{3.6-51a}$$

where

$$a_n = \frac{1}{2\pi}\int_0^{2\pi} \cos n\theta \, \cosh\left(2\cos\theta\right) d\theta \tag{3.6-51b}$$

The function $\cosh\left(z + \frac{1}{z}\right)$ is analytic for all non-zero finite values of $z$. Therefore it can be expanded in a Laurent series at any point $z$ about the origin in the region $0 < |z| < \infty$. Equation (3.6-35a) becomes

$$\cosh\left(z + \frac{1}{z}\right) = \sum_{n=0}^{\infty} a_n z^n + \sum_{n=1}^{\infty} b_n\left(\frac{1}{z}\right)^n \tag{3.6-52}$$

where $a_n$ and $b_n$ are given by Equations (3.6-35b, c).

To evaluate $a_n$ and $b_n$, we choose a common circle C, the unit circle centered at the origin. We note also that on interchanging $z$ and $\frac{1}{z}$, the function remains unchanged and

$$a_n = b_n \tag{3.6-53}$$

On C, the unit circle, we have

$$z = e^{i\theta} \tag{3.5-54a}$$

$$dz = i\, e^{i\theta}\, d\theta \tag{3.6-54b}$$

The coefficients $a_n$ are given by

$$a_n = \frac{1}{2\pi i} \int_0^{2\pi} \frac{\cosh\left(e^{i\theta} + e^{-i\theta}\right)}{e^{i(n+1)\theta}} i\, e^{i\theta}\, d\theta \tag{3.6-55a}$$

$$= \frac{1}{2\pi} \int_0^{2\pi} \cosh\left(2\cos\theta\right) e^{-in\theta}\, d\theta \tag{3.6-55b}$$

$$= \frac{1}{2\pi} \int_0^{2\pi} \cosh\left(2\cos\theta\right) \left[\cos n\theta - i\sin n\theta\right] d\theta \tag{3.6-55c}$$

To evaluate the second term on the right side of Equation (3.6-55c), we divide the region of integration and write

$$\int_0^{2\pi} \cosh\left(2\cos\theta\right) \sin n\theta\, d\theta = \int_0^{\pi} \cosh\left(2\cos\theta\right) \sin n\theta\, d\theta + \int_{\pi}^{2\pi} \cosh\left(2\cos\theta\right) \sin n\theta\, d\theta \tag{3.6-56a}$$

$$= \int_0^{\pi} \cosh\left(2\cos\theta\right) \sin n\theta\, d\theta + \int_{\pi}^{0} \cosh\left[2\cos(2\pi-\phi)\right] \sin\left[n(2\pi-\phi)\right] (-d\phi) \tag{3.6-56b}$$

$$= \int_0^{\pi} \cosh\left(2\cos\theta\right) \sin n\theta\, d\theta + \int_0^{\pi} \cosh\left(2\cos\phi\right) \left[-\sin n\phi\right] d\phi \tag{3.6-56c}$$

$$= 0 \tag{3.6-56d}$$

Combining Equations (3.6-55c, 56d) yields

$$a_n = \frac{1}{2\pi} \int_0^{2\pi} \cosh\left(2\cos\theta\right)\cos n\theta \ d\theta \tag{3.6-51b}$$

***Example 3.6-8.*** Deduce the complex potential for a two-dimensional, irrotational, incompressible flow past an infinite stationary circular cylinder of radius a.

The centre of the cylinder is taken to be the origin. In Example 3.3-6, we have shown that an analytic function can be a suitable complex potential for an irrotational, incompressible flow. There is no flow in the region $|z| \le a$ and singularities may be present in this region. In the region $a \le |z| < \infty$, there is no singularity and the complex potential $\Phi(z)$ is analytic and can be represented by a Laurent series. We start by choosing the simplest Laurent series given by

$$\Phi(z) = a_1 z + \frac{b_1}{z} \tag{3.6-57}$$

We have shown in Example 3.3-6 that the potential $\phi$ and the stream function $\psi$ are given by the real and imaginary parts of $\Phi(z)$ respectively. Separating $\Phi(z)$ into its real and imaginary parts, we obtain

$$\phi + i\,\psi = a_1\,(x + i\,y) + \frac{b_1\,(x - i\,y)}{\left(x^2 + y^2\right)} \tag{3.6-58a}$$

$$= a_1 x + \frac{b_1 x}{x^2 + y^2} + i\,y\left(a_1 - \frac{b_1}{x^2 + y^2}\right) \tag{3.6-58b}$$

From Equation (3.6-58b), we deduce

$$\phi = a_1 x + \frac{b_1 x}{x^2 + y^2} \tag{3.6-59a}$$

$$\psi = y\left(a_1 - \frac{b_1}{x^2 + y^2}\right) \tag{3.6-59b}$$

The cylinder is a streamline and we can assume that $\psi$ is zero on the cylinder. This implies that $\psi = 0$ for $x^2 + y^2 = a^2$, for all values of y and, from Equation (3.6-59b), we obtain

$$0 = \left(a_1 - \frac{b_1}{a^2}\right) \tag{3.6-60}$$

Far away from the cylinder, the flow is undisturbed by the cylinder and we assume the velocity distribution to be

$$\underline{v} = \left(v_\infty, 0\right) \tag{3.6-61}$$

From Equations (3.3-62a, 6-59a), we obtain as $(x^2 + y^2)$ tends to infinity

$$v_\infty = a_1 \tag{3.6-62}$$

From Equations (3.6-60, 62), we deduce

$$b_1 = a^2 \, v_\infty \tag{3.6-63}$$

Equation (3.6-57) can be written as

$$\Phi = v_\infty \left(z + \frac{a^2}{z}\right) \tag{3.6-64}$$

The function $\Phi$ is analytic in the annulus $a \leq |z| < \infty$ and satisfies all the boundary conditions and is the complex potential for the present flow.

## 3.7 RESIDUE THEORY

We have defined the singular point (singularity) $z_0$ of $f(z)$ to be the point at which $f(z)$ ceases to be analytic. If $z_0$ is a singular point but $f(z)$ is analytic in the region $0 < |z - z_0| < R$ for some positive R, $z_0$ is an **isolated singular point**. The function $1/(z - 2)$ has a singular point at $z = 2$ and is analytic in the region $0 < |z - 2| < R$. That is to say, $f(z)$ is analytic in a region in which the point $z = 2$ has been removed. The point $z = 2$ is an isolated singular point. The function $\ell n \, z$ is singular at the origin but also along the negative part of the real axis as illustrated in Figure 3.4-1. The origin is not an isolated singular point. There are many other singular points near the origin.

If $z_0$ is an isolated singular point in the annulus $0 < |z - z_0| < R$, $f(z)$ has a Laurent series representation which can be written as

$$f(z) = \sum_{n=0}^{\infty} a_n \left(z - z_0\right)^n + \sum_{n=1}^{\infty} b_n \left(z - z_0\right)^{-n} \tag{3.6-35a}$$

where the coefficients $a_n$, $b_n$ are given by Equations (3.6-35b, c).

We consider three types of singularities

(i)     If all the coefficients $b_n$ are zero, the Laurent series reduces to the Taylor series. The singular point is a **removable singular point**. We define

$$\lim_{z \to z_0} f(z) = a_0 \qquad\qquad (3.7\text{-}1)$$

For example, the function $\frac{\sin z}{z}$ is not defined at the origin. The function has a series representation

$$f(z) = \frac{\sin z}{z} = 1 - \frac{z^2}{3!} + \frac{z^4}{5!} \quad .... \qquad\qquad (3.7\text{-}2)$$

The series in Equation (3.7-2) is convergent and we define $f(0)$ to be one. We can also obtain this result using l'Hôpital's rule. The origin is a removable singularity.

(ii)    If all but a finite number of the coefficients $b_n$ are zero, $z_0$ is a **pole**. The singular point $z_0$ is a **pole of order m** if $(b_1, \ldots , b_m)$ are non-zero and the coefficients $b_n$ vanish for all $n > m$.

If $m$ is one, $z_0$ is a **simple pole**. The coefficient $b_1$ is the **residue** and is denoted as Res $(z_0)$ or Res $[f(z), z_0]$

(iii)   If all the coefficients $b_n$ do not vanish, $z_0$ is an **essential singular point**.

***Example 3.7-1.*** Discuss the nature of the singularities of

(i) $\dfrac{\cos z - 1}{z^2}$,              (ii) $e^{1/z^2}$,                    (iii) $\dfrac{\sinh z}{z^5}$

at the origin.

(i)     The expansion of $\cos z$ in powers of $z$ is known; the expansion of $(\cos z - 1)/z^2$ can be deduced to be

$$\frac{\cos z - 1}{z^2} = -\frac{1}{2} + \frac{z^2}{4!} - \frac{z^4}{6!} + ... \qquad\qquad (3.7\text{-}3)$$

We define the value of $(\cos z - 1)/z^2$ at the origin to be $(-1/2)$. In so doing, the function is analytic everywhere. The origin is a removable singular point. Note that the function is not defined at the origin. We can also use l'Hôpital's rule to deduce that its value at the origin is $(-1/2)$.

(ii) The expansion of the exponential function is

$$e^{1/z^2} = 1 + \frac{1}{z^2} + \frac{1}{2z^4} + \dots + \frac{1}{n!(z^{2n})} + \dots \tag{3.7-4}$$

We have an infinite series, the origin is an essential singular point.

(iii) The expansion of $(\sinh z)/z^5$ is obtained by dividing the expansion of $\sinh z$ by $z^5$ to obtain

$$\frac{\sinh z}{z^5} = \frac{1}{z^4} + \frac{1}{3!\,z^2} + \frac{1}{5!} + \frac{z^2}{7!} + \dots \tag{3.7-5}$$

The origin is a pole of order four.

●

The **zero** of a function $f(z)$ is the point at which the value of $f$ is zero. The Taylor series expansion of $f(z)$ about any point $z_0$ (not a singular point) can be written as

$$f(z) = \sum_{n=0}^{\infty} a_n (z - z_0)^n \tag{3.7-6a}$$

$$a_n = \frac{f^{(n)}(z_0)}{n!} \tag{3.7-6b}$$

If $a_0$ is zero and the other coefficients $a_n$ do not vanish, $f(z_0)$ is zero and $z_0$ is a **simple zero**. If $a_0, a_1, \dots, a_{m-1}$ are all zero and $a_n$ $(n \geq m)$ are non-zero, $f(z)$ has a **zero of order m** at $z_0$. That is to say, the first $(m-1)$ derivatives of $f(z)$ at $z_0$ vanish. Equation (3.7-6a) becomes

$$f(z) = a_m (z - z_0)^m + a_{m+1} (z - z_0)^{m+1} + \dots \tag{3.7-7a}$$

$$= (z - z_0)^m [a_m + a_{m+1} (z - z_0) + \dots] \tag{3.7-7b}$$

$$= (z - z_0)^m \sum_{n=m}^{\infty} a_n (z - z_0)^{n-m} \tag{3.7-7c}$$

$$= (z - z_0)^m g(z) \tag{3.7-7d}$$

where

$$g(z) = \sum_{n=m}^{\infty} a_n (z - z_0)^{n-m} = (z - z_0)^{-m} f(z) \tag{3.7-7e,f}$$

$$g(z_0) = a_m \neq 0 \tag{3.7-7g,h}$$

If $h(z)$ has a pole of order $m$, its Laurent series can be written as

$$h(z) = \sum_{n=0}^{\infty} a_n (z - z_0)^n + \frac{b_1}{(z - z_0)} + \dots + \frac{b_m}{(z - z_0)^m} \tag{3.7-8a}$$

$$= (z - z_0)^{-m} \left[ \sum_{n=0}^{\infty} a_n (z - z_0)^{n+m} + b_1 (z - z_0)^{m-1} + \dots + b_m \right] \tag{3.7-8b}$$

$$= (z - z_0)^{-m} \, \ell \, (z) \tag{3.7-8c}$$

where

$$\ell \, (z) = \sum_{n=0}^{\infty} a_n (z - z_0)^{n+m} + b_1 (z - z_0)^{m-1} + \dots + b_m = (z - z_0)^m \, h(z) \tag{3.7-8d,e}$$

$$\ell \, (z_0) = b_m \neq 0 \tag{3.7-8f,g}$$

If $f(z)$ has a zero of order $m$ at $z_0$, $\dfrac{1}{f(z)}$ has a pole of order $m$ at $z_0$.

If $h(z)$ has a pole of order $m$ at $z_0$, $\dfrac{1}{h(z)}$ has a removable singularity at $z_0$.

If $f(z)$ and $g(z)$ have poles of order $k$ and $m$ respectively at $z_0$, their product $fg$ has a pole of order $k + m$ at $z_0$.

***Example 3.7-2.*** Determine the order of the pole of $(2 \cos z - 2 + z^2)^{-2}$ at the origin.

Instead of looking at the poles of $(2 \cos z - 2 + z^2)^{-2}$, it is easier to consider the zeros of $(2 \cos z - 2 + z^2)^2$. Expanding the function $f(z)$, we have

$$f(z) = (2 \cos z - 2 + z^2)^2 = \left[ 2 \left( 1 - \frac{z^2}{2!} + \frac{z^4}{4!} - \frac{z^6}{6!} + \dots \right) - 2 + z^2 \right]^2 \tag{3.7-9a,b}$$

$$= \left[ \frac{2z^4}{4!} - \frac{2z^6}{6!} + \dots \right]^2 \tag{3.7-9c}$$

$$= z^8 \left[ \frac{2}{4!} - \frac{2z^2}{6!} + \dots \right]^2 \tag{3.7-9d}$$

The function $f(z)$ has a zero of order 8. Therefore, the function $1/f(z)$ has a pole of order 8.

●

If all the singularities of $f(z)$ in the finite complex plane are poles, $f(z)$ is a **meromorphic** function.

## Cauchy's Residue Theorem

Let $C$ be a simple closed contour, positively oriented, within and on which a function $f(z)$ is analytic except at a finite number of singular points $z_1, z_2, \ldots, z_n$ contained in the interior of $C$. **Cauchy's residue theorem** states that

$$\int_C f(z)\,dz = 2\pi i \sum_{n=1}^{k} \operatorname{Res}\left[f(z), z_n\right] \tag{3.7-10}$$

Proof. Let $C_1, C_2, \ldots, C_k$ be $k$ circles each positively oriented, with their centers at the isolated singular points $z_1, z_2, \ldots, z_k$ respectively, as shown in Figure 3.7-1. Each circle $C_j$ $(j = 1, \ldots, k)$ lies inside $C$ and exterior to the other circles. From Equation (3.5-33), we have

$$\int_C f(z)\,dz = \sum_{j=1}^{k} \int_{C_j} f(z)\,dz \tag{3.7-11}$$

Consider the integral $\int_{C_i} f(z)\,dz$ for a fixed value of i. The residue $b_1$ as defined by Equation (3.6-35c) is given by

$$b_1 = \frac{1}{2\pi i} \int_{C_i} f(z)\,dz \tag{3.7-12a}$$

$$2\pi i\, b_1 = \int_{C_i} f(z)\,dz \tag{3.7-12b}$$

Equation (3.7-10) is obtained by combining Equations (3.7-11, 12b).

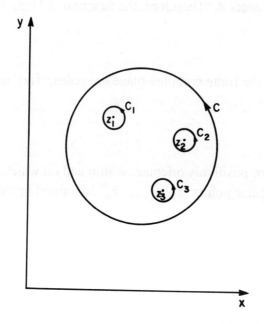

**FIGURE 3.7-1    Illustration for the proof of Cauchy's residue theorem**

**Some Methods of Evaluating the Residues**

*(i)    Simple pole at $z_i$*

The Laurent series of $f(z)$ about $z_i$ can be written as

$$f(z) = \sum_{n=0}^{\infty} a_n (z - z_i)^n + \frac{b_1}{(z - z_i)} \tag{3.7-13}$$

From Equation (3.7-13), we deduce that $b_1$ is given by

$$b_1 = \lim_{z \to z_i} (z - z_i) f(z) \tag{3.7-14}$$

If the function $f(z)$ is given in the form of a rational function, that is to say

$$f(z) = \frac{g(z)}{h(z)} \tag{3.7-15}$$

$f(z)$ has a simple pole at $z_i$ which implies that $h(z)$ has a simple zero at $z_i$ [$h(z_i) = 0$]. Equation (3.7-14) can be written as

$$b_1 = \lim_{z \to z_i} \left[ \frac{(z - z_i)\, g(z)}{h(z) - h(z_i)} \right] \qquad (3.7\text{-}16a)$$

$$= \lim_{z \to z_i} \left\{ \frac{g(z)}{[h(z) - h(z_i)]/(z - z_i)} \right\} \qquad (3.7\text{-}16b)$$

$$= \frac{g(z_i)}{h'(z_i)} \qquad (3.7\text{-}16c)$$

The derivative $h'(z_i)$ is non-zero because $h(z)$ has only a simple zero at $z_i$.

## (ii) *Pole of order* $m \geq 2$

The Laurent series of $f(z)$ about $z_i$ is

$$f(z) = \sum_{n=0}^{\infty} a_n (z - z_i)^n + \frac{b_1}{(z - z_i)} + \frac{b_2}{(z - z_i)^2} + \ldots + \frac{b_m}{(z - z_i)^m} \qquad (3.7\text{-}17a)$$

$$= (z - z_i)^{-m}\, g(z) \qquad (3.7\text{-}17b)$$

where $g(z)$ is analytic and non-zero at $z_i$ (see Equations 3.7-8a to g).

Since $g(z)$ is analytic, it has a Taylor series which can be written as

$$g(z) = \sum_{n=0}^{\infty} \frac{g^{(n)}(z_i)}{n!} (z - z_i)^n \qquad (3.7\text{-}18)$$

Comparing Equations (3.7-17a, b, 18), we find that the residue $b_1$ is given by

$$b_1 = \frac{g^{(m-1)}(z_i)}{(m - 1)!} \qquad (3.7\text{-}19a)$$

$$= \lim_{z \to z_i} \frac{d^{m-1}}{dz^{m-1}} \frac{\left[ (z - z_i)^m\, f(z) \right]}{(m - 1)!} \qquad (3.7\text{-}19b)$$

If $f(z)$ is as given by Equation (3.7-15), we can expand $g(z)$ and $h(z)$ in their Taylor series. By comparing coefficients of powers of $(z - z_i)$ with Equation (3.7-17a), we can determine $b_1$.

In many cases, it is possible to decompose $f(z)$ into its partial fractions.

### *(iii)*   *Essential singular point*

In this case, we have to expand $f(z)$ in powers of $(z - z_i)$, including negative powers, and obtain $b_1$.

We illustrate the method of calculating residues by evaluating contour integrals.

*Example 3.7-3.* Use the residue theorem to evaluate the following integrals

(i)  $\displaystyle\int_{C_1} \frac{(z-2)\,dz}{z(z-1)}$ ,       (ii)  $\displaystyle\int_{C_2} \frac{5\,dz}{z(2+z-z^2)}$ ,       (iii)  $\displaystyle\int_{C_3} z\,e^{2/z}\,dz$

where $C_1$, $C_2$ and $C_3$ are circles with center at the origin and of radii 2, 1/2 and 1 respectively.

(i)      The function $\dfrac{(z-2)}{z(z-1)}$ has two simple poles, one at the origin and the other at $z = 1$, both are inside $C_1$.

$$\int_{C_1} \frac{(z-2)\,dz}{z(z-1)} = 2\pi i\left\{ \mathrm{Res}\left[f(z),\,0\right] + \mathrm{Res}\left[f(z),\,1\right] \right\} \tag{3.7-20}$$

The residue at the origin is obtained by using Equation (3.7-14)

$$b_1 = \lim_{z\to 0} \frac{z(z-2)}{z(z-1)} = 2 \tag{3.7-21a,b}$$

Similarly, the residue at $z = 1$ is given by

$$b_1 = \lim_{z\to 1} \frac{(z-1)(z-2)}{z(z-1)} = -1 \tag{3.7-22a,b}$$

Using Equations (3.7-20, 21a, b, 22a, b), we obtain

$$\int_{C_1} \frac{(z-2)\,dz}{z(z-1)} = 2\pi i[2-1] = 2\pi i \tag{3.7-23a,b}$$

(ii)      We rewrite the function as

$$\frac{1}{z(2+z-z^2)} = \frac{1}{z(2+z)(1-z)} \tag{3.7-24}$$

The function has singularities at the origin, at $z = -2$ and at $z = 1$, of which only the origin is interior to $C_2$.

The residue at the origin is given by

$$b_1 = \lim_{z \to 0} \frac{z}{z(2+z)(1-z)} = \frac{1}{2} \qquad (3.7\text{-}25a,b)$$

The integral is

$$\int_{C_2} \frac{5\,dz}{z(2+z-z^2)} = 2\pi i \left[\frac{5}{2}\right] = 5\pi i \qquad (3.7\text{-}26a,b)$$

(iii)    The function $z\,e^{2/z}$ has an essential singularity at the origin. We expand the exponential function and obtain

$$z\,e^{2/z} = z \sum_{n=0}^{\infty} \frac{1}{n!} \left(\frac{2}{z}\right)^n \qquad (3.7\text{-}27a)$$

$$= z\left[1 + \frac{2}{z} + \frac{1}{2}\left(\frac{2}{z}\right)^2 + \frac{1}{3!}\left(\frac{2}{z}\right)^3 + \ldots\right] \qquad (3.7\text{-}27b)$$

$$= z + 2 + \frac{2}{z} + \frac{4}{3z^2} + \ldots \qquad (3.7\text{-}27c)$$

From Equation (3.7-27c), we find

$$b_1 = 2 \qquad (3.7\text{-}28)$$

$$\int_{C_3} z\,e^{2/z}\,dz = 2\pi i\,(2) = 4\pi i \qquad (3.7\text{-}29a,b)$$

***Example 3.7-4.*** Evaluate the following

(i)  $\displaystyle\int_C \frac{e^{2z}}{\cosh \pi z}\,dz,$          (ii)  $\displaystyle\int_C \frac{\pi \cot(\pi z)}{z^2}\,dz$

where $C$ is a unit circle with center at the origin.

(i)    The zeros of $\cosh \pi z$ which lie interior to $C$ are

$$z_1 = \frac{i}{2}, \qquad z_2 = -\frac{i}{2} \qquad (3.7\text{-}30a,b)$$

Using Equation (3.7-16c), the residue at $i/2$ and $-i/2$ are respectively given by

$$b_1 = \left. \frac{e^{2z}}{\pi \sinh \pi z} \right|_{z=i/2} = \frac{e^i}{\pi (i \sin \pi/2)} = -\frac{i e^i}{\pi} \qquad (3.7\text{-}31a,b,c)$$

$$b_1 = \left. \frac{e^{2z}}{\pi \sinh \pi z} \right|_{z=-i/2} = \frac{e^{-i}}{\pi [i \sin (-\pi/2)]} = +\frac{i e^{-i}}{\pi} \qquad (3.7\text{-}32a,b,c)$$

The integral is given by

$$\int_C \frac{e^{2z}}{\cosh \pi z} \, dz = 2\pi i \left[ -\frac{i e^i}{\pi} + \frac{i e^{-i}}{\pi} \right] = 2 \left[ e^i - e^{-i} \right] = 4i \sin 1 \qquad (3.7\text{-}33a,b,c)$$

(ii)     We write the function as

$$\frac{\pi \cot (\pi z)}{z^2} = \frac{\pi \cos (\pi z)}{z^2 \sin (\pi z)} \qquad (3.7\text{-}34)$$

The only zero of $\sin \pi z$ which is inside $C$ is the origin. The function $\pi \cot (\pi z)/z^2$ has a pole of order three at the origin. The residue $b_1$ can be determined using Equation (3.7-19b) which gives

$$b_1 = \lim_{z \to 0} \frac{d^2}{dz^2} \left[ \frac{z^3 \pi \cos \pi z}{2! \, z^2 \sin \pi z} \right] = \frac{\pi}{2} \lim_{z \to 0} \frac{d^2}{dz^2} \left[ \frac{z \cos \pi z}{\sin \pi z} \right] \qquad (3.7\text{-}35a,b)$$

$$= \pi^2 \lim_{z \to 0} \left[ \frac{\pi z \cos \pi z - \sin \pi z}{\sin^3 \pi z} \right] \qquad (3.7\text{-}35c)$$

To evaluate the limit in Equation (3.7-35c), we use l'Hôpital's rule and we obtain

$$b_1 = \pi^2 \lim_{z \to 0} \left[ \frac{-\pi^2 z \sin \pi z}{3\pi \sin^2 \pi z \cos \pi z} \right] = \pi^2 \left[ -\frac{1}{3} \right] \qquad (3.7\text{-}36a,b)$$

The integral is given by

$$\int_C \frac{\pi \cot (\pi z)}{z^2} \, dz = 2\pi i \left[ -\frac{\pi^2}{3} \right] = -\frac{2 i \pi^3}{3} \qquad (3.7\text{-}37a,b)$$

**Example 3.7-5.** The force $\underline{F}$ per unit length exerted on a cylinder of infinite length in a steady irrotational flow can be determined by complex variable methods. Blasius (1908) has shown that if $F_x$ and $F_y$ are the $x$ and $y$ components of $\underline{F}$

$$F_x - i\,F_y = \frac{i\rho}{2} \int_C \left(\frac{d\Phi}{dz}\right)^2 dz \qquad (3.7\text{-}38)$$

where $\rho$ is the density of the fluid, $\Phi$ is the complex potential, and $C$ is the curve defining the surface of the cylinder.

Calculate the force if

$$\Phi = -v_\infty (z + a^2/z) \qquad (3.7\text{-}39)$$

In Example 3.6-8, we have shown that Equation (3.7-39) represents the complex potential for a flow past a circular cylinder of radius $a$ with center at the origin.

Differentiating Equation (3.7-39) and using Equation (3.7-38), we obtain

$$F_x - i\,F_y = \frac{i\rho}{2} \int_C v_\infty^2 \left(1 - \frac{2a^2}{z^2} + \frac{a^4}{z^4}\right) dz \qquad (3.7\text{-}40a)$$

$$= i\rho \frac{v_\infty^2}{2} 2\pi i \sum \text{Res} \qquad (3.7\text{-}40b)$$

The residue is zero and this means that there is no force acting on the cylinder. This surprising result is known as **d'Alembert's paradox** and is discussed at length by Batchelor (1967).

●

The integral of real variables can be evaluated using the residue theorem. We consider a few examples.

**Triginometric Integrals**

To evaluate integrals of the form

$$I = \int_0^{2\pi} F\left(\cos\theta,\ \sin\theta\right) d\theta \qquad (3.7\text{-}41)$$

where $F(\cos\theta,\ \sin\theta)$ is a rational function of $\cos\theta$ and $\sin\theta$, we write

$$z = e^{i\theta} \qquad (3.7\text{-}42)$$

The trigonometric functions $\cos\theta$ and $\sin\theta$ can be written as

$$\cos\theta = \frac{1}{2}\left(z+\frac{1}{z}\right), \qquad \sin\theta = \frac{1}{2i}\left(z-\frac{1}{z}\right) \qquad\qquad (3.7\text{-}43a,b)$$

Substituting Equations (3.7-42, 43a, b) into Equation (3.7-41), we obtain

$$I = \int_C f(z)\,dz = 2\pi i \sum \text{Res}\,(f(z), z_n) \qquad\qquad (3.7\text{-}44a,b)$$

where C is the unit circle, $z_n$ are the poles of $f(z)$ inside the unit circle.

Note that from Equation (3.7-42) the limits of the integration 0 to $2\pi$ correspond to the unit circle and the integral is taken in the anticlockwise direction.

***Example 3.7-6.*** Show that

$$I = \int_0^{2\pi} \frac{d\theta}{a+b\cos\theta} = \frac{2\pi}{\sqrt{a^2-b^2}}, \qquad a>b>0 \qquad\qquad (3.7\text{-}45a,b)$$

Substituting Equations (3.7-42, 43a) into Equation (3.7-45a), we obtain

$$I = \int_C \frac{-i\,z^{-1}\,dz}{a+\frac{b}{2}\left(z+z^{-1}\right)} = -2i \int_C \frac{dz}{b\,z^2+2az+b} \qquad\qquad (3.7\text{-}46a,b)$$

The poles of the function are the zeros of

$$f(z) = bz^2 + 2az + b = 0 \qquad\qquad (3.7\text{-}47a,b)$$

They are

$$z_1 = -\frac{a+\sqrt{a^2-b^2}}{b} \qquad\qquad (3.7\text{-}48a)$$

$$z_2 = -\frac{a-\sqrt{a^2-b^2}}{b} \qquad\qquad (3.7\text{-}48b)$$

Since $a>b>0$, $|z_2|<1$ and the only pole inside C is $z_2$.

We evaluate the residue $b_1$ using Equation (3.7-14) and we have

$$b_1 = \lim_{z\to z_2} \frac{(z-z_2)}{b(z-z_1)(z-z_2)} = \frac{1}{2\sqrt{a^2-b^2}} \qquad\qquad (3.7\text{-}49a,b)$$

Combining Equations (3.7-44b, 46b, 49b) results in

$$I = \frac{(2\pi i)(-2i)}{2\sqrt{a^2 - b^2}} = \frac{2\pi}{\sqrt{a^2 - b^2}} \qquad (3.7\text{-}50a,b)$$

## Improper Integrals of Rational Functions

We recall from the theory of functions of real variables that if $f(x)$ is a continuous function on $0 \leq x < \infty$, the improper integral of $f$ over $[0, \infty)$ is defined by

$$\int_0^\infty f(x)\, dx = \lim_{a \to \infty} \int_0^a f(x)\, dx \qquad (3.7\text{-}51a)$$

provided the limit exists. Similarly if $f(x)$ is continuous on $(-\infty, 0]$, we have

$$\int_{-\infty}^0 f(x)\, dx = \lim_{a \to -\infty} \int_a^0 f(x)\, dx \qquad (3.7\text{-}51b)$$

When both limits exist, we write, for an integrable function on the whole real line $(-\infty, \infty)$

$$\int_{-\infty}^\infty f(x)\, dx = \lim_{a \to -\infty} \int_a^0 f(x)\, dx + \lim_{a \to \infty} \int_0^a f(x)\, dx \qquad (3.7\text{-}52a)$$

$$= \int_{-\infty}^0 f(x)\, dx + \int_0^\infty f(x)\, dx \qquad (3.7\text{-}52b)$$

The value of the improper integral is computed as

$$\int_{-\infty}^\infty f(x)\, dx = \lim_{a \to \infty} \int_{-a}^a f(x)\, dx \qquad (3.7\text{-}53)$$

It may happen that the limit in Equation (3.7-53) exists, but the limits on the right side of Equation (3.7-52a) may not exist. Consider, for example, the function $f(x)$ defined by

$$f(x) = x \qquad (3.7\text{-}54)$$

The limit

$$\lim_{a \to \infty} \int_0^a x \, dx = \lim_{a \to \infty} \frac{a^2}{2} \tag{3.7-55}$$

does not exist.  But

$$\lim_{a \to \infty} \int_{-a}^a x \, dx = 0 \tag{3.7-56}$$

This leads us to define the **Cauchy principal value (p.v.)** of an integral over the interval $(-\infty, \infty)$.  This is defined by Equation (3.7-53) if the limit exist.  From now on, the principal value of $\int_{-\infty}^{\infty} f(x) \, dx$ is implied whenever the integral appears.

Consider the integral $\int_{-\infty}^{\infty} f(x) \, dx$ where

(i)      $f(x) = P(x) / Q(x)$ \hfill (3.7-57)

(ii)     $P(x)$ and $Q(x)$ are polynomials.

(iii)    $Q(x)$ has no real zeros.

(iv)     The degree of $P(x)$ is at least two less than that of $Q(x)$.

The integral is then given by

$$\int_{-\infty}^{\infty} f(x) \, dx = 2\pi i \sum_{n=1}^{k} \text{Res}\left[f(z), z_n\right] \tag{3.7-58}$$

where $z_1, z_2, \ldots, z_k$ are the poles of $f(z)$ that lie in the upper half plane.

Note that the integral on the left side involves the real line.  The contour associated with the complex integration must therefore also include the real line.

The contour we choose is a semi-circle, centered at the origin with radius $R$ in the upper half plane as shown in Figure 3.7-2.  We denote the semi-circle by $\Gamma$ and we choose $R$ to be large enough so that the semi-circle encloses all the poles of $f(z)$.  By the residue theorem, we have

$$\int_{-R}^{R} f(x) \, dx + \int_{\Gamma} f(z) \, dz = 2\pi i \sum_{n=1}^{k} \text{Res}\left[f(z), z_n\right] \tag{3.7-59}$$

Equation (3.7-58) implies the integral along $\Gamma$ to be zero and this will be established next.

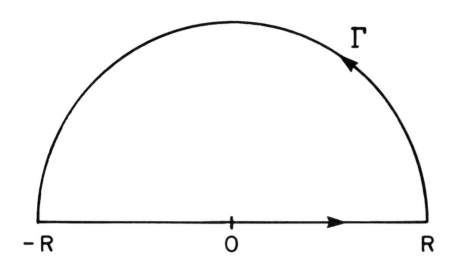

**FIGURE 3.7-2    Contour integral for evaluating infinite integrals**

From condition (iv), we deduce that for sufficiently large $R$

$$|\, z\, f(z)\,| \; < \; \varepsilon \tag{3.7-60}$$

for all points of $\Gamma$. By substituting

$$z \; = \; R\, e^{i\theta} \tag{3.7-61}$$

into the second integral on the left side of Equation (3.7-59), we obtain

$$\left| \int_{\Gamma} f(z)\, dz \right| \; = \; \left| \int_{0}^{\pi} f\!\left(R\, e^{i\theta}\right) i\, R\, e^{i\theta}\, d\theta \right| \tag{3.7-62a}$$

$$< \; \varepsilon \int_{0}^{\pi} d\theta \tag{3.7-62b}$$

$$< \; \varepsilon\, \pi \tag{3.7-62c}$$

Equation (3.7-62b) is obtained by using inequality (3.7-60).

As $R \longrightarrow \infty$, $\varepsilon \longrightarrow 0$ and the integral around $\Gamma$ is zero. Equation (3.7-59) reduces to Equation (3.7-58).

***Example 3.7-7.*** Show that

$$\int_{-\infty}^{\infty} \frac{dx}{x^4 + a^4} = \frac{\sqrt{2}\,\pi}{2a^3}, \qquad a > 0 \tag{3.7-63}$$

The function

$$f(x) = \frac{1}{x^4 + a^4} \tag{3.7-64}$$

is in the form of Equation (3.7-57) and using Equation (3.7-58), we have

$$\int_{-\infty}^{\infty} \frac{dx}{x^4 + a^4} = 2\pi i \sum_{n=1}^{k} \text{Res}\left[\frac{1}{z^4 + a^4}, z_n\right] \tag{3.7-65}$$

To determine the zeros of $z^4 + a^4$, we write

$$z^4 = -a^4 = a^4 \, e^{i\pi (1 + 2n)} \tag{3.7-66}$$

Using De Moivre's theorem, we find that the zeros are

$$z_1 = a\,e^{i\pi/4}, \qquad z_2 = a\,e^{3i\pi/4}, \qquad z_3 = a\,e^{5i\pi/4}, \qquad z_4 = a\,e^{7i\pi/4} \tag{3.7-67a-d}$$

Of the four zeros only $z_1$ and $z_2$ are in the upper half plane. We denote any one of them by $\alpha$. From Equations (3.7-66), we find that

$$\alpha^4 = -a^4 \tag{3.7-68}$$

The residue at $z = \alpha$ is given by

$$\lim_{z \to \alpha} (z - \alpha)\, f(z) = \lim_{z \to \alpha} \frac{(z - \alpha)}{z^4 - \alpha^4} \tag{3.7-69a}$$

$$= \lim_{z \to \alpha} \frac{(z - \alpha)}{(z - \alpha)(z + \alpha)\left(z^2 + \alpha^2\right)} \tag{3.7-69b}$$

$$= \frac{1}{4\alpha^3} = -\frac{\alpha}{4a^4} \tag{3.7-69c,d}$$

Substituting Equation (3.7-69d) into Equation (3.7-65), we obtain

$$\int_{-\infty}^{\infty} \frac{dx}{x^4 + a^4} = -\frac{2\pi i}{4a^4} \left[ a\,e^{i\pi/4} + a\,e^{3i\pi/4} \right] \tag{3.7-70a}$$

$$= -\frac{i\pi}{2a^3} \left[ \cos\frac{\pi}{4} + \cos\frac{3\pi}{4} + i\left( \sin\frac{\pi}{4} + \sin\frac{3\pi}{4} \right) \right] \tag{3.7-70b}$$

$$= \frac{\sqrt{2}\,\pi}{2a^3} \tag{3.7-70c}$$

## Evaluating Integrals Using Jordan's Lemma

Let $f(x)$ be of the form given by Equation (3.7-57), $P(x)$ and $Q(x)$ satisfy conditions (ii) and (iii). Condition (iv) is replaced by the condition that the degree of $P(x)$ is at least one less than that of $Q(x)$. If $\Gamma$ is the semi-circle shown in Figure 3.7-2, **Jordan's lemma** states

$$\int_{\Gamma} e^{imz} f(z)\, dz \to 0 \quad \text{as} \quad R \to \infty \tag{3.7-71}$$

where $m$ is a positive integer.

To evaluate the integral $\int_{-\infty}^{\infty} e^{imx} f(x)\, dx$, we integrate around the contour shown in Figure 3.7-2 and we have

$$\int_{-R}^{R} e^{imx} f(x)\, dx + \int_{\Gamma} e^{imz} f(z)\, dz = 2\pi i \sum_{n=1}^{k} \text{Res}\left[ e^{imz} f(z), z_n \right] \tag{3.7-72}$$

Using Jordan's lemma and letting $R \longrightarrow \infty$, Equation (3.7-72) reduces to

$$\int_{-\infty}^{\infty} e^{imx} f(x)\, dx = 2\pi i \sum_{n=1}^{k} \text{Res}\left[ e^{imz} f(z), z_n \right] \tag{3.7-73}$$

By separating Equation (3.7-73) into its real and imaginary parts, we can evaluate $\int_{-\infty}^{\infty} f(x) \cos mx\, dx$ and $\int_{-\infty}^{\infty} f(x) \sin mx\, dx$.

***Example 3.7-8.*** Show that

$$\int_{-\infty}^{\infty} \frac{\cos x \, dx}{\left(x^2 + a^2\right)\left(x^2 + b^2\right)} = \frac{\pi}{a^2 - b^2}\left(\frac{e^{-b}}{b} - \frac{e^{-a}}{a}\right), \quad a > b > 0 \tag{3.7-74}$$

From Equation (3.7-73), we have

$$\int_{-\infty}^{\infty} \frac{\cos x \, dx}{\left(x^2 + a^2\right)\left(x^2 + b^2\right)} = \text{Re}\left\{2\pi i \sum_{n=1}^{k} \text{Res}\left[\frac{e^{iz}}{\left(z^2 + a^2\right)\left(z^2 + b^2\right)}, z_n\right]\right\} \tag{3.7-75}$$

The four poles of the function are

$$z_1 = i\,a, \qquad z_2 = -i\,a, \qquad z_3 = i\,b, \qquad z_4 = -i\,b \tag{3.7-76a-d}$$

Of the four poles, only $z_1$ and $z_3$ are in the upper half plane.

The residue $b_1$ at $z_1$ is given by

$$b_1 = \lim_{z \to i\,a} \frac{(z - i\,a)\, e^{iz}}{(z - i\,a)(z + i\,a)(z - i\,b)(z + i\,b)} \tag{3.7-77a}$$

$$= \frac{e^{-a}}{2i\,a\left(b^2 - a^2\right)} \tag{3.7-77b}$$

Similarly the residue $b_1$ at $z_3$ is

$$b_1 = \frac{e^{-b}}{2i\,b\left(a^2 - b^2\right)} \tag{3.7-78}$$

Substituting Equations (3.7-77b, 78) into Equation (3.7-75), we obtain

$$\int_{-\infty}^{\infty} \frac{\cos x \, dx}{\left(x^2 + a^2\right)\left(x^2 + b^2\right)} = \text{Re}\left\{2\pi i\left[\frac{e^{-a}}{2i\,a\left(b^2 - a^2\right)} + \frac{e^{-b}}{2i\,b\left(a^2 - b^2\right)}\right]\right\} \tag{3.7-79a}$$

$$= \frac{\pi}{\left(a^2 - b^2\right)}\left[\frac{e^{-b}}{b} - \frac{e^{-a}}{a}\right] \tag{3.7-79b}$$

## Poles on the Real Axis

So far, we have assumed that the poles of $f(z)$ were not on the real axis, since we chose $Q(x) \neq 0$. If some of the poles are on the real axis, we indent the contour shown in Figure 3.7-2 by making small semi-circles in the upper half plane to remove the poles from the real axis. Suppose $f(z)$ has a pole at $z = a$, where $a$ is real. The contour is indented by making a semi-circle of radius $\varepsilon$, centered at $z = a$ in the upper half plane as shown in Figure 3.7-3. We denote the small semi-circle by $\gamma$. The contour integral given in Equation (3.7-59) is now modified as follows

$$\int_{-R}^{a-\varepsilon} f(x)\, dx + \int_{\gamma} f(z)\, dz + \int_{a+\varepsilon}^{R} f(x)\, dx + \int_{\Gamma} f(z)\, dz \;=\; 2\pi i \sum_{n=1}^{k} \mathrm{Res}\left[f(z), z_n\right] \qquad (3.7\text{-}80)$$

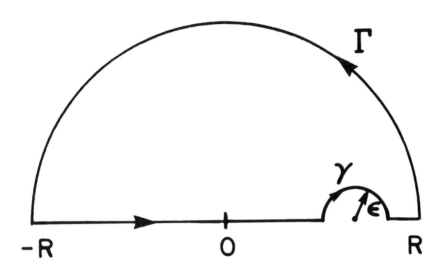

**FIGURE 3.7-3    Contour integral with a pole on the real axis**

It was shown earlier that in the limit as $R \longrightarrow \infty$, $\displaystyle\int_{\Gamma} f(z)\, dz \longrightarrow 0$. We now consider the integral around $\gamma$. On $\gamma$, we have

$$z \;=\; a + \varepsilon\, e^{i\theta} \qquad\qquad\qquad\qquad\qquad (3.7\text{-}81)$$

and the contour integral becomes

$$\int_{\gamma} f(z)\, dz \;=\; \int_{\pi}^{0} f(a + \varepsilon e^{i\theta})\, \varepsilon\, i\, e^{i\theta}\, d\theta \qquad\qquad (3.7\text{-}82)$$

Since $f(z)$ has a simple pole at $z = a$, we can write

$$f(z) = g(z)/(z-a) \tag{3.7-83}$$

where $g(z)$ is analytic and can be expanded as a Taylor series about the point $a$. The function $g(z)$ can be approximated by

$$g(z) = g(a) + 0(\varepsilon) \tag{3.7-84}$$

Substituting Equations (3.7-83, 84) into Equation (3.7-82), we obtain, as $\varepsilon \longrightarrow 0$,

$$\int_{\gamma} f(z)\, dz \ \rightarrow \ i \int_{\pi}^{0} g(a)\, d\theta \tag{3.7-85a}$$

$$\rightarrow \ -i\, \pi\, g(a) \tag{3.7-85b}$$

We note from Equations (3.7-14, 83) that $g(a)$ is the residue of $f(z)$ at $a$ and is denoted by $R_a$. In the limit, as $R \longrightarrow \infty$ and $\varepsilon \longrightarrow 0$, Equation (3.7-80) becomes

$$\int_{-\infty}^{\infty} f(x)\, dx \ = \ 2\pi i \sum_{n=1}^{k} \text{Res}\left[f(z), z_n\right] + \pi\, i\, R_a \tag{3.7-86}$$

If there is more than one simple pole on the real axis and if we denote these poles by $a_1, a_2, ... , a_\ell$, we replace $R_a$ in Equation (3.7-86) by the sum of $R_{a_i}$, that is to say, the sum of the residues at $a_1$, $a_2, ... , a_\ell$. The same modification can be applied to Equation (3.7-72). We illustrate this by considering an example.

***Example 3.7-9***.  Show that

$$\int_{-\infty}^{\infty} \frac{\cos x \, dx}{a^2 - x^2} \ = \ \frac{\pi \sin a}{a}, \quad a \text{ is real} \tag{3.7-87}$$

We consider the integral $\displaystyle\int_{C} \frac{e^{iz}}{a^2 - z^2}\, dz$, where $C$ is the indented contour to be defined later. The function $f(z) \ [= 1/(a^2 - z^2)]$ has two poles on the real axis, at $-a$ and $a$. We indent the contour by making two small semi-circles of radius $\varepsilon$, one at $z = -a$ denoted by $\gamma_1$, and the other at $z = a$ denoted by $\gamma_2$. The contour $C$ is shown in Figure 3.7-4.

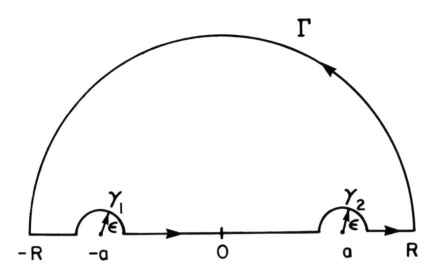

**FIGURE 3.7-4**     **Contour integral with two poles on the real axis**

Equation (3.7-72) is modified to

$$\int_{-R}^{-a-\varepsilon} \frac{e^{ix}}{a^2-x^2}\,dx + \int_{\gamma_1} \frac{e^{iz}}{a^2-z^2}\,dz + \int_{-a+\varepsilon}^{a-\varepsilon} \frac{e^{ix}}{a^2-x^2}\,dx + \int_{\gamma_2} \frac{e^{iz}}{a^2-z^2}\,dz + \int_{a+\varepsilon}^{R} \frac{e^{ix}}{a^2-x^2}\,dx$$

$$+ \int_{\Gamma} \frac{e^{iz}}{a^2-z^2}\,dz \; = \; 2\,\pi i \sum_{n=1}^{k} \text{Res}\left[\frac{e^{iz}}{a^2-z^2}, \, z_n\right] \tag{3.7-88}$$

Using Jordan's lemma, we have as $R \longrightarrow \infty$

$$\int_{\Gamma} \frac{e^{iz}}{a^2-z^2}\,dz \; = \; 0 \tag{3.7-89}$$

To evaluate the integral around $\gamma_1$, we write

$$z \; = \; -a + \varepsilon\,e^{i\theta} \tag{3.7-90}$$

$$\int_{\gamma_1} \frac{e^{iz}}{a^2-z^2}\,dz \; = \; \int_{\pi}^{0} \frac{\exp\!\left(-ia + i\varepsilon\,e^{i\theta}\right)}{\varepsilon\,e^{i\theta}\left(2a - \varepsilon\,e^{i\theta}\right)}\, i\,\varepsilon\,e^{i\theta}\,d\theta \tag{3.7-91}$$

In the limit as $\varepsilon \longrightarrow 0$, Equation (3.7-91) simplifies to

$$\int_{\gamma_1} \frac{e^{iz}}{a^2 - z^2} \, dz \;\rightarrow\; \frac{i\,e^{-ia}}{2a} \int_{\pi}^{0} d\theta \tag{3.7-92a}$$

$$\rightarrow\; -\frac{i\,\pi\,e^{-ia}}{2a} \tag{3.7-92b}$$

Using Equation (3.7-14), we find that the residue $R_{-a}$ at $z = -a$ is given by

$$R_{-a} \;=\; \lim_{z \to -a} \frac{(z+a)\,e^{i\,z}}{(a-z)(a+z)} \tag{3.7-93a}$$

$$=\; \frac{e^{-ia}}{2a} \tag{3.7-93b}$$

From Equations (3.7-92b, 93b), we obtain

$$\lim_{\varepsilon \to 0} \int_{\gamma_1} \frac{e^{iz}}{a^2 - z^2} \, dz \;=\; -i\,\pi\,R_{-a} \tag{3.7-94}$$

Equation (3.7-94) is a special case of Equation (3.7-85b).

Similarly we have

$$\lim_{\varepsilon \to 0} \int_{\gamma_2} \frac{e^{iz}}{a^2 - z^2} \, dz \;=\; -i\,\pi\,R_{a} \tag{3.7-95}$$

where $R_a$ is the residue at $z = a$.

The function $e^{iz} / (a^2 - z^2)$ has no other poles except at $-a$ and $a$, so the right side of Equation (3.7-88) is zero. In the limiting case $R \longrightarrow \infty$, $\varepsilon \longrightarrow 0$, Equation (3.7-88) simplifies to

$$\int_{-\infty}^{\infty} \frac{e^{ix}}{a^2 - x^2} \, dx \;=\; i\,\pi \left[ R_a + R_{-a} \right] \tag{3.7-96}$$

The residue $R_a$ is given by

$$R_a \;=\; \lim_{z \to a} \frac{(z-a)\,e^{i\,z}}{(a-z)(a+z)} \tag{3.7-97a}$$

$$= -\frac{e^{ia}}{2a} \tag{3.7-97b}$$

Substituting the values of $R_a$ and $R_{-a}$ into Equation (3.7-96), we obtain

$$\int_{-\infty}^{\infty} \frac{e^{ix}}{a^2 - x^2} \, dx = \frac{i\pi}{2a} \left[ e^{-ia} - e^{ia} \right] \tag{3.7-98a}$$

$$= \frac{\pi}{a} (\sin a) \tag{3.7-98b}$$

Separating Equation (3.7-98b) into its real and imaginary parts, we have

$$\int_{-\infty}^{\infty} \frac{\cos x}{a^2 - x^2} \, dx = \frac{\pi \sin a}{a} \tag{3.7-99a}$$

$$\int_{-\infty}^{\infty} \frac{\sin x}{a^2 - x^2} \, dx = 0 \tag{3.7-99b}$$

## 3.8   CONFORMAL MAPPING

In this section, we consider the function f defined by Equation (3.3-5) to be a mapping from one subset of a complex plane to another. From Equations (3.3-6a, b), we regard one complex plane to be z [= (x, y)] and the other to be w [= (u, v)] and f establishes the correspondence between the (x, y) plane and the (u, v) plane. We examine the geometric properties of this **mapping**. If to each point in the (u, v) plane there corresponds one and only one point in the (x, y) plane and vice versa, the mapping is **one to one**.

In Section 3.3, we have defined a curve in the (x, y) plane by introducing a parameter t. The equation of a curve C in the (x, y) plane is written as

$$z(t) = x(t) + i\, y(t) \tag{3.8-1}$$

The curve C is transformed to a curve $\Gamma$ in the (u, v) plane. The equation of $\Gamma$ is written as

$$w(t) = f(z) = u(t) + i\, v(t) \tag{3.8-2a,b}$$

Suppose the curve C passes through a point P at which t takes the value $t_0$. The tangent to C at P is given by

$$\frac{dz}{dt}\bigg|_{t_0} = \frac{dx}{dt}\bigg|_{t_0} + i\frac{dy}{dt}\bigg|_{t_0} \qquad (3.8\text{-}3)$$

The tangent to the curve $\Gamma$ in the (u, v) plane at the point $[u(t_0), v(t_0)]$ is

$$\frac{dw}{dt}\bigg|_{t_0} = \frac{du}{dt}\bigg|_{t_0} + i\frac{dv}{dt}\bigg|_{t_0} \qquad (3.8\text{-}4)$$

Using the chain rule $\frac{du}{dt}$ and $\frac{dv}{dt}$ can be written, in matrix form, as

$$\begin{bmatrix} \dfrac{du}{dt} \\[3mm] \dfrac{dv}{dt} \end{bmatrix} = \begin{bmatrix} \dfrac{\partial u}{\partial x} & \dfrac{\partial u}{\partial y} \\[3mm] \dfrac{\partial v}{\partial x} & \dfrac{\partial v}{\partial y} \end{bmatrix} \begin{bmatrix} \dfrac{dx}{dt} \\[3mm] \dfrac{dy}{dt} \end{bmatrix} \qquad (3.8\text{-}5)$$

Equation (3.8-5) expresses the **transformation** of tangent vectors from the z-plane to the w-plane. The condition for a unique solution is that the **determinant**

$$J = \begin{vmatrix} \dfrac{\partial u}{\partial x} & \dfrac{\partial u}{\partial y} \\[3mm] \dfrac{\partial v}{\partial x} & \dfrac{\partial v}{\partial y} \end{vmatrix} \neq 0 \qquad (3.8\text{-}6)$$

The determinant $J$ is the **Jacobian**. If it is non-zero, it ensures that non-zero tangent vectors in one plane are transformed to non-zero tangent vectors in the other plane.

Using the Cauchy-Riemann conditions [Equations (3.3-29a, b)], $J$ can be written as

$$J = \frac{\partial u}{\partial x}\frac{\partial v}{\partial y} - \frac{\partial u}{\partial y}\frac{\partial v}{\partial x} \qquad (3.8\text{-}7a)$$

$$= \left(\frac{\partial u}{\partial x}\right)^2 + \left(\frac{\partial v}{\partial x}\right)^2 \qquad (3.8\text{-}7b)$$

$$= \left|\frac{\partial u}{\partial x} + i\frac{\partial v}{\partial x}\right|^2 \qquad (3.8\text{-}7c)$$

$$= \left|\frac{df}{dz}\right|^2 \qquad (3.8\text{-}7d)$$

The condition that the determinant be non-zero is equivalent to the condition that $\frac{df}{dz}$ be non-zero.

The curves $C_1$ and $C_2$ in the z-plane intersect at $(x_0, y_0)$. Let $\Delta\theta$ be the angle between the tangent vectors to $C_1$ and $C_2$ at $(x_0, y_0)$, measured from $C_1$ to $C_2$ as illustrated in Figure 3.8-1. Suppose $C_1$ and $C_2$ are mapped by f to $\Gamma_1$ and $\Gamma_2$ respectively in the w-plane. The curves $\Gamma_1$ and $\Gamma_2$ intersect at $(u_0, v_0)$ and let $\Delta\alpha$ be the angle between the tangent vectors to $\Gamma_1$ and $\Gamma_2$ at $(u_0, v_0)$, measured from $\Gamma_1$ and $\Gamma_2$, as shown in Figure 3.8-1. If $\Delta\theta$ is equal to $\Delta\alpha$, the mapping is **conformal**. If the magnitude of $\Delta\theta$ is equal to the magnitude $\Delta\alpha$, but the sense is not the same, the mapping is **isogonal**. Thus, in a conformal mapping, both the magnitude and sense of angles are preserved. The conditions that f represents a conformal mapping are

(a)     f is analytic;

(b)     f is single valued;

(c)     $\dfrac{df}{dz}$ is non-zero.

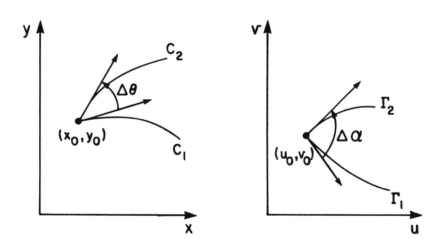

**FIGURE 3.8-1     Conformal mapping**

The Jacobian J introduced earlier plays an important role in the theory of transformations. Here, we briefly review the case of two variables. Further discussions are given in Chapter 4. Let $u_1$ and $u_2$ be two differentiable functions of two variables $x_1$ and $x_2$. The Jacobian J is defined as

$$J = \frac{\partial(u_1, u_2)}{\partial(x_1, x_2)} = \begin{vmatrix} \dfrac{\partial u_1}{\partial x_1} & \dfrac{\partial u_1}{\partial x_2} \\[3mm] \dfrac{\partial u_2}{\partial x_1} & \dfrac{\partial u_2}{\partial x_2} \end{vmatrix} \qquad (3.8\text{-}8a,b)$$

$J$ is also known as the **functional determinant**.

For the inverse function to exist, that is to say, for the possibility to write

$$u_1 = u_1 (x_1, x_2), \qquad\qquad u_2 = u_2 (x_1, x_2) \qquad (3.8\text{-}9a,b)$$

we require that the Jacobian to be non-zero.

The points at which $J$ is zero are **singular points**.

By differentiating Equations (3.8-9a, b) partially with respect to $x_1$ and $x_2$, and solving for the partial derivatives $\dfrac{\partial x_i}{\partial u_j}$, we obtain

$$\frac{\partial x_1}{\partial u_1} = \frac{1}{J} \frac{\partial u_2}{\partial x_2} , \qquad\qquad \frac{\partial x_1}{\partial u_2} = -\frac{1}{J} \frac{\partial u_1}{\partial x_2} \qquad (3.8\text{-}10a,b)$$

$$\frac{\partial x_2}{\partial u_1} = -\frac{1}{J} \frac{\partial u_2}{\partial x_1} , \qquad\qquad \frac{\partial x_2}{\partial u_2} = \frac{1}{J} \frac{\partial u_1}{\partial x_1} \qquad (3.8\text{-}10c,d)$$

We can also write $1/J$ as

$$\frac{1}{J} = \begin{vmatrix} \dfrac{\partial x_1}{\partial u_1} & \dfrac{\partial x_1}{\partial u_2} \\[5mm] \dfrac{\partial x_2}{\partial u_1} & \dfrac{\partial x_2}{\partial u_2} \end{vmatrix} \qquad (3.8\text{-}11)$$

If the Jacobian vanishes, $u_1$ and $u_2$ are not independent. That is to say, there exists a relationship between $u_1$ and $u_2$ and

$$f (u_1, u_2) = 0 \qquad (3.8\text{-}12)$$

In Equation (3.8-12), $x_1$ and $x_2$ do not occur explicitly. This is analogous to the case in linear algebra where two vectors $\underline{a} (a_1, a_2)$ and $\underline{b} (b_1, b_2)$ are linearly dependent if the determinant

$$\begin{vmatrix} a_1 & a_2 \\ b_1 & b_2 \end{vmatrix} = 0 \tag{3.8-13}$$

If $x_1$ and $x_2$ are also functions of $y_1$ and $y_2$, we have

$$\frac{\partial(u_1, u_2)}{\partial(y_1, y_2)} = \frac{\partial(u_1, u_2)}{\partial(x_1, x_2)} \frac{\partial(x_1, x_2)}{\partial(y_1, y_2)} \tag{3.8-14}$$

If a closed region $R$ in the $(x_1, x_2)$ plane is mapped into a closed region $R'$ in the $(u_1, u_2)$ plane, the double integral of any function $\phi$ over $R$ is given by

$$\iint\limits_{R} \phi(x_1, x_2) \, dx_1 \, dx_2 = \iint\limits_{R'} \phi[x_1(u_1, u_2), x_2(u_1, u_2)] \, |J| \, du_1 \, du_2 \tag{3.8-15}$$

If $\phi$ is unity, we obtain the area of the closed region. The Jacobian $J$ gives the magnification of the area due to the transformation.

We now consider some transformations in the complex plane.

**Linear Transformation**

Consider the transformation

$$w = A z + B \tag{3.8-16}$$

where $A$ and $B$ are complex constants.

If we take $A$ to be unity, Equation (3.8-16) becomes

$$u + i v = (x + i y) + (B_1 + i B_2) \tag{3.8-17}$$

where $B$ is written as $B_1 + i B_2$, $B_1$ and $B_2$ are real constants.

Separating Equation (3.8-17) into its real and imaginary parts, we obtain

$$(u, v) = (x + B_1, y + B_2) \tag{3.8-18}$$

Equation (3.8-18) shows that a point $(u, v)$ in the w-plane is mapped to a point $(x + B_1, y + B_2)$ in the z-plane. This corresponds to a displacement of $(x, y)$ by $(B_1, B_2)$ as shown in Figure 3.8-2.

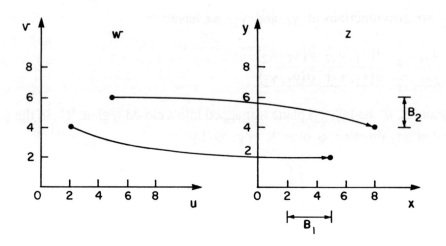

**FIGURE 3.8-2    Translation**

If we take B to be zero and we write w, A and z in their polar form as

$$w = \rho\, e^{i\phi}, \qquad A = a\, e^{i\alpha}, \qquad z = r\, e^{i\theta} \qquad\qquad (3.8\text{-}19a,b,c)$$

Equation (3.8-16) becomes

$$\rho\, e^{i\phi} = a\, r\, e^{i(\alpha+\theta)} \qquad\qquad (3.8\text{-}20)$$

We deduce

$$\rho = ar, \qquad\qquad \phi = \alpha + \theta \qquad\qquad (3.8\text{-}21a,b)$$

Equation (3.8-21a) represents a contraction (a < 1) or an expansion (a > 1) of the radius vector r by the factor a. Equation (3.8-21b) is a rotation through an angle $\alpha$. This transformation is illustrated in Figure 3.8-3. Combining these two cases, we deduce that the linear transformation given by Equation (3.8-16) represents a displacement, a magnification, and a rotation. The shape of all the figures is preserved.

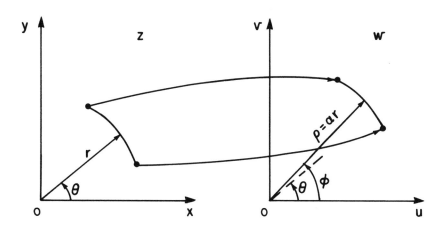

**FIGURE 3.8-3    Rotation and magnification**

***Example 3.8-1***. A rectangular region R in the z-plane is bounded by

$$x = 0 , \qquad x = 1 , \qquad y = 0 , \qquad y = 2 \qquad\qquad (3.8\text{-}22a\text{-}d)$$

Determine the region $R'$ of the w-plane into which R is mapped under the transformation

$$w = (1 + i) z + (1 - 2i) \qquad\qquad (3.8\text{-}23)$$

Equation (3.8-23) can be written as

$$u + i v = (1 + i) (x + i y) + (1 - 2i) \qquad\qquad (3.8\text{-}24a)$$

$$= (x - y + 1) + i (x + y - 2) \qquad\qquad (3.8\text{-}24b)$$

It follows that

$$u = x - y + 1, \qquad\qquad v = x + y - 2 \qquad\qquad (3.8\text{-}25a,b)$$

From Equation (3.8-25a, b), we find that the points (0, 0), (1, 0), (1, 2) and (0, 2) in the z-plane are mapped into (1, – 2), (2, – 1), (0, 1) and (– 1, 0) respectively in the w-plane. The lines

$$y = 0, \qquad x = 1, \qquad y = 2, \qquad x = 0 \qquad\qquad (3.8\text{-}26a\text{-}d)$$

in the z-plane are mapped into

$$u - v = 3, \qquad u + v = 1, \qquad u - v = -1, \qquad u + v = -1 \tag{3.8-27a-d}$$

respectively in the w-plane.

The rectangle R in the z-plane is mapped into rectangle R' in the w-plane, as shown in Figure 3.8-4. We note that all the points in the z-plane are displaced by $(1, -2)$ in the w-plane, the distance between two points has been magnified by the factor $\sqrt{2}$ and all the lines have undergone a rotation of $\pi/4$.

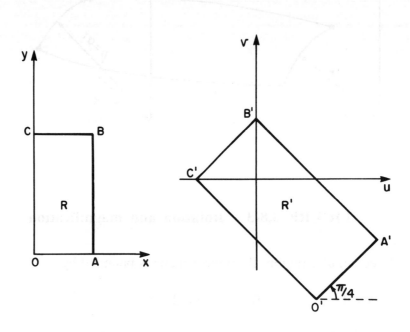

**FIGURE 3.8-4     Mapping of a rectangle in the z-plane
to another rectangle in the w-plane**

The transformation can also be deduced by comparing Equations (3.8-16, 23). We identify A and B to be

$$A = (1 + i) = \sqrt{2}\, e^{i\pi/4} \tag{3.8-28a,b}$$

$$B = (1 - 2i) \tag{3.8-28c}$$

The transformation is as described earlier.

**Reciprocal Transformation**

Consider the transformation

$$w = \frac{1}{z} \tag{3.8-29}$$

Writing $z$ in its polar form, Equation (3.8-29) becomes

$$w = \frac{1}{r} e^{-i\theta} \qquad\qquad (3.8\text{-}30)$$

From Equation (3.8-30), we deduce that the **reciprocal transformation** consists of an **inversion** with respect to the unit circle and a reflection about the real axis. This is illustrated in Figure 3.8-5. The process of inversion was introduced in Example 3.5-7. We recall that if $z_0$ is a point in the complex plane, its inverse point $z_1$, with respect to the unit circle, is given by

$$z_1 = \frac{1}{\overline{z}_0} = \frac{1}{r} e^{i\theta} \qquad\qquad (3.8\text{-}31a,b)$$

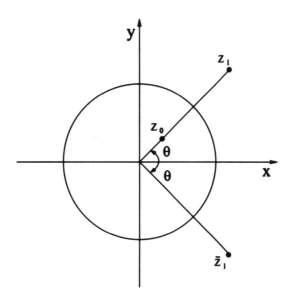

**FIGURE 3.8-5    Reciprocal transformation**

On reflecting about the real axis, $z_1$ is transformed to $w$.

We note that, in this case, the inverse mapping is also a reciprocal transformation. That is to say, from Equation (3.8-29), we obtain

$$z = \frac{1}{w} \qquad\qquad (3.8\text{-}32)$$

On applying the transformation twice in succession, we obtain the identity transformation. The reciprocal transformation maps circles or straight lines into straight lines and circles. The equation

$$a\,(x^2 + y^2) + b_1 x + b_2 y + c = 0 \tag{3.8-33}$$

represents a circle if a is non-zero and a straight line if a is zero.

Equation (3.8-33) can also be written as

$$a z \overline{z} + b \overline{z} + \overline{b} z + c = 0 \tag{3.8-34}$$

where a and c are real constants and b is a complex constant.

If a is zero, we have a straight line; if a is non-zero and c is zero, we have a circle passing through the origin.

Substituting Equation (3.8-29) into Equation (3.8-34), we obtain

$$c w \overline{w} + \overline{b}\,\overline{w} + b w + a = 0 \tag{3.8-35}$$

From Equations (3.8-34, 35), we deduce the following

(i)     the straight line not passing through the origin in the z-plane (a = 0, c ≠ 0) is mapped into a circle passing through the origin in the w-plane;

(ii)    the circle passing through the origin in the z-plane (a ≠ 0, c = 0) is mapped into a straight line not passing through the origin in the w-plane;

(iii)   the straight line passing through the origin in the z-plane (a = c = 0) is mapped into a straight line passing through the origin in the w-plane;

(iv)    the circle not passing through the origin in the z-plane (a ≠ 0, c ≠ 0) is mapped into a circle not passing through the origin in the w-plane.

Note that the unit circle, center at the origin in the z-plane, is mapped into a unit circle, center at the origin in the w-plane. But the points inside the circle in the z-plane are mapped to points outside the circle in the w-plane.

## Bilinear Transformation

The **bilinear** (Möbius) **transformation** is given by

$$w = \frac{az + b}{cz + d} \tag{3.8-36}$$

where a, b, c, d are complex constants and (ad − bc) is non-zero.

Equation (3.8-36) can be written as

$$w = \frac{a}{c} + \frac{(bc - ad)}{c(cz + d)} \tag{3.8-37a}$$

$$= \frac{a}{c} + \frac{(bc - ad)}{c} z^* \tag{3.8-37b}$$

$$z^* = \frac{1}{w^*} \tag{3.8-37c}$$

$$w^* = cz + d \tag{3.8-37d}$$

The bilinear transformation is a combination of a translation, a rotation, a magnification, and an inversion. If $c$ is zero, Equation (3.8-36) reduces to Equation (3.8-16).

Simplifying Equation (3.8-36) yields

$$czw + dw - az - b = 0 \tag{3.8-38}$$

Equation (3.8-38) is bilinear in $z$ and $w$ (linear in both $z$ and $w$).

From Equation (3.8-38), we can obtain the inverse transformation

$$z = + \frac{b - dw}{cw - a} \tag{3.8-39}$$

which is again a bilinear transformation.

Let $z_1$, $z_2$, $z_3$ and $z_4$ be any four points in the z-plane and $w_1$, $w_2$, $w_3$ and $w_4$ be the corresponding points in the w-plane. Using Equation (3.8-36), we have

$$w_r - w_s = \frac{az_r + b}{cz_r + d} - \frac{az_s + b}{cz_s + d} \tag{3.8-40a}$$

$$= \frac{(ad - bc)(z_r - z_s)}{(cz_r + d)(cz_s + d)}, \qquad r, s = 1, 2, 3, 4 \tag{3.8-40b}$$

From Equation (3.8-40b), it follows that

$$\frac{(w_1 - w_4)(w_3 - w_2)}{(w_1 - w_2)(w_3 - w_4)} = \frac{(z_1 - z_4)(z_3 - z_2)}{(z_1 - z_2)(z_3 - z_4)} \tag{3.8-41}$$

The ratio in Equation (3.8-41) is the **cross ratio** of the four points and the cross ratio is an invariant under a bilinear transformation. If three points $(z_1, z_2, z_3)$ in the z-plane are known to map to three points in the w-plane, we obtain via Equation (3.8-41) a unique relationship between $w$ $(= w_4)$ and $z$ $(= z_4)$.

***Example 3.8-2***. Determine the bilinear transformation that maps the points $(1 + i, -i, 2 - i)$ in the z-plane into $(0, 1, i)$ in the w-plane.

From Equation (3.8-41), we obtain

$$\frac{(-w)(i-1)}{(-1)(i-w)} = \frac{(1+i-z)(2)}{(1+2i)(2-i-z)} \tag{3.8-42}$$

On simplifying Equation (3.8-42), we obtain

$$w = \frac{2(1-i+iz)}{(5-3i)-z(1+i)} \tag{3.8-43}$$

***Example 3.8-3***. Find the bilinear transformation that maps the upper half plane $\mathrm{Im}\,(z) \geq 0$ into the unit circle $|w| \leq 1$.

From Equation (3.8-36), we deduce

(i)      the point $z = -b/a$ corresponds to $w = 0$;

(ii)     the point $z = -d/c$ corresponds to $w = \infty$.

We denote $-b/a$ by $z_0$ and $-d/c$ by $\bar{z}_0$, with the condition that $\mathrm{Im}\,(z_0)$ is positive. The point $z_0$ is then mapped to the origin which is inside the unit circle, and the lower plane $(\bar{z}_0)$ is mapped into the outside of the unit circle. The bilinear transformation can be written as

$$w = \alpha\,\frac{(z-z_0)}{(z-\bar{z}_0)} \tag{3.8-44}$$

The origin in the z-plane is mapped into a point on the unit circle in the w-plane. Using Equation (3.8-44), we have

$$|w| = |\alpha| = 1 \tag{3.8-45a,b}$$

Combining Equations (3.8-44, 45a, b), we obtain

$$w = \frac{e^{i\theta}(z-z_0)}{(z-\bar{z}_0)} \tag{3.8-46}$$

The real axis of the z-plane is mapped into

$$w = \frac{e^{i\theta}(x-z_0)}{(x-\bar{z}_0)} \tag{3.8-47}$$

From Equation (3.8-47), we obtain

$$w \, \overline{w} \; = \; \frac{e^{i\theta}(x - z_0)}{(x - \overline{z}_0)} \; \frac{e^{-i\theta}(x - \overline{z}_0)}{(x - z_0)} \; = \; 1 \tag{3.8-48}$$

Thus the real axis of the z-plane is mapped into the unit circle in the w-plane. It has already been shown that a point on the upper z-plane is mapped into a point inside the unit circle in the w-plane. Equation (3.8-46) is the required transformation.

## Schwarz-Christoffel Transformation

Suppose we have a polygon with interior angles $\alpha_1$, $\alpha_2$ ... and we wish to map its boundary into the real axis ($v = 0$) of the w-plane. The **Schwarz-Christoffel transformation** is the required transformation and is given by

$$\frac{dw}{dz} \; = \; K \, (w - a_1)^{1 - \alpha_1/\pi} \, (w - a_2)^{1 - \alpha_2/\pi} \; ... \tag{3.8-49}$$

where $K$ is a constant and $a_1$, $a_2$ are the real values of $w \, (= u)$ corresponding to the vertices of the polygon.

If one vertex of the polygon is at infinity, its interior angle is zero and we may take the corresponding point $a$ in the w-plane to be infinity. The factor $(w - a_\infty)$ in Equation (3.8-49) may be considered to be a constant which can be absorbed in $K$.

***Example 3.8-4***. An ideal fluid is bounded by a semi-infinite rectangle given by

$$x \geq 0, \qquad 1 \geq y \geq 0 \tag{3.8-50}$$

The motion of the ideal fluid is due to a source of strength $m$ at $\dfrac{(1 + i)}{2}$. Determine the complex potential $\Phi$, which would allow the determination of the velocity and of the streamlines, as shown in Example 3.3-6.

The semi-rectangular channel ABCD is shown in Figure 3.8-6. We map the vertices A, B, C, and D to the points $(-\infty, 0)$, $(-1, 0)$, $(1, 0)$, and $(\infty, 0)$ respectively in the w-plane. In this example, Equation (3.8-49) becomes

$$\frac{dw}{dz} \; = \; K \, (w + 1)^{1/2} \, (w - 1)^{1/2} \tag{3.8-51}$$

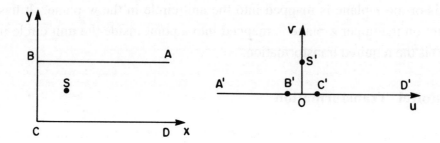

**FIGURE 3.8-6   Mapping a rectangle into the upper half plane**

Integrating Equation (3.8-51), we have

$$z = \frac{1}{K} \int \frac{dw}{\sqrt{w^2 - 1}} \tag{3.8-52a}$$

$$= \frac{1}{K} \cosh^{-1} w + K_1 \tag{3.8-52b}$$

where $K_1$ is a constant.

Inverting Equation (3.8-52b) yields

$$w = \cosh (K z - K K_1) \tag{3.8-53}$$

The mapping of the point B $(= i)$ to B' $(= -1)$ and C $(= 0)$ to C' $(= 1)$ implies

$$-1 = \cosh (i K - K K_1) \tag{3.8-54a}$$

$$1 = \cosh (- K K_1) \tag{3.8-54b}$$

From Equations (3.8-54a, b), we obtain

$$K_1 = 0, \qquad K = \pi \tag{3.8-55a,b}$$

Substituting Equations (3.8-55a, b) into Equation (3.8-53), we have

$$w = \cosh(\pi z) \tag{3.8-56}$$

The source in the z-plane at $\dfrac{(1+i)}{2}$ is mapped into a source at

$$w = \cosh\left[\frac{\pi}{2}(1+i)\right] \tag{3.8-57a}$$

$$= \cosh\frac{\pi}{2}\cosh\frac{i\pi}{2} + \sinh\frac{\pi}{2}\sinh\frac{i\pi}{2} \tag{3.8-57b}$$

$$= i\sinh\frac{\pi}{2} \tag{3.8-57c}$$

The source at $S\left[=\dfrac{(1+i)}{2}\right]$ in the z-plane is mapped into a source at $S'\left[= i\sinh\dfrac{\pi}{2}\right]$ in the w-plane, as shown in Figure 3.8-6.

The complex potential $\Phi$ due to a source of strength $m$ at $i\sinh\dfrac{\pi}{2}$ and a wall along the real axis is [Milne-Thomson, 1965, p. 210]

$$\Phi = -m\,\ell n\left(w - i\sinh\frac{\pi}{2}\right) - m\,\ell n\left(w + i\sinh\frac{\pi}{2}\right) \tag{3.8-58a}$$

$$= -m\,\ell n\left(w^2 + \sinh^2\frac{\pi}{2}\right) \tag{3.8-58b}$$

$$= -m\,\ell n\left(\cosh^2\pi z + \sinh^2\frac{\pi}{2}\right) \tag{3.8-58c}$$

### Joukowski Transformation

The **Joukowski transformation** plays an important role in aerodynamics. It transforms an aerofoil into a circle. The transformation can be written as

$$w = z + \frac{1}{z} \tag{3.8-59}$$

The singular points of the transformation are the points at which $\dfrac{dw}{dz}$ is zero. From Equation (3.8-59), we find that the singular points are

$$z = \pm 1 \tag{3.8-60}$$

Consider a circle of radius $a$ with center at the origin in the z-plane. In the w-plane, we have

$$w = a\,e^{i\theta} + \frac{e^{-i\theta}}{a} \tag{3.8-61}$$

Separating Equation (3.8-61) into its real and imaginary parts, we obtain

$$u = \frac{(a^2 + 1)}{a} \cos\theta, \qquad v = \frac{(a^2 - 1)}{a} \sin\theta \qquad\qquad (3.8\text{-}62a,b)$$

Eliminating $\theta$ from Equations (3.8-62a, b) yields

$$\frac{a^2 u^2}{(a^2 + 1)} + \frac{a^2 v^2}{(a^2 - 1)} = 1 \qquad\qquad (3.8\text{-}63)$$

A circle in the z-plane is mapped into an ellipse in the w-plane, as long as the radius of the circle is not one, which is equivalent to saying that the circle does not pass through the singular points. If the radius of the circle is one ($a = 1$), Equations (3.8-62a, b) become

$$u = 2 \cos\theta, \qquad v = 0 \qquad\qquad (3.8\text{-}64a,b)$$

The circle passing through both singular points in the z-plane is mapped into a segment of the real axis in the w-plane, as shown in Figure 3.8-7. A circle passing through one singular point and enclosing the other singular point in the z-plane is mapped into an aerofoil in the w-plane.

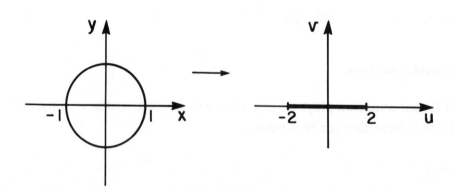

**FIGURE 3.8-7    Mapping of a unit circle into a segment of the real line**

***Example 3.8-5.*** Discuss the transformation of the circles $C_1$, $C_2$, and $C_3$ in the z-plane into the w-plane under the Joukowski transformation [Equation (3.8-59)]. The center of $C_1$ and $C_3$ is at the origin and their radii are one and two respectively. The center of $C_2$ is on the real axis at $(1/2, 0)$ and its radius is $3/2$.

The circles $C_1$, $C_2$, and $C_3$ are shown in Figure 3.8-8. The circle $C_2$ passes through one singular point $(z = -1)$ and encloses the other singular point $(z = 1)$. It also intersects $C_1$ at the singular point A and $C_3$ at the point B $[= (2, 0)]$.

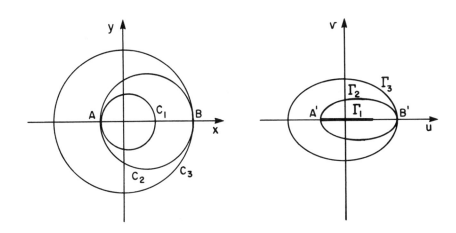

**FIGURE 3.8-8    Mapping of the circles $C_1$, $C_2$, and $C_3$ from the z-plane to the w-plane**

The unit circle $C_1$ is mapped into $\Gamma_1$, a segment of the real axis in the w-plane [Equation (3.8-64a, b)] and can be written as

$$-2 \leq u \leq 2, \qquad v = 0 \tag{3.8-65a,b}$$

The circle $C_3$ is mapped into $\Gamma_3$, an ellipse [Equation (3.8-63)] and, in this example, can be written as

$$\frac{u^2}{25} + \frac{v^2}{9} = \frac{1}{4} \tag{3.8-66}$$

The equation for $C_2$ is given by

$$z = \frac{1}{2} + \frac{3}{2} e^{i\theta} \tag{3.8-67}$$

Substituting Equation (3.8-67) into Equation (3.8-59) yields

$$w = \frac{1}{2}\left[(1 + 3\cos\theta) + 3i\sin\theta\right] + \frac{2\left[(1 + 3\cos\theta) - 3i\sin\theta\right]}{10 + 6\cos\theta} \qquad (3.8\text{-}68)$$

Separating $w$ into its real and imaginary parts, we obtain

$$u = (1 + 3\cos\theta)\left(\frac{1}{2} + \frac{1}{5 + 3\cos\theta}\right) \qquad (3.8\text{-}69a)$$

$$v = 3\sin\theta\left(\frac{1}{2} - \frac{1}{5 + 3\cos\theta}\right) \qquad (3.8\text{-}69b)$$

The circle $C_2$ in the z-plane is transformed into an aerofoil $\Gamma_2$ in the w-plane. From Equations (3.8-69a, b), it is seen that $\Gamma_2$ is symmetrical about the real axis.

The points A and B in the z-plane are mapped into the points A' and B' in the w-plane. The mapping of the three circles from the z-plane to the w-plane is shown in Figure 3.8-8.

●

In Table 3.8-1, we list some useful transformations which map a domain in the z-plane conformally to a domain in the w-plane. A more extensive table is given in Kober (1952).

## TABLE 3.8-1

### Some useful transformations

| z-plane | w-plane | Transformation |
|---------|---------|----------------|
| Upper half plane $y \geq 0$ | unit circle $\lvert w \rvert \leq 1$ | $w = \dfrac{i - z}{i + z}$ |
| Infinite strip of finite width $-\infty \leq y \leq \infty,\ 0 \leq x \leq \pi$ | unit circle $\lvert w \rvert \leq 1$ | $w = \dfrac{1 + i\,e^{iz}}{1 - i\,e^{iz}}$ |
| Region outside the ellipse $\dfrac{x^2}{a^2} + \dfrac{y^2}{b^2} = 1$ | unit circle $\lvert w \rvert \leq 1$ | $z = \dfrac{1}{2}\left[(a - b)w + \dfrac{(a + b)}{w}\right]$ |
| Sector $\lvert z \rvert \leq 1,\ 0 \leq \theta \leq \dfrac{\pi}{n}$ | upper half plane $v \geq 0$ | $w = \left(\dfrac{z^n + 1}{z^n - 1}\right)^2$ |
| Strip $0 \leq y \leq \pi$ | upper half plane $v \geq 0$ | $w = e^z$ |

In Section 3.3, we have mentioned the importance of Laplace's equation and have shown that an analytic solution is a solution of Laplace's equation. In all physical problems, the solution has to satisfy certain boundary conditions. If the geometry of the region of interest is complicated, the problem of imposing the boundary conditions can be demanding. By using a conformal transformation, we can map the complicated region into a simpler region, such as the unit circle. Let $\phi$ be a harmonic function satisfying conditions at the boundary, such as

$$\phi = \phi_0 \quad \text{or} \quad \frac{d\phi}{dn} = 0 \qquad\qquad (3.8\text{-}70\text{a,b})$$

where $\phi_0$ is a real constant and $\dfrac{d\phi}{dn}$ is the derivative of $\phi$ with respect to the normal to the boundary curve.

Using a conformal transformation to the w-plane, the transformed function $\kappa(u, v)$ $\{= \phi [x(u, v), y(u, v)]\}$ is also harmonic satisfying the same boundary conditions [Equations (3.8-70a, b)] at the transformed boundary. If we need to solve Laplace's equation subject to a boundary condition in a complicated domain D, we can transform the domain to a simpler one D' and solve the problem in D'. We can then invert back to D. The main problem is to determine the suitable transformation from D to D'. Riemann has shown that, for any simply connected domain which is not the whole complex plane, there exists a transformation that can map it into a unit disk ($|z| < 1$). Unfortunately it is not shown how the transformation can be obtained.

We have applied this technique in Example 3.8-4. In Example 3.5-7, we have shown that if the value of a harmonic function is given on a circle of radius R, its value at any other point is known. In theory, the solution of Laplace's equation is known for any domain, since we can map the domain into a unit circle.


## PROBLEMS

1a.    Write $z = 4\sqrt{2} + 4\sqrt{2}\, i$ in polar form. Determine all three values of $z^{1/3}$ and plot them in the Argand diagram.

2a.    The complex viscosity $\eta^*$ of a linear viscoelastic liquid is given by

$$\eta^* = \int_0^\infty f(s)\, e^{-i\omega s}\, ds$$

where f, $\omega$, and s are real.

If we write

$$\eta^* = \eta' - i\eta''$$

$$f(s) = K e^{-s/\lambda}$$

where $K$ and $\lambda$ are constants, calculate $\eta'$ and $\eta''$. The quantities $\eta'$ and $\eta''$ are the dynamic viscosity and dynamic rigidity respectively.

Answer: $\dfrac{\lambda K}{1 + \lambda^2 \omega^2}$ , $\dfrac{\omega K \lambda^2}{1 + \lambda^2 \omega^2}$

3b.    Determine and plot the curve given by

$$\left| \frac{z+1}{z-1} \right| = \sqrt{2}$$

4a.    Discuss the continuity and differentiability of the following functions

   (i)     $f(z) = z^2$                          (iii)    $f(z) = z^2 + \dfrac{1}{z}$

   (ii)    $f(z) = \dfrac{1}{z}$                          (iv)    $f(z) = |z|$

5a.    Starting form the definition of $f'(z_0)$ given by Equation (3.3-26), use the Cauchy-Riemann conditions to obtain two other expressions for $f'(z_0)$, one in terms of $u$ only and the other in terms of $v$ only.

6b.    Show that the equation

$$u(x, y) = \ln \sqrt{x^2 + y^2}$$

is a harmonic function.

Obtain the harmonic conjugate $v(x, y)$ and the function $f(z)$. If $f(z)$ is the complex potential $\Phi$, identify the potential $\phi$ and the stream function $\psi$. The stream lines are given by $\psi = $ constant, what are the stream lines in this case? Do they represent the flow due to a source at the origin?

Answer: $\arctan(y/x)$

7b.    If $f(z)$ is an analytic function, show that

$$\left( \frac{\partial^2}{\partial x^2} + \frac{\partial^2}{\partial y^2} \right) |f(z)|^2 = 4|f'(z)|^2$$

8a.     Determine the real and imaginary parts of

$$f(z) = \frac{e^z}{z}$$

Are the Cauchy-Riemann conditions satisfied?  Compute $f'(z)$.  Is the origin a singular point?

9a.     If $f(z, \bar{z})$ is an analytic function, show that $\frac{df}{d\bar{z}} = 0$.

10a.    Calculate $\lim\limits_{z \to 0} \frac{\sin z}{z}$ along the real and imaginary axes.

11b.    Evaluate $\displaystyle\int_{-1}^{1} \bar{z}\, dz$ along the semi-circles $C_1$ and $C_2$.

C$_1$ is the upper semi-circle of unit radius, centered at the origin traversed in the clockwise direction from $(-1, 0)$ to $(1, 0)$ and $C_2$ is the image of $C_1$ about the real axis.  This is shown in Figure 3.P-11b.

If C is the closed curve $(C_1 - C_2)$, what is the value of $\displaystyle\int_C \bar{z}\, dz$ ?  Is the integral zero?  If not, what condition(s) in the Cauchy's theorem is (are) not satisfied?

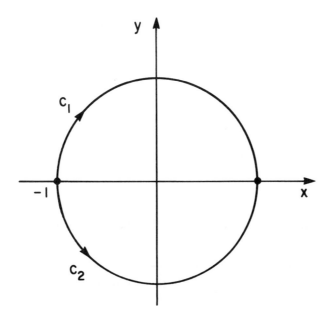

**FIGURE 3.P-11b     Integration around semi-circles**

12a.   Evaluate the integral $\displaystyle\int_C \frac{\sin z}{(z - z_0)^2}\, dz$ , where C is a simple closed curve for the following

cases

(i)      $z_0$ is not enclosed by C                                                                     Answer: 0

(ii)     $z_0$ is enclosed by C                                                                        Answer: $2\pi i \cos z_0$

13b.   The circulation $\Gamma$ round a closed curve C in a two-dimensional flow is given by

$$\Gamma = \text{Re} \int_C \frac{d\Phi}{dz}\, dz$$

where $\Phi$ is the complex potential.

If C is the unit circle with center at the origin, for which of the following $\Phi$ is $\Gamma$ zero?

(i)      $\Phi = z$

(ii)     $\Phi = i \log z$

(iii)    $\Phi = \dfrac{1}{z - 2}$

If $\Gamma$ is not zero, evaluate $\Gamma$.

What is the value of $\Gamma$ if the radius of C is changed from 1 to $3/2$?

14a.   Evaluate the integral

$$\int_C \frac{(\cos^2 z + 2\cosh^4 z)^4}{z}\, dz$$

where C is any simple closed curve that encloses the origin.

Answer: $162\pi i$

15b.   Show that the expansion of $\dfrac{1}{1 - z}$ about the origin is

$$\frac{1}{1 - z} = 1 + z + \dots + z^n + \dots , \quad |z| < 1$$

Expand the same function about the point (0, 1/2). Note that this point lies inside the circle of convergence and is not on the line joining the origin to the singular point. Determine the radius of convergence of the new series.

Sketch the circles of convergence of the two series. Do they coincide? If not, are there points at which the series about the point (0, 1/2) is convergent while the series about the origin is not convergent? This process of extending the region of convergence is called the analytic continuation of the function.

16a.   Find the Taylor series of $\sin z$ about the point $z = \pi/2$.

17a.   Obtain the Laurent series of $\dfrac{e^{2z}}{(z-1)^3}$ about the point $z = 1$. What is its radius of convergence?

18a.   Find the Laurent series of $\dfrac{z}{(z-1)(z-3)}$

   (i)     in powers of $z$,

   (ii)    in powers of $(z-1)$.

19a.   Expand $\mathrm{Ln}\left(\dfrac{z}{1-z}\right)$ for $|z| > 1$.

20a.   Locate the poles of the following functions and specify their order

   (i)     $\dfrac{z(z-2)}{(z+1)^3(z^2+4)}$                    (ii)     $\dfrac{e^z}{\cos^2 z}$

21a.   Locate the singularities of the following functions and discuss their nature

   (i)     $e^{1/z}$                                                  (iii)    $z\sin(1/z)$

   (ii)    $e^z/z^3$                                                (iv)     $(\cos z)/(z-\pi)^2$

22a.   The function $\csc(1/z)$ has singularities at the origin and at points on the real axis given by $z = 1/n\pi$. Is the origin an isolated singular point?

23b.   To describe the behavior of a function $f(z)$ at infinity, we make the transformation $\zeta = 1/z$ and the point at infinity in the z-plane is mapped to the origin in the $\zeta$-plane. Discuss the behavior of the following functions at infinity

   (i)     $e^z$                                                     (iii)    $z\sin(1/z)$

   (ii)    $z^{1/2}$                                                (iv)     $z/[(z-1)(z-3)]$

24a.  Find the residues at the poles of the following functions

   (i)      $1 / (1 - z)$                                    (iv)   $e^z / \sin z$

   (ii)     $1 / (1 - z^2)$                                  (v)    $\cos z / z^3$

   (iii)    $\cos z / \sin z$                                (vi)   $1 / [z^2 (z - 1) (z - 2)]$

                                             Answer:  (i)    $-1$          (iv)  $1$
                                                      (ii)   $-1/2, -1/2$  (v)   $-1/2$
                                                      (iii)  $1$           (vi)  $3/4, -1, 1/4$

25a.  Evaluate the following integrals by means of the Cauchy residue theorem

   (i)    $\displaystyle\int_C \frac{z}{(2z - 1)} \, dz$,  C is the unit circle centered at the origin          Answer: $i\pi/2$

   (ii)   $\displaystyle\int_C \frac{(z^2 + 5z + 11)}{z (z - 1)^2 (z - 3)} \, dz$,  C is the circle $|z| = 2$          Answer: $-35 i \pi / 6$

   (iii)  $\displaystyle\int_C \frac{z}{1 - 2e^{-iz}} \, dz$,  C is any closed contour in the upper plane          Answer: $0$

26b.  The complex potential $\Phi$ of a two-dimensional motion of a fluid is given by

$$\Phi = v_\infty (z + a^2/z) + i k \, \ell n \, (z/a)$$

where $v_\infty$, $k$, and $a$ are constants (real).

Determine the velocity components and show that there are two stagnation points ($\underline{v} = \underline{0}$) on C, the circle of radius $a$ centered at the origin if $v_\infty > k / (2a)$. Verify that the stream function is zero on C. Calculate the components $F_x$ and $F_y$ of the force acting on C.

                                                            Answer: $0$ , $2 \pi \rho \, v_\infty k$

27a.  Show, by the method of contour integration, that

   (i)    $\displaystyle\int_0^{2\pi} \frac{d\theta}{3 - \cos \theta} = \frac{\pi}{\sqrt{2}}$

(ii)    $\displaystyle\int_0^{2\pi} \frac{\sin^2\theta \, d\theta}{2 - \cos\theta} = \pi\,(4 - 2\sqrt{3}\,)$

(iii)    $\displaystyle\int_0^{2\pi} \frac{\cos^3\theta \, d\theta}{1 - 2p\cos\theta + p^2} = \frac{\pi\,(1 - p + p^2)}{1 - p}, \quad 0 < p < 1$

28a.    Evaluate, using a semi-circle in the upper half-plane, the following integrals

(i)    $\displaystyle\int_{-\infty}^{\infty} \frac{x^2 \, dx}{1 + x^4}$      (ii)    $\displaystyle\int_{-\infty}^{\infty} \frac{x^2 \, dx}{(x^2 + 4)\,(x^2 + 9)}$      (iii)    $\displaystyle\int_{-\infty}^{\infty} \frac{x^2 \, dx}{(1 + x^2)^2}$

Answer:    (i) $\pi\sqrt{2}\,/2$,    (ii) $\pi/5$,    (iii) $\pi/2$

29b.    Show that

$$\int_{-\infty}^{\infty} \frac{dx}{ax^2 + bx + c} = \frac{2\pi}{\sqrt{(4ac - b^2)}}$$

For what values of a, b, and c is the result valid?

What modifications need to be made to the contour if the conditions on a, b, and c are not satisfied?

30a.    Use Jordan's lemma to calculate

$$\int_{-\infty}^{\infty} \frac{e^{i\alpha x} \, dx}{(x^2 + a^2)}$$

Deduce

(i)    $\displaystyle\int_{-\infty}^{\infty} \frac{\cos \alpha x \, dx}{x^2 + a^2}$      (ii)    $\displaystyle\int_{-\infty}^{\infty} \frac{\sin \alpha \, dx}{x^2 + a^2}$

Answer:    (i) $\pi\,e^{-\alpha a}/a$,    (ii) 0

# CHAPTER 4

# VECTOR AND TENSOR ANALYSIS

## 4.1 INTRODUCTION

In this chapter, we will be dealing mainly with vectors and second order tensors. A **vector** is defined as a quantity that has both magnitude and direction. In a three-dimensional space, it is described by three numbers (components). For example, velocity is a vector and, when referred to rectangular Cartesian axes, it is specified by its three components $(v_x, v_y, v_z)$. A vector is also defined as a **tensor of order one**. A **scalar** has only magnitude and is completely characterized by one number. It is a **tensor of order zero** and temperature is an example of a scalar. A **second order tensor**, in a three-dimensional space, is represented by nine numbers (components). The extra stress tensor in fluid mechanics is an example of a second order tensor and can be denoted by $\tau_{xx}$, $\tau_{xy}$, $\tau_{xz}$, $\tau_{yx}$, $\tau_{yy}$, $\tau_{yz}$, $\tau_{zx}$, $\tau_{zy}$, and $\tau_{zz}$.. Third and fourth order tensors will also be introduced.

## 4.2 VECTORS

In this section, we briefly review some of the known and useful results of vector algebra and of vector calculus. In Table 4.2-1, $\underline{v}$, $\underline{a}$ and $\underline{q}$ are vectors, $c_1$ and $c_2$ are scalars and $\varphi$ and $\psi$ are differentiable scalar functions of position x, y, z, or time t. $\underline{i}$, $\underline{j}$ and $\underline{k}$ are unit vectors along the x, y, z axes respectively.

### TABLE 4.2-1

### Results of vector analysis

---

$$\underline{a} + \underline{v} = \underline{v} + \underline{a} \tag{4.2-1}$$

$$\underline{v} + (\underline{a} + \underline{q}) = (\underline{v} + \underline{a}) + \underline{q} \tag{4.2-2}$$

$$c_1 \underline{v} = \underline{v} c_1 \tag{4.2-3}$$

$$(c_1 + c_2) \underline{v} = c_1 \underline{v} + c_2 \underline{v} \tag{4.2-4}$$

$$\underline{a} \cdot \underline{v} = \underline{v} \cdot \underline{a} \quad \text{(dot or scalar product)} \tag{4.2-5}$$

$$\underline{a} \cdot \underline{v} = |\underline{a}| |\underline{v}| \cos \theta \quad (\theta \text{ is the angle between } \underline{a} \text{ and } \underline{v}) \tag{4.2-6}$$

$$\underline{v} \bullet (\underline{a} + \underline{q}) = \underline{v} \bullet \underline{a} + \underline{v} \bullet \underline{q} \tag{4.2-7}$$

$$c_1 (\underline{v} \bullet \underline{a}) = (c_1 \underline{v}) \bullet \underline{a} = \underline{v} \bullet (c_1 \underline{a}) \tag{4.2-8a,b}$$

$$\underline{a} \times \underline{v} = -\underline{v} \times \underline{a} \quad \text{(cross or vector product)} \tag{4.2-9}$$

$$\underline{a} \times \underline{v} = |\underline{a}| \, |\underline{v}| \, (\sin \theta) \, \underline{n}, \quad \underline{n} \text{ is a unit vector orthogonal to } \underline{v} \text{ and } \underline{a} \tag{4.2-10}$$

$$\underline{a} \times \underline{v} = \begin{vmatrix} \underline{i} & \underline{j} & \underline{k} \\ a_x & a_y & a_z \\ v_x & v_y & v_z \end{vmatrix} \tag{4.2-11}$$

$$\underline{v} \times (\underline{a} + \underline{q}) = \underline{v} \times \underline{a} + \underline{v} \times \underline{q} \tag{4.2-12}$$

$$c_1 (\underline{v} \times \underline{a}) = (c_1 \underline{v}) \times \underline{a} = \underline{v} \times (c_1 \underline{a}) \tag{4.2-13a,b}$$

$$\underline{v} \bullet (\underline{a} \times \underline{q}) = \underline{a} \bullet (\underline{q} \times \underline{v}) = \underline{q} \bullet (\underline{v} \times \underline{a}) \tag{4.2-14a,b}$$

$$\underline{v} \bullet (\underline{a} \times \underline{q}) = \begin{vmatrix} v_x & v_y & v_z \\ a_x & a_y & a_z \\ q_x & q_y & q_z \end{vmatrix} \tag{4.2-15}$$

$$\frac{d}{dt} (\underline{v} + \underline{a}) = \frac{d\underline{v}}{dt} + \frac{d\underline{a}}{dt} \tag{4.2-16}$$

$$\frac{d}{dt} (\underline{v} \bullet \underline{a}) = \underline{v} \bullet \frac{d\underline{a}}{dt} + \frac{d\underline{v}}{dt} \bullet \underline{a} \tag{4.2-17}$$

$$\frac{d}{dt} (\underline{v} \times \underline{a}) = \underline{v} \times \frac{d\underline{a}}{dt} + \frac{d\underline{v}}{dt} \times \underline{a} \tag{4.2-18}$$

$$\frac{d}{dt} (\varphi \, \underline{v}) = \varphi \frac{d\underline{v}}{dt} + \frac{d\varphi}{dt} \, \underline{v} \tag{4.2-19}$$

$$\frac{d}{dt} (\underline{v} \bullet \underline{a} \times \underline{q}) = \underline{v} \bullet \underline{a} \times \frac{d\underline{q}}{dt} + \underline{v} \bullet \frac{d\underline{a}}{dt} \times \underline{q} + \frac{d\underline{v}}{dt} \bullet \underline{a} \times \underline{q} \tag{4.2-20}$$

$$\frac{d}{dt} [\underline{v} \times (\underline{a} \times \underline{q})] = \underline{v} \times \left( \underline{a} \times \frac{d\underline{q}}{dt} \right) + \underline{v} \times \left( \frac{d\underline{a}}{dt} \times \underline{q} \right) + \frac{d\underline{v}}{dt} \times (\underline{a} \times \underline{q}) \tag{4.2-21}$$

$$\underline{\nabla} = \underline{i} \frac{\partial}{\partial x} + \underline{j} \frac{\partial}{\partial y} + \underline{k} \frac{\partial}{\partial z} \tag{4.2-22}$$

$$\underline{\nabla} \varphi = \underline{i} \frac{\partial \varphi}{\partial x} + \underline{j} \frac{\partial \varphi}{\partial y} + \underline{k} \frac{\partial \varphi}{\partial z} = \text{grad } \varphi \tag{4.2-23a,b}$$

$$\underline{\nabla} \cdot \underline{v} = \frac{\partial v_x}{\partial x} + \frac{\partial v_y}{\partial y} + \frac{\partial v_z}{\partial z}$$

$$= \text{divergence } \underline{v} \text{ or } (\text{div } \underline{v})$$

(4.2-24a,b)

$$\underline{\nabla} \times \underline{v} = \begin{vmatrix} \underline{i} & \underline{j} & \underline{k} \\ \dfrac{\partial}{\partial x} & \dfrac{\partial}{\partial y} & \dfrac{\partial}{\partial x} \\ v_x & v_y & v_z \end{vmatrix}$$

$$= \text{curl } \underline{v} \text{ or } (\text{rot } \underline{v})$$

(4.2-25a,b)

$$\underline{\nabla} (\varphi + \psi) = \underline{\nabla} \varphi + \underline{\nabla} \psi$$

(4.2-26)

$$\underline{\nabla} \cdot (\underline{v} + \underline{a}) = \underline{\nabla} \cdot \underline{v} + \underline{\nabla} \cdot \underline{a}$$

(4.2-27)

$$\underline{\nabla} \times (\underline{v} + \underline{a}) = \underline{\nabla} \times \underline{v} + \underline{\nabla} \times \underline{a}$$

(4.2-28)

$$\underline{\nabla} \times (\underline{v} \times \underline{a}) = (\underline{a} \cdot \underline{\nabla}) \underline{v} - \underline{a} (\underline{\nabla} \cdot \underline{v}) - (\underline{v} \cdot \underline{\nabla}) \underline{a} + \underline{v} (\underline{\nabla} \cdot \underline{a})$$

(4.2-29)

$$\underline{\nabla} \cdot (\underline{\nabla} \varphi) \equiv \nabla^2 \varphi = \frac{\partial^2 \varphi}{\partial x^2} + \frac{\partial^2 \varphi}{\partial y^2} + \frac{\partial^2 \varphi}{\partial z^2}$$

(4.2-30a,b)

$$\nabla^2 = \frac{\partial^2}{\partial x^2} + \frac{\partial^2}{\partial y^2} + \frac{\partial^2}{\partial z^2} \quad \text{is the Laplacian operator}$$

(4.2-31)

---

***Example 4.2-1.*** Show that $\underline{\nabla} \cdot \underline{v} \neq \underline{v} \cdot \underline{\nabla}$

$\underline{\nabla} \cdot \underline{v}$ is a scalar quantity and is given by

$$\underline{\nabla} \cdot \underline{v} = \frac{\partial v_x}{\partial x} + \frac{\partial v_y}{\partial y} + \frac{\partial v_z}{\partial z}$$

(4.2-32)

On the other hand $\underline{v} \cdot \underline{\nabla}$ is a scalar operator and is given by

$$\underline{v} \cdot \underline{\nabla} = v_x \frac{\partial}{\partial x} + v_y \frac{\partial}{\partial y} + v_z \frac{\partial}{\partial z}$$

(4.2-33)

Thus $\underline{\nabla} \cdot \underline{v} \neq \underline{v} \cdot \underline{\nabla}$.

***Example 4.2-2.*** If $\underline{r} = (x, y, z) = \underline{i} x + \underline{j} y + \underline{k} z$, $r = |\underline{r}|$ and $\underline{\omega}$ is a constant vector, show that

(i)        $\underline{\nabla}\,\varphi(r) = \dfrac{1}{r}\,\dfrac{d\varphi}{dr}\,\underline{r}$                                                                       (4.2-34)

(ii)       $\underline{\nabla}\cdot(\underline{\omega}\times\underline{r}) = 0$                                                                              (4.2-35)

(iii)      $\underline{\nabla}\times(\underline{\omega}\times\underline{r}) = 2\underline{\omega}$                                                                          (4.2-36)

(i)        $\underline{\nabla}\,\varphi(r) \;=\; \underline{i}\,\dfrac{\partial\varphi}{\partial x} + \underline{j}\,\dfrac{\partial\varphi}{\partial y} + \underline{k}\,\dfrac{\partial\varphi}{\partial z}$                                                         (4.2-37a)

$$= \underline{i}\,\dfrac{d\varphi}{dr}\,\dfrac{\partial r}{\partial x} + \underline{j}\,\dfrac{d\varphi}{dr}\,\dfrac{\partial r}{\partial y} + \underline{k}\,\dfrac{d\varphi}{dr}\,\dfrac{\partial r}{\partial z}$$                                   (4.2-37b)

$$\dfrac{\partial r}{\partial x} = \dfrac{\partial}{\partial x}\left(x^2 + y^2 + z^2\right)^{1/2}$$                                                         (4.2-38a)

$$= \left(x^2 + y^2 + z^2\right)^{-1/2} x$$                                                                  (4.2-38b)

$$= \dfrac{x}{r}$$                                                                                         (4.2-38c)

We have similar expressions for $\dfrac{\partial r}{\partial y}$ and $\dfrac{\partial r}{\partial z}$ and, on substituting these expressions into Equation (4.2-37b), we obtain

$$\underline{\nabla}\,\varphi(r) \;=\; \dfrac{d\varphi}{dr}\left(\dfrac{x}{r}\,\underline{i} + \dfrac{y}{r}\,\underline{j} + \dfrac{z}{r}\,\underline{k}\right)$$                                                   (4.2-39a)

$$= \dfrac{1}{r}\,\dfrac{d\varphi}{dr}\,\underline{r}$$                                                                         (4.2-39b)

(ii)       $\underline{\nabla}\cdot(\underline{\omega}\times\underline{r}) \;=\; \dfrac{\partial}{\partial x}(z\omega_y - y\omega_z) + \dfrac{\partial}{\partial y}(x\omega_z - z\omega_x) + \dfrac{\partial}{\partial z}(y\omega_x - x\omega_y)$              (4.2-40a)

$$= 0$$                                                                                             (4.2-40b)

(iii)      From Equation (4.2-29), we have

$$\underline{\nabla}\times(\underline{\omega}\times\underline{r}) \;=\; (\underline{r}\cdot\underline{\nabla})\,\underline{\omega} - \underline{r}\,(\underline{\nabla}\cdot\underline{\omega}) - (\underline{\omega}\cdot\underline{\nabla})\,\underline{r} + \underline{\omega}\,(\underline{\nabla}\cdot\underline{r})$$           (4.2-41)

The first two terms on the right side of Equation (4.2-41) are zero since $\underline{\omega}$ is a constant.

$$(\underline{\omega}\cdot\underline{\nabla})\,\underline{r} \;=\; \left(\omega_x\dfrac{\partial}{\partial x} + \omega_y\dfrac{\partial}{\partial y} + \omega_z\dfrac{\partial}{\partial z}\right)\underline{r}$$                                         (4.2-42a)

$$= (\underline{i}\,\omega_x + \underline{j}\,\omega_y + \underline{k}\,\omega_z) = \underline{\omega}$$                                                       (4.2-42b,c)

$$\underline{\nabla} \cdot \underline{r} = \left( \underline{i} \frac{\partial}{\partial x} + \underline{j} \frac{\partial}{\partial y} + \underline{k} \frac{\partial}{\partial z} \right) \cdot (x\underline{i} + y\underline{j} + z\underline{k}) \tag{4.2-43a}$$

$$= 1 + 1 + 1 \tag{4.2-43b}$$

$$= 3 \tag{4.2-43c}$$

Combining Equations (4.2-41, 42c, 43c) yields

$$\underline{\nabla} \times (\underline{\omega} \times \underline{r}) = -\underline{\omega} + 3\underline{\omega} \tag{4.2-44a}$$

$$= 2\underline{\omega} \tag{4.2-44b}$$

## 4.3   LINE, SURFACE AND VOLUME INTEGRALS

### Line Integral of a Scalar Function

In the definite integral defined by Equation (1.3-2), the integration is along the x-axis. We now consider the integral of a scalar function $\varphi(x, y, z)$ along a curve C from the point A to the point B as illustrated in Figure 4.3-1.

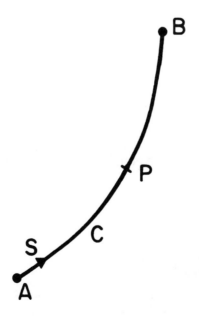

**FIGURE 4.3-1    Integral along a curve C**

Let the curve  C  be represented parametrically by

$$\underline{r}\,(s) = x(s)\,\underline{i} + y(s)\underline{j} + z(s)\,\underline{k} \tag{4.3-1}$$

where  s  is the arc length of  C  and the points  A  and  B  correspond to  s = a  and  s = b respectively. We assume that  $\underline{r}$  (s)  is continuous and has continuous first derivatives for all values of s  under consideration. This means that  C  has a unique tangent at each point and such a curve is referred to as a **smooth curve**. We shall consider only smooth curves. The **line or curvilinear integral** of  φ  along the curve  C  from  s = a  to  s = b  is defined as

$$I = \int_{a}^{b} \varphi[x(s),\, y(s),\, z(s)]\, ds \tag{4.3-2a}$$

$$= \int_{AB} \varphi(s)\, ds \tag{4.3-2b}$$

$$= \int_{C} \varphi(s)\, ds \tag{4.3-2c}$$

If the curve  C  is closed, then  A  coincides with  B  and the line integral around a closed curve  C  is usually denoted by

$$I = \oint \varphi(s)\, ds \tag{4.3-3}$$

where a modified integral sign is introduced.

Unless otherwise specified, the integral is taken along the **positive direction**. The positive direction along a closed curve is the direction such that as we move around the curve, the region enclosed is to our left.

The properties of ordinary definite integrals are equally valid for line integrals. Thus

$$\int_{a}^{b} \varphi(s)\, ds = -\int_{b}^{a} \varphi(s)\, ds \tag{4.3-4a}$$

$$\int_{AB} \varphi(s)\, ds = \int_{AP} \varphi(s)\, ds + \int_{PB} \varphi(s)\, ds \tag{4.3-4b}$$

where  P  is a point on the curve between  A  and  B.

The evaluation of the line integrals is done by writing them as ordinary integrals. The representation of the curve  C  by the arc length  s  is not always simple for integration. It might be more convenient to use a new parameter  t  instead of  s. Then in Equation (4.3-1), the variable is  t  and not  s. The line element  ds  is then given by

$$ds = \sqrt{d\underline{r} \cdot d\underline{r}} \tag{4.3-5a}$$

$$= \sqrt{\frac{d\underline{r}}{dt} \cdot \frac{d\underline{r}}{dt}} \; dt \tag{4.3-5b}$$

$$= \sqrt{\left(\frac{dx}{dt}\right)^2 + \left(\frac{dy}{dt}\right)^2 + \left(\frac{dz}{dt}\right)^2} \; dt \tag{4.3-5c}$$

***Example 4.3-1***. Evaluate the line integral

$$I = \int_C \varphi \, [x(s), \, y(s), \, z(s)] \; ds \tag{4.3-6}$$

where $\varphi = x^2 + y^2$, and $C$ is the triangle OAB with O the origin, A the point $(1, 0, 0)$ and B the point $(0, 2, 0)$ as shown in Figure 4.3-2.

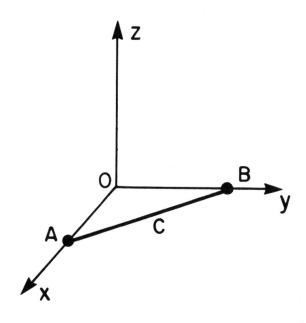

**FIGURE 4.3-2    Integral of $\varphi$ around the closed curve C**

From Figure 4.3-2, it can be seen that the integral $I$ may be written as

$$I = \int_{OA} \varphi \; ds + \int_{AB} \varphi \; ds + \int_{BO} \varphi \; ds \tag{4.3-7}$$

Along OA, $y = z = 0$, so

$$\int_{OA} \varphi \, ds = \int_0^1 x^2 \, dx = \left[ \frac{x^3}{3} \right]_0^1 = \frac{1}{3} \tag{4.3-8a,b,c}$$

The line AB is given by

$$y = -2 (x - 1) \tag{4.3-9a}$$

$$z = 0 \tag{4.3-9b}$$

Equations (4.3-9a, b) may be written in parametric form as

$$x = 1 - t \tag{4.3-10a}$$

$$y = 2t \tag{4.3-10b}$$

$$z = 0 \tag{4.3-10c}$$

where the points A and B correspond to $t = 0$ and $t = 1$ respectively.

If s is the arc length of AB, then

$$\left( \frac{ds}{dt} \right) = \sqrt{ \left( \frac{dx}{dt} \right)^2 + \left( \frac{dy}{dt} \right)^2 + \left( \frac{dz}{dt} \right)^2 } \tag{4.3-11a}$$

$$= \sqrt{ (-1)^2 + (2)^2 } \tag{4.3-11b}$$

$$= \sqrt{5} \tag{4.3-11c}$$

$$\int_{AB} \varphi \, ds = \int_0^1 \left[ (1-t)^2 + 4t^2 \right] \frac{ds}{dt} \, dt \tag{4.3-12a}$$

$$= \sqrt{5} \int_0^1 \left[ 1 - 2t + 5t^2 \right] dt \tag{4.3-12b}$$

$$= \frac{5\sqrt{5}}{3} \tag{4.3-12c}$$

Along BO, $x = z = 0$, so

$$\int_{BO} \varphi \, ds = \int_2^0 y^2 \, dy \tag{4.3-13a}$$

$$= -\frac{8}{3} \tag{4.3-13b}$$

Combining Equations (4.3-7, 8a, b, c, 12c, 13b), we obtain

$$I = \frac{1}{3} + \frac{5\sqrt{5}}{3} - \frac{8}{3} \tag{4.3-14a}$$

$$= \frac{5\sqrt{5} - 7}{3} \tag{4.3-14b}$$

**Line Integral of a Vector Function**

Let $\underline{v}$ [x (s), y (s), z (s)] be a vector field defined at all points of a smooth curve C and $\underline{T}$ be the **unit tangent** to C. The tangent to C at the point P is by definition the line joining P to a neighboring point P' as P' approaches P. Thus $\underline{T}$ is by definition given by

$$\underline{T} = \lim_{\Delta s \to 0} \frac{\underline{r}(s + \Delta s) - \underline{r}(s)}{\Delta s} \tag{4.3-15a}$$

$$= \frac{d\underline{r}}{ds} \tag{4.3-15b}$$

If C is defined in terms of t instead of s, then

$$\underline{T} = \frac{d\underline{r}}{dt} \Big/ \left| \frac{d\underline{r}}{dt} \right| \tag{4.3-16}$$

The **scalar line integral** of $\underline{v}$ along C is then defined as

$$I = \int_C \underline{v} \cdot \underline{T} \, ds \tag{4.3-17a}$$

$$= \int_C \underline{v} \cdot d\underline{r} \tag{4.3-17b}$$

where $d\underline{r} = \underline{T} \, ds$.

If $\underline{v}$ is a velocity and C is a closed curve then I, in fluids mechanics, is known as the circulation around C. If $\underline{F}$ (= $\underline{v}$) is a force, then I represents the work done by $\underline{F}$ in moving a particle along C.

***Example 4.3-2.*** Evaluate

$$I = \int_A^B \underline{v} \cdot d\underline{r} \tag{4.3-18}$$

where $\underline{v} = (2x, 0, 2x + 2y + 2z) = \underline{i}\,(2x) + \underline{k}\,(2x + 2y + 2z)$, and along the arc of the circle $x^2 + y^2 = 1$, $z = 0$ joining $A = (1, 0, 0)$ to $B = (0, 1, 0)$.

The path of integration is shown in Figure 4.3-3. The vector position $\underline{r}$ of any point on C which is part of a circle can be given by

$$\underline{r} = \cos \theta \, \underline{i} + \sin \theta \, \underline{j} + 0 \underline{k} \qquad\qquad (4.3\text{-}19)$$

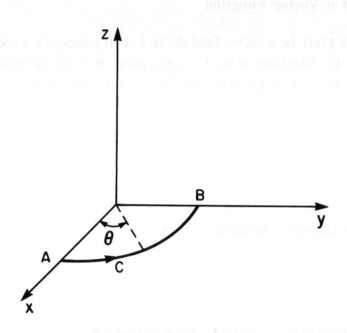

**FIGURE 4.3-3    Line integral along curve C**

$$\int_A^B \underline{v} \bullet d\underline{r} = \int_0^{\pi/2} [2 \cos \theta \, \underline{i} + (2 \cos \theta + 2 \sin \theta) \underline{k}\,] \bullet [-\sin \theta \, \underline{i} + \cos \theta \underline{j}\,]\, d\theta \qquad (4.3\text{-}20a)$$

$$= \int_0^{\pi/2} -2 \cos \theta \sin \theta \, d\theta \qquad\qquad (4.3\text{-}20b)$$

$$= \frac{1}{2} \left[\cos 2\,\theta\right]_0^{\pi/2} \qquad\qquad (4.3\text{-}20c)$$

$$= -1 \qquad\qquad (4.3\text{-}20d)$$

***Example 4.3-3***.  Calculate the work done by the force $\underline{F} = (x, -z, 2y)$ in displacing a particle along the parabola $y = 2x^2$, $z = 2$ from the point $(0, 0, 2)$ to $(1, 2, 2)$.

The path of integration is shown in Figure 4.3-4 with point A = (0, 0, 2) and point B = (1, 2, 2).

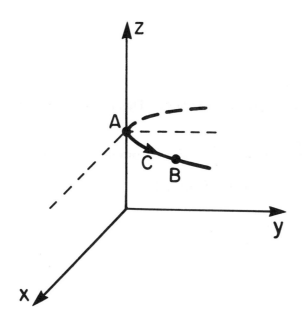

**FIGURE 4.3-4    Line integral along a parabola**

The vector position $\underline{r}$ of any point on the parabola may be written as

$$\underline{r} = t\,\underline{i} + 2t^2\,\underline{j} + 2\,\underline{k} \tag{4.3-21}$$

This has been obtained by choosing $x = t$. It then follows that $y = 2t^2$ since $y = 2x^2$. Such parametrization is usually done via educated guessing. The more practice one has, the luckier one gets.

The point A corresponds to $t = 0$ and the point B corresponds to $t = 1$.

$$\int_A^B \underline{F} \cdot d\underline{r} = \int_0^1 [t\,\underline{i} - 2\underline{j} + 4t^2\underline{k}] \cdot [\underline{i} + 4t\underline{j}]\, dt \tag{4.3-22a}$$

$$= \int_0^1 [t - 8t]\, dt = -\frac{7}{2} \tag{4.3-22b,c}$$

●

The evaluation of **vector line integrals**, such as

$$\underline{I} = \int_C \underline{v} \, ds \tag{4.3-23a}$$

or

$$\underline{I} = \int_C \underline{v} \times d\underline{r} \tag{4.3-23b}$$

can be done by integrating each component separately. This is illustrated in the next example.

***Example 4.3-4.*** Evaluate $\int_C \underline{r} \, ds$ and $\int_C \underline{r} \times d\underline{r}$ from the point (a, 0, 0) to the point (a, 0, 2πb) on the circular helix illustrated in Figure 4.3-5, given by

$$\underline{r} = (x, y, z) = (a \cos t, \, a \sin t, \, bt) \tag{4.3-24}$$

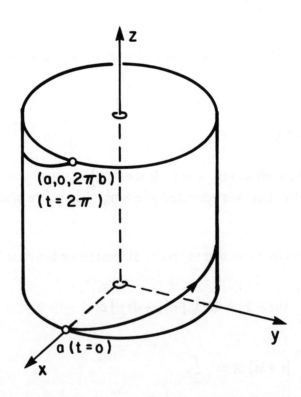

**FIGURE 4.3-5   Line integral along a helix**

We note that the point (a, 0, 0) corresponds to $t = 0$ and the point (a, 0, $2\pi$b) corresponds to $t = 2\pi$. The line element ds is given, via Equation (4.3-5c), by

$$ds = \sqrt{(-a \sin t)^2 + (a \cos t)^2 + b^2} \; dt \tag{4.3-25a}$$

$$= \sqrt{a^2 + b^2} \; dt \tag{4.3-25b}$$

$$\int_C \underline{r} \; ds = \int_0^{2\pi} [(a \cos t) \, \underline{i} + (a \sin t) \, \underline{j} + (bt) \, \underline{k}] \sqrt{a^2 + b^2} \; dt \tag{4.3-26a}$$

$$= \sqrt{a^2 + b^2} \left[ (a \sin t) \, \underline{i} - (a \cos t) \underline{j} + \left( \frac{bt^2}{2} \right) \underline{k} \right]_0^{2\pi} \tag{4.3-26b}$$

$$= \sqrt{a^2 + b^2} \left[ 2 \pi^2 b \, \underline{k} \right] \tag{4.3-26c}$$

$$\int_C \underline{r} \times d\underline{r} = \int_0^{2\pi} \left[ a \cos t \, \underline{i} + a \sin t \, \underline{j} + bt \, \underline{k} \right] \times \left[ -a \sin t \, \underline{i} + a \cos t \, \underline{j} + b \, \underline{k} \right] dt \tag{4.3-27a}$$

$$= \int_0^{2\pi} \left[ ab (\sin t - t \cos t) \, \underline{i} - ab (t \sin t + \cos t) \underline{j} + a^2 \, \underline{k} \right] dt \tag{4.3-27b}$$

$$= \left[ ab (-t \sin t - 2 \cos t) \, \underline{i} - ab (-t \cos t + 2 \sin t) \underline{j} + a^2 t \, \underline{k} \right]_0^{2\pi} \tag{4.3-27c}$$

$$= \left[ 2\pi \, ab \, \underline{j} + 2\pi \, a^2 \, \underline{k} \right] \tag{4.3-27d}$$

## Repeated Integrals

In Equation (1.8-3), we have defined a function $I(x)$ by integrating a function $f(x, y)$ with respect to y between $y = u(x)$ to $y = v(x)$. If we now integrate $I(x)$ with respect to x between the limits $x = a$ and $x = b$, we have

$$\int_a^b I(x) \; dx = \int_a^b \left[ \int_{u(x)}^{v(x)} f(x, y) \; dy \right] dx \tag{4.3-28a}$$

$$= \int_a^b \int_{u(x)}^{v(x)} f(x, y) \; dy \; dx \tag{4.3-28b}$$

The above integral is an example of a **repeated integral** and is evaluated by integrating in the order given in Equation (4.3-28a). If $u(x) = c$ and $v(x) = d$ where c and d are constants, then the order of integration is not important. If further $f(x, y)$ is **separable** and may be written as a product of a function of $x$ and a function of $y$, then

$$\int_a^b \int_c^d f(x, y) \, dy \, dx = \int_a^b \int_c^d \varphi(x) \, \psi(y) \, dy \, dx \qquad (4.3\text{-}29a)$$

$$= \left[ \int_a^b \varphi(x) \, dx \right] \left[ \int_c^d \psi(y) \, dy \right] \qquad (4.3\text{-}29b)$$

Thus in this case the repeated integral becomes the product of two single integrals.

In a single integral the integration is taken along a curve. In a double integral the integration is taken over an area. Thus, in Equation (4.3-28b), the area $A$ over which the integration is to be performed is the area bounded by the curves $y = u(x)$, $y = v(x)$, $x = a$ and $x = b$. Thus an alternative notation is

$$\int_a^b \int_{u(x)}^{v(x)} f(x, y) \, dy \, dx = \iint_A f(x, y) \, dy \, dx \qquad (4.3\text{-}30)$$

If $f(x, y) = 1$, then the double integral in Equation (4.3-30) is the area $A$. If $z = f(x, y)$ then the double integral is the volume of the cylinder formed by lines parallel to the z-axis and bounded by $z = 0$ and $z = f$. The area of the base of the cylinder is $A$.

We can extend the process of double integration to triple or higher integration. Thus a **triple integral** can be defined as

$$I^* = \int_a^b \left[ \int_{u(x)}^{v(x)} \left\{ \int_{p(x,y)}^{q(x,y)} f(x, y, z) \, dz \right\} dy \right] dx \qquad (4.3\text{-}31)$$

In Equation (4.3-31), we integrate $f(x, y, z)$ with respect to $z$ between the limits $z = p(x, y)$ to $z = q(x, y)$ resulting in a function of $x$ and $y$, say $g(x, y)$. The triple integral is thus reduced to a double integral and can be evaluated as shown before.

***Example 4.3-5.*** A thin plate is bounded by the parabola $y = 2x - x^2$ and $y = 0$. Determine its mass if the density at any point $(x, y)$ is $1/(1 + x)$.

The equation of the parabola may be written as

$$y = 1 - x^2 + 2x - 1 \qquad (4.3\text{-}32a)$$

$$= 1 - (x - 1)^2 \qquad (4.3\text{-}32b)$$

The parabola intersects $y = 0$ at

$$x = 0 \tag{4.3-33a}$$

$$x = 2 \tag{4.3-33b}$$

The shape of the plate is shown in Figure 4.3-6. The mass $M$ of the plate of unit thickness is given by

$$M = \int_0^2 \left[ \int_{y=0}^{2x-x^2} \frac{1}{1+x} \, dy \right] dx \tag{4.3-34}$$

Note that the double integral results from the fact that we consider a plate of unit thickness.

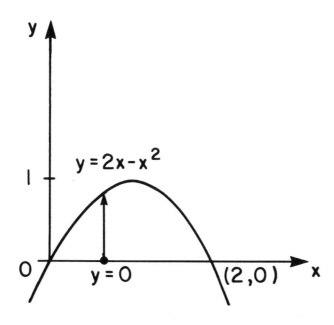

**FIGURE 4.3-6    Shape of the flat plate. The integral with respect to y is indicated by the arrow**

In Equation (4.3-34), we integrate with respect to $y$ first from $y = 0$ to $y = 2x - x^2$ as can be seen from Figure 4.3-6. Then we integrate with respect to $x$ from $x = 0$ to $x = 2$. Since the density is independent of $y$, we obtain on integrating with respect to $y$

$$M = \int_0^2 \left[ \frac{y}{1+x} \right]_0^{2x-x^2} dx \tag{4.3-35a}$$

$$= \int_0^2 \frac{x\,(2-x)}{(1+x)}\ dx \tag{4.3-35b}$$

$$= \int_0^2 \left[ -x + \frac{3x}{1+x} \right]\ dx \tag{4.3-35c}$$

$$= \left[ -\frac{x^2}{2} + 3x - 3\ell n\,(1+x) \right]_0^2 \tag{4.3-35d}$$

$$= 4 - 3\ell n\,3 \tag{4.3-35e}$$

***Example 4.3-6***. Find the area bounded by the curves $y = x^2$, $y = 0$, $x = 0$ and $x = 1$.

Figure 4.3-7 illustrates the area. If we integrate with respect to $y$ first, then the area $A$ is given by

$$A = \int_0^1 \left[ \int_0^{x^2} dy \right] dx \tag{4.3-36a}$$

$$= \int_0^1 x^2\ dx = \left[ \frac{x^3}{3} \right]_0^1 = \frac{1}{3} \tag{4.3-36b,c,d}$$

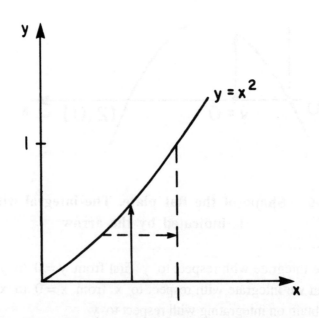

**FIGURE 4.3-7    Sketch of the area**

We can interchange the order of integration, then A is given by

$$A = \int_0^1 \left[ \int_{\sqrt{y}}^1 dx \right] dy \qquad\qquad (4.3\text{-}37a)$$

$$= \int_0^1 (1 - \sqrt{y}) \, dy \qquad\qquad (4.3\text{-}37b)$$

$$= \left[ y - \frac{2}{3} y^{3/2} \right]_0^1 = \frac{1}{3} \qquad\qquad (4.3\text{-}37c,d)$$

In this example we have seen that it is easier to integrate with respect to y first and then with respect to x. The limits of integration can be inferred from the diagram. It is always useful to sketch the region of integration.

●

A change of variables may sometimes make the evaluation of a double integral easier. If u and v are two new variables such that

$$x = x(u, v), \qquad y = y(u, v) \qquad\qquad (4.3\text{-}38a,b)$$

then

$$\iint_A f(x, y) \, dx \, dy = \iint_{A'} f[x(u, v), y(u, v)] \, |J| \, du \, dv \qquad\qquad (4.3\text{-}39)$$

where A is the area of integration in the xy-plane and A' is the corresponding area in the uv-plane. The Jacobian J is given by

$$J = \frac{\partial(x, y)}{\partial(u, v)} = \begin{vmatrix} \dfrac{\partial x}{\partial u} & \dfrac{\partial x}{\partial v} \\[2mm] \dfrac{\partial y}{\partial u} & \dfrac{\partial y}{\partial v} \end{vmatrix} \qquad\qquad (4.3\text{-}40a,b)$$

Similarly for a triple integral, we have

$$\iiint_V f(x, y, z) \, dx \, dy \, dz = \iiint_{V'} \left[ f(x(u, v, w), y(u, v, w), z(u, v, w)) \, |J| \, du \, dv \, dw \right] \qquad (4.3\text{-}41)$$

where V is the region of integration in the xyz-space and V' is the corresponding region in the uvw-space. The variables u, v, w are defined by

$$x = x(u, v, w), \qquad y = y(u, v, w), \qquad z = z(u, v, w) \qquad \text{(4.3-42a,b,c)}$$

The Jacobian J is

$$J = \frac{\partial(x, y, z)}{\partial(u, v, w)} = \begin{vmatrix} \dfrac{\partial x}{\partial u} & \dfrac{\partial x}{\partial v} & \dfrac{\partial x}{\partial w} \\[2mm] \dfrac{\partial y}{\partial u} & \dfrac{\partial y}{\partial v} & \dfrac{\partial y}{\partial w} \\[2mm] \dfrac{\partial z}{\partial u} & \dfrac{\partial z}{\partial v} & \dfrac{\partial z}{\partial w} \end{vmatrix} \qquad \text{(4.3-43a,b)}$$

**Example 4.3-7.** Evaluate the triple integral, by a suitable change of variables.

$$\iiint\limits_V \left[ x^2 + y^2 + z^2 \right] dx\, dy\, dz$$

where V is the region bounded by the ellipsoid

$$\frac{x^2}{a^2} + \frac{y^2}{b^2} + \frac{z^2}{c^2} = 1 \qquad \text{(4.3-44)}$$

The ellipsoid given by Equation (4.3-44) can be written in a parametric form as

$$x = ar \sin \theta \cos \phi \qquad \text{(4.3-45a)}$$

$$y = br \sin \theta \sin \phi \qquad \text{(4.3-45b)}$$

$$x = cr \cos \theta \qquad \text{(4.3-45c)}$$

The relations between $(x, y, z)$ and $(r, \theta, \phi)$ are illustrated in Figure 4.3-8.

It can be seen from Figure 4.3-8 that the range of r, $\theta$, $\phi$ is

$$0 \le r \le 1, \qquad 0 \le \theta \le \pi, \qquad 0 \le \phi \le 2\pi \qquad \text{(4.3-46a,b,c)}$$

The Jacobian J is given by

$$J = \begin{vmatrix} a \sin \theta \cos \phi & ar \cos \theta \cos \phi & -ar \sin \theta \sin \phi \\ b \sin \theta \sin \phi & br \cos \theta \sin \phi & br \sin \theta \cos \phi \\ c \cos \theta & -cr \sin \theta & 0 \end{vmatrix} \qquad \text{(4.3-47a)}$$

$$= abc\, r^2 \sin \theta \qquad \text{(4.3-47b)}$$

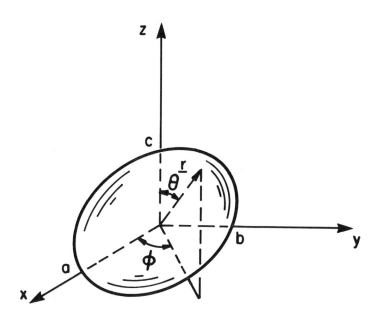

**FIGURE 4.3-8    Relations between (x, y, z) and (r, θ, φ) for an ellipsoid**

$$\iiint\limits_{V} \left[x^2 + y^2 + z^2\right] dx\, dy\, dz$$

$$= \int_{\theta=0}^{\pi} \int_{\phi=0}^{2\pi} \int_{r=0}^{1} \left[r^2(a^2\sin^2\theta\,\cos^2\phi + b^2\sin^2\theta\,\sin^2\phi + c^2\cos^2\theta\,)\,abcr^2\sin\theta\,dr\,d\phi\,d\theta\right] \quad (4.3\text{-}48a)$$

$$= abc\left[\frac{r^5}{5}\right]_0^1 \int_{\theta=0}^{\pi} \int_{\phi=0}^{2\pi} \left[a^2\sin^3\theta\,\cos^2\phi + b^2\sin^3\theta\,\sin^2\phi + c^2\cos^2\theta\,\sin\theta\right] d\phi\,d\theta \quad (4.3\text{-}48b)$$

$$= \frac{abc}{5}\int_0^{\pi} \left\{ a^2\sin^3\theta\left[\frac{\phi}{2} + \frac{1}{4}\sin 2\phi\right]_0^{2\pi} + b^2\sin^3\theta\left[\frac{\phi}{2} - \frac{1}{4}\sin 2\phi\right]_0^{2\pi} \right.$$

$$\left. + c^2\cos^2\theta\,\sin\theta\left[\phi\right]_0^{2\pi} \right\} d\theta \quad (4.3\text{-}48c)$$

$$= \frac{abc}{5}\int_0^{\pi} \left\{ \sin^3\theta\,(a^2\pi + b^2\pi) + (c^2\cos^2\theta\,\sin\theta)\,(2\pi) \right\} d\theta \quad (4.3\text{-}48d)$$

$$= \frac{\pi\, abc}{5} \left\{ (a^2 + b^2) \left[ -\frac{1}{3} (\cos\theta \sin^2\theta + 2 \cos\theta) \right]_0^\pi + 2c^2 \left[ -\frac{\cos^3\theta}{3} \right]_0^\pi \right\} \qquad (4.3\text{-}48e)$$

$$= \frac{4\,\pi\, abc}{15} (a^2 + b^2 + c^2) \qquad (4.3\text{-}48f)$$

## Surfaces

The equation of a surface S may be written as

$$\varphi\, (x, y, z) = \text{constant} \qquad (4.3\text{-}49a)$$

or

$$z = f\, (x, y) \qquad (4.3\text{-}49b)$$

or in parametric form as

$$x = x\, (u, v), \qquad y = y\, (u, v), \qquad z = z\, (u, v) \qquad (4.3\text{-}49c,d,e)$$

Equation (4.3-49c, d, e) defines a mapping (projection) of a region A in the xyz-space into a region A' in the uv-plane.

Thus the equation of the surface of a sphere of radius a and center at the origin may be written as

$$x^2 + y^2 + z^2 = a^2 \qquad (4.3\text{-}50a)$$

$$z = \pm \sqrt{a^2 - x^2 - y^2} \qquad (4.3\text{-}50b)$$

$$x = a \sin\theta \cos\phi, \qquad y = a \sin\theta \sin\phi, \qquad z = a \cos\theta \qquad (4.3\text{-}50c,d,e)$$

We note that the surface of a sphere of radius a, in the xyz-space is mapped into a rectangle, $0 \le \theta \le \pi$, $0 \le \phi \le 2\pi$, in the $\theta\phi$-plane.

Let P be a point with vector position $\underline{r}$ on a general surface S given by Equations (4.3-49c to e). If we keep the value of v fixed and let u vary, then P will trace out a curve $C_u$ as shown in Figure 4.3-9. Similarly by fixing the value of u and letting v vary, a curve $C_v$ will be traced out. The tangent to $C_u$ is given by $\frac{\partial r}{\partial u}$ $(= \underline{r}_u)$ and the tangent to $C_v$ is $\frac{\partial r}{\partial v}$ $(= \underline{r}_v)$. A **unit normal** $\underline{n}$ to the surface S is a vector which is perpendicular to the tangents to $C_u$ and $C_v$ and is given via Equation (4.2-10) by

$$\underline{n} = \frac{(\underline{r}_u \times \underline{r}_v)}{|\underline{r}_u \times \underline{r}_v|} \qquad (4.3\text{-}51)$$

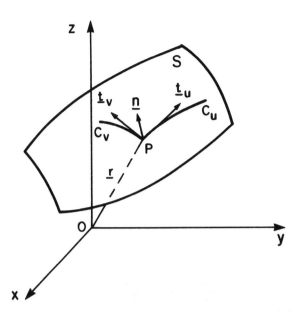

**FIGURE 4.3-9    Curves $C_u$ and $C_v$ on surface S**

The curve $C_u$ can also be given by Equation (4.3-1) and since $C_u$ lies on the surface S, [x (s), y (s), z (s)] satisfy Equation (4.3-49a). On differentiating with respect to s and using the chain rule we obtain

$$\frac{\partial\varphi}{\partial x}\frac{dx}{ds}+\frac{\partial\varphi}{\partial y}\frac{dy}{ds}+\frac{\partial\varphi}{\partial z}\frac{dz}{ds}=0 \qquad (4.3\text{-}52)$$

From the definition of grad $\varphi$ and the tangent to $C_u$, we find that Equation (4.3-52) can be written as

$$\text{grad } \varphi \cdot \underline{T}_u = 0 \qquad (4.3\text{-}53)$$

where $\underline{T}_u$ is the unit tangent to $C_u$.

Similarly we can deduce that

$$\text{grad } \varphi \cdot \underline{T}_v = 0 \qquad (4.3\text{-}54)$$

where $\underline{T}_v$ is the unit tangent to $C_v$.

From Equations (4.3-53, 54), we deduce that grad $\varphi$ is perpendicular to both $\underline{T}_u$ and $\underline{T}_v$ and thus is a normal to the surface S. That is to say, $\underline{n}$ is also given by

$$\underline{n} = \frac{\text{grad } \varphi}{|\text{ grad } \varphi |} \qquad (4.3\text{-}55)$$

If $\underline{r}_u$ and $\underline{r}_v$ are interchanged in Equation (4.3-51), the sign of $\underline{n}$ is reversed. Therefore, we need to establish a convention to label one side of S to be positive. If S is a closed surface, $\underline{n}$ is chosen to be positive if it points outwards.

The surfaces we shall consider are two-sided surfaces. However, there are surfaces which have one side only. A **Möbius strip** illustrated in Figure 4.3-10 is an example of a one-sided surface. It is obtained by twisting a strip of paper once and gluing the ends together. An insect can crawl on that strip and reach all points on the strip without ever having to cross an edge! Thus the strip has only one side. In such a case we cannot designate a positive side. If we cut the strip along the centre line, we obtain one circle.

**FIGURE 4.3-10    Möbius  strip**

A surface is said to be **smooth** if its unit normal exists and is continuous everywhere on the surface. The union of a finite number of smooth surfaces forms a **simple surface**.

*Example 4.3-8*. Find the unit normal $\underline{n}$ to the surface of a sphere of radius a, centered at the origin, using Equations (4.3-51, 55).

The equation of the surface of the sphere is given by Equations (4.3-50c to e)

$$\frac{\partial \underline{r}}{\partial \theta} = a \cos \theta \cos \phi \, \underline{i} + a \cos \theta \sin \phi \, \underline{j} - a \sin \theta \, \underline{k} \qquad (4.3\text{-}56a)$$

$$\frac{\partial \underline{r}}{\partial \phi} = -a \sin \theta \sin \phi \underline{i} + a \sin \theta \cos \phi \underline{j} \tag{4.3-56b}$$

$$\underline{n} = \frac{(\underline{r}_\theta \times \underline{r}_\phi)}{|\underline{r}_\theta \times \underline{r}_\phi|} \tag{4.3-57a}$$

$$= \frac{(a^2 \sin^2\theta \cos \phi, \; a^2 \sin^2\theta \sin \phi, \; a^2 \sin \theta \cos \theta)}{a^2 \sin \theta} \tag{4.3-57b}$$

$$= (\sin \theta \cos \phi, \; \sin \theta \sin \phi, \; \cos \theta) \tag{4.3-57c}$$

$$= \frac{\underline{r}}{a} \tag{4.3-57d}$$

The equation of the surface of the sphere is also given by Equation (4.3-50a) and $\varphi$ is given by

$$\varphi = x^2 + y^2 + z^2 = a^2 = \text{constant} \tag{4.3-58}$$

$$\text{grad } \varphi = 2x \underline{i} + 2y \underline{j} + 2z \underline{k} \tag{4.3-59a}$$

$$= 2 \underline{r} \tag{4.3-59b}$$

From Equation (4.3-55) we have

$$\underline{n} = \frac{2 \underline{r}}{|2 \underline{r}|} \tag{4.3-60a}$$

$$= \frac{\underline{r}}{a} \tag{4.3-60b}$$

since $|\underline{r}| = a$.

## Surface and Volume Integrals

The **surface integral** is an extension of the double integral to an integration over a surface S. If $\varphi(x, y, z)$ is a scalar function, then the surface integral of $\varphi$ over S is denoted by

$$I = \iint_S \varphi(x, y, z) \, dS \tag{4.3-61}$$

In Equation (4.3-61), z is not an independent variable, it has to satisfy the equation of the surface. That is to say, it is given by Equation (4.3-49b). If S is given in parametric form, then the surface element dS is given by

$$dS = |\underline{r}_u \times \underline{r}_v| \, du \, dv \tag{4.3-62}$$

The surface integral $I$ given by Equation (4.3-61) becomes

$$I = \iint_{S'} \varphi[x(u, v), y(u, v), z(u, v)] \; |\underline{r}_u \times \underline{r}_v| \; du \; dv \tag{4.3-63}$$

where $S'$ is the region in the uv-plane corresponding to $S$ in the xyz-space.

If $\underline{v}$ is a vector field, then from Equations (4.3-51, 62), we deduce the following relations

$$\iint_{S} \underline{v} \cdot \underline{n} \; dS = \iint_{S'} \underline{v}(u, v) \cdot (\underline{r}_u \times \underline{r}_v) \; du \; dv \tag{4.3-64}$$

$$\iint_{S} \underline{v} \; dS = \iint_{S} (v_x \underline{i} + v_y \underline{j} + v_z \underline{k}) \; dS \tag{4.3-65a}$$

$$= \iint_{S'} (v_x \underline{i} + v_y \underline{j} + v_z \underline{k}) |\underline{r}_u \times \underline{r}_v| \; du \; dv \tag{4.3-65b}$$

$$\iint_{S} \varphi \; \underline{n} \; dS = \iint_{S'} \varphi(u, v) \, (\underline{r}_u \times \underline{r}_v) \; du \; dv \tag{4.3-66}$$

***Example 4.3-9***. Evaluate $\iint_{S} \underline{v} \cdot \underline{n} \; dS$ where $\underline{v}$ is the vector $(4x, y, z)$ and $S$ is the plane $2x + y + 2z = 6$ in the positive octant.

The region $S$ over which the integration is carried out is shown in Figure 4.3-11a. On $S$

$$z = 3 - x - y/2 \tag{4.3-67}$$

The original inclined plane in the xyz-space is replaced by its projection in the uv-plane and we write

$$u = x, \qquad y = v \tag{4.3-68a,b}$$

It follows from Equation (4.3-67) that

$$z = 3 - u - v/2 \tag{4.3-68c}$$

$S'$, the corresponding region of $S$ in the uv-plane, is shown in Figure 4.3-11b. The vector position $\underline{r}$ of any point on $S$ is

$$\underline{r} = u \, \underline{i} + v \, \underline{j} + (3 - u - v/2) \, \underline{k} \tag{4.3-69}$$

$$\underline{r}_u = \underline{i} - \underline{k} \tag{4.3-70a}$$

$$\underline{r}_v = \underline{j} - \frac{1}{2}\underline{k} \tag{4.3-70b}$$

$$\underline{r}_u \times \underline{r}_v = \underline{i} + \frac{1}{2}\underline{j} + \underline{k} \tag{4.3-70c}$$

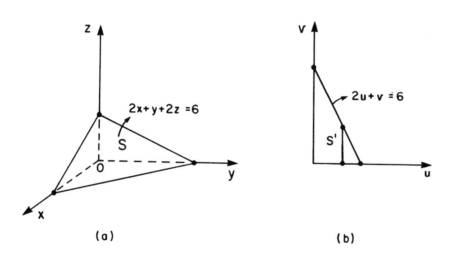

**FIGURE 4.3-11**  (a) Region S in the xyz-space, (b) Region S' in the uv-plane

Using Equation (4.3-65), we have

$$\iint_S \underline{v} \bullet \underline{n} \, dS = \iint_{S'} \left(4\,u,\, v,\, 3 - u - \frac{v}{2}\right) \bullet \left(1,\, \frac{1}{2},\, 1\right) dv \, du \tag{4.3-71a}$$

$$= \int_{u=0}^{3} \int_{v=0}^{6-2u} 3(u + 1)\, dv \, du \tag{4.3-71b}$$

$$= 3 \int_0^3 \left[uv + v\right]_0^{6-2u} du \tag{4.3-71c}$$

$$= 3 \int_0^3 \left[-2u^2 + 4u + 6\right] du = 54 \tag{4.3-71d,e}$$

If $\varphi(x, y, z)$ is a scalar function defined throughout a volume $V$, then the **volume integral** of $\varphi$ over $V$ is a triple integral of $\varphi$ and is written as

$$\iiint\limits_{V} \varphi(x, y, z) \, dx \, dy \, dz = \iiint\limits_{V} \varphi(x, y, z) \, dV \qquad (4.3\text{-}72)$$

Some authors use only a single integral sign $\left( \int\limits_{V} \varphi(x, y, z) \, dV \right)$, instead of three, for the volume integral as written on the right side of Equation (4.3-72).

Evaluation of volume integrals can sometimes be simplified by a transformation of variables. Thus if we make the transformation from the xyz-space to the uvw-space, then the vector position $\underline{r}$ of any point can be written as

$$\underline{r} = x(u, v, w)\,\underline{i} + y(u, v, w)\,\underline{j} + z(u, v, w)\,\underline{k} \qquad (4.3\text{-}73)$$

The tangent to the $C_u$ curve, the curve generated by allowing $u$ to vary while keeping $v$ and $w$ fixed, is $\underline{r}_u$. Similarly, one defines the tangents $\underline{r}_v$ and $\underline{r}_w$. Thus the volume element $dV$ is given by

$$dV = \left| (\underline{r}_u \times \underline{r}_v) \bullet \underline{r}_w \right| du \, dv \, dw \qquad (4.3\text{-}74)$$

The integral given in Equation (4.3-72) becomes

$$\iiint\limits_{V} \varphi(x, y, z) \, dx \, dy \, dz = \iiint\limits_{V'} \varphi[x(u, v, w), y(u, v, w), z(u, v, w)] \left| (\underline{r}_u \times \underline{r}_v) \bullet \underline{r}_w \right| du \, dv \, dw$$

$$(4.3\text{-}75)$$

where $V'$ is the volume in uvw-space corresponding to $V$ in xyz-space. Equation (4.3-75) is identical to Equation (4.3-41).

If $\underline{v}(x, y, z)$ is a vector function, then the volume integral of $\underline{v}$ is evaluated by decomposing $\underline{v}$ into its components and evaluating each component separately. Thus

$$\iiint\limits_{V} \underline{v} \, dV = \iiint\limits_{V} [\underline{i} \, v_x + \underline{j} \, v_y + \underline{k} \, v_z] \, dx \, dy \, dz \qquad (4.3\text{-}76a)$$

$$= \iiint\limits_{V'} [\underline{i} \, v_x + \underline{j} \, v_y + \underline{k} \, v_z] \left| (\underline{r}_u \times \underline{r}_v) \bullet \underline{r}_w \right| du \, dv \, dw \qquad (4.3\text{-}76b)$$

Examples of volume integrals are given in the next section.

## 4.4 RELATIONS BETWEEN LINE, SURFACE AND VOLUME INTEGRALS

Two important theorems in vector analysis are **Gauss' theorem** and **Stokes' theorem**. We shall state these theorems without proof. They are also known by other names and the origins of these theorems were summarized by Ericksen (1960).

### Gauss' (divergence) Theorem

If a vector field $\underline{v}$ and its divergence are defined throughout a volume $V$ bounded by a simple closed surface $S$, then

$$\iint\limits_{S} \underline{v} \cdot \underline{n} \, dS = \iiint\limits_{V} \text{div} \, \underline{v} \, dV = \iiint\limits_{V} \underline{\nabla} \cdot \underline{v} \, dV \qquad (4.4\text{-}1a,b)$$

where $\underline{n}$ is the unit outward normal to $S$.

***Example 4.4-1.*** Verify Gauss' theorem if $\underline{v} = (x, y, z)/r^2$ and $V$ is the volume enclosed by the spheres $x^2 + y^2 + z^2 = 1$ and $x^2 + y^2 + z^2 = \varepsilon^2$, $\varepsilon < 1$.

$$\text{div} \, \underline{v} = \frac{\partial}{\partial x}\left(\frac{x}{r^2}\right) + \frac{\partial}{\partial y}\left(\frac{y}{r^2}\right) + \frac{\partial}{\partial z}\left(\frac{z}{r^2}\right) \qquad (4.4\text{-}2a)$$

$$= \frac{1}{r^2} - \frac{2x^2}{r^4} + \frac{1}{r^2} - \frac{2y^2}{r^4} + \frac{1}{r^2} - \frac{2z^2}{r^4} = \frac{1}{r^2} \qquad (4.4\text{-}2b,c)$$

We make a change of variables from $(x, y, z)$ to $(r, \theta, \phi)$ and these two sets of variables are related by an equation similar to Equations (4.3-50c to e), which is

$$x = r \sin \theta \cos \phi, \qquad y = r \sin \theta \sin \phi, \qquad z = r \cos \theta \qquad (4.4\text{-}3a,b,c)$$

The tangent vectors are

$$\underline{r}_r = \sin \theta \cos \phi \, \underline{i} + \sin \theta \sin \phi \, \underline{j} + \cos \theta \, \underline{k} \qquad (4.4\text{-}4a)$$

$$\underline{r}_\theta = r \cos \theta \cos \phi \, \underline{i} + r \cos \theta \sin \phi \, \underline{j} - r \sin \theta \, \underline{k} \qquad (4.4\text{-}4b)$$

$$\underline{r}_\phi = -r \sin \theta \sin \phi \, \underline{i} + r \sin \theta \cos \phi \, \underline{j} \qquad (4.4\text{-}4c)$$

$$dV = \left| (\underline{r}_r \times \underline{r}_\theta) \cdot \underline{r}_\phi \right| dr \, d\theta \, d\phi = r^2 \sin \theta \, dr \, d\theta \, d\phi \qquad (4.4\text{-}5a,b)$$

The volume $V$ is enclosed by two spheres, $S_1$ with radius $1$ and $S_\varepsilon$ with radius $\varepsilon$, as shown in Figure 4.4-1.

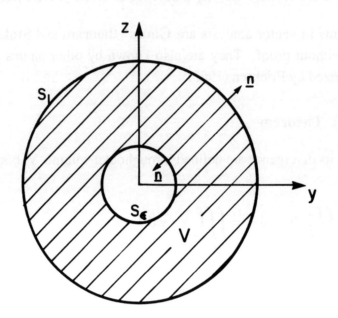

**FIGURE 4.4-1    Shaded volume V enclosed by sphere $S_1$ and $S_\epsilon$**

$$\iiint\limits_V \text{div } \underline{v} \, dV = \int_{\theta=0}^{\pi} \left[ \int_{\phi=0}^{2\pi} \left\{ \int_{r=\epsilon}^{1} \frac{1}{r^2} \, r^2 \sin\theta \, dr \right\} d\phi \right] d\theta \qquad (4.4\text{-}6a)$$

$$= \int_{\theta=0}^{\pi} \left[ \int_{\phi=0}^{2\pi} (1-\epsilon) \sin\theta \, d\phi \right] d\theta \qquad (4.4\text{-}6b)$$

$$= \int_{0}^{\pi} 2\pi \, (1-\epsilon) \sin\theta \, d\theta \qquad (4.4\text{-}6c)$$

$$= 4\pi \, (1-\epsilon) \qquad (4.4\text{-}6d)$$

The unit normal to a sphere is given by Equation (4.3-60b).  We also note that $\underline{v}$ may be written as

$$\underline{v} = \frac{\underline{r}}{r^2} \qquad (4.4\text{-}7)$$

Using Equation (4.3-64), we obtain

$$\underset{S_1}{\iint} \underline{v} \cdot \underline{n} \, dS = \int_{\theta=0}^{\pi} \left[ \int_{\phi=0}^{2\pi} \underline{r} \cdot \underline{r} \, \sin \theta \, d\phi \right] d\theta \qquad\qquad (4.4\text{-}8a)$$

$$= \int_0^{\pi} 2\pi \, \sin \theta \, d\theta \qquad\qquad (4.4\text{-}8b)$$

$$= 4\pi \qquad\qquad (4.4\text{-}8c)$$

Note that in Equation (4.4-8b), $\underline{r} \cdot \underline{r} = 1$ since $S_1$ has radius 1.

On $S_\epsilon$, the outward normal is towards the origin as shown in Figure 4.4-1. In this case

$$\underline{n} = -\frac{\underline{r}}{\epsilon} \qquad\qquad (4.4\text{-}9)$$

$$\underset{S_\epsilon}{\iint} \underline{v} \cdot \underline{n} \, dS = \int_{\theta=0}^{\pi} \left[ \int_{\phi=0}^{2\pi} \frac{\underline{r}}{\epsilon^2} \cdot \frac{-\underline{r}}{\epsilon} \, \epsilon^2 \sin \theta \, d\phi \right] d\theta \qquad\qquad (4.4\text{-}10a)$$

$$= \int_{\theta=0}^{\pi} \left[ \int_{\phi=0}^{2\pi} -\epsilon \, \sin \theta \, d\phi \right] d\theta \qquad\qquad (4.4\text{-}10b)$$

$$= -4\pi \, \epsilon \qquad\qquad (4.4\text{-}10c)$$

From Equations (4.4-6d, 8c, 10c), it can be seen that

$$\underset{V}{\iiint} \operatorname{div} \underline{v} \, dV = \underset{S_1}{\iint} \underline{v} \cdot \underline{n} \, dS + \underset{S_\epsilon}{\iint} \underline{v} \cdot \underline{n} \, dS \qquad\qquad (4.4\text{-}11a)$$

$$= \underset{S}{\iint} \underline{v} \cdot \underline{n} \, dS \qquad\qquad (4.4\text{-}11b)$$

Thus the divergence theorem is verified.

In this example, $\operatorname{div} \underline{v}$ $(= 1/r^2)$ is not defined at the origin and so we cannot apply Gauss' theorem in a region that includes the origin. We circumvent this problem by enclosing the origin with a sphere of radius $\epsilon$. To obtain the volume integral of $\operatorname{div} \underline{v}$ in a region that includes the origin, we let $\epsilon \to 0$. In this example, we obtain from Equation (4.4-6d) with $\epsilon \to 0$

$$\iiint\limits_{V} \text{div } \underline{v} \ dV \ = \ 4\pi \tag{4.4-12}$$

●

We can deduce the following from Gauss' theorem.

(i)     $$\iint\limits_{S} \varphi \ \underline{n} \ dS \ = \ \iiint\limits_{V} \underline{\nabla} \ \varphi \ dV \tag{4.4-13}$$

where, as usual,  S  is the surface that encloses volume  V.

Let  $\underline{q}$  be an arbitrary constant vector.

$$\underline{\nabla} \bullet (\varphi \ \underline{q}) \ = \ \frac{\partial}{\partial x} \ (\varphi q_x) + \frac{\partial}{\partial y} \ (\varphi q_y) + \frac{\partial}{\partial z} \ (\varphi q_z) \tag{4.4-14a}$$

$$= \ q_x \ \frac{\partial \varphi}{\partial x} \ + \ q_y \ \frac{\partial \varphi}{\partial y} \ + \ q_z \ \frac{\partial \varphi}{\partial z} \tag{4.4-14b}$$

$$= \ \underline{q} \bullet \underline{\nabla} \ \varphi \tag{4.4-14c}$$

From Gauss' theorem, we have

$$\iint\limits_{S} \varphi \ \underline{q} \bullet \underline{n} \ dS \ = \ \iiint\limits_{V} \text{div } (\varphi \ \underline{q}) \ dV \tag{4.4-15a}$$

$$= \ \iiint\limits_{V} \underline{q} \bullet \underline{\nabla} \ \varphi \ dV \tag{4.4-15b}$$

Equation (4.4-15b) may be written as

$$\underline{q} \bullet \left[ \iint\limits_{S} \varphi \ \underline{n} \ dS \ - \ \iiint\limits_{V} \underline{\nabla} \ \varphi \ dV \right] \ = \ \underline{q} \bullet \underline{F} = 0 \tag{4.4-16a,b}$$

Since  $\underline{q}$  is arbitrary, it is not necessarily perpendicular to  $\underline{F}$.  However, since the right side of Equation (4.4-16a) is zero, it follows that  $\underline{F} = \underline{0}$,  verifying Equation (4.4-13).

(ii)     $$\iint\limits_{S} (\underline{n} \times \underline{v}) \ dS \ = \ \iiint\limits_{V} (\underline{\nabla} \times \underline{v}) \ dV \tag{4.4-17}$$

From Equations (4.2-14a, b), we can deduce that

$$\underline{\nabla} \cdot (\underline{v} \times \underline{q}) = \underline{q} \cdot (\underline{\nabla} \times \underline{v}) \tag{4.4-18}$$

where $\underline{q}$ is an arbitrary constant vector.

Gauss' theorem yields

$$\iint_S (\underline{v} \times \underline{q}) \cdot \underline{n} \, dS = \iiint_V \underline{\nabla} \cdot (\underline{v} \times \underline{q}) \, dV \tag{4.4-19}$$

Using Equations (4.2-14a, b, 4-18), Equation (4.4-19) becomes

$$\underline{q} \cdot \iint_S (\underline{n} \times \underline{v}) \, dS = \underline{q} \cdot \iiint_V (\underline{\nabla} \times \underline{v}) \, dV \tag{4.4-20}$$

It follows as in (i) that Equation (4.4-17) is true.

(iii) $$\iint_S \varphi \frac{\partial \psi}{\partial n} \, dS = \iiint_V (\varphi \, \nabla^2 \psi + \underline{\nabla} \varphi \cdot \underline{\nabla} \psi) \, dV \tag{4.4-21}$$

where $\dfrac{\partial}{\partial n}$ is the directional derivative and is defined in Equation (1.5-10).

Equation (4.4-21) can be derived as in cases (i) and (ii) by considering a vector $\varphi \, \underline{\nabla} \psi$ and applying Gauss' theorem. Equation (4.4-21) is the mathematical statement of **Green's first theorem**. **Green's second theorem** is obtained by interchanging $\varphi$ and $\psi$ in Equation (4.4-21) and subtracting the resulting expression from Equation (4.4-21). The result is

$$\iint_S \left( \varphi \frac{\partial \psi}{\partial n} - \psi \frac{\partial \varphi}{\partial n} \right) dS = \iiint_V (\varphi \, \nabla^2 \psi - \psi \, \nabla^2 \varphi) \, dV \tag{4.4-22}$$

(iv) $$\iint_A \left( \frac{\partial v_x}{\partial x} + \frac{\partial v_y}{\partial y} \right) dx \, dy = \oint_C (v_x \, dy - v_y \, dx) \tag{4.4-23}$$

where C is the curve enclosing the area A in the xy-plane.

Equation (4.4-23) is the statement of Gauss' theorem in two-dimensions.

***Example 4.4-2.*** Apply Gauss' theorem to a right circular cylinder bounded by the planes $z = 0$ and $z = h$, as shown in Figure 4.4-2. The vector field $\underline{v}$ is a function of $x$ and $y$ only and $v_z = 0$.

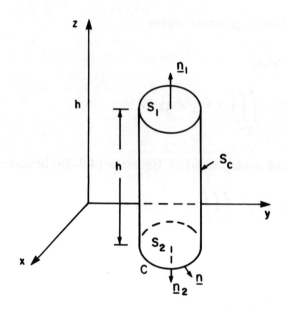

**FIGURE 4.4-2     Right circular cylinder bounded by planes $S_1$, $S_2$, and curved surface $S_c$**

The three surfaces that enclose the volume $V$ are $S_1$ (the plane $z = h$), $S_2$ (the plane $z = 0$) and the curved surface $S_c$. The unit outward normal to $S_1$, $\underline{n}_1$ is in the positive z-direction whereas $\underline{n}_2$, the unit outward normal to $S_2$ is in the opposite direction.

$$\iint_S \underline{v} \cdot \underline{n} \ dS = \iint_{S_1} \underline{v} \cdot \underline{n}_1 \ dS + \iint_{S_2} \underline{v} \cdot \underline{n}_2 \ dS + \iint_{S_c} \underline{v} \cdot \underline{n} \ dS \qquad (4.4\text{-}24)$$

Since $\underline{n}_1$ and $\underline{n}_2$ are of opposite sign and $\underline{v}$ is independent of $z$, the first two integrals on the right side of Equation (4.4-24) cancel each other. Applying Gauss' theorem yields

$$\iint_{S_c} \underline{v} \cdot \underline{n} \ dS = \iiint_V \text{div} \ \underline{v} \ dV \qquad (4.4\text{-}25a)$$

$$= \iiint_V \left( \frac{\partial v_x}{\partial x} + \frac{\partial v_y}{\partial y} \right) dV \qquad (4.4\text{-}25b)$$

Since we are considering a right circular cylinder of height $h$,

$$dV = h \ dS_2 \qquad (4.4\text{-}26a)$$

$$dS_c = h \, ds \qquad (4.4\text{-}26b)$$

where $ds$ is the line element along the circle that encloses the surface $S_2$.

Substituting Equations (4.4-26a, b) into Equation (4.4-25b), we obtain

$$h \int_C \underline{v} \cdot \underline{n} \, ds \;=\; h \iint_{S_2} \left( \frac{\partial v_x}{\partial x} + \frac{\partial v_y}{\partial y} \right) dx \, dy \qquad (4.4\text{-}27)$$

The normal $\underline{n}$ to the curve C is given by

$$\underline{n} \;=\; \left( \frac{\partial y}{\partial s} , -\frac{\partial x}{\partial s} , 0 \right) \qquad (4.4\text{-}28)$$

Combining Equations (4.4-27, 28) and dividing by $h$, we obtain

$$\int_C \left( v_x \, dy - v_y \, dx \right) \;=\; \iint_{S_2} \left( \frac{\partial v_x}{\partial x} + \frac{\partial v_y}{\partial y} \right) dx \, dy \qquad (4.4\text{-}29)$$

$$= \iint_A \left( \frac{\partial v_x}{\partial x} + \frac{\partial v_y}{\partial y} \right) dx \, dy \qquad (4.4\text{-}23)$$

where we have replaced $S_2$ by A. We have thus deduced Gauss' theorem in two dimensions [Equation (4.4-23)] for this simple geometry.

## Stokes' Theorem

If the vector field $\underline{v}$ and curl $\underline{v}$ are defined everywhere on a simple open surface S bounded by a curve C, then

$$\oint_C \underline{v} \cdot \underline{T} \, ds \;=\; \iint_S \mathrm{curl}\, \underline{v} \cdot \underline{n} \, dS \qquad (4.4\text{-}30)$$

where $\underline{T}$ is a unit tangent to C and $\underline{n}$ a unit outward normal to the surface S.

If the surface S is enclosed by two curves C and $C_1$, then at a point A on C we make a cut and draw a curve from A to a point B on $C_1$. We then go around $C_1$ in the positive direction once and leave $C_1$ at B to rejoin C at A and proceed along C in the positive direction until completion of the circuit. This is illustrated in Figure 4.4-3. The direction of AB from C to $C_1$ is opposite to the direction from $C_1$ to C. Therefore, the line integrals along AB will cancel. Thus we only need to evaluate the integrals along C and $C_1$. We have to ensure that the direction of integration is chosen properly. The same technique of making appropriate cuts can be extended to surfaces enclosed by more than two simple curves.

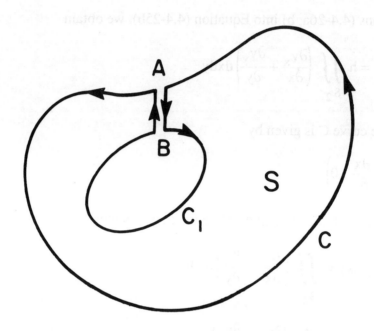

**FIGURE 4.4-3    Surface S enclosed by two curves C and $C_1$**

***Example 4.4-3***. Verify Stokes' theorem in the case where $\underline{v} = (-y^3, x^3, 0)$ and $S$ is the circular disk $x^2 + y^2 = 1$ with $z = 0$.

Equations (4.2-25a, b) define curl $\underline{v}$ as

$$\text{curl } \underline{v} = \begin{vmatrix} \underline{i} & \underline{j} & \underline{k} \\[6pt] \dfrac{\partial}{\partial x} & \dfrac{\partial}{\partial y} & \dfrac{\partial}{\partial z} \\[8pt] -y^3 & x^3 & 0 \end{vmatrix} \tag{4.4-31a}$$

$$= \left[ 0, 0, 3\,(x^2 + y^2) \right] \tag{4.4-31b}$$

$$\underline{n} = (0, 0, 1) \tag{4.4-32}$$

$$\iint\limits_{S} \text{curl } \underline{v} \cdot \underline{n} \, dS = \iint\limits_{S} 3\,(x^2 + y^2) \, dS \tag{4.4-33a}$$

$$= 3 \int_{0}^{2\pi} \int_{0}^{1} r^2 \, r \, dr \, d\theta \qquad (4.4\text{-}33b)$$

$$= \frac{3\pi}{2} \qquad (4.4\text{-}33c)$$

The curve $C$ enclosing $S$ is the unit circle ($r = 1$). In parametric form, the unit tangent $\underline{T}$ is given by

$$\underline{T} = \left( \frac{dx}{d\theta}, \frac{dy}{d\theta} \right) \qquad (4.4\text{-}34a)$$

$$= (-\sin\theta, \cos\theta) \qquad (4.4\text{-}34b)$$

$$\oint_{C} \underline{v} \cdot \underline{T} \, ds = \int_{0}^{2\pi} \left( -\sin^3\theta, \cos^3\theta, 0 \right) \cdot \left( -\sin\theta, \cos\theta, 0 \right) d\theta \qquad (4.4\text{-}35a)$$

$$= \int_{0}^{2\pi} \left( \sin^4\theta + \cos^4\theta \right) d\theta \qquad (4.4\text{-}35b)$$

$$= \left[ -\frac{\cos\theta \sin^3\theta}{4} + \frac{\sin\theta \cos^3\theta}{4} \right]_{0}^{2\pi} + \frac{3}{4} \left[ \int_{0}^{2\pi} \left( \sin^2\theta + \cos^2\theta \right) d\theta \right] \qquad (4.4\text{-}35c)$$

$$= \frac{3\pi}{2} \qquad (4.4\text{-}35d)$$

Thus Equation (4.4-30) is verified.

The following can be deduced from Stokes' theorem.

(i)      If the line integral $\displaystyle\int_{C} \underline{v} \cdot \underline{T} \, ds$ is independent of the path of integration and depends only on its end points, then the line integral of $\underline{v}$ around a closed curve $C$ (a curve with coincident end points) is zero. Such a vector $\underline{v}$ is called a **conservative vector**. Since the line integral is zero, it follows, from Stokes' theorem, that curl $\underline{v} = \underline{0}$. $\underline{v}$ is then said to be **irrotational**.

(ii)     
$$\iint_{S} \underline{n} \, dS = \frac{1}{2} \oint_{C} \underline{r} \times d\underline{r} \qquad (4.4\text{-}36)$$

Let $\underline{q}$ be an arbitrary constant vector. From Equation (4.2-29), we have

$$\text{curl} \left( \underline{q} \times \underline{r} \right) = (\underline{r} \cdot \nabla) \underline{q} - \underline{r} (\nabla \cdot \underline{q}) - (\underline{q} \cdot \nabla) \underline{r} + \underline{q} (\nabla \cdot \underline{r}) \qquad (4.4\text{-}37a)$$

$$= - \underline{q} + 3 \, \underline{q} \tag{4.4-37b}$$

$$= 2 \, \underline{q} \tag{4.4-37c}$$

From Stokes' theorem

$$\iint\limits_{S} \text{curl} \left( \underline{q} \times \underline{r} \right) \cdot \underline{n} \, dS = \oint\limits_{C} \left( \underline{q} \times \underline{r} \right) \cdot d\underline{r} \tag{4.4-38}$$

From Equations (4.4-37c, 2-14), we deduce that Equation (4.4-38) may be written as

$$2 \, \underline{q} \cdot \iint\limits_{S} \underline{n} \, dS = \underline{q} \cdot \oint\limits_{C} \underline{r} \times d\underline{r} \tag{4.4-39}$$

Since $\underline{q}$ is an arbitrary vector, Equation (4.4-36) is satisfied.

(iii)     Stokes' theorem in two-dimensions is given by

$$\oint\limits_{C} \left( v_x \, dx + v_y \, dy \right) = \iint\limits_{A} \left( \frac{\partial v_y}{\partial x} - \frac{\partial v_x}{\partial y} \right) dx \, dy \tag{4.4-40}$$

where A is the area in the xy-plane enclosed by curve C.

## 4.5     APPLICATIONS

### Conservation of Mass

Consider a fixed volume V of a fluid of density $\rho$ enclosed by a surface S. The mass m of the fluid is

$$m = \iiint\limits_{V} \rho \, dV \tag{4.5-1}$$

The rate of influx of mass

$$Q = \frac{dm}{dt} = \iiint\limits_{V} \frac{\partial \rho}{\partial t} \, dV \tag{4.5-2a,b}$$

where $Q = - \iint\limits_{S} \rho \, \underline{v} \cdot \underline{n} \, dS$ \hfill (4.5-3) and $\underline{v}$ is the velocity of the fluid.

The minus sign indicates that $\underline{n}$ is an outward normal.

Combining Equations (4.5-2a, b, 3) yields

$$\iiint_V \frac{\partial \rho}{\partial t} dV = -\iint_S \rho \underline{v} \cdot \underline{n} \, dS \qquad (4.5\text{-}4a)$$

$$= -\iiint_V \text{div} \, (\rho \, \underline{v}) \, dV \qquad (4.5\text{-}4b)$$

We have applied Gauss' theorem to the right side of Equation (4.5-4a) to transform the surface integral to the volume integral and as a result we obtain Equation (4.5-4b) which can be written as

$$\iiint_V \left[ \frac{\partial \rho}{\partial t} + \text{div} \, (\rho \, \underline{v}) \right] dV = 0 \qquad (4.5\text{-}5)$$

Since Equation (4.5-5) is valid for any arbitrary volume $V$ it must true at every point. (This is similar to $\int_a^b f(x) \, dx = 0$ for any limits $a$ and $b$, $f(x)$ must be zero). Therefore

$$\frac{\partial \rho}{\partial t} + \text{div} \, (\rho \, \underline{v}) = 0 \qquad (4.5\text{-}6)$$

Expanding $\text{div} \, (\rho \, \underline{v})$, we have

$$\text{div} \, (\rho \, \underline{v}) = \frac{\partial}{\partial x} (\rho \, v_x) + \frac{\partial}{\partial y} (\rho \, v_y) + \frac{\partial}{\partial z} (\rho \, v_z) \qquad (4.5\text{-}7a)$$

$$= \rho \left[ \frac{\partial v_x}{\partial x} + \frac{\partial v_y}{\partial y} + \frac{\partial v_z}{\partial z} \right] + v_x \frac{\partial \rho}{\partial x} + v_y \frac{\partial \rho}{\partial y} + v_z \frac{\partial \rho}{\partial z} \qquad (4.5\text{-}7b)$$

$$= \rho \, \text{div} \, \underline{v} + \underline{v} \cdot \text{grad} \, \rho \qquad (4.5\text{-}7c)$$

Combining Equations (4.5-6, 7c) yields

$$\frac{\partial \rho}{\partial t} + \underline{v} \cdot \text{grad} \, \rho + \rho \, \text{div} \, \underline{v} = 0 \qquad (4.5\text{-}8a)$$

which can be written as

$$\frac{D\rho}{Dt} + \rho \, \text{div} \, \underline{v} = 0 \qquad (4.5\text{-}8b)$$

where $\dfrac{D}{Dt} = \dfrac{\partial}{\partial t} + \underline{v} \bullet \text{grad}$   is the material or substantial derivative.

Equation (4.5-8b) is the equation of continuity in fluid dynamics, which for incompressible fluids ($\rho$ = constant) simplifies to

$$\text{div } \underline{v} = \underline{\nabla} \bullet \underline{v} = 0 \qquad\qquad (4.5\text{-}9a,b)$$

## Solution of Poisson's Equation

In many applications, such as electrostatics, the potential $\varphi$ satisfies the equation

$$\nabla^2 \varphi = \rho \,(x, y, z) \qquad\qquad (4.5\text{-}10)$$

Equation (4.5-10) is known as **Poisson's equation** and if $\rho = 0$, we have **Laplace's equation**, which is a homogeneous equation.

We propose to solve the inhomogeneous Equation (4.5-10) using the method of Green's functions which was introduced in Section 1.18 for ordinary differential equations.

We want to determine the value of $\varphi$ at a point P, inside a volume V enclosed by a surface S.

We assume the boundary condition to be

$$\varphi = f\,(x, y, z) \text{ on } S \qquad\qquad (4.5\text{-}11)$$

The boundary condition given by Equation (4.5-11) which gives the value of $\varphi$ on S is known as the **Dirichlet condition**, which is sufficient to ensure a unique solution to Equation (4.5-10).  In some problems, $\dfrac{\partial \varphi}{\partial n}$ is given at the boundary and this condition is the **Neumann condition**.

We recall from Section 1.18 that we need to construct a Green's function G that satisfies the homogeneous equation, in the present case, $\nabla^2 G = 0$, everywhere except at P.  We choose P to be the origin.  We also assume that G satisfies the homogeneous boundary condition, that is to say

$$G = 0 \quad \text{on } S \qquad\qquad (4.5\text{-}12)$$

Replacing $\psi$ by G in Equation (4.4-22) we have

$$\iint\limits_{S} \left(\varphi \frac{\partial G}{\partial n} - G \frac{\partial \varphi}{\partial n}\right) dS = \iiint\limits_{V} \left(\varphi \nabla^2 G - G \nabla^2 \varphi\right) dV \qquad\qquad (4.5\text{-}13)$$

Since G and $\nabla^2 G$ are not defined at the origin, we need to isolate the origin by enclosing it with a sphere of radius $\varepsilon$, surface $S_\varepsilon$ and volume $V_\varepsilon$ as in Example 4.4-1.

In $V - V_\varepsilon$, the region enclosed by $S$ and $S_\varepsilon$, G satisfies Laplace's equation, Equation (4.5-12) and $\varphi$ satisfies Equations (4.5-10, 11). Equation (4.5-13) becomes

$$\iint\limits_{S} f \frac{\partial G}{\partial n}\, dS + \iint\limits_{S_\varepsilon} \left[ \varphi \frac{\partial G}{\partial n} - G \frac{\partial \varphi}{\partial n} \right] dS = - \iiint\limits_{V-V_\varepsilon} G \rho\, dV \tag{4.5-14}$$

Near the origin we assume $\frac{1}{r}$ to be the dominant term of G.

G is singular at the origin. As in Example 4.4-1, we work in terms of spherical coordinates. On $S_\varepsilon$ we have

$$r = \varepsilon \tag{4.5-15a}$$

$$\frac{\partial}{\partial n} = -\frac{\partial}{\partial r} \tag{4.5-15b}$$

$$G \approx -\frac{1}{r} \tag{4.5-15c}$$

$$\iint\limits_{S_\varepsilon} \left[ \varphi \frac{\partial G}{\partial n} - G \frac{\partial \varphi}{\partial n} \right] dS = -\int_o^\pi \left[ \int_o^{2\pi} \left\{ \left(\frac{1}{\varepsilon^2}\right) \varphi + \frac{1}{\varepsilon} \frac{\partial \varphi}{\partial r} \right\} \varepsilon^2 \sin\theta\, d\phi \right] d\theta \tag{4.5-15d}$$

$$= -\int_o^\pi \left[ \int_o^{2\pi} \left\{ \varphi + \varepsilon \frac{\partial \varphi}{\partial r} \right\} \sin\theta\, d\phi \right] d\theta \tag{4.5-15e}$$

$$= -4\pi \, \varphi\,(0), \quad \text{as } \varepsilon \to 0 \tag{4.5-15f}$$

To evaluate the right side of Equation (4.5-14) as $\varepsilon \to 0$, we write

$$\iiint\limits_{V-V_\varepsilon} G \rho\, dV = \iiint\limits_{V} G \rho\, dV - I_\varepsilon \tag{4.5-16}$$

where $I_\varepsilon$ is the contribution from $V_\varepsilon$.

To evaluate $I_\varepsilon$ we note that $\rho$ is finite everywhere and let its upper bound in $V_\varepsilon$ be M. Then in $V_\varepsilon$ we have

$$|\rho| \le M \tag{4.5-17a}$$

$$V_\varepsilon = \frac{4}{3}\pi\,\varepsilon^3 \tag{4.5-17b}$$

$$G \approx \frac{1}{\varepsilon} \tag{4.5-17c}$$

$$I_\varepsilon \le M\frac{4}{3}\pi\,\varepsilon^3\frac{1}{\varepsilon} \tag{4.5-17d}$$

$$\le \frac{4}{3}\pi\,M\,\varepsilon^2 \tag{4.5-17e}$$

Thus as $\varepsilon \to 0$, $I_\varepsilon \to 0$. $\tag{4.5-18}$

Combining Equations (4.5-14, 15f, 16, 18), we obtain

$$\varphi(0) = \frac{1}{4\pi}\left[\iiint\limits_V G\,\rho\,dV + \iint\limits_S f\frac{\partial G}{\partial n}\,dS\right] \tag{4.5-19}$$

Equation (4.5-19) gives the value of $\varphi$ at P in the form of an integral involving the Green's function G. We will be in a position to construct G, following Chapters 5 and 6 on partial differential equations.

## Non-Existence of Periodic Solutions

Many dynamical systems to be covered in Chapter 10 are governed by a **non-linear autonomous** system

$$\frac{dx}{dt} = f(x, y) \tag{4.5-20a}$$

$$\frac{dy}{dt} = g(x, y) \tag{4.5-20b}$$

where $f(x, y)$ and $g(x, y)$ are continuous functions with continuous partial derivatives. The system is said to be **autonomous** because f and g do not depend explicitly on time t. If we can solve Equations (4.5-20a, b), we obtain x and y as functions of t. On eliminating t between them, we generate a relationship between x and y. We can plot x versus y and the obtained curve is the **path** or **trajectory**. The xy-plane is known as the **phase plane**. If x and y are periodic and of period T, then

$$x(t + T) = x(t) \tag{4.5-21a}$$

$$y(t + T) = y(t) \tag{4.5-21b}$$

The path in the phase plane will then be a closed curve. The closed curve will be traversed once as t increases from $t_0$ to $t_0 + T$, for every $t_0$. Thus a periodic solution corresponds to a closed path in the phase plane. If in Equation (4.4-23) we set $v_x = f$ and $v_y = g$, then we can write

$$\iint\limits_A \left( \frac{\partial f}{\partial x} + \frac{\partial g}{\partial y} \right) dx \, dy = \oint\limits_C (f \, dy - g \, dx) \qquad (4.5\text{-}22a)$$

$$= \int_{t_0}^{t_0+T} \left( f \frac{dy}{dt} - g \frac{dx}{dt} \right) dt \qquad (4.5\text{-}22b)$$

$$= 0 \qquad (4.5\text{-}22c)$$

Equation (4.5-22c) follows from using Equations (4.5-20a, b). Equation (4.5-22b) is obtained by introducing the variable t. Integrating once around the closed curve C corresponds to integrating from $t = t_0$ to $t = t_0 + T$.

If $\frac{\partial f}{\partial x} + \frac{\partial g}{\partial y}$ is of one sign only in the phase plane, that is $\frac{\partial f}{\partial x} + \frac{\partial g}{\partial y}$ is either positive or negative throughout the phase plane, then the left side of Equation (4.5-22c) cannot be zero. So this leads to a contradiction which implies that there is no closed curve. In other words there is no periodic solution. Thus if $\frac{\partial f}{\partial x} + \frac{\partial g}{\partial y}$ does not change sign in the phase space, then the system given by Equations (4.5-20a, b) does not have a periodic solution. This is **Bendixson's negative criterion**. Other criteria for determining the existence or non-existence of periodic solutions are given in Cesari (1971).

It should be pointed out that the non-linear Equations (4.5-20a, b) are usually difficult to solve exactly and only approximate solutions can be obtained. Thus it is of interest to have analytical criteria to determine the existence of periodic solutions.

## Maxwell's Equations

Faraday discovered that if a closed circuit is being moved across a magnetic field or if the circuit is placed in a varying magnetic field, a current is generated in the loop. This experimental observation is usually stated as **Neumann's law** and **Lenz's law**.

(i)      Neumann's law: if the magnetic flux N through a closed circuit varies with time, then an additional electromotive force (e.m.f.) is set up in the circuit and is of magnitude $\frac{dN}{dt}$.

(ii)      Lenz's law: the current induced in the circuit opposes the change in N.

Thus laws (i) and (ii) can be expressed as

$$\text{e.m.f.} = -\frac{dN}{dt} \qquad (4.5\text{-}23)$$

provided both the e.m.f. and $N$ are measured in the same system of units. Since the e.m.f. around a closed circuit is equal to the change in potential in going around the circuit once, we have

$$\text{e.m.f.} = \oint_C \underline{E} \cdot \underline{T} \, ds \qquad (4.5\text{-}24)$$

where $\underline{E}$ is the electric field.

The total normal magnetic induction $N$ across any surface $S$ is

$$N = \iint_S \underline{B} \cdot \underline{n} \, dS \qquad (4.5\text{-}25)$$

where $\underline{B}$ is the magnetic induction. Since $S$ is a fixed surface, we obtain by combining Equations (4.5-23 to 25)

$$\oint_C \underline{E} \cdot \underline{T} \, ds = -\iint_S \frac{\partial \underline{B}}{\partial t} \cdot \underline{n} \, dS \qquad (4.5\text{-}26)$$

Applying Stokes' theorem, we write

$$\iint_S \underline{n} \cdot \text{curl} \, \underline{E} \, dS = -\iint_S \frac{\partial \underline{B}}{\partial t} \cdot \underline{n} \, dS \qquad (4.5\text{-}27)$$

Since Equation (4.5-27) holds for every $S$, it follows that

$$\text{curl} \, \underline{E} = -\frac{\partial \underline{B}}{\partial t} \qquad (4.5\text{-}28)$$

Equation (4.5-28) is one of Maxwell's equations in electromagnetic field theory.

## 4.6   GENERAL CURVILINEAR COORDINATE SYSTEMS AND HIGHER ORDER TENSORS

### Cartesian Vectors and Summation Convention

A vector $\underline{v}$ is represented by its components which depend on the choice of the coordinate system. If we change the coordinate system, the components will generally change. The vector $\underline{v}$ however is independent of the coordinate system. Thus there is a relationship between the components of $\underline{v}$ in one coordinate system and the components of the same vector $\underline{v}$ in another coordinate system. It is the object of this section to establish such relationships.

We start by considering a transformation from one Cartesian coordinate system (x, y, z) to another Cartesian coordinate system $(\overline{x}, \overline{y}, \overline{z})$ obtained by rotating the (x, y, z) system as shown in Figure 4.6-1. Vector $\underline{v}$ has components $v_x$, $v_y$, $v_z$ in the (x, y, z) system and components $\overline{v}_x$, $\overline{v}_y$, $\overline{v}_z$ in the $(\overline{x}, \overline{y}, \overline{z})$ system. We wish to relate those components, and this can be achieved as follows.

$$\underline{v} = v_x \underline{i} + v_y \underline{j} + v_z \underline{k} = \overline{v}_x \overline{\underline{i}} + \overline{v}_y \overline{\underline{j}} + \overline{v}_z \overline{\underline{k}} \qquad (4.6\text{-}1a,b)$$

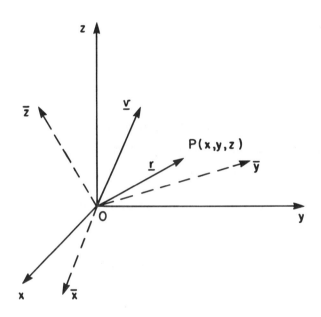

**FIGURE 4.6-1    Vectors $\underline{v}$ and $\underline{r}$ in two coordinate systems**

The component $v_x$ is obtained by forming the dot product of Equations (4.6-1a, b) with the unit vector $\underline{i}$.

$$v_x = \overline{v}_x \overline{\underline{i}} \cdot \underline{i} + \overline{v}_y \overline{\underline{j}} \cdot \underline{i} + \overline{v}_z \overline{\underline{k}} \cdot \underline{i} \qquad (4.6\text{-}2a)$$

Similarly, $v_y$ and $v_z$ are obtained through forming dot products with unit vectors $\underline{j}$ and $\underline{k}$ respectively.

$$v_y = \overline{v}_x \overline{\underline{i}} \cdot \underline{j} + \overline{v}_y \overline{\underline{j}} \cdot \underline{j} + \overline{v}_z \overline{\underline{k}} \cdot \underline{j} \qquad (4.6\text{-}2b)$$

$$v_z = \overline{v}_x \overline{\underline{i}} \cdot \underline{k} + \overline{v}_y \overline{\underline{j}} \cdot \underline{k} + \overline{v}_z \overline{\underline{k}} \cdot \underline{k} \qquad (4.6\text{-}2c)$$

Note that in the rectangular Cartesian system

$$\underline{i} \cdot \underline{j} = \underline{j} \cdot \underline{k} = \underline{i} \cdot \underline{k} = 0 \qquad\qquad (4.6\text{-}3a,b,c)$$

$$\underline{i} \cdot \underline{i} = \underline{j} \cdot \underline{j} = \underline{k} \cdot \underline{k} = 1 \qquad\qquad (4.6\text{-}3d,e,f)$$

Inverting the set of Equations (4.6-2a to c), we obtain

$$\begin{pmatrix} \overline{v}_x \\ \overline{v}_y \\ \overline{v}_z \end{pmatrix} = \begin{pmatrix} \ell_{11} & \ell_{12} & \ell_{13} \\ \ell_{21} & \ell_{22} & \ell_{23} \\ \ell_{31} & \ell_{32} & \ell_{33} \end{pmatrix} \begin{pmatrix} v_x \\ v_y \\ v_z \end{pmatrix} \qquad\qquad (4.6\text{-}4)$$

where $\ell_{11} = \overline{\underline{i}} \cdot \underline{i} = |1|\,|1| \cos\left(\overline{\underline{i}}, \underline{i}\right)$ $\qquad\qquad (4.6\text{-}5a,b)$

$\qquad\quad \ell_{21} = \overline{\underline{j}} \cdot \underline{i} = |1|\,|1| \cos\left(\overline{\underline{j}}, \underline{i}\right)$ $\qquad\qquad (4.6\text{-}5c,d)$

$\cos\left(\overline{\underline{i}}, \underline{i}\right)$ is the cosine of the angle between the unit vectors $\overline{\underline{i}}$ and $\underline{i}$. The nine quantities $\ell_{11}$, $\ell_{21}$, ... are the **direction cosines** and can be represented in a more compact form by $\ell_{mn}$, where both indices $m$ and $n$ take the values 1, 2 and 3.

In Equation (4.6-4), the indices in $\ell_{mn}$ are written as numbers instead of as x, y, z and this notation has the following advantages

(i)     in the case of an extension to an n dimensional space, where n can be greater than 26, we would run out of letters. Indeed, we can extend it to an infinite dimensional space;

(ii)    the notation is more compact.

Similarly a vector $\underline{v}$ with components $(v_x, v_y, v_z)$ can be represented by $v_m$, where m takes the values 1, 2 and 3. In this notation, $v_1 = v_x$, $v_2 = v_y$ and $v_3 = v_z$. The components $(\overline{v}_x, \overline{v}_y, \overline{v}_z)$ will be denoted by $(\overline{v}_1, \overline{v}_2, \overline{v}_3)$, the unit vectors $(\underline{i}, \underline{j}, \underline{k})$ and $(\overline{\underline{i}}, \overline{\underline{j}}, \overline{\underline{k}})$ will be denoted by $(\underline{\delta}_1, \underline{\delta}_2, \underline{\delta}_3)$ and $(\overline{\underline{\delta}}_1, \overline{\underline{\delta}}_2, \overline{\underline{\delta}}_3)$ respectively. The coordinates (x, y, z) and $(\overline{x}, \overline{y}, \overline{z})$ will be relabelled as $(x^1, x^2, x^3)$ and $(\overline{x}^1, \overline{x}^2, \overline{x}^3)$ respectively. Note that the indices in $(x^1, x^2, x^3)$ are written as superscripts and the reason for this notation will be explained later. Thus $x^2$ is not x squared and we shall denote $x^2$ squared as $(x^2)^2$ with a bracket round $x^2$.

The three equations given in matrix form in Equation (4.6-4) can now be written as

$$v_m = \sum_{n=1}^{3} \overline{v}_n \, \ell_{nm} \qquad\qquad (4.6\text{-}6a)$$

$$\quad = \overline{v}_n \, \ell_{nm} \qquad\qquad (4.6\text{-}6b)$$

In Equation (4.6-6a), the index m is known as the **free index** and it can take any of its possible values. In our case, m can take the values of 1, 2 or 3. Once a value of m is chosen, we must apply the same value to m wherever m occurs. Thus the three equations represented by Equation (4.6-6a) are obtained by assigning the value of m = 1, 2 and 3 in turn. The index n is called the **dummy** (repeated) index and the right side of Equation (4.6-6a) is a summation over all the possible values of n as indicated by the $\Sigma$ sign. The summation sign occurs so frequently that it is useful to adopt a convention, known as the **Einstein summation convention.** According to this convention, whenever an index occurs twice and twice only in an expression it implies summation over all the possible values of that index, unless stated otherwise. Thus in Equation (4.6-6b), the index n is a dummy index and the expression on the right side of Equation (4.6-6b) implies summation over all possible values of n. Since n is a dummy suffix, we can replace it by other letter, p say, and Equation (4.6-6b) can equally well be written as

$$v_m = \bar{v}_p \ell_{pm} \tag{4.6-7}$$

The right side of Equation (4.6-7) is a summation over all the possible values of p, which is the same as the summation over all the possible values of n as implied by Equation (4.6-6b). To obey the summation convention, we should not replace the dummy index n (or p) by the free index m. Similarly the free index m in Equation (4.6-7) can be replaced by any other letter except n (or p).

The components $(\bar{v}_1, \bar{v}_2, \bar{v}_3)$ can be expressed in terms of $(v_1, v_2, v_3)$ by inverting Equation (4.6-7), making use of the properties of $\ell_{pm}$, or by repeating the process used in obtaining Equation (4.6-7). We shall adopt the second procedure to demonstrate the elegance and conciseness of the summation convention. Equations (4.6-1a,b) can be written as

$$\underline{v} = v_m \underline{\delta}_m = \bar{v}_n \underline{\bar{\delta}}_n \tag{4.6-1a,b}$$

Forming the dot product with $\underline{\bar{\delta}}_r$ (we cannot use $\underline{\bar{\delta}}_n$ or $\underline{\bar{\delta}}_m$ due to the summation convention), we obtain

$$v_m \underline{\delta}_m \bullet \underline{\bar{\delta}}_r = \bar{v}_n \underline{\bar{\delta}}_n \bullet \underline{\bar{\delta}}_r \tag{4.6-8}$$

Since the $\left(\bar{x}^1, \bar{x}^2, \bar{x}^3\right)$ coordinate system is an orthonormal coordinate system

$$\underline{\bar{\delta}}_n \bullet \underline{\bar{\delta}}_r = \delta_{nr} \tag{4.6-9}$$

where $\delta_{nr}$ is the Kronecker delta and is equal to 1 if n = r, and is equal to 0 if n ≠ r.

From the definition of $\ell_{mn}$, Equations (4.6-5a,b), we have

$$\underline{\delta}_m \bullet \underline{\bar{\delta}}_r = \ell_{rm} \tag{4.6-10}$$

Note that the first index in $\ell_{rm}$, which is $r$, is associated with $\bar{\delta}_r$ and the second index, $m$, is associated with $\underline{\delta}_m$ of the unbarred coordinate system.

Substituting Equations (4.6-9, 10) into Equation (4.6-8), we obtain

$$\ell_{rm} \, v_m = \bar{v}_n \, \delta_{nr} \qquad\qquad\qquad\qquad (4.6\text{-}11a)$$

$$= \bar{v}_r \qquad\qquad\qquad\qquad (4.6\text{-}11b)$$

On summing the right side of Equation (4.6-11a) for all possible values of $n$, we find, due to the definition of $\delta_{nr}$, that the only surviving term is $\bar{v}_r$.

If $P$ is any point in space, and the coordinates of $P$ are $(x^1, x^2, x^3)$ and $\left(\bar{x}^{\,1}, \bar{x}^{\,2}, \bar{x}^{\,3}\right)$ relative to the $O\,x^1\,x^2\,x^3$ and $O\bar{x}^{\,1}\,\bar{x}^{\,2}\,\bar{x}^{\,3}$ coordinate systems respectively, then the components of the vector position $\underline{OP}$ $(= \underline{r})$ will transform from the barred to the unbarred system or vice-versa, according to Equations (4.6-7, 11b).  That is to say

$$x^m = \bar{x}^{\,p}\,\ell_{pm} \qquad\qquad\qquad\qquad (4.6\text{-}12a)$$

$$\ell_{rm}\,x^m = \bar{x}^{\,r} \qquad\qquad\qquad\qquad (4.6\text{-}12b)$$

Here we have freely written the indices as superscripts and subscripts.  Although this is permissible in our present coordinate transformation, it is not generally permissible to do so in a general coordinate transformation, as we shall discuss in the next section.

Because the direction cosines $\ell_{rm}$ are constants, the transformation, as defined by Equations (4.6-12a,b), from $O\,x^1\,x^2\,x^3$ to $O\bar{x}^{\,1}\,\bar{x}^{\,2}\,\bar{x}^{\,3}$ is a linear transformation.  Further we note from Equation (4.6-12b) that

$$\ell_{rm} = \frac{\partial \bar{x}^{\,r}}{\partial x^m} \qquad\qquad\qquad\qquad (4.6\text{-}13a)$$

Note that $\ell_{rm}$ can also be obtained via Equation (4.6-12a) by replacing the dummy index $p$ by $r$, to yield

$$\ell_{rm} = \frac{\partial x^m}{\partial \bar{x}^{\,r}} \qquad\qquad\qquad\qquad (4.6\text{-}13b)$$

Equations (4.6-13a, b) imply that

$$\frac{\partial \bar{x}^{\,r}}{\partial x^m} = \frac{\partial x^m}{\partial \bar{x}^{\,r}} \qquad\qquad\qquad\qquad (4.6\text{-}13c)$$

This is only true in the case of a Cartesian transformation.

Substituting Equations (4.6-13a, b) into Equation (4.6-11b), we obtain

$$
\overline{v}_r = \begin{cases} \dfrac{\partial x^m}{\partial \overline{x}^r} \, v_m \\[4mm] \dfrac{\partial \overline{x}^r}{\partial x^m} \, v_m \end{cases} \qquad\qquad (4.6\text{-}14a,b)
$$

Equations (4.6-14a, b) describe two types of transformation laws. Components that transform according to Equation (4.6-14a) are called **covariant components** $(v_m)$ and the indices are written as subscripts; components that transform according to Equation (4.6-14b) are known as **contravariant components** and the indices are written as superscripts. Thus in proper tensor notation, Equation (4.6-14b) should be written as

$$
\overline{v}^{\,r} = \frac{\partial \overline{x}^r}{\partial x^m} \, v^m \qquad\qquad (4.6\text{-}15)
$$

Note the symmetry in the notation. In Equation (4.6-14a), $m$ is a dummy index, it occurs once as a superscript in $\dfrac{\partial x^m}{\partial \overline{x}^r}$ (in the expression $\dfrac{\partial x^m}{\partial \overline{x}^r}$, we regard $m$ as a superscript and $r$ as a subscript) and once as a subscript in $v_m$. We can consider them as "cancelling" each other. On the right side, we are left with a subscript $r$ and on the left side we also have only a subscript r. The superscript $m$ is associated with $O x^1 x^2 x^3$ and the subscript $r$ with $O \overline{x}^{\,1} \overline{x}^{\,2} \overline{x}^{\,3}$. Similarly in Equation (4.6-15), the index $m$ "cancels" and we have the superscript $r$ on both sides of the equation. Thus in a general coordinate transformation, we need to distinguish between covariant and contravariant components. But for Cartesian components, both laws of transformation are equivalent [Equations (4.6-14a, b)] and thus there is no need to make a distinction between subscripts (covariant components) and superscripts (contravariant components).

## General Curvilinear Coordinate Systems

The rectangular Cartesian coordinate system is not always the most convenient coordinate system to use in solving problems. The laminar flow of a fluid in a circular pipe is solved using a cylindrical polar coordinate system. The velocity is then a function of the radial position $r$ only and not of two variables $x^1$ and $x^2$. Similarly the spherical polar coordinate system is chosen for solving flow past a sphere. The choice of coordinate system depends on the geometry of the problem.

We now consider a **general curvilinear coordinate** system $(q^1, q^2, q^3)$ as shown in Figure 4.6-2.

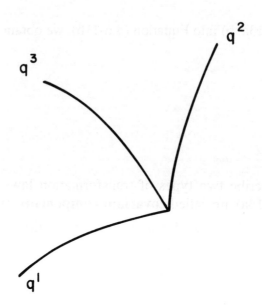

**FIGURE 4.6-2     Generalized coordinate system**

This coordinate system is related to the rectangular Cartesian coordinate system $(x^1, x^2, x^3)$ by

$$x^1 = x^1 (q^1, q^2, q^3) \tag{4.6-16a}$$

$$x^2 = x^2 (q^1, q^2, q^3) \tag{4.6-16b}$$

$$x^3 = x^3 (q^1, q^2, q^3) \tag{4.6-16c}$$

or in concise notation

$$x^m = x^m (q^n) \tag{4.6-17}$$

We assume that the transformation from $(x^1, x^2, x^3)$ to $(q^1, q^2, q^3)$ is one-to-one and it can be inverted as

$$q^n = q^n (x^m) \tag{4.6-18}$$

The **base vectors** $\underline{\delta}_m$ of the orthonormal coordinate system can be defined as being tangent to the coordinate axes $x^m$. Likewise, we define a set of base vectors $\underline{g}_m$ of the generalized curvilinear coordinate system as being tangent to the coordinate axes $q^m$, as illustrated in Figure 4.6-3. The vector connecting the origin to point P is $\underline{r}$.

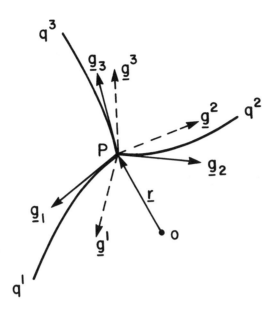

**FIGURE 4.6-3    Generalized coordinate system and base vectors**

The base vectors $\underline{g}_m$ are given by

$$\underline{g}_m = \frac{\partial \underline{r}}{\partial q^m} = \frac{\partial x^p}{\partial q^m} \underline{\delta}_p \qquad\qquad (4.6\text{-}19a,b)$$

Base vectors defined as tangents to coordinate curves are **covariant base vectors**.

The base vectors $\underline{\delta}_m$ can also be considered as being normal to the planes $x^m = $ constant. For example the base vector $\underline{\delta}_3$ is normal to the surface $x^3 = $ constant, that is to say, it is parallel to the $x^3$-axis. In the particular case of the rectangular Cartesian coordinate system, the tangent and the normal coincide, which explains why Equation (4.6-13c) is valid. This is not generally the case. Base vectors defined as normal to coordinate surfaces $q^n = $ constant are **contravariant base vectors** and are denoted by $\underline{g}^n$ as illustrated in Figure 4.6-3. They are given by Equation (4.3-55) and, in the present notation, they are written as

$$\underline{g}^n = \underline{\nabla} q^n = \text{grad } q^n = \frac{\partial q^n}{\partial x^\ell} \underline{\delta}_\ell \qquad\qquad (4.6\text{-}20a,b,c)$$

Note that both $\underline{g}_m$ and $\underline{g}^n$ are not necessarily unit vectors.

***Example 4.6-1.*** Obtain the covariant and contravariant base vectors for the spherical polar coordinate system.

The transformation from the rectangular Cartesian coordinate system $(x^1, x^2, x^3)$ to the spherical polar coordinates system $(r, \theta, \phi)$ is

$$x^1 = r \sin \theta \cos \phi, \qquad x^2 = r \sin \theta \sin \phi, \qquad x^3 = r \cos \theta \qquad \text{(4.6-21a,b,c)}$$

(In our notation, $q^1 = r$, $q^2 = \theta$, $q^3 = \phi$).

The covariant base vectors are given by Equation (4.6-19b)

$$\underline{g}_r = \frac{\partial x^1}{\partial r} \underline{\delta}_1 + \frac{\partial x^2}{\partial r} \underline{\delta}_2 + \frac{\partial x^3}{\partial r} \underline{\delta}_3 \qquad \text{(4.6-22a)}$$

$$= \sin \theta \cos \phi \, \underline{\delta}_1 + \sin \theta \sin \phi \, \underline{\delta}_2 + \cos \theta \, \underline{\delta}_3 \qquad \text{(4.6-22b)}$$

$$\underline{g}_\theta = \frac{\partial x^1}{\partial \theta} \underline{\delta}_1 + \frac{\partial x^2}{\partial \theta} \underline{\delta}_2 + \frac{\partial x^3}{\partial \theta} \underline{\delta}_3 \qquad \text{(4.6-22c)}$$

$$= r \cos \theta \cos \phi \, \underline{\delta}_1 + r \cos \theta \sin \phi \, \underline{\delta}_2 - r \sin \theta \, \underline{\delta}_3 \qquad \text{(4.6-22d)}$$

$$\underline{g}_\phi = \frac{\partial x^1}{\partial \phi} \underline{\delta}_1 + \frac{\partial x^2}{\partial \phi} \underline{\delta}_2 + \frac{\partial x^3}{\partial \phi} \underline{\delta}_3 \qquad \text{(4.6-22e)}$$

$$= - r \sin \theta \sin \phi \, \underline{\delta}_1 + r \sin \theta \cos \phi \, \underline{\delta}_2 \qquad \text{(4.6-22f)}$$

To obtain the contravariant base vectors $\underline{g}^n$ we have to invert Equations (4.6-21a, b, c). We then obtain

$$r = \sqrt{\left(x^1\right)^2 + \left(x^2\right)^2 + \left(x^3\right)^2} \qquad \text{(4.6-23a)}$$

$$\theta = \arctan\left(\sqrt{\frac{\left(x^1\right)^2 + \left(x^2\right)^2}{\left(x^3\right)^2}}\right) \qquad \text{(4.6-23b)}$$

$$\phi = \arctan\left(\frac{x^2}{x^1}\right) \qquad \text{(4.6-23c)}$$

We calculate $\underline{g}^n$ from Equation (4.6-20c)

$$\underline{g}^r = \frac{\partial r}{\partial x^1} \underline{\delta}_1 + \frac{\partial r}{\partial x^2} \underline{\delta}_2 + \frac{\partial r}{\partial x^3} \underline{\delta}_3 \qquad \text{(4.6-24a)}$$

$$= \sin \theta \cos \phi \, \underline{\delta}_1 + \sin \theta \sin \phi \, \underline{\delta}_2 + \cos \theta \, \underline{\delta}_3 \qquad \text{(4.6-24b)}$$

$$\underline{g}^{\theta} = \frac{\partial \theta}{\partial x^1} \underline{\delta}_1 + \frac{\partial \theta}{\partial x^2} \underline{\delta}_2 + \frac{\partial \theta}{\partial x^3} \underline{\delta}_3 \tag{4.6-24c}$$

$$= \frac{\cos \phi \cos \theta}{r} \underline{\delta}_1 + \frac{\sin \phi \cos \theta}{r} \underline{\delta}_2 - \frac{\sin \theta}{r} \underline{\delta}_3 \tag{4.6-24d}$$

$$\underline{g}^{\phi} = -\frac{\sin \phi}{r \sin \theta} \underline{\delta}_1 + \frac{\cos \phi}{r \sin \theta} \underline{\delta}_2 \tag{4.6-24e}$$

Note that

$$\underline{g}^r = \underline{g}_r \tag{4.6-25a}$$

$$\underline{g}^{\theta} = \frac{1}{r^2} \underline{g}_{\theta} \tag{4.6-25b}$$

$$\underline{g}^{\phi} = \frac{1}{r^2 \sin^2 \theta} \underline{g}_{\phi} \tag{4.6-25c}$$

Thus $\underline{g}_{\theta}$, $\underline{g}_{\phi}$ are parallel to $\underline{g}^{\theta}$, $\underline{g}^{\phi}$ respectively, but they are not generally equal in magnitude.

●

We now consider the transformation from one generalized coordinate system $q^m$ to another generalized coordinate system $\overline{q}^n$, where as usual, the indices $m$ and $n$ can take the values 1, 2 and 3.

The relation between these coordinate systems is

$$q^m = q^m (\overline{q}^n) \tag{4.6-26}$$

or $\quad \overline{q}^n = \overline{q}^n (q^m) \tag{4.6-27}$

The base vectors will be given by

$$\overline{\underline{g}}_m = \frac{\partial \underline{r}}{\partial \overline{q}^m} \tag{4.6-28}$$

and $\quad \overline{\underline{g}}^n = \frac{\partial \overline{q}^n}{\partial x^{\ell}} \underline{\delta}_{\ell} = \underline{\nabla} \overline{q}^n \tag{4.6-29a,b}$

We will now establish the relationship between the components of a vector $\underline{v}$ in the two generalized curvilinear coordinate systems $q^m$ and $\overline{q}^n$. Any vector $\underline{v}$ may be written as

$$\underline{v} = v^m \underline{g}_m = \overline{v}^n \overline{\underline{g}}_n \tag{4.6-30a,b}$$

Since in the general case, covariant and contravariant base vectors are different, one has to pay special attention to the position of the indices. As far as the summation convention is concerned, repeated indices should appear as subscript-superscript pairs.

In the general case, we can no longer take dot products, as in Equation (4.6-8), to establish the relation between components $v^m$ and $\bar{v}^m$. General coordinate systems are not necessarily orthogonal and $\underline{g}_1 \cdot \underline{g}_2$ is not necessarily zero.

Those relationships are obtained via the definitions [Equations (4.6-19, 20)] and the chain rule, as follows

$$\underline{g}_m = \frac{\partial \underline{r}}{\partial \bar{q}^n} \frac{\partial \bar{q}^n}{\partial q^m} \tag{4.6-31a}$$

$$= \bar{\underline{g}}_n \frac{\partial \bar{q}^n}{\partial q^m} \tag{4.6-31b}$$

Substitution in Equation (4.6-30a, b) yields

$$\underline{v} = v^m \frac{\partial \bar{q}^n}{\partial q^m} \bar{\underline{g}}_n = \bar{v}^n \bar{\underline{g}}_n \tag{4.6-32a,b}$$

It now follows that

$$\bar{v}^n = \frac{\partial \bar{q}^n}{\partial q^m} v^m \tag{4.6-33}$$

Components of vectors which transform according to Equation (4.6-33) are **contravariant components**. See also Equation (4.6-15).

***Example 4.6-2***. Obtain the law of transformation of the velocity components $v^i$.

The velocity components $v^m$, in the $q^m$ coordinate system, are defined as

$$v^m = \lim_{\Delta t \to o} \frac{\Delta q^m}{\Delta t} \tag{4.6-34a}$$

$$= \frac{\partial q^m}{\partial t} \tag{4.6-34b}$$

On transforming to the $\bar{q}^s$ coordinate system, the components $\bar{v}^s$ are defined as

$$\bar{v}^s = \lim_{\Delta t \to o} \frac{\Delta \bar{q}^s}{\Delta t} \tag{4.6-35a}$$

$$= \frac{\partial \overline{q}^s}{\partial t} \tag{4.6-35b}$$

According to the chain rule

$$\Delta \overline{q}^s = \frac{\partial \overline{q}^s}{\partial q^m} \Delta q^m \tag{4.6-36}$$

Substituting Equation (4.6-36) into (4.6-35a), we obtain

$$\overline{v}^s = \lim_{\Delta t \to o} \frac{\partial \overline{q}^s}{\partial q^m} \frac{\Delta q^m}{\Delta t} \tag{4.6-37a}$$

$$= \frac{\partial \overline{q}^s}{\partial q^m} v^m \tag{4.6-37b}$$

Thus the components $\overline{v}^s$ transform as contravariant components.

●

We further note from Equation (4.6-36) that the components $\Delta q^m$ also transform as contravariant components. Although the coordinates $q^m$ are not components of a vector, $\Delta q^m$ are contravariant components and the indices are written as superscripts. The transformation from $q^m$ to $\overline{q}^n$ is arbitrary, as can be seen from Equation (4.6-26, 27), but we still write the indices as superscripts because $\Delta q^m$ are contravariant components.

In Equations (4.6-30a, b), we have expressed the vector $\underline{v}$ in terms of the covariant base vectors $\underline{g}_m$. We could equally have expressed $\underline{v}$ in terms of the contravariant base vectors $\underline{g}^n$. That is

$$\underline{v} = v_m \underline{g}^m = \overline{v}_n \overline{\underline{g}}^n \tag{4.6-38}$$

The relation between $\overline{\underline{g}}^n$ and $\underline{g}^m$ can be deduced, using the chain rule, as follows

$$\underline{g}^m = \underline{\nabla} q^m \tag{4.6-39}$$

$$= \frac{\partial q^m}{\partial \overline{q}^n} \underline{\nabla} \overline{q}^n \tag{4.6-40a}$$

$$= \frac{\partial q^m}{\partial \overline{q}^n} \overline{\underline{g}}^n \tag{4.6-40b}$$

Substituting Equation (4.6-40b) into Equation (4.6-38), we obtain

$$\frac{\partial q^m}{\partial \overline{q}^n} v_m \overline{\underline{g}}^n = \overline{v}_n \overline{\underline{g}}^n \tag{4.6-41}$$

We then deduce

$$\bar{v}_n = \frac{\partial q^m}{\partial \bar{q}^n} \, v_m \qquad\qquad (4.6\text{-}42)$$

Components of vectors which transform according to Equation (4.6-42) are known as **covariant components** [see, also Equation (4.6-14a)].

***Example 4.6-3.*** Show that the components of $\underline{\nabla} \, \varphi$, where $\varphi$ is a scalar function of $q^1$, $q^2$ and $q^3$, transform as covariant components of a vector.

Let

$$\underline{u} = \underline{\nabla} \, \varphi \qquad\qquad (4.6\text{-}43a)$$

then we define $u_m$ as

$$u_m = \frac{\partial \varphi}{\partial q^m} \qquad\qquad (4.6\text{-}43b)$$

In the $\left(\bar{q}^{\,1}, \bar{q}^{\,2}, \bar{q}^{\,3}\right)$ coordinate system, the components $\bar{u}_n$ are given by

$$\bar{u}_n = \frac{\partial \varphi}{\partial \bar{q}^n} \qquad\qquad (4.6\text{-}44)$$

Applying the chain rule to Equation (4.6-44), we have

$$\bar{u}_n = \frac{\partial \varphi}{\partial q^m} \frac{\partial q^m}{\partial \bar{q}^n} \qquad\qquad (4.6\text{-}45a)$$

$$= \frac{\partial q^m}{\partial \bar{q}^n} \, u_m \qquad\qquad (4.6\text{-}45b)$$

Thus the components $\bar{u}_n$ transform as covariant components.

**Tensors of Arbitrary Order**

So far we have considered only scalars and vectors. A scalar is a tensor of order zero and its numerical value at a point remains invariant when the coordinate system is transformed. A vector has both a magnitude and a direction and its components transform according to Equations (4.6-33, 42) when the coordinate system is transformed. One index is sufficient to specify its components. It is a tensor of order one. In Example 4.6-3, we have seen that $\underline{\nabla} \, \varphi$ is a vector but $\varphi$ is a scalar. Thus the quantity $\underline{\nabla} \, \underline{v}$ is a tensor of order two and is known as the velocity gradient. In fluid mechanics, the rate of deformation tensor, which is a tensor of order two, is equal to the sum of the velocity gradient and its transpose. The stress tensor, which is another tensor of order two, maps linearly the surface force on the surface of a deformable continuum to the unit normal on the surface. Thus a tensor of order two transforms a vector $\underline{u}$ linearly to another vector $\underline{v}$. This can be written as

$$u_m = T_{mn} v^n \tag{4.6-46}$$

In Equation (4.6-46), we have adopted the summation convention by writing the dummy index as a subscript-superscript pair. The components of $\underline{u}$ are written as covariant components and the components of $\underline{v}$ are written in contravariant form. The quantities $T_{mn}$ are the components of a second order tensor which we denote as $\underline{\underline{T}}$. In a three-dimensional space, m and n can take the values 1, 2 and 3 and $\underline{\underline{T}}$ has nine components. We need two indices to specify the components of a second order tensor.

In the $\left(\bar{q}^{\,1}, \bar{q}^{\,2}, \bar{q}^{\,3}\right)$ coordinate system, Equation (4.6-46) becomes

$$\bar{u}_m = \bar{T}_{mn} \bar{v}^{\,n} \tag{4.6-47}$$

Using Equations (4.6-33, 42), Equation (4.6-47) becomes

$$\frac{\partial q^r}{\partial \bar{q}^{\,m}} u_r = \bar{T}_{mn} \frac{\partial \bar{q}^{\,n}}{\partial q^s} v^s \tag{4.6-48}$$

Multiplying both sides of Equation (4.6-48) by $\dfrac{\partial \bar{q}^{\,m}}{\partial q^p}$, we obtain

$$\frac{\partial \bar{q}^{\,m}}{\partial q^p} \frac{\partial q^r}{\partial \bar{q}^{\,m}} u_r = \bar{T}_{mn} \frac{\partial \bar{q}^{\,m}}{\partial q^p} \frac{\partial \bar{q}^{\,n}}{\partial q^s} v^s \tag{4.6-49}$$

Note that $\dfrac{\partial q^r}{\partial q^p}$ can, according to the chain rule, be written as $\dfrac{\partial q^r}{\partial \bar{q}^{\,m}} \dfrac{\partial \bar{q}^{\,m}}{\partial q^p}$. One recognizes the quantity $\dfrac{\partial q^r}{\partial q^p}$ (in the unbarred coordinate system) as representing the Kronecker delta. Similarly, $\dfrac{\partial \bar{q}^{\,s}}{\partial \bar{q}^{\,t}}$ would represent $\delta^s_{\,t}$, while a quantity such as $\dfrac{\partial q^i}{\partial \bar{q}^{\,\ell}}$ is not in general $\delta^i_{\,\ell}$. It thus follows that

$$\bar{T}_{mn} \frac{\partial \bar{q}^{\,m}}{\partial q^p} \frac{\partial \bar{q}^{\,n}}{\partial q^s} v^s = \delta^r_p u_r = u_p \tag{4.6-50a,b}$$

In Equation (4.6-46), we are at liberty to change the free index m to p and the dummy index n to s, we then obtain

$$u_p = T_{ps} v^s \tag{4.6-51}$$

Comparing Equations (4.6-50b, 51), we deduce

$$T_{ps} = \frac{\partial \overline{q}^{\,m}}{\partial q^p} \frac{\partial \overline{q}^{\,n}}{\partial q^s} \overline{T}_{mn}$$

(4.6-52)

Interchanging $\left(q^1, q^2, q^3\right)$ and $\left(\overline{q}^{\,1}, \overline{q}^{\,2}, \overline{q}^{\,3}\right)$ in Equation (4.6-52), it becomes

$$\overline{T}_{ps} = \frac{\partial q^m}{\partial \overline{q}^{\,p}} \frac{\partial q^n}{\partial \overline{q}^{\,s}} T_{mn}$$

(4.6-53)

Components that transform according to Equation (4.6-52) are known as **covariant components**. Note the similarity to Equation (4.6-42).

**Contravariant components** $\overline{T}^{\,mn}$ transform according to

$$\overline{T}^{\,mn} = \frac{\partial \overline{q}^{\,m}}{\partial q^r} \frac{\partial \overline{q}^{\,n}}{\partial q^s} T^{rs}$$

(4.6-54)

For second order tensors, in addition to covariant and contravariant components, we can have **mixed components** $\overline{T}^{\,m}_{\;n}$ which transform according to

$$\overline{T}^{\,m}_{\;n} = \frac{\partial \overline{q}^{\,m}}{\partial q^r} \frac{\partial q^s}{\partial \overline{q}^{\,n}} T^{\,r}_{\;s}$$

(4.6-55)

Tensors of order higher than two also exist and are frequently used. An example of a tensor of order three is the permutation tensor, which will be defined in the next section. A tensor that maps linearly a second order tensor to another second order tensor is a **tensor of order four**. The constitutive equation of a linear elastic material may be written as

$$\tau_{ij} = c_{ijk\ell} \, \gamma^{k\ell}$$

(4.6-56)

where $\tau_{ij}$, $\gamma^{k\ell}$ and $c_{ijk\ell}$ are the stress tensor, the infinitesimal strain tensor and the elastic tensor respectively. The components $c_{ijk\ell}$ need four indices and are the components of a fourth order tensor.

The law of transformation for a fourth order tensor can be obtained by generalizing Equations (4.6-53 to 55) as follows

$$T_{prst} = \frac{\partial \overline{q}^{\,i}}{\partial q^p} \frac{\partial \overline{q}^{\,j}}{\partial q^r} \frac{\partial \overline{q}^{\,k}}{\partial q^s} \frac{\partial \overline{q}^{\,\ell}}{\partial q^t} \overline{T}_{ijk\ell}$$

(4.6-57a)

$$T^{prst} = \frac{\partial q^p}{\partial \overline{q}^i} \frac{\partial q^r}{\partial \overline{q}^j} \frac{\partial q^s}{\partial \overline{q}^k} \frac{\partial q^t}{\partial \overline{q}^\ell} \overline{T}^{ijk\ell} \qquad (4.6\text{-}57b)$$

$$T^p_{rst} = \frac{\partial q^p}{\partial \overline{q}^i} \frac{\partial \overline{q}^j}{\partial q^r} \frac{\partial \overline{q}^k}{\partial q^s} \frac{\partial \overline{q}^\ell}{\partial q^t} \overline{T}^i_{jk\ell} \qquad (4.6\text{-}57c)$$

A second order tensor may also be defined as the juxtaposition of two vectors $\underline{u}$ and $\underline{v}$. The second order tensor $\underline{\underline{T}}$ may be defined as the **dyadic product** of vectors $\underline{u}$ and $\underline{v}$.

$$\underline{\underline{T}} = \underline{u}\,\underline{v} = \underline{u} \otimes \underline{v} \qquad (4.6\text{-}58a,b)$$

Equation (4.6-58) can also be written as

$$\underline{\underline{T}} = u^i \underline{g}_i \, v^j \underline{g}_j = u_m \underline{g}^m v_n \underline{g}^n \qquad (4.6\text{-}59a,b)$$

$$= u^i v^j \underline{g}_i \underline{g}_j = u_m v_n \underline{g}^m \underline{g}^n \qquad (4.6\text{-}59c,d)$$

$$= T^{ij} \underline{g}_i \underline{g}_j = T_{mn} \underline{g}^m \underline{g}^n \qquad (4.6\text{-}59e,f)$$

A second order tensor $\underline{\underline{T}}$ is also known as a **dyad** and the notation adopted in Equations (4.6-59a to f) is called the dyadic notation. The juxtaposition of the two vectors $\underline{u}$ and $\underline{v}$ is also known as the **outer product** of the two vectors. The **commutative law** does not hold and $\underline{g}_i \underline{g}_j$ is in general not equal to $\underline{g}_j \underline{g}_i$. The component $T_{ij}$ is not necessarily equal to the component $T_{ji}$. If they are equal, $\underline{\underline{T}}$ is a **symmetric tensor**. Tensors of higher order can likewise be defined.

## Metric and Permutation Tensors

Equation (4.6-19a) defines the covariant base vector $\underline{g}_m$. If P and Q are two neighboring points with vector positions $\underline{r}$ and $\underline{r} + d\underline{r}$ with coordinates $(q^i)$ and $(q^i + dq^i)$ respectively, then the square of the distance, $ds^2$, between P and Q is

$$ds^2 = d\underline{r} \cdot d\underline{r} \qquad (4.6\text{-}60a)$$

$$= \frac{\partial \underline{r}}{\partial q^m} \cdot \frac{\partial \underline{r}}{\partial q^n} dq^m dq^n \qquad (4.6\text{-}60b)$$

$$= \underline{g}_m \cdot \underline{g}_n \, dq^m dq^n \qquad (4.6\text{-}60c)$$

$$= g_{mn} \, dq^m dq^n \qquad (4.6\text{-}60d)$$

The second order tensor $g_{mn}$ is known as the **metric tensor** and in general is a function of $(q^i)$. In an orthonormal coordinate system, the metric tensor simplifies to the Kronecker delta. In an orthogonal coordinate system

$$g_{mn} = \begin{cases} = 0 & \text{if } m \neq n \\ \neq 0 & \text{if } m = n \end{cases} \qquad (4.6\text{-}61\text{a,b})$$

Since the dot product is commutative, $g_{mn}$ is symmetric.

The dual (conjugate or associate) of $g_{mn}$ which is denoted by $g^{rs}$ is defined as

$$g^{rs} = \underline{g}^r \cdot \underline{g}^s \qquad (4.6\text{-}62\text{a})$$

$$= \underline{\nabla} q^r \cdot \underline{\nabla} q^s \qquad (4.6\text{-}62\text{b})$$

From Equation (4.6-19b), we have

$$\underline{g}_m \cdot \underline{g}_j = \frac{\partial x^p}{\partial q^m} \, \underline{\delta}_p \cdot \frac{\partial x^\ell}{\partial q^j} \, \underline{\delta}_\ell \qquad (4.6\text{-}63\text{a})$$

$$= \frac{\partial x^p}{\partial q^m} \, \frac{\partial x^\ell}{\partial q^j} \, \delta_{p\ell} \qquad (4.6\text{-}63\text{b})$$

$$= \frac{\partial x^p}{\partial q^m} \, \frac{\partial x^p}{\partial q^j} \qquad (4.6\text{-}63\text{c})$$

Equation (4.6-63b) shows that the components of the metric tensor $g_{mj}$ transform as covariant components, hence the subscript notation.

Starting from Equation (4.6-20c), we can deduce an expression for $g^{jn}$ in terms of $(x^i)$ as follows

$$\underline{g}^j \cdot \underline{g}^n = \frac{\partial q^j}{\partial x^\ell} \, \underline{\delta}_\ell \cdot \frac{\partial q^n}{\partial x^r} \, \underline{\delta}_r \qquad (4.6\text{-}64\text{a})$$

$$= \frac{\partial q^j}{\partial x^\ell} \, \frac{\partial q^n}{\partial x^r} \, \delta_{\ell r} \qquad (4.6\text{-}64\text{b})$$

$$= \frac{\partial q^j}{\partial x^r} \, \frac{\partial q^n}{\partial x^r} \qquad (4.6\text{-}64\text{c})$$

It can be seen from Equation (4.6-64b) that $g^{jn}$ components transform as contravariant components. The relationship between $g_{mj}$ and $g^{jn}$ can be established by combining Equations (4.6-63c, 64c) and the result is

$$g_{mj}\, g^{jn} = \frac{\partial x^p}{\partial q^m}\, \frac{\partial x^p}{\partial q^j}\, \frac{\partial q^j}{\partial x^r}\, \frac{\partial q^n}{\partial x^r} \tag{4.6-65a}$$

$$= \frac{\partial x^p}{\partial q^m}\, \frac{\partial x^p}{\partial x^r}\, \frac{\partial q^n}{\partial x^r} \tag{4.6-65b}$$

$$= \frac{\partial x^p}{\partial q^m}\, \frac{\partial q^n}{\partial x^r}\, \delta_{pr} \tag{4.6-65c}$$

$$= \frac{\partial x^r}{\partial q^m}\, \frac{\partial q^n}{\partial x^r} \tag{4.6-65d}$$

$$= \delta^n_m \tag{4.6-65e}$$

In Equation (4.6-65e), we have written the Kronecker delta as a mixed tensor so as to conform to the convention that an index appearing as a subscript (superscript) on one side of the equation should also appear as a subscript (superscript) on the other side of the equation. The fundamental reason for writing the Kronecker delta as a mixed tensor is because it transforms as a mixed second order tensor. The convention is framed so as to be compatible with the rules of transformation.

The permutation tensor is an example of a third order tensor which in an orthonormal coordinate system is denoted by $e_{ijk}$, and is defined as

$$e_{ijk} = \begin{cases} 0, & \text{if any two indices are equal} \\ 1, & \text{if the indices 1, 2, 3 appear in the clockwise direction} \\ -1, & \text{if the indices 1, 2, 3 appear in the anticlockwise direction} \end{cases} \tag{4.6-66a,b,c}$$

Thus, for example,

$$e_{112} = e_{122} = 0 \tag{4.6-67a,b}$$

$$e_{123} = e_{312} = 1 \tag{4.6-67c,d}$$

$$e_{321} = e_{213} = -1 \tag{4.6-67e,f}$$

Let $\underline{w}$ be the vector product of two vectors $\underline{u}$ and $\underline{v}$. Then in an orthonormal coordinate system, $w_i$ is given by

$$w_i = e_{ijk}\, u_j\, v_k \tag{4.6-68}$$

If $\underline{\underline{D}}$ is a $(3 \times 3)$ matrix, with elements $d_{ij}$, then the determinant of $\underline{\underline{D}}$ can be written as

$$|\underline{\underline{D}}| = e_{ijk}\, d_{i1}\, d_{j2}\, d_{k3} \tag{4.6-69}$$

We extend the definition of the permutation tensor to a general curvilinear coordinate system and we denote it by $\varepsilon_{\ell mn}$. We put a bar over the components of the vectors in the present coordinate system, so as to avoid confusion with the Cartesian components. Thus $\overline{w}_\ell$ is expressed in a general barred coordinate system as

$$\overline{w}_\ell = \varepsilon_{\ell mn}\, \overline{u}^{\,m}\, \overline{v}^{\,n} \tag{4.6-70}$$

Transforming $\left(\overline{w}_\ell,\, \overline{u}^{\,m} \text{ and } \overline{v}^{\,n}\right)$ to the orthonormal coordinate system using Equations (4.6-33, 42), identifying the coordinates $q^m$ as $x^m$ and $\overline{q}^{\,n}$ as $q^n$, Equation (4.6-70) becomes

$$\frac{\partial x^r}{\partial q^\ell}\, w_r = \varepsilon_{\ell mn}\, \frac{\partial q^m}{\partial x^s}\, u^s\, \frac{\partial q^n}{\partial x^t}\, v^t \tag{4.6-71a}$$

$$= \varepsilon_{\ell mn}\, \frac{\partial q^m}{\partial x^s}\, \frac{\partial q^n}{\partial x^t}\, u^s\, v^t \tag{4.6-71b}$$

Multiplying both sides of Equation (4.6-71b) by $\dfrac{\partial q^\ell}{\partial x^p}$, we obtain

$$\frac{\partial q^\ell}{\partial x^p}\, \frac{\partial x^r}{\partial q^\ell}\, w^r = \varepsilon_{\ell mn}\, \frac{\partial q^\ell}{\partial x^p}\, \frac{\partial q^m}{\partial x^s}\, \frac{\partial q^n}{\partial x^t}\, u^s\, v^t \tag{4.6-72}$$

Using the chain rule on the left side of the equation and noting that in an orthonormal coordinate system, we do not distinguish between covariant and contravariant components, Equation (4.6-72) becomes

$$w_p = \varepsilon_{\ell mn}\, \frac{\partial q^\ell}{\partial x^p}\, \frac{\partial q^m}{\partial x^s}\, \frac{\partial q^n}{\partial x^t}\, u_s\, v_t \tag{4.6-73a}$$

$$= e_{pst}\, u_s\, v_t \tag{4.6-73b}$$

From Equations (4.6-73a, b), we deduce that

$$e_{pst} = \varepsilon_{\ell mn}\, \frac{\partial q^\ell}{\partial x^p}\, \frac{\partial q^m}{\partial x^s}\, \frac{\partial q^n}{\partial x^t} \tag{4.6-74}$$

which is the law of transformation of covariant components. Alternatively $\varepsilon_{\ell mn}$ may be written as

$$\varepsilon_{\ell mn} = \sqrt{g}\, e_{\ell mn} \tag{4.6-75}$$

where $g$ is the determinant of the metric tensor $g_{mn}$. The contravariant form of the permutation tensor is

$$\varepsilon^{ijk} = \frac{1}{\sqrt{g}} \, e_{ijk} \tag{4.6-76}$$

***Example 4.6-4.*** Calculate the metric tensor $g_{ij}$, its dual $g^{rs}$ and its determinant $g$ for the spherical polar coordinate system.

The base vectors have been obtained in Example 4.6-1.

The components of the metric tensor are given by

$$g_{rr} = \underline{g}_r \cdot \underline{g}_r = 1 \tag{4.6-77a,b}$$

$$g_{\theta\theta} = \underline{g}_\theta \cdot \underline{g}_\theta = r^2 \tag{4.6-77c,d}$$

$$g_{\phi\phi} = \underline{g}_\phi \cdot \underline{g}_\phi = r^2 \sin^2 \theta \tag{4.6-77e,f}$$

All the other $g_{ij}$ are zero since the spherical polar coordinate system is orthogonal.

$$g^{rr} = \underline{g}^r \cdot \underline{g}^r = 1 \tag{4.6-78a,b}$$

$$g^{\theta\theta} = \underline{g}^\theta \cdot \underline{g}^\theta = \frac{1}{r^2} \tag{4.6-78c,d}$$

$$g^{\phi\phi} = \underline{g}_\phi \cdot \underline{g}_\phi = \frac{1}{r^2 \sin^2 \theta} \tag{4.6-78e,f}$$

The determinant $g$ is given by

$$g = r^4 \sin^2 \theta \tag{4.6-79}$$

## Covariant, Contravariant and Physical Components

It has been noted that the covariant and contravariant base vectors do not in general have the same dimensions and it is not surprising that the covariant and contravariant components of a vector also do not have the same dimensions. Via the metric tensor it is possible to establish a relationship between these two types of components and this is done as follows

$$\underline{v} = v_m \, \underline{g}^m = v^n \, \underline{g}_n \tag{4.6-80a,b}$$

Forming the dot product with $\underline{g}_r$, we have

$$v_m \, \underline{g}^m \cdot \underline{g}_r = v^n \, \underline{g}_n \cdot \underline{g}_r \tag{4.6-81a}$$

$$v_m \, \delta_r^m = v^n \, g_{nr} \tag{4.6-81b}$$

$$v_r = g_{nr} \, v^n \tag{4.6-81c}$$

Equation (4.6-81c) shows the transformation from contravariant components to covariant components. Similarly for higher order tensors we have

$$T_{rs} = g_{nr} \, g_{ms} \, T^{nm} \tag{4.6-82a}$$

$$T^{nm} = g^{nr} \, g^{ms} \, T_{rs} \tag{4.6-82b}$$

$$T^{nm}_{k\ell} = g^{nr} \, g^{ms} \, g_{pk} \, g_{q\ell} \, T^{pq}_{rs} \tag{4.6-82c}$$

In a space in which a metric is defined, it is possible to transform covariant components to contravariant components and vice versa. This process is known as **lowering and raising indices**. It was pointed out in Example 4.6-1 that $\underline{g}_r$ and $\underline{g}_\theta$ do not have the same magnitude and they have different dimensions. This makes it necessary to define the so called physical components. The **physical components** of a vector are expressed in terms of normalized base vectors. The normalized covariant base vectors are given by

$$\underline{g}_{(n)} = \frac{\underline{g}_n}{|\underline{g}_n|} = \frac{\underline{g}_n}{\sqrt{g_{nn}}} \tag{4.6-83a,b}$$

Note that in Equations (4.6-83a, b) there is no summation and the index $n$ is contained in brackets to designate physical components. A vector $\underline{v}$ may be written as

$$\underline{v} = v^m \, \underline{g}_m = v_{(n)} \, \underline{g}_{(n)} \tag{4.6-84a,b}$$

$$= \frac{v_{(n)} \, \underline{g}_n}{\sqrt{g_{nn}}} \tag{4.6-84c}$$

Combining Equations (4.6-84a, c), noting that $m$ is a dummy index, yields

$$\left( v^n - \frac{v_{(n)}}{\sqrt{g_{nn}}} \right) \underline{g}_n = \underline{0} \tag{4.6-85}$$

Since the vectors $\underline{g}_n$ are base vectors and are linearly independent, Equation (4.6-85) implies that for each n,

$$v^n = \frac{v_{(n)}}{\sqrt{g_{nn}}} \tag{4.6-86a}$$

or $\quad v_{(n)} = \sqrt{g_{nn}} \; v^n$ (4.6-86b)

That is to say, the physical components $v_{(n)}$ can be obtained via the metric tensor. Note that in Equation (4.6-86a, b) there is no summation over n. The index n occurs three times!

The contravariant component $v^n$ can be written as

$$v^n = g^{nm} \; v_m \tag{4.6-87}$$

Substituting Equation (4.6-87) into Equation (4.6-86b), we obtain

$$v_{(n)} = \sqrt{g_{nn}} \; g^{nm} \; v_m \tag{4.6-88}$$

In an orthogonal coordinate system, combining Equations (4.6-61a, b, 88), we obtain

$$\sqrt{g^{nn}} \; v_n = v_{(n)} \tag{4.6-89}$$

Similarly the physical components of higher order tensors are defined. For an orthogonal coordinate system we have

$$T_{(mr)} = \sqrt{g_{mm}} \; \sqrt{g_{rr}} \; T^{mr} \tag{4.6-90a}$$

$$= \sqrt{g^{mm}} \; \sqrt{g^{rr}} \; T_{mr} \tag{4.6-90b}$$

$$= \sqrt{g^{mm}} \; \sqrt{g_{rr}} \; T^r_m \tag{4.6-90c}$$

So far, we have defined the physical components, in terms of normalized covariant base vectors. We could equally have chosen to define physical components via normalized contravariant components. In the framework of an orthogonal coordinate system, both are identical. In the case of a non-orthogonal coordinate system, they may not be identical since Equation (4.6-89) could not be obtained from Equation (4.6-88). Indeed, for a non-orthogonal system, $g^{nm}$ in Equation (4.6-88) represents three non-zero components for each value of n. In Equation (4.6-89), there is no summation over n.

***Example 4.6-5.*** Obtain the contravariant, covariant and physical components of the velocity vector $\underline{v}$ of a particle in the spherical polar coordinate system.

In Example 4.6-2, we have shown that $\dfrac{\partial q^m}{\partial t}$ transforms as contravariant components. The contravariant components of $\underline{v}$ are

$$v^r = \frac{dr}{dt} = \dot{r} \tag{4.6-91a,b}$$

$$v^\theta = \frac{d\theta}{dt} = \dot\theta \tag{4.6-91c,d}$$

$$v^\phi = \frac{d\phi}{dt} = \dot\phi \tag{4.6-91e,f}$$

Using Equation (4.6-81c), we have

$$v_r = g_{rr} v^r = \dot r \tag{4.6-92a,b}$$

$$v_\theta = g_{\theta\theta} v^\theta = r^2 \dot\theta \tag{4.6-92c,d}$$

$$v_\phi = g_{\phi\phi} v^\phi = r^2 \left(\sin^2\theta\right) \dot\phi \tag{4.6-92e,f}$$

The physical components are given by Equation (4.6-86b)

$$v_{(r)} = \sqrt{g_{rr}} \, v^r = \dot r \tag{4.6-93a,b}$$

$$v_{(\theta)} = \sqrt{g_{\theta\theta}} \, v^\theta = r \, \dot\theta \tag{4.6-93c,d}$$

$$v_{(\phi)} = \sqrt{g_{\phi\phi}} \, v^\phi = r \left(\sin\theta\right) \dot\phi \tag{4.6-93e,f}$$

***Example 4.6-6.*** Calculate the covariant, contravariant and physical components of $\underline\nabla f$ in the spherical polar coordinate system.

In Example 4.6-3, we have shown that $\dfrac{\partial\varphi}{\partial q^m}$ is a covariant component. If we denote $\underline\nabla f$ by $\underline u$, then

$$u_r = \frac{\partial f}{\partial r} \tag{4.6-94a}$$

$$u_\theta = \frac{\partial f}{\partial\theta} \tag{4.6-94b}$$

$$u_\phi = \frac{\partial f}{\partial\phi} \tag{4.6-94c}$$

The contravariant components are

$$u^r = g^{rr} u_r = \frac{\partial f}{\partial r} \tag{4.6-95a,b}$$

$$u^\theta = g^{\theta\theta} u_\theta = \frac{1}{r^2} \frac{\partial f}{\partial \theta} \qquad (4.6\text{-}95\text{c,d})$$

$$u^\phi = g^{\phi\phi} u_\phi = \frac{1}{r^2 \sin^2\theta} \frac{\partial f}{\partial \phi} \qquad (4.6\text{-}95\text{e,f})$$

The physical components are

$$u_{(r)} = \sqrt{g^{rr}} \, u_r = \frac{\partial f}{\partial r} \qquad (4.6\text{-}96\text{a,b})$$

$$u_{(\theta)} = \sqrt{g^{\theta\theta}} \, u_\theta = \frac{1}{r} \frac{\partial f}{\partial \theta} \qquad (4.6\text{-}96\text{c,d})$$

$$u_{(\phi)} = \sqrt{g^{\phi\phi}} \, u_\phi = \frac{1}{r \sin\theta} \frac{\partial f}{\partial \phi} \qquad (4.6\text{-}96\text{e,f})$$

In Examples 4.6-5 and 6, we note that the r covariant and contravariant components do not have the same dimension as the $\theta$, $\phi$ components, whereas the r, $\theta$ and $\phi$ physical components, all have the same dimension. In a space in which a metric tensor exists, a tensor can be represented in terms of covariant, contravariant or physical components. They can be transformed from one to the other by the process of raising or lowering the indices, as shown in Examples 4.6-5 and 6.

●

The laws of physics are independent of the coordinate system and they should be written in tensorial form. The quantities that enter into the equations expressing these laws should be in covariant or contravariant components. Each expression in the equation should be a tensor component of the same kind and order. That is to say, if on the right side of the equation we have a mixed component which is covariant of order m and contravariant of order n, then on the left side we must also have a mixed component, covariant of order m and contravariant of order n.

Finally the components of a tensor have to be measured in terms of certain units and it is desirable to express all the components in terms of the same physical dimensions.

In the rectangular Cartesian coordinate system, there is no distinction between covariant, contravariant and physical components. The metric tensor is the Kronecker delta. Many authors omit the word physical when referring to physical components of a tensor. Thus it is safe to assume that unless otherwise stated, the components of a tensor refer to physical components.

## 4.7   COVARIANT DIFFERENTIATION

We have shown in Example 4.6-3 that if $\varphi$ is a scalar function, $\dfrac{\partial \varphi}{\partial q^i}$ is a covariant component of a

vector. We shall presently show that if $v^i$ is the contravariant component of a vector $\underline{v}$, $\dfrac{\partial v^i}{\partial q^j}$ does

not transform as tensor.

Consider the transformation given by Equations (4.6-26, 27). On transforming from $(q^1, q^2, q^3)$ to $(\bar{q}^{\,1}, \bar{q}^{\,2}, \bar{q}^{\,3})$, we have

$$\frac{\partial v^i}{\partial q^j} = \frac{\partial}{\partial \bar{q}^{\,s}} \left( \frac{\partial q^i}{\partial \bar{q}^{\,\ell}} \, \bar{v}^{\,\ell} \right) \frac{\partial \bar{q}^{\,s}}{\partial q^j} \qquad\qquad (4.7\text{-}1a)$$

$$= \frac{\partial \bar{v}^{\,\ell}}{\partial \bar{q}^{\,s}} \frac{\partial q^i}{\partial \bar{q}^{\,\ell}} \frac{\partial \bar{q}^{\,s}}{\partial q^j} + \bar{v}^{\,\ell} \frac{\partial^2 q^i}{\partial \bar{q}^{\,s} \partial \bar{q}^{\,\ell}} \frac{\partial \bar{q}^{\,s}}{\partial q^j} \qquad\qquad (4.7\text{-}1b)$$

Comparing Equations (4.6-55, 7-1b), we note that $\dfrac{\partial v^i}{\partial q^j}$ transforms as a mixed component if

$\dfrac{\partial^2 q^i}{\partial \bar{q}^{\,s} \partial \bar{q}^{\,\ell}} = 0$. That is to say, $\dfrac{\partial v^i}{\partial q^j}$ transforms as a component of a tensor only if the transformation

given by Equation (4.6-26) is a linear transformation. In general the partial deviative of the components of a vector is not a tensor. This is because the base vectors are in general not constants, but functions of $(q^1, q^2, q^3)$. On taking the partial deviative of a vector there is a contribution from the base vectors and this has to be taken into account.

From Equation (4.6-30a), we obtain

$$\frac{\partial \underline{v}}{\partial q^j} = \frac{\partial v^m}{\partial q^j} \, \underline{g}_m + v^m \frac{\partial \underline{g}_m}{\partial q^j} \qquad\qquad (4.7\text{-}2)$$

Using Equation (4.6-19b), we find

$$\frac{\partial \underline{g}_m}{\partial q^j} = \frac{\partial^2 x^p}{\partial q^m \partial q^j} \, \underline{\delta}_p \qquad\qquad (4.7\text{-}3)$$

Note that $\underline{\delta}_p$, the base vector in the rectangular Cartesian coordinate system is a constant. The transformation of the covariant base vector $\underline{\delta}_p$ to $\underline{g}_\ell$ is given by Equation (4.6-31b)

$$\frac{\partial \underline{v}}{\partial q^j} = \frac{\partial v^m}{\partial q^j} \underline{g}_m + v^m \frac{\partial^2 x^p}{\partial q^m \partial q^j} \frac{\partial q^\ell}{\partial x^p} \underline{g}_\ell \tag{4.7-5a}$$

$$= \left[ \frac{\partial v^\ell}{\partial q^j} + v^m \frac{\partial^2 x^p}{\partial q^m \partial q^j} \frac{\partial q^\ell}{\partial x^p} \right] \underline{g}_\ell \tag{4.7-5b}$$

$$= \left[ \frac{\partial v^\ell}{\partial q^j} + \left\{ {\ell \atop m\ \ j} \right\} v^m \right] \underline{g}_\ell \tag{4.7-5c}$$

$$= \left[ v^\ell{}_{,j} \right] \underline{g}_\ell \tag{4.7-5d}$$

In going from Equation (4.7-5a to b), we have replaced the dummy index $m$ in the first term on the right side of the equation by $\ell$, allowing us to factor out $\underline{g}_\ell$. The quantity

$$\left\{ {\ell \atop m\ \ j} \right\} = \frac{\partial^2 x^p}{\partial q^m \partial q^j} \frac{\partial q^\ell}{\partial x^p} \tag{4.7-6}$$

is known as the **Christoffel symbol of the second kind** and is also denoted by $\Gamma^\ell{}_{mj}$. The **covariant derivative** of $v^\ell$ with respect to $q^j$ is denoted by $v^\ell{}_{,j}$ or $v^\ell|_j$. From Equations (4.7-5c, d), it is given by

$$v^\ell{}_{,j} = \frac{\partial v^\ell}{\partial q^j} + \left\{ {\ell \atop m\ \ j} \right\} v^m \tag{4.7-7}$$

The notation correctly suggests that $v^\ell{}_{,j}$ represents a mixed component of a second order tensor. $\frac{\partial v^\ell}{\partial q^j}$ and $\left\{ {\ell \atop m\ \ j} \right\}$ do not transform as tensors.

The **Christoffel symbol of the first kind**, denoted by [rs, t] (or $\Gamma_{rst}$), is defined as

$$[rs,\ t] = g_{\ell t} \left\{ {\ell \atop r\ \ s} \right\} \tag{4.7-8}$$

The covariant derivative of the covariant component $v_\ell$ is given by

$$v_{\ell,j} = \frac{\partial v_\ell}{\partial q^j} - \left\{ {m \atop \ell\ \ j} \right\} v_m \tag{4.7-9}$$

The covariant derivative of higher order tensors are

$$T^{\ell m}{}_{,j} = \frac{\partial T^{\ell m}}{\partial q^j} + \begin{Bmatrix} \ell \\ s\ j \end{Bmatrix} T^{sm} + \begin{Bmatrix} m \\ t\ j \end{Bmatrix} T^{\ell t} \qquad (4.7\text{-}10a)$$

$$T_{\ell m,\, j} = \frac{\partial T_{\ell m}}{\partial q^j} - \begin{Bmatrix} s \\ \ell\ j \end{Bmatrix} T_{sm} - \begin{Bmatrix} t \\ m\ j \end{Bmatrix} T_{\ell t} \qquad (4.7\text{-}10b)$$

$$T^{\ell}{}_{m,\, j} = \frac{\partial T^{\ell}{}_{m}}{\partial q^j} + \begin{Bmatrix} \ell \\ s\ j \end{Bmatrix} T^{s}{}_{m} - \begin{Bmatrix} t \\ m\ j \end{Bmatrix} T^{\ell}{}_{t} \qquad (4.7\text{-}10c)$$

$$T^{\ell m}{}_{n,\, j} = \frac{\partial T^{\ell m}{}_{n}}{\partial q^j} + \begin{Bmatrix} \ell \\ s\ j \end{Bmatrix} T^{sm}{}_{n} + \begin{Bmatrix} m \\ t\ j \end{Bmatrix} T^{\ell t}{}_{n} - \begin{Bmatrix} u \\ n\ j \end{Bmatrix} T^{\ell m}{}_{u} \qquad (4.7\text{-}10d)$$

$$T^{\ell}{}_{mn,\, j} = \frac{\partial T^{\ell}{}_{mn}}{\partial q^j} + \begin{Bmatrix} \ell \\ s\ j \end{Bmatrix} T^{s}{}_{mn} - \begin{Bmatrix} t \\ m\ j \end{Bmatrix} T^{\ell}{}_{tn} - \begin{Bmatrix} u \\ n\ j \end{Bmatrix} T^{\ell}{}_{mu} \qquad (4.7\text{-}10e)$$

The covariant derivative of tensors of arbitrary order can be written down by observing the pattern shown in Equations (4.7-10a to e).

## Properties of Christoffel Symbols

(i)     The Christoffel symbol of the second kind can be calculated in terms of the metric tensor and is given by

$$\begin{Bmatrix} \ell \\ m\ j \end{Bmatrix} = \frac{1}{2}\, g^{\ell k} \left[ \frac{\partial g_{jk}}{\partial q^m} + \frac{\partial g_{mk}}{\partial q^j} - \frac{\partial g_{mj}}{\partial q^k} \right] \qquad (4.7\text{-}11)$$

(ii)    $$\begin{Bmatrix} \ell \\ m\ j \end{Bmatrix} = \begin{Bmatrix} \ell \\ j\ m \end{Bmatrix} \qquad \text{(symmetry)} \qquad (4.7\text{-}12)$$

(iii)   If the coordinate system is orthogonal

$$\begin{Bmatrix} \ell \\ m\ j \end{Bmatrix} = 0, \quad \text{if } \ell,\ m,\ \text{and } j \text{ are all different} \qquad (4.7\text{-}13a)$$

$$\begin{Bmatrix} \ell \\ \ell\ j \end{Bmatrix} = \frac{1}{2}\, g^{\ell\ell}\, \frac{\partial g_{\ell\ell}}{\partial q^j} \qquad (4.7\text{-}13b)$$

$$\begin{Bmatrix} \ell \\ j\ j \end{Bmatrix} = -\frac{1}{2}\, g^{\ell\ell}\, \frac{\partial g_{jj}}{\partial q^\ell} \qquad (4.7\text{-}13c)$$

$$\left\{ \begin{matrix} \ell \\ \ell \ \ell \end{matrix} \right\} = \frac{1}{2} g^{\ell\ell} \frac{\partial g_{\ell\ell}}{\partial q^\ell} \tag{4.7-13d}$$

In Equations (4.7-13a to d), no summation is implied.

(iv)  On performing the transformation given by Equations (4.6-26, 27), the Christoffel symbol transforms as

$$\left\{ \begin{matrix} \overline{\ell} \\ m \ j \end{matrix} \right\} = \left\{ \begin{matrix} r \\ s \ t \end{matrix} \right\} \frac{\partial \overline{q}^\ell}{\partial q^r} \frac{\partial q^s}{\partial \overline{q}^m} \frac{\partial q^t}{\partial \overline{q}^j} + \frac{\partial \overline{q}^\ell}{\partial q^t} \frac{\partial^2 q^t}{\partial \overline{q}^m \partial \overline{q}^j} \tag{4.7-14}$$

Note that only for the linear transformation $\left( \dfrac{\partial^2 q^t}{\partial \overline{q}^m \partial \overline{q}^j} = 0 \right)$, does the Christoffel symbol

transform as a mixed third order tensor. In the rectangular Cartesian coordinate system, all the Christoffel symbols are zero, since the metric tensors are constants.

## Rules of Covariant Differentiation

(i)  The covariant derivative of the sum (or difference) of two tensors is the sum (or difference) of their covariant derivatives

$$(T_{ij} + S_{ij})_{,k} = T_{ij,k} + S_{ij,k} \tag{4.7-15}$$

(ii)  The covariant derivative of a dot (or outer) product of two tensors is equal to the sum of the two terms obtained by the dot (or outer) product of each tensor with the covariant derivative of the other tensor

$$\left( T_{i\ell} S^\ell_{\ j} \right)_{,k} = T_{i\ell} S^\ell_{\ j,k} + \left( T_{i\ell,k} \right) S^\ell_{\ j} \tag{4.7-16a}$$

$$\left( T_{i\ell} S_{mj} \right)_{,k} = T_{i\ell} S_{mj,k} + \left( T_{i\ell,k} \right) S_{mj} \tag{4.7-16b}$$

Note that in Equation (4.7-16a), $\ell$ is a dummy index and the dot product of the two second order tensors is another second order tensor. In Equation (4.7-16b), we have formed the outer product of two second order tensors resulting in a fourth order tensor.

(iii)  The metric tensors $g_{ij}$, $g^{\ell m}$ and the Kronecker delta $\delta^r_{\ s}$ are constants with respect to covariant differentiation

$$g_{ij,k} = g^{\ell m}_{\ \ ,k} = \delta^r_{\ s,k} = 0 \tag{4.7-17a,b,c}$$

A consequence of this result is that the metric and Kronecker tensors can be put outside the covariant differentiation sign. That is to say

$$\left( g_{ij} \, T^j{}_{\ell m} \right)_{,k} = \left( T^j{}_{\ell m,k} \right) g_{ij} \tag{4.7-18}$$

(iv)      The covariant derivative of $v_\ell$ with respect to $q^j$ is a second order tensor and we can take its covariant derivative with respect to $q^k$. Combining Equations (4.7-9, 10b), we obtain

$$\left( v_{\ell,j} \right)_{,k} = \frac{\partial}{\partial q^k} \left( v_{\ell,j} \right) - \begin{Bmatrix} s \\ \ell \;\; k \end{Bmatrix} v_{s,j} - \begin{Bmatrix} s \\ j \;\; k \end{Bmatrix} v_{\ell,s} \tag{4.7-19a}$$

$$= \frac{\partial}{\partial q^k} \left[ \frac{\partial v_\ell}{\partial q^j} - \begin{Bmatrix} m \\ \ell \;\; j \end{Bmatrix} v_m \right] - \begin{Bmatrix} s \\ \ell \;\; k \end{Bmatrix} \left[ \frac{\partial v_s}{\partial q^j} - \begin{Bmatrix} m \\ s \;\; j \end{Bmatrix} v_m \right] - \begin{Bmatrix} s \\ j \;\; k \end{Bmatrix} \left[ \frac{\partial v_\ell}{\partial q^s} - \begin{Bmatrix} m \\ \ell \;\; s \end{Bmatrix} v_m \right]$$
$$\tag{4.7-19b}$$

$$= \frac{\partial^2 v_\ell}{\partial q^k \, \partial q^j} - \left[ \frac{\partial}{\partial q^k} \begin{Bmatrix} m \\ \ell \;\; j \end{Bmatrix} \right] v_m - \begin{Bmatrix} m \\ \ell \;\; j \end{Bmatrix} \frac{\partial v_m}{\partial q^k} - \begin{Bmatrix} s \\ \ell \;\; k \end{Bmatrix} \frac{\partial v_s}{\partial q^j} - \begin{Bmatrix} s \\ j \;\; k \end{Bmatrix} \frac{\partial v_\ell}{\partial q^s}$$

$$+ \begin{Bmatrix} s \\ \ell \;\; k \end{Bmatrix} \begin{Bmatrix} m \\ s \;\; j \end{Bmatrix} v_m + \begin{Bmatrix} s \\ j \;\; k \end{Bmatrix} \begin{Bmatrix} m \\ \ell \;\; s \end{Bmatrix} v_m = v_{\ell,jk} \tag{4.7-19c, d}$$

If we interchange the order of differentiation, Equation (4.7-19c) becomes

$$v_{\ell,kj} = \frac{\partial^2 v_\ell}{\partial q^j \, \partial q^k} - \left[ \frac{\partial}{\partial q^j} \begin{Bmatrix} m \\ \ell \;\; k \end{Bmatrix} \right] v_m - \begin{Bmatrix} m \\ \ell \;\; k \end{Bmatrix} \frac{\partial v_m}{\partial q^j} - \begin{Bmatrix} s \\ \ell \;\; j \end{Bmatrix} \frac{\partial v_s}{\partial q^k} - \begin{Bmatrix} s \\ k \;\; j \end{Bmatrix} \frac{\partial v_\ell}{\partial q^s}$$

$$+ \begin{Bmatrix} s \\ \ell \;\; j \end{Bmatrix} \begin{Bmatrix} m \\ s \;\; k \end{Bmatrix} v_m + \begin{Bmatrix} s \\ k \;\; j \end{Bmatrix} \begin{Bmatrix} m \\ \ell \;\; s \end{Bmatrix} v_m \tag{4.7-19e}$$

On the right side of Equation (4.7-19e), $m$ is a dummy index in the third term and it can be replaced by $s$, likewise in the fourth term $s$ can be replaced by $m$. Noting that the Christoffel symbol is symmetric (Equation 4.7-12), and assuming that $v_\ell$ is continuous with continuous second partial derivatives, it follows from Equations (4.7-19c, d, e) that

$$v_{\ell,jk} - v_{\ell,kj} = \left[ \frac{\partial}{\partial q^j} \begin{Bmatrix} m \\ \ell \;\; k \end{Bmatrix} - \frac{\partial}{\partial q^k} \begin{Bmatrix} m \\ \ell \;\; j \end{Bmatrix} + \begin{Bmatrix} s \\ \ell \;\; k \end{Bmatrix} \begin{Bmatrix} m \\ s \;\; j \end{Bmatrix} - \begin{Bmatrix} s \\ \ell \;\; j \end{Bmatrix} \begin{Bmatrix} m \\ s \;\; k \end{Bmatrix} \right] v_m \tag{4.7-20a}$$

$$= R^m{}_{\ell jk} \, v_m \tag{4.7-20b}$$

The left side of Equation (4.7-20b) is a covariant component of a third order tensor. $R^m{}_{\ell jk}$ is a mixed fourth order tensor, contravariant of order one and covariant of order three. The dot product $R^m{}_{\ell jk} \, v_m$ is a covariant component of a third order tensor. Thus the terms on both sides of Equation (4.7-20b) are covariant components of third order tensors. The fourth order tensor $R^m{}_{\ell jk}$ is the **Riemann-Christoffel tensor**.

We can interchange the order of covariant differentiation if $R^m{}_{\ell jk} = 0$. The Riemann-Christoffel tensor is a property of the space and is independent of the vector $\underline{v}$. In a Euclidean space, we can always set up a rectangular Cartesian coordinate system and the Christoffel symbols are zero, consequently the Riemann Christoffel tensor is zero. $R^m{}_{\ell jk}$ is a fourth order tensor and transforms according to Equation (4.6-57c) and is thus zero in all coordinate systems. We conclude that in a Euclidean space, $R^m{}_{\ell jk}$ is zero and the order of covariant differentiation is not important as long as the components of the tensor have continuous second partial derivatives.

***Example 4.7-1.*** Calculate the Christoffel symbols of the second kind for the spherical polar coordinate system $(r, \theta, \phi)$.

Since the coordinate system is orthogonal, we make use of Equations (4.7-13a to d). The metric tensors are given by Equations (4.6-77a to f, 78a to f). The only non-zero Christoffel symbols are

$$\left\{ \begin{matrix} & r & \\ \theta & & \theta \end{matrix} \right\} = -\frac{1}{2} \, g^{rr} \, \frac{\partial}{\partial r} \, (g_{\theta\theta}) = -r \tag{4.7-21a,b}$$

$$\left\{ \begin{matrix} & r & \\ \phi & & \phi \end{matrix} \right\} = -\frac{1}{2} \, g^{rr} \, \frac{\partial}{\partial r} \, (g_{\phi\phi}) = -r \sin^2 \theta \tag{4.7-21c,d}$$

$$\left\{ \begin{matrix} & \theta & \\ \theta & & r \end{matrix} \right\} = \frac{1}{2} \, g^{\theta\theta} \, \frac{\partial}{\partial r} \, (g_{\theta\theta}) = \frac{1}{r} \tag{4.7-21e,f}$$

$$\left\{ \begin{matrix} & \theta & \\ \phi & & \phi \end{matrix} \right\} = -\frac{1}{2} \, g^{\theta\theta} \, \frac{\partial}{\partial \theta} \, (g_{\phi\phi}) = -\sin \theta \cos \theta \tag{4.7-21g,h}$$

$$\left\{ \begin{matrix} & \phi & \\ \phi & & r \end{matrix} \right\} = \frac{1}{2} \, g^{\phi\phi} \, \frac{\partial}{\partial r} \, (g_{\phi\phi}) = \frac{1}{r} \tag{4.7-21i,j}$$

$$\left\{ \begin{matrix} & \phi & \\ \phi & & \theta \end{matrix} \right\} = \frac{1}{2} \, g^{\phi\phi} \, \frac{\partial}{\partial \theta} \, (g_{\phi\phi}) = \cot \theta \tag{4.7-21k,l}$$

***Example 4.7-2.*** Let $v^i$ be the contravariant components of the velocity vector $\underline{v}$. Obtain $v^i{}_{,i}$ in spherical polar coordinate system. That is to say, calculate $v^r{}_{,r} + v^\theta{}_{,\theta} + v^\phi{}_{,\phi}$. Rewrite this expression in physical components.

From Equations (4.7-7, 21a to l), we have

$$v^r{}_{,r} = \frac{\partial v^r}{\partial r} + \begin{Bmatrix} r \\ m\ \ r \end{Bmatrix} v^m \tag{4.7-22a}$$

$$= \frac{\partial v^r}{\partial r} + \begin{Bmatrix} r \\ r\ \ r \end{Bmatrix} v^r + \begin{Bmatrix} r \\ \theta\ \ r \end{Bmatrix} v^\theta + \begin{Bmatrix} r \\ \phi\ \ r \end{Bmatrix} v^\phi \tag{4.7-22b}$$

$$= \frac{\partial v^r}{\partial r} \tag{4.7-22c}$$

$$v^\theta{}_{,\theta} = \frac{\partial v^\theta}{\partial \theta} + \begin{Bmatrix} \theta \\ r\ \ \theta \end{Bmatrix} v^r + \begin{Bmatrix} \theta \\ \theta\ \ \theta \end{Bmatrix} v^\theta + \begin{Bmatrix} \theta \\ \phi\ \ \theta \end{Bmatrix} v^\phi \tag{4.7-23a}$$

$$= \frac{\partial v^\theta}{\partial \theta} + \frac{1}{r}\, v^r \tag{4.7-23b}$$

$$v^\phi{}_{,\phi} = \frac{\partial v^\phi}{\partial \phi} + \begin{Bmatrix} \phi \\ r\ \ \phi \end{Bmatrix} v^r + \begin{Bmatrix} \phi \\ \theta\ \ \phi \end{Bmatrix} v^\theta + \begin{Bmatrix} \phi \\ \phi\ \ \phi \end{Bmatrix} v^\phi \tag{4.7-24a}$$

$$= \frac{\partial v^\phi}{\partial \phi} + \frac{1}{r}\, v^r + \cot\theta\, v^\theta \tag{4.7-24b}$$

$$v^r{}_{,r} + v^\theta{}_{,\theta} + v^\phi{}_{,\phi} = \frac{\partial v^r}{\partial r} + \frac{\partial v^\theta}{\partial \theta} + \frac{\partial v^\phi}{\partial \phi} + \frac{2}{r}\, v^r + \cot\theta\, v^\theta \tag{4.7-25}$$

Rewriting in physical components, using Equation (4.6-86b), we have

$$v^r = v_{(r)}\,, \qquad v^\theta = \frac{1}{r}\, v_{(\theta)}\,, \qquad v^\phi = \frac{1}{r\sin\theta}\, v_{(\phi)} \tag{4.7-26a,b,c}$$

Substituting Equations (4.7-26a to c) into Equation (4.7-25), we obtain

$$v^r{}_{,r} + v^\theta{}_{,\theta} + v^\phi{}_{,\phi} = \frac{\partial}{\partial r}\left(v_{(r)}\right) + \frac{\partial}{\partial \theta}\left(\frac{1}{r} v_{(\theta)}\right) + \frac{\partial}{\partial \phi}\left(\frac{1}{r\sin\theta} v_{(\phi)}\right) + \frac{2\, v_{(r)}}{r} + \frac{\cot\theta}{r}\, v_{(\theta)} \tag{4.7-27}$$

We note that on the right side of Equation (4.7-27) every term has the same physical dimension, namely the reciprocal of time. Equation (4.7-27) expresses the divergence of the velocity vector $\underline{v}$ and is a scalar, which is shown in the next example.

***Example 4.7-3.*** Show that $v^i{}_{,i}$ is a scalar.

The component $v^i_{,j}$ is a mixed tensor and will transform according to Equation (4.6-55)

$$\overline{v}^{\,i}_{\;,j} = \frac{\partial \overline{q}^{\,i}}{\partial q^r} \frac{\partial q^s}{\partial \overline{q}^{\,j}} v^r_{\;,s} \qquad\qquad (4.7\text{-}28)$$

Setting $j = i$ and summing

$$\overline{v}^{\,i}_{\;,i} = \frac{\partial \overline{q}^{\,i}}{\partial q^r} \frac{\partial q^s}{\partial \overline{q}^{\,i}} v^r_{\;,s} \qquad\qquad (4.7\text{-}29a)$$

$$= \delta^s_{\;r} v^r_{\;,s} \qquad\qquad (4.7\text{-}29b)$$

$$= v^s_{\;,s} \qquad\qquad (4.7\text{-}29c)$$

$$= v^i_{\;,i} \qquad\qquad (4.7\text{-}29d)$$

Equations (4.7-29b, c) are obtained by using the usual chain rule, and the property of the Kronecker delta respectively. The indices $i$ and $s$ are dummy and can be interchanged freely. The sum $v^i_{\;,i}$ is independent of the coordinate system and is a scalar or an invariant. That is to say, we obtain the same value in the barred and in the unbarred coordinate system.

## Grad, Div, and Curl

The operator $\underline{\nabla}$ in a general curvilinear coordinate system can be defined as

$$\underline{\nabla} = \underline{g}^r \frac{\partial}{\partial q^r} \qquad\qquad (4.7\text{-}30)$$

If $\varphi$ is a scalar function

$$\text{grad } \varphi = \underline{\nabla} \varphi = \underline{g}^r \frac{\partial \varphi}{\partial q^r} \qquad\qquad (4.7\text{-}31a,b)$$

In Example 4.6-3, we have shown that $\dfrac{\partial \varphi}{\partial q^r}$ transforms as a covariant tensor.

If $\underline{v}$ is a vector

$$\underline{\nabla}\, \underline{v} = \underline{g}^r \frac{\partial}{\partial q^r} v_s\, \underline{g}^s = \underline{g}^r \frac{\partial}{\partial q^r} v^t\, \underline{g}_t \qquad\qquad (4.7\text{-}32a,b)$$

$$= \underline{g}^r\, \underline{g}^s\, v_{s,r} = \underline{g}^r\, \underline{g}_t\, v^t_{\;,r} \qquad\qquad (4.7\text{-}32c,d)$$

The components $v_{s,r}$ and $v^t_{,r}$ are the covariant and mixed components respectively of a second order tensor. This confirms the statement made earlier that the operation of taking the grad of a tensor raises the order of the tensor.

In Example 4.7-3, we have calculated the divergence of a vector, which in the $\underline{\nabla}$ notation is written as

$$\underline{\nabla} \cdot \underline{v} = \underline{g}^r \frac{\partial}{\partial q^r} \cdot v^s \underline{g}_s \tag{4.7-33a}$$

$$= v^s_{,r} \underline{g}^r \cdot \underline{g}_s \tag{4.7-33b}$$

$$= v^s_{,r} \delta^r_s \tag{4.7-33c}$$

$$= v^s_{,s} \tag{4.7-33d}$$

The **divergence of a second order tensor** is defined as

$$\underline{\nabla} \cdot \underline{\underline{T}} = \underline{g}^r \frac{\partial}{\partial q^r} \cdot T^{st} \underline{g}_s \underline{g}_t \tag{4.7-34a}$$

$$= T^{st}_{,r} \underline{g}^r \cdot \underline{g}_s \underline{g}_t \tag{4.7-34b}$$

$$= T^{st}_{,r} \delta^r_s \underline{g}_t \tag{4.7-34c}$$

$$= T^{st}_{,s} \underline{g}_t \tag{4.7-34d}$$

The component $T^{st}_{,s}$ is the contravariant component of a vector, since $t$ is the only free index.

The divergence of a tensor of order $n$ is a tensor of order $n-1$. Thus the divergence of a vector is a scalar and the divergence of a tensor of order two is a vector.

The **Laplacian** of a scalar function $\varphi$ is given by

$$\nabla^2 \varphi = \text{div} \ (\text{grad} \ \varphi) = \underline{\nabla} \cdot (\underline{\nabla} \varphi) \tag{4.7-35a,b}$$

If we denote grad $\varphi$ by $\underline{v}$, then

$$v_i = \frac{\partial \varphi}{\partial q^i} \tag{4.7-36}$$

$\nabla^2 \varphi$ can be written as

$$\nabla^2 \varphi = \underline{\nabla} \cdot \underline{v} = v^s{}_{,s} \tag{4.7-37a,b}$$

$$= \left( g^{si} \frac{\partial \varphi}{\partial q^i} \right)_{,s} \tag{4.7-37c}$$

$$= g^{si} \left( \frac{\partial \varphi}{\partial q^i} \right)_{,s} \tag{4.7-37d}$$

$$= g^{si} \left[ \frac{\partial^2 \varphi}{\partial q^i \, \partial q^s} - \begin{Bmatrix} j \\ i \ s \end{Bmatrix} \frac{\partial \varphi}{\partial q^j} \right] \tag{4.7-37e}$$

Note that $\underline{\nabla} \cdot \underline{v}$ is defined as the covariant derivative of contravariant components. In Equation (4.7-37c), we have used Equation (4.6-81c) to transform the covariant components $v_i$ to contravariant components $v^s$. The metric tensor is a constant with respect to the covariant derivative and Equation (4.7-37d) follows from Equation (4.7-37c). If the coordinate system is orthogonal, Equation (4.7-37e) simplifies to

$$\nabla^2 \varphi = g^{ii} \left[ \frac{\partial^2 \varphi}{\partial q^i \, \partial q^i} - \begin{Bmatrix} j \\ i \ i \end{Bmatrix} \frac{\partial \varphi}{\partial q^j} \right] \tag{4.7-38}$$

If $\underline{u}$ is curl $\underline{v}$

$$\underline{u} = \underline{g}^r \frac{\partial}{\partial q^r} \times v_s \, \underline{g}^s \tag{4.7-39a}$$

$$= v_{s,r} \, \underline{g}^r \times \underline{g}^s \tag{4.7-39b}$$

$$= v_{s,r} \, \varepsilon^{rst} \, \underline{g}_t \tag{4.7-39c}$$

$$= \left[ \frac{\partial v_s}{\partial q^r} - \begin{Bmatrix} p \\ s \ r \end{Bmatrix} v_p \right] \frac{e_{rst}}{\sqrt{g}} \, \underline{g}_t \tag{4.7-39d}$$

Note that $\underline{u}$ is expressed in terms of contravariant components and Equation (4.7-39d) may be written as

$$u^t = \left[ \frac{\partial v_s}{\partial q^r} - \begin{Bmatrix} p \\ s \ r \end{Bmatrix} v_p \right] \frac{e_{rst}}{\sqrt{g}} \tag{4.7-40}$$

If the coordinate system is orthogonal, then $\begin{Bmatrix} p \\ s \ r \end{Bmatrix}$ is zero if all three suffixes are different but if $r = s$, the permutation tensor $e_{rst}$ is zero. The only possibility of non-zero contributions from the Christoffel symbol arises when $p = s$ and $p = r$. The Christoffel symbol is symmetric [Equation

(4.7-12)], $e_{rst} = -e_{srt}$, and the contributions from the Christoffel symbol cancel out. To clarify this point further, consider the case $t = 1$. The only non-zero contribution from $e_{rst}$ arises when $r = 2$, $s = 3$ and when $r = 3$, $s = 2$. Thus Equation (4.7-40) becomes

$$u^1 = \frac{e_{231}}{\sqrt{g}} \left[ \frac{\partial v_3}{\partial q^2} - \begin{Bmatrix} 3 \\ 3\ 2 \end{Bmatrix} v_3 - \begin{Bmatrix} 2 \\ 3\ 2 \end{Bmatrix} v_2 \right] + \frac{e_{321}}{\sqrt{g}} \left[ \frac{\partial v_2}{\partial q^3} - \begin{Bmatrix} 2 \\ 2\ 3 \end{Bmatrix} v_2 - \begin{Bmatrix} 3 \\ 2\ 3 \end{Bmatrix} v_3 \right] \qquad (4.7\text{-}41)$$

From the definition of the permutation tensor

$$e_{231} = 1 , \qquad e_{321} = -1 \qquad\qquad\qquad\qquad\qquad\qquad\qquad (4.7\text{-}42a,b)$$

Using Equations (4.7-42a, b), we find that the contributions from the Christoffel symbols cancel out and Equation (4.7-41) reduces to

$$u^1 = \frac{1}{\sqrt{g}} \left[ \frac{\partial v_3}{\partial q^2} - \frac{\partial v_2}{\partial q^3} \right] \qquad\qquad\qquad\qquad\qquad\qquad\qquad (4.7\text{-}43)$$

In an orthogonal coordinate system, Equation (4.7-40) simplifies to

$$u^t = \frac{e_{rst}}{\sqrt{g}} \frac{\partial v_s}{\partial q^r} \qquad\qquad\qquad\qquad\qquad\qquad\qquad\qquad (4.7\text{-}44)$$

***Example 4.7-4.*** If $f$ is a scalar function of position, write down $\nabla^2 f$ in the spherical polar coordinate system $(r, \theta, \phi)$.

Since the spherical polar coordinate system is orthogonal, we can then use Equation (4.7-38) and we have

$$\nabla^2 f = g^{rr} \left[ \frac{\partial^2 f}{\partial r^2} - \begin{Bmatrix} j \\ r\ r \end{Bmatrix} \frac{\partial f}{\partial q^j} \right] + g^{\theta\theta} \left[ \frac{\partial^2 f}{\partial \theta^2} - \begin{Bmatrix} j \\ \theta\ \theta \end{Bmatrix} \frac{\partial f}{\partial q^j} \right] + g^{\phi\phi} \left[ \frac{\partial^2 f}{\partial \phi^2} - \begin{Bmatrix} j \\ \phi\ \phi \end{Bmatrix} \frac{\partial f}{\partial q^j} \right]$$

$$(4.7\text{-}45)$$

Using Equations (4.6-78a to f, 7-21a to l), Equation (4.7-45) becomes

$$\nabla^2 f = \frac{\partial^2 f}{\partial r^2} + \frac{1}{r^2} \left[ \frac{\partial^2 f}{\partial \theta^2} + r \frac{\partial f}{\partial r} \right] + \frac{1}{r^2 \sin^2 \theta} \left[ \frac{\partial^2 f}{\partial \phi^2} + r \sin^2 \theta \frac{\partial f}{\partial r} + \sin \theta \cos \theta \frac{\partial f}{\partial \theta} \right]$$

$$(4.7\text{-}46a)$$

$$= \frac{\partial^2 f}{\partial r^2} + \frac{1}{r^2} \frac{\partial^2 f}{\partial \theta^2} + \frac{1}{r^2 \sin^2 \theta} \frac{\partial^2 f}{\partial \phi^2} + \frac{2}{r} \frac{\partial f}{\partial r} + \frac{\cot \theta}{r^2} \frac{\partial f}{\partial \theta} \qquad\qquad (4.7\text{-}46b)$$

***Example 4.7-5.*** Calculate the physical components of curl $\underline{v}$ in the spherical polar coordinate system.

If we denote curl $\underline{v}$ by $\underline{u}$, we have from Equations (4.6-79, 7-44)

$$u^r = \frac{1}{r^2 \sin \theta} \left( \frac{\partial v_\phi}{\partial \theta} - \frac{\partial v_\theta}{\partial \phi} \right) \tag{4.7-47a}$$

$$u^\theta = \frac{1}{r^2 \sin \theta} \left( \frac{\partial v_r}{\partial \phi} - \frac{\partial v_\phi}{\partial r} \right) \tag{4.7-47b}$$

$$u^\phi = \frac{1}{r^2 \sin \theta} \left( \frac{\partial v_\theta}{\partial r} - \frac{\partial v_r}{\partial \theta} \right) \tag{4.7-47c}$$

Transforming all the components to physical components via Equations (4.6-86b, 89), we obtain

$$u_{(r)} = \frac{1}{r^2 \sin \theta} \left[ \frac{\partial}{\partial \theta} \left( r \sin \theta \, v_{(\phi)} \right) - \frac{\partial}{\partial \phi} \left( r \, v_{(\theta)} \right) \right] \tag{4.7-48a}$$

$$= \frac{1}{r \sin \theta} \left[ \sin \theta \, \frac{\partial v_{(\phi)}}{\partial \theta} + \cos \theta \, v_{(\phi)} - \frac{\partial v_{(\theta)}}{\partial \phi} \right] \tag{4.7-48b}$$

$$u_{(\theta)} = \frac{r}{r^2 \sin \theta} \left[ \frac{\partial}{\partial \phi} \, v_{(r)} - \frac{\partial}{\partial r} \left( r \sin \theta \, v_{(\phi)} \right) \right] \tag{4.7-48c}$$

$$= \frac{1}{r \sin \theta} \left[ \frac{\partial v_{(r)}}{\partial \phi} - r \sin \theta \, \frac{\partial v_{(\phi)}}{\partial r} - \sin \theta \, v_{(\phi)} \right] \tag{4.7-48d}$$

$$u_{(\phi)} = \frac{r \sin \theta}{r^2 \sin \theta} \left[ \frac{\partial}{\partial r} \left( r \, v_{(\theta)} \right) - \frac{\partial}{\partial \theta} \left( v_{(r)} \right) \right] \tag{4.7-48e}$$

$$= \frac{1}{r} \left[ r \frac{\partial v_{(\theta)}}{\partial r} + v_{(\theta)} - \frac{\partial v_{(r)}}{\partial \theta} \right] \tag{4.7-48f}$$

***Example 4.7-6.*** The equation of motion for slow flows may be written as

$$\text{grad } p = \text{div } \underline{\tau} \tag{4.7-49}$$

where $p$ is a scalar (pressure) and $\underline{\tau}$ is the stress tensor. Write down Equation (4.7-49) in component form for the spherical polar coordinate system.

Assume that $p$ and $\underline{\tau}$ are functions of $r$ and $\theta$ only, and that $\underline{\tau}$ is symmetric. Equation (4.7-49) is written in the so-called coordinate free form. For a general curvilinear coordinate system $(q^1, q^2, q^3)$, the equation can be written in component form as

$$\frac{\partial p}{\partial q^i} = \tau_i{}^j{}_{,j} \tag{4.7-50}$$

From Example 4.6-2, it is known that $\dfrac{\partial p}{\partial q^i}$ is a covariant component of order one, as is the right side of Equation (4.7-50).

Expanding Equation (4.7-50), we have

$$\frac{\partial p}{\partial q^1} = \tau_1{}^1{}_{,1} + \tau_1{}^2{}_{,2} + \tau_1{}^3{}_{,3} \tag{4.7-51a}$$

$$= \frac{\partial \tau_1{}^1}{\partial q^1} + \left\{ {1 \atop s\ 1} \right\} \tau_1{}^s - \left\{ {t \atop 1\ 1} \right\} \tau_t{}^1 + \frac{\partial \tau_1{}^2}{\partial q^2} + \left\{ {2 \atop s\ 2} \right\} \tau_1{}^s - \left\{ {t \atop 1\ 2} \right\} \tau_t{}^2$$

$$+ \frac{\partial \tau_1{}^3}{\partial q^3} + \left\{ {3 \atop s\ 3} \right\} \tau_1{}^s - \left\{ {t \atop 1\ 3} \right\} \tau_t{}^3 \tag{4.7-51b}$$

$$\frac{\partial p}{\partial q^2} = \tau_2{}^1{}_{,1} + \tau_2{}^2{}_{,2} + \tau_2{}^3{}_{,3} \tag{4.7-51c}$$

$$= \frac{\partial \tau_2{}^1}{\partial q^1} + \left\{ {1 \atop s\ 1} \right\} \tau_2{}^s - \left\{ {t \atop 2\ 1} \right\} \tau_t{}^1 + \frac{\partial \tau_2{}^2}{\partial q^2} + \left\{ {2 \atop s\ 2} \right\} \tau_2{}^s - \left\{ {t \atop 2\ 2} \right\} \tau_t{}^2$$

$$+ \frac{\partial \tau_2{}^3}{\partial q^3} + \left\{ {3 \atop s\ 3} \right\} \tau_2{}^s - \left\{ {t \atop 2\ 3} \right\} \tau_t{}^3 \tag{4.7-51d}$$

$$\frac{\partial p}{\partial q^3} = \tau_3{}^1{}_{,1} + \tau_3{}^2{}_{,2} + \tau_3{}^3{}_{,3} \tag{4.7-51e}$$

$$= \frac{\partial \tau_3{}^1}{\partial q^1} + \left\{ {1 \atop s\ 1} \right\} \tau_3{}^s - \left\{ {t \atop 3\ 1} \right\} \tau_t{}^1 + \frac{\partial \tau_3{}^2}{\partial q^2} + \left\{ {2 \atop s\ 2} \right\} \tau_3{}^s - \left\{ {t \atop 3\ 2} \right\} \tau_t{}^2$$

$$+ \frac{\partial \tau_3{}^3}{\partial q^3} + \left\{ {3 \atop s\ 3} \right\} \tau_3{}^s - \left\{ {t \atop 3\ 3} \right\} \tau_t{}^3 \tag{4.7-51f}$$

For the spherical polar coordinate system we identify

$$q^1 = r, \qquad q^2 = \theta, \qquad q^3 = \phi \tag{4.7-52}$$

Making use of Equations (4.7-21a to l), Equations (4.7-51b, d, f) become

$$\frac{\partial p}{\partial r} = \frac{\partial \tau_r{}^r}{\partial r} + \frac{\partial \tau_r{}^\theta}{\partial \theta} + \frac{\tau_r{}^r}{r} - \frac{\tau_\theta{}^\theta}{r} + \frac{\tau_r{}^r}{r} + \cot\theta\,\tau_r{}^\theta - \frac{\tau_\phi{}^\phi}{r} \qquad (4.7\text{-}53\text{a})$$

$$\frac{\partial p}{\partial \theta} = \frac{\partial \tau_\theta{}^r}{\partial r} - \frac{\tau_\theta{}^r}{r} + \frac{\partial \tau_\theta{}^\theta}{\partial \theta} + \frac{\tau_\theta{}^r}{r} + r\,\tau_r{}^\theta + \frac{\tau_\theta{}^r}{r} + \cot\theta\left(\tau_\theta{}^\theta - \tau_\phi{}^\phi\right) \qquad (4.7\text{-}53\text{b})$$

$$0 = \frac{\partial \tau_\phi{}^r}{\partial r} - \frac{\tau_\phi{}^r}{r} + \frac{\partial \tau_\phi{}^\theta}{\partial \theta} + \frac{\tau_\phi{}^r}{r} - \cot\theta\,\tau_\phi{}^\theta + \frac{\tau_\phi{}^r}{r} + \cot\theta\,\tau_\phi{}^\phi$$

$$+ r\sin^2\theta\,\tau_r{}^\phi + \sin\theta\cos\theta\,\tau_\theta{}^\phi \qquad (4.7\text{-}53\text{c})$$

Transforming all the covariant and mixed components to physical components, we obtain

$$\frac{\partial p}{\partial r} = \frac{\partial}{\partial r}\left(\tau_{(rr)}\right) + \frac{\partial}{\partial \theta}\left(\frac{\tau_{(r\theta)}}{r}\right) + \left(\frac{2\,\tau_{(rr)} - \tau_{(\theta\theta)} - \tau_{(\phi\phi)}}{r}\right) + \frac{\tau_{(r\theta)}\cot\theta}{r} \qquad (4.7\text{-}54\text{a})$$

$$\frac{1}{r}\frac{\partial p}{\partial \theta} = \frac{1}{r}\left[\frac{\partial}{\partial r}\left(r\,\tau_{(\theta r)}\right) + \frac{\partial(\tau_{\theta\theta})}{\partial \theta} + 2\,\tau_{(\theta r)} + \cot\theta\left(\tau_{(\theta\theta)} - \tau_{(\phi\phi)}\right)\right] \qquad (4.7\text{-}54\text{b})$$

$$0 = \frac{1}{r\sin\theta}\left[\frac{\partial}{\partial r}\left(r\sin\theta\,\tau_{(\phi r)}\right) + \frac{\partial}{\partial \theta}\left(\sin\theta\,\tau_{(\phi\theta)}\right) + 2\sin\theta\,\tau_{(\phi r)} + \cos\theta\,\tau_{(\theta\phi)}\right] \qquad (4.7\text{-}54\text{c})$$

Note that in Equations (4.7-54a to c) every term has the same dimension.

## 4.8   INTEGRAL TRANSFORMS

The divergence theorem which transforms a volume integral to a surface integral and Stokes' theorem which transforms a surface integral to a line integral can be extended to higher order tensors and higher dimensional spaces. In this section, we state the divergence theorem and Stokes' theorem for a first and second order tensor in a generalized coordinate system.

We recall that the divergence theorem for a vector $\underline{u}$ is

$$\int_V \operatorname{div}\underline{u}\,dV = \int_S \underline{u}\cdot\underline{n}\,dS \qquad (4.8\text{-}1\text{a})$$

or

$$\int_V u^j{}_{,j}\,dV = \int_S u^j n_j\,dS \qquad (4.8\text{-}1\text{b})$$

where $V$ is the volume enclosed by the surface $S$ whose outward unit normal is $\underline{n}$.

To extend the theorem to a second order tensor $\underline{\underline{T}}$, we may replace $\underline{u}$ by $\underline{\underline{T}}$ in Equation (4.8-1a) and, in component form, we obtain

$$\int_V T^{ij},_j \ dV = \int_S T^{ij} n_j \ dS \tag{4.8-2}$$

We note that Equation (4.8-2) is a vector equation. Both sides, contain one free index (i) and it represents three equations for $i = 1, 2$ and 3. Equation (4.8-1b) is one scalar equation, containing no free index.

Stokes' theorem may be written as

$$\int_S \left( \varepsilon^{ijk} u_{k,j} \right) n_i \ dS = \oint_C u_i \ dq^i \tag{4.8-3}$$

where, as usual, $C$ is the closed curve bounding the surface $S$.

For a second order tensor $\underline{\underline{T}}$, Equation (4.8-3) becomes

$$\int_S \left( \varepsilon^{ijk} T_{k\ell,j} \right) n_i \ dS = \oint_C T_{k\ell} \ dq^k \tag{4.8-4}$$

Again, Equation (4.8-4) represents three equations ($\ell = 1, 2$ and 3).

***Example 4.8-1.*** Show that

$$\int_V \text{curl} \ \underline{u} \ dV = \int_S \underline{n} \times \underline{u} \ dS \tag{4.8-5}$$

Let

$$T^{ij} = \varepsilon^{ijk} u_k \tag{4.8-6}$$

On substituting Equation (4.8-6) into Equation (4.8-2) and noting that $\varepsilon^{ijk}$ is a constant with respect to covariant differentiation, we obtain the following

$$\int_V \left( \varepsilon^{ijk} u_k \right),_j \ dV = \int_S \varepsilon^{ijk} u_k n_j \ dS \tag{4.8-7a}$$

$$\int_V \varepsilon^{ijk} u_{k,j} \, dV = \int_S \varepsilon^{ijk} n_j u_k \, dS \tag{4.8-7b}$$

Equation (4.8-7b) is the component form of Equation (4.8-5).

***Example 4.8-2.*** Applying the law of conservation of momentum to a volume of a continuous medium in motion, Bird et al. (1987) have obtained the equation

$$\frac{d}{dt} \int_V \rho \, \underline{v} \, dV = -\int_S [\underline{n} \cdot \rho \, \underline{v} \, \underline{v}] \, dS - \int_S \underline{n} \cdot \underline{\underline{\pi}} \, dS + \int_V \rho \, \underline{g} \, dV \tag{4.8-8}$$

where $\underline{v}$ is the velocity, $\rho$ is the density, $\underline{\underline{\pi}}$ is the stress tensor and $\underline{g}$ is the gravity force.

Write down the equation in component form and hence deduce the equation of motion at each point in space.

We choose to write the components as contravariant components and hence Equation (4.8-8) becomes

$$\frac{d}{dt} \int_V \rho \, v^i \, dV = -\int_S n_j \, \rho \, v^j \, v^i \, dS - \int_S n_j \, \pi^{ji} \, dS + \int_V \rho \, g^i \, dV \tag{4.8-9}$$

To obtain the equation of motion at each point, we need to use the divergence theorem to transform the surface integrals to volume integrals. Since the volume $V$ is fixed in space, we may include the time derivative inside the volume $V$. Equation (4.8-9) now becomes

$$\int_V \frac{d}{dt} \left( \rho \, v^i \right) dV = -\int_V \left( \rho \, v^j \, v^i \right)_{,j} \, dV - \int_V \pi^{ji}_{,j} \, dV + \int_V \rho \, g^i \, dV \tag{4.8-10}$$

Since $V$ is an arbitrary volume, Equation (4.8-10) holds at every point and we obtain

$$\frac{\partial}{\partial t} \left( \rho \, v^i \right) + \left( \rho \, v^j \, v^i \right)_{,j} = -\pi^{ji}_{,j} + \rho \, g^i \tag{4.8-11}$$

Expanding the left side of Equation (4.8-11), we obtain

$$\rho \frac{\partial v^i}{\partial t} + v^i \frac{\partial \rho}{\partial t} + \rho \, v^j \, v^i_{,j} + \rho \, v^j_{,j} \, v^i + \rho_{,j} \, v^j \, v^i = \rho \left( \frac{\partial v^i}{\partial t} + v^j \, v^i_{,j} \right) + v^i \left( \frac{\partial \rho}{\partial t} + v^j \, \rho_{,j} + \rho \, v^j_{,j} \right) \tag{4.8-12}$$

From the mass balance, we obtain the equation of continuity which may be written as

$$\frac{\partial \rho}{\partial t} + \left( \rho \, v^j \right)_{,j} = 0 \tag{4.8-13}$$

Combining Equations (4.8-12, 13), Equation (4.8-11) becomes

$$\rho\left(\frac{\partial v^i}{\partial t} + v^j\, v^i_{,j}\right) = -\pi^{ji}_{,j} + \rho\, g^i \qquad (4.8\text{-}14)$$

The left side of Equation (4.8-14) is often written as $\dfrac{Dv^i}{Dt}\left(\dfrac{dv^i}{dt}\right)$ where $\dfrac{D}{Dt}$ is known as the **substantial (material) derivative**. It is the time derivative following a material element.

## 4.9   ISOTROPIC, OBJECTIVE TENSORS AND TENSOR-VALUED FUNCTIONS

### Isotropic Tensors

Many materials are **isotropic**, that is to say, their properties are independent of direction. Thus if these properties are described by tensors, the components of these tensors are identical in all rectangular Cartesian coordinate systems. To find out if a tensor is isotropic or not, we express its components in a rectangular Cartesian coordinate system $(x^1, x^2, x^3)$ and rotate the axes to obtain a new coordinate system $(\bar{x}^1, \bar{x}^2, \bar{x}^3)$. If in the new coordinate system the components are identical, the tensor is isotropic. Below we list the isotropic tensors of order zero to four. Here we consider only Cartesian coordinate systems.

i)      All tensors of order zero are isotropic. Since tensors of order zero are scalars and are independent of direction, they are isotropic.

ii)     $\underline{0}$ is the only isotropic tensor of order one. If $(u_1, u_2, u_3)$ are the components of a tensor of order one (a vector) in the $(x^i)$ coordinate system, and $(\bar{u}^i)$ are the components of the same vector in the $(\bar{x}^i)$ coordinate system, we can write

$$\bar{u}_m = \ell_{mn}\, u_n \qquad (4.9\text{-}1)$$

Let $(\bar{x}^i)$ be the coordinate axes obtained by rotating the $(x^i)$ system through $\pi$ rad about the $x^3$-axis, then

$$\ell_{11} = -1, \qquad \ell_{22} = -1, \qquad \ell_{33} = 1, \qquad \text{the other } \ell_{ij} = 0. \qquad (4.9\text{-}2a,b,c,d)$$

Combining Equations (4.9-1, 2a to d) yields

$$\bar{u}_1 = -u_1, \qquad \bar{u}_2 = -u_2, \qquad \bar{u}_3 = u_3 \qquad (4.9\text{-}3a,b,c)$$

From Equations (4.9-3a, b), we deduce that for $\underline{u}$ to be isotropic

$$\bar{u}_1 = u_1 = 0, \qquad \bar{u}_2 = u_2 = 0 \qquad (4.9\text{-}4a,b,c,d)$$

By rotating the axes about the $x^1$-axis through $\pi$ radians, we obtain

$$u_3 = 0 \qquad\qquad\qquad (4.9\text{-}5)$$

Thus $\underline{0}$ is the only isotropic tensor, that is to say, there is no non-zero isotropic vector.

iii)    The Kronecker delta $\delta_{ij}$ is isotropic. If $T_{ij}$ is a component of a second order tensor in the $(x^i)$ coordinate system and $\overline{T}_{rs}$ is a component of the same tensor in the $(\overline{x}^{\,i})$ coordinate system, then

$$\overline{T}_{rs} = \ell_{ri}\,\ell_{sj}\,T_{ij} \qquad\qquad\qquad (4.9\text{-}6)$$

Letting $T_{ij}$ be the Kronecker delta, Equation (4.9-6) becomes

$$\overline{T}_{rs} = \ell_{ri}\,\ell_{sj}\,\delta_{ij} \qquad\qquad\qquad (4.9\text{-}7\text{a})$$

$$= \ell_{ri}\,\ell_{si} = \delta_{rs} \qquad\qquad\qquad (4.9\text{-}7\text{b,c})$$

Thus the Kronecker delta transforms into itself and is thus an isotropic tensor.

In fluid mechanics, the isotropic part of the stress tensor $\pi_{ij}^{(0)}$ can be written as

$$\pi_{ij}^{(0)} = -p\,\delta_{ij} \qquad\qquad\qquad (4.9\text{-}8)$$

where $p$ is a scalar.

Any second order isotropic tensor can be expressed in terms of the Kronecker delta.

iv)    The permutation tensor $e_{ijk}$ is an isotropic tensor of third order.

A useful relation between $e_{ijk}$ and $\delta_{rs}$ is

$$e_{ijk}\,e_{rsk} = \delta_{ir}\,\delta_{js} - \delta_{is}\,\delta_{jr} \qquad\qquad\qquad (4.9\text{-}9)$$

v)    Any fourth order isotropic tensor $c_{ijk\ell}$ may be expressed as

$$c_{ijk\ell} = \lambda\,\delta_{ij}\,\delta_{k\ell} + \mu\,\delta_{ik}\,\delta_{j\ell} + \nu\,\delta_{i\ell}\,\delta_{jk} \qquad\qquad\qquad (4.9\text{-}10)$$

where $\lambda$, $\mu$ and $\nu$ are scalars.

***Example 4.9-1.*** For a Newtonian fluid, the deviatoric stress tensor $\underset{=}{\tau}$ depends linearly on the rate-of-deformation $\underset{=}{\dot\gamma}$. Obtain the constitutive equation of an isotropic, incompressible Newtonian fluid.

The constitutive equation of a Newtonian fluid may be written as

$$\tau_{ij} = \pm c_{ijk\ell}\,\dot{\gamma}_{k\ell} \tag{4.9-11}$$

Since the fluid is isotropic, we obtain using Equation (4.9-10)

$$\tau_{ij} = \pm\left[\lambda\,\delta_{ij}\delta_{k\ell} + \mu\,\delta_{ik}\delta_{j\ell} + \nu\,\delta_{i\ell}\delta_{jk}\right]\dot{\gamma}_{k\ell} \tag{4.9-12a}$$

$$= \pm\left[\lambda\,\dot{\gamma}_{kk}\delta_{ij} + \mu\,\dot{\gamma}_{i\ell}\delta_{j\ell} + \nu\,\dot{\gamma}_{ki}\delta_{jk}\right] \tag{4.9-12b}$$

$$= \pm\left[\lambda\,\dot{\gamma}_{kk}\delta_{ij} + \mu\,\dot{\gamma}_{ij} + \nu\,\dot{\gamma}_{ji}\right] \tag{4.9-12c}$$

$$= \pm(\nu + \mu)\,\dot{\gamma}_{ij} \tag{4.9-12d}$$

Equation (4.9-12d) follows from Equation (4.9-12c) since the fluid is incompressible. That is to say, $\dot{\gamma}_{kk}$ is zero. Also $\dot{\gamma}_{ij}$ is symmetric. The coefficient $(\nu + \mu)$ is known as the viscosity of the fluid. Some authors adopt the positive sign in Equation (4.9-12a) and others (Bird et al., 1987) adopt the negative sign.

## Objective Tensors

The constitutive equation of a material should be independent of the motion of the material. Alternatively we may state that the constitutive equation should be the same for all observers, irrespective whether they are at rest or in motion. Quantities which are indifferent to the motion of the observers are known as **objective quantities**.

Consider two observers, one at rest (coordinate system $\underline{x} = x^i$), and the other in relative motion (coordinate system $x^{*i} = \underline{x}^*$). Since the second observer is both translating and rotating relative to the first one, we can relate these two systems by

$$\underline{x}^* = \underline{c}(t) + \underline{\underline{Q}}(t)\bullet\underline{x} \tag{4.9-13}$$

The vector $\underline{c}(t)$ in Equation (4.9-13) denotes the translation of the second observer relative to the first observer. The matrix $\underline{\underline{Q}}(t)$ denotes the rotation of the second observer relative to the first one and the elements of $\underline{\underline{Q}}$ are the $\ell_{ij}$, the direction cosines of the axes $x^{*i}$ relative to $x^i$. Note that in the present transformation, both $\underline{c}$ and $\underline{\underline{Q}}$ are functions of time $t$. Such transformation is known as a **transformation of frames of reference**. $\underline{\underline{Q}}$ is orthogonal at all times.

Objective tensors are thus tensors which are invariant under a change of frame of reference. If we denote the components of a vector $\underline{u}$ relative to the $x^i$ coordinate system as $u^i$ and the components of the same vector $\underline{u}$ relative to the $x^{*i}$ coordinate system as $u^{*i}$, then if

$$\underline{u}^* = \underset{=}{Q} \bullet \underline{u} \qquad\qquad (4.9\text{-}14)$$

$\underline{u}$ is an objective tensor (vector).

Note that due to the rotation of the axes, the components $u^{*i}$ transform to components $u^i$ under the usual tensor transformation laws.

Equally a second order tensor $\underset{=}{T}$ is an objective tensor if

$$\underset{=}{T}^* = \underset{=}{Q} \bullet \underset{=}{T} \bullet \underset{=}{Q}^\dagger \qquad\qquad (4.9\text{-}15)$$

We now examine the objectivity of some tensors.

i)      Velocity vector $\underline{v}$

Differentiating Equation (4.9-13), we obtain

$$\underline{v}^* = \underset{=}{\dot{\underline{c}}}(t) + \underset{=}{Q}(t) \bullet \underline{v} + \underset{=}{\dot{Q}}(t) \bullet \underline{x} \qquad\qquad (4.9\text{-}16)$$

Since $\dot{\underline{c}}(t)$ and $\underset{=}{\dot{Q}}(t)$ do not vanish at all times, $\underline{v}$ is not an objective vector. This observation is a common experience. Sitting in a moving bus and watching the passenger sitting opposite to us, we seem to be at rest, but to an observer standing on the road we are travelling at a finite velocity.

ii)     The line element $(ds)^2$

From Equation (4.9-13), we have

$$d\underline{x}^* = \underset{=}{Q} \bullet d\underline{x} \qquad\qquad (4.9\text{-}17)$$

$$\left(ds^*\right)^2 = d\underline{x}^{*\dagger}\, d\underline{x}^* = d\underline{x}^* \bullet d\underline{x}^* \qquad\qquad (4.9\text{-}18a,b)$$

$$= d\underline{x}^\dagger\, \underset{=}{Q}^\dagger \bullet \underset{=}{Q}\, d\underline{x} \qquad\qquad (4.9\text{-}18c)$$

$$= d\underline{x}^\dagger\, d\underline{x} = (ds)^2 \qquad\qquad (4.9\text{-}18d,e)$$

Equation (4.9-18d) follows since $\underline{\underline{Q}}$ is orthogonal (i.e. $\underline{\underline{Q}}^\dagger \underline{\underline{Q}} = \underline{\underline{I}}$ ).

Thus $(ds)^2$ is an objective quantity. The length of an object in non-relativistic mechanics does not depend on the motion of the observer.

iii)      The rate-of-deformation $\underline{\underline{\dot\gamma}}$

Let

$$\underline{\underline{L}}^* = \frac{\partial \underline{v}^*}{\partial \underline{x}^*} , \qquad \underline{\underline{L}} = \frac{\partial \underline{v}}{\partial \underline{x}} \qquad\qquad (4.9\text{-}19a,b)$$

Using Equation (4.9-16), $\underline{\underline{L}}^*$ becomes

$$\underline{\underline{L}}^* = \frac{\partial}{\partial \underline{x}} \left[ \underline{\dot c}(t) + \underline{\underline{Q}} \cdot \underline{v} + \underline{\underline{\dot Q}} \cdot \underline{x} \right] \frac{\partial \underline{x}}{\partial \underline{x}^*} \qquad\qquad (4.9\text{-}20)$$

Inverting Equation (4.9-13), we obtain

$$\underline{\underline{Q}}^\dagger \cdot \underline{x}^* - \underline{\underline{Q}}^\dagger \cdot \underline{c} = \underline{x} \qquad\qquad (4.9\text{-}21)$$

Note that $\dfrac{\partial \underline{x}}{\partial \underline{x}^*} = \underline{\underline{Q}}^\dagger$.

Combining Equations (4.9-20, 21) yields

$$\underline{\underline{L}}^* = \underline{\underline{Q}} \cdot \underline{\underline{L}} \cdot \underline{\underline{Q}}^\dagger + \underline{\underline{\dot Q}} \cdot \underline{\underline{Q}}^\dagger \qquad\qquad (4.9\text{-}22)$$

$\underline{\underline{L}}^*$ is not an objective tensor since $\underline{\underline{\dot Q}} \cdot \underline{\underline{Q}}^\dagger$ is not zero at all times.

The rate-of-deformation $\underline{\underline{\dot\gamma}}^*$ is defined as

$$\underline{\underline{\dot\gamma}}^* = \underline{\underline{L}}^* + \underline{\underline{L}}^{*\dagger} \qquad\qquad (4.9\text{-}23a)$$

$$= \underline{\underline{Q}} \cdot \underline{\underline{L}} \cdot \underline{\underline{Q}}^\dagger + \underline{\underline{\dot Q}} \cdot \underline{\underline{Q}}^\dagger + \underline{\underline{Q}} \cdot \underline{\underline{L}}^\dagger \cdot \underline{\underline{Q}}^\dagger + \underline{\underline{Q}} \cdot \underline{\underline{\dot Q}}^\dagger \qquad\qquad (4.9\text{-}23b)$$

$$= \underline{\underline{Q}} \cdot \left( \underline{\underline{L}} + \underline{\underline{L}}^\dagger \right) \cdot \underline{\underline{Q}}^\dagger + \left( \underline{\underline{\dot Q}} \cdot \underline{\underline{Q}}^\dagger + \underline{\underline{Q}} \cdot \underline{\underline{\dot Q}}^\dagger \right) \qquad\qquad (4.9\text{-}23c)$$

$$= \underset{=}{Q} \cdot \underset{=}{\dot{\gamma}} \cdot \underset{=}{Q}^{\dagger} + \frac{d}{dt}\left(\underset{=}{Q} \cdot \underset{=}{Q}^{\dagger}\right) \tag{4.9-23d}$$

$$= \underset{=}{Q} \cdot \underset{=}{\dot{\gamma}} \cdot \underset{=}{Q}^{\dagger} \tag{4.9-23e}$$

Equation (4.9-23e) follows since $\underset{=}{Q}$ is orthogonal. Thus $\underset{=}{\dot{\gamma}}$ is an objective tensor and is an admissible quantity in a constitutive equation.

Not all equations in physics are objective. The equation of motion is not objective, because velocity and acceleration are not objective quantities. The equation of motion holds only relative to an inertial frame of reference. If the motion of the earth can be neglected then a frame of reference fixed on the surface of the earth can be considered to be an inertial frame.

## Tensor-Valued Functions

As indicated earlier, the **invariants of a tensor** are scalar quantities which remain unchanged when the coordinate system is transformed. They have an important role in tensor-valued functions. The scalar product of $\underset{\sim}{u}$ and $\underset{\sim}{u}$ ($u^i u_i$) is an invariant. For second order tensors, we have three principal invariants. These invariants arise naturally when we consider the eigenvalues and eigenvectors of a second order tensor. A non-zero vector $\underset{\sim}{u}$ is said to be an eigenvector of a second order tensor $\underset{=}{T}$ if the product $\underset{=}{T} \cdot \underset{\sim}{u}$ is parallel to $\underset{\sim}{u}$. This can be expressed as

$$\underset{=}{T} \cdot \underset{\sim}{u} = \lambda \underset{\sim}{u} \tag{4.9-24}$$

where $\lambda$ is a scalar and is an eigenvalue of $\underset{=}{T}$.

The condition for the existence of a non-zero solution to Equation (4.9-24) is

$$\det\left[\underset{=}{T} - \lambda \underset{=}{I}\right] = 0 \tag{4.9-25}$$

On expanding the determinant, we obtain

$$-\lambda^3 + I_1 \lambda^2 - I_2 \lambda + I_3 = 0 \tag{4.9-26}$$

where $I_1 = \text{tr } \underset{=}{T}$, $\quad I_2 = \frac{1}{2}\left[\left(\text{tr } \underset{=}{T}\right)^2 - \text{tr}\left(\underset{=}{T}^2\right)\right]$, $\quad I_3 = \det \underset{=}{T}$.

The trace of tensor $\underset{=}{T}$ ($\text{tr } \underset{=}{T}$) is the sum of the diagonal elements.

The functions $I_1$, $I_2$ and $I_3$ are known as the **principal invariants** of $\underset{=}{T}$, and Equation (4.9-26) is its **characteristic equation**.

Another set of invariants is defined as

$$I = \text{tr}\ \underline{\underline{T}} = I_1 \qquad\qquad (4.9\text{-}27a,b)$$

$$II = \text{tr}\ \underline{\underline{T}}^{\,2} = I_1^2 - 2\,I_2 \qquad\qquad (4.9\text{-}27c,d)$$

$$III = \text{tr}\ \underline{\underline{T}}^{\,3} = \tfrac{1}{2}\left(6\,I_3 + 2\,I_1^3 - 6\,I_1\,I_2\right) \qquad\qquad (4.9\text{-}27e,f)$$

If $\underline{\underline{T}}$ is a symmetric tensor, then its eigenvalues are real and it is diagonalisable.  That is to say, if $\lambda_1,\ \lambda_2$ and $\lambda_3$ are its eigenvalues, $\underline{\underline{T}}$ can be transformed to a diagonal matrix with $\lambda_1,\ \lambda_2$ and $\lambda_3$ as its diagonal elements.  Then

$$I = \lambda_1 + \lambda_2 + \lambda_3 \qquad\qquad (4.9\text{-}28a)$$

$$I_2 = \lambda_1\lambda_2 + \lambda_1\lambda_3 + \lambda_2\lambda_3 \qquad\qquad (4.9\text{-}28b)$$

$$I_3 = \lambda_1\lambda_2\lambda_3 \qquad\qquad (4.9\text{-}28c)$$

Further if $\varphi$ is a scalar function of $\underline{\underline{T}}$, then $\varphi$ is a function of the invariants of $\underline{\underline{T}}$, which in turn is a function of $\lambda_1,\ \lambda_2$ and $\lambda_3$ in the case of a symmetric tensor $\underline{\underline{T}}$.

We also need to consider tensor-valued functions of $\underline{\underline{T}}$ and we write

$$\underline{\underline{S}} = \underline{\underline{F}}\left(\underline{\underline{T}}\right) \qquad \text{or} \qquad S_{ij} = F_{ij}\left(T_{k\ell}\right) \qquad\qquad (4.9\text{-}29a,b)$$

If $\underline{\underline{S}}$ can be expressed as a polynomial in $\underline{\underline{T}}$, then the Cayley-Hamilton theorem can be used to simplify the representation of $\underline{\underline{S}}$.

The **Cayley-Hamilton theorem** can be stated as follows.

Every matrix satisfies its own characteristic equation.

That is to say, $\underline{\underline{T}}$ satisfies Equation (4.9-26) and substituting $\underline{\underline{T}}$ for $\lambda$ yields

$$-\underline{\underline{T}}^{\,3} + I_1\,\underline{\underline{T}}^{\,2} - I_2\,\underline{\underline{T}} + I_3\,\underline{\underline{I}} = 0 \qquad\qquad (4.9\text{-}30)$$

Expanding $\underline{\underline{F}}$ as a polynomial in $\underline{\underline{T}}$, we have

$$\underline{\underline{S}} = \alpha_0\,\underline{\underline{I}} + \alpha_1\,\underline{\underline{T}} + \alpha_2\,\underline{\underline{T}}^{\,2} + \alpha_3\,\underline{\underline{T}}^{\,3} + \ldots + \alpha_n\,\underline{\underline{T}}^{\,n} \qquad\qquad (4.9\text{-}31)$$

where $\alpha_0,\ \alpha_1,\ \ldots,\ \alpha_n$ are constants.

Using the Cayley-Hamilton theorem [Equation (4.9-30)], we find that $\underset{=}{T}^3$ can be replaced by $\underset{=}{T}^2$, $\underset{=}{T}$, $\underset{=}{I}$ and $I_1$, $I_2$ and $I_3$. Similarly, all powers of $\underset{=}{T}$ higher than two in Equation (4.9-31) can be replaced by $\underset{=}{T}^2$, $\underset{=}{T}$, $\underset{=}{I}$, and the three invariants of $\underset{=}{T}$. Thus Equation (4.9-31) simplifies to

$$\underset{=}{S} = \beta_0 \underset{=}{I} + \beta_1 \underset{=}{T} + \beta_2 \underset{=}{T}^2 \tag{4.9-32}$$

where $\beta_0$, $\beta_1$, and $\beta_2$ are functions of the invariants of $\underset{=}{T}$.

In continuum mechanics, it is not uncommon to restrict attention to isotropic materials and $\underset{=}{F}$ is then an isotropic function; that is to say, $\underset{=}{F}$ is the same in all Cartesian coordinate systems. Thus, on rotating the axes, we have

$$\overline{\underset{=}{S}} = \underset{=}{Q} \underset{=}{S} \underset{=}{Q}^\dagger = \underset{=}{Q} \underset{=}{F}\left(\underset{=}{T}\right) \underset{=}{Q}^\dagger \tag{4.9-33a,b}$$

$$\overline{\underset{=}{T}} = \underset{=}{Q} \underset{=}{T} \underset{=}{Q}^\dagger \tag{4.9-33c}$$

where $\underset{=}{Q}$ is the orthogonal matrix defined in Equation (4.9-13).

Since $\underset{=}{F}$ is the same in both coordinate systems,

$$\overline{\underset{=}{S}} = \underset{=}{F}\left(\overline{\underset{=}{T}}\right) \tag{4.9-34}$$

Combining the two sets of equations, we deduce

$$\underset{=}{Q} \underset{=}{F}\left(\underset{=}{T}\right) \underset{=}{Q}^\dagger = \underset{=}{F}\left(\underset{=}{Q} \underset{=}{T} \underset{=}{Q}^\dagger\right) \tag{4.9-35}$$

Equation (4.9-35) defines an **isotropic second order tensor-valued function** $\underset{=}{F}$.

If $\underset{=}{F}$ is isotropic, and $\underset{=}{S}$ and $\underset{=}{T}$ are symmetric, Equation (4.9-32) is a representation of $\underset{=}{S}$ without requiring that it can be expressed as a polynomial in $\underset{=}{T}$.

We note that $\underset{=}{T}^2$ may be written in terms of $\underset{=}{T}$, $\underset{=}{I}$ and the inverse of $\underset{=}{T}$ $\left(\underset{=}{T}^{-1}\right)$ if it exists. On premultiplying Equation (4.9-30) by $\underset{=}{T}^{-1}$, we obtain

$$-\underset{=}{T}^2 + I_1 \underset{=}{T} - I_2 \underset{=}{I} + I_3 \underset{=}{T}^{-1} = 0 \tag{4.9-36}$$

Thus an alternative representation of $\underset{=}{S}$ is

$$\underset{=}{S} = \gamma_0 \underset{=}{I} + \gamma_1 \underset{=}{T} + \gamma_{-1} \underset{=}{T}^{-1} \tag{4.9-37}$$

where $\gamma_0$, $\gamma_1$ and $\gamma_{-1}$ are functions of $I_1$, $I_2$ and $I_3$.

The function $\underline{\underline{F}}$ can be a function of more than one tensor. In the case where $\underline{\underline{F}}$ is an isotropic function of two symmetric tensors, $\underline{\underline{T}}_1$ and $\underline{\underline{T}}_2$,

$$\underline{\underline{S}} = \underline{\underline{F}}\left(\underline{\underline{T}}_1, \underline{\underline{T}}_2\right) \tag{4.9-38a}$$

$$= \psi_0 \underline{\underline{I}} + \psi_1 \underline{\underline{T}}_1 + \psi_2 \underline{\underline{T}}_2 + \psi_3 \underline{\underline{T}}_1^2 + \psi_4 \underline{\underline{T}}_2^2 + \psi_5 \left(\underline{\underline{T}}_1 \underline{\underline{T}}_2 + \underline{\underline{T}}_2 \underline{\underline{T}}_1\right)$$

$$+ \psi_6 \left(\underline{\underline{T}}_1^2 \underline{\underline{T}}_2 + \underline{\underline{T}}_2 \underline{\underline{T}}_1^2\right) + \psi_7 \left(\underline{\underline{T}}_1 \underline{\underline{T}}_2^2 + \underline{\underline{T}}_2^2 \underline{\underline{T}}_1\right) + \psi_8 \left(\underline{\underline{T}}_1^2 \underline{\underline{T}}_2^2 + \underline{\underline{T}}_2^2 \underline{\underline{T}}_1^2\right) \tag{4.9-38b}$$

The quantities $\psi_0, \ldots, \psi_8$ are functions of the invariants of $\underline{\underline{T}}_1$, $\underline{\underline{T}}_2$ and their products.

The ten principal invariants are

$$\mathrm{tr}\left(\underline{\underline{T}}_i\right), \quad \mathrm{tr}\left(\underline{\underline{T}}_i^2\right), \quad \mathrm{tr}\left(\underline{\underline{T}}_i^3\right), \quad (i = 1, 2)$$

$$\mathrm{tr}\left(\underline{\underline{T}}_1 \underline{\underline{T}}_2\right), \quad \mathrm{tr}\left(\underline{\underline{T}}_1^2 \underline{\underline{T}}_2\right), \quad \mathrm{tr}\left(\underline{\underline{T}}_1 \underline{\underline{T}}_2^2\right), \quad \mathrm{tr}\left(\underline{\underline{T}}_1^2 \underline{\underline{T}}_2^2\right)$$

In extending from one tensor to two tensors, we have increased the number of functions from three to eight. The number of arguments of each function has increased from three to ten.

The results for a function of an arbitrary number of tensors can be found in Truesdell and Noll (1965).

**Example 4.9-2.** A Stokesian fluid is a fluid whose deviatoric stress tensor $\underline{\underline{\tau}}$ depends on the rate-of-deformation $\underline{\underline{\dot{\gamma}}}$. Obtain a representation for $\underline{\underline{\tau}}$.

$\underline{\underline{\tau}}$ is a tensor-valued function of $\underline{\underline{\dot{\gamma}}}$ and if we assume that $\underline{\underline{\tau}}$ can be expanded as a power series of $\underline{\underline{\dot{\gamma}}}$ we obtain Equation (4.9-32) which, in this case, is written as

$$\underline{\underline{\tau}} = \beta_1 \underline{\underline{\dot{\gamma}}} + \beta_2 \underline{\underline{\dot{\gamma}}}^2 \tag{4.9-39}$$

The term $\beta_0 \underline{\underline{I}}$ has been dropped since we are considering the deviatoric stress.

A Stokesian fluid is isotropic. Since both $\underline{\underline{\tau}}$ and $\underline{\underline{\dot{\gamma}}}$ are symmetric, Equation (4.9-39) is an exact representation of $\underline{\underline{\tau}}$. If the fluid is incompressible, the first invariant of $\underline{\underline{\dot{\gamma}}}$ is zero because the equation of continuity has to be satisfied. In some flows, such as shear flows, the third invariant is also zero. Thus, in shear flows, $\beta_1$ and $\beta_2$ can be functions of the second invariant of $\dot{\underset{=}{\gamma}}$.

As a matter of historical interest, we observe that Stokes proposed his constitutive equation in 1845, Hamilton stated the Cayley-Hamilton theorem in 1853 for a special class of matrices, generalized by Cayley in 1858, but it was only in 1945 that Reiner combined both results to obtain Equation (4.9-39). Rivlin obtained Equation (4.9-39) two years later without requiring the polynomial approximation. It is thus not surprising that Equation (4.9-39) is known as the constitutive equation of a Reiner-Rivlin fluid.

It is perhaps appropriate to close this chapter by observing that Lord Kelvin (Crowe, 1967) did not believe vectors would be of the slightest use to any creature. Can we imagine a present-day physicist not using vectors at all? Many engineers believe tensors are of no use. Are they better prophets than Lord Kelvin?

## PROBLEMS

1a.    If the magnitude of a vector $\underline{a}\,(t)$ is a constant, show that $\underline{a}\,(t)$ is perpendicular to $\dfrac{d\underline{a}}{dt}$.

2a.    If $\underline{v} \times \underline{a} = \underline{q} \times \underline{a}$, determine the relation between $\underline{v}$ and $\underline{q}$.

3a.    Determine $\underline{\nabla}\,\varphi$ in each of the following cases

      (i)     $\varphi = ax + by + cz$

      (ii)    $\varphi = ax^2 + 2bxy + 2cz^2$

      (iii)   $\varphi = f(r), \ r^2 = x^2 + y^2 + z^2$                                    Answer: $\underline{r}\,f'(r)/r$

4b.    Show that

$$\underline{\nabla}\,(\underline{v} \bullet \underline{v}) = 2\,(\underline{v} \bullet \underline{\nabla})\,\underline{v} + 2\underline{v} \times \operatorname{curl} \underline{v}$$

If $\underline{v} = \underline{\nabla}\,\varphi$, deduce that $\operatorname{curl} \underline{v} = 0$, that is to say, $\underline{v}$ is a conservative field. If $\underline{v}$ is a conservative field, show that the acceleration defined by

$$\underline{a} = \frac{\partial \underline{v}}{\partial t} + (\underline{v} \bullet \underline{\nabla})\,\underline{v}$$

is also a conservative field. Determine the acceleration potential $\varphi$.

5a.    If $\underline{v} = 2yz\underline{i} - x^2 y\,\underline{j} + xz^2\underline{k}$ and $\varphi = xyz$, calculate the following

      (i)  $(\underline{v} \bullet \underline{\nabla})\,\varphi,$      (ii)  $\underline{v} \bullet (\underline{\nabla}\,\varphi),$      (iii)  $(\underline{v} \times \underline{\nabla})\,\varphi,$      (iv)  $(\underline{v} \bullet \underline{\nabla})\,\underline{\nabla}\,\varphi$

6a.     Evaluate the following

(i)     $\displaystyle\int_0^A (x+y+z)\,ds$

along the line joining the origin 0 to the point A (1, 1, 1).                    Answer: $3\sqrt{3}/2$

(ii)    $\displaystyle\int_C \left(x^2+y^2\right)ds$

along the semi-circle $x^2+y^2=1,\ y\geq 0,\ z=0$.                              Answer: $\pi$

(iii)   $\displaystyle\oint_C xy\,ds$

along the sides of the square of unit length.  That is to say, along the 4 lines given by
(a) $y=0, 0\leq x\leq 1$,  (b) $x=1, 0\leq y\leq 1$,  (c) $y=1, 0\leq x\leq 1$,  (d) $x=0, 0\leq y\leq 1$.

Answer: 0

7a.     If $\underline{v}=(x,y,z)$, evaluate the following

(i)     $\displaystyle\int_0^A \underline{v}\bullet d\underline{r}$                                                              Answer: 42

(ii)    $\displaystyle\int_0^A \underline{v}\times d\underline{r}$                                                       Answer: $\dfrac{32}{5}\underline{i}-8\underline{j}+\dfrac{8}{3}\underline{k}$

along the curve given in parametric form by  $x=t,\ y=t^2,\ z=t^3$  from the origin 0 to point A (2, 4, 8).

8b.     By inverting the order of integration, evaluate

(i)     $\displaystyle\int_0^1 \int_x^{2-x} \frac{x}{y}\,dy\,dx$                                                Answer: $2\,\ell n\,2-1$

(ii)    $\displaystyle\int_0^1 \int_y^1 \frac{x^3}{(x^2+y^2)}\,dx\,dy$                                       Answer: $\pi/12$

[Hint: sketch the area of integration, split the area into two parts, if necessary.]

9b.     By means of the transformation

$$x=r\cos\theta,\qquad y=r\sin\theta$$

evaluate

$$\iint_A e^{-\left(x^2 + y^2\right)} dx\, dy$$

where $A$ is the positive quadrant ($x \geq 0$, $y \geq 0$) of the xy-plane.

Hence, deduce the value of

(i) $\displaystyle\int_0^\infty e^{-x^2} dx$,        (ii) $\displaystyle\int_0^\infty e^{-\alpha x^2} dx$

where $\alpha$ is a constant.        Answer: (i) $\dfrac{\sqrt{\pi}}{2}$,  (ii) $\dfrac{1}{2}\sqrt{\dfrac{\pi}{\alpha}}$

10b.    Evaluate

$$\iiint_V \underline{v}\, dV$$

where $\underline{v} = 2xz\underline{i} - x\underline{j} + y^2\underline{k}$ and $V$ is the region bounded by the surfaces $x = 0$, $y = 0$, $y = 6$, $z = x^2$, $z = 4$.

[Hint: $V$ is shown in Figure 4.P-10b.]        Answer: $(128, -24, 384)$

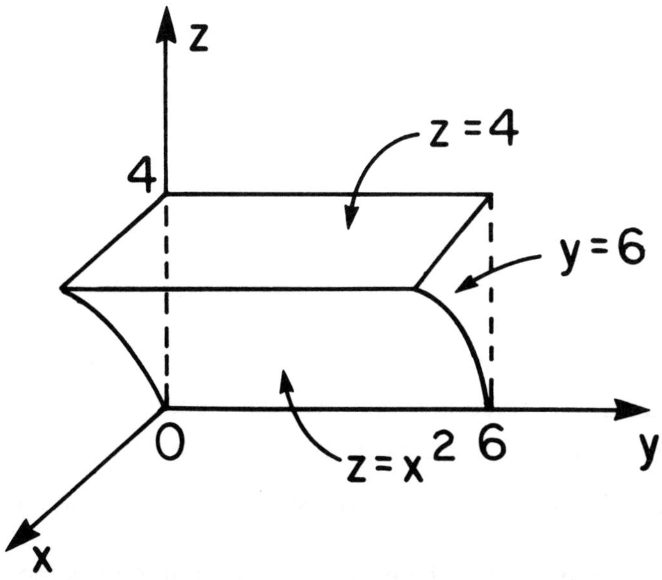

**FIGURE 4.P-10b    Volume of integration**

11a.    $S$ is the complete surface of a cube whose sides are of length $2a$. Evaluate

(i)     $\displaystyle\iint_S \left(x^2 + y^2\right) dS$

(ii)    $\displaystyle\iint_S \left(x\underline{i} + y\underline{j} + z\underline{k}\right) \bullet \underline{n}\ dS$

where $\underline{n}$ is the unit normal.                          Answer: (i) $\dfrac{80\ a^4}{3}$, (ii) $24\ a^3$

12a.    Use the divergence theorem to evaluate

$$\iint_S \underline{v} \bullet \underline{n}\ dS$$

where $\underline{v} = x^3\underline{i} + y^3\underline{j} + z^3\underline{k}$ and S is the surface of a sphere of radius a, $\underline{n}$ is the unit

outward normal to S.                                                        Answer: $\dfrac{12\ a^5}{5}$

13b.    Verify the divergence theorem in each of the following cases, by evaluating both the volume
        and surface integrals.

(i)     $\underline{v} = x\underline{i} + 4y\underline{j} + z\underline{k}$, the volume V is bounded by the coordinate planes and the plane
        $2x + y + 2z = 6$ in the positive octant.

(ii)    $\underline{v} = x\underline{i} + y\underline{j} + \left(z^2 - 1\right)\underline{k}$, V is the region occupied by the circular cylinder $x^2 + y^2 \leq 1$,
        $z = \pm 1$.

14a.    Show that Equation (4.3-75) is identical to Equation (4.3-41).

15a.    Show that Equation (4.4-40) is equivalent to Equation (4.4-23).

16b.    Bird et al. (1960) obtained the equation of continuity for a binary mixture by considering a
        volume element $\Delta x\ \Delta y\ \Delta z$ fixed in space. We can equally consider an arbitrary volume V
        fixed in space. The volume V is enclosed by a surface S. The time rate of change of mass of
        A in V is $\dfrac{\partial}{\partial t}\displaystyle\int_V \rho_A\ dV$, where $\rho_A$ is the density of A. The output of A across the surface
        is $\displaystyle\int_S \underline{n}_A \bullet \underline{n}\ dS$, $\underline{n}_A (= \rho_A\ \underline{v}_A)$ is the mass flux vector and $\underline{n}$ is the unit outward normal to
        S. The rate of production of A by chemical reaction in volume V is $\displaystyle\int_V r_A\ dV$. Using the
        mass balance and the divergence theorem to transform the surface integral to a volume integral,
        deduce the equation

$$\frac{\partial \rho_A}{\partial t} + \underline{\nabla} \cdot \underline{n}_A = r_A$$

17a. The set of axes $O\overline{x}\,\overline{y}\,\overline{z}$ is obtained by rotating the set of axes $Oxyz$ through an angle of 60° about the z-axis, the direction of rotation being from the x-axis to the y-axis. Write down the set of direction cosines which corresponds to this rotation. Hence, show that

$$\begin{bmatrix} \overline{x} \\ \overline{y} \\ \overline{z} \end{bmatrix} = \begin{bmatrix} \frac{1}{2} & \frac{\sqrt{3}}{2} & 0 \\ -\frac{\sqrt{3}}{2} & \frac{1}{2} & 0 \\ 0 & 0 & 1 \end{bmatrix} \begin{bmatrix} x \\ y \\ z \end{bmatrix}$$

If the coordinates of a point $P$ are $x = 1$, $y = 2$, $z = 3$, find $\overline{x}$, $\overline{y}$, $\overline{z}$. Calculate $\sqrt{x^2 + y^2 + z^2}$ and $\sqrt{\overline{x}^2 + \overline{y}^2 + \overline{z}^2}$. Explain your result.

18a. The set of axes $Oxyz$, $O\overline{x}\,\overline{y}\,\overline{z}$ are as defined in Problem 17a. If the vectors $\underline{u}$ and $\underline{v}$ have components $(2, 3, 4)$ and $(4, 5, 6)$ when referred to $Oxyz$, obtain their components when referred to $O\overline{x}\,\overline{y}\,\overline{z}$. Calculate the sum of the products $u_i v_i$ and $\overline{u}_i \overline{v}_i$. Verify that $u_i v_i = \overline{u}_i \overline{v}_i$.

19a. The transformation from the rectangular Cartesian coordinate system $(x^1, x^2, x^3)$ to the elliptical coordinate system $(\xi, \eta, z)$ is

$$x^1 = \cosh \xi \cos \eta, \quad x^2 = \sinh \xi \sin \eta, \quad x^3 = z \; ; \quad 0 \le \xi \le \infty, \quad -\pi < \eta \le \pi, \quad -\infty < z < \infty.$$

Determine the geometrical shapes of the $\xi$ and $\eta$ coordinate curves.

Calculate the covariant and contravariant base vectors of the elliptical coordinate system. Is it an orthogonal coordinate system?

20b. Calculate the metric tensor $g_{ij}$ and its dual $g^{rs}$ for the $(\xi, \eta, z)$ coordinate system considered in Problem 19a, by using the results obtained in Problem 19a.

Also obtain $g_{ij}$ by using the transformation given by Equation (4.6-53) and the fact that in a rectangular Cartesian coordinate system the metric tensor is the Kronecker delta $\delta_{ij}$.

21a. Let $m = n$ in Equation (4.6-55). Using the summation convention, show that $T^n{}_n$ is a scalar. This process of setting a superscript equal to a subscript is known as the process of contraction.

22a. Let $T_{ij}$ be a covariant tensor of order two. $v_{<j>}$ is a quantity whose tensorial properties are not known. But the inner product $T_{ij} v_{<j>}$ is known to be a covariant vector (tensor of order one).  Show that $v_{<j>}$ is a contravariant vector.  That is to say,  $v_{<j>}$ transforms as given by Equation (4.6-33).

This method of ascertaining whether a quantity is a tensor or not is known as the quotient law.

23a.     The Taylor (1934) four-roll mill consists of four infinitely long cylinders immersed in a fluid.  The axes of the cylinders pass through the corners of a square as shown in Figure 4.P-23a.  Each cylinder rotates in the direction opposite to the two adjacent cylinders.  The resulting flow field is a point of stagnation at O; the fluid is drawn in at  A  and  C  and is expelled at  B  and D.  The stagnation point  O  is taken to be the origin of a rectangular Cartesian coordinate system  $Ox^1 x^2$, as shown in Figure 4.P-23a.  The velocity components are

$$v_1 = -cx^1, \qquad v_2 = cx^2$$

where  c  is a constant.

The transformation from the polar coordinate system $(r, \theta)$ to the Cartesian coordinate system is given by

$$x^1 = r \cos \theta, \qquad x^2 = r \sin \theta.$$

Obtain the physical components of the velocity in the polar coordinate system.

Answer:  $[cr (\sin^2 \theta - \cos^2 \theta), cr \sin \theta \cos \theta]$

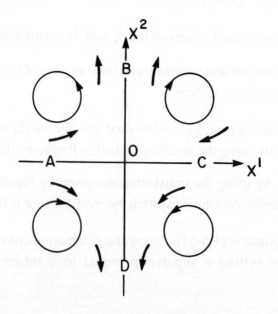

**FIGURE  4.P-23a     Four-roll  mill**

24b. In Bird et al. (1987), Appendix E, it is required to evaluate the volume integral of the dyad $\underline{u}\,\underline{u}$, where $\underline{u}$ is a unit vector, over a unit sphere. In the sperical polar coordinate system (see Example 4.4-1), the volume integral may be written as

$$\int_0^1 \int_0^\pi \int_0^{2\pi} \underline{u}\,\underline{u}\, r^2 \sin\theta\, dr\, d\theta\, d\phi$$

Show that the volume integral is equal to $\frac{4\pi}{9}\,\underline{\underline{I}}$.

25a. Show that the only non-zero Christoffel symbols of the second kind corresponding to the cylindrical polar coordinate system $(r, \theta, z)$ are $\left\{\begin{matrix} r \\ \theta\ \ \theta \end{matrix}\right\}$ and $\left\{\begin{matrix} \theta \\ r\ \ \theta \end{matrix}\right\}$. Determine them.

The transformation from the Cartesian coordinate system $(x^1, x^2, x^3)$ to the cylindrical coordinate system $(r, \theta, z)$ is

$$x^1 = r \cos\theta, \qquad x^2 = r \sin\theta, \qquad x^3 = z.$$

The $(r, \theta, z)$ system is orthogonal and the metric tensors are $g_{rr} = g_{zz} = 1$, $g_{\theta\theta} = r^2$.

26b. In modelling a vibrating jet, Chan Man Fong et al. (1993) assumed that the physical components of the velocity field, in the cylindrical polar coordinate system $(r, \theta, z)$, may be written as

$$v_{(r)} = \frac{r}{2}\frac{dw}{dz} + u_1(r, z)\cos\theta$$

$$v_{(\theta)} = -\frac{\partial}{\partial r}(r\,u_1)\sin\theta$$

$$v_{(z)} = -w(z).$$

The covariant component of the rate-of-strain tensor $\dot{\gamma}_{ij}$ is defined as

$$\dot{\gamma}_{ij} = v_{i,j} + v_{j,i}$$

Show that the physical components of $\dot{\underline{\underline{\gamma}}}$ (denoted by $\underline{\underline{A}}_1$ in the original paper) are

$$\begin{bmatrix} \dfrac{dw}{dz} + 2\dfrac{\partial u_1}{\partial r}\cos\theta & -(r\dfrac{\partial^2 u_1}{\partial r^2} + \dfrac{\partial u_1}{\partial r})\sin\theta & \dfrac{r}{2}\dfrac{d^2 w}{dz^2} + \dfrac{\partial u_1}{\partial z}\cos\theta \\[3ex] -(r\dfrac{\partial^2 u_1}{\partial r^2} + \dfrac{\partial u_1}{\partial r})\sin\theta & \dfrac{dw}{dz} - 2\dfrac{\partial u_1}{\partial r}\cos\theta & -(r\dfrac{\partial^2 u_1}{\partial r\partial z} + \dfrac{\partial u_1}{\partial z})\sin\theta \\[3ex] \dfrac{r}{2}\dfrac{d^2 w}{dz^2} + \dfrac{\partial u_1}{\partial z}\cos\theta & -(r\dfrac{\partial^2 u_1}{\partial r\partial z} + \dfrac{\partial u_1}{\partial z})\sin\theta & -2\dfrac{dw}{dz} \end{bmatrix}$$

27a.    Starting from the definition of div $\underline{v}$ given by Equation (4.7-33d), show that for an orthogonal coordinate system, div $\underline{v}$ may be written as

$$\text{div }\underline{v} \;=\; \frac{1}{h_1 h_2 h_3}\left[\frac{\partial}{\partial q^1}(h_2 h_3 v_{(1)}) + \frac{\partial}{\partial q^2}(h_1 h_3 v_{(2)}) + \frac{\partial}{\partial q^3}(h_1 h_2 v_{(3)})\right]$$

where $g_{ii} = h_i^2$.

28a.    Show that for any vector $\underline{v}$

(i)      $v_{i,j} - v_{j,i} \;=\; \dfrac{\partial v_i}{\partial q^j} - \dfrac{\partial v_j}{\partial q^i}$

(ii)     $\varepsilon^{ijk}(v_{i,j} - v_{j,i})$ is a vector.  Is it a covariant or a contravariant vector?

29a.    Write the covariant derivatives with respect to $q^k$ of each of the following

(i) $g_{ij} v^j$,          (ii) $g^{ij} v_j$,          (iii) $T_{ij} v^j$

30a.    Calculate the physical components of grad $\varphi$ in the cylindrical polar coordinate system (see Problem 25a).

31a.    Find $\nabla^2 \varphi$ in the cylindrical polar coordinate system.

32a.    If $T^{ij}$ is a symmetric tensor, show that $\varepsilon_{ijk} T^{ij} = 0$.

33b.    The symmetric stress tensor $\pi_{ij}$ satisfies the equation

$$\pi_{\ni j,\, j} = 0.$$

Show that

$$\frac{1}{2} \int_V \pi_{ij} \, \dot{\gamma}^{ij} \, dV = \int_S v^i \, \pi_{ij} \, n^j \, dS$$

where V is the volume enclosing the surface S, $v^i$ are the velocity components and

$$\dot{\gamma}_{ij} = v_{i,j} + v_{j,i} \,.$$

34b. By considering the transformation of coordinates given in Problem 17a, show that if $T_{ij}$ is an isotropic tensor, it is necessarily of the form $\lambda \, \delta_{ij}$.

35b. The rate-of-deformation tensor $\underset{=}{\dot{\gamma}}$ has been shown to be objective. Show that

(i)     $\dfrac{D\dot{\gamma}^{ij}}{Dt} = \dfrac{\partial \dot{\gamma}^{ij}}{\partial t} + v^s \, \dot{\gamma}^{ij}{}_{,s}$ is not objective;

(ii)    $\dot{\gamma}^{ij}{}_{(2)} = \dfrac{\partial \dot{\gamma}^{ij}}{\partial t} + v^s \dfrac{\partial \dot{\gamma}^{ij}}{\partial x^s} - \dfrac{\partial v^i}{\partial x^s} \, \dot{\gamma}^{sj} - \dfrac{\partial v^j}{\partial x^s} \, \dot{\gamma}^{ij}$ is objective.

[Note that in (ii) we have ordinary partial derivatives and not covariant derivatives as in (i). Verify that the Christoffel symbols cancel. The derivative defined in (ii) is known as the Oldroyd contravariant upper convected derivative.]

36a. The rate-of-deformation tensor $\underset{=}{\dot{\gamma}}$ for (i) a simple shear flow and (ii) an uniaxial elongational flow are given respectively by

(i)     $\underset{=}{\dot{\gamma}} = \dot{\gamma} \begin{bmatrix} 0 & 1 & 0 \\ 1 & 0 & 0 \\ 0 & 0 & 0 \end{bmatrix}$

(ii)    $\underset{=}{\dot{\gamma}} = \dot{\varepsilon} \begin{bmatrix} -1 & 0 & 0 \\ 0 & -1 & 0 \\ 0 & 0 & 2 \end{bmatrix}$

Compute the principal invariants of $\underset{=}{\dot{\gamma}}$ for the two flows.      Answer: (i) $2 \, \dot{\gamma}^2$, (ii) $6 \, \dot{\varepsilon}^2$

37b. The scalar function $\phi \, (I_1, I_2)$ is a function of the first two principal invariants of a second order tensor $\underline{T}$. Working with Cartesian components, show that

$$\frac{\partial \phi}{\partial T_{rs}} = \frac{\partial \phi}{\partial I_1} \, \delta_{rs} + \frac{\partial \phi}{\partial I_2} \left( I_1 \, \delta_{rs} - T_{rs} \right)$$

# CHAPTER 5

# PARTIAL DIFFERENTIAL EQUATIONS I

## 5.1    INTRODUCTION

In the first two chapters, we have considered **ordinary differential equations** (O.D.E.).  An O.D.E. is one in which there is only one independent variable and all derivatives are ordinary derivatives.  A **partial differential equation** (P.D.E.) is one in which there are two or more independent variables and the derivatives that occur in it are partial derivatives.

Most processes that are of interest to engineers and scientists take place in a two or three-dimensional space.  Frequently, time may also be involved.  The number of independent variables is usually more than one and the equations governing these processes are partial differential equations.

In Chapter 3, we have seen that to determine the velocity field of an incompressible and irrotational flow, we have to solve Laplace's equation, which is a partial differential equation.  The equation describing the vibrations of a string can be written as

$$\frac{\partial^2 y}{\partial t^2} = c^2 \frac{\partial^2 y}{\partial x^2} \tag{5.1-1}$$

where  $y$  is the displacement of the string from its equilibrium position,  $t$  is the time,  $x$  is the coordinate of a point on the string, and  $c$  is a constant.  The diffusion of a material in a homogeneous medium is governed by the equation

$$\frac{\partial c}{\partial t} = \alpha^2 \frac{\partial^2 c}{\partial x^2} \tag{5.1-2}$$

where  $\alpha$  is a constant,  $c$  is the concentration,  $t$  and  $x$  are time and position as in Equation (5.1-1). In Chapter 10, we shall see that, in mechanics, Hamilton's equations of motion are given by a set of partial differential equations.  Other examples of P.D.E. and their solutions will be given in this and the next chapter.

For simplicity, we consider only two independent variables , which we denote by  $x_1$  and  $x_2$ .  We denote the unknown function by  $u(x_1, x_2)$ .  A P.D.E. can be written as

$$f\left(x_1, x_2; u, \frac{\partial u}{\partial x_1}, \frac{\partial u}{\partial x_2}, \frac{\partial^2 u}{\partial x_1^2}, \dots \right) = 0 \qquad (5.1\text{-}3)$$

where f is a function.

The **order** of the equation is defined to be the order of the highest order partial derivative in the equation. If f is **linear** in u and its derivatives, that is to say, the coefficients of u and its derivatives are only functions of the independent variables $x_1$ and $x_2$, the equation is **linear**. Equations (5.1-1, 2) are of second order and are linear. If f is linear only in the highest order partial derivatives, the equation is **quasi-linear**. An example is

$$\frac{\partial u}{\partial x_1} + u \frac{\partial u}{\partial x_2} = 0 \qquad (5.1\text{-}4)$$

If the equation is neither linear nor quasi-linear, it is **non-linear**. An example of a non-linear equation is

$$\frac{\partial u}{\partial x_1} + \left(\frac{\partial u}{\partial x_2}\right)^2 = 0 \qquad (5.1\text{-}5)$$

Just as for ordinary differential equations, a **homogeneous** P.D.E. is an equation in which there is no term which is a function of the independent variables only. A first order linear **non-homogeneous** partial differential equation can be written as

$$a_1(x_1, x_2) \frac{\partial u}{\partial x_1} + a_2(x_1, x_2) \frac{\partial u}{\partial x_2} + a_3(x_1, x_2) u = g(x_1, x_2) \qquad (5.1\text{-}6)$$

If g is zero, the equation is homogeneous.

## 5.2   FIRST ORDER EQUATIONS

In this section, we present several methods of solving first order partial differential equations.

**Method of Characteristics**

Consider the linear homogeneous equation which is written as

$$a_1(x_1, x_2) \frac{\partial u}{\partial x_1} + a_2(x_1, x_2) \frac{\partial u}{\partial x_2} = 0 \qquad (5.2\text{-}1)$$

where $a_1$ and $a_2$ are given functions.

The solution of Equation (5.2-1) is a function $u$ that satisfies Equation (5.2-1) in the $(x_1, x_2)$ plane. Geometrically, we regard $u$ as a surface in the $(x_1, x_2)$ plane and the surface is generated by a family of curves, $\Gamma_1, \Gamma_2, \ldots$ Let $\Gamma_i$ be a curve in the $(x_1, x_2)$ plane, defined in parametric form by [see Equations (4.3-1, 15)]

$$\frac{dx_1}{ds} = a_1(x_1, x_2) \tag{5.2-2a}$$

$$\frac{dx_2}{ds} = a_2(x_1, x_2) \tag{5.2-2b}$$

Along $\Gamma_i$, $u$ is a function of one variable $s$ and we can write

$$\frac{du}{ds} = \frac{\partial u}{\partial x_1}\frac{dx_1}{ds} + \frac{\partial u}{\partial x_2}\frac{dx_2}{ds} \tag{5.2-3}$$

Combining Equations (5.2-2a, b, 3) yields

$$\frac{du}{ds} = 0 \tag{5.2-4}$$

The solution of Equation (5.2-4) is

$$u = C \tag{5.2-5}$$

where $C$ is a constant.

Along $\Gamma_i$, $u$ is a constant and $\Gamma_i$ is a **characteristic curve**. The characteristics are obtained by solving Equations (5.2-2a, b) which, by eliminating $s$, can be written as

$$\frac{dx_1}{a_1} = \frac{dx_2}{a_2} \tag{5.2-6}$$

The solution has an arbitrary constant and the characteristics form a one-parameter family of curves. To obtain a solution, additional conditions need to be prescribed. For a first order ordinary differential equation, we need to impose an initial condition, that is to say, we need to know the value of the function at a given point. For a partial differential equation, we prescribe the value of the function $u$ along a curve.

***Example 5.2-1.*** Solve the equation

$$\frac{\partial u}{\partial x_1} + x_1\frac{\partial u}{\partial x_2} = 0 \tag{5.2-7}$$

given that

$$u\,(0,\,x_2) \;=\; 1 + x_2^2, \qquad 0 < x_2 < 2 \tag{5.2-8}$$

The equations for the characteristics are [Equations (5.2-2a, b)]

$$\frac{dx_1}{ds} \;=\; 1, \qquad\qquad \frac{dx_2}{ds} \;=\; x_1 \tag{5.2-9a,b}$$

Combining Equations (5.2-9a, b) yields

$$\frac{dx_2}{dx_1} \;=\; x_1 \tag{5.2-10}$$

The solution is

$$x_1^2 - 2x_2 \;=\; K \tag{5.2-11}$$

where $K$ is a constant.

The characteristics are parabolas and are shown in Figure 5.2-1. On each of the parabolas, the value of $u$ is constant. The values of $u$ are given along the $x_2$-axis in the interval $0 < x_2 < 2$ and we have denoted this line segment by $\gamma$ in Figure 5.2-1. The values of $u$ in the region bounded by the characteristics passing through the origin and $(0, 2)$ can be determined. We denote this region by $R$ which is shaded in Figure 5.2-1. Let $(\bar{x}_1, \bar{x}_2)$ be any point in $R$. The equation of the characteristic that passes through $(\bar{x}_1, \bar{x}_2)$ is given by Equation (5.2-11), which is

$$x_1^2 - 2x_2 \;=\; \bar{x}_1^{\,2} - 2\bar{x}_2 \;=\; K \tag{5.2-12}$$

On this parabola, $u$ is a constant and we write

$$u \;=\; C \tag{5.2-13}$$

The parabola will intersect $\gamma$ at the point

$$x_1 = 0, \qquad x_2 = \frac{1}{2}\left(2\bar{x}_2 - \bar{x}_1^{\,2}\right) \tag{5.2-14a,b}$$

The value of $u$ along $\gamma$ is given by Equation (5.2-8) and combining Equations (5.2-8, 13, 14b) yields

$$C \;=\; 1 + \frac{1}{4}\left(2\bar{x}_2 - \bar{x}_1^{\,2}\right)^2 \tag{5.2-15}$$

Since $(\bar{x}_1, \bar{x}_2)$ is any point in $R$, we can replace it by $(x_1, x_2)$ and the solution in $R$ is

$$u \;=\; 1 + \frac{1}{4}\left(2x_2 - x_1^2\right)^2 \tag{5.2-16}$$

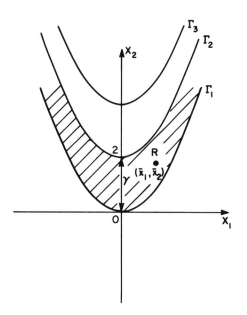

**FIGURE 5.2-1**   **Characteristics $\Gamma$ of the differential equation and the line $\gamma$ along which initial data are given**

●

Note that $\gamma$ is not a segment of a characteristic $\Gamma$. If $\gamma$ is part of $\Gamma$, the conditions must be consistent. That is to say, since $u$ is constant on $\Gamma$, $u$ must also be constant on $\gamma$, otherwise there is no solution.

This method can be applied to a **quasi-linear** equation which can be written as

$$f_1(x_1, x_2, u) \frac{\partial u}{\partial x_1} + f_2(x_1, x_2, u) \frac{\partial u}{\partial x_2} = g(x_1, x_2, u) \qquad (5.2\text{-}17)$$

Suppose the solution can be written in an implicit form as

$$\Phi(x_1, x_2, u) = 0 \qquad (5.2\text{-}18)$$

Differentiating with respect to $x_1$ and $x_2$, we obtain

$$\frac{\partial \Phi}{\partial x_1} + \frac{\partial \Phi}{\partial u} \frac{\partial u}{\partial x_1} = 0 \qquad (5.2\text{-}19a)$$

$$\frac{\partial \Phi}{\partial x_2} + \frac{\partial \Phi}{\partial u} \frac{\partial u}{\partial x_2} = 0 \qquad (5.2\text{-}19b)$$

Assuming that $\dfrac{\partial \Phi}{\partial u}$ is not identically zero, we deduce

$$\frac{\partial u}{\partial x_1} = -\frac{\partial \Phi}{\partial x_1} \Big/ \frac{\partial \Phi}{\partial u} \tag{5.2-20a}$$

$$\frac{\partial u}{\partial x_2} = -\frac{\partial \Phi}{\partial x_2} \Big/ \frac{\partial \Phi}{\partial u} \tag{5.2-20b}$$

Substituting Equations (5.2-20a, b) into Equation (5.2-17), we obtain

$$f_1 \frac{\partial \Phi}{\partial x_1} + f_2 \frac{\partial \Phi}{\partial x_2} + g \frac{\partial \Phi}{\partial u} = 0 \tag{5.2-21}$$

Equation (5.2-21) is now a linear equation in $\Phi$. However, it now involves three variables $x_1$, $x_2$, and $u$. We consider all three as independent variables (in the original problem, only $x_1$ and $x_2$ are independent variables and $u$ is the dependent variable). The characteristics of Equation (5.2-21) are given by

$$\frac{dx_1}{f_1} = \frac{dx_2}{f_2} = \frac{du}{g} \tag{5.2-22a,b}$$

The characteristics of Equation (5.2-17) are obtained by solving Equation (5.2-22a). Solving Equation (5.2-22b) yields the values of $u$ along the characteristics and Equation (5.2-17) is solved.

Note that, in this case, $f_1$ and $f_2$ are also functions of $u$ and that the characteristics depend on $u$. The initial values of $u$ are given along the curve $\gamma$ and the solution $(u)$ is defined along the characteristics that pass through $\gamma$, as shown in Figure 5.2-2.

The characteristics can also be given in parametric form and Equations (5.2-22a, b) can be written as

$$\frac{dx_1}{ds} = f_1(x_1, x_2, u) \tag{5.2-23a}$$

$$\frac{dx_2}{ds} = f_2(x_1, x_2, u) \tag{5.2-23b}$$

$$\frac{du}{ds} = g(x_1, x_2, u) \tag{5.2-23c}$$

Note that in this case $\dfrac{du}{ds}$ is not zero, unless $g$ is zero. If $g$ is zero, $u$ is a constant along the characteristics.

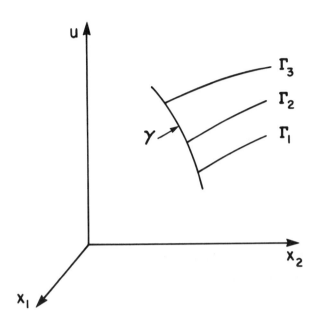

**FIGURE 5.2-2**    **Characteristics $\Gamma_i$ of the differential equation and the initial data curve $\gamma$**

*Example 5.2-2*. Model the movement of cars at a traffic light

We model the automobile traffic on the road by considering the flow to be one-dimensional and the road to be the x-axis. We denote the local density (number of automobiles per unit length) by $\rho\,(x)$ and the velocity by $v_x$. The principle of the conservation of the number of cars (Equation A.I-1) implies that

$$\frac{\partial \rho}{\partial t} + \frac{\partial}{\partial x}(\rho\, v_x) = 0 \qquad\qquad (5.2\text{-}24)$$

Our driving experience leads us to believe that $v_x$ is a decreasing function of $\rho$. If $\rho$ is zero, we can travel at the maximum allowable speed $v_m$ and if $\rho$ reaches a critical value $\rho_c$, we have a traffic jam and $v_x$ reduces to zero. A simple function that describes this situation is

$$v_x = v_m\,(1 - \rho/\rho_c) \qquad\qquad (5.2\text{-}25)$$

We non-dimensionalize Equations (5.2-24, 25) by introducing

$$u = \frac{v_x}{v_m}, \qquad \sigma = \frac{\rho}{\rho_c}, \qquad \tau = \frac{v_m t}{L}, \qquad \xi = \frac{x}{L} \qquad\qquad (5.2\text{-}26\text{a,b,c,d})$$

where $L$ is a typical length, which can be the distance between two traffic lights (i.e. a block).

Equations (5.2-24, 25) now become

$$\frac{\partial \sigma}{\partial \tau} + \frac{\partial}{\partial \xi} (\sigma u) = 0 \qquad (5.2\text{-}27\text{a})$$

$$u = (1 - \sigma) \qquad (5.2\text{-}27\text{b})$$

Combining Equations (5.2-27a, b) yields

$$\frac{\partial \sigma}{\partial \tau} + (1 - 2\sigma) \frac{\partial \sigma}{\partial \xi} = 0 \qquad (5.2\text{-}28)$$

We now assume that at the origin ($\xi = 0$) there is a traffic light which turns green at $\tau = 0$. The road in front of the traffic light is free of vehicles and behind the traffic light the automobiles are bumper-to-bumper. The density $\sigma$ initially can be written as

$$\sigma (\xi, 0) = \begin{cases} 1, & \xi < 0 \\ 0, & \xi > 0 \end{cases} \qquad (5.8\text{-}29\text{a,b})$$

The characteristics of Equation (5.2-28) are given by

$$\frac{d\tau}{1} = \frac{d\xi}{1 - 2\sigma} \qquad (5.2\text{-}30\text{a})$$

or $\quad \dfrac{d\xi}{d\tau} = 1 - 2\sigma \qquad (5.2\text{-}30\text{b})$

We note that the slopes of the characteristics change sign at $\sigma$ equals to a half. Also along each characteristic, $\sigma$ is a constant (g = 0). The factor $(1 - 2\sigma)$ can be interpreted as the speed of the propagation of the state $\sigma$. This can be shown by differentiating $\sigma$ along a characteristic curve. We have

$$d\sigma = 0 = \frac{\partial \sigma}{\partial \tau} d\tau + \frac{\partial \sigma}{\partial \xi} d\xi \qquad (5.2\text{-}31\text{a,b})$$

The speed of propagation c of the state $\sigma$ is given by

$$c = \frac{d\xi}{d\tau} \qquad (5.2\text{-}32)$$

From Equations (5.2-31a, b), we deduce

$$c = -\frac{\partial \sigma}{\partial \tau} \Big/ \frac{\partial \sigma}{\partial \xi} \tag{5.2-33a}$$

$$= (1 - 2\sigma) \tag{5.2-33b}$$

Equation (5.2-33b) is obtained from Equation (5.2-28).

Comparing Equations (5.2-27b, 33b), we find that the velocity of the vehicles is greater than the speed of propagation of $\sigma$. That is to say, $u > c$.

We now draw the characteristics which are shown in Figure 5.2-3. The initial conditions are given along the $\xi$-axis. For positive values of $\xi$, $\sigma$ is zero and we deduce from Equation (5.2-30b) that the slopes of the characteristics are one. Let $\Gamma_1$ be the characteristic that passes through the origin. For negative values of $\xi$, the slopes are –1 and we denote by $\Gamma_2$ the characteristic that passes through the origin. The characteristics $\Gamma_1$ and $\Gamma_2$ divide the $(\xi, \tau)$ upper half plane into three regions which we denote by I, II, and III as shown in Figure 5.2-3. On the characteristics, the value of $\sigma$ is a constant; in I and III, the values of $\sigma$ are zero and one respectively. In II, the slopes of the characteristics vary from one to minus one and this implies (Equation 5.2-30b) that the values of $\sigma$ vary from zero to one. The $\tau$-axis corresponds to a value of $\sigma$ equals to one half.

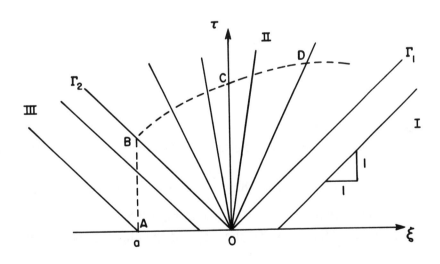

**FIGURE 5.2-3**    **Characteristics and path of vehicles (...)**
**for the continuum traffic model**

Let  D  be any point on a characteristic in II, as illustrated in Figure 5.2-3.  Integrating Equation (5.2-30b) and imposing the condition that the characteristic passes through the origin yields

$$\xi = (1 - 2\sigma)\,\tau \qquad (5.2\text{-}34)$$

From Equation (5.2-34), we deduce that the value of  $\sigma$  on the characteristic through  D  is

$$\sigma = \frac{1}{2}\left(1 - \frac{\xi}{\tau}\right) \qquad (5.2\text{-}35)$$

The equations for  $\Gamma_1\,(\sigma = 0)$  and  $\Gamma_2\,(\sigma = 1)$  are respectively

$$\xi = \tau \qquad (5.2\text{-}36a)$$

$$\xi = -\tau \qquad (5.2\text{-}36b)$$

Having obtained a description of  $\sigma$, we now consider the path of an individual automobile.  In this model, the car at the origin takes off at maximum speed since the state  $\sigma$  associated with it is always zero (along  $\Gamma_1$).  The car initially at  A  $(\xi = -a)$  in Figure 5.2-3 is associated with the condition  $\sigma = 1$.  That is to say, it cannot move.  It has to wait a finite time for the car density  $\sigma$  to become less than one, before it can move.  This is associated with point  B  in the figure.  The time it has to wait can be obtained from Equation (5.2-36b) and is

$$\tau = a \qquad (5.2\text{-}37)$$

From now on as it moves forward in space and time, the value of  $\sigma$  is given by Equation (5.2-35).  The speed  u  of the car is given by Equation (5.2-27b) and is written as

$$u = \frac{d\xi}{d\tau} = \left[1 - \frac{1}{2}\left(1 - \frac{\xi}{\tau}\right)\right] \qquad (5.2\text{-}38a,b)$$

Note that, in Equation (5.2-30b), $\dfrac{d\xi}{d\tau}$ refers to the slope of the characteristics which are defined by the P.D.E. [Equation (5.2-28)].  In Equation (5.2-38b), $\dfrac{d\xi}{d\tau}$ refers to the speed of the car and is defined by the trajectory of the car, as shown by the broken line in Figure 5.2-3.  Note also that, in the case of the first car,  $\sigma = 0$  and  $u = c$.  That is to say, the slope of the characteristic is also the velocity of the car.

Integrating Equation (5.2-38b) yields

$$\xi = \tau + K\sqrt{\tau} \qquad (5.2\text{-}39)$$

where  K  is an arbitrary constant.

Imposing the condition that the vehicle passes through B [= (−a, a)], Equation (5.2-39) becomes

$$\xi = \tau - 2\sqrt{a\tau} \qquad (5.2\text{-}40)$$

Equation (5.2-39) is the equation of a parabola and is the path of the vehicle shown as a broken line in Figure 5.2-3. The time required for the vehicle to reach the traffic light ($\xi = 0$) from B can be obtained from Equation (5.2-40) and is given by

$$\tau = 4a \qquad (5.2\text{-}41)$$

The last vehicle in the block ($\xi = -1$) crosses the traffic light after five units of time (5a) and if the light changes in less than five units of time, the car has to wait for the next change.

The typical path of a vehicle starting at point A can be summarized as follows. After the light turns green, it remains stationary (AB) for a finite time (a), after which it starts to accelerate to cross the traffic light after a further 4a units of time at point C. It continues to accelerate until it reaches its maximum speed asymptotically.

●

The quasi-linear equation considered in this example can be used to describe the movement of glaciers and floods. Further details are given in Rhee et al. (1986); see Problem 3b.

## Lagrange's Method

Lagrange's method is similar to the method of characteristics. Equation (5.2-17) is now also solved via Equations (5.2-22a, b). The latter equations are now referred to as the **auxiliary** (or subsidiary) equations. They can also be written as

$$\frac{dx_2}{dx_1} = \frac{f_2}{f_1}, \qquad \frac{du}{dx_1} = \frac{g}{f_1} \qquad (5.2\text{-}42a,b)$$

Instead of defining the solution along the characteristics, in Lagrange's method the solutions of Equations (5.2-42a, b) are written in implicit form as

$$\Phi_1\,(x_1,\,x_2,\,u) = C_1 \qquad (5.2\text{-}43a)$$

$$\Phi_2\,(x_1,\,x_2,\,u) = C_2 \qquad (5.2\text{-}43b)$$

where $C_1$ and $C_2$ are the arbitrary constants of integration.

The functions $\Phi_1$ and $\Phi_2$ are integrals of the auxiliary equations and are solutions of Equation (5.2-17). This can be shown by differentiating $\Phi_1$ to yield

$$0 = \frac{\partial \Phi_1}{\partial x_1}\, dx_1 + \frac{\partial \Phi_1}{\partial x_2}\, dx_2 + \frac{\partial \Phi_1}{\partial u}\, du \tag{5.2-44}$$

Using Equations (5.2-42a, b), Equation (5.2-44) becomes

$$f_1 \frac{\partial \Phi_1}{\partial x_1} + f_2 \frac{\partial \Phi_1}{\partial x_2} + g \frac{\partial \Phi_1}{\partial u} = 0 \tag{5.2-45}$$

From Equations (5.2-20a, b), we deduce that $\Phi_1$ is a solution of Equation (5.2-17). Similarly, $\Phi_2$ is a solution and a general solution can be written as

$$F(\Phi_1, \Phi_2) = 0 \tag{5.2-46}$$

where F is an arbitrary function.

The system of Equations (5.2-22 a , b) might not be easy to integrate. A possible method of solving the equations involves writing Equations (5.2-22 a, b) as follows

$$\frac{dx_1}{f_1} = \frac{dx_2}{f_2} = \frac{du}{g} = \frac{\lambda\, dx_1 + \mu\, dx_2 + \nu\, du}{\lambda f_1 + \mu f_2 + \nu g} \tag{5.2-47a,b,c}$$

where we used the following relation

$$\frac{a}{b} = \frac{c}{d} = \frac{\lambda a + \mu c}{\lambda b + \mu d} \tag{5.2-48a,b}$$

We now choose the functions $\lambda$, $\mu$, and $\nu$ such that

$$\lambda f_1 + \mu f_2 + \nu g = 0 \tag{5.2-49}$$

Since we are now dividing by zero, we have to require that

$$\lambda\, dx_1 + \mu\, dx_2 + \nu\, du = 0 \tag{5.2-50}$$

Comparing the exact differentials [Equations (5.2-44, 50)], we identify

$$\lambda = \frac{\partial \Phi_1}{\partial x_1}, \quad \mu = \frac{\partial \Phi_1}{\partial x_2}, \quad \nu = \frac{\partial \Phi_1}{\partial u} \tag{5.2-51a,b,c}$$

Since $\lambda$, $\mu$ and $\nu$ are known functions of $x_1$, $x_2$ and u, we can integrate Equations (5.2-51a to c) to obtain $\Phi_1$. A different choice for the set $\{\lambda, \mu, \nu\}$ in Equation (5.2-49) will provide a different function $\Phi_2$.

The general solution will be given by Equation (5.2-46) which implicitly determines $u$ as a function of $x_1$ and $x_2$. The arbitrary function $F$ is determined by the initial conditions. Different choices of $\Phi_i$ will generate different functions F, but the initial conditions will ensure that the solution is unique.

***Example 5.2-3.*** Chromatography is widely used as a method of separation of chemical species and references are given in Rhee et al. (1986). The simplest case is the chromatography of a single solute. Suppose that the solution flows along the z-direction through a bed with void fraction $\in$ at a constant superficial velocity $v_0$. If $c$ is the concentration in the fluid phase and $n$ is the concentration in the stationary phase, a mass balance [Rhee et al. (1986) or Bird et al. (1960)] yields

$$v_0 \frac{\partial c}{\partial z} + \in \frac{\partial c}{\partial t} + \left(1 - \in\right) \frac{\partial n}{\partial t} = 0 \qquad (5.2\text{-}52)$$

We assume a linear relation between $n$ and $c$ and write

$$n = Kc \qquad (5.2\text{-}53)$$

where $K$ is a constant.

Combining Equations (5.2-52, 53) yields

$$v_0 \frac{\partial c}{\partial z} + \left[\in + K\left(1 - \in\right)\right] \frac{\partial c}{\partial t} = 0 \qquad (5.2\text{-}54)$$

The initial and boundary conditions are assumed to be

$$c\,(z, 0) = 0, \qquad c\,(0, t) = c_0\,H(t) \qquad (5.2\text{-}55a,b)$$

where $c_0$ is a constant and $H(t)$ is the Heaviside function. [Equation (1.17-8a, b)].

The auxiliary equations are

$$\frac{dz}{v_0} = \frac{dt}{\alpha} = \frac{dc}{0} \qquad (5.2\text{-}56a,b)$$

where $\alpha = \in + K\,(1 - \in)$                                                   (5.2-56c)

From Equation (5.2-56a), we have

$$\frac{dt}{dz} = \frac{\alpha}{v_0} = \beta \qquad (5.2\text{-}57a,b)$$

On integrating, we obtain

$$t - \beta z = C_1 \qquad (5.2\text{ -}58)$$

where $C_1$ is a constant.

From Equation (5.2-56b), we deduce that

$$\frac{dc}{dt} = 0 \quad \Rightarrow \quad c = C_2 \tag{5.2-59a,b}$$

The general solution is

$$F(t - \beta z, c) = 0 \tag{5.2-60a}$$

or     $c = f(t - \beta z)$ \hfill (5.2-60b)

Our aim is now to determine the form of the function $f$. That is to say, we want to know if $f$ is an exponential function, a trigonometric function, etc. To determine the form of the function, we use Equations (5.2-55a, b). From Equation (5.2-55b), we determine that the unknown function $f$ has to be a Heaviside function. The solution is given by Equation (5.2-61). The reader should verify that this equation satisfies the other boundary condition.

From Equations (5.2-55a, b), we deduce that

$$c = c_0 H(t - \beta z) \tag{5.2-61}$$

## Transformation Method

We illustrate this method by considering a first order partial differential equation which can be written as

$$a_1(x_1, x_2)\frac{\partial u}{\partial x_1} + a_2(x_1, x_2)\frac{\partial u}{\partial x_2} + a_3(x_1, x_2)u = f(x_1, x_2) \tag{5.2-62}$$

We make a change of independent variables and write

$$\xi = \xi(x_1, x_2), \qquad \eta = \eta(x_1, x_2) \tag{5.2-63a,b}$$

Using the chain rule, we obtain

$$\frac{\partial u}{\partial x_1} = \frac{\partial u}{\partial \xi}\frac{\partial \xi}{\partial x_1} + \frac{\partial u}{\partial \eta}\frac{\partial \eta}{\partial x_1} \tag{5.2-64a}$$

$$\frac{\partial u}{\partial x_2} = \frac{\partial u}{\partial \xi}\frac{\partial \xi}{\partial x_2} + \frac{\partial u}{\partial \eta}\frac{\partial \eta}{\partial x_2} \tag{5.2-64b}$$

Substituting Equations (5.2-64a, b) into Equation (5.2-62) yields

$$\left( a_1 \frac{\partial \xi}{\partial x_1} + a_2 \frac{\partial \xi}{\partial x_2} \right) \frac{\partial u}{\partial \xi} + \left( a_1 \frac{\partial \eta}{\partial x_1} + a_2 \frac{\partial \eta}{\partial x_2} \right) \frac{\partial u}{\partial \eta} + a_3 (\xi, \eta) \, u = f(\xi, \eta) \qquad (5.2\text{-}65)$$

We now choose $\xi$ and $\eta$ such that we reduce the P.D.E. to an equation which involves only one derivative. One possibility is to choose

$$\xi = x_1 \qquad (5.2\text{-}66a)$$

and to require the coefficient of $\frac{\partial u}{\partial \eta}$ to be zero. That is to say

$$a_1 \frac{\partial \eta}{\partial x_1} + a_2 \frac{\partial \eta}{\partial x_2} = 0 \qquad (5.2\text{-}66b)$$

This allows us to obtain $\eta$ by solving

$$\frac{dx_1}{dx_2} = \frac{a_1}{a_2} \qquad (5.2\text{-}67)$$

Along each of the characteristics given by Equation (5.2-67), $\eta$ is constant.

Equation (5.2-65) simplifies to

$$a_1 (\xi, \eta) \frac{\partial u}{\partial \xi} + a_3 (\xi, \eta) \, u = f(\xi, \eta) \qquad (5.2\text{-}68a)$$

or $\qquad \dfrac{\partial u}{\partial \xi} + \dfrac{a_3}{a_1} u = \dfrac{f}{a_1} \qquad (5.2\text{-}68b)$

Equation (5.2-68b) can be solved by introducing an integrating factor $I$ defined as

$$I = \exp \left[ \int \frac{a_3}{a_1} \, d\xi \right] \qquad (5.2\text{-}69)$$

Equation (5.2-68b) can now be written as

$$\frac{\partial}{\partial \xi} (u \, I) = \frac{f}{a_1} I \qquad (5.2\text{-}70)$$

On integrating, we obtain the solution to the problem.

$$u = \frac{1}{I} \left[ \int \frac{f}{a_1} I \, d\xi + F(\eta) \right] \qquad (5.2\text{-}71)$$

where $F(\eta)$ is an arbitrary function of $\eta$, which is determined from the initial (boundary) conditions. The solution in terms of $x_1$ and $x_2$ is obtained via Equations (5.2-66a, b).

The transformation from $(x_1, x_2)$ to $(\xi, \eta)$ is valid as long as the Jacobian

$$J = \begin{vmatrix} \dfrac{\partial \xi}{\partial x_1} & \dfrac{\partial \xi}{\partial x_2} \\[3mm] \dfrac{\partial \eta}{\partial x_1} & \dfrac{\partial \eta}{\partial x_2} \end{vmatrix} \tag{5.2-72}$$

is non-zero.

***Example 5.2-4***. Solve the problem considered in Example 5.2-3 by the present method. Equations (5.2-54) can be written as

$$\alpha \, \frac{\partial c}{\partial t} + v_0 \, \frac{\partial c}{\partial z} = 0 \tag{5.2-73}$$

We identify $a_1$ to be $\alpha$, $a_2$ to be $v_0$, $\xi$ to be $t$, and $\eta$ is obtained by solving

$$\frac{dt}{dz} = \frac{\alpha}{v_0} \tag{5.2-74}$$

Solving Equation (5.2-74), we find that $\eta$ is given by

$$\eta = \left( t - \frac{\alpha z}{v_0} \right) = t - \beta z = \text{constant} \tag{5.2-75a,b,c}$$

Equation (5.2-73), in terms of $\xi$ and $\eta$, becomes

$$\frac{\partial c}{\partial \xi} = 0 \tag{5.2-76}$$

The solution is

$$c = F(\eta) \tag{5.2-77}$$

where $F$ is an arbitrary function.

Combining Equations (5.2-55a, b, 75b, 77) yields

$$c = c_0 \, H(t - \beta z) \tag{5.2-78}$$

which is the solution obtained by Lagrange's method.

***Example 5.2-5.*** Solve the following equation using the three methods described in this section

$$x_1 \frac{\partial u}{\partial x_1} + x_2 \frac{\partial u}{\partial x_2} = 2x_1 x_2 \qquad (5.2\text{-}79)$$

subject to the condition

$$u = 2 \quad \text{on} \ x_2 = x_1^2 \qquad (5.2\text{-}80)$$

In this example, Equations (5.2-22a, b) are

$$\frac{dx_1}{x_1} = \frac{dx_2}{x_2} = \frac{du}{2x_1 x_2} \qquad (5.2\text{-}81a,b)$$

From Equation (5.2-81a), we deduce that the characteristics are

$$x_2 = K x_1 \qquad (5.2\text{-}82)$$

where $K$ is a constant.

The characteristics are straight lines through the origin, with slope $K$. Combining Equations (5.2-81b, 82), we obtain

$$\frac{dx_1}{x_1} = \frac{du}{2K x_1^2} \qquad (5.2\text{-}83)$$

Solving Equation (5.2-83), we obtain along each characteristic

$$u = K x_1^2 + C \qquad (5.2\text{-}84)$$

where $C$ is a constant.

The constant $C$ depends on the characteristic; that is to say, it is a function of $K \ (= x_2/x_1)$. Using Equation (5.2-82), Equation (5.2-84) can be written as

$$u = x_1 x_2 + f(x_2/x_1) \qquad (5.2\text{-}85)$$

Imposing the condition given by Equation (5.2-80) yields

$$2 = x_1^3 + f(x_1) \qquad (5.2\text{-}86)$$

We deduce

$$f(x_1) = 2 - x_1^3 \qquad (5.2\text{-}87)$$

Substituting f into Equation (5.2-85), we obtain

$$u = x_1 x_2 + 2 - (x_2/x_1)^3 \tag{5.2-88}$$

We now solve Equation (5.2-79) by Lagrange's method.

From Equations (5.2-47a to c, 81a, b), we note that by choosing

$$\lambda = x_2, \qquad \mu = x_1, \qquad \text{and } \nu = -1 \tag{5.2-89a,b,c}$$

we obtain

$$x_2 \, dx_1 + x_1 \, dx_2 - du = 0 \tag{5.2-90}$$

Equation (5.2-90) is an exact differential and can be written as

$$d (x_1 x_2 - u) = 0 \tag{5.2-91}$$

Equation (5.2-91) implies that

$$x_1 x_2 - u = \text{constant} \tag{5.2-92}$$

The two solutions of the auxiliary equations are given by Equations (5.2-82, 92). The general solution [Equation 5.2-46)] can be written as

$$x_1 x_2 - u = g (x_2/x_1) \tag{5.2-93}$$

where g is an arbitrary function.

Using Equation (5.2-80), we find, as in the previous method, that g is given by

$$g (x_2/x_1) = (x_2/x_1)^3 - 2 \tag{5.2-94}$$

It follows that Equation (5.2-88) is the solution of Equations (5.2-79, 80).

We now use the transformation method. We introduce two new variables $\xi$ and $\eta$ and choose $\xi$ to be $x_1$. The variable $\eta$ is given by Equation (5.2-66b) and in this example, it is

$$x_1 \frac{\partial \eta}{\partial x_1} + x_2 \frac{\partial \eta}{\partial x_2} = 0 \tag{5.2-95}$$

The characteristics of Equation (5.2-95) are

$$\frac{dx_1}{x_1} = \frac{dx_2}{x_2} \tag{5.2-96}$$

The solution of Equation (5.2-96) is

$$\frac{x_2}{x_1} = \text{constant} \tag{5.2-97}$$

The variable $\eta$ is given by

$$\eta = x_2/x_1 \tag{5.2-98}$$

Equation (5.2-79) becomes, in terms of $\xi$ and $\eta$,

$$\xi \frac{\partial u}{\partial \xi} = 2\xi^2 \eta \tag{5.2-99}$$

Solving Equation (5.2-99), we obtain

$$u = \xi^2 \eta + h(\eta) \tag{5.2-100}$$

where $h$ is an arbitrary function of $\eta$.

Equation (5.2-80) now becomes

$$u = 2 \quad \text{on } \xi = \eta \tag{5.2-101a, b}$$

Combining Equations (5.2-100, 101a, b) yields

$$h(\eta) = 2 - \eta^3 \tag{5.2-102}$$

It follows that the solution is

$$u = 2 + \xi^2 \eta - \eta^3 \tag{5.2-103}$$

Reverting to the original variables (replacing $\xi$ by $x_1$ and $\eta$ by $x_2/x_1$), Equation (5.2-103) becomes Equation (5.2-71). As expected, all three methods yield the same solution.

## 5.3   SECOND ORDER LINEAR EQUATIONS

Second order linear partial differential equations are frequently encountered in the applications of mathematics. They are classified into three types: **hyperbolic, parabolic, and elliptic**. Each type describes a different physical phenomenon. Hyperbolic equations describe wave phenomena, parabolic equations describe diffusion processes, and elliptic equations describe equilibrium conditions.

We next deduce the canonical form of each type via a transformation of independent variables.

### Classification

The general second order linear partial differential equation in two independent variables $(x_1, x_2)$ can be written as

$$a_{11} \frac{\partial^2 u}{\partial x_1^2} + a_{12} \frac{\partial^2 u}{\partial x_1 \partial x_2} + a_{22} \frac{\partial^2 u}{\partial x_2^2} + b_1 \frac{\partial u}{\partial x_1} + b_2 \frac{\partial u}{\partial x_2} + cu = f(x_1, x_2) \qquad (5.3\text{-}1)$$

where $a_{11}$, $a_{12}$, $a_{22}$, $b_1$, $b_2$, and $c$ are functions of $x_1$ and $x_2$.

We simplify Equation (5.3-1) by changing the variables $(x_1, x_2)$ to $(\xi, \eta)$, in an attempt to reduce the number of higher derivatives. This transformation is given by Equations (5.2-63a, b). From Equation (5.2-64a), we have

$$\frac{\partial^2 u}{\partial x_1^2} = \frac{\partial^2 \xi}{\partial x_1^2} \frac{\partial u}{\partial \xi} + \frac{\partial \xi}{\partial x_1} \left[ \frac{\partial}{\partial \xi} \left( \frac{\partial u}{\partial \xi} \right) \frac{\partial \xi}{\partial x_1} + \frac{\partial}{\partial \eta} \left( \frac{\partial u}{\partial \xi} \right) \frac{\partial \eta}{\partial x_1} \right] + \frac{\partial^2 \eta}{\partial x_1^2} \frac{\partial u}{\partial \eta}$$

$$+ \frac{\partial \eta}{\partial x_1} \left[ \frac{\partial}{\partial \xi} \left( \frac{\partial u}{\partial \eta} \right) \frac{\partial \xi}{\partial x_1} + \frac{\partial}{\partial \eta} \left( \frac{\partial u}{\partial \eta} \right) \frac{\partial \eta}{\partial x_1} \right] \qquad (5.3\text{-}2a)$$

$$= \left( \frac{\partial \xi}{\partial x_1} \right)^2 \frac{\partial^2 u}{\partial \xi^2} + 2 \frac{\partial \xi}{\partial x_1} \frac{\partial \eta}{\partial x_1} \frac{\partial^2 u}{\partial \xi \partial \eta} + \left( \frac{\partial \eta}{\partial x_1} \right)^2 \frac{\partial^2 u}{\partial \eta^2} + \frac{\partial^2 \xi}{\partial x_1^2} \frac{\partial u}{\partial \xi} + \frac{\partial^2 \eta}{\partial x_1^2} \frac{\partial u}{\partial \eta}$$

$$(5.3\text{-}2b)$$

Similarly, we can compute $\dfrac{\partial^2 u}{\partial x_1 \partial x_2}$ and $\dfrac{\partial^2 u}{\partial x_2^2}$. Equation (5.3-1) now becomes

$$\frac{\partial^2 u}{\partial \xi^2}\left[a_{11}\left(\frac{\partial \xi}{\partial x_1}\right)^2 + a_{12}\frac{\partial \xi}{\partial x_1}\frac{\partial \xi}{\partial x_2} + a_{22}\left(\frac{\partial \xi}{\partial x_2}\right)^2\right] + \frac{\partial^2 u}{\partial \xi \partial \eta}\left[2a_{11}\frac{\partial \xi}{\partial x_1}\frac{\partial \eta}{\partial x_1} + a_{12}\left(\frac{\partial \xi}{\partial x_1}\frac{\partial \eta}{\partial x_2}\right.\right.$$

$$\left.\left. + \frac{\partial \xi}{\partial x_2}\frac{\partial \eta}{\partial x_1}\right) + 2a_{22}\frac{\partial \xi}{\partial x_2}\frac{\partial \eta}{\partial x_2}\right] + \frac{\partial^2 u}{\partial \eta^2}\left[a_{11}\left(\frac{\partial \eta}{\partial x_1}\right)^2 + a_{12}\frac{\partial \eta}{\partial x_1}\frac{\partial \eta}{\partial x_2} + a_{22}\left(\frac{\partial \eta}{\partial x_2}\right)^2\right]$$

$$+ \frac{\partial u}{\partial \xi}\left[a_{11}\frac{\partial^2 \xi}{\partial x_1^2} + a_{12}\frac{\partial^2 \xi}{\partial x_1 \partial x_2} + a_{22}\frac{\partial^2 \xi}{\partial x_2^2} + b_1\frac{\partial \xi}{\partial x_1} + b_2\frac{\partial \xi}{\partial x_2}\right] + \frac{\partial u}{\partial \eta}\left[a_{11}\frac{\partial^2 \eta}{\partial x_1^2} + a_{12}\frac{\partial^2 \eta}{\partial x_1 \partial x_2}\right.$$

$$\left. + a_{22}\frac{\partial^2 \eta}{\partial x_2^2} + b_1\frac{\partial \eta}{\partial x_1} + b_2\frac{\partial \eta}{\partial x_2}\right] + cu = f \tag{5.3-3}$$

The transformation $(x_1, x_2)$ to $(\xi, \eta)$ is not yet defined. As mentioned earlier, the aim of the transformation is to simplify Equation (5.3-1). We examine the possibility of defining a transformation such that the coefficient of $\dfrac{\partial^2 u}{\partial \xi^2}$ vanishes. This does not affect the generality of the equation. That is to say, Equation (5.3-3) with the coefficient of $\dfrac{\partial^2 u}{\partial \xi^2} = 0$ will be equivalent to Equation (5.3-1). An example is provided by Equations (5.4-6b, 21).

That is to say, we seek the conditions under which it is possible to have

$$a_{11}\left(\frac{\partial \xi}{\partial x_1}\right)^2 + a_{12}\frac{\partial \xi}{\partial x_1}\frac{\partial \xi}{\partial x_2} + a_{22}\left(\frac{\partial \xi}{\partial x_2}\right)^2 = 0 \tag{5.3-4}$$

The solution of Equation (5.3-4) provides the desired transformation for $\xi$.

Assuming that $\dfrac{\partial \xi}{\partial x_2}$ is non-zero, Equation (5.3-4) can be written as

$$a_{11}\left(\frac{\partial \xi}{\partial x_1}\bigg/\frac{\partial \xi}{\partial x_2}\right)^2 + a_{12}\left(\frac{\partial \xi}{\partial x_1}\bigg/\frac{\partial \xi}{\partial x_2}\right) + a_{22} = 0 \tag{5.3-5}$$

The solution is

$$\left(\frac{\partial \xi}{\partial x_1}\bigg/\frac{\partial \xi}{\partial x_2}\right) = \frac{-a_{12} \pm \sqrt{a_{12}^2 - 4a_{11}a_{22}}}{2a_{11}} \tag{5.3-6}$$

Consider the following cases.

(i)     $a_{12}^2 > 4a_{11}a_{22}$

In this case, we have two distinct roots and this is the **hyperbolic case**. Consider the curve

$$\xi = \text{constant} \tag{5.3-7}$$

On differentiating, we obtain

$$\frac{\partial \xi}{\partial x_1} \, dx_1 + \frac{\partial \xi}{\partial x_2} \, dx_2 = 0 \tag{5.3-8}$$

It follows that

$$\frac{\partial \xi}{\partial x_1} \Big/ \frac{\partial \xi}{\partial x_2} = -\frac{dx_2}{dx_1} \tag{5.3-9}$$

Substituting Equation (5.3-9) into Equation (5.3-6) yields

$$\frac{dx_2}{dx_1} = \frac{a_{12} \mp \sqrt{a_{12}^2 - 4a_{11}a_{22}}}{2a_{11}} \tag{5.3-10}$$

We have two possible solutions; that is to say, we have two possible curves ($\xi$ = constant), one corresponding to the positive root and the other to the negative root. Noting that the coefficient of $\dfrac{\partial^2 u}{\partial \eta^2}$ is identical to that of $\dfrac{\partial^2 u}{\partial \xi^2}$ if we replace $\xi$ by $\eta$, we choose one root for the curve $\xi$ (= constant) and the other root for $\eta$ (= constant), since $\eta$ is also a function of $x_1$ and $x_2$. These two curves are the **characteristics** of the partial differential equation. For different values of the constants, these characteristics span a two-dimensional space. For a hyperbolic equation, we have two real characteristics ($\xi$ = constant, $\eta$ = constant) and they are obtained by solving the equations

$$\frac{dx_2}{dx_1} = \frac{a_{12} + \sqrt{a_{12}^2 - 4a_{11}a_{22}}}{2a_{11}} \tag{5.3-11a}$$

$$\frac{dx_2}{dx_1} = \frac{a_{12} - \sqrt{a_{12}^2 - 4a_{11}a_{22}}}{2a_{11}} \tag{5.3-11b}$$

respectively.

With the coefficients of $\dfrac{\partial^2 u}{\partial \xi^2}$ and $\dfrac{\partial^2 u}{\partial \eta^2}$ equal to zero, we obtain the **canonical form** of a hyperbolic equation which is written as

$$\frac{\partial^2 u}{\partial \xi \partial \eta} + \beta_1 \frac{\partial u}{\partial \xi} + \beta_2 \frac{\partial u}{\partial \eta} + \gamma u = \sigma \qquad (5.3\text{-}12)$$

where $\beta_1$, $\beta_2$, $\gamma$, and $\sigma$ are functions of $\xi$ and $\eta$.

Note that Equation (5.1-1) is hyperbolic.

In Equation (5.3-6), we have assumed $a_{11}$ to be non-zero; but if $a_{11}$ is zero, we can still solve for $\dfrac{\partial \xi}{\partial x_1} \Big/ \dfrac{\partial \xi}{\partial x_2}$ from Equation (5.3-5) and finally obtain Equation (5.3-12). If both $a_{11}$ and $a_{22}$ are zero, the equation is already in the canonical form.

(ii)    $a_{12}^2 = 4 a_{11} a_{22}$

This is the **parabolic** case and the two characteristics are coincident. We choose $\xi$ ($=$ constant) to be a characteristic and the coefficient of $\dfrac{\partial^2 u}{\partial \xi^2}$ to be zero. The coefficient of $\dfrac{\partial^2 u}{\partial \xi \partial \eta}$, which we denote by $B$, can be written as

$$B = \left\{ 2 a_{11} \frac{\partial \eta}{\partial x_1} \left( \frac{\partial \xi}{\partial x_1} \Big/ \frac{\partial \xi}{\partial x_2} \right) + a_{12} \left[ \frac{\partial \eta}{\partial x_2} \left( \frac{\partial \xi}{\partial x_1} \Big/ \frac{\partial \xi}{\partial x_2} \right) + \frac{\partial \eta}{\partial x_1} \right] + 2 a_{22} \frac{\partial \eta}{\partial x_2} \right\} \frac{\partial \xi}{\partial x_2} \qquad (5.3\text{-}13)$$

Substituting Equation (5.3-6) into Equation (5.3-13), we have

$$B = \left\{ 2 a_{11} \frac{\partial \eta}{\partial x_1} \left( -\frac{a_{12}}{2 a_{11}} \right) + a_{12} \left[ \frac{\partial \eta}{\partial x_2} \left( -\frac{a_{12}}{2 a_{11}} \right) + \frac{\partial \eta}{\partial x_1} \right] + 2 a_{22} \frac{\partial \eta}{\partial x_2} \right\} \frac{\partial \xi}{\partial x_2} \qquad (5.3\text{-}14a)$$

$$= \left[ \frac{\partial \eta}{\partial x_2} \left( -\frac{a_{12}^2}{2 a_{11}} + 2 a_{22} \right) \right] \frac{\partial \xi}{\partial x_2} \qquad (5.3\text{-}14b)$$

$$= 0 \qquad (5.3\text{-}14c)$$

The canonical form of a parabolic equation is

$$\frac{\partial^2 u}{\partial \eta^2} + \beta_3 \frac{\partial u}{\partial \xi} + \beta_4 \frac{\partial u}{\partial \eta} + \gamma_1 u = \sigma_1 \qquad (5.3\text{-}15)$$

where $\beta_3$, $\beta_4$, $\gamma_1$, and $\sigma_1$ are functions of $\xi$ and $\eta$.

Equation (5.1-2) is a parabolic equation.

(iii)    $a_{12}^2 < 4 a_{11} a_{22}$

This is the **elliptic** case and the characteristics are complex. Since the coefficients $a_{ij}$ are real, the characteristics are complex conjugates [Equations (5.3-11a, b)] and we write

$$\xi = \alpha + i\beta \qquad (5.3\text{-}16a)$$

$$\eta = \alpha - i\beta \qquad (5.3\text{-}16b)$$

where $\alpha$ and $\beta$ are real.

Writing Equation (5.3-4) in terms of $\alpha$ and $\beta$ and equating the real and imaginary parts to zero yields

$$a_{11} \left( \frac{\partial \alpha}{\partial x_1} \right)^2 + a_{12} \frac{\partial \alpha}{\partial x_1} \frac{\partial \alpha}{\partial x_2} + a_{22} \left( \frac{\partial \alpha}{\partial x_2} \right)^2 = a_{11} \left( \frac{\partial \beta}{\partial x_1} \right)^2 + a_{12} \frac{\partial \beta}{\partial x_1} \frac{\partial \beta}{\partial x_2} + a_{22} \left( \frac{\partial \beta}{\partial x_2} \right)^2$$

$$(5.3\text{-}17a)$$

$$2 a_{11} \frac{\partial \alpha}{\partial x_1} \frac{\partial \beta}{\partial x_1} + a_{12} \left( \frac{\partial \alpha}{\partial x_1} \frac{\partial \beta}{\partial x_2} + \frac{\partial \alpha}{\partial x_2} \frac{\partial \beta}{\partial x_1} \right) + 2 a_{22} \frac{\partial \alpha}{\partial x_2} \frac{\partial \beta}{\partial x_2} = 0 \qquad (5.3\text{-}17b)$$

Rewriting Equations (5.3-3) in terms of $\alpha$ and $\beta$ and using Equations (5.3-17a, b), we find that the coefficient of $\dfrac{\partial^2 u}{\partial \alpha \, \partial \beta}$ is zero and the coefficients of $\dfrac{\partial^2 u}{\partial \alpha^2}$ and $\dfrac{\partial^2 u}{\partial \beta^2}$ are equal. The canonical form of an elliptic equation is

$$\frac{\partial^2 u}{\partial \alpha^2} + \frac{\partial^2 u}{\partial \beta^2} + \beta_5 \frac{\partial u}{\partial \alpha} + \beta_6 \frac{\partial u}{\partial \beta} + \gamma_2 u = \sigma_2 \qquad (5.3\text{-}18)$$

where $\beta_5$, $\beta_6$, $\gamma_2$, and $\sigma_2$ are functions of $\alpha$ and $\beta$.

Laplace's equation is an elliptic equation.

As usual, we require the transformations from $(x_1, x_2)$ to $(\xi, \eta)$ to be non-singular; that is to say, the Jacobian $J$, given by Equation (5.2-72), is non-zero. In the elliptic case, we replace the complex functions $\xi$ and $\eta$ by the real quantities $\alpha$ and $\beta$ and this is implied in all subsequent discussions.

Let the coefficients of $\dfrac{\partial^2 u}{\partial \xi^2}$, $\dfrac{\partial^2 u}{\partial \xi \, \partial \eta}$, and $\dfrac{\partial^2 u}{\partial \eta^2}$ be denoted by $\alpha_{11}$, $\alpha_{12}$, and $\alpha_{22}$ respectively. By direct computation from Equations (5.2-72, 3-3), we obtain

$$\alpha_{12}^2 - 4\alpha_{11}\alpha_{22} = J^2 \left(a_{12}^2 - 4a_{11}a_{22}\right) \tag{5.3-19}$$

The Jacobian $J$ is real, its square is positive, and Equation (5.3-19) implies that the sign of $a_{12}^2 - 4a_{11}a_{22}$ is preserved on a transformation of coordinates. That is to say, the type (hyperbolic, parabolic, or elliptic) of a P.D.E. is independent of the coordinate system. The quantity $a_{12}^2 - 4a_{11}a_{22}$ can be a function of $(x_1, x_2)$, its sign can depend on the region of the $(x_1, x_2)$ plane. The type of an equation may change on moving from one region to another.

***Example 5.3-1.*** Reduce the equation

$$\frac{\partial^2 u}{\partial x_1^2} + x_2 \frac{\partial^2 u}{\partial x_2^2} = 0 \tag{5.3-20}$$

to its canonical form.

Equation (5.3-20) is elliptic in the region $x_2 > 0$, hyperbolic in the region $x_2 < 0$, and parabolic along the $x_1$-axis.

In this example, Equations (5.3-11a, b) become

$$\frac{dx_2}{dx_1} = \sqrt{-x_2} \tag{5.3-21a}$$

$$\frac{dx_2}{dx_1} = -\sqrt{-x_2} \tag{5.3-21b}$$

We consider the regions ($x_2 > 0$ and $x_2 < 0$) separately.

In the upper half plane ($x_2 > 0$), the characteristics are complex. On integrating Equations (5.3-21a, b), we obtain

$$x_1 = 2i x_2^{1/2} + \text{constant} \tag{5.3-22a}$$

$$x_1 = -2i x_2^{1/2} + \text{constant} \qquad (5.3\text{-}22b)$$

The solutions are

$$\xi = x_1 - 2i x_2^{1/2} = \text{constant} \qquad (5.3\text{-}23a,b)$$

$$\eta = x_1 + 2i x_2^{1/2} = \text{constant} \qquad (5.3\text{-}23c,d)$$

We deduce that

$$\alpha = x_1, \qquad \beta = 2 x_2^{1/2} \qquad (5.3\text{-}24a,b)$$

Using Equations (5.3-24a, b), Equation (5.3-20) becomes

$$\frac{\partial^2 u}{\partial \alpha^2} + \frac{\partial^2 u}{\partial \beta^2} - \frac{1}{\beta} \frac{\partial u}{\partial \beta} = 0 \qquad (5.3\text{-}25)$$

In the region $x_2 < 0$, the characteristics are real and the solutions of Equations (5.3-21a, b) are

$$\xi = x_1 - 2 x_2^{1/2} = \text{constant} \qquad (5.3\text{-}26a,b)$$

$$\eta = x_1 + 2 x_2^{1/2} = \text{constant} \qquad (5.3\text{-}26c,d)$$

On transforming to $\xi$ and $\eta$, Equation (5.3-20) becomes

$$\frac{\partial^2 u}{\partial \xi \, \partial \eta} + \frac{1}{2(\eta - \xi)} \left( \frac{\partial u}{\partial \eta} - \frac{\partial u}{\partial \xi} \right) = 0 \qquad (5.3\text{-}27)$$

Equations (5.3-26a, c) show that the characteristics are parabolas.

On the $x_1$-axis ($x_2 = 0$), the equation is parabolic and Equation (5.3-20) reduces to the canonical form

$$\frac{\partial^2 u}{\partial x_1^2} = 0 \qquad (5.3\text{-}28)$$

We note that Equations (5.3-25, 27) have singularities at

$$\beta = 0, \qquad \eta = \xi \qquad (5.3\text{-}29a,b)$$

In both cases, the singularities correspond to the $x_1$-axis ($x_2 = 0$). That is to say, both hyperbolic and elliptic equations have singularities on the $x_1$-axis and it is across this axis that Equation (5.3-20) changes from elliptic to hyperbolic. In a given problem, the boundary and initial conditions will have to change according to the region in the $(x_1, x_2)$ plane. This will be addressed later.

## 5.4   METHOD OF SEPARATION OF VARIABLES

This method relies on the possibility of writing the unknown function $u(x_1, x_2)$ as the product of two functions $f_1(x_1) f_2(x_2)$. In many cases, the partial differential equation reduces to two ordinary differential equations for $f_1$ and $f_2$. We can extend this method to more than two variables. We illustrate this method by solving the following equations.

### Wave Equation

An elastic string is tied at its ends $(x = 0, x = L)$. The displacement from the equilibrium position is $y$, as shown in Figure 5.4-1. We assume that the tension $T$ in the string is uniform and the density of the string is $\rho$. Consider a small element $PQ$ of the string of length $\delta s$. At equilibrium, it is at $P_0 Q_0$, as shown in Figure 5.4-1. The tension $T$ acts along the tangents at $P$ and $Q$.

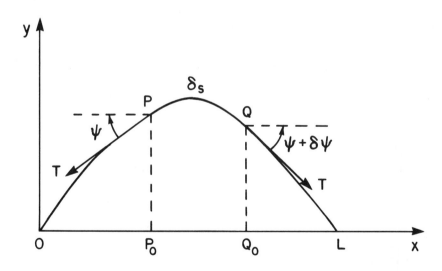

**FIGURE 5.4-1    Vibrating string**

The angles these tangents make with the x-axis are $\psi$ and $\psi + \delta\psi$. The y component of the equation of motion is

$$\rho\ \delta s\ \frac{\partial^2 y}{\partial t^2}\ =\ T \sin(\psi + \delta \psi) - T \sin \psi \tag{5.4-1a}$$

$$=\ T(\sin \psi \cos \delta \psi + \cos \psi \sin \delta \psi) - T \sin \psi \tag{5.4-1b}$$

$$=\ T\,\delta\psi \cos \psi + 0\,(\delta\psi)^2 \tag{5.4-1c}$$

By definition

$$\frac{\partial y}{\partial x}\ =\ \tan \psi \tag{5.4-2}$$

On differentiating, we obtain

$$\frac{\partial^2 y}{\partial x^2}\ \delta x\ =\ \sec^2 \psi\ \delta\psi \tag{5.4-3a}$$

$$=\ \left[1 + \left(\frac{\partial y}{\partial x}\right)^2\right] \delta\psi \tag{5.4-3b}$$

$$\approx\ \delta\psi \tag{5.4-3c}$$

Equation (5.4-3c) is obtained by assuming that $\frac{\partial y}{\partial x}$ is small.  Its square and higher powers can then be neglected.  Similarly

$$\cos \psi\ =\ \left[1 + \left(\frac{\partial y}{\partial x}\right)^2\right]^{-1/2} \tag{5.4-4a}$$

$$\approx\ 1 \tag{5.4-4b}$$

$$\delta s\ =\ \delta x \left[1 + \left(\frac{\partial y}{\partial x}\right)^2\right]^{1/2} \tag{5.4-5a}$$

$$\approx\ \delta x \tag{5.4-5b}$$

Substituting Equations (5.4-3c, 4b, 5b) into Equation (5.4-1c) yields

$$\frac{\partial^2 y}{\partial t^2}\ =\ \frac{T}{\rho}\ \frac{\partial^2 y}{\partial x^2}\ =\ c^2\ \frac{\partial^2 y}{\partial x^2} \tag{5.4-6a,b}$$

Note that Equation (5.4-6b) is identical to Equation (5.1-1).

To complete the problem, we need to impose additional conditions. The string is tied at its ends, so there is no displacement at these two points. This condition is written as

$$y(0, t) = y(L, t) = 0 \tag{5.4-7a,b}$$

The vibrations depend on the initial shape of the string and the speed at which the string is released. These are expressed as

$$y(x, 0) = f(x), \qquad \frac{\partial y}{\partial t}\bigg|_{t=0} = g(x) \tag{5.4-8a,b}$$

where $f(x)$ and $g(x)$ are given functions of $x$.

The conditions given by Equations (5.4-7a, b) are the **boundary conditions** and those given by Equations (5.4-8a, b) are the **initial (Cauchy) conditions**.

We now seek a solution of the form

$$y = X(x) T(t) \tag{5.4-9}$$

Differentiating and substituting into Equation (5.4-6b), we obtain

$$X \frac{d^2 T}{dt^2} = c^2 \frac{d^2 X}{dx^2} T \tag{5.4-10}$$

The function $XT$ is not identically zero; dividing Equation (5.4-10) by $XT$ yields

$$\frac{1}{T} \frac{d^2 T}{dt^2} = \frac{c^2}{X} \frac{d^2 X}{dx^2} = \text{constant} \tag{5.4-11}$$

Since the left side of Equation (5.4-11) is a function of $t$ only and the right side a function of $x$ only, where $x$ and $t$ are independent variables, each side is equal to a constant. We choose the constant to be $-n^2$, so as to satisfy the boundary conditions.

Equation (5.4-11) represents two ordinary linear equations which can be written as follows

$$\frac{d^2 T}{dt^2} + n^2 T = 0 \tag{5.4-12a}$$

$$\frac{d^2 X}{dx^2} + \frac{n^2}{c^2} X = 0 \tag{5.4-12b}$$

The P.D.E. in two independent variables has been transformed to a pair of O.D.E.'s. Equations (5.4-12b, 7a, b) form a Sturm-Liouville problem which is discussed in Chapter 2. The solutions of Equations (5.4-12a, b) depend on the constant n and can be written as

$$T_n = A_n \cos (nt) + B_n \sin (nt) \tag{5.4-13a}$$

$$X_n = C_n \cos \left(\frac{nx}{c}\right) + D_n \sin \left(\frac{nx}{c}\right) \tag{5.4-13b}$$

where $A_n$, $B_n$, $C_n$, and $D_n$ are arbitrary constants.

Equations (5.4-7a, b) imply

$$C_n = 0 \tag{5.4-14a}$$

$$\sin \left(\frac{nL}{c}\right) = 0 \tag{5.4-14b}$$

Note that we do not impose both $C_n$ and $D_n$ to be zero because we are not interested in the trivial solution (y = 0). From Equation (5.4-14b), we deduce

$$\left(\frac{nL}{c}\right) = s\pi \qquad (s = 1, 2, ...) \tag{5.4-15}$$

where s is an integer.

Using the principle of superposition, we express the solution as

$$y = \sum_{s=1}^{\infty} \left[A_s \cos \left(\frac{s\pi ct}{L}\right) + B_s \sin \left(\frac{s\pi ct}{L}\right)\right] \sin \left(\frac{s\pi x}{L}\right) \tag{5.4-16}$$

Assuming term-by-term differentiation to be permissible, we have

$$\frac{\partial y}{\partial t} = \sum_{s=1}^{\infty} \frac{s\pi c}{L} \left[-A_s \sin \left(\frac{s\pi ct}{L}\right) + B_s \cos \left(\frac{s\pi ct}{L}\right)\right] \sin \left(\frac{s\pi x}{L}\right) \tag{5.4-17}$$

Imposing the initial conditions [Equations (5.4-8a, b)] yields

$$f(x) = \sum_{s=1}^{\infty} A_s \sin \left(\frac{s\pi x}{L}\right) \tag{5.4-18a}$$

$$g(x) = \sum_{s=1}^{\infty} \frac{s\pi c}{L} B_s \sin \left(\frac{s\pi x}{L}\right) \tag{5.4-18b}$$

The right sides of Equations (5.4-18a, b) are the Fourier series of $f(x)$ and $g(x)$ and the Fourier coefficients are given by

$$A_s = \frac{2}{L} \int_0^L f(x) \sin\left(\frac{s\pi x}{L}\right) dx \tag{5.4-19a}$$

$$B_s = \frac{2}{s\pi c} \int_0^L g(x) \sin\left(\frac{s\pi x}{L}\right) dx \tag{5.4-19b}$$

The solution $y$ is periodic both in $x$ and $t$. Its period in $x$ is $(2L)$ and in $t$, it is $\left(\frac{2L}{c}\right)$. If the string is released from rest $[g(x) = 0]$, the coefficients $B_s$ are zero for all $s$. We note that we first impose the boundary conditions and then the initial conditions. This procedure is necessary because the initial conditions are given as functions and not as constants. The P.D.E. is decomposed in two O.D.E.'s and on integrating, we obtain constants and not arbitrary functions. It is necessary to use the superposition principle, that is to say, to expand the functions given in the initial conditions in terms of appropriate eigenfunctions. The coefficients of the expansion are then determined. In this case, the eigenfunctions are the trigonometric functions and we obtain a Fourier series.

## D'Alembert's Solution

D'Alembert proposed a solution to the wave equation by transforming Equation (5.4-6b) to its canonical form [Equation (5.3-12)]. This is accomplished by introducing the characteristics, which are [Equations (5.3-11a, b)]

$$\xi = x + ct, \qquad \eta = x - ct \tag{5.4-20a,b}$$

From Equation (5.3-3) with $x_1 (= x)$, $x_2 (= t)$, we find that Equation (5.4-6b) becomes

$$\frac{\partial^2 y}{\partial \xi \, \partial \eta} = 0 \tag{5.4-21}$$

The solution is

$$y = F(\xi) + G(\eta) \tag{5.4-22a}$$

$$= F(x + ct) + G(x - ct) \tag{5.4-22b}$$

where $F$ and $G$ are arbitrary functions.

On differentiating $y$ partially with respect to $t$, we obtain

$$\frac{\partial y}{\partial t} = c\, F'(x+ct) - c\, G'(x-ct) \tag{5.4-23}$$

where the prime ($'$) denotes differentiation with respect to the argument $[(x+ct)$ or $(x-ct)]$.

The initial conditions [Equations (5.4-8a, b)] imply

$$f(x) = F(x) + G(x) \tag{5.4-24a}$$

$$g(x) = c\, F'(x) - c\, G'(x) \tag{5.4-24b}$$

Integrating Equation (5.4-24b) yields

$$\frac{1}{c} \int_{x_0}^{x} g(\zeta)\, d\zeta = F(x) - G(x) \tag{5.4-24c}$$

where $x_0$ is arbitrary.

From Equations (5.4-24a, c), we obtain

$$F(x) = \frac{1}{2} \left[ f(x) + \frac{1}{c} \int_{x_0}^{x} g(\zeta)\, d\zeta \right] \tag{5.4-25a}$$

$$G(x) = \frac{1}{2} \left[ f(x) - \frac{1}{c} \int_{x_0}^{x} g(\zeta)\, d\zeta \right] \tag{5.4-25b}$$

Substituting Equations (5.4-25a, b) into Equation (5.4-22b) yields

$$y = \frac{1}{2} \left[ f(x+ct) + f(x-ct) + \frac{1}{c} \int_{x_0}^{x+ct} g(\zeta)\, d\zeta - \frac{1}{c} \int_{x_0}^{x-ct} g(\zeta)\, d\zeta \right] \tag{5.4-26a}$$

$$= \frac{1}{2} \left[ f(x+ct) + f(x-ct) + \frac{1}{c} \int_{x_0}^{x+ct} g(\zeta)\, d\zeta + \frac{1}{c} \int_{x-ct}^{x_0} g(\zeta)\, d\zeta \right] \tag{5.4-26b}$$

$$= \frac{1}{2} \left[ f(x+ct) + f(x-ct) + \frac{1}{c} \int_{x-ct}^{x+ct} g(\zeta)\, d\zeta \right] \tag{5.4-26c}$$

This solution is **d'Alembert's solution** and is equivalent to Equation (5.4-16). We note in this case that, on integrating Equation (5.4-21), we obtain arbitrary functions and these functions are determined by using the initial conditions. The functions $f(x)$ and $g(x)$ are defined only in the interval

$$0 \leq x \leq L \tag{5.4-27}$$

We need to extend the range of the interval when $(x + ct)$ exceeds $L$ and when $(x - ct)$ is negative. To do this, we need to impose the boundary conditions [Equations (5.4-7a, b)] and from Equation (5.4-22b), we deduce

$$F(ct) + G(-ct) = 0 \tag{5.4-28a}$$

$$F(L + ct) + G(L - ct) = 0 \tag{5.4-28b}$$

If we replace $L + ct$ by $\sigma$, Equation (5.4-28b) becomes

$$F(\sigma) + G(2L - \sigma) = 0 \tag{5.4-29}$$

The value of $t$ is not fixed and Equations (5.4-28a, 29) are valid for any $t$ and $\sigma$. We deduce that, for any $\sigma$,

$$G(-\sigma) = G(2L - \sigma) \tag{5.4-30}$$

Equation (5.4-30) shows that $G$ is periodic with period $2L$. Similarly, $F$ is also periodic with period $2L$. The solution can now be extended for all values of $(x - ct)$ and $(x + ct)$.

To examine further the physical significance of the solution, we consider the simpler case where the string is released from rest. From Equation (5.4-8b), it is seen that this implies that $g$ is zero.

Equation (5.4-26c) is then written as

$$y = \frac{1}{2}\left[f(x + ct) + f(x - ct)\right] \tag{5.4-31}$$

Consider a function $f(\eta)$. For a fixed value of $\eta$, $f(\eta)$ is constant. The transformation $\eta$ $[= (x - ct)]$ represents a displacement to the right at a speed $c$. That is to say, if initially $(t = 0)$ the point $x_1$ corresponds to a constant value $\eta$, then at time $t$ the point $x$ that corresponds to the same value of $\eta$ is $x_1 + ct$. Thus $f(x - ct)$ is a pattern moving to the right at a speed $c$ without changing shape. From Equation (5.4-31), we observe that initially the shape of the string is given by $f(x)$. Afterwards $(t > 0)$, the original pattern breaks up into two similar patterns but only half the size (amplitude) of the initial one. One pattern moves to the right and the other $[f(\xi)]$ to the left as shown in Figure 5.4-2. Equation (5.4-31) represents the superposition of two waves, one moving to the right and the other to the left, and at the end points $(x = 0$ and $x = L)$, they are reflected. They are periodic and of period $2L$.

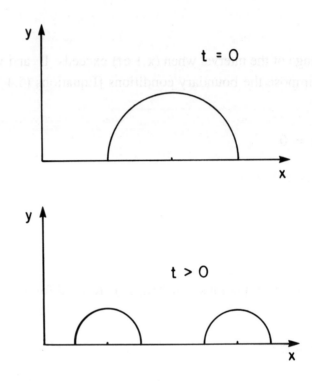

**FIGURE 5.4-2   Displacement of an initial pattern**

Next we consider the case where f is zero and g is non-zero. Equation (5.4-26c) simplifies to

$$y(x, t) = \frac{1}{2c} \int_{x-ct}^{x+ct} g(\zeta) \, d\zeta \tag{5.4-32}$$

The value of y at (x, t) depends only on the interval (x − c t) and (x + c t). This is the **domain of dependence** and it is shown in Figure 5.4-3. It can be seen from Equation (5.4-26c) that the domain of dependence is the same in the case f ≠ 0.

Motivated by physical considerations, boundary and initial conditions [Equations (5.4-7a, b, 8a, b)] have been imposed. These conditions are sufficient and necessary to determine a unique solution.

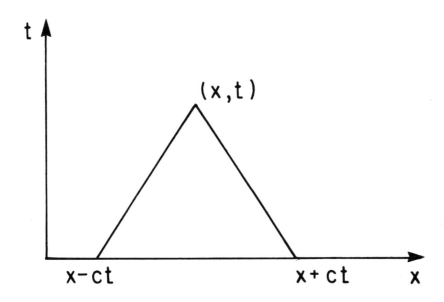

**FIGURE 5.4-3    Domaine of influence**

## Diffusion Equation

Diffusion is a process by which matter is transported from one part of a system to another. Consider the diffusion of chemical species A in a binary system of A and B. Under appropriate conditions, Equation (A.IV-1) simplifies to

$$\frac{\partial c_A}{\partial t} = \mathcal{D}_{AB} \frac{\partial^2 c_A}{\partial x^2} \tag{5.4-33}$$

where $c_A$ is the concentration of A, $\mathcal{D}_{AB}$ is the diffusivity, t is the time and x is the position.

For simplicity, we write Equation (5.4-33) as Equation (5.1-2) which is reproduced here for convenience

$$\frac{\partial c}{\partial t} = \alpha^2 \frac{\partial^2 c}{\partial x^2} \tag{5.1-2}$$

Equation (5.1-2) also describes the conduction of heat in an isotropic medium. In this case, c is the temperature and $\alpha^2$ is the thermal diffusivity.

We consider the diffusion out of a plane sheet of finite thickness L. Initially the distribution of the diffusing substance is $f(x)$. The surfaces are kept at zero concentration at all times. The initial and boundary conditions can be written as

$$c(x, 0) = f(x), \qquad\qquad\qquad 0 < x < L \qquad\qquad\qquad\qquad (5.4\text{-}34a)$$

$$c(0, t) = 0, \qquad c(L, t) = 0, \qquad t > 0 \qquad\qquad\qquad\qquad (5.4\text{-}34b,c)$$

We seek a solution of the form

$$c(x, t) = X(x)\, T(t) \qquad\qquad\qquad\qquad\qquad\qquad\qquad (5.4\text{-}35)$$

Proceeding as in the case of the wave equation, we find that the partial differential equation leads to two ordinary differential equations. They are

$$\frac{dT}{dt} + n^2 T = 0 \qquad\qquad\qquad\qquad\qquad\qquad\qquad\qquad (5.4\text{-}36a)$$

$$\frac{d^2 X}{dx^2} + \frac{n^2}{\alpha^2} X = 0 \qquad\qquad\qquad\qquad\qquad\qquad\qquad (5.4\text{-}36b)$$

where $n^2$ is a positive constant which is determined by the boundary conditions.

Equations (5.4-36b, 34b, c) constitute an eigenvalue problem. The solution is

$$X_n = A_n \cos\frac{n\,x}{\alpha} + B_n \sin\frac{n\,x}{\alpha} \qquad\qquad\qquad\qquad\qquad (5.4\text{-}37)$$

Imposing the boundary conditions, we deduce

$$A_n = 0, \qquad n = \frac{s\,\pi\,\alpha}{L}, \qquad s = 1, 2, \dots \qquad\qquad\qquad (5.4\text{-}38a,b,c)$$

The solution of Equation (5.4-36a) is

$$T_n = C_n\, e^{-n^2 t} \qquad\qquad\qquad\qquad\qquad\qquad\qquad\qquad (5.4\text{-}39)$$

where $C_n$ is a constant.

Using the principle of superposition, the general solution of Equation (5.1-2) subject to Equations (5.4-34b, c) is

$$c = \sum_{s=1}^{\infty} B_s \exp\left(-s^2 \pi^2 \alpha^2 t / L^2\right) \sin\frac{s\,\pi\,x}{L} \qquad\qquad\qquad (5.4\text{-}40)$$

The Fourier coefficients $B_s$ are obtained from the initial conditions and they are given by

$$B_s = \frac{2}{L} \int_0^L f(x) \sin \frac{s\pi x}{L} \, dx \qquad (5.4-41)$$

The solution is sinusoidal in $x$, as in the case of a vibrating string, decaying exponentially in time, unlike the case of a vibrating string. It decays more rapidly with increasing $\alpha$ in agreement with physical expectation.

## Laplace's Equation

We consider the steady temperature in a semi-infinite rectangular solid. Historically, this is the first problem considered in detail by Fourier, as pointed out in Carslaw and Jaeger (1973). The solid is bounded by the planes

$$x = 0, \qquad x = L, \qquad y = 0, \qquad y \longrightarrow \infty \qquad (5.4\text{-}42a,b,c,d)$$

The two dimensional heat equation is given by (see Appendix III)

$$\frac{\partial T}{\partial t} = \alpha^2 \left( \frac{\partial^2 T}{\partial x^2} + \frac{\partial^2 T}{\partial y^2} \right) \qquad (5.4\text{-}43)$$

where $T$ is the temperature and $\alpha^2$ is the thermal diffusivity.

At steady state, Equation (5.4-43) reduces to

$$\frac{\partial^2 T}{\partial x^2} + \frac{\partial^2 T}{\partial y^2} = 0 \qquad (5.4\text{-}44)$$

The walls ($x = 0$, $x = L$) are kept at zero temperature and the wall ($y = 0$) is kept at a temperature $f(x)$. As $y$ tends to infinity, the temperature tends to zero.

The boundary conditions are

$$T(0, y) = T(L, y) = 0 \qquad (5.4\text{-}45a,b)$$

$$T(x, 0) = f(x), \qquad T \longrightarrow 0 \text{ as } y \longrightarrow \infty \qquad (5.4\text{-}45c,d)$$

We employ the method of separation of variables and write

$$T = X(x)\, Y(y) \qquad (5.4\text{-}46)$$

Differentiating, substituting the resulting expression into Equation (5.4-44), and dividing by $XY$ ($\neq 0$) yields

$$\frac{d^2X}{dx^2} + n^2 X = 0 \tag{5.4-47a}$$

$$\frac{d^2Y}{dy^2} - n^2 Y = 0 \tag{5.4-47b}$$

The solution of Equation (5.4-47a) satisfying Equations (5.4-45a, b) is, as in the previous two cases,

$$X_n = \sin nx, \qquad n = \frac{s\pi}{L} \tag{5.4-48a,b}$$

where $s$ is an integer.

The solution of Equation (5.4-47b) is

$$Y_n = A_n e^{ny} + B_n e^{-ny} \tag{5.4-49}$$

To satisfy Equation (5.4-45d), we require $A_n$ to be zero. Using the principle of superposition, $T$ is given by

$$T = \sum_{s=1}^{\infty} B_s e^{-s\pi y/L} \sin \frac{s\pi x}{L} \tag{5.4-50}$$

Combining Equations (5.4-45c, 50) yields

$$f(x) = \sum_{s=1}^{\infty} B_s \sin \frac{s\pi x}{L} \tag{5.4-51}$$

From Equation (5.4-51), we deduce

$$B_s = \frac{2}{L} \int_0^L f(x) \sin \frac{s\pi x}{L} \, dx \tag{5.4-52}$$

Substituting Equation (5.4-52) into Equation (5.4-50), we obtain

$$T = \frac{2}{L} \sum_{s=1}^{\infty} \int_0^L \left( f(\xi) \sin \frac{s\pi \xi}{L} \right) d\xi \, e^{-s\pi y/L} \sin \frac{s\pi x}{L} \tag{5.4-53}$$

The temperature is sinusoidal in $x$ and decays exponentially to zero with increasing $y$ as required by the boundary conditions.

We have so far considered the dependent variable to be a function of two independent variables. In a three-dimensional space, the wave equation, the diffusion equation, and the Laplace equation are written respectively as

$$\nabla^2 u = \frac{1}{c^2} \frac{\partial^2 u}{\partial t^2} \qquad\qquad (5.4\text{-}54a)$$

$$\nabla^2 u = \frac{1}{\alpha^2} \frac{\partial u}{\partial t} \qquad\qquad (5.4\text{-}54b)$$

$$\nabla^2 u = 0 \qquad\qquad (5.4\text{-}54c)$$

where $\nabla^2$ is the Laplacian operator [Equation 4.2-31)] and $u$ is the dependent variable.

We note that at steady state, both the wave and diffusion equations reduce to Laplace's equation. The rectangular Cartesian coordinate system is not always the most suitable coordinate system to refer to. In many cases, it is more appropriate to choose the cylindrical polar or the spherical polar coordinate system. We next solve Laplace's equation in these two coordinate systems.

## 5.5  CYLINDRICAL AND SPHERICAL POLAR COORDINATE SYSTEMS

### Cylindrical Polar Coordinate System

In Chapter 4 (Problem 25a), we have defined the cylindrical polar coordinate system $(r, \theta, z)$. Using Equation (4.7-38), we deduce (Chapter 4, Problem 31a) that Laplace's equation can be written as

$$\nabla^2 u = \frac{\partial^2 u}{\partial r^2} + \frac{1}{r} \frac{\partial u}{\partial r} + \frac{1}{r^2} \frac{\partial^2 u}{\partial \theta^2} + \frac{\partial^2 u}{\partial z^2} = 0 \qquad\qquad (5.5\text{-}1a,b)$$

We assume that $u$ is of the form

$$u = F(r, \theta) Z(z) \qquad\qquad (5.5\text{-}2)$$

Differentiating, substituting the resulting expression into Equation (5.5-1b), and dividing by $FZ$ ($\neq 0$), we obtain

$$\frac{1}{F} \frac{\partial^2 F}{\partial r^2} + \frac{1}{rF} \frac{\partial F}{\partial r} + \frac{1}{r^2 F} \frac{\partial^2 F}{\partial \theta^2} = -\frac{1}{Z} \frac{d^2 Z}{dz^2} = n^2 \qquad\qquad (5.5\text{-}3a,b)$$

where $n^2$ is a constant.

Note that the left side is a function of r and θ and the right side is a function of z, and since r, θ, and z are independent variables, each side must be equal to a constant, which we denote by $n^2$. From Equations (5.5-3a, b), we obtain two equations which can be written as

$$\frac{d^2 Z}{dz^2} + n^2 Z = 0 \tag{5.5-4}$$

$$\frac{\partial^2 F}{\partial r^2} + \frac{1}{r} \frac{\partial F}{\partial r} + \frac{1}{r^2} \frac{\partial^2 F}{\partial \theta^2} - n^2 F = 0 \tag{5.5-5}$$

We now write F as

$$F = R(r) \, \Theta(\theta) \tag{5.5-6}$$

Proceeding in the usual manner, we obtain

$$\frac{r^2}{R} \frac{d^2 R}{dr^2} + \frac{r}{R} \frac{dR}{dr} - n^2 r^2 = -\frac{d^2 \Theta}{d\theta^2} = m^2 \tag{5.5-7a,b}$$

Equations (5.5-7a, b) can be written as

$$r^2 \frac{d^2 R}{dr^2} + r \frac{dR}{dr} - \left(n^2 r^2 + m^2\right) R = 0 \tag{5.5-8a}$$

$$\frac{d^2 \Theta}{d\theta^2} + m^2 \Theta = 0 \tag{5.5-8b}$$

Solving Equation (5.5-1b) implies solving Equations (5.5-4, 8a, b). We consider the simpler case where u is not a function of z. That is to say, we consider the two-dimensional case. The constant $n^2$ is zero and Equation (5.5-8a) simplifies to

$$r^2 \frac{d^2 R}{dr^2} + r \frac{dR}{dr} - m^2 R = 0 \tag{5.5-9}$$

Equation (5.5-8b) remains unchanged and its solution is

$$\Theta = A \cos m\theta + B \sin m\theta, \qquad m \neq 0 \tag{5.5-10}$$

where A and B are constants.

The function Θ must be single valued, that is to say

$$\cos m\theta = \cos m (\theta + 2 s \pi) \qquad (5.5\text{-}11)$$

where $s$ is an integer.

This implies that $m$ must be an integer. Equation (5.5-9) is the Euler (unidimensional) equation and its solution is

$$R = Cr^m + Dr^{-m}, \qquad m \neq 0 \qquad (5.5\text{-}12)$$

where $C$ and $D$ are constants.

If $m$ is zero, $\Theta$ and $R$ are given by

$$\Theta = A_0\theta + B_0 \qquad (5.5\text{-}13a)$$

$$R = C_0 \ln r + D_0 \qquad (5.5\text{-}13b)$$

where $A_0$, $B_0$, $C_0$, and $D_0$ are constants.

The solution $\ln r$ is the **fundamental solution** and is singular at the origin. If the function has no singularity and is periodic in $\theta$, the constants $D$, $A_0$, and $C_0$ are zero. Using the principle of superposition, $F$ is given by

$$F = E_0 + \sum_{m=1}^{\infty} \left[ A_m \cos m\theta + B_m \sin m\theta \right] r^m \qquad (5.5\text{-}14)$$

where $E_0$, $A_m$, and $B_m$ are constants.

In the three-dimensional case $(m \neq 0)$, we have to solve Equations (5.5-4, 8, 9). The solution of Equation (5.5-4) is

$$Z = H \cos n z + G \sin n z \qquad (5.5\text{-}15)$$

where $H$ and $G$ are constants.

To solve Equation (5.5-8a), we make a change of variable and write

$$\sigma = i n r \qquad (i = \sqrt{-1}) \qquad (5.5\text{-}16)$$

Equation (5.5-8a) becomes

$$\sigma^2 \frac{d^2R}{d\sigma^2} + \sigma \frac{dR}{d\sigma} + \left( \sigma^2 - m^2 \right) R = 0 \qquad (5.5\text{-}17)$$

Equation (5.5-17) is the Bessel equation with complex argument. Its solution is

$$R = L\, I_m\,(nr) + M\, K_m\,(nr) \qquad\qquad (5.5\text{-}18)$$

where $I_m$ and $K_m$ are the modified Bessel functions of the first and second kind respectively. L and M are constants.

The function $K_m$ has a singularity at the origin. We assume that the function remains finite at the origin. The constant M must then be zero. If the radius of the cylinder is a, there is no loss of generality in assuming that R(a) is one and Equation (5.5-18) becomes

$$R = \frac{I_m\,(nr)}{I_m\,(na)} \qquad\qquad (5.5\text{-}19)$$

The constants A, B, H, K, and n are to be determined from the boundary conditions. To complete the problem, we assume the boundary conditions to be

$$u\,(r,\,\theta,\,0) = u\,(r,\,\theta,\,\pi) = 0 \qquad\qquad (5.5\text{-}20a,b)$$

$$u\,(a,\,\theta,\,z) = f\,(\theta,\,z) \qquad\qquad (5.5\text{-}20c)$$

Equations (5.5-20a, b) imply that H is zero and that n is an integer.

The general solution is

$$u = \sum_{n=1}^{\infty} G_n \sin nz\, \frac{I_0\,(nr)}{I_0\,(na)} + \sum_{n=1}^{\infty}\sum_{m=1}^{\infty} G_n \sin nz \Big[A_m \cos m\theta + B_m \sin m\theta\Big]\frac{I_m\,(nr)}{I_m\,(na)}$$

$$(5.5\text{-}21a)$$

$$= \frac{1}{2}\sum_{n=1}^{\infty} A_{n0} \sin nz\, \frac{I_0\,(nr)}{I_0\,(na)} + \sum_{m,n=1}^{\infty} \Big[A_{nm} \cos m\theta + B_{nm} \sin m\theta\Big]\sin nz\, \frac{I_m\,(nr)}{I_m\,(na)}$$

$$(5.5\text{-}21b)$$

Applying Equation (5.5-20c) to Equation (5.5-21b) yields

$$f\,(\theta,\,z) = \frac{1}{2}\sum_{n=1}^{\infty} A_{n0} \sin nz + \sum_{m,n=1}^{\infty} \Big[A_{nm} \cos m\theta + B_{nm} \sin m\theta\Big]\sin nz \qquad (5.5\text{-}22)$$

The right side of Equation (5.5-22) is the (double) **Fourier series** of $f\,(\theta,\,z)$. The coefficients $A_{nm}$ and $B_{nm}$ are given by

$$A_{nm} = \frac{2}{\pi^2} \int_0^\pi \int_0^{2\pi} f(\theta, z) \sin nz \cos m\theta \, d\theta \, dz \qquad (5.5\text{-}23a)$$

$$B_{nm} = \frac{2}{\pi^2} \int_0^\pi \int_0^{2\pi} f(\theta, z) \sin nz \sin m\theta \, d\theta \, dz \qquad (5.5\text{-}23b)$$

We have assumed the height of the cylinder to be $\pi$ for convenience. If the height is $L$, it can be scaled to $\pi$ by introducing $\bar{z} \left(= \frac{z\pi}{L}\right)$. If the cylinder is semi-infinite, and $u$ tends to zero as $z$ tends to infinity, the solution cannot be sinusoidal in $z$ as given here. The sign of $n^2$ in Equations (5.5-3a, b) has to be changed. Equations (5.5-4, 8) become

$$\frac{d^2 Z}{dz^2} - n^2 Z = 0 \qquad (5.5\text{-}24a)$$

$$r^2 \frac{d^2 R}{dr^2} + r \frac{dR}{dr} + \left(n^2 r^2 - m^2\right) R = 0 \qquad (5.5\text{-}24b)$$

Equation (5.5-8b) remains unchanged. The solution of Equation (5.5-24a) that satisfies the condition $Z$ tends to zero as $z$ tends to infinity is

$$Z = \overline{K} e^{-nz} \qquad (5.5\text{-}25)$$

where $\overline{K}$ is a constant.

Equation (5.5-24b) is the Bessel equation and the solution that remains finite at the origin is

$$R = \overline{L} J_m(nr) \qquad (5.5\text{-}26)$$

where $J_m$ is the Bessel function of the first kind, and $\overline{L}$ is a constant.

We have already assumed that $u$ is finite as $r$ tends to zero, and $u$ tends to zero as $z$ tends to infinity. The remaining conditions are assumed to be

$$u(a, \theta, z) = 0, \qquad u(r, \theta, 0) = f(r, \theta) \qquad (5.5\text{-}27a,b)$$

Equation (5.5-27a) implies that $na$ are the zeros of $J_m$. There are an infinite number of them and we denote them as $\lambda_{nm}$ $(\lambda_{n1}, \lambda_{n2}, ...)$

$$J_m(na) = J_m(\lambda_{nm}) = 0 \qquad (5.5\text{-}28a,b)$$

The general solution is given by

$$u = \sum_{n=1}^{\infty} \overline{K}_n \exp\left(-\frac{\lambda_{n0} z}{a}\right) J_0\left(\frac{\lambda_{n0} r}{a}\right)$$

$$+ \sum_{n=1}^{\infty} \sum_{m=1}^{\infty} \overline{K}_n \exp\left(-\frac{\lambda_{nm} z}{a}\right) J_m\left(\frac{\lambda_{nm} r}{a}\right) \left[A_m \cos m\theta + B_m \sin m\theta\right] \qquad (5.5\text{-}29a)$$

$$= \frac{1}{2} \sum_{n=1}^{\infty} A_{0n} \exp\left(-\frac{\lambda_{n0} z}{a}\right) J_0\left(\frac{\lambda_{n0} r}{a}\right)$$

$$+ \sum_{m,n=1}^{\infty} \left[A_{nm} \cos m\theta + B_{nm} \sin m\theta\right] \exp\left(-\frac{\lambda_{nm} z}{a}\right) J_m\left(\frac{\lambda_{nm} r}{a}\right) \qquad (5.5\text{-}29b)$$

Equation (5.5-27b) implies that

$$f(r, \theta) = \frac{1}{2} \sum_{n=1}^{\infty} A_{0n} J_0\left(\frac{\lambda_{n0} r}{a}\right) + \sum_{m,n=1}^{\infty} \left[A_{nm} \cos m\theta + B_{nm} \sin m\theta\right] J_m\left(\frac{\lambda_{nm} r}{a}\right) \qquad (5.5\text{-}30)$$

Equation (5.5-30) is the (double) Fourier-Bessel series expression of $f(r, \theta)$.

We recall that the Bessel functions are orthogonal and their properties are discussed in Chapter 2. The coefficients $A_{nm}$ and $B_{nm}$ are given by

$$A_{nm} = \frac{2}{\pi^2 a^2 J_{m+1}^2(\lambda_{nm})} \int_0^a \int_0^{2\pi} f(r, \theta) J_m\left(\frac{\lambda_{nm} r}{a}\right) \cos m\theta \; r \, dr \, d\theta \qquad (5.5\text{-}31a)$$

$$B_{nm} = \frac{2}{\pi^2 a^2 J_{m+1}^2(\lambda_{nm})} \int_0^a \int_0^{2\pi} f(r, \theta) J_m\left(\frac{\lambda_{nm} r}{a}\right) \sin m\theta \; r \, dr \, d\theta \qquad (5.5\text{-}31b)$$

## Spherical Polar Coordinate System

Laplace's equation in the spherical polar coordinate system $(r, \theta, \phi)$ was deduced in Example 4.7-4 and can be written as

$$\frac{\partial^2 u}{\partial r^2} + \frac{2}{r} \frac{\partial u}{\partial r} + \frac{1}{r^2} \frac{\partial^2 u}{\partial \theta^2} + \frac{\cot \theta}{r^2} \frac{\partial u}{\partial \theta} + \frac{1}{r^2 \sin^2 \theta} \frac{\partial^2 u}{\partial \phi^2} = 0 \qquad (5.5\text{-}32)$$

We assume that $u$ can be written as

$$u = F(r, \theta) \, \Phi(\phi) \tag{5.5-33}$$

In terms of $F$ and $\Phi$, Equation (5.5-32) becomes

$$\frac{r^2 \sin^2 \theta}{F} \left( \frac{\partial^2 F}{\partial r^2} + \frac{2}{r} \frac{\partial F}{\partial r} + \frac{1}{r^2} \frac{\partial^2 F}{\partial \theta^2} + \frac{\cot \theta}{r^2} \frac{\partial F}{\partial \theta} \right) = -\frac{1}{\Phi} \frac{d^2 \Phi}{d\phi^2} = m^2 \tag{5.5-34a,b}$$

where $m^2$ is a constant.

Equations (5.5-34a, b) can be written as

$$\frac{d^2 \Phi}{d\phi^2} + m^2 \Phi = 0 \tag{5.5-35a}$$

$$r^2 \sin^2 \theta \left( \frac{\partial^2 F}{\partial r^2} + \frac{2}{r} \frac{\partial F}{\partial r} + \frac{1}{r^2} \frac{\partial^2 F}{\partial \theta^2} + \frac{\cot \theta}{r^2} \frac{\partial F}{\partial \theta} \right) - m^2 F = 0 \tag{5.5-35b}$$

We now write $F$ as

$$F = R(r) \, \Theta(\theta) \tag{5.5-36}$$

Equation (5.5-35b) can be written as two ordinary differential equations. They are

$$r^2 \frac{d^2 R}{dr^2} + 2r \frac{dR}{dr} - \lambda R = 0 \tag{5.5-37a}$$

$$\frac{d^2 \Theta}{d\theta^2} + \cot \theta \frac{d\Theta}{d\theta} - \left( \frac{m^2}{\sin^2 \theta} - \lambda \right) \Theta = 0 \tag{5.5-37b}$$

In many physical situations, $u$ is independent of $\phi$, and we first consider this case. This implies that $m$ is zero and we need to solve Equation (5.5-37a) and

$$\frac{d^2 \Theta}{d\theta^2} + \cot \theta \frac{d\Theta}{d\theta} + \lambda \Theta = 0 \tag{5.5-37c}$$

We transform Equation (5.5-37c) into the standard form of a Legendre equation by writing $x$ as $\cos \theta$ and we obtain

$$\left(1 - x^2\right) \frac{d^2\Theta}{dx^2} - 2x \frac{d\Theta}{dx} + \lambda\Theta = 0 \qquad\qquad (5.5\text{-}37d)$$

By writing $\lambda$ as $n(n+1)$, Equation (5.5-37d) becomes the standard Legendre equation discussed in Chapter 2. The solution is

$$\Theta = A\,P_n(x) + B\,Q_n(x) \qquad\qquad (5.5\text{-}38)$$

where $A$ and $B$ are constants and $P_n$ and $Q_n$ are the Legendre functions of the first and second kind respectively.

The singular points of the equation are at $x \ (= \pm 1)$ which corresponds to $\theta \ (= 0, \pi)$. If the solution is finite for all values of $\theta$, $B$ must be zero, $P_n$ are the Legendre polynomials, and $n$ is an integer.

Equation (5.5-37a) is the Euler equation and its solution is

$$R = Cr^n + \frac{D}{r^{n+1}} \qquad\qquad (5.5\text{-}39)$$

Note that we have replaced $\lambda$ by $n(n+1)$ and $n$ is an integer. If $R$ is finite at the origin, $D$ must be zero. The general solution is

$$u = \sum_{n=0}^{\infty} A_n\, r^n\, P_n(\cos\theta) \qquad\qquad (5.5\text{-}40)$$

Suppose the boundary condition on the sphere of radius $a$ is specified as

$$u(a, \theta) = f(\theta) \qquad\qquad (5.5\text{-}41)$$

Combining Equations (5.5-40, 41) yields

$$f(\theta) = \sum_{n=0}^{\infty} A_n\, a^n\, P_n(\cos\theta) \qquad\qquad (5.5\text{-}42)$$

Equation (5.5-42) shows that $f(\theta)$ is expressed as a Fourier-Legendre series and the coefficients $A_n$ can be determined using the orthogonal properties of Legendre polynomials. The coefficients $A_n$ are given by

$$A_n = \frac{(2n+1)}{2a^n} \int_0^{\pi} f(\theta)\, P_n(\cos\theta) \sin\theta \, d\theta \qquad\qquad (5.5\text{-}43)$$

We have now obtained the solution inside the sphere. If the region of interest is outside the sphere, the condition at the origin cannot be imposed and D is not necessarily zero. The solution is then given as

$$u = \sum_{n=0}^{\infty} \left( C_n r^n + \frac{D_n}{r^{n+1}} \right) P_n(\cos\theta) \qquad (5.5\text{-}44)$$

The coefficients $C_n$ and $D_n$ are determined from the boundary conditions. The region can be enclosed by surfaces of spheres of radii a and b with the possibility that b tends to infinity. When n is zero, the solution is $\frac{1}{r}$ and it is the **fundamental solution**. In the general case, u is a function of all three variables, and we have to solve Equations (5.5-35a, 37a, b). The solution of Equation (5.5-35a) involves $\sin m\phi$ and $\cos m\phi$ and we require the function to be single valued. Therefore, m has to be an integer. The solution of Equation (5.5-37b) is the associated Legendre function. The problem can be completed by imposing appropriate boundary conditions.

The method of separation of variables consists of writing the unknown function of two or more independent variables as a product of two or more functions, such that each of these functions is a function of one variable only. The partial differential equation is transformed to two or more ordinary differential equations. Together with the boundary conditions, the differential equations form eigenvalue problems which generate eigenfunctions. The initial conditions are expanded in terms of these eigenfunctions in the form of Fourier series. The constants generated by the solutions of the ordinary differential equations are identified as the coefficients of the Fourier series and they can be determined. In general, the solution of a partial differential equation involves undetermined functions rather than undetermined constants. In the method of separation of variables, the undetermined functions are expressed as Fourier series. Possible base functions for the series are trigonometric functions, hyperbolic functions, and orthogonal functions considered in Chapter 2. The functions to be chosen depend on the coordinate system.

## 5.6   BOUNDARY AND INITIAL CONDITIONS

The three types of second order linear partial differential equations have been derived in a physical context and the prescribed boundary and initial conditions have been based on physical considerations. Each type of equation requires different conditions. These are shown in Figure 5.6-1. For convenience, we denote the dependent variable by u. For a hyperbolic equation, both u and $\frac{\partial u}{\partial t}$ are given initially and at the boundaries only u is given. In the case of a parabolic equation, only u needs to be known initially. The initial value of $\frac{\partial u}{\partial t}$ can be determined from Equation (5.1-2). In addition, the values of u are given at the boundaries. These two types of equations describe evolutionary processes and, from the initial conditions, we can integrate forward to determine the values of u at a later stage. Laplace's equation characterizes the steady state and u is prescribed on the surfaces that enclose the region of interest. In Example 3.5-7, we have shown that the solution of Laplace's equation in two-dimensions is determined by the values of the function on the boundary curve. The conditions to be specified have to be compatible with the particular type of equation. We elaborate the importance of imposing the appropriate boundary and initial conditions by considering two examples.

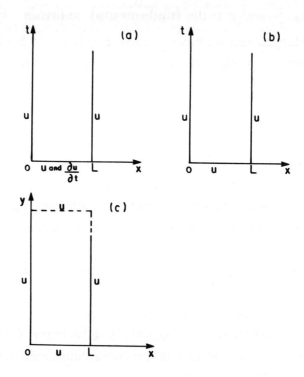

**FIGURE 5.6-1**     **Prescribed conditions for (a) hyperbolic, (b) parabolic, and (c) elliptic equations**

***Example 5.6-1***.  Solve the equation

$$\frac{\partial^2 u}{\partial x^2} + \frac{\partial^2 u}{\partial y^2} = 0 \tag{5.6-1}$$

subject to

$$u(x, 0) = 0, \qquad \left. \frac{\partial u}{\partial y} \right|_{y=0} = \frac{\sin nx}{n} \tag{5.6-2a,b}$$

This example was first considered by Hadamard.

Note that we have to solve an elliptic equation and that we did not impose conditions on the curve enclosing the region required. Instead, we have imposed initial conditions; that is to say, conditions at $y = 0$. Using the method of separation of variables and writing $u$ as $X(x) Y(y)$, Equation (5.6-1) is transformed to

$$X'' + n^2 X = 0 \tag{5.6-3a}$$

$$Y'' - n^2 Y = 0 \tag{5.6-3b}$$

The solution that satisfies Equations (5.6-2a, b) is

$$u = \frac{\sinh ny \, \sin nx}{n^2} \tag{5.6-4}$$

As $n$ tends to infinity, both $u$ and $\dfrac{\partial u}{\partial y}$ tend to zero but the solution $u$ oscillates with increasing amplitude. The solution does not approach zero as the initial conditions tend to zero, as a result of having imposed the wrong type of boundary conditions. In most cases, we expect that a small change in the initial conditions leads to a small change to the corresponding solution. This example violates this rule and the problem is not **well posed**.

***Example 5.6-2.*** Solve the equation

$$\frac{\partial^2 u}{\partial x \, \partial t} = 0 \tag{5.6-5}$$

subject to

$$u(x, 0) = f_1(x), \qquad u(x, 1) = f_2(x) \tag{5.6-6a,b}$$

$$u(0, t) = g_1(t), \qquad u(1, t) = g_2(t) \tag{5.6-6c,d}$$

Equation (5.6-5) is a hyperbolic equation and we have prescribed boundary conditions, as opposed to the required initial conditions.

The solution of Equation (5.6-5) is

$$u = F(x) + G(t) \tag{5.6-7}$$

Imposing Equations (5.6-6a to d) yields

$$F(x) + G(0) = f_1(x) \tag{5.6-8a}$$

$$F(x) + G(1) = f_2(x) \qquad\qquad\qquad\qquad (5.6\text{-}8b)$$

$$F(0) + G(t) = g_1(t) \qquad\qquad\qquad\qquad (5.6\text{-}8c)$$

$$F(1) + G(t) = g_2(t) \qquad\qquad\qquad\qquad (5.6\text{-}8d)$$

From Equations (5.6-8a, b), we deduce that both $f_1(x) - F(x)$ and $f_2(x) - F(x)$ are constants and this implies that $f_1(x)$ and $f_2(x)$ can differ only by a constant, and are therefore not arbitrary. In most physical problems, the boundary conditions can be imposed arbitrarily, that is to say, the condition imposed at one end $(x = 0)$ is independent of the condition at the other end $(x = 1)$. This is another example of an ill posed problem.

Hadamard has proposed the following conditions for a **well posed problem**:

(i)      the existence of a solution;

(ii)     the solution is unique;

(iii)    small changes in initial and boundary conditions as well as in the coefficients of the equations lead to small changes in the solution.

The equations solved in Section 5.4 satisfy Hadamard's conditions. In Section 4.5, we have defined **Dirichlet's problem** ($u$ is given on the boundary) and **Neumann's problem** ($\frac{\partial u}{\partial n}$ is given on the boundary). If $u$ is given on part of the boundary and $\frac{\partial u}{\partial n}$ on the remaining part of the boundary, we have **Robin's problem**.

## 5.7   NON-HOMOGENEOUS PROBLEMS

The method of separation of variables was used to solve homogeneous partial differential equations with homogeneous boundary conditions. For some non-homogeneous problems, we can introduce an auxiliary function and the problem can be reduced to a homogeneous problem. We illustrate this method by considering a few examples.

***Example 5.7-1.*** Chan Man Fong et al. (1993) considered the transient flow of a thin layer of a Maxwell fluid on a rotating disk. In dimensionless form, the equations governing the motion are

$$\frac{\partial f}{\partial t_1} - 1 = \frac{\partial \tau}{\partial \bar{z}} \qquad\qquad\qquad\qquad (5.7\text{-}1a)$$

$$\tau + \lambda_1 \frac{\partial \tau}{\partial t_1} = \frac{\partial f}{\partial \bar{z}} \qquad\qquad\qquad\qquad (5.7\text{-}1b)$$

where $t_1$ and $\bar{z}$ are the dimensionless time and height respectively; $\lambda_1$ is a dimensionless number characterizing the relaxation time of the fluid. The functions $f$ and $\tau$ are related to the dimensionless radial velocity $u_0$ and the dimensionless shear stress $\tau_{(rz)}$ by

$$u_0 = r_1 \, f(\bar{z}, t_1), \qquad \tau_{(rz)} = r_1 \, \tau(\bar{z}, t_1) \qquad\qquad (5.7\text{-}2a,b)$$

where $r_1$ is the dimensionless radial distance.

The initial and boundary conditions are

$$f(\bar{z}, 0) = \tau(\bar{z}, 0) = f(0, t_1) = \tau(1, t_1) = 0 \qquad\qquad (5.7\text{-}3a,b,c,d)$$

Combining Equations (5.7-1a, b) yields

$$\lambda_1 \frac{\partial^2 f}{\partial t_1^2} + \frac{\partial f}{\partial t_1} - \frac{\partial^2 f}{\partial \bar{z}^2} = 1 \qquad\qquad (5.7\text{-}4)$$

Equation (5.7-4) is non-homogeneous and the method of separation of variables cannot be applied directly. We introduce another function and write

$$f(\bar{z}, t_1) = f_1(\bar{z}) + g(\bar{z}, t_1) \qquad\qquad (5.7\text{-}5)$$

Note that this choice introduces one extra degree of freedom. That is to say, one function on the left side is replaced by two functions on the right side.

Substituting Equation (5.7-5) into Equation (5.7-4) and separating the resulting equation into an equation for $f_1$ and another one for $g$, we obtain

$$\frac{d^2 f_1}{d\bar{z}^2} = -1 \qquad\qquad (5.7\text{-}6a)$$

$$\lambda_1 \frac{\partial^2 g}{\partial t_1^2} + \frac{\partial g}{\partial t_1} - \frac{\partial^2 g}{\partial \bar{z}^2} = 0 \qquad\qquad (5.7\text{-}6b)$$

As a result of having introduced an extra degree of freedom, we are allowed to choose Equation (5.7-6a) so as to remove the inhomogeneity.

Equations (5.7-1b, 3d) imply

$$\frac{\partial f}{\partial \bar{z}}(1, t_1) = 0 \qquad\qquad (5.7\text{-}7)$$

We assume that $f_1$ satisfies boundary conditions (5.7-3c, 7), namely

$$f_1(0) = \frac{df_1}{d\overline{z}}\bigg|_{\overline{z}=0} = 0 \tag{5.7-8a,b}$$

The solution of Equation (5.7-6a) subject to Equations (5.7-8a, b) is

$$f_1 = -\frac{\overline{z}^2}{2} + \overline{z} \tag{5.7-9}$$

The boundary conditions that $g$ has to satisfy are deduced from Equations (5.7-3c, 7) and are

$$g(0, t_1) = \frac{\partial g}{\partial \overline{z}}(1, t_1) = 0 \tag{5.7-10a,b}$$

The method of separation of variables can now be used to solve the boundary value problem defined by Equations (5.7-6b, 10a, b) and we write

$$g(\overline{z}, t_1) = Z(\overline{z}) \, T(t_1) \tag{5.7-11}$$

Substituting Equation (5.7-11) into Equation (5.7-6b) and separating the resulting expression into functions of $\overline{z}$ and $t_1$, we obtain

$$\lambda_1 \frac{T''}{T} + \frac{T'}{T} = \frac{Z''}{Z} = -\alpha^2 \tag{5.7-12a,b}$$

Equations (5.7-12a, b) can be written as

$$\lambda_1 T'' + T' + \alpha^2 T = 0 \tag{5.7-13a}$$

$$Z'' + \alpha^2 Z = 0 \tag{5.7-13b}$$

The solutions of Equations (5.7-13a, b) are respectively

$$T = K_1 e^{-at_1} + K_2 e^{-bt_1} \tag{5.7-14a}$$

$$Z = K_3 \cos \alpha \overline{z} + K_4 \sin \alpha \overline{z} \tag{5.7-14b}$$

where the $K_i$ are constants and where $a$ and $b$ are given by

$$a = \left[1 - \sqrt{1 - 4\lambda_1 \alpha^2}\right] / 2\lambda_1 \tag{5.7-14c}$$

$$b = \left[1 + \sqrt{1 - 4\lambda_1 \alpha^2}\right] / 2\lambda_1 \tag{5.7-14d}$$

Imposing Equations (5.7-10a, b) yields

$$K_3 = 0 \tag{5.7-15a}$$

$$\alpha = \frac{(2n+1)\pi}{2}, \quad n \text{ is an integer} \tag{5.7-15b}$$

From Equations (5.7-9, 11, 14a to d, 15a, b) and using the principle of superposition, we find that f can be written as

$$f = -\frac{\bar{z}^2}{2} + \bar{z} + \sum_{n=0}^{\infty} \left[ A_n e^{-a_n t_1} + B_n e^{-b_n t_1} \right] \sin \alpha_n \bar{z} \tag{5.7-16}$$

where $A_n$ and $B_n$ are constants (Fourier coefficients), $\alpha_n$ is given by Equation (5.7-15b), and $a_n$ and $b_n$ are given by Equations (5.7-14c, d) with $\alpha$, $a$, and $b$ being replaced by $\alpha_n$, $a_n$, and $b_n$.

Substituting Equation (5.7-16) into Equation (5.7-1b) and solving the resulting equation subject to Equation (5.7-3d), we obtain

$$\tau = 1 - \bar{z} + \sum_{n=0} \left[ \frac{a_n A_n e^{-a_n t} + b_n B_n e^{-b_n t}}{\alpha_n} \right] \cos \alpha_n \bar{z} \tag{5.7-17}$$

Applying the initial conditions given by Equations (5.7-3a, b) on f and $\tau$ yields

$$\frac{\bar{z}^2}{2} - \bar{z} = \sum_{n=0} \left[ A_n + B_n \right] \sin \alpha_n \bar{z} \tag{5.7-18a}$$

$$\bar{z} - 1 = \sum_{n=0} \left[ \frac{a_n A_n + b_n B_n}{\alpha_n} \right] \cos \alpha_n \bar{z} \tag{5.7-18b}$$

The Fourier coefficients $A_n$ and $B_n$ are determined in the usual manner, that is to say, by multiplying both sides of Equations (5.7-18a, b) by $\sin \alpha_n \bar{z}$ and $\cos \alpha_n \bar{z}$ and integrating with respect to $\bar{z}$ between 0 and 1. The coefficients $A_n$ and $B_n$ are found to be

$$A_n = \frac{2\left(b_n - \alpha_n^2\right)}{\alpha_n^3 (a_n - b_n)}, \qquad B_n = \frac{-2\left(a_n - \alpha_n^3\right)}{\alpha_n^3 (a_n - b_n)} \tag{5.7-19a,b}$$

We can now substitute $A_n$ and $B_n$ into Equations (5.7-16, 17) and f and $\tau$ are given in the form of an infinite series.

We next consider an example where the boundary condition is non-homogeneous.

***Example 5.7-2.*** The diffusion of oxygen into blood is important in surgery. Hershey et al. (1967) have modelled this process as a wetted-wall column. It is illustrated in Figure 5.7-1. A film of blood of thickness L flows down along the z-axis and is in direct contact with a rising stream of oxygen.

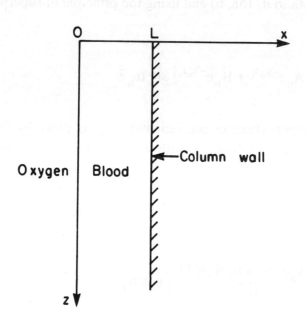

**FIGURE 5.7-1     Cross section of a wetted-wall column**

The diffusion equation can be written as [Equation (A.IV-1)]

$$\frac{Dc}{Dt} = \mathcal{D}_{0B} \nabla^2 c \tag{5.7-20a}$$

where $\mathcal{D}_{0B}$ is the diffusion coefficient which is assumed to be constant, c is the concentration of oxygen in the blood, and $\frac{D}{Dt}$ is the **substantial (material) derivative**. It is given by

$$\frac{Dc}{Dt} = \frac{\partial c}{\partial t} + \underline{v} \cdot \text{grad}\, c \tag{5.7-20b}$$

where $\underline{v}$ is the velocity of the blood.

In this problem, the following assumptions are made:

(i)      the system is at steady state $\left(\frac{\partial}{\partial t} = 0\right)$,

(ii)    negligible diffusion in the z-direction,

(iii)   $v_z$ is constant,

(iv)    no change in concentration in the y-direction.

Equations (5.7-20a, b) can be simplified to yield

$$v_z \frac{\partial c}{\partial z} = \mathcal{D}_{0B} \frac{\partial^2 c}{\partial x^2}$$

(5.7-21)

The appropriate boundary and initial conditions are

$$c(0, z) = c_f, \qquad c(x, 0) = c_i, \qquad \left. \frac{\partial c}{\partial x} \right|_{x=L} = 0$$

(5.7-22a,b,c)

where $c_f$ is the concentration at the blood/oxygen interface, $c_i$ is the concentration at the inlet.

Note that Equations (5.7-22a, b) are non-homogeneous.  To reduce Equation (5.7-22a) to a homogeneous equation, we write c as

$$c(x, z) = f(x, z) + g(x)$$

(5.7-23)

Substituting Equation (5.7-23) into Equation (5.7-21), we write

$$v_z \frac{\partial f}{\partial z} = \mathcal{D}_{0B} \frac{\partial^2 f}{\partial x^2}$$

(5.7-24a)

$$\frac{d^2 g}{dx^2} = 0$$

(5.7-24b)

Again, since we have introduced an extra function g, we are allowed to impose an extra condition. This condition is given by Equation (5.7-24b) and g also satisfies

$$g(0) = c_f, \qquad \left. \frac{dg}{dx} \right|_{x=L} = 0$$

(5.7-25a,b)

The solution of Equation (5.7-24b) subject to Equations (5.7-25a, b) is

$$g = c_f$$

(5.7-25c)

Combining Equations (5.7-22a to c, 23, 25a, b) yields the boundary conditions on f and they are given by

$$f(0, z) = 0, \qquad f(x, 0) = c_i - c_f, \qquad \left.\frac{df}{dx}\right|_{x=L} = 0 \qquad (5.7\text{-}26a,b,c)$$

We can now use the method of separation of variables to solve Equation (5.7-24a) subject to Equations (5.7-26a to c). We write f as

$$f = X(x) Z(z) \qquad (5.7\text{-}27)$$

Substituting Equation (5.7-27) into Equation (5.7-24a) and separating the variables, we obtain

$$\frac{Z'}{Z} = \frac{\mathcal{D}_{0B} X''}{v_z X} = -n^2 \qquad (5.7\text{-}28a,b)$$

where $n^2$ is the separation constant.

Equations (5.7-28a, b) can be written as

$$Z' + n^2 Z = 0 \qquad (5.7\text{-}29a)$$

$$X'' + \frac{n^2 v_z}{\mathcal{D}_{0B}} X = 0 \qquad (5.7\text{-}29b)$$

The solutions of Equations (5.7-29a, b) are respectively

$$Z = A\, e^{-n^2 z} \qquad (5.7\text{-}30a)$$

$$X = B \cos n \sqrt{\frac{v_z}{\mathcal{D}_{0B}}}\, x + C \sin n \sqrt{\frac{v_z}{\mathcal{D}_{0B}}}\, x \qquad (5.7\text{-}30b)$$

where A, B and C are constants.

Imposing conditions (5.7-26a, c) yields

$$B = 0, \qquad n = \frac{(2s + 1)\pi}{2L} \sqrt{\frac{\mathcal{D}_{0B}}{v_z}} \qquad (5.7\text{-}31a,b)$$

where s (= 0, 1, 2, ...) is an integer.

Combining Equations (5.7-23, 25c, 27, 30a, b, 31a, b) and applying the principle of superposition, we obtain

$$c = \sum_{s=0}^{\infty} A_s \exp\left[\frac{-(2s+1)^2\pi^2 \mathcal{D}_{0B}\, z}{4L^2 v_z}\right] \sin\frac{(2s+1)\pi x}{2L} + c_f \tag{5.7-32}$$

The boundary condition (5.7-22b) implies

$$c_i - c_f = \sum_{s=0}^{\infty} A_s \sin\frac{(2s+1)\pi x}{2L} \tag{5.7-33}$$

The Fourier coefficients $A_s$ are given by

$$A_s = \frac{2(c_i - c_f)}{L} \int_0^L \sin\frac{(2s+1)\pi x}{2L}\, dx \tag{5.7-34a}$$

$$= \frac{4(c_i - c_f)}{(2s+1)\pi} \tag{5.7-34b}$$

Substituting Equation (5.7-34b) into Equation (5.7-32), we obtain

$$c = \frac{4(c_i - c_f)}{\pi} \sum_{s=0}^{\infty} \frac{1}{(2s+1)} \exp\left[\frac{-(2s+1)^2\pi^2 \mathcal{D}_{0B}\, z}{4L^2 v_z}\right] \sin\frac{(2s+1)\pi x}{2L} + c_f \tag{5.7-35}$$

The average concentration across the film at a point $z$ is defined as

$$\langle c \rangle = \frac{\int_0^L c(x,z)\, dx}{L} \tag{5.7-36a}$$

$$= \frac{8(c_i - c_f)}{\pi^2} \sum_{s=0}^{\infty} \frac{1}{(2s+1)^2} \exp\left[\frac{-(2s+1)^2\pi^2 \mathcal{D}_{0B}\, z}{4L^2 v_z}\right] + c_f \tag{5.7-36b}$$

Equation (5.3-36b) is obtained by integrating Equation (5.3-35) and is identical to Equation (6) of Hershey et al. (1967), except for a difference in notation.

●

In the two examples so far considered, the non-homogeneous terms are constants. The method can be extended to the case where the non-homogeneous terms are functions of one of the independent variables. We illustrate this method in the following example.

***Example 5.7-3.*** Wilson (1979) solved the problem of film boiling on a sphere in forced convection. We simplify the problem by considering the steady state case. The equation of energy [Equation (A.III-3)] can be simplified to yield

$$\frac{\partial^2 T}{\partial r^2} + \frac{2}{r}\frac{\partial T}{\partial r} + \frac{1}{r^2 \sin\theta}\frac{\partial}{\partial\theta}\left(\sin\theta\,\frac{\partial T}{\partial\theta}\right) = 0 \tag{5.7-37}$$

where $T(r, \theta)$ is the temperature.

The boundary conditions are: on the surface of the sphere $(r = 1)$

$$\frac{\partial T}{\partial r} = f(\theta) = \frac{3}{4}\sin^2\theta\left(1 - \frac{3}{2}\cos\theta + \frac{1}{2}\cos^3\theta\right)^{-1/2} \tag{5.7-38a,b}$$

$T$ is finite at the origin.

Note that there is a typographical error in the expression for $f(\theta)$ in Wilson (1979).

We use the method of separation of variables and write $T$ as

$$T = R(r)\,\Theta(\theta) \tag{5.7-39}$$

Substituting $T$ and its derivatives into Equation (5.7-37), we obtain

$$r^2\frac{d^2 R}{dr^2} + 2r\frac{dR}{dr} - n(n+1)R = 0 \tag{5.7-40a}$$

$$\frac{d^2\Theta}{d\theta^2} + \cot\theta\,\frac{d\Theta}{d\theta} + n(n+1)\Theta = 0 \tag{5.7-40b}$$

The solution of Equations (5.7-40a, b) are given in Section 5 [Equations (5.5-40)]. The general solution which is finite at the origin is

$$T = A_0 + \sum_{n=1}^{\infty} A_n r^n P_n(\cos\theta) \tag{5.7-41}$$

where $P_n(\cos\theta)$ are the Legendre polynomials.

We note that Equation (5.7-40b) generates Legendre polynomials and it is appropriate, as discussed in Section 6, to expand $f(\theta)$ as a series expansion in Legendre polynomials. We express $f(\theta)$ as

$$f(\theta) = \sum_{n=1}^{\infty} B_n P_n(\cos \theta) \tag{5.7-42}$$

The coefficients $B_n$ are determined using the orthogonal properties of $P_n(\cos \theta)$. We recall that

$$\int_{-1}^{1} P_m(\cos \theta) P_n(\cos \theta) \, d(\cos \theta) = \begin{cases} 0 & \text{if } m \neq n \\ \dfrac{2}{2n+1} & \text{if } m = n \end{cases} \tag{5.7-43a,b}$$

Multiplying both sides of Equation (5.7-42) by $P_m(\cos \theta)$ and using Equations (5.7-43a, b), we deduce

$$B_n = \frac{(2n+1)}{2} \int_{-1}^{1} f(\theta) P_n(\cos \theta) \, d(\cos \theta) \tag{5.7-44a}$$

$$= \frac{(2n+1)}{2} \int_{\pi}^{0} f(\theta) P_n(\cos \theta) (-\sin \theta) \, d\theta \tag{5.7-44b}$$

$$= \frac{(2n+1)}{2} \int_{0}^{\pi} f(\theta) P_n(\cos \theta) (\sin \theta) \, d\theta \tag{5.7-44c}$$

The function $f(\theta)$ is given, $P_n(\cos \theta)$ are known in terms of powers of $\cos \theta$ and the integral can be evaluated. We calculate $B_0$ in detail. We know that $P_0$ is one, $f(\theta)$ is given by Equation (5.7-38b), and $B_0$ can be written as

$$B_0 = \frac{1}{2} \int_{0}^{\pi} \frac{3 (\sin^2 \theta) \sin \theta \, d\theta}{4 \sqrt{1 - \frac{3}{2} \cos \theta + \frac{1}{2} \cos^3 \theta}} \tag{5.7-45a}$$

$$= \frac{3}{8} \int_{-1}^{1} \frac{(1 - x^2) \, dx}{\sqrt{1 - \frac{3}{2} x + \frac{1}{2} x^3}} \tag{5.7-45b}$$

$$= \frac{3}{8} \int_{0}^{2} \frac{2 \, dy}{3 \sqrt{y}} \tag{5.7-45c}$$

$$= \frac{\sqrt{2}}{2} \tag{5.7-45d}$$

Equations (5.7-45b, c) are obtained by making the following substitutions respectively

$$x = \cos \theta \qquad\qquad (5.7\text{-}46a)$$

$$y = 1 - \frac{3}{2} x + \frac{1}{2} x^3 \qquad\qquad (5.7\text{-}46b)$$

Similarly $B_1$, $B_2$, ... can be calculated. The boundary condition given by Equation (5.7-38a) can now be written as

$$\frac{\partial T}{\partial r}\bigg|_{r=1} = \sum_{n=0}^{\infty} B_n P_n (\cos \theta) \qquad\qquad (5.7\text{-}47)$$

Differentiating $T$ from Equation (5.7-41) and using Equation (5.7-47), we obtain

$$\sum_{n=0}^{\infty} n A_n P_n (\cos \theta) = \sum_{n=0}^{\infty} B_n P_n (\cos \theta) \qquad\qquad (5.7\text{-}48)$$

We deduce that

$$A_n = \frac{B_n}{n} \qquad\qquad (5.7\text{-}49)$$

for all $n \geq 1$.

The coefficients $B_n$ are known and so the $A_n$ ($n \geq 1$) can be determined. The constant $A_0$ is arbitrary. This is not surprising because, for a Neumann problem, the solution is arbitrary to the extent of a constant.

The method of **integral transforms**, introduced next, can be used to solve partial differential equations, including non-homogeneous problems.

## 5.8    LAPLACE TRANSFORMS

In Chapter 1, we have used **Laplace transforms** to simplify an ordinary differential equation to an algebraic equation. It is equally possible to use Laplace transforms to reduce a partial differential equation in two independent variables to an ordinary differential equation. We recall that the Laplace transform of a function $f(t)$ was defined by Equation (1.17-1).

Likewise, if $u(x, t)$ is a function of two independent variables $x$ and $t$, we define the Laplace transform of $u$ with respect to $t$ by

$$L[u(x, t)] = U(s, x) = \int_0^{\infty} e^{-st} u(x,t) \, dt \qquad\qquad (5.8\text{-}1)$$

Other properties of Laplace transforms are listed in Chapter 1. We illustrate the method of Laplace transforms by solving a few partial differential equations.

***Example 5.8-1.*** Wimmers et al. (1984) studied the diffusion of a gas from a bubble into a liquid in which a chemical reaction occurs. If the bubble is at rest and has a constant radius $R$, the concentration of the absorbed gas in the liquid phase is found, by simplifying Equation (A.IV-3), to be

$$\frac{\partial c_A}{\partial t} = \mathcal{D}_{AB} \left[ \frac{1}{r^2} \frac{\partial}{\partial r} \left( r^2 \frac{\partial c_A}{\partial r} \right) \right] + R_A \tag{5.8-2}$$

where $c_A$ is the concentration of gas $A$ in the liquid, $\mathcal{D}_{AB}$ is the diffusion coefficient of solute $A$ in solvent $B$, $R_A$ is the production rate of $A$, $t$ is the time, and $r$ is the radial distance.

We assume that the rate equation is first order and $R_A$ is given by

$$R_A = -k\, c_A \tag{5.8-3}$$

where $k$ is the rate constant.

Combining Equations (5.8-2, 3) yields

$$\frac{\partial c_A}{\partial t} = \mathcal{D}_{AB} \left[ \frac{1}{r^2} \frac{\partial}{\partial r} \left( r^2 \frac{\partial c_A}{\partial r} \right) \right] - k\, c_A \tag{5.8-4}$$

The boundary and initial conditions are

$$c_A(R, t) = c_0, \qquad c_A(r, 0) = 0 \ (r > R), \qquad c_A \longrightarrow 0 \text{ as } r \longrightarrow \infty \text{ for all } t \tag{5.8-5a,b,c}$$

We denote the Laplace transform of $c_A$ by $C$ $\left( = \int_0^\infty e^{-st} c_A(r, t)\, dt \right)$. Taking the Laplace transform of Equation (5.8-4) and using the initial conditions [Equation (5.8-5b)], we obtain

$$(s + k)\, C = \mathcal{D}_{AB} \left[ \frac{1}{r^2} \frac{d}{dr} \left( r^2 \frac{dC}{dr} \right) \right] = \mathcal{D}_{AB} \left[ \frac{d^2 C}{dr^2} + \frac{2}{r} \frac{dC}{dr} \right] \tag{5.8-6a,b}$$

We make the substitution

$$C = \frac{\overline{C}}{r} \tag{5.8-7}$$

and Equation (5.8-6b) reduces to the following equation with constant coefficients

$$\frac{d^2\overline{C}}{dr^2} - \alpha^2\,\overline{C} = 0 \tag{5.8-8a}$$

$$\alpha^2 = (s + k)\big/\mathcal{D}_{AB} \tag{5.8-8b}$$

The solution of Equation (5.8-8a) satisfying Equation (5.8-5c) is

$$\overline{C} = A\,e^{-\alpha r} \tag{5.8-9}$$

From Equation (5.8-7), we obtain

$$C = \frac{A}{r}\,e^{-\alpha r} \tag{5.8-10}$$

The constant $A$ is obtained by using Equation (5.8-5a). The Laplace transform of $c_A\,(R, t)$ is $c_0/s$. Substituting this into Equation (5.8-10) yields

$$\frac{c_0}{s} = \frac{A}{R}\,e^{-\alpha R} \tag{5.8-11a}$$

$$A = \frac{R\,c_0}{s}\,e^{\alpha R} \tag{5.8-11b}$$

The solution can now be written as

$$C = \frac{R\,c_0}{r\,s}\,e^{-\alpha(r-R)} \tag{5.8-12a}$$

$$= \frac{R\,c_0}{r\,s}\,\exp\left[-\left(\frac{s+k}{\mathcal{D}_{AB}}\right)^{1/2}(r-R)\right] \tag{5.8-12b}$$

$$= \frac{R\,c_0}{r}\left(\frac{1}{\sigma-k}\right)\exp\left[-\sqrt{\sigma}\,\frac{(r-R)}{\sqrt{\mathcal{D}_{AB}}}\right] \tag{5.8-12c}$$

Equation (5.8-12c) is in a standard form and is obtained by substituting

$$\sigma = s + k \tag{5.8-13}$$

From the table of Laplace transforms given in Table 1.17-1, we find that $c_A$ is given by

$$c_A = \frac{R\,c_0}{2r}\,e^{kt}\left\{e^{-x\sqrt{k}}\,\operatorname{erfc}\left[\frac{x}{2\sqrt{t}} - \sqrt{k\,t}\right] + e^{x\sqrt{k}}\,\operatorname{erfc}\left[\frac{x}{2\sqrt{t}} + \sqrt{k\,t}\right]\right\} \tag{5.8-14a}$$

$$x = (r - R)\big/\sqrt{\mathcal{D}_{AB}} \tag{5.8-14b}$$

where **erfc** is the complement of the **error function** (erf). These functions are defined by

$$\operatorname{erf} x = \frac{2}{\sqrt{\pi}} \int_0^x e^{-u^2} \, du \qquad (5.8\text{-}15a)$$

$$\operatorname{erfc} x = 1 - \operatorname{erf} x = \frac{2}{\sqrt{\pi}} \int_x^\infty e^{-u^2} \, du \qquad (5.8\text{-}15b)$$

(To show that $\dfrac{2}{\sqrt{\pi}} \displaystyle\int_0^\infty e^{-u^2} \, du$ is one, see Problem 9b in Chapter 4.)

If no chemical reaction occurs, k is zero and Equation (5.8-12b) reduces to

$$C = \frac{R\, c_0}{r\, s} \exp\left[ -\sqrt{\frac{s}{\mathcal{D}_{AB}}} \,(r - R) \right] \qquad (5.8\text{-}16)$$

From the table of Laplace transforms, we obtain

$$c_A = \frac{c_0 R}{r} \operatorname{erfc}\left( \frac{r - R}{2\sqrt{\mathcal{D}_{AB} t}} \right) \qquad (5.8\text{-}17)$$

●

In using the method of Laplace transforms, we need to invert the transform $F(s)$ so as to obtain the solution $f(t)$. The method we have used is to read the inversion from the table of transforms. There is a general inversion formula which can be written as

$$f(t) = L^{-1}\left[F(s)\right] = \frac{1}{2\pi i} \int_{\gamma - i\infty}^{\gamma + i\infty} e^{st} \, F(s) \, ds \qquad (5.8\text{-}18)$$

where s $(= x + iy)$ is a complex variable and the path of integration is along a straight line L in the complex s-plane, as shown in Figure 5.8-1. The constant $\gamma$ is chosen so that all singularities of $F(s)$ are to the left of L. The integration is usually done by taking L to be a finite line, closing the contour, then letting the contour tend to infinity, and evaluating the integral using the Cauchy residue theorem [Equation (3.7-10)]. One method of obtaining a closed contour C is by joining the line L to an arc of a circle centered at the origin and of radius R. This is illustrated in Figure 5.8-2.

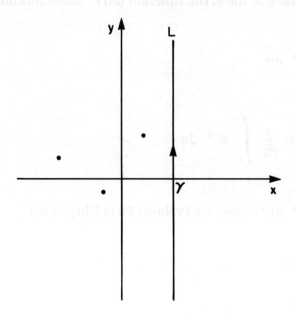

**FIGURE 5.8-1    Path of integration (L) in the inversion of Laplace transforms. The dots (•) are singularities of F(s)**

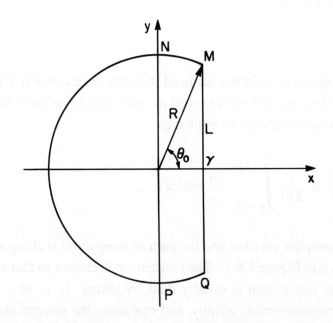

**FIGURE 5.8-2    Closed contour in the inversion of Laplace transforms**

From Equation (3.7-10), we have

$$\frac{1}{2\pi i} \int_C e^{st} F(s) \, ds = \sum_n \text{Res}\left[e^{st} F(s), s_n\right] \tag{5.8-19}$$

where $s_n$ are the singularities of $e^{st} F(s)$.

The integral around $C$ can be decomposed as

$$\int_C e^{st} F(s) \, ds = \int_L e^{st} F(s) \, ds + \int_{MN} e^{st} F(s) \, ds + \int_{NP} e^{st} F(s) \, ds + \int_{PQ} e^{st} F(s) \, ds$$

$$\tag{5.8-20}$$

On the semicircle $NP$, we have

$$s = R e^{i\theta}, \qquad \frac{\pi}{2} < \theta < \frac{3\pi}{2} \tag{5.8-21a,b}$$

$$\int_{NP} \left|e^{st} F(s) \, ds\right| = \int_{\pi/2}^{3\pi/2} \left|\exp\left[R t \left(\cos\theta + i \sin\theta\right)\right] F\left(R e^{i\theta}\right) i R e^{i\theta} \, d\theta\right| \tag{5.8-22a}$$

$$= \int_{\pi/2}^{3\pi/2} \left|\exp\left(R t \cos\theta\right) \exp\left(i R t \sin\theta\right) F\left(R e^{i\theta}\right) i R e^{i\theta} \, d\theta\right| \tag{5.8-22b}$$

$$= \int_{\pi/2}^{3\pi/2} e^{R t \cos\theta} \left|F\left(R e^{i\theta}\right)\right| R \, d\theta \tag{5.8-22c}$$

since $\left|e^{i a}\right| = 1$ for all real $a$.

We assume that, on the arc of the circle, $F$ satisfies the inequality

$$\left|F(s)\right| < \frac{\alpha}{R^m} \tag{5.8-23}$$

where $\alpha$ and $m$ are positive constants. That is to say, when $R \longrightarrow \infty$, $F(s) \longrightarrow 0$.

From Equations (5.8-22a, b, c, 23), we deduce that, because $\cos\theta$ is negative, the integral tends to zero as $R$ tends to infinity.

On the arc $MN$, we have the following inequalities

$$\left| \int_{MN} e^{st} F(s) \, ds \right| < \int_{MN} \left| e^{st} F(s) \, ds \right| \tag{5.8-24a}$$

$$< \int_{\theta_0}^{\pi/2} e^{Rt\cos\theta} \frac{\alpha}{R^m} R \, d\theta \tag{5.8-24b}$$

$$< \frac{\alpha}{R^{m-1}} \int_{\theta_0}^{\pi/2} e^{Rt\cos\theta} \, d\theta \tag{5.8-24c}$$

$$< \frac{\alpha}{R^{m-1}} \int_{\theta_0}^{\pi/2} e^{Rt\gamma/R} \, d\theta \tag{5.8-24d}$$

$$< \frac{\alpha \, e^{\gamma t}}{R^{m-1}} \left( \frac{\pi}{2} - \theta_0 \right) \tag{5.8-24e}$$

where

$$\theta_0 = \cos^{-1}(\gamma/R) \tag{5.8-24f}$$

If we let $R$ tend to infinity, $(\gamma/R)$ will tend to zero. Expanding $\cos^{-1}(\gamma/R)$ about zero in a Taylor series yields

$$\cos^{-1}(\gamma/R) = \frac{\pi}{2} - \frac{\gamma}{R} + \dots \tag{5.8-25}$$

From inequality (5.8-24e) and Equation (5.8-25), we deduce

$$\left| \int_{MN} e^{st} F(s) \, ds \right| < \frac{\alpha \, e^{\gamma t}}{R^m} \gamma \tag{5.8-26}$$

The integral $\int_{MN} e^{st} F(s) \, ds$ tends to zero as $R$ tends to infinity. Similarly, the $\int_{PQ} e^{st} F(s) \, ds$ tends to zero as $R$ tends to infinity.

If $F(s)$ satisfies inequality (5.8-23), Equation (5.8-19) becomes

$$\frac{1}{2\pi i} \int_L e^{st} F(s) \, ds = \sum_n \text{Res}\left[ e^{st} F(s), s_n \right] \tag{5.8-27}$$

Combining Equations (5.8-18, 27) yields

$$f(t) = \sum_n \text{Res}\left[e^{st}\, F(s),\, s_n\right] \tag{5.8-28}$$

We have assumed that $e^{st}\, F(s)$ has no branch points inside the region enclosed by C. If $e^{st}\, F(s)$ has branch points, the appropriate cuts should be made as discussed in Chapter 3. We illustrate this point by the following example.

***Example 5.8-2.*** We consider the heat conduction problem of a semi-infinite region $x > 0$. The governing equation, obtained by simplifying Equation (A.III-1), is

$$\frac{\partial T}{\partial t} = k^* \frac{\partial^2 T}{\partial x^2} \tag{5.8-29}$$

where $k^*$ $(= k/\rho \hat{C}_p)$ is the thermal diffusivity and T is the temperature.

The initial and boundary conditions are

$$T(x, 0) = 0, \quad T(0, t) = c_0 \sin(\omega t + \varepsilon), \quad T \longrightarrow 0 \text{ as } x \longrightarrow \infty \tag{5.8-30a,b,c}$$

where $c_0$, $\omega$, and $\varepsilon$ are constants.

Taking the Laplace transform of Equation (5.8-29) and using Equation (5.8-30a), we obtain

$$s\overline{T} = k^* \frac{d^2 \overline{T}}{dx^2} \tag{5.8-31}$$

where $\overline{T}$ is the Laplace transform of T.

The solution of Equation (5.8-31) is

$$\overline{T} = A \exp\left[\left(\sqrt{s/k^*}\,\right) x\right] + B \exp\left[-\left(\sqrt{s/k^*}\,\right) x\right] \tag{5.8-32}$$

where A and B are constants.

The condition that $T \longrightarrow 0$ as $x \longrightarrow \infty$ implies that $\overline{T} \longrightarrow 0$ as $x \longrightarrow \infty$. The constant A has to be zero. The condition at the origin $(x = 0)$ implies that $\overline{T}$ at the origin is given by

$$\overline{T} = \int_0^\infty e^{-st} c_0 \sin(\omega t + \varepsilon)\, dt \tag{5.8-33a}$$

$$= \frac{c_0}{s^2 + \omega^2} \left(s \sin\varepsilon + \omega \cos \varepsilon\right) \tag{5.8-33b}$$

Applying the boundary conditions on $\bar{T}$, we obtain

$$\bar{T} = \frac{c_0}{s^2 + \omega^2} \left(s \sin\varepsilon + \omega \cos \varepsilon\right) \exp\left[-x \sqrt{s/k^*}\right] \tag{5.8-34}$$

The inversion formula [Equation (5.8-18)] gives

$$T(x, t) = \frac{1}{2\pi i} \int_{\gamma - i\infty}^{\gamma + i\infty} e^{st} \frac{c_0}{s^2 + \omega^2} \left(s \sin\varepsilon + \omega \cos \varepsilon\right) \exp\left[-x \sqrt{s/k^*}\right] ds \tag{5.8-35}$$

To evaluate the integral, we transform it to a closed contour integral as explained earlier. We note that the integrand has a branch point at the origin ($s = 0$) and simple poles at

$$s_1 = i\,\omega, \qquad s_2 = -i\,\omega \tag{5.8-36a,b}$$

The contour shown in Figure 5.8-2 is no longer appropriate. We have to remove the origin and the negative part of the real axis. The resulting contour is illustrated in Figure 5.8-3. The integrand is a single-valued function of $s$ on and inside contour C. Contour C consists of the line L, the arc MN, the line $L_1$, the small circle $C_1$, the line $L_2$, and finally the arc PQ. On letting the radius R tend to infinity, the integrals along the arcs MN and PQ are zero, as shown previously. Similarly, as $\varepsilon$ tends to zero, the integral around $C_1$ tends to zero. On $L_1$, the argument of $s$ is $\pi$ and on $L_2$, it is $-\pi$. On $L_1$ and $L_2$, we write

$$s = \rho\, e^{i\pi}, \qquad s = \rho\, e^{-i\pi} \tag{5.8-37a,b}$$

where $\rho$ is real and positive.

The square roots of $s$ on $L_1$ and $L_2$ are given respectively by

$$\sqrt{s} = i \sqrt{\rho}, \qquad \sqrt{s} = -i \sqrt{\rho} \tag{5.8-38a,b}$$

We now evaluate the integrand of Equation (5.8-35) along $L_1$ and $L_2$. We denote the integrals by $I_1$ and $I_2$. From Equations (5.8-35, 37a, b, 38a, b), we obtain

$$I_1 = \int_\infty^0 e^{-\rho t} \frac{c_0}{\rho^2 + \omega^2} \left(-\rho \sin\varepsilon + \omega \cos \varepsilon\right) \exp\left[-i\left(\sqrt{\rho/k^*}\right)x\right] (-d\rho) \tag{5.8-39a}$$

$$= \int_0^\infty e^{-\rho t} \frac{c_0}{\rho^2 + \omega^2} (-\rho \sin\varepsilon + \omega \cos \varepsilon) \exp\left[-i\left(\sqrt{\rho/k^*}\right)x\right] d\rho \qquad (5.8\text{-}39b)$$

$$I_2 = \int_0^\infty e^{-\rho t} \frac{c_0}{\rho^2 + \omega^2} (-\rho \sin\varepsilon + \omega \cos \varepsilon) \exp\left[i\left(\sqrt{\rho/k^*}\right)x\right] (-d\rho) \qquad (5.8\text{-}39c)$$

Combining Equations (5.8-39b, c) yields

$$I = I_1 + I_2 = \int_0^\infty e^{-\rho t} \frac{c_0}{\rho^2 + \omega^2} (-\rho \sin\varepsilon + \omega \cos \varepsilon) \left[e^{-i\sqrt{\rho/k^*}\,x} - e^{i\sqrt{\rho/k^*}\,x}\right] d\rho \qquad (5.8\text{-}40a,b)$$

$$= \int_0^\infty e^{-\rho t} \frac{c_0}{\rho^2 + \omega^2} (-\rho \sin\varepsilon + \omega \cos \varepsilon) (-2i) \sin\left[\left(\sqrt{\rho/k^*}\right)x\right] d\rho \qquad (5.8\text{-}40c)$$

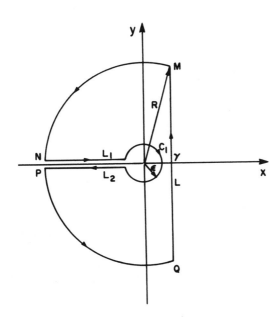

**FIGURE 5.8-3　Contour integral with a branch point at the origin**

From Equation (3.7-14), we find that the residue $R_1$ at $s_1$ is given by

$$R_1 = \lim_{s \to i\omega} \frac{(s - i\omega) e^{st} c_0}{(s - i\omega)(s + i\omega)} (s \sin\varepsilon + \omega \cos \varepsilon) \exp\left[-\left(\sqrt{s/k^*}\right)x\right] \qquad (5.8\text{-}41a)$$

$$= \frac{c_0 \, e^{i\omega t}}{2i\omega} \, (i\omega \sin\varepsilon + \omega \cos\varepsilon) \exp\left[-\left(\sqrt{i\omega/k^*}\right) x\right] \qquad (5.8\text{-}41\text{b})$$

$$= \frac{c_0 \, e^{i\omega t}}{2i} \, (\cos\varepsilon + i \sin\varepsilon) \exp\left[-\frac{(1+i)}{\sqrt{2}} \left(\sqrt{\omega/k^*}\right) x\right] \qquad (5.8\text{-}41\text{c})$$

$$= \frac{c_0 \, e^{i\omega t}}{2i} \, e^{i\varepsilon} \, e^{-\left(\sqrt{\omega/2k^*}\right)x} \, \exp\left[-i\left(\sqrt{\omega/2k^*}\right) x\right] \qquad (5.8\text{-}41\text{d})$$

Similarly, the residue $R_2$ at $s_2$ is found to be

$$R_2 = -\frac{c_0 \, e^{-i\omega t}}{2i} \, e^{-i\varepsilon} \, e^{-\left(\sqrt{\omega/2k^*}\right)x} \, \exp\left[i\left(\sqrt{\omega/2k^*}\right) x\right] \qquad (5.8\text{-}41\text{e})$$

The sum of residues $R$ is given by

$$R = R_1 + R_2 = \frac{c_0 \left[\exp i \left(\omega t + \varepsilon - x\sqrt{\omega/2k^*}\right) - \exp -i \left(\omega t + \varepsilon - x\sqrt{\omega/2k^*}\right)\right]}{2i} \, e^{-x\sqrt{\omega/2k^*}}$$

$$(5.8\text{-}42\text{a,b})$$

$$= c_0 \sin\left(\omega t + \varepsilon - x\sqrt{\omega/2k^*}\right) e^{-x\sqrt{\omega/2k^*}} \qquad (5.8\text{-}42\text{c})$$

Using Cauchy's residue theorem [Equation (3.7-10)], we can write

$$\int_C = \int_L + \int_{MN} + \int_{L_1} + \int_{C_1} + \int_{L_2} + \int_{PQ} \qquad (5.8\text{-}43\text{a})$$

$$= 2\pi i c_0 \sin\left(\omega t + \varepsilon - x\sqrt{\omega/2k^*}\right) e^{-x\sqrt{\omega/2k^*}} \qquad (5.8\text{-}43\text{b})$$

We have shown that $\int_{MN}$, $\int_{C_1}$, and $\int_{PQ}$ are equal to zero. The function $T$ given in Equation (5.8-35) is $\frac{1}{2\pi i} \int_L$. From Equations (5.8-40c, 43b), we deduce that $T$ is given by

$$T = c_0 \sin\left(\omega t + \varepsilon - x\sqrt{\omega/2k^*}\right) e^{-x\sqrt{\omega/2k^*}}$$

$$+ \frac{c_0}{\pi} \int_0^\infty \frac{e^{-\rho t}}{\rho^2 + \omega^2} \, (-\rho \sin\varepsilon + \omega \cos\varepsilon) \sin\left(x\sqrt{\rho/k^*}\right) d\rho \qquad (5.8\text{-}44)$$

We note that as t tends to infinity, the integrand on the right side of Equation (5.8-44) tends to zero. For sufficiently large t, the solution is given approximately by the first term on the right side of Equation (5.8-44).

***Example 5.8-3***. Model the smoke dispersion from a high chimney and solve it by the method of Laplace transforms.

Smoke diffuses from a chimney of height h into the atmosphere. Suppose the prevailing wind is in one direction only (x-direction) and has a constant velocity V. The vertical axis is the z-axis and the base of the chimney is at the origin, as shown in Figure 5.8-4. The average cross-sectional concentration c (x, z), obtained by simplifying Equation (A.IV-1), is

$$V \frac{\partial c}{\partial x} = D \left[ \left( \frac{\partial^2 c}{\partial x^2} \right) + \left( \frac{\partial^2 c}{\partial z^2} \right) \right] \qquad (5.8\text{-}45)$$

where $v_x = V$, $R_A = 0$, and the diffusion coefficient $\mathscr{D}_{AB} = D$ (a constant).

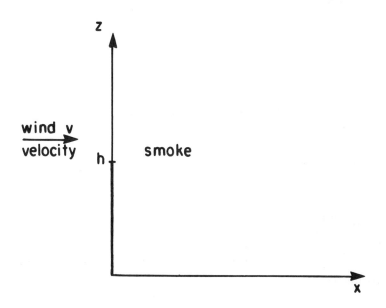

**FIGURE 5.8-4    Smoke dispersion from a chimney**

We further assume that the derivative with respect to x is small compared to the derivative with respect to z. Then Equation (5.8-45) simplifies to

$$V \frac{\partial c}{\partial x} = D \frac{\partial^2 c}{\partial z^2} \qquad (5.8\text{-}46)$$

The inversion height (where there is no transport of smoke) is at infinity and the rate of deposit of smoke at ground level is negligible. Under these assumptions, the boundary conditions are

$$\text{at } z = 0, \quad \frac{\partial c}{\partial z} = 0 \; ; \qquad \qquad \text{at } z \longrightarrow \infty, \quad \frac{\partial c}{\partial z} = 0 \qquad (5.8\text{-}47a,b)$$

$$\text{at } x = 0, \quad c = (Q/V) \, \delta (z - h) \; ; \qquad \text{at } x \longrightarrow \infty, \quad c = 0 \qquad (5.8\text{-}48a,b)$$

where $Q$ is the source strength (flowrate) emanating from the chimney and $\delta$ is the Dirac delta function.

We introduce the dimensionless quantities $\bar{x}$, $\bar{z}$, and $\bar{c}$ by writing

$$\bar{x} = \frac{Dx}{Vh^2}, \qquad \bar{z} = \frac{z}{h}, \qquad \bar{c} = \frac{Vc}{Q} \qquad (5.8\text{-}49a,b,c)$$

Equations (5.8-46 to 48b) become

$$\frac{\partial \bar{c}}{\partial \bar{x}} = \frac{\partial^2 \bar{c}}{\partial \bar{z}^2} \qquad (5.8\text{-}50)$$

$$\left. \frac{\partial \bar{c}}{\partial \bar{z}} \right|_{\bar{z}=0} = 0, \qquad \left. \frac{\partial \bar{c}}{\partial \bar{z}} \right|_{\bar{z} \longrightarrow \infty} = 0, \qquad \bar{c} \, \big|_{\bar{x}=0} = \delta (\bar{z} - 1) \qquad (5.8\text{-}51a,b,c)$$

The Laplace transform $C$ is defined by

$$C = \int_0^\infty e^{-s\bar{x}} \, \bar{c} \, (\bar{x}, \bar{z}) \, d\bar{x} \qquad (5.8\text{-}52)$$

Equation (5.8-50) transforms to

$$\frac{d^2 \bar{C}}{d\bar{z}^2} - s C = -\delta (\bar{z} - 1) \qquad (5.8\text{-}53)$$

This is a second order equation which we solve for $\bar{z} < 1$ and for $\bar{z} > 1$ separately because the delta function introduces a singularity at $\bar{z} = 1$. For both $\bar{z} < 1$ and $\bar{z} > 1$, $\delta (\bar{z} - 1) = 0$. The solutions are given in terms of exponential functions, keeping in mind that a second order equation is associated with two arbitrary constants.

The solutions satisfying Equations (5.8-51a to c) are

$$C = A\,e^{-\bar{z}\sqrt{s}}, \qquad\qquad \bar{z} > 1 \qquad\qquad (5.8\text{-}54a)$$

$$= B\,(e^{\bar{z}\sqrt{s}} + e^{-\bar{z}\sqrt{s}}), \qquad \bar{z} < 1 \qquad\qquad (5.8\text{-}54b)$$

where A and B are constants.

We further assume that C is continuous at $\bar{z} = 1$ and this implies that

$$A = B\,(e^{2\sqrt{s}} + 1) \qquad\qquad (5.8\text{-}55)$$

The derivative of C at $\bar{z} = 1$ is not continuous and to evaluate it, we integrate Equation (5.8-53) and obtain

$$\left[\frac{dC}{d\bar{z}}\right]_{1-\varepsilon}^{1+\varepsilon} - s \int_{1-\varepsilon}^{1+\varepsilon} C\,d\bar{z} = -\int_{1-\varepsilon}^{1+\varepsilon} \delta\,(\bar{z}-1)\,d\bar{z} = -1 \qquad\qquad (5.8\text{-}56)$$

Combining Equations (5.8-54a, b, 56) and letting $\varepsilon \longrightarrow 0$, we deduce that

$$\sqrt{s}\,[A\,e^{-\sqrt{s}} + B\,(e^{\sqrt{s}} - e^{-\sqrt{s}})] = 1 \qquad\qquad (5.8\text{-}57)$$

Note that the second integral on the left side of Equation (5.8-56) is zero because C is continuous at $\bar{z} = 1$.

The values of A and B are found to be

$$A = \frac{1}{2\sqrt{s}}\,(e^{\sqrt{s}} + e^{-\sqrt{s}}), \qquad B = \frac{e^{-\sqrt{s}}}{2\sqrt{s}} \qquad\qquad (5.8\text{-}58a,b)$$

To determine the concentration at ground level $(\bar{z} = 0)$, we determine $C\,(0, s)$ and this is given by

$$C\,(0, s) = e^{-\sqrt{s}}/\sqrt{s} \qquad\qquad (5.8\text{-}59)$$

Using Equation (5.8-18), we invert C and obtain

$$\bar{c}\,(\bar{x}, 0) = \frac{1}{2\pi i} \int_{\gamma-i\infty}^{\gamma+i\infty} \frac{e^{s\bar{x}}\,e^{-\sqrt{s}}}{\sqrt{s}}\,ds \qquad\qquad (5.8\text{-}60)$$

We note that the integrand has a branch point at $s = 0$ and we proceed to evaluate the integral as in Example 5.8-2. The only non-zero contribution to the integral is along $L_1$ and $L_2$ (Figure 5.8-3). On $L_1$ and $L_2$, the square roots of s can respectively be written as

$$\sqrt{s} = i\rho, \qquad \sqrt{s} = -i\rho \tag{5.8-61a,b}$$

Substituting Equations (5.8-61a, b) into Equation (5.8-60) yields

$$\bar{c}(\bar{x}, 0) = \frac{1}{\pi} \int_{-\infty}^{0} \exp\left[-(\rho^2\bar{x} + i\rho)\right] d\rho - \frac{1}{\pi} \int_{0}^{-\infty} \exp\left[-(\rho^2\bar{x} - i\rho)\right] d\rho \tag{5.8-62a}$$

$$= \frac{1}{\pi} \int_{-\infty}^{\infty} \exp\left[-(\rho^2\bar{x} + i\rho)\right] d\rho \tag{5.8-62b}$$

$$= \frac{1}{\pi} \int_{-\infty}^{\infty} \exp\left[-(\rho\sqrt{\bar{x}} + i/2\sqrt{\bar{x}})^2 - 1/4\bar{x}\right] d\rho \tag{5.8-62c}$$

$$= \frac{\exp(-1/4\bar{x})}{\pi} \int_{-\infty}^{\infty} \exp\left[-(\rho\sqrt{\bar{x}} + i/2\sqrt{\bar{x}})^2\right] d\rho \tag{5.8-62d}$$

$$= \frac{\exp(-1/4\bar{x})}{\sqrt{\pi\bar{x}}} \tag{5.8-62e}$$

The maximum concentration $\bar{c}_{max}$ is given by

$$\frac{d\bar{c}}{d\bar{x}} = \frac{\exp(-1/4\bar{x})}{2\sqrt{\pi\bar{x}^3}}\left[-1 + \frac{1}{2\bar{x}}\right] = 0 \tag{5.8-63a,b}$$

The solution is

$$\bar{x} = 1/2, \qquad \bar{c}_{max} = \sqrt{\frac{2}{\pi e}} \tag{5.8-64a,b}$$

In dimensional variables, the maximum concentration is $(Q/V)(2/\pi e)^{1/2}$ at a distance $Vh^2/2D$ from the chimney. From Equations (5.8-49a, b), it is seen that the assumption $\partial/\partial z \gg \partial/\partial x$ requires that $Vh/D \gg 1$.

## 5.9    FOURIER TRANSFORMS

From Section 2.8, we deduce that if $f(x)$ is an **odd periodic function of period 2L**, its **Fourier series** can be written as

$$f(x) = \sum_{n=1}^{\infty} b_n \sin\frac{n\pi x}{L} \tag{5.9-1a}$$

$$b_n = \frac{2}{L} \int_0^L f(x) \sin \frac{n\pi x}{L} \, dx \qquad\qquad (5.9\text{-}1b)$$

We can regard $b_n$ as the **finite Fourier sine transform** of $f(x)$. In keeping with the notation used in Laplace transforms, we write

$$F_{sn} = \frac{2}{L} \int_0^L f(x) \sin \frac{n\pi x}{L} \, dx \qquad\qquad (5.9\text{-}2)$$

From Equation (5.9-1a), we deduce that the inverse is

$$f(x) = \sum_{n=1}^{\infty} F_{sn} \sin \frac{n\pi x}{L} \qquad\qquad (5.9\text{-}3)$$

Similarly if $f(x)$ is an **even periodic function of period 2L**, we can define its **finite Fourier cosine transform** as

$$F_{cn} = \frac{2}{L} \int_0^L f(x) \cos \frac{n\pi x}{L} \, dx \qquad\qquad (5.9\text{-}4)$$

Its inverse is given by

$$f(x) = \frac{1}{2} F_{c0} + \sum_{n=1}^{\infty} F_{cn} \cos \frac{n\pi x}{L} \qquad\qquad (5.9\text{-}5)$$

The Fourier sine and cosine transforms can be used to solve ordinary and partial differential equations. The procedure is similar to that employed when Laplace transforms are adopted. We solve an electrostatic problem using the method of Fourier transforms.

***Example 5.9-1.*** In the semi-infinite space bounded by the planes $x = 0$, $x = \pi$, and $y = 0$ as shown in Figure 5.9-1, there is a uniform distribution of charge of density $(\rho/4\pi)$. If $u$ is the potential, $u$ satisfies the equation

$$\frac{\partial^2 u}{\partial x^2} + \frac{\partial^2 u}{\partial y^2} = -\rho \qquad\qquad (5.9\text{-}6)$$

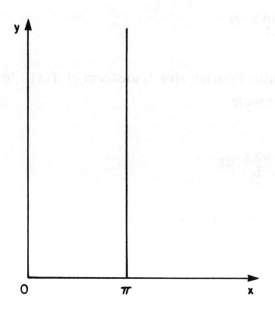

**FIGURE 5.9-1    Semi-infinite region**

The boundary conditions are

$$u\,(0, y) = 0\,, \qquad u\,(\pi, y) = 1\,, \qquad u\,(x, 0) = 0 \qquad\qquad (5.9\text{-}7a,b,c)$$

We assume $u$ to be bounded throughout the region.

The region is bounded in $x$ and we use the finite sine transform method to solve Equation (5.9-6).
We multiply both sides of Equation (5.9-6) by $\dfrac{2}{\pi} \sin nx$ and integrate from $0$ to $\pi$.

$$\frac{2}{\pi}\int_0^\pi \frac{\partial^2 u}{\partial x^2}\sin nx\, dx + \frac{2}{\pi}\int_0^\pi \frac{\partial^2 u}{\partial y^2}\sin nx\, dx = -\frac{2\rho}{\pi}\int_0^\pi \sin nx\, dx \qquad (5.9\text{-}8)$$

The first integral can be integrated by parts and we obtain

$$\frac{2}{\pi}\int_0^\pi \frac{\partial^2 u}{\partial x^2}\sin nx\, dx = \frac{2}{\pi}\left\{\left[\frac{\partial u}{\partial x}\sin nx\right]_0^\pi - n\int_0^\pi \frac{\partial u}{\partial x}\cos nx\, dx\right\} \qquad (5.9\text{-}9a)$$

$$= \frac{2}{\pi}\left\{-n\,[u\cos nx]_0^\pi - n^2\int_0^\pi u\sin nx\, dx\right\} \qquad (5.9\text{-}9b)$$

$$= -\frac{2n}{\pi}(-1)^n - n^2 U_{sn} \tag{5.9-9c}$$

where $U_{sn}$ is the finite sine transform of u.

Substituting Equation (5.9-9c) into Equation (5.9-8) and carrying out the remaining integration yields

$$(-1)^{n+1}\frac{2n}{\pi} - n^2 U_{sn} + \frac{d^2 U_{sn}}{dy^2} = 2\rho\frac{[(-1)^n - 1]}{n\pi} \tag{5.9-10}$$

The solution of Equation (5.9-10) is

$$U_{sn} = A e^{ny} + B e^{-ny} - 2\rho\frac{[(-1)^n - 1]}{\pi n^3} - \frac{2(-1)^n}{n\pi} \tag{5.9-11}$$

The condition that u is bounded implies that A is zero. From Equation (5.9-7c), we deduce

$$y = 0, \qquad U_{sn} = 0 \tag{5.9-12a,b}$$

From Equations (5.9-11, 12a, b), we obtain

$$B = 2\rho\frac{[(-1)^n - 1]}{\pi n^3} + \frac{2(-1)^n}{n\pi} \tag{5.9-13}$$

Equation (5.9-11) simplifies to

$$U_{sn} = \frac{2}{\pi}\left\{\frac{\rho[(-1)^n - 1]}{n^3} + \frac{(-1)^n}{n}\right\}\left\{e^{-ny} - 1\right\} \tag{5.9-14}$$

Using Equation (5.9-3), we express u as

$$u(x,y) = \sum_{n=1}^{\infty} U_{sn}\sin nx \tag{5.9-15a}$$

$$= \frac{2}{\pi}\sum_{n=1}^{\infty}\left\{\frac{\rho[(-1)^n - 1]}{n^3} + \frac{(-1)^n}{n}\right\}\left\{e^{-ny} - 1\right\}\sin nx \tag{5.9-15b}$$

●

So far we have assumed that the function is defined in a finite interval. In many physical problems, the region can be considered to be infinite. The **Fourier series** is then extended to a **Fourier integral**. The function $f(x)$ can be represented in complex form as

$$f(x) = \frac{1}{2\pi} \int_{-\infty}^{\infty} F(\alpha) e^{i\alpha x} d\alpha \qquad\qquad (5.9\text{-}16a)$$

$$F(\alpha) = \int_{-\infty}^{\infty} f(x) e^{-i\alpha x} dx \qquad\qquad (5.9\text{-}16b)$$

We define $F(\alpha)$ as the **Fourier transform** of $f(x)$ and its inverse is given by Equation (5.9-16a). An alternative form of writing Equation (5.9-16a) is

$$f(x) = \frac{1}{\pi} \int_{0}^{\infty} [A(\alpha) \cos \alpha x + B(\alpha) \sin \alpha x] d\alpha \qquad\qquad (5.9\text{-}17a)$$

$$A(\alpha) = \int_{-\infty}^{\infty} f(x) \cos \alpha x \, dx \qquad\qquad (5.9\text{-}17b)$$

$$B(\alpha) = \int_{-\infty}^{\infty} f(x) \sin \alpha x \, dx \qquad\qquad (5.9\text{-}17c)$$

If $f(x)$ is odd, $A(\alpha)$ is zero and Equations (5.9-17a, c) become

$$f(x) = \frac{2}{\pi} \int_{0}^{\infty} B(\alpha) \sin \alpha x \, d\alpha \qquad\qquad (5.9\text{-}18a)$$

$$B(\alpha) = \int_{0}^{\infty} f(x) \sin \alpha x \, dx \qquad\qquad (5.9\text{-}18b)$$

We define $B(\alpha)$ as the **Fourier sine transform** of $f(x)$ and Equations (5.9-18a, b) can be written as

$$f(x) = \frac{2}{\pi} \int_{0}^{\infty} F_s(\alpha) \sin \alpha x \, d\alpha \qquad\qquad (5.9\text{-}19a)$$

$$F_s(\alpha) = \int_{0}^{\infty} f(x) \sin \alpha x \, dx \qquad\qquad (5.9\text{-}19b)$$

The inverse of the Fourier sine transform $F_s(\alpha)$ is given by Equation (5.9-19a). Similarly the **Fourier cosine transform** and its inverse are defined by

$$F_c(\alpha) = \int_0^\infty f(x) \cos \alpha x \, dx \tag{5.9-20a}$$

$$f(x) = \frac{2}{\pi} \int_0^\infty F_c(\alpha) \cos \alpha x \, d\alpha \tag{5.9-20b}$$

The definitions given by some authors differ from those given here by having different factors in front of the integral. For the sake of symmetry, some authors define the Fourier cosine transform and its inverse as

$$F_c(\alpha) = \sqrt{\frac{2}{\pi}} \int_0^\infty f(x) \cos \alpha x \, dx \tag{5.9-21a}$$

$$f(x) = \sqrt{\frac{2}{\pi}} \int_0^\infty F_c(\alpha) \cos \alpha x \, d\alpha \tag{5.9-21b}$$

Note that both sets of Equations (5.9-20a, b, 21a, b) lead to

$$f(x) = \frac{2}{\pi} \int_0^\infty \left\{ \int_0^\infty f(x) \cos \alpha x \, dx \right\} \cos \alpha x \, d\alpha \tag{5.9-22}$$

Similarly the Fourier transform [Equation (5.9-16b)] and the Fourier sine transform [Equation (5.9-19b)] can be defined with the factor $\sqrt{\frac{2}{\pi}}$ in front of the integral.

***Example 5.9-2.*** Sedahmed (1986) studied the rate of mass transfer at a cathode. The mass diffusion equation is determined by simplifying Equation (A.IV-1) with $R_A = 0$ to yield

$$\frac{\partial c}{\partial t} = \mathcal{D} \frac{\partial^2 c}{\partial y^2} \tag{5.9-23}$$

where $c$ is the concentration and $\mathcal{D}$ is the diffusivity.

The initial and boundary conditions are

$$c = c_i \qquad \text{for } y = 0 \text{ at } t > 0 \tag{5.9-24a}$$

$$c \longrightarrow c_b \qquad \text{as} \quad y \longrightarrow \infty \quad \text{at} \ t > 0 \tag{5.9-24b}$$

$$c = c_b \qquad \text{for} \ y > 0 \ \text{at} \ t = 0 \tag{5.9-24c}$$

For simplicity, we write

$$c^* = (c_b - c) / (c_b - c_i) \tag{5.9-25}$$

Equations (5.9-23, 24a to c) become

$$\frac{\partial c^*}{\partial t} = \mathcal{D} \frac{\partial^2 c^*}{\partial y^2} \tag{5.2-26a}$$

$$c^*(0, t) = 1 , \quad c^* \longrightarrow 0 \ \text{as} \ y \longrightarrow \infty , \quad c^*(y, 0) = 0 \ \text{for} \ y > 0 \tag{5.2-26b,c,d}$$

We take the Fourier sine transform of Equation (5.9-26a); that is to say, we multiply both sides of the equation by $\sin \alpha y$ and integrate with respect to y from 0 to infinity.  The right side is now

$$\mathcal{D} \int_0^\infty \frac{\partial^2 c^*}{\partial y^2} \sin \alpha y \ dy = \mathcal{D} \left\{ \left[ \frac{\partial c^*}{\partial y} \sin \alpha y \right]_0^\infty - \int_0^\infty \alpha \frac{\partial c^*}{\partial y} \cos \alpha y \ dy \right\} \tag{5.9-27a}$$

$$= \mathcal{D} \left\{ -\alpha \left[ c^* \cos \alpha y \right]_0^\infty - \alpha^2 \int_0^\infty c^* \sin \alpha y \ dy \right\} \tag{5.9-27b}$$

$$= \mathcal{D} \left[ \alpha - \alpha^2 C_s \right] \tag{5.9-27c}$$

where $C_s$ is the sine transform of $c^*$.

Using Equation (5.9-27c), Equation (5.9-26a) becomes

$$\frac{dC_s}{dt} + \alpha^2 \mathcal{D} C_s = \alpha \mathcal{D} \tag{5-9-28}$$

The solution is

$$C_s = \frac{1}{\alpha} + A \exp \left[ -(\alpha^2 t \, \mathcal{D}) \right] \tag{5.9-29}$$

where A is an arbitrary constant.

From Equation (5.9-26d), we deduce that

$$\text{at} \ t = 0, \ C_s = 0 \tag{5.9-30}$$

Applying Equation (5.9-30), we find that $C_s$ is given by

$$C_s = \frac{1}{\alpha} (1 - e^{-\alpha^2 t \mathscr{D}}) \tag{5.9-31}$$

To obtain $c^*$, we can either look up in a table of sine transforms or use the inversion formula [Equation (5.9-19a)]. Using the inversion formula, we obtain

$$c^* (y, t) = \frac{2}{\pi} \int_0^\infty \frac{(1 - e^{-\alpha^2 t \mathscr{D}})}{\alpha} \sin \alpha y \, d\alpha \tag{5.9-32}$$

From the table of Fourier sine transforms (see Table 5.9-1), we find that the inverse of $C_s$ is

$$c^* = \mathrm{erfc} \, (y/2\sqrt{\mathscr{D} t}) \tag{5.9-33}$$

which is the solution quoted by Sedahmed (1986).

●

In Examples 5.9-1 and 2, we have used the sine transform because the values of u and c are given at the origin. We choose the cosine transform if the first derivative is given at the origin. If the region of interest is the whole infinite space ($-\infty$ to $\infty$), the Fourier transform [Equations (5.9-16a, b)] is used. This is shown in the next example.

## TABLE 5.9-1

### Fourier sine transform

| $f(x)$ | $F_s(\alpha)$ |
| --- | --- |
| $\begin{cases} 1, & 0 < x < a \\ 0, & a < x < \infty \end{cases}$ | $\frac{1}{\alpha} [1 - \cos(a\alpha)]$ |
| $e^{-x}$ | $\alpha/(1 + \alpha^2)$ |
| $x e^{-x}$ | $2\alpha/(1 + \alpha^2)^2$ |
| $\mathrm{erfc} \, ax$ | $\frac{1}{\alpha} [1 - \exp(-\alpha^2/4a^2)]$ |
| $x^{-a}, \quad 0 < a < 1$ | $\frac{1}{\alpha^{1-a}} \cos\left(\frac{\pi a}{2}\right) \Gamma(1 - a)$ |
| | $\Gamma$ is the gamma function |

***Example 5.9-3***.  Solve the wave equation

$$\frac{\partial^2 u}{\partial t^2} - c^2 \frac{\partial^2 u}{\partial x^2} = 0, \quad -\infty < x < \infty \qquad (5.9\text{-}34)$$

subject to the initial conditions

$$u(x, 0) = f(x) \qquad (5.9\text{-}35a)$$

$$\left. \frac{\partial u}{\partial t} \right|_{t=0} = 0 \qquad (5.9\text{-}35b)$$

Taking the Fourier transform of Equation (5.9-34) and integrating by parts, as before, we obtain

$$\frac{d^2 U}{dt^2} + c^2 \alpha^2 U = 0 \qquad (5.9\text{-}36)$$

where $U \left( = \int_{-\infty}^{\infty} u(x, t) e^{-i\alpha x} dx \right)$ is the Fourier transform of $u$.

The solution is

$$U = A e^{i c \alpha t} + B e^{-i c \alpha t} \qquad (5.9\text{-}37)$$

where $A$ and $B$ are constants.

Differentiating $U$ with respect to $t$ yields

$$\frac{dU}{dt} = i \alpha c (A e^{i c \alpha t} - B e^{-i c \alpha t}) \qquad (5.9\text{-}38)$$

The initial conditions imply

$$F(\alpha) = A + B \qquad (5.9\text{-}39a)$$

$$0 = A - B \qquad (5.9\text{-}39b)$$

where

$$F(\alpha) = \int_{-\infty}^{\infty} f(x) e^{-i\alpha x} dx \qquad (5.9\text{-}39c)$$

Solving Equations (5.9-39a, b), we obtain

$$A = B = \frac{1}{2} F(\alpha) \tag{5.9-40a,b}$$

Equation (5.9-37) becomes

$$U = \frac{1}{2} F(\alpha) [e^{ic\alpha t} + e^{-ic\alpha t}] \tag{5.9-41}$$

Using the inversion formula [Equation (5.9-16a)], we obtain

$$u = \frac{1}{2\pi} \cdot \frac{1}{2} \int_{-\infty}^{\infty} F(\alpha) [e^{ic\alpha t} + e^{-ic\alpha t}] e^{i\alpha x} d\alpha \tag{5.9-42a}$$

$$= \frac{1}{4\pi} \int_{-\infty}^{\infty} F(\alpha) [e^{i\alpha(x+ct)} + e^{i\alpha(x-ct)}] d\alpha \tag{5.9-42b}$$

Equation (5.9-16a) allows one to write Equation (5.9-42b) as

$$u = \frac{1}{2} [f(x + ct) + f(x - ct)] \tag{5.9-43}$$

Equation (5.9-43) is d'Alembert's solution.

## 5.10  HANKEL AND MELLIN TRANSFORMS

We have seen that the solution of Laplace's equation in cylindrical polar coordinates involves Bessel functions. In problems formulated in cylindrical polar coordinates and if the radial distance $r$ is from 0 to infinity, the derivatives with respect to $r$ can be removed by the application of the Hankel transform. We define the **Hankel transform** of $f(r)$ by

$$F_n(\alpha) = \int_0^{\infty} f(r) \, r \, J_n(\alpha r) \, dr \tag{5.10-1}$$

where $J_n(\alpha r)$ is the Bessel function of the first kind of order $r$.

The **inverse** is given by

$$f(r) = \int_0^{\infty} \alpha \, F_n(\alpha) \, J_n(\alpha r) \, d\alpha \tag{5.10-2}$$

The **Mellin transform** is defined by

$$F\,(m)\;=\;\int_0^{\infty} x^{m-1}\,f\,(x)\,dx \tag{5.10-3}$$

Its **inverse** is given by

$$f\,(x)\;=\;\frac{1}{2\pi i}\int_{\gamma-i\infty}^{\gamma+i\infty} x^{-m}\,F\,(m)\,dm \tag{5.10-4}$$

The exponent $m$ in Equation (5.10-3) is chosen such that the integral exists. The contour integral in Equation (5.10-4) is similar to the contour used in the inversion formula for the Laplace transform [Equation (5.8-18)]. As an example of the use of the Hankel transform, we consider the problem of a charged disk.

*Example 5.10-1.* The potential $u$ due to a flat circular disk of unit radius satisfies Laplace's equation and the following boundary conditions

$$\frac{\partial^2 u}{\partial r^2}+\frac{1}{r}\frac{\partial u}{\partial r}+\frac{\partial^2 u}{\partial z^2}=0 \tag{5.10-5a}$$

$$\text{on } z=0,\quad u=u_0\,,\;\;0\le r\le 1\,;\quad \frac{\partial u}{\partial z}=0\,,\;r>1 \tag{5.10-5b,c}$$

$$u\longrightarrow 0\quad\text{as}\quad z\longrightarrow\infty \tag{5.10-5d}$$

Equation (5.10-5a) suggests that we take the Bessel function of order zero. Multiplying every term of Equation (5.10-5a) by $r\,J_0(\alpha r)$ and integrating with respect to $r$ from zero to infinity, we obtain

$$\int_0^{\infty} r\left(\frac{\partial^2 u}{\partial r^2}+\frac{1}{r}\frac{\partial u}{\partial r}\right)J_0(\alpha r)\,dr+\frac{d^2}{dz^2}\int_0^{\infty} r\,u\,(r,\,z)\,J_0(\alpha r)\,dr=0 \tag{5.10-6}$$

The first integral can be written as

$$\int_0^{\infty} r\left(\frac{\partial^2 u}{\partial r^2}+\frac{1}{r}\frac{\partial u}{\partial r}\right)J_0(\alpha r)\,dr=\int_0^{\infty}\frac{\partial}{\partial r}\left(r\frac{\partial u}{\partial r}\right)J_0(\alpha r)\,dr \tag{5.10-7a}$$

$$= \left[ r \frac{\partial u}{\partial r} J_0(\alpha r) \right]_0^\infty - \int_0^\infty \alpha r \frac{\partial u}{\partial r} J_0'(\alpha r) \, dr \qquad (5.10\text{-}7b)$$

$$= -\left[ \alpha r u J_0'(\alpha r) \right]_0^\infty + \int_0^\infty \alpha u \, [J_0'(\alpha r) + r\alpha \, J_0''(\alpha r)] \, dr$$
$$(5.10\text{-}7c)$$

$$= -\int_0^\infty \alpha^2 r u \, J_0(\alpha r) \, dr \qquad (5.10\text{-}7d)$$

$$= -\alpha^2 U \qquad (5.10\text{-}7e)$$

The terms inside the square brackets are zero because of the boundary conditions. Equation (5.10-7d) is obtained by recognizing that $J_0(\alpha r)$ satisfies the following equation

$$\alpha r \, J_0'' + J_0' + \alpha r \, J_0 = 0 \qquad (5.10\text{-}8)$$

Equation (5.10-6) becomes

$$\frac{d^2 U}{dz^2} - \alpha^2 U = 0 \qquad (5.10\text{-}9)$$

The solution that satisfies Equation (5.10-5d) is

$$U = A \, e^{-\alpha z} \qquad (5.10\text{-}10)$$

where A is a constant.

To obtain u, we use Equation (5.10-2) which leads to

$$u = \int_0^\infty \alpha \, A(\alpha) \, e^{-\alpha z} J_0(\alpha r) \, d\alpha \qquad (5.10\text{-}11)$$

Differentiating with respect to z yields

$$\frac{\partial u}{\partial z} = -\int_0^\infty \alpha^2 \, A(\alpha) \, e^{-\alpha z} J_0(\alpha r) \, d\alpha \qquad (5.10\text{-}12)$$

Imposing the boundary condition on z = 0, we have

$$\int_0^\infty \alpha\, A(\alpha)\, J_0(\alpha r)\, d\alpha \;=\; u_0 \,, \quad 0 \le r < 1 \tag{5.10-13a}$$

$$\int_0^\infty \alpha^2\, A(\alpha)\, J_0(\alpha r)\, d\alpha \;=\; 0 \,, \quad r > 1 \tag{5.10-13b}$$

We make the substitution

$$\alpha^2 A(\alpha) \;=\; \frac{2}{\pi}\, u_0\, B(\alpha) \tag{5.10-14}$$

Equations (5.10-13a, b) become

$$\int_0^\infty \alpha^{-1}\, B(\alpha)\, J_0(\alpha r)\, d\alpha \;=\; \frac{\pi}{2} \,, \quad 0 \le r < 1 \tag{5.10-15a}$$

$$\int_0^\infty B(\alpha)\, J_0(\alpha r)\, d\alpha \;=\; 0 \,, \quad r > 1 \tag{5.10-15b}$$

From the integral properties of Bessel functions [see Watson (1966), p. 405], we note that

$$\int_0^\infty \alpha^{-1}\, J_0(\alpha r)\, \sin\alpha\, d\alpha \;=\; \frac{\pi}{2} \,, \quad 0 \le r < 1 \tag{5.10-16a}$$

$$\int_0^\infty J_0(\alpha r)\, \sin\alpha\, d\alpha \;=\; 0 \,, \quad r > 1 \tag{5.10-16b}$$

Comparing Equations (5.10-15a, b, 16a, b), we deduce that

$$B(\alpha) \;=\; \sin\alpha \tag{5.10-17}$$

From Equations (5.10-11, 14, 17), we obtain

$$u \;=\; \frac{2 u_0}{\pi} \int_0^\infty \alpha^{-1}\, \sin\alpha\, e^{-\alpha z}\, J_0(\alpha r)\, d\alpha \tag{5.10-18}$$

Note that in this example, the boundary conditions are of the Robin type ($u$ is given on a section of the boundary and $\dfrac{\partial u}{\partial z}$ on the remaining part of the boundary) and we determine $A(\alpha)$ after the inversion.

If the boundary conditions are of the Dirichlet or Neumann type, we determine $A(\alpha)$ and then carry out the inversion.

***Example 5.10-2.*** The steady heat flow problem in a wedge, as shown in Figure 5.10-1, consists of solving Laplace's equation, which can be obtained from Equation (A.III-2)

$$r^2 \frac{\partial^2 T}{\partial r^2} + r \frac{\partial T}{\partial r} + \frac{\partial^2 T}{\partial \theta^2} = 0 \tag{5.10-19}$$

where $T$ is the temperature, subject to

$$\text{on } \theta = \pm\theta_0, \quad T = 1, \ 0 < r < a; \quad T = 0, \ r > a \tag{5.10-20a,b}$$

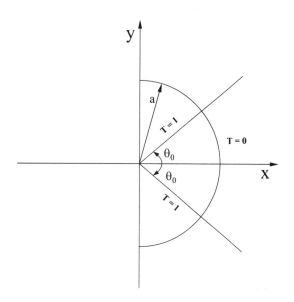

**FIGURE 5.10-1    Heat flow in a wedge**

Taking the Mellin transform of Equation (5.10-19), we obtain

$$\int_0^\infty r^{m-1} \left( r^2 \frac{\partial^2 T}{\partial r^2} + r \frac{\partial T}{\partial r} \right) dr + \frac{d^2}{d\theta^2} \int_0^\infty r^{m-1} T \, dr = 0 \tag{5.10-21}$$

We integrate the first integral by parts and Equation (5.10-21) becomes

$$\frac{d^2\overline{T}}{d\theta^2} + m^2\overline{T} = 0 \tag{5.10-22}$$

where $\overline{T}$ is the Mellin transform of $T$.

From Equations (5.10-20a, b) we deduce that $T$ and $\overline{T}$ are even functions of $\theta$.

The appropriate solution of Equation (5.10-22) is

$$\overline{T} = A \cos m\theta \tag{5.10-23}$$

On $\theta = \pm\theta_0$, $\overline{T}$ is given by

$$\overline{T} = \int_0^a r^{m-1}\, dr = \frac{1}{m}\Big[r^m\Big]_0^a = \frac{a^m}{m} \tag{5.10-24a,b,c}$$

Substituting Equations (5.10-24a to c) into Equation (5.10-23), we obtain

$$\frac{a^m}{m} = A \cos m\theta_0 \tag{5.10-25}$$

Substituting the value of $A$ into Equation (5.10-23) yields

$$\overline{T} = \frac{a^m \cos m\theta}{m \cos m\theta_0} \tag{5.10-26}$$

Using the inversion formula [Equation 5.10-4)], $T$ is expressed as

$$T = \frac{1}{2\pi i}\int_{\gamma-i\infty}^{\gamma+i\infty} \frac{r^{-m}\, a^m \cos m\theta}{m \cos m\theta_0}\, dm \tag{5.10-27}$$

The integrand has a simple pole at the origin. We choose the line integral to be along the imaginary axis of $m$ with an indenture at the origin, as shown in Figure 5.10-2. The indenture is in the form of a semi-circle of radius $\varepsilon$. Equation (5.10-27) can be written as

$$T = \frac{1}{2\pi i}\left[\int_{-i\infty}^{-i\varepsilon}\left(\frac{a}{r}\right)^m \frac{\cos m\theta}{m\cos m\theta_0}\, dm + \int_{-i\varepsilon}^{i\varepsilon}\left(\frac{a}{r}\right)^m \frac{\cos m\theta}{m\cos m\theta_0}\, dm + \int_{i\varepsilon}^{i\infty}\left(\frac{a}{r}\right)^m \frac{\cos m\theta}{m\cos m\theta_0}\, dm\right] \tag{5.10-28}$$

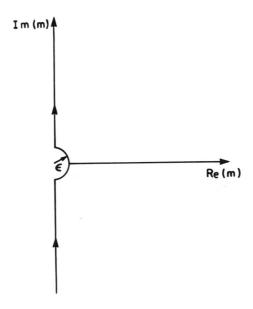

**FIGURE 5.10-2   Contour integral for the evaluation of the inverse
of Mellin transform**

To evaluate the middle integral, we write

$$m = \varepsilon\, e^{i\theta} \tag{5.10-29}$$

$$\lim_{\varepsilon \to 0} \int_{-i\varepsilon}^{i\varepsilon} \left(\frac{a}{r}\right)^m \frac{\cos m\theta}{m \cos m\theta_0}\, dm = \lim_{\varepsilon \to 0} \int_{-\pi/2}^{\pi/2} \frac{i\varepsilon e^{i\theta}}{\varepsilon e^{i\theta}}\, d\theta \tag{5.10-30a}$$

$$= i\,\pi \tag{5.10-30b}$$

The first and last integrals are evaluated by substituting

$$m = i\,y \tag{5.10-31}$$

$$\lim_{\varepsilon \to 0} \int_{-i\infty}^{-i\varepsilon} \left(\frac{a}{r}\right)^m \frac{\cos m\theta}{m \cos m\theta_0}\, dm = \int_{-\infty}^{0} \left(\frac{a}{r}\right)^{iy} \frac{\cos (iy\,\theta)}{iy \cos (iy\,\theta_0)}\, i\, dy \tag{5.10-32a}$$

$$= \int_{-\infty}^{0} \exp\left[iy \ln (a/r)\right] \frac{\cosh y\theta}{y \cosh y\theta_0}\, dy \tag{5.10-32b}$$

$$= -\int_0^\infty \exp[-i\,y\,\ln(a/r)]\,\frac{\cosh(y\,\theta)}{y\,\cosh(y\,\theta_0)}\,dy \qquad (5.10\text{-}32c)$$

Equation (5.10-32b) is obtained by using Equation (3.4-27) noting that cosh is an even function.

Similarly, we find

$$\lim_{\varepsilon \to 0}\int_{i\varepsilon}^{i\infty}\left(\frac{a}{r}\right)^m \frac{\cos m\theta}{m\,\cos m\theta_0}\,dm \;=\; \int_0^\infty \exp[i\,y\,\ln(a/r)]\,\frac{\cosh y\,\theta}{y\,\cosh y\,\theta_0}\,dy \qquad (5.10\text{-}33)$$

Substituting Equations (5.10-30b, 32c, 33) into Equation (5.10-28) yields

$$T = \frac{1}{2\pi i}\left\{ i\,\pi + \int_0^\infty \frac{\cosh y\,\theta}{y\,\cosh y\,\theta_0}\left[e^{i\,y\,\ln(a/r)} - e^{-i\,y\,\ln(a/r)}\right]dy \right\} \qquad (5.10\text{-}34a)$$

$$= \frac{1}{2\pi i}\left\{ i\,\pi + \int_0^\infty \frac{\cosh y\,\theta}{y\,\cosh y\,\theta_0}\,[2i\,\sin(y\,\ln(a/r))]\,dy \right\} \qquad (5.10\text{-}34b)$$

$$= \frac{1}{2} + \frac{1}{\pi}\int_0^\infty \frac{\cosh y\,\theta}{y\,\cosh y\,\theta_0}\,[\sin(y\,\ln(a/r))]\,dy \qquad (5.10\text{-}34c)$$

We have the solution in the form of an integral.

●

So far, we have considered partial differential equations in two independent variables. Applying the integral transform method, we have reduced a partial differential equation to an ordinary differential equation. Similarly, if we start with a partial differential equation involving three independent variables, we need to apply two transforms to reduce it to an ordinary differential equation. In general, an equation in $n$ independent variables needs $(n-1)$ transforms to reduce it to an ordinary differential equation. We need to invert $(n-1)$ times so as to obtain the solution. This process is illustrated in the next example.

*Example 5.10-3.* Stastna et al. (1991) considered the diffusion of a cylindrical drop through a membrane. A circular cylindrical drop $K_1$ of radius $a$ and height $h$ diffuses into a circular membrane $K_2$ of infinite radius and of thickness $\ell$ as shown in Figure 5.10-3. It is assumed that $K_1$ is filled with a penetrant of concentration $c_1$ which is a function of $z$ and $t$ only. In the membrane $K_2$, the concentration is $c_2$ and it is a function of $z$, $r$, and $t$. Symmetry is assumed and there is no dependence on $\theta$. The diffusivities are assumed to be constant and we denote them by

$\mathcal{D}_1$ and $\mathcal{D}_2$ in $K_1$ and $K_2$ respectively. The relevant equations are obtained via Equation (A.IV-2).

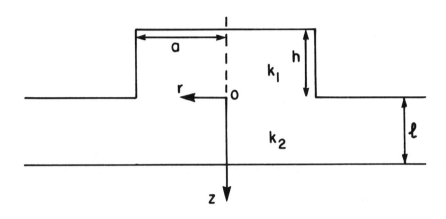

**FIGURE 5.10-3   Cylinder $K_1$ and membrane $K_2$**

In $K_1$, we have

$$\frac{\partial c_1}{\partial t} = \mathcal{D}_1 \frac{\partial^2 c_1}{\partial z^2} \qquad (5.10\text{-}35a)$$

$$c_1\big|_{t=0} = c_1^0 \qquad (5.10\text{-}35b)$$

$$\frac{\partial c_1}{\partial z}\bigg|_{z=-h} = 0 \, , \qquad \frac{\partial c_1}{\partial z}\bigg|_{z=0} = K\,f(t) \qquad (5.10\text{-}35c,d)$$

where $c_1^0$ is a constant, $K\ (= \mathcal{D}_2/\mathcal{D}_1)$ is the ratio of the diffusivities.

In $K_2$, we have

$$\frac{\partial c_2}{\partial t} = \mathcal{D}_2 \left[ \frac{1}{r}\frac{\partial}{\partial r}\left( r\frac{\partial c_2}{\partial r} \right) + \frac{\partial^2 c_2}{\partial z^2} \right] \qquad (5.10\text{-}36a)$$

$$c_2\big|_{t=0} = 0 \qquad (5.10\text{-}36b)$$

$$\frac{\partial c_2}{\partial z}\bigg|_{z=0} = \begin{cases} f(t), & 0 < r < a \\ 0, & r > a \end{cases} \qquad (5.10\text{-}36c,d)$$

$$c_2\big|_{z=\ell} = 0 \qquad (5.10\text{-}36e)$$

Taking the Laplace transform with respect to $t$ of Equations (5.10-35a to d), we obtain

$$s\,C_1 - c_1^0 = \mathcal{D}_1\,\frac{d^2C_1}{dz} \qquad (5.10\text{-}37a)$$

$$\frac{dC_1}{dz}\bigg|_{z=-h} = 0, \qquad \frac{dC_1}{\partial z}\bigg|_{z=0} = K\,F(s) \qquad (5.10\text{-}37b,c)$$

where $C_1 \left(= \int_0^\infty e^{-st}\,c_1\,(z,\,t)\,dt\right)$ is the Laplace transform of $c_1$. Likewise $F(s)$ is the Laplace transform of $f(t)$.

The solution of Equation (5.10-37a) is

$$C_1 = A\,\cosh\left(z\,\sqrt{\frac{s}{\mathcal{D}_1}}\,\right) + B\,\sinh\left(z\,\sqrt{\frac{s}{\mathcal{D}_1}}\,\right) + \frac{c_1^0}{s} \qquad (5.10\text{-}38)$$

where $A$ and $B$ are constants.

Imposing conditions (5.10-37b, c), we obtain

$$C_1 = K\,F\,\sqrt{\frac{\mathcal{D}_1}{s}}\;\frac{\cosh\,(z+h)\,\sqrt{s/\mathcal{D}_1}}{\sinh\,(h\,\sqrt{s/\mathcal{D}_1}\,)} + \frac{c_1^0}{s} \qquad (5.10\text{-}39)$$

From the table of the Laplace transforms, we find that $c_1$ is given by

$$c_1 = \frac{\mathcal{D}_2}{h}\int_0^t f(\tau)\left[1 + 2\sum_{n=1}^\infty (-1)^n\,\exp\left\{-\left(\frac{n\pi}{h}\right)^2 (t-\tau)\,\mathcal{D}_1\right\}\cos\frac{n\pi\,(z+h)}{h}\right]d\tau + c_1^0$$
$$(5.10\text{-}40)$$

To solve Equation (5.10-36a), we first take its Hankel transform. Proceeding as in Example 5.10-1, we obtain

$$\frac{\partial C_2}{\partial t} = \mathcal{D}_2\left[-\alpha^2 C_2 + \frac{\partial^2 C_2}{\partial z^2}\right] \qquad (5.10\text{-}41)$$

where $C_2 \left( = \int_0^\infty r \, c_2 \, J_0(\alpha r) \, dr \right)$ is the Hankel transform of $c_2$. Note that $C_2$ is a function of $z$ and $t$.

The Hankel transforms of Equations (5.10-36b to e) are given by

$$C_2 \Big|_{t=0} = 0, \qquad C_2 \Big|_{z=\ell} = 0 \tag{5.10-42a,b}$$

$$\frac{\partial C_2}{\partial z} \bigg|_{z=0} = \int_0^a r \, J_0(\alpha r) \, f(t) \, dr \tag{5.10-42c}$$

$$= f(t) \int_0^a r \, J_0(\alpha r) \, dr \tag{5.10-42d}$$

$$= \frac{f(t) \, a \, J_1(\alpha a)}{\alpha} \tag{5.10-42e}$$

Equation (5.10-42e) is obtained by using the properties of Bessel functions. To remove the time derivative in Equation (5.10-41), we take its Laplace transform. The resulting equation is

$$s \, \overline{C}_2 = \mathcal{D}_2 \left[ -\alpha^2 \overline{C}_2 + \frac{d^2 \overline{C}_2}{dz^2} \right] \tag{5.10-43}$$

where $\overline{C}_2 \left( = \int_0^\infty e^{-st} \, C_2 \, dt \right)$ is the Laplace transform of $C_2$.

Equations (5.10-42b, e) become, on taking their Laplace transforms,

$$\overline{C}_2 \Big|_{z=\ell} = 0, \qquad \frac{d\overline{C}_2}{dz} \bigg|_{z=0} = \frac{F(s) \, a \, J_1(\alpha a)}{\alpha} \tag{5.10-44a,b}$$

where $F(s)$ is the Laplace transform of $f(t)$.

The solution of Equation (5.10-43) is

$$\overline{C}_2 = A \cosh \beta z + B \sinh \beta z \tag{5.10-45}$$

$$\beta = \sqrt{\alpha^2 + s/\mathcal{D}_2} \tag{5.10-46}$$

where $A$ and $B$ are constants.

Using Equations (5.10-44a, b), we obtain

$$\overline{C}_2 = \frac{F(s)\, a\, J_1(\alpha a)}{\alpha \beta \cosh \beta \ell} \sinh \beta (z - \ell) \tag{5.10-47}$$

From the table of Laplace transforms, we find that

$$C_2 = \frac{2a\, J_1(\alpha a)\, \mathcal{D}_2}{\alpha \ell} \int_0^t f(\tau) \sum_{n=1}^{\infty} (-1)^{n-1} \exp\left\{-\left[\alpha^2 + \pi^2\left(\frac{2n-1}{2\ell}\right)^2\right](t-\tau)\, \mathcal{D}_2\right\}$$

$$\sin \frac{(2n-1)(z-\ell)\pi}{2\ell}\, d\tau \tag{5.10-48}$$

Using Equation (5.10-2), we obtain

$$c_2 = \int_0^{\infty} \alpha\, C_2\, J_0(\alpha r)\, d\alpha \tag{5.10-49a}$$

$$= \frac{2a\mathcal{D}_2}{\ell} \int_0^{\infty} J_1(\alpha a)\, J_0(\alpha r) \int_0^t f(\tau) \sum_{n=1}^{\infty} (-1)^{n-1}$$

$$\exp\left\{-\left[\alpha^2 + \pi^2\left(\frac{2n-1}{2\ell}\right)^2\right](t-\tau)\, \mathcal{D}_2\right\} \sin \frac{(2n-1)(z-\ell)\pi}{2\ell}\, d\tau\, d\alpha \tag{5.10-49b}$$

which is the solution given by Stastna et al. (1991).

If $f(t)$ is specified, then $c_1$ and $c_2$ can be determined from Equations (5.10-40, 49) respectively.

## 5.11  SUMMARY

The method of separation of variables is suitable for homogeneous equations with homogeneous boundary conditions. The solution is often expressed as an infinite series. If the equations are non-homogeneous, the method of separation of variables cannot be applied directly. We need to introduce an auxiliary function and the method can be complicated.

The method of integral transforms can be applied to linear differential equations (ordinary or partial) with arbitrary boundary and initial conditions. The problems solved by the method of separation of variables can be solved by the method of integral transforms. The various transforms $F(s)$ of a function $f(t)$ introduced in this chapter can be written as

$$F(s) = \int_a^b K(t, s)\, f(t)\, dt \tag{5.11-1}$$

where $K(t, s)$ is the **kernel**.

Table 5.11-1 lists the kernel corresponding to various transformations.

The inverse process, that is to say, finding $f(t)$ given $F(s)$, can be complicated. If $f(t)$ can be read from a table, the solution is obtained immediately. Otherwise, it might involve evaluating an integral which cannot be written in a closed form. Numerical methods can be used to evaluate the integral. Extensive tables of integral transforms are given by Erdelyi et al. (1954). Other kernels than those discussed in this chapter have been considered by Latta in the handbook edited by Pearson (1974).

The methods of separation of variables and of integral transforms are applicable to linear equations only. In the next chapter, we shall consider methods which can be used to solve non-linear equations.

## TABLE 5.11-1

### Kernels of various transforms

| Transforms | a | b | $K(t, s)$ |
|---|---|---|---|
| Laplace | 0 | $\infty$ | $e^{-st}$ |
| Fourier | $-\infty$ | $\infty$ | $e^{-ist}$ |
| Fourier sine | 0 | $\infty$ | $\sin(st)$ |
| Fourier cosine | 0 | $\infty$ | $\cos(st)$ |
| Fourier finite sine | 0 | L | $\sin \dfrac{s\pi t}{L}$ |
| Fourier finite cosine | 0 | L | $\cos \dfrac{s\pi t}{L}$ |
| Hankel | 0 | $\infty$ | $t\,J_n(st)$ |
| Mellin | 0 | $\infty$ | $t^{s-1}$ |

## PROBLEMS

1a.   Solve the following equations

(i)   $\dfrac{\partial u}{\partial x_1} + \dfrac{\partial u}{\partial x_2} = 0$

given that $u(x_1, 0) = x_1^2$                                   Answer: $(x_1 - x_2)^2$

(ii)   $x_1 \dfrac{\partial u}{\partial x_1} + x_2 \dfrac{\partial u}{\partial x_2} = u$

given that $u = 5$ on $x_2 = x_1^2$                              Answer: $5 x_1^2 / x_2$

(iii)   $x_2 \dfrac{\partial u}{\partial x_1} - 2 x_1 x_2 \dfrac{\partial u}{\partial x_2} = 2 x_1 u$

given that $u(0, x_2) = (\sinh x_2)/x_2$              Answer: $\dfrac{1}{x_2} \sinh(x_1^2 + x_2)$

2b.   The equations governing one-dimensional compressible flow are the equations of continuity and motion given in Appendix I and II. They can be written, in the absence of a pressure gradient, as

$$\frac{\partial \rho}{\partial t} + \frac{\partial}{\partial x}(\rho v_x) = 0$$

$$\frac{\partial v_x}{\partial t} + v_x \frac{\partial v_x}{\partial x} = 0$$

where $\rho$ is the density and $v_x$ is the velocity.

Obtain $\rho$ and $v_x$ if initially $(t = 0)$, $\rho$ and $v_x$ are given by

$$\rho = \rho_0 (1 + x), \qquad v_x = v_0 (1 + x)$$

where $\rho_0$ and $v_0$ are constants.                 Answer: $v_0 (1 + x) / (1 + t v_0)$

$$\rho_0 (1 + x) / (1 + t v_0)^2$$

3b.   We model the downhill movement of a glacier as the sliding of a block of height $h$ down a plane surface of constant slope $\alpha$. The direction of flow is taken to be the x-axis and the flow velocity $v_x$ is assumed to be a function of $x$ and $t$. As a first approximation, the shear stress $\tau$ between the glacier and its bed is given by

$$\tau = \rho g h \alpha$$

where $\rho$ is the density and $g$ is the gravitational acceleration.

It is further assumed that the relationship between $\tau$ and $v_x$ is

$$v_x = k\tau^n$$

where $k$ and $n$ are constants.

The equation of continuity [Equation(A.I-1)] can be written as

$$\frac{\partial h}{\partial t} + \frac{\partial}{\partial x}(hv_x) = 0$$

Assuming that $h$ is a function of $x$ and $t$, obtain $h$.

4a.   Show that the non-homogeneous equation

$$x_1 \frac{\partial u}{\partial x_1} + x_2 \frac{\partial u}{\partial x_2} + u = x_1^2 + x_2^2$$

reduces to

$$x_1 \frac{\partial v}{\partial x_1} + x_2 \frac{\partial v}{\partial x_2} + v = 0$$

if we assume

$$u = v + \frac{1}{3}(x_1^2 + x_2^2)$$

Obtain $u$ if on $x = 1$, $u$ is given by

$$u = e^{-x_2^2} \cos x_2 + \frac{1}{3}(1 + x_2^2)$$

$$\text{Answer: } \frac{1}{3}(x_1^2 + x_2^2) + \frac{\exp(-x_2^2/x_1^2)\cos(x_2/x_1)}{x_1}$$

5a.   Determine the regions where the following equations are hyperbolic, parabolic, and elliptic.

(i)   $$\frac{\partial^2 u}{\partial x_1^2} + 4\frac{\partial^2 u}{\partial x_1 \partial x_2} + 4\frac{\partial^2 u}{\partial x_2^2} = 0$$

(ii)  $$\frac{\partial^2 u}{\partial x_1^2} + x_1\frac{\partial^2 u}{\partial x_1 \partial x_2} + x_2\frac{\partial^2 u}{\partial x_2^2} = 0$$

(iii)    $x_1 \dfrac{\partial^2 u}{\partial x_1^2} + x_2 \dfrac{\partial^2 u}{\partial x_2^2} + x_1 x_2 \dfrac{\partial u}{\partial x_2} + u = 0$

6a.    Find the characteristic coordinates $\xi$, $\eta$ for

$$(\cos^2 x_2 - \sin^2 x_2) \dfrac{\partial^2 u}{\partial x_1^2} + 2 \cos x_2 \dfrac{\partial^2 u}{\partial x_1 \partial x_2} + \dfrac{\partial^2 u}{\partial x_2^2} + u = 0$$

Rewrite the equation in terms of $\xi$ and $\eta$.

Answer:  $\sin x_2 - \cos x_2 - x_1$

$\sin x_2 + \cos x_2 - x_1$

7a.    Use d'Alembert's method and the method of separation of variables to solve the wave equation [Equation (5.4-6b)] given that the initial conditions are

$$y(x, 0) = \sin x, \qquad \left.\dfrac{\partial y}{\partial t}\right|_{t=0} = 0$$

Answer:  $\sin x \cos c t$

8a.    Solve the equation

$$\dfrac{\partial u}{\partial x_1} + \dfrac{\partial u}{\partial x_2} = 0$$

by the method of separation of variables.

Compare the solution with the solution obtained in Problem 1a (i).

What condition(s) has (have) to be imposed to satisfy

$$u(x_1, 0) = x_1^2$$

9a.    Solve the following equations, subject to the given conditions

(i)    $\dfrac{\partial^2 u}{\partial t^2} = \dfrac{\partial^2 u}{\partial x^2}$,          $0 < x < \pi, \ t > 0$

$u(0, t) = u(\pi, t) = 0 ; \qquad \left.\dfrac{\partial u}{\partial t}\right|_{t=0} = 0 ; \qquad u(x, 0) = x(\pi - x)$

Answer:  $\dfrac{8}{\pi} \displaystyle\sum_{s=0}^{\infty} \dfrac{\cos(2s+1)t \ \sin(2s+1)x}{(2s+1)^3}$

(ii)    $\dfrac{\partial u}{\partial t} = k \dfrac{\partial^2 u}{\partial x^2}$,        $0 < x < \pi,\ t > 0,\ k$ is a constant

       $u(0, t) = u(\pi, t) = 0$ ;    $u(x, 0) = \sin 3x$        Answer: $e^{-9kt} \sin 3x$

(iii)    $\dfrac{\partial^2 u}{\partial x^2} + \dfrac{\partial^2 u}{\partial y^2} = 0$ ,      $0 < x < \pi,\ 0 < y < \pi,$

       $u(0, y) = u(\pi, y) = u(x, \pi) = 0$ ;    $u(x, 0) = \sin^2 x$

$$\text{Answer: } -\frac{8}{\pi} \sum_{s=1}^{\infty} \frac{\sinh(2s-1)(\pi-y)\ \sin(2s-1)x}{(4s^2-1)(2s-3)\sinh(2s-1)\pi}$$

10b.    In deriving the wave equation, we have assumed that the frictional resistance is negligible. If the resistance is assumed to be proportional to the velocity, the wave equation becomes

$$\frac{\partial^2 y}{\partial t^2} + k \frac{\partial y}{\partial t} = c^2 \frac{\partial^2 y}{\partial x^2}$$

where $k$ is a constant.

Obtain the solution if $y$ satisfies the following conditions

$$y(0, t) = y(1, t) = y(x, 0) = 0 ; \quad \left. \frac{\partial y}{\partial t} \right|_{t=0} = v_0\, \delta\!\left(x - \frac{1}{2}\right)$$

where $v_0$ is a constant and $\delta$ is the Dirac function.

$$\text{Answer: } \sum_{s=0}^{\infty} \frac{2v_0}{q_{s+1}}\, e^{-kt/2} \sin q_{2s+1}t\ \sin(2s+1)\pi x$$

$$q_{2s+1} = c^2(2s+1)^2 \pi^2 - k^2/4$$

11a.    The one-dimensional heat equation can be written as (see Appendix III)

$$\frac{\partial T}{\partial t} = \alpha \frac{\partial^2 T}{\partial x^2}$$

where $T(x, t)$ is the temperature and $\alpha$ is the thermal diffusivity.

Solve the heat equation for a rod of unit length. The boundary and initial conditions are

$$T(0, t) = T(1, t) = 0; \quad T(x, 0) = \begin{cases} 2x, & 0 \le x \le \dfrac{1}{2} \\ 2(1-x), & \dfrac{1}{2} \le x \le 1 \end{cases}$$

Answer: $\displaystyle 8 \sum_{s=0}^{\infty} \frac{(-1)^s \, e^{-(2s+1)^2 \alpha \pi^2 t} \sin(2s+1)\pi x}{\pi^2 (2s+1)^2}$

12a.  Hershey et al. (1967) considered the transport of oxygen into blood down a wetted wall column. A film of blood of uniform thickness falls with velocity $v_z$. The blood is in direct contact with a rising stream of oxygen, as shown in Figure (5.7-1). The equation governing the diffusion process is (see Appendix IV)

$$v_z \frac{\partial c_A}{\partial z} = \mathscr{D}_{AB} \frac{\partial^2 c_A}{\partial x^2}$$

where $c_A$ is the concentration of oxygen in the blood and $\mathscr{D}_{AB}$ is the diffusion coefficient.

The appropriate boundary conditions are

$$c_A(0, z) = c_f, \qquad c_A(x, 0) = c_0$$

and at the wall $x = L$,  $\quad \dfrac{\partial c_A}{\partial x} = 0$

Assuming that the oxygen does not penetrate very far into the blood, the velocity $v_z$ may be approximated by the velocity at the interface which we denote by $v_m$, which is the maximum velocity. By writing

$$c = \frac{c_A - c_f}{c_0 - c_f}$$

obtain the partial differential equation governing the process of diffusion. Deduce the boundary conditions. Note that the boundary conditions are now homogeneous at $x = 0$ and at $x = L$.

Answer: $\displaystyle \frac{4(c_0 - c_f)}{\pi} \sum_{s=0}^{\infty} \frac{\exp[-(2s+1)^2 \pi^2 \mathscr{D}_{AB} z / 4L^2 v_z]}{(2s+1)} \sin \frac{(2s+1)\pi x}{2L} + c_f$

13a.  The steady temperature $T(x, y)$ in a square $0 < x < a$, $0 < y < a$ is given by

$$\frac{\partial^2 T}{\partial x^2} + \frac{\partial^2 T}{\partial y^2} = 0$$

Obtain the temperature $T$ for the following boundary conditions:

(i)     $T(0, y) = T(a, y) = T(x, 0) = 0$,     $T(x, a) = x(a - x)$

$$\text{Answer: } \frac{8a^2}{\pi^3} \sum_{s=0}^{\infty} \frac{\sin(2s+1)\pi x/a \ \sinh(2s+1)\pi y/a}{(2s+1)^3 \sinh(2s+1)\pi}$$

(ii)     $T(x, 0) = \left.\frac{\partial T}{\partial x}\right|_{x=0} = \left.\frac{\partial T}{\partial x}\right|_{x=a} = 0$,     $T(x, a) = x(a - x)$

$$\text{Answer: } \frac{8a^3}{\pi^2} \sum_{s=0}^{\infty} \frac{\cos(2s+1)\pi x/a \ \sinh(2s+1)\pi y/a}{(2s+1)^2 \sinh(2s+1)\pi}$$

14b.    In Chapter 3, it is shown that the potential $\varphi$ associated with the flow of an incompressible inviscid fluid satisfies Laplace's equation. Consider the flow past a circular cylinder of radius $a$ with its centre at the origin. In the polar coordinate system, Laplace's equation is written as

$$\frac{\partial^2 \varphi}{\partial r^2} + \frac{1}{r}\frac{\partial \varphi}{\partial r} + \frac{1}{r^2}\frac{\partial^2 \varphi}{\partial \theta^2} = 0$$

The appropriate boundary conditions are

$$r = a, \quad \frac{\partial \varphi}{\partial r} = 0; \qquad r \longrightarrow \infty, \quad \varphi \longrightarrow r\, v_x \cos\theta$$

where $v_x$ is constant (velocity far from the cylinder).

Obtain $\varphi$ and compare your answer with that given in Example 3.6-8.

$$\text{Answer: } v_x \left(r + a^2/r\right)\cos\theta$$

15a.    Determine the potential $\varphi$ for the flow of an inviscid, incompressible fluid past a sphere of radius $a$. Laplace's equation in the spherical polar coordinate system, assuming spherical symmetry, is written as

$$\frac{1}{r^2}\frac{\partial}{\partial r}\left(r^2\frac{\partial \varphi}{\partial r}\right) + \frac{1}{r^2 \sin\theta}\frac{\partial}{\partial \theta}\left(\sin\theta \frac{\partial \varphi}{\partial \theta}\right) = 0$$

Choosing the centre of the sphere to be the origin of the coordinate system, the boundary conditions are

$$\left.\frac{\partial \varphi}{\partial r}\right|_{r=a} = 0, \qquad \varphi \longrightarrow v_\infty\, r \cos \theta \;\; \text{as} \;\; r \longrightarrow \infty$$

Answer: $v_\infty\, (r - a^3/2r^2) \cos \theta$

16a.   To find the temperature inside the earth, we consider a very simple model. We assume the earth to be a sphere with constant thermal properties. There is no source of heat. The temperature is at steady state. The temperature distribution on the surface of the earth has a maximum at the equator and a minimum at the poles. For such a model, the temperature $T$ satisfies the following equations

$$\frac{1}{r^2} \frac{\partial}{\partial r}\left(r^2 \frac{\partial T}{\partial r}\right) + \frac{1}{r^2 \sin \theta} \frac{\partial}{\partial \theta}\left(\sin \theta \frac{\partial T}{\partial \theta}\right) = 0$$

$T$ is finite at the centre of the earth ($r = 0$). $T = \sin^2 \theta$ on the surface of the earth ($r = 1$).

Determine the temperature at the centre of the earth.

Answer: $2/3$

17b.   The molecular diffusion of species $A$ inside a spherical drop is described by (see Appendix IV)

$$\frac{\partial c_A}{\partial t} = \mathcal{D}_{AB}\left[\frac{1}{r^2} \frac{\partial}{\partial r}\left(r^2 \frac{\partial c_A}{\partial r}\right)\right]$$

where $c_A\,(r, t)$ is the concentration, $\mathcal{D}_{AB}$ is the diffusivity.

The initial and boundary conditions are

$$t = 0, \quad c_A = c_0; \quad \text{on the surface } (r = 1),\; c_A = c_s;$$

$c_A$ is finite at the origin. The quantities $c_0$ and $c_s$ are constants.

By writing $u = r c_A$, show that the equation is reduced to one with constant coefficients. Determine the initial and boundary conditions on $u$. Apply the appropriate substitution such that the boundary condition at $r = 1$ is homogeneous. Determine $c_A$.

Answer: $\displaystyle 2\,(c_0 - c_s) \sum_{s=1}^{\infty} \frac{(-1)^{s+1}}{s\pi r}\, e^{-s^2\pi^2 \mathcal{D}_{AB} t} \sin s\pi r + c_s$

18b.    The equation governing the diffusion of species  A  in a medium  B  and reacting with it
        irreversibly according to a first order reaction is given by (see Appendix IV)

$$\frac{\partial c_A}{\partial t} = \mathcal{D}_{AB} \nabla^2 c_A - k_1 c_A$$

where  $c_A$  is the concentration,  $\mathcal{D}_{AB}$  is the diffusivity, and  $k_1$  is the rate constant for the
reaction.

The initial and boundary conditions are

$$t = 0, \quad c_A = 0 ; \quad \text{at the surface,} \ c_A = c_s$$

Verify that if  f  is the solution of

$$\frac{\partial f}{\partial t} = \mathcal{D}_{AB} \nabla^2 f$$

then  $c_A$  is given by

$$c_A = k_1 \int_0^t f e^{-k_1 t'} \, dt' + f e^{-k_1 t}$$

Show that  $c_A$  and  f  satisfy the same initial and boundary conditions.

19b.    If heat is produced at a rate proportional to the temperature  T,  then  T  satisfies the equation

$$\frac{\partial T}{\partial t} = \alpha \frac{\partial^2 T}{\partial x^2} + k_1 T$$

where  $\alpha$  and  $k_1$  are constants.

Assume that the initial and boundary conditions are

$$T(x, 0) = 0, \quad T(0, t) = 0, \quad T(1, t) = T_1$$

Solve the differential equation, subject to the given initial and boundary conditions, for the case
$k_1 = 0$.  Then, use the result given in Problem 18b to obtain the solution for non-zero values
of  $k_1$.  Verify that the solution satisfies the differential equation, the initial and the boundary
conditions.

20b.    Can you use the method of separation of variables to solve the equation

$$\frac{\partial^2 u}{\partial x^2} + \frac{\partial^2 u}{\partial y^2} + (x^2 + y^2)^2 u = 0$$

Transform to the polar coordinate system and show that the equation can be written as

$$\frac{1}{r} \frac{\partial}{\partial r} \left( r \frac{\partial u}{\partial r} \right) + \frac{1}{r^2} \frac{\partial^2 u}{\partial \theta^2} + r^4 u = 0$$

By writing $u = R(r) \Theta(\theta)$, deduce that $F$ and $\Theta$ are solutions of

$$r \frac{d}{dr} \left( r \frac{dR}{dr} \right) + (r^6 - n^2) R = 0, \qquad \frac{d^2 \Theta}{d\theta^2} + n^2 \Theta = 0$$

where $n^2$ is the separation constant.

Obtain a series solution for $R$ (see Chapter 2).

If the region of interest is $0 < r < 0.5$ and given that

$$u(0.5, \theta) = \sin 2\theta$$

obtain an approximate solution for $u$.

Answer: $4r^2 (1 - r^6/60) \sin 2\theta$

21b.    The velocity component $v_z$ of a Newtonian fluid flowing in a long circular pipe of unit radius under a pressure gradient is given by [see Equation (1.19-20)]

$$v_z = v_m (1 - r^2)$$

where $v_m$ is the maximum velocity (velocity along the centre line).

If the pressure gradient is stopped, then the transient velocity $v_{zt}$ is given by

$$\frac{\partial v_{zt}}{\partial t} = \frac{v}{r} \frac{\partial}{\partial r} \left( r \frac{\partial v_{zt}}{\partial r} \right)$$

where $v$ is the kinematic viscosity.

The initial and boundary conditions are

$$v_{zt}(r, 0) = v_m (1 - r^2), \qquad v_{zt}(1, t) = 0, \qquad v_{zt} \text{ is finite at the origin}$$

Solve for $v_{zt}$. By considering only the first term of the infinite series, estimate the time required for the velocity at the centre line to be reduced by 90%.

$$\text{Answer:} \quad \frac{4v_m \, J_2(2.4) \, e^{-(2.4)^2 \nu t} \, J_0(0)}{[2.4 \, J_1(2.4)]^2}$$

22b.  Chiappetta and Sobel (1987) have modeled the combustion-gas sampling probe as a hemisphere whose flat surface is kept at a constant temperature $T_C$. Gas at constant temperature $T_G > T_C$ is flowing over the probe. The system is steady and the temperature $T$ at any point inside the hemisphere satisfies Laplace's equation, which in spherical polar coordinate, can be written as (see Appendix III)

$$\frac{\partial}{\partial r}\left(r^2 \frac{\partial T}{\partial r}\right) + \frac{1}{\sin \theta} \frac{\partial}{\partial \theta}\left(\sin \theta \frac{\partial T}{\partial \theta}\right) = 0, \qquad 0 \le r \le a, \ -\frac{\pi}{2} \le \theta \le \frac{\pi}{2}$$

where $a$ is the radius of the hemisphere.

The boundary conditions are

on the flat surface:         $T\left(r, \frac{\pi}{2}\right) = T\left(r, -\frac{\pi}{2}\right) = T_C$;

on the curved surface:      $k \dfrac{\partial T}{\partial r}(a, \theta) = h\,[T_G - T(a, \theta)]$

where $k$ is the thermal conductivity and $h$ is the heat transfer coefficient; $T$ is finite everywhere in the hemisphere.

Note that the boundary conditions are not homogeneous and we introduce $u$ defined by

$$u = T - T_C$$

Solve for $u$ and then, obtain $T$.

23b.  Burgers (1948) has proposed the equation

$$\frac{\partial v}{\partial t} + v\frac{\partial v}{\partial x} = \nu \frac{\partial^2 v}{\partial x^2}$$

where $\nu$ is a positive constant, as a model for a turbulent flow in a channel.

Show that the transformation (Cole-Hopf transformation)

$$v(x, t) = A \frac{\partial}{\partial x} [\ln \theta (x, t)]$$

transforms Burger's equation to the diffusion equation for a suitable choice of constant A.

Determine A and solve for $\theta$ by the method of separation of variables. Obtain $v(x, t)$ if the initial and boundary conditions are

$$v(x, 0) = v_0, \quad 0 \le x \le a; \quad v(0, t) = v(a, t) = 0$$

Answer: $\displaystyle\sum_{s=0}^{\infty} A_s e^{-s^2 \pi^2 v \, t/a^2} \cos \pi s x / a$

$$A_s = \frac{v_0}{av} [1/(v_0^2/4v^2 + s^2\pi^2/a^2)] [1 - (-1)^s e^{-av_0/2v}]$$

24b.  By introducing a suitable auxiliary function, solve the following equations by the method of separation of variables, subject to the given conditions:

(i)    $$\frac{\partial^2 u}{\partial t^2} = c^2 \frac{\partial^2 u}{\partial x^2} + x, \quad 0 < x < 1$$

$$u(0, t) = u(1, t) = 0; \quad u(x, 0) = \left.\frac{\partial u}{\partial t}\right|_{t=0} = 0$$

Answer: $\displaystyle\frac{x(1 - x^2)}{6c^2} + \sum_{s=1}^{\infty} \frac{(-1)^s 2}{(s\pi)^3} \cos s\pi ct \, \sin s\pi x$

(ii)   $$\frac{\partial u}{\partial t} = \frac{\partial^2 u}{\partial x^2} + e^{-t} \sin 3x$$

$$u(0, t) = 0, \quad u(\pi, t) = 1, \quad u(x, 0) = x/\pi$$

Answer: $\displaystyle\frac{x}{\pi} + \frac{1}{8} (e^{-t} - e^{-9t}) \sin 3x$

25a.  Use the method of Laplace transforms to solve the following equations:

(i)    $$\frac{\partial u}{\partial x} + x \frac{\partial u}{\partial t} = 0; \quad u(x, 0) = 0, \quad u(0, t) = t$$

Answer:  $0, \qquad 0 < t < x^2/2$
$t - x^2/2, \quad t > x^2/2$

(ii) $\quad \dfrac{\partial u}{\partial x} + 2x \dfrac{\partial u}{\partial t} = 2x ; \quad u(x, 0) = u(0, t) = 1$

$$\text{Answer: } \begin{aligned} &t + 1, && 0 < t < x^2 \\ &(t + 1) - (t - x^2), && t > x^2 \end{aligned}$$

(iii) $\quad \dfrac{\partial^2 u}{\partial t^2} = c^2 \dfrac{\partial^2 u}{\partial x^2} - g ; \quad u(x, 0) = \left. \dfrac{\partial u}{\partial t} \right|_{t=0} = 0, \quad x \ge 0$

$$u(0, t) = 0, \quad \lim_{x \to \infty} \frac{\partial u}{\partial x} = 0, \quad t \ge 0 ; \quad c \text{ and } g \text{ are constants.}$$

$$\text{Answer: } \begin{aligned} &-g t^2 / 2, && 0 < t < x/c \\ &-\frac{g}{2} [t^2 - (t - x/c)^2], && t > x/c \end{aligned}$$

26b. A composite solid is made of two layers $K_1$ and $K_2$. Layer $K_1$ $(0 \le x \le a)$ is initially kept at a temperature $T_0$ and is in perfect thermal contact with layer $K_2$ $(a < x \infty)$ which is initially kept at zero temperature. The face $x = 0$ is insulated. The thermal conductivity and diffusivity in $K_1$ and $K_2$ are respectively $k_1$, $\alpha_1$ and $k_2$, $\alpha_2$. The problem can be expressed as (see Appendix III)

$$\frac{\partial T}{\partial t} = \alpha_1 \frac{\partial^2 T}{\partial x^2}, \quad 0 < x < a, \ t > 0$$

$$\frac{\partial T}{\partial t} = \alpha_2 \frac{\partial^2 T}{\partial x^2}, \quad x > a, \ t > 0$$

$$T(x, 0) = T_0, \ 0 < x < a ; \quad T(x, 0) = 0, \ x > a$$

$$\left. \frac{\partial T}{\partial x} \right|_{x=0} = 0, \quad \lim_{x \to \infty} T(x, t) = 0$$

$T(a - 0, t) = T(a + 0, t)$ (continuity of temperature at the interface)

$$k_1 \left. \frac{\partial T}{\partial x} \right|_{x=a_-} = k_2 \left. \frac{\partial T}{\partial x} \right|_{x=a_+} \quad \text{(continuity of heat flux at the interface)}$$

Solve for $T$ using the method of Laplace transform.

$$\text{Answer: } T_0 - \frac{T_0 (1 - \lambda)}{2} \sum_{s=0}^{\infty} \lambda^n \left\{ \text{erfc} \left[ \frac{(2s + 1) a - x}{2 \sqrt{\alpha_1 t}} \right] + \text{erfc} \left[ \frac{(2s + 1) a + x}{2 \sqrt{\alpha_1 t}} \right] \right\}, \quad 0 < x < a$$

$$\frac{T_0\,(1+\lambda)}{2} \sum_{s=0}^{\infty} \lambda^n \left\{ \mathrm{erfc}\left[\frac{2sa+\mu\,(x-a)}{2\,\sqrt{\alpha_1 t}}\right] - \mathrm{erfc}\left[\frac{(2s+2)\,a+\mu\,(x-a)}{2\,\sqrt{\alpha_1 t}}\right] \right\},\quad x > a$$

$$\mu = \sqrt{\alpha_1/\alpha_2}\,,\qquad \lambda = \frac{k_1\sqrt{\alpha_2}-k_2\sqrt{\alpha_1}}{k_1\sqrt{\alpha_2}+k_2\sqrt{\alpha_1}}$$

27b.    Show that the solution of

$$\frac{\partial^2 u}{\partial t^2} = \frac{\partial^2 u}{\partial x^2} + \frac{\partial^4 u}{\partial x^2 \partial t^2}\,,\qquad x > 0\,,\;\; t > 0$$

subject to

$$u = \frac{\partial u}{\partial t} = 0\,,\qquad t = 0\,,\;\; x > 0$$

$$u\,(0,t) = 1\,,\qquad u \longrightarrow 0 \;\;\text{as}\;\; x \longrightarrow \infty \;\;\text{for all}\;\; t > 0$$

can be written as

$$u = \frac{1}{2\pi i} \int_{\gamma-i\infty}^{\gamma+i\infty} \frac{1}{s}\,\exp\left(-sx/\sqrt{1+s^2} + st\right)\,ds$$

Deduce that

$$\left.\frac{\partial u}{\partial x}\right|_{x=0} = -J_0\,(t)$$

[Hint:  Use the table of Laplace transforms.]

28b.    Use the method of Fourier transform to solve Poisson's equation

$$\frac{\partial^2 u}{\partial x^2} + \frac{\partial^2 u}{\partial y^2} = -\rho\,,\qquad 0 < x < \pi\,,\;\; y > 0$$

$$u\,(0,y) = 0\,,\qquad u\,(\pi,y) = 1\,,\qquad u\,(x,0) = 0$$

u  is bounded throughout the region.

Answer: $\dfrac{2}{\pi} \sum B_s\,(e^{-sy} - 1)\,\sin sx$

$$B_s = \frac{1}{s^2}\left\{(-1)^s\,s + \rho\,[1-(-1)^s]\,/\,s\right\}$$

29b.   Show that the Fourier transform of $e^{-a x^2}$ is $\sqrt{\dfrac{\pi}{a}}\ e^{-\alpha^2/4a}$.

[Hint:  Complete the square in the exponential term and use the appropriate contour.]

Solve by the method of Fourier transform the equation

$$\frac{\partial^2 u}{\partial x^2} + \frac{\partial^2 u}{\partial y^2} - u = 0, \qquad -\infty < x < \infty, \ \ 0 < y < 1$$

subject to

$$\frac{\partial u}{\partial y}\bigg|_{y=0} = 0, \quad u(x, 1) = e^{-x^2}, \quad u(x, y) \longrightarrow 0 \ \text{as} \ x \longrightarrow \pm\infty$$

Leave your answer in integral form.

# CHAPTER 6

# PARTIAL DIFFERENTIAL EQUATIONS II

## 6.1 INTRODUCTION

In Chapter 5, we have used the method of separation of variables and transform methods to solve partial differential equations. In this chapter, we introduce additional methods which can be used to solve P.D.E.'s. We consider also non-linear equations.

P.D.E.'s can be classified in three types, **hyperbolic**, **elliptic**, and **parabolic**, and for each type we discuss one method. Here, the hyperbolic equations are solved by the **method of characteristics**, a method which has been encountered in Chapter 5. The method of **Green's function** which was employed in Chapter 1 to solve non-homogeneous O.D.E.'s is now extended to solve elliptic P.D.E.'s, and a **similarity variable** is introduced to solve parabolic equations.

The equations we considered in Chapter 5, namely the wave equation, the diffusion (heat) equation, and Laplace's equation are all important equations in classical physics. In this chapter, we seek the solution of **Schrödinger's equation** which is one of the most important equations in quantum mechanics.

## 6.2 METHOD OF CHARACTERISTICS

In Chapter 5, we have used the method of characteristics to solve first order partial differential equations and the wave equation (d'Alembert's solution). We recall that the **characteristics** are the curves along which information is propagated. Discontinuous initial data are carried along the characteristics. It is also shown that hyperbolic equations have real characteristics. We now describe **Riemann's method** of solving hyperbolic equations.

The canonical form of a linear hyperbolic equation is

$$L(u) = \frac{\partial^2 u}{\partial x \, \partial y} + a(x, y) \frac{\partial u}{\partial x} + b(x, y) \frac{\partial u}{\partial y} + c(x, y) \, u = \rho(x, y) \qquad (6.2\text{-}1a,b)$$

We assume that the prescribed conditions are on a curve $\gamma$ and that they can be written as

$$u = g(x, y), \qquad \frac{\partial u}{\partial n} = h(x, y) \qquad (6.2\text{-}2a,b)$$

where $\dfrac{\partial u}{\partial n}$ is the normal derivative of u on γ.

We also assume that γ is not a characteristic of Equations (6.2-1a, b), as shown in Figure 6.2-1.

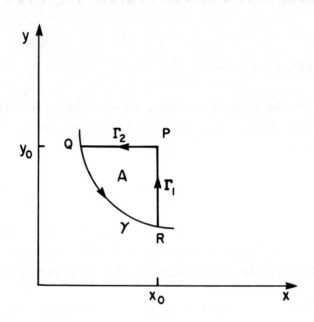

**FIGURE  6.2-1**     **Characteristics $\Gamma_1$ and $\Gamma_2$ and curve γ**
**along which conditions are prescribed**

The characteristics are given by [Equations (5.3-11a, b)]

$$\Gamma_1 : \quad x = \text{constant} , \qquad \Gamma_2 : \quad y = \text{constant} \tag{6.2-3a,b}$$

We now determine the value of u at a point $P(x_0, y_0)$ which is a point of intersection of two characteristics. The solution at P is determined by the domain of dependence A (see also Figure 5.4-3).

The **adjoint operator** $L^*$ of the operator L defined by Equations (6.2-1a, b) is [see Equation (6.5-1) for the general case]

$$L^*(v) = \frac{\partial^2 v}{\partial x \, \partial y} - \frac{\partial}{\partial x}(av) - \frac{\partial}{\partial y}(bv) + cv \tag{6.2-4}$$

If $L^* = L$, L is **self-adjoint**.

From Equations (6.2-1a, b, 4), we obtain

$$v\, L\,(u) - u\, L^*\,(v) = \frac{\partial}{\partial x}\left(auv - u\frac{\partial v}{\partial y}\right) + \frac{\partial}{\partial y}\left(buv + v\frac{\partial u}{\partial x}\right) \tag{6.2-5}$$

Applying Gauss' two-dimensional theorem [Equation (4.4-23)] to Equation (6.2-5), we obtain

$$\iint\limits_{A} [v\, L\,(u) - u\, L^*\,(v)]\, dx\, dy = \oint\limits_{C} \left[\left(auv - u\frac{\partial v}{\partial y}\right) dy - \left(buv + v\frac{\partial u}{\partial x}\right) dx\right] \tag{6.2-6}$$

where $C$ is the closed curve formed by $\gamma$, $\Gamma_2$, and $\Gamma_1$, as shown in Figure 6.2-1. Noting that on $\Gamma_1$, $dx = 0$ and on $\Gamma_2$, $dy = 0$, it follows that

$$\oint\limits_{C} \left[\left(auv - u\frac{\partial v}{\partial y}\right) dy - \left(buv + v\frac{\partial u}{\partial x}\right) dx\right] = \int\limits_{\gamma} \left[\left(auv - u\frac{\partial v}{\partial y}\right) dy - \left(buv + v\frac{\partial u}{\partial x}\right) dx\right]$$

$$+ \int_{R}^{P} \left(auv - u\frac{\partial v}{\partial y}\right) dy - \int_{P}^{Q} \left(buv + v\frac{\partial u}{\partial x}\right) dx \tag{6.2-7}$$

Integrating $\displaystyle\int_{P}^{Q} v\,\frac{\partial u}{\partial x}\, dx$ by parts yields

$$\int_{P}^{Q} v\,\frac{\partial u}{\partial x}\, dx = \left[vu\right]_{P}^{Q} - \int_{P}^{Q} u\,\frac{\partial v}{\partial x}\, dx \tag{6.2-8}$$

Combining Equations (6.2-6 to 8), we obtain

$$\iint\limits_{A} [v\, L\,(u) - u\, L^*\,(v)]\, dx\, dy = \int\limits_{\gamma} \left[\left(auv - u\frac{\partial v}{\partial y}\right) dy - \left(buv + v\frac{\partial u}{\partial x}\right) dx\right]$$

$$+ \int_{R}^{P} \left(auv - u\frac{\partial v}{\partial y}\right) dy - v(Q)\,u(Q) + v(P)\,u(P) - \int_{P}^{Q} \left(buv - u\frac{\partial v}{\partial x}\right) dx \tag{6.2-9}$$

Equation (6.2-9) expresses the value of $u$ at $P$ in terms of $v$ and the sum of integrals. By choosing $v$ appropriately, we can reduce some of the integrals to zero. Following Riemann, we impose on $v$ (the **Riemann function**) the following conditions

(a)     v satisfies the adjoint equation, that is to say

$$L^*(v) = 0 \qquad\qquad (6.2\text{-}10a)$$

(b)     along $x = \text{constant}\ (\Gamma_1)$

$$\frac{\partial v}{\partial y} = av \qquad\qquad (6.2\text{-}10b)$$

(c)     along $y = \text{constant}\ (\Gamma_2)$

$$\frac{\partial v}{\partial x} = bv \qquad\qquad (6.2\text{-}10c)$$

(d)     at the point P

$$v(x_0, y_0) = 1 \qquad\qquad (6.2\text{-}10d)$$

Substituting Equations (6.2-10a to d) into Equation (6.2-9) yields

$$u(P) = v(Q)\,u(Q) - \int_\gamma \left[\left(auv - u\frac{\partial v}{\partial y}\right)dy - \left(buv + v\frac{\partial u}{\partial x}\right)dx\right] + \iint_A v\rho\,dx\,dy \qquad (6.2\text{-}11)$$

Equation (6.2-11) is not symmetrical in the sense that it only involves the point Q and the derivative $\frac{\partial u}{\partial x}$ but not R and $\frac{\partial u}{\partial y}$. To obtain a more symmetric expression, we consider the identity

$$\int_Q^R \left[\frac{\partial}{\partial x}(uv)\,dx + \frac{\partial}{\partial y}(uv)\,dy\right] = \int_Q^R d(uv) = u(R)\,v(R) - u(Q)\,v(Q) \qquad (6.2\text{-}12a,b)$$

Substituting $u(Q)\,v(Q)$ from Equations (6.2-12a, b) into Equation (6.2-9) and using conditions (6.2-10a to d), we obtain

$$u(P) = u(R)\,v(R) - \int_\gamma \left[\left(auv - u\frac{\partial v}{\partial y}\right)dy - \left(buv + v\frac{\partial u}{\partial x}\right)dx\right]$$

$$- \int_\gamma \left[\frac{\partial}{\partial x}(uv)\,dx + \frac{\partial}{\partial y}(uv)\,dy\right] + \iint_A v\rho\,dx\,dy \qquad (6.2\text{-}13a)$$

$$= u(R)\,v(R) - \int_\gamma \left[\left(auv + v\frac{\partial u}{\partial y}\right)dy - \left(buv - u\frac{\partial v}{\partial x}\right)dx\right] + \iint_A v\rho\,dx\,dy \qquad (6.2\text{-}13b)$$

Adding Equations (6.2-11, 13b) yields

$$u\,(P) \;=\; \frac{1}{2}\,[u\,(R)\,v\,(R) + u\,(Q)\,v\,(Q)] + \iint\limits_{A} v\,\rho\;dx\,dy$$

$$-\,\frac{1}{2}\int\limits_{\gamma}\;[(2\,auv - u\frac{\partial v}{\partial y} + v\frac{\partial u}{\partial y})\,dy - (2\,buv + v\frac{\partial u}{\partial x} - u\frac{\partial v}{\partial x})\,dx] \qquad\qquad (6.2\text{-}14)$$

The function $u$ and its first partial derivatives on $\gamma$ are given by Equations (6.2-2a, b), $R$ and $Q$ are on $\gamma$; that is, $u\,(R)$ and $u\,(Q)$ are known, $\rho$ is given in $A$, and if the Riemann function $v$ is known, $u\,(P)$ can be evaluated. The problem of obtaining a solution to Equations (6.2-1a, b, 2a, b) has now been transformed to finding the Riemann function $v$ satisfying Equations (6.2-10a to d). Once $v$ is found, $u\,(P)$ can be obtained from Equation (6.2-14). Note that Equation (6.2-14) can be written in a general form as

$$u(P) \;=\; \int K(\xi,\,P)\,\rho(\xi)\,d\xi \qquad\qquad (6.2\text{-}15)$$

where the kernel $K$ depends on the operator and the boundary conditions, and is independent of $\rho$. The Riemann method is similar to the Green's function method [see Equation (1.18-1)].

Note that we have assumed that each characteristic intersects $\gamma$ at no more than one point. This condition is necessary, otherwise the problem might have no solution.

**Example 6.2-1.** Solve the following equation using Riemann's method

$$\frac{\partial^2 u}{\partial x\,\partial y} - \frac{1}{2x}\frac{\partial u}{\partial y} \;=\; 0 \qquad\qquad (6.2\text{-}16)$$

subject to

$$u\,(x,\,y) = f\,(x,\,y)\,, \qquad\qquad \frac{\partial u}{\partial n} = h\,(x,\,y) \qquad\qquad (6.2\text{-}17a,b)$$

on the line $y = -x + c$, where $c$ is a positive constant.

The characteristics are given by Equations (6.2-3a, b) and the adjoint operator [Equation (6.2-4)] is

$$\frac{\partial^2 v}{\partial x\,\partial y} + \frac{1}{2x}\frac{\partial v}{\partial y} \;=\; 0 \qquad\qquad (6.2\text{-}18)$$

On integrating with respect to y, we obtain

$$\frac{\partial v}{\partial x} + \frac{v}{2x} = F(x) \tag{6.2-19}$$

where $F(x)$ is an arbitrary function.

The integrating factor of Equation (6.2-19) is $\sqrt{x}$ and the solution is

$$\sqrt{x} \ v = \int \sqrt{x} \ F(x) \, dx + G(y) \tag{6.2-20a}$$

or     $v = F_1(x) + G(y)/\sqrt{x} \tag{6.2-20b}$

where $F_1$ and $G$ are arbitrary functions.

The function $v$ has to satisfy Equations (6.2-10b to d) and this implies that

on $x = $ constant ,                     $G'(y)/\sqrt{x} = 0 \tag{6.2-21a}$

on $y = $ constant ,             $\dfrac{dF_1}{dx} - \dfrac{1}{2} G(y)/x^{3/2} = -v/(2x) \tag{6.2-21b}$

at the point $P(x_0, y_0)$ ,             $v(x_0, y_0) = 1 \tag{6.2-21c}$

From Equations (6.2-21a, b), we deduce that $G$ is a constant and

$$\frac{dF_1}{dx} + \frac{F_1}{2x} = 0 \tag{6.2-22}$$

The solution is

$$F_1 = K_0/\sqrt{x} \tag{6.2-23}$$

where $K_0$ is a constant.

Combining Equations (6.2-20b, 21c, 23) yields

$$v = \sqrt{x_0/x} \tag{6.2-24}$$

The solution $u(x_0, y_0)$ is given by Equation (6.2-14) and, in this example, it simplifies to

$$u(x_0, y_0) = \frac{1}{2} [u(R) \, v(R) + u(Q) \, v(Q)] + \frac{1}{2} \int_\gamma \left( v \frac{\partial u}{\partial y} + \frac{2uv}{x} + v \frac{\partial u}{\partial x} - u \frac{\partial v}{\partial x} \right) dx \tag{6.2-25}$$

where $\gamma$ is the line $y = -x + c$ and on $\gamma$, $dx = -dy$.

From, Figure 6.2-1, we deduce that $Q$ and $R$ have coordinates $(c - y_0, y_0)$ and $(x_0, c - x_0)$ respectively. From Equations (6.2-17a, 24), we obtain

$$u(R) = f(x_0, c - x_0), \qquad u(Q) = f(c - y_0, y_0) \tag{6.2-26a,b}$$

$$v(R) = 1, \qquad v(Q) = \sqrt{x_0/(c - y_0)} \tag{6.2-26c,d}$$

Using Equation (1.5-13), Equation (6.2-17b) can be written as

$$\frac{\partial u}{\partial x} + \frac{\partial u}{\partial y} = \sqrt{2}\, h(x, y) \tag{6.2-27}$$

Combining Equations (6.2-17a,b, 24, 25, 26a to d, 27) yields

$$u(x_0, y_0) = \frac{1}{2}\left\{ f(x_0, c-x_0) + \sqrt{x_0/(c-y_0)}\, f(c-y_0, y_0) + \int_{c-y_0}^{x_0} [(\sqrt{2x_0/x})\, h(x, c-x) \right.$$

$$\left. + \frac{5}{2}\, (\sqrt{x_0/x^3})\, f(x, c-x)]\, dx \right\} \tag{6.2-28}$$

The functions $f$ and $h$ are given and the value of $u$ at any point $(x_0, y_0)$ can be calculated.

***Example 6.2-2.*** Solve the telegraph equation

$$\frac{\partial^2 u}{\partial t^2} - c^2 \frac{\partial^2 u}{\partial x^2} + \alpha \frac{\partial u}{\partial t} + \beta u = 0 \tag{6.2-29}$$

subject to

$$u(x, 0) = f_1(x), \qquad \left.\frac{\partial u}{\partial t}\right|_{t=0} = f_2(x) \tag{6.2-30a,b}$$

by Riemann's method.

Equation (6.2-29) arises in the transmission of electrical impulses in a long cable with distributed capacitance, inductance, and resistance. Note that if $\alpha = \beta = 0$ (no loss of current and no resistance), Equation (6.2-26) reduces to the wave equation [Equation (5.1.1)].

We can simplify Equation (6.2-29) by assuming that $u$ is of the form

$$u(x, t) = \theta(t) \, w(x, t) \tag{6.2-31}$$

In terms of $\theta$ and $w$, Equation (6.2-29) becomes

$$\theta\left(\frac{\partial^2 w}{\partial t^2} - c^2 \frac{\partial^2 w}{\partial x^2}\right) + \frac{\partial w}{\partial t}\left(2\frac{d\theta}{dt} + \alpha\theta\right) + w\left(\frac{d^2\theta}{dt^2} + \alpha\frac{d\theta}{dt} + \beta\theta\right) = 0 \tag{6.2-32}$$

Replacing $u(x, t)$ by two functions $\theta$ and $w$ in Equation (6.2-31) allows us to impose an additional condition on say $\theta$ so as to simplify Equation (6.2-32).

We choose to set the coefficient of $\dfrac{\partial w}{\partial t}$ to zero. That is to say,

$$2\frac{d\theta}{dt} + \alpha\theta = 0 \tag{6.2-33}$$

The solution of Equation (6.2-33) is

$$\theta = C_0 \, e^{-\alpha t/2} \tag{6.2-34}$$

where $C_0$ is a constant.

Substituting Equation (6.2-34) into Equation (6.2-32) yields

$$\frac{\partial^2 w}{\partial t^2} - c^2 \frac{\partial^2 w}{\partial x^2} + K'w = 0 \tag{6.2-35a}$$

where

$$K' = \beta - \alpha^2/4 \tag{6.2-35b}$$

The characteristics of Equation (6.2-35a) are [Equations 5.4-20a, b]

$$\xi = x + ct = \text{constant}, \qquad \eta = x - ct = \text{constant} \tag{6.2-36a,b,c,d}$$

Using $\xi$ and $\eta$ as the new variables, Equations (6.2-30a, b, 35a) become

$$\frac{\partial^2 w}{\partial \xi \, d\eta} + Kw = 0, \qquad K = K'/4c^2 \tag{6.2-37a,b}$$

$$w\,\big|_{\xi=\eta} = f_1(\xi) \tag{6.2-38a}$$

$$c \left( \frac{\partial w}{\partial \xi} - \frac{\partial w}{\partial \eta} \right) \bigg|_{\xi=\eta} = f_2(\xi) + \frac{\alpha}{2} f_1(\xi) = g_1(\xi) \qquad (6.2\text{-}38b,c)$$

We note from Equations (6.2-36a, c) that $t = 0$ corresponds to $\xi = \eta$.

We evaluate the value of $w$ at a point $P(\xi_0, \eta_0)$. The point $P$, the characteristics, and the curve $\gamma$ on which $w$ and its derivatives are prescribed by Equations (6.2-38a to c) are shown in Figure 6.2-2.

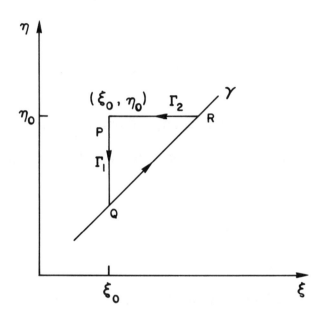

**FIGURE 6.2-2**     **Characteristics $\Gamma_1$ and $\Gamma_2$ of the telegraph equation**

The Riemann function $v$ has to satisfy Equations (6.2-10a to d) and the operator $L$ defined by Equation (6.2-37) is self adjoint. The function $v$ is a function of $\xi$ and $\eta$ and depends also on $\xi_0$ and $\eta_0$. We write $v$ as $v(\xi, \eta; \xi_0, \eta_0)$. The characteristics $\Gamma_1$ and $\Gamma_2$ are given by

$$\Gamma_1: \quad \xi = \xi_0 \qquad ; \qquad \Gamma_2: \quad \eta = \eta_0 \qquad (6.2\text{-}39a,b)$$

The conditions that $v$ has to satisfy can be written as

$$\frac{\partial^2 v}{\partial \xi \, d\eta} + Kv = 0 \qquad (6.2\text{-}40a)$$

$$\frac{\partial}{\partial \xi} [v(\xi, \eta_0; \xi_0, \eta_0)] = 0 \qquad (6.2\text{-}40b)$$

$$\frac{\partial}{\partial \eta} [v (\xi_0, \eta; \xi_0, \eta_0)] = 0 \qquad\qquad (6.2\text{-}40c)$$

$$v (\xi_0, \eta_0; \xi_0, \eta_0) = 1 \qquad\qquad (6.2\text{-}40d)$$

Rather than solving Equation (6.2-40a) right away, we first examine the form that $v$ might take. We expand $v$ about $(\xi_0, \eta_0)$ in a Taylor series and we obtain

$$v (\xi, \eta; \xi_0, \eta_0) = 1 + (\xi - \xi_0) \frac{\partial v}{\partial \xi} + (\eta - \eta_0) \frac{\partial v}{\partial \eta} + \frac{1}{2} (\xi - \xi_0)^2 \frac{\partial^2 v}{\partial \xi^2}$$

$$+ \frac{1}{2} (\eta - \eta_0)^2 \frac{\partial^2 v}{\partial \eta^2} + (\xi - \xi_0) (\eta - \eta_0) \frac{\partial^2 v}{\partial \xi \, d\eta} + \dots \qquad\qquad (6.2\text{-}41)$$

All the derivatives are evaluated at the point $(\xi_0, \eta_0)$. From Equations (6.2-40b, c), we deduce that $\dfrac{\partial^n v}{\partial \eta^n}$ and $\dfrac{\partial^n v}{\partial \xi^n}$ are zero at $(\xi_0, \eta_0)$. From Equations (6.2-40a, d), we obtain

$$\frac{\partial^2 v}{\partial \xi \, \partial \eta}\bigg|_{\xi_0, \eta_0} = -Kv\big|_{\xi_0, \eta_0} = -K \qquad\qquad (6.2\text{-}42a,b)$$

We deduce that $\dfrac{\partial^{2n} v}{\partial \xi^n \, \partial \eta^n}\bigg|_{\xi_0, \eta_0}$ are the only non-zero terms and, from Equation (6-2-41), we deduce that $v$ is of the form

$$v = f [(\xi - \xi_0) (\eta - \eta_0)] = f (s) \qquad\qquad (6.2\text{-}43a,b)$$

On differentiating, we have

$$\frac{\partial v}{\partial \xi} = \frac{df}{ds} \frac{ds}{\partial \xi} = (\eta - \eta_0) \frac{df}{ds} \qquad\qquad (6.2\text{-}44a,b)$$

$$\frac{\partial^2 v}{\partial \xi \, \partial \eta} = \frac{df}{ds} + (\eta - \eta_0) (\xi - \xi_0) \frac{d^2 f}{ds^2} = \frac{df}{ds} + s \frac{d^2 f}{ds^2} \qquad\qquad (6.2\text{-}44c,d)$$

Substituting Equations (6.2-44a to d) into Equation (6.2-40a), we obtain the following ordinary differential equation

$$s \frac{d^2 f}{ds^2} + \frac{df}{ds} + Kf = 0 \qquad\qquad (6.2\text{-}45)$$

We make a change of variable and write

$$\lambda = (4 \, K \, s)^{1/2} \qquad\qquad (6.2\text{-}46)$$

Using the chain rule, we have

$$\frac{df}{ds} = \frac{df}{d\lambda} \frac{d\lambda}{ds} = 2 \, K \, (4 \, K \, s)^{-1/2} \frac{df}{d\lambda} \qquad\qquad (6.2\text{-}47a,b)$$

$$\frac{d^2 f}{ds^2} = -4 \, K^2 \, (4 \, K \, s)^{-3/2} \frac{df}{d\lambda} + 4 \, K^2 \, (4 \, K \, s)^{-1} \frac{d^2 f}{d\lambda^2} \qquad\qquad (6.2\text{-}47c)$$

Equation (6.2-45) becomes

$$\frac{d^2 f}{d\lambda^2} + \frac{1}{\lambda} \frac{df}{d\lambda} + f = 0 \qquad\qquad (6.2\text{-}48)$$

Equation (6.2-48) is Bessel's equation of order zero and the solution which is regular at the origin $(\lambda = 0)$ is $J_0(\lambda)$. The Riemann function $v$ is

$$v = J_0(\lambda) = J_0 \left[ \sqrt{4K \, (\xi - \xi_0) \, (\eta - \eta_0)} \right] \qquad\qquad (6.2\text{-}49a,b)$$

We use Equation (6.2-14) to determine $w(P)$. We note that $\rho$, $a$, and $b$ are zero. On $\gamma$ $(\xi = \eta)$, we find that $w(P)$ is given by

$$w(P) = \frac{1}{2} \, [w(R) \, v(R) + w(Q) \, v(Q)] - \frac{1}{2} \int_{Q}^{R} \left[ v \left( \frac{\partial w}{\partial \eta} - \frac{\partial w}{\partial \xi} \right) + w \left( \frac{\partial v}{\partial \xi} - \frac{\partial v}{\partial \eta} \right) \right] d\xi \qquad (6.2\text{-}50)$$

The points $Q$ and $R$ are respectively $(\xi_0, \xi_0)$ and $(\eta_0, \eta_0)$. The functions $v$ and $w$ required in Equation (6.2-50) are given respectively by Equations (6.2-38a to c, 49a, b). The required values of $v$, $w$, and their derivatives are

$$v(Q) = v(R) = J_0(0) = 1 \qquad\qquad (6.2\text{-}51a,b,c)$$

$$w(Q) = f_1(\xi_0), \qquad w(R) = f_1(\eta_0) \qquad\qquad (6.2\text{-}52a,b)$$

$$\left. \left( \frac{\partial w}{\partial \xi} - \frac{\partial w}{\partial \eta} \right) \right|_{\xi = \eta} = \frac{1}{c} \, g_1(\xi) \tag{6.2-53}$$

$$\left. \frac{\partial v}{\partial \xi} \right|_{\xi = \eta} = \left[ \frac{\sqrt{K} \, (\eta - \eta_0)}{\sqrt{(\xi - \xi_0)(\eta - \eta_0)}} \, J_0'(\lambda) \right]_{\eta = \xi} \tag{6.2-54}$$

$$\left. \frac{\partial v}{\partial \eta} \right|_{\xi = \eta} = \left[ \frac{\sqrt{K} \, (\xi - \xi_0)}{\sqrt{(\xi - \xi_0)(\eta - \eta_0)}} \, J_0'(\lambda) \right]_{\eta = \xi} \tag{6.2-55}$$

$$\left. \left[ \frac{\partial v}{\partial \xi} - \frac{\partial v}{\partial \eta} \right] \right|_{\xi = \eta} = \left[ \frac{\sqrt{K} \, (\xi_0 - \eta_0)}{\sqrt{(\xi - \xi_0)(\eta - \eta_0)}} \, J_0'(\lambda) \right]_{\eta = \xi} \tag{6.2-56}$$

Substituting Equations (6.2-51a to 56) into Equation (6.2-50) and replacing the integration variable $\xi$ by $z$, to avoid confusion, yields

$$w(P) = \frac{1}{2} \left[ f_1(\xi_0) + f_1(\eta_0) \right] + \frac{1}{2c} \int_{\xi_0}^{\eta_0} J_0(\lambda_1) \, g_1(z) \, dz$$

$$- \frac{1}{2} \int_{\xi_0}^{\eta_0} \frac{f_1(z) \sqrt{K} \, (\xi_0 - \eta_0) J_0'(\lambda_1)}{\sqrt{(z - \xi_0)(z - \eta_0)}} \, dz \tag{6.2-57a}$$

where

$$\lambda_1 = \sqrt{4K \, (z - \xi_0)(z - \eta_0)} \tag{6.2-57b}$$

We revert to the $x$ and $t$ variables, $\xi_0$ and $\eta_0$ are replaced by

$$\xi_0 = x + ct, \qquad \eta_0 = x - ct \tag{6.2-58a,b}$$

The quantity $\lambda_1$ in Equation (6.2-57b) is now written as

$$\lambda_1 = \sqrt{4K \, (z - x - ct)(z - x + ct)} = \sqrt{4K \, [(z - x)^2 - c^2 t^2]} \tag{6.2-59a,b}$$

Substituting Equations (6.2-58a to 59b) into Equation (6.2-57a), we obtain

$$w(P) = \frac{1}{2} \left[ f_1(x + ct) + f_1(x - ct) \right] + \frac{1}{2} \int_{x + ct}^{x - ct} F(x, t, z) \, dz \tag{6.2-60a}$$

where

$$F(x, t, z) = \frac{1}{c} J_0(\lambda_1) g_1(z) - [\sqrt{K} (2ct) f_1(z) J_0'(\lambda_1)] / [(z-x)^2 - c^2 t^2] \qquad (6.2\text{-}60\text{b})$$

The solution $u$ is obtained from Equations (6.2-31, 34) and is

$$u = e^{-\alpha t/2} w \qquad (6.2\text{-}61)$$

Comparing Equations (6.2-61, 5.4-31, 32), we realize that, in the present example, $u$ is damped.

●

An extension of the method described here to higher order problems is given in Courant and Hilbert (1966). Abbott (1966) has described and solved several problems of engineering interest by the method of characteristics. In Chapter 8, a numerical method for solving hyperbolic equations based on characteristics is presented.

## 6.3   SIMILARITY SOLUTIONS

In many physical problems, the solution does not depend on the independent variables separately but on a combination of the independent variables. In this case, a **similarity solution** exists and the existence of a similarity solution is usually associated with the process of diffusion. We describe the method of obtaining a similarity solution by considering a few examples.

***Example 6.3-1***. Look for a similarity solution for the diffusion problem obtained by considering a generalized form of Equation (A.IV-1)

$$\frac{\partial c}{\partial t} = \frac{\partial}{\partial x} \left[ D(c) \frac{\partial c}{\partial x} \right], \qquad x \geq 0, \ t \geq 0 \qquad (6.3\text{-}1)$$

$$c(0, t) = c_0, \qquad \lim_{x \to \infty} c(x, t) = c_1, \qquad c(x, 0) = c_2 \qquad (6.3\text{-}2a,b,c)$$

We introduce a **similarity variable** by writing

$$\eta = x^\alpha t^\beta \qquad (6.3\text{-}3)$$

where $\alpha$ and $\beta$ are constants.

We assume $c$ to be a function of $\eta$ only. Its derivatives are given by

$$\frac{\partial c}{\partial t} = \frac{dc}{d\eta} \frac{\partial \eta}{\partial t} = \frac{dc}{d\eta} \beta x^\alpha t^{\beta-1} \qquad (6.3\text{-}4a,b)$$

$$\frac{\partial}{\partial x} \left( D \frac{\partial c}{\partial x} \right) = \alpha (\alpha - 1) x^{\alpha-2} t^\beta D \frac{dc}{d\eta} + \alpha^2 x^{2\alpha-2} t^{2\beta} \frac{d}{d\eta} \left( D \frac{dc}{d\eta} \right) \qquad (6.3\text{-}4c)$$

Substituting Equations (6.3-4a to c) into Equation (6.3-1) yields

$$\beta\, x^{\alpha}\, t^{\beta-1} \frac{dc}{d\eta} = \alpha\,(\alpha-1)\, x^{\alpha-2}\, t^{\beta} D\, \frac{dc}{d\eta} + \alpha^2\, x^{2\alpha-2}\, t^{2\beta} \frac{d}{d\eta}\left(D\, \frac{dc}{d\eta}\right) \tag{6.3-5}$$

We now substitute, wherever possible, the combination of $x$ and $t$ by $\eta$ and Equation (6.3-5) becomes

$$\beta\, \eta\left(\frac{x^2}{t}\right) \frac{dc}{d\eta} = \alpha\,(\alpha-1)\, \eta\, D\, \frac{dc}{d\eta} + \alpha^2\, \eta^2 \frac{d}{d\eta}\left(D\, \frac{dc}{d\eta}\right) \tag{6.3-6}$$

Both sides of Equation (6.3-6) are functions of $\eta$ only if $(x^2/t)$ is a function of $\eta$. We choose $\eta$ to be

$$\eta = x/\sqrt{t} \tag{6.3-7}$$

This choice implies that

$$\alpha = 1, \qquad \beta = -\frac{1}{2} \tag{6.3-8a,b}$$

Substituting Equations (6.3-8a, b) into Equation (6.3-6) yields

$$-\frac{1}{2}\, \eta\, \frac{dc}{d\eta} = \frac{d}{d\eta}\left(D\, \frac{dc}{d\eta}\right) \tag{6.3-9}$$

Note that the value of $\alpha$ has been chosen so that the first term on the right side of Equation (6.3-6) vanishes.

We now check that the initial and boundary conditions are consistent with the choice of $\eta$. From Equation (6.3-7), we deduce

$$x = 0 \Rightarrow \eta = 0, \quad x \longrightarrow \infty \Rightarrow \eta \longrightarrow \infty, \quad t = 0 \Rightarrow \eta \longrightarrow \infty \tag{6.3-10a to f}$$

Equations (6.3-2a to c, 10a to f) imply that a similarity solution exists if

$$c_1 = c_2 \tag{6.3-11}$$

If $D$ is a function of $c$, Equation (6.3-9) is a non-linear equation and we can solve it numerically. It is generally easier to solve an O.D.E. than a P.D.E. Thus, looking for a similarity solution can simplify the problem. On the curves $(\eta = x/\sqrt{t} = \text{constant})$, $c(\eta)$ is a constant, hence the name **similarity solution**.

To simplify the problem, we assume $D$ to be constant and

$$c_1 = c_2 = 0 \tag{6.3-12a,b}$$

Equation (6.3-9) becomes

$$\frac{d^2 c}{d\eta^2} + \frac{1}{2D} \eta \frac{dc}{d\eta} = 0 \tag{6.3-13}$$

The solution is

$$c = c_3 \int_0^\eta \exp(-\xi^2/4D)\, d\xi + c_4 \tag{6.3-14}$$

where $c_3$ and $c_4$ are constants.

Note that the lower limit of integration in Equation (6.3-14) is arbitrary. That is to say, had we integrated between the limits 1 and $\eta$, the constants $c_3$ and $c_4$ would have been adjusted accordingly. However, since the boundary conditions refer to $\eta = 0$, the obvious choice for the lower limit is zero.

Imposing the initial and boundary conditions [Equations (6.3-2a to c, 10a to f, 12a, b)] yields

$$c_4 = c_0, \qquad c_3 = -c_0 \Big/ \int_0^\infty \exp(-\xi^2/4D)\, d\xi = -c_0 \sqrt{\frac{1}{\pi D}} \tag{6.3-15a,b,c}$$

Equation (6.3-14) can now be written as

$$c = c_0 \left[ 1 - \sqrt{\frac{1}{\pi D}} \int_0^{x/\sqrt{t}} \exp(-\xi^2/4D)\, d\xi \right] \tag{6.3-16}$$

***Example 6.3-2.*** Obtain similarity solutions for the two-dimensional boundary layer flow over a flat plate of a Newtonian fluid. The two-dimensional boundary layer equation of a Newtonian fluid can be written via Equations (A.I-1, II-1) as

$$\frac{\partial v_x}{\partial x} + \frac{\partial v_y}{\partial y} = 0 \tag{6.3-17}$$

$$v_x \frac{\partial v_x}{\partial x} + v_y \frac{\partial v_x}{\partial y} = v_\infty \frac{dv_\infty}{dx} + v \frac{\partial^2 v_x}{\partial y^2} \tag{6.3-18}$$

where $v_x$ and $v_y$ are the x and y components of the velocity, $v_\infty$ is the free stream velocity, and $v$ is the kinematic viscosity (see also Rosenhead, 1963). We define a stream function $\psi(x, y)$ by

$$v_x = \frac{\partial \psi}{\partial y}, \qquad v_y = -\frac{\partial \psi}{\partial x} \qquad\qquad (6.3\text{-}19a,b)$$

Equation (6.3-17) is satisfied automatically and Equation (6.3-18) becomes

$$\frac{\partial \psi}{\partial y}\frac{\partial^2 \psi}{\partial x \partial y} - \frac{\partial \psi}{\partial x}\frac{\partial^2 \psi}{\partial y^2} = v_\infty \frac{dv_\infty}{dx} + v\frac{\partial^3 \psi}{\partial y^3} \qquad\qquad (6.3\text{-}20)$$

The appropriate boundary conditions are

$$y = 0, \qquad \frac{\partial \psi}{\partial x} = \frac{\partial \psi}{\partial y} = 0, \qquad \lim_{y \to \infty}\frac{\partial \psi}{\partial y} = v_\infty \qquad\qquad (6.3\text{-}21a,b,c,d)$$

We assume $\eta$ to be of a more general form than Equation (6.3-3) and write

$$\eta = y/\xi(x) \qquad\qquad (6.3\text{-}22)$$

where $\xi$ is to be specified later. Equation (6.3-20) suggests that $\psi$ can be written as

$$\psi = v_\infty(x)\,\xi(x)\,f(\eta) \qquad\qquad (6.3\text{-}23)$$

Computing the derivatives of $\psi$, we have

$$\frac{\partial \psi}{\partial x} = \frac{dv_\infty}{dx}\xi f + v_\infty \frac{d\xi}{dx}f + v_\infty \xi \frac{df}{d\eta}\frac{\partial \eta}{\partial x} = \frac{dv_\infty}{dx}\xi f + v_\infty \frac{d\xi}{dx}f - v_\infty \eta \frac{d\xi}{dx}\frac{df}{d\eta} \qquad (6.3\text{-}24a,b)$$

$$\frac{\partial \psi}{\partial y} = v_\infty \frac{df}{d\eta}, \qquad \frac{\partial^2 \psi}{\partial x \partial y} = \frac{dv_\infty}{dx}\frac{df}{d\eta} - \frac{v_\infty \eta}{\xi}\frac{d\xi}{dx}\frac{d^2 f}{d\eta^2} \qquad (6.3\text{-}24c,d)$$

$$\frac{\partial^2 \psi}{\partial y^2} = \frac{v_\infty}{\xi}\frac{d^2 f}{d\eta^2}, \qquad \frac{\partial^3 \psi}{\partial y^3} = \frac{v_\infty}{\xi^2}\frac{d^3 f}{d\eta^3} \qquad (6.3\text{-}24e,f)$$

Substituting Equations (6.3-24a to f) into Equation (6.3-20), we obtain after some simplification

$$\frac{d^3 f}{d\eta^3} + \left[\frac{\xi}{v}\frac{d}{dx}(\xi v_\infty)\right]f\frac{d^2 f}{d\eta^2} + \left(\frac{\xi^2}{v}\frac{dv_\infty}{dx}\right)\left[1 - \left(\frac{df}{d\eta}\right)^2\right] = 0 \qquad (6.3\text{-}25)$$

For the existence of similarity solutions, we require that

$$\frac{\xi}{v}\frac{d}{dx}(\xi v_\infty) = \alpha \qquad\qquad (6.3\text{-}26a)$$

$$\frac{\xi^2}{v} \frac{dv_\infty}{dx} = \beta \qquad\qquad (6.3\text{-}26b)$$

where $\alpha$ and $\beta$ are constants.

Equation (6.3-25) can now be written as

$$\frac{d^3f}{d\eta^3} + \alpha f \frac{d^2f}{d\eta^2} + \beta \left[ 1 - \left( \frac{df}{d\eta} \right)^2 \right] = 0 \qquad\qquad (6.3\text{-}27)$$

Equation (6.3-27) is the **Falker Skan equation**. Various values of $\alpha$ and $\beta$ are associated with various outer flow situations. The following cases have been examined.

**Blasius Flow**

In this case, we choose

$$\alpha = \frac{1}{2}, \qquad \beta = 0 \qquad\qquad (6.3\text{-}28a,b)$$

From Equations (6.3-26a, b, 28 a, b), we deduce

$$v_\infty = \text{constant}, \qquad \xi = (v\, x / v_\infty)^{1/2} \qquad\qquad (6.3\text{-}29a,b)$$

This flow corresponds to the flow over a flat plate and Equation (6.3-27) simplifies to

$$\frac{d^3f}{d\eta^3} + \frac{1}{2} f \frac{d^2f}{d\eta^2} = 0 \qquad\qquad (6.3\text{-}30)$$

The boundary conditions [Equations (6.3-21a to d)] are

$$f(0) = f'(0) = 0, \qquad f' \longrightarrow 1 \ \text{ as } \ \eta \longrightarrow \infty \qquad\qquad (6.3\text{-}31a,b,c)$$

where the prime denotes differentiation with respect to $\eta$.

Equation (6.3-30) is a third order non-linear O.D.E. and its solution, subject to the three boundary conditions, can be obtained numerically.

**Stagnation Flow**

Substituting the values of

$$\alpha = 1, \qquad \beta = 1 \qquad\qquad (6.3\text{-}32a,b)$$

into Equations (6.3-26a, b) leads to

$$v_\infty \frac{d\xi}{dx} = 0 \; , \qquad \frac{\xi^2}{v} \frac{dv_\infty}{dx} = 1 \tag{6.3-33a,b}$$

The solution can be written as

$$\xi = \sqrt{\frac{v}{c}} \; , \qquad v_\infty = c \, x \tag{6.3-34a,b}$$

where c is a constant.

Equation (6.3-27) and the boundary conditions are now

$$\frac{d^3 f}{d\eta^3} + f \frac{d^2 f}{d\eta^2} - \left(\frac{df}{d\eta}\right)^2 + 1 = 0 \tag{6.3-35}$$

and    $f(0) = f'(0) = 0 \; , \quad f' \longrightarrow 1 \quad \text{as} \quad \eta \longrightarrow \infty$ $\tag{6.3-36a,b,c}$

The system defined by Equations (6.3-35, 36a to c) can be solved numerically.

The solution of various flows is discussed in Rosenhead (1963).

***Example 6.3-3*. Stefan problems** are concerned with the phase change across a moving boundary. We consider the phase change from a solid to a liquid. The boundary between the solid [x < s (t)] and the liquid [x > s (t)] is given by

$$x = s(t) \tag{6.3-37}$$

and is shown in Figure 6.3-1.

We assume that there is no motion and that the temperature T, in both phases, satisfies a simplified form of Equation (A.III-1)

$$\rho \, c \, \frac{\partial T}{\partial t} = k \frac{\partial^2 T}{\partial x^2} \tag{6.3-38}$$

where $\rho$ is the density, c is the specific heat, and k is the thermal conductivity. We assume that they are constants and have the same values in both phases.

At the interface, the temperature is at the melting point $T_m$.

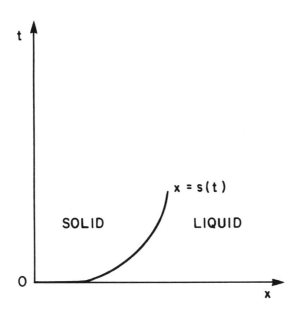

**FIGURE 6.3-1    Solid and liquid phase**

An energy balance yields the Stefan condition

$$\left[-k \frac{\partial T}{\partial x}\right]_{x=s_-}^{x=s_+} = L \frac{ds}{dt} \tag{6.3-39}$$

where $L$ is the latent heat.

We use **Birkhoff's method** to obtain the similarity variables.

A parameter $a$ is introduced and all variables are written as

$$\overline{t} = a^{\alpha_1} t , \qquad \overline{x} = a^{\alpha_2} x , \qquad \overline{T} = a^{\alpha_3} T \tag{6.3-40a,b,c}$$

where $\alpha_1$, $\alpha_2$, and $\alpha_3$ are constants.

In terms of the new variables, Equation (6.3-38) becomes

$$\rho \, c \, a^{\alpha_1 - \alpha_3} \frac{\partial \overline{T}}{\partial \overline{t}} = k \, a^{2\alpha_2 - \alpha_3} \frac{\partial^2 \overline{T}}{\partial \overline{x}^2} \tag{6.3-41}$$

The equation should be independent of $a$ and this implies that

$$\alpha_1 - \alpha_3 = 2\alpha_2 - \alpha_3 \tag{6.3-42}$$

We now combine the variables $x$, $t$, and $T$ in such a way as to reduce the number of variables.

We introduce a set of new variables defined by

$$\eta = \frac{x}{t^{\alpha_2/\alpha_1}}, \qquad \tau = \frac{T}{t^{\alpha_3/\alpha_1}} \qquad\qquad (6.3\text{-}43\text{a,b})$$

The ratios $\alpha_2/\alpha_1$ and $\alpha_3/\alpha_1$ can be obtained from Equation (6.3-42), leading to

$$1 - \frac{\alpha_3}{\alpha_1} = 2\frac{\alpha_2}{\alpha_1} - \frac{\alpha_3}{\alpha_1} \qquad\qquad (6.3\text{-}44)$$

One possible solution is

$$\frac{\alpha_3}{\alpha_1} = 0, \qquad \frac{\alpha_2}{\alpha_1} = \frac{1}{2} \qquad\qquad (6.3\text{-}45\text{a,b})$$

Combining Equations (6.3-43a, b, 45a, b), we obtain

$$\eta = \frac{x}{\sqrt{t}}, \qquad \tau = T \qquad\qquad (6.3\text{-}46\text{a,b})$$

Note that the same $\eta$ was obtained in Example 6.3-1.

We define

$$f(\eta) = \tau - T_m \qquad\qquad (6.3\text{-}47)$$

Equation (6.3-38) is reduced to an O.D.E. and is written as

$$\frac{d^2f}{d\eta^2} + \frac{1}{2D}\eta\frac{df}{d\eta} = 0 \qquad\qquad (6.3\text{-}48)$$

where $D = k/\rho c$.

For the interface, we assume that $s$ is given by

$$s = s_0\sqrt{t} \qquad\qquad (6.3\text{-}49)$$

where $s_0$ is a constant.

The interface ($x = s$) is now given by

$$\eta = s_0 \qquad\qquad (6.3\text{-}50)$$

The conditions at the interface are

$$f(s_0) = 0, \qquad \left[\frac{df}{d\eta}\right]_{s_{0_-}}^{s_{0_+}} = -\frac{L\, s_0}{2k} \qquad\qquad (6.3\text{-}51a,b)$$

where $s_{0_-}$ is in the solid phase and $s_{0_+}$ is in the liquid phase.

To complete the problem, we need to define the domain which we assume to be semi-infinite $(0 < x < \infty)$. The additional boundary conditions we impose are

$$f(0) = -1, \qquad \lim_{\eta \to \infty} f(\eta) = 0 \qquad\qquad (6.3\text{-}52a,b)$$

Equations (6.2-52a, b) imply that, at $x = 0$, the temperature is one degree below the melting point and beyond the interface, we have the liquid phase and the temperature is at the melting point ($f = 0$).

The solution of Equation (6.3-48) is

$$f = C_1 + C_0 \int_0^\eta e^{-\xi^2/4D}\, d\xi \qquad\qquad (6.3\text{-}53)$$

The constants $C_0$ and $C_1$ are determined from Equations (6.3-51a, 52a) and $f$ becomes

$$f = -1 + \frac{\displaystyle\int_0^\eta e^{-\xi^2/4D}\, d\xi}{\displaystyle\int_0^{s_0} e^{-\xi^2/4D}\, d\xi}, \qquad 0 < \eta < s_0 \qquad\qquad (6.3\text{-}54a)$$

$$f = 0, \qquad\qquad\qquad\qquad \eta > s_0 \qquad\qquad (6.3\text{-}54b)$$

Imposing Equation (6.3-51b) yields

$$\frac{L s_0}{2k}\, e^{s_0^2/4D} \int_0^{s_0} e^{-\xi^2/4D}\, d\xi = 1 \qquad\qquad (6.3\text{-}55)$$

The interface ($s_0$) is obtained by solving Equation (6.3-55).

Crank (1984) has discussed Stefan problems in detail.

In this section, we have introduced three methods of obtaining similarity variables. These methods can be extended to more than two independent variables. Further details can be found in Ames (1972) and in Bluman and Cole (1974).

## 6.4   GREEN'S FUNCTIONS

### Dirichlet Problems

Green's functions were used to solve boundary value problems in Chapter 1. This method is widely used to solve elliptic partial differential equations. In Chapter 4, it is shown, as an application of Green's theorem, that the solution of Poisson's equation is known if the Green's function for that problem is available. In particular, the Dirichlet problem defined by

$$\nabla^2 u = \rho\,(x, y, z)\,, \qquad \text{in } V \tag{6.4-1a}$$

$$u = f\,(x, y, z)\,, \qquad \text{on } S \tag{6.4-1b}$$

where $S$ is the surface enclosing $V$. The value of $u$ at a point $P$ is given by

$$u\,(P) = \frac{1}{4\pi}\left[\iiint\limits_V G\,\rho\;dV + \iint\limits_S f\,\frac{\partial G}{\partial n}\;dS\right] \tag{6.4-2}$$

where $G$ is the Green's function for the problem.

Similarly, in a plane, if

$$\nabla^2 u = \rho\,(x, y), \qquad \text{in } A \tag{6.4-3a}$$

$$u = f\,(x, y), \qquad \text{on } C \tag{6.4-3b}$$

where $C$ is the curve enclosing the domain $A$,

$$u\,(P) = \frac{1}{2\pi}\left[\iint\limits_A G\,\rho\;dA + \int\limits_C f\,\frac{\partial G}{\partial n}\;ds\right] \tag{6.4-4}$$

$G$ is the two-dimensional Green's function.

Note that some authors include the factors $(1/4\pi, 1/2\pi)$ in the definition of $G$. These factors are then absent in Equations (6.4-2, 4).

We now have to compute $G$. The conditions which $G$ has to satisfy in the case of ordinary differential equations are stated in Chapter 1. These conditions can be extended to partial differential equations. In Chapter 4, it is stated that $G$ has to satisfy the following conditions

(a)     $\nabla^2 G = 0$,             everywhere in $V$ except at $P$                                    (6.4-5a)

(b)     $G = 0$,                on S                                   (6.4-5b)

(c)     $G \approx -1/r$,        near P                                (6.4-5c)

where $r$ is the radial distance from P.

A simple physical interpretation of G is that it is the potential due to placing a unit charge at P and earthing the surface S. We can rewrite Equation (6.4-5a) in terms of the Dirac delta function. The two-dimensional delta function $\delta(x - x_0, y - y_0)$ is written as a product of two delta functions in the following manner

$$\delta(x - x_0, y - y_0) = \delta(x - x_0)\,\delta(y - y_0) \tag{6.4-6}$$

Using the definition of the delta function [Equation (1.17-18d)], we have for any differentiable function $f(x, y)$

$$\iint\limits_{-\infty}^{\infty} f(x, y)\,\delta(x - x_0, y - y_0)\,dx\,dy = \iint\limits_{-\infty}^{\infty} f(x, y)\,\delta(x - x_0)\,\delta(y - y_0)\,dx\,dy \tag{6.4-7a}$$

$$= \int_{-\infty}^{\infty} f(x_0, y)\,\delta(y - y_0)\,dy = f(x_0, y_0) \tag{6.4-7b,c}$$

Similarly, in the three-dimensional case,

$$\delta(x - x_0, y - y_0, z - z_0) = \delta(x - x_0)\,\delta(y - y_0)\,\delta(z - z_0) \tag{6.4-8a}$$

$$\iiint\limits_{-\infty}^{\infty} f(x, y, z)\,\delta(x - x_0)\,\delta(y - y_0)\,\delta(z - z_0)\,dx\,dy\,dz = f(x_0, y_0, z_0) \tag{6.4-8b}$$

If P is the point $(x_0, y_0, z_0)$, Equation (6.4-5a) can be written as

(a')    $\nabla^2 G = 4\pi\,\delta(x - x_0)\,\delta(y - y_0)\,\delta(z - z_0)$                    (6.4-5a')

In the two-dimensional case, G satisfies the following conditions

(a)     $\nabla^2 G = 2\pi\,\delta(x - x_0)\,\delta(y - y_0)$                              (6.4-9a)

(b)     $G = 0$,                on C                                   (6.4-9b)

(c)     $G \approx \ln r$,        near P                                (6.4-9c) where $r$ is the radial distance from $P$.

Note that near $P$, $G$ behaves as the fundamental solution of Laplace's equation [Equations (5.5-13b, 39)] and $G$ satisfies Equations (6.4-5a' or 9a).

Condition (c) for the three and two-dimensional cases can respectively be replaced by

$$\iint_{S_\varepsilon} \frac{\partial G}{\partial n} \, dS = 4\pi \qquad\qquad (6.4\text{-}5c')$$

$$\int_{C_\varepsilon} \frac{\partial G}{\partial n} \, ds = 2\pi \qquad\qquad (6.4\text{-}9c')$$

where $S_\varepsilon$ and $C_\varepsilon$ are respectively the spherical surface and circumference that enclose the point $P$. Each is of radius $\varepsilon$ with $\varepsilon \longrightarrow 0$.

Since the Laplacian is **self-adjoint**, $G$ is **symmetric**, that is

$$G\,(x, y, z; x_0, y_0, z_0) = G\,(x_0, y_0, z_0; x, y, z) \qquad\qquad (6.4\text{-}10)$$

This can be deduced from physical considerations. The potential at $(x, y, z)$ due to a unit charge at $(x_0, y_0, z_0)$ is the same as the potential at $(x_0, y_0, z_0)$ due to a unit charge at $(x, y, z)$.

It is generally not easy to calculate $G$, except for simple domains such as a half space or a circle. We next deduce $G$ for simple geometries by the method of images.

### Upper half plane

Consider the upper half plane $z \geq 0$. The function $G$ is the potential of a unit charge placed at $(x_0, y_0, z_0)$ and the plane $z = 0$ is kept at zero potential. To maintain the plane $z = 0$ at zero potential, we place a unit charge of opposite sign at $(x_0, y_0, -z_0)$. The Green's function $G\,(x, y, z; x_0, y_0, z_0)$ is given by

$$G = \frac{1}{r'} - \frac{1}{r} \qquad\qquad (6.4\text{-}11a)$$

$$r = \sqrt{(x - x_0)^2 + (y - y_0)^2 + (z - z_0)^2} \qquad\qquad (6.4\text{-}11b)$$

$$r' = \sqrt{(x - x_0)^2 + (y - y_0)^2 + (z + z_0)^2} \qquad\qquad (6.4\text{-}11c)$$

In the two-dimensional case ($y \geq 0$), $G$ is given by

$$G = \frac{1}{2}\ln\left[(x - x_0)^2 + (y - y_0)^2\right] - \frac{1}{2}\ln\left[(x - x_0)^2 + (y + y_0)^2\right] \qquad (6.4\text{-}12)$$

### *Sphere and circle*

Consider a sphere of radius $a$ with a unit negative charge placed at P. The point P is at a distance $b$ from O, the centre of the sphere, as shown in Figure 6.4-1. The image system that produces a zero potential on the surface of the sphere is a charge $a/b$ placed at the inverse point P' of P. The **inverse point** P' is defined such that

$$OP \cdot OP' = a^2 \qquad (6.4\text{-}13)$$

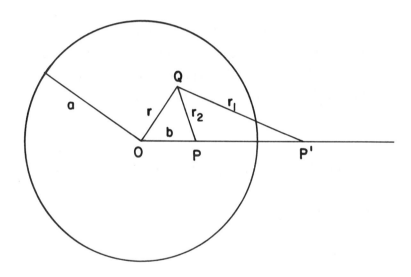

**FIGURE 6.4-1    Image system for a sphere and a circle**

If Q is any point, G is given by

$$G = -\frac{1}{r_2} + \frac{a}{br_1} \qquad (6.4\text{-}14)$$

where $r_1$ and $r_2$ are the distances from Q to P' and P respectively.

On introducing spherical polar coordinates $(r, \theta, \phi)$, the vector positions of Q, P, and P' can be written respectively as

$$\overrightarrow{OQ} = r\left[\sin\theta\cos\phi, \ \sin\theta\sin\phi, \ \cos\theta\right] \qquad (5.4\text{-}15a)$$

$$\overrightarrow{OP} = b\left[\sin\theta_0 \cos\phi_0, \ \sin\theta_0 \sin\phi_0, \ \cos\theta_0\right] \qquad (6.4\text{-}15b)$$

$$\overrightarrow{OP}\,' = \frac{a^2}{b}\left[\sin\theta_0 \cos\phi_0, \ \sin\theta_0 \sin\phi_0, \ \cos\theta_0\right] \qquad (6.4\text{-}15c)$$

It follows that

$$r_1^2 = (\overrightarrow{OQ} - \overrightarrow{OP}\,') \cdot (\overrightarrow{OQ} - \overrightarrow{OP}\,') = r^2 + a^4/b^2 - 2\,\overrightarrow{OQ} \cdot \overrightarrow{OP}\,' \qquad (6.4\text{-}15d,e)$$

$$= r^2 + a^4/b^2 - 2\,(a^2 r/b)\cos\Theta \qquad (6.4\text{-}15f)$$

$$r_2^2 = (\overrightarrow{OQ} - \overrightarrow{OP}) \cdot (\overrightarrow{OQ} - \overrightarrow{OP}) = r^2 + b^2 - 2r\,b\cos\Theta \qquad (6.4\text{-}15g,h)$$

$$\cos\Theta = \sin\theta\,\sin\theta_0 \cos(\phi - \phi_0) + \cos\theta\,\cos\theta_0 \qquad (6.4\text{-}15i)$$

In the two-dimensional case, we suppose the circle to be of radius a with a unit charge placed at P. Again, the point P is at a distance b from O, the centre of the circle. The image system is a charge of opposite sign placed at P', the inverse of P. Equation (6.4-13) defines the point P'. The appropriate G at any point Q which satisfies G = 0 on the circle is

$$G = \ell n \ ar_2/(br_1) \qquad (6.4\text{-}16)$$

where $r_1$ and $r_2$ are the distances from Q to P' and P respectively.

In polar coordinates $(r, \theta)$, $r_1$ and $r_2$ are given by

$$r_1^2 = r^2 + a^4/b^2 - 2\,(a^2 r/b)\cos(\theta - \theta_0) \qquad (6.4\text{-}17a)$$

$$r_2^2 = r^2 + b^2 - 2\,r b\cos(\theta - \theta_0) \qquad (6.4\text{-}17b)$$

where $\theta$ and $\theta_0$ are the polar angles of OQ and OP (OP') respectively.

The method of images is simple and the image system for other simple boundaries can be found in Ferraro (1956). There are other techniques of obtaining G. Among them are splitting G into the sum of two functions $G_1$ and $G_2$. We can expand G in terms of eigenfunctions. We illustrate these methods by considering the following examples.

***Example 6.4-1.*** Determine the Green's function for the Dirichlet problem in the unit circle.

We need to find a function G that satisfies Equations (6.4-9a to c). We write G as a sum of two functions

$$G = G_1 + G_2 \qquad (6.4\text{-}18)$$

Let $G_1$ be the fundamental solution. It satisfies Equations (6.4-9a, c). If $P$ is the point $(x_0, y_0)$, $G_1$ is given by

$$G_1 = \ln r_1 \tag{6.4-19a}$$

where

$$r_1 = \sqrt{(x - x_0)^2 + (y - y_0)^2} \tag{6.4-19b}$$

The function $G_1$ does not satisfy the boundary condition. We choose $G_2$ such that the boundary condition is satisfied. That is to say, on $C$, the unit circle, we have

$$G_2 = -G_1 \tag{6.4-20}$$

In addition, $G_2$ has to satisfy Laplace's equation and should not have a singularity inside the unit circle. In Chapter 5, using the method of separation of variables, we have found that $G_2$ is given by Equation (5.5-14) and, for convenience, it is reproduced here.

$$G_2 = E_0 + \sum_{m=1}^{\infty} [A_m \cos m\theta + B_m \sin m\theta] \, r^m \tag{5.5-14}$$

Using the polar coordinate system, we denote $(x_0, y_0)$ by $(b \cos\theta_0, b \sin\theta_0)$ and $r_1$ is given by

$$r_1 = \sqrt{(r\cos\theta - b\cos\theta_0)^2 + (r\sin\theta - b\sin\theta_0)^2} \tag{6.4-21a}$$

$$= \sqrt{r^2 + b^2 - 2rb\cos(\theta - \theta_0)} \tag{6.4-21b}$$

Combining Equations (6.4-19a, 20, 21a, b) yields

$$E_0 + \sum_{m=1}^{\infty} [A_m \cos m\theta + B_m \sin m\theta] = -\frac{1}{2} \ln [1 + b^2 - 2b\cos(\theta - \theta_0)] \tag{6.4-22a}$$

$$= -\frac{1}{2} \ln [(1 - be^{i(\theta-\theta_0)})(1 - be^{-i(\theta-\theta_0)})] \tag{6.4-22b}$$

$$= -\frac{1}{2} [\ln(1 - be^{i(\theta-\theta_0)}) + \ln(1 - be^{-i(\theta-\theta_0)})] \tag{6.4-22c}$$

$$= \frac{1}{2} \left[ \sum_{m=1}^{\infty} \frac{b^m}{m} (e^{im(\theta-\theta_0)} + e^{-im(\theta-\theta_0)}) \right] \tag{6.4-22d}$$

$$= \sum_{m=1}^{\infty} \frac{b^m}{m} \cos m(\theta - \theta_0) \tag{6.4-22e}$$

$$= \sum_{m=1}^{\infty} \frac{b^m}{m} (\cos m\theta \, \cos m\theta_0 + \sin m\theta \, \sin m\theta_0) \quad (6.4\text{-}22f)$$

Comparing the coefficients of $\cos m\theta$ and $\sin m\theta$, we obtain

$$E_0 = 0, \quad A_m = (b^m \cos m\theta_0)/m, \quad B_m = (b^m \sin m\theta_0)/m \quad (6.4\text{-}23a,b,c)$$

Combining Equations (5.5-14, 6.4-23a to c) yields

$$G_2 = \sum_{m=1}^{\infty} \frac{b^m}{m} [\cos m\theta_0 \, \cos m\theta + \sin m\theta_0 \, \sin m\theta] \, r^m \quad (6.4\text{-}24a)$$

$$= \sum_{m=1}^{\infty} \frac{(br)^m}{m} [\cos m\theta_0 \, \cos m\theta + \sin m\theta_0 \, \sin m\theta] \quad (6.4\text{-}24b)$$

$$= -\frac{1}{2} \ell n \, [1 + b^2 r^2 - 2br \cos(\theta - \theta_0)] \quad (6.4\text{-}24c)$$

From Equations (6.4-18, 19a, 21b, 24c), we deduce that $G$ is given by

$$G = \frac{1}{2} \ell n \left[ \frac{r^2 + b^2 - 2rb \cos(\theta - \theta_0)}{1 + b^2 r^2 - 2br \cos(\theta - \theta_0)} \right] \quad (6.4\text{-}25)$$

We have obtained $G$ inside the unit circle, and we note that $G$ obtained by this method [Equation (6.4-25)] is identical to the one obtained by the method of images [Equations (6.4-16, 17a, b)]. We can equally derive $G$ for the region outside the unit circle. In this case, Equation (5.5-14) is no longer appropriate and the expansion in Equation (6.4-22c) has to be modified.

*Example 6.4-2*. Solve Poisson's equation

$$\nabla^2 u = \rho \qquad \text{inside the unit circle} \quad (6.4\text{-}26a)$$

$$u = f \qquad \text{on the unit circle} \quad (6.4\text{-}26b)$$

Equation (6.4-4) enables us to write the value of $u$ at a point $P$ as follows

$$u(P) = \frac{1}{2\pi} \left[ \int_0^1 \int_0^{2\pi} \rho G r \, d\theta \, dr + \int_0^{2\pi} f \frac{\partial G}{\partial r} \, d\theta \right] \quad (6.4\text{-}27)$$

From Equation (6.4-25), we obtain

$$\left.\frac{\partial G}{\partial r}\right|_{r=1} = \frac{1-b^2}{1-2b\cos(\theta-\theta_0)+b^2} \tag{6.4-28}$$

The combination of Equations (6.4-25, 27, 28) yields the value of u at $(b\cos\theta_0, b\sin\theta_0)$. If $\rho$ is zero, u reduces to

$$u\,(b\cos\theta_0, b\sin\theta_0) = \frac{1}{2\pi} \int_0^{2\pi} \frac{f\,(1-b^2)}{1-2b\cos(\theta-\theta_0)+b^2}\,d\theta \tag{6.4-29}$$

Equation (6.4-29) is **Poisson's integral** which is the solution of Laplace's equation satisfying Equation (6.4-26b).

***Example 6.4-3.*** Obtain the Green's function G inside the rectangle $0 \le x \le a$, $0 \le y \le b$ for the Dirichlet problem.

We expand $G\,(x, y\,; x_0, y_0)$ as

$$G\,(x, y; x_0, y_0) = \sum_{\substack{m=1 \\ n=1}}^{\infty \atop \infty} c_{mn}(x_0, y_0)\, V_{mn}(x, y) \tag{6.4-30}$$

where $V_{mn}$ are the eigenfunctions defined by

$$\nabla^2 V_{mn} = \lambda_{mn} V_{mn} \tag{6.4-31a}$$

$$V_{mn}\,(0, y) = V_{mn}\,(a, y) = V_{mn}\,(x, 0) = V_{mn}\,(x, b) = 0 \tag{6.4-31b,c,d,e}$$

By the method of separation of variables or by inspection, the eigenfunctions $V_{mn}$ and the eigenvalues $\lambda_{mn}$ are given by

$$V_{mn} = C \sin\left(\frac{m\pi x}{a}\right) \sin\left(\frac{n\pi y}{b}\right) \tag{6.4-32a}$$

$$\lambda_{mn} = -\pi^2\left(\frac{m^2}{a^2} + \frac{n^2}{b^2}\right) \tag{6.4-32b}$$

where C is a constant, m and n are integers.

We normalize the eigenfunctions, that is to say, we require

$$\int_0^a \int_0^b V_{mn}^2 \, dy \, dx = 1 = C^2 \int_0^a \int_0^b \sin^2 \frac{m\pi x}{a} \sin^2 \frac{n\pi y}{b} \, dy \, dx = C^2 \frac{ab}{4} \qquad (6.4\text{-}33a,b,c)$$

The normalized eigenfunctions are then

$$V_{mn} = \frac{2}{\sqrt{ab}} \sin\left(\frac{m\pi x}{a}\right) \sin\left(\frac{n\pi y}{b}\right) \qquad (6.4\text{-}34)$$

From Equations (6.4-9a, 30, 31a, 34), we obtain

$$\nabla^2 G = 2\pi \, \delta(x - x_0) \, \delta(y - y_0) = \frac{2}{\sqrt{ab}} \sum_{m,n} c_{mn} \lambda_{mn} \sin\left(\frac{m\pi x}{a}\right) \sin\left(\frac{n\pi y}{b}\right) \qquad (6.4\text{-}35a,b)$$

The Fourier coefficients $c_{mn}$ are given by

$$c_{mn} = \frac{4\pi}{\lambda_{mn} \sqrt{ab}} \int_0^a \int_0^b \sin\left(\frac{m\pi x}{a}\right) \sin\left(\frac{n\pi y}{b}\right) \delta(x - x_0) \, \delta(y - y_0) \, dx \, dy \qquad (6.4\text{-}36a)$$

$$= \frac{4\pi}{\lambda_{mn} \sqrt{ab}} \sin\left(\frac{m\pi x_0}{a}\right) \sin\left(\frac{n\pi y_0}{b}\right) \qquad (6.4\text{-}36b)$$

Substituting $c_{mn}$ into Equation (6.4-30), we obtain

$$G = -\frac{8\pi}{ab} \sum_{m,n} \frac{\sin(m\pi x/a) \sin(m\pi x_0/a) \sin(n\pi y/b) \sin(n\pi y_0/b)}{(m\pi/a)^2 + (n\pi/b)^2} \qquad (6.4\text{-}37)$$

## Neumann Problems

In this case, it is the normal derivative $\partial u / \partial n$ which is given on the surface. It is tempting to extend the theory developed for Dirichlet problems to this case, by simply changing the boundary condition $G = 0$ to $\partial G / \partial n = 0$. This is not as straightforward as it may seem. We examine some of the causes.

From Equation (4.4-22), we deduce by setting $\varphi = 1$ and $\psi = G$

$$\iint_S \frac{\partial G}{\partial n} \, dS = \iiint_V \nabla^2 G \, dV \qquad (6.4\text{-}38)$$

$$\iint\limits_{S} \frac{\partial G}{\partial n}\, dS = \iiint\limits_{V} \nabla^2 G\, dV \qquad (6.4\text{-}38)$$

Let P be the singular point. By definition $\nabla^2 G = 0$ everywhere except at P. Enclose P by a sphere of radius $\varepsilon$, denote the surface of this sphere by $S_\varepsilon$ and the volume by $V_\varepsilon$. In the volume $V - V_\varepsilon$, $\nabla^2 G$ is zero and, from Equation (6.4-38), we obtain

$$\iint\limits_{S} \frac{\partial G}{\partial n}\, dS + \iint\limits_{S_\varepsilon} \frac{\partial G}{\partial n}\, dS = 0 \qquad (6.4\text{-}39)$$

Imposing the condition $\partial G/\partial n = 0$ on S, implies

$$\iint\limits_{S_\varepsilon} \frac{\partial G}{\partial n}\, dS = 0 \qquad (6.4\text{-}40)$$

Equation (6.4-5c') contradicts Equation (6.4-40). Note that if G satisfies Laplace's equation everywhere in V, the boundary condition must satisfy the left side of Equation (6.4-38). One possible physical interpretation is to consider the problem of heat conduction. At steady state, the temperature satisfies Equation (5.4-44) which is Laplace's equation. To maintain a steady temperature, the heat flux across the boundary must be zero and this is expressed by the left side of Equation (6.4-38) where G is the temperature.

In Example 5.7-3, it is stated that the solution to a Neumann problem is arbitrary to the extent of a constant. It follows that, if we use the method adopted in Example 6.4-1 to determine G, the function $G_2$ will not be uniquely defined. Also, in Example 6-4-3, for a Neumann problem the eigenfunctions $V_{mn}$ are of the form $\cos m\pi x \cos n\pi y$ and one of the eigenvalues $\lambda_{mn}$ can be zero (m = 0, n = 0) leading to the impossibility of determining $c_{mn}$. Thus, modifying the boundary condition is not sufficient and we need to impose additional conditions. One possibility is to modify condition (a) [Equation (6.4-5a or 9a)]. We impose the following conditions on the Green's function for the **Neumann problem** (N)

(a) $\quad \nabla^2 N = A\ (\text{constant})\,,\quad$ everywhere in V except at P $\qquad (6.4\text{-}41a)$

$$= A + 4\pi\,\delta\,(\underline{r} - \underline{r}_0) \qquad (6.4\text{-}41b)$$

(b) $\quad \dfrac{\partial N}{\partial n} = 0\,,$ on S $\qquad (6.4\text{-}41c)$

(c) $\quad N \approx -1/r\,,$ near P $\qquad (6.4\text{-}41d)$

In Equation (6.4-41b), $\underline{r}$ is the vector position of any point and $\underline{r}_0$ is the vector position of P. From Equations (6.4-5c', 38, 41a, c), we deduce that

$$\iiint\limits_V A\, dV = A\, V = 4\pi \qquad\qquad\qquad (6.4\text{-}42a,b)$$

The **two-dimensional Neumann function** satisfies the following conditions

(a) $\qquad \nabla^2 N = A_1 + 2\pi\, \delta\,(\underline{r} - \underline{r}_0) \qquad\qquad\qquad (6.4\text{-}43a)$

(b) $\qquad \dfrac{\partial N}{\partial n} = 0\,,\ \text{on}\ C \qquad\qquad\qquad (6.4\text{-}43b)$

(c) $\qquad N \approx \ell n\, r\,,\ \text{near}\ P \qquad\qquad\qquad (6.4\text{-}43c)$

where $A_1$ is a constant.

Poisson's equation along with the Neumann condition are written as

$$\nabla^2 u = \rho\,(x, y, z)\,, \qquad \text{in}\ V\ (\text{or}\ S\ \text{in a plane}) \qquad (6.4\text{-}44a)$$

$$\frac{\partial u}{\partial n} = h\,(x, y, z)\,, \qquad \text{on}\ S\ (\text{or}\ C\ \text{in a plane}) \qquad (6.4\text{-}44b)$$

Substituting Equations (6.4-41b, c, 44a, b) into Equation (4.4-22) and setting $\psi = N$ and $\varphi = u$ yields

$$-\iint\limits_S Nh\, dS = \iiint\limits_V [Au + 4\pi u\, \delta\,(\underline{r} - \underline{r}_0) - N\rho]\, dV \qquad (6.4\text{-}45a)$$

$$= A \iiint\limits_V u\, dV + 4\pi\, u\,(P) - \iiint\limits_V N\rho\, dV \qquad (6.4\text{-}45b)$$

From Equation (6.4-45b), we obtain

$$u\,(P) = \frac{1}{4\pi} \left[ \iiint\limits_V N\rho\, dV - \iint\limits_S Nh\, dS \right] + \text{constant} \qquad (6.4\text{-}46)$$

where the constant term is $A \iiint\limits_V u \, dV$. This term can be ignored since the solution to a Neumann problem is arbitrary to the extent of a constant. The solution of Equations (6.4-44a, b) is given by Equation (6.4-46) and can be calculated once $N$ is known.

The analog of Equation (6.4-46) in two-dimensions is

$$u(P) = \frac{1}{2\pi} \left[ \iint\limits_S N\rho \, dS - \int\limits_C Nh \, ds \right] \tag{6.4-47}$$

Neumann problems are usually associated with hydrodynamics where the usual boundary condition is that the normal velocity ($\partial u / \partial n$, u is the potential) at the boundary is zero. We can determine $N$ by the method of images. We illustrate this method in the next example.

***Example 6.4-4.*** Find the Neumann function $N$ for the plane $y \geq 0$.

A source is placed at $(x_0, y_0)$. In order for the fluid not to cross the line $y = 0$, we need to place another source of equal strength at $(x_0, -y_0)$. If $N$ is the potential, $N$ is given by (Milne-Thomson, 1967)

$$N = \frac{1}{2} \ln \left[ (x - x_0)^2 + (y - y_0)^2 \right] + \frac{1}{2} \ln \left[ (x - x_0)^2 + (y + y_0)^2 \right] \tag{6.4-48a}$$

$$= \frac{1}{2} \ln \left[ (x - x_0)^2 + (y - y_0)^2 \right] \left[ (x - x_0)^2 + (y + y_0)^2 \right] \tag{6.4-48b}$$

On differentiating partially with respect to $y$, we obtain

$$\frac{\partial N}{\partial y} = \frac{(y - y_0)}{(x - x_0)^2 + (y - y_0)^2} + \frac{(y + y_0)}{(x - x_0)^2 + (y + y_0)^2} \tag{6.4-49}$$

It follows that

$$\left. \frac{\partial N}{\partial y} \right|_{y=0} = 0 \tag{6.4-50}$$

●

Note that, in hydrodynamics, the image of a source is another source and not a sink. This is the difference between hydrodynamics and electrostatics. Further discussions on the image system in hydrodynamics can be found in Milne-Thomson (1967). Neumann's functions can also be constructed by the method of eigenfunction expansion as explained in the next example.

*Example 6.4-5.* Find Neumann's function $N$ inside the rectangle $0 \leq x \leq a$, $0 \leq y \leq b$.

In Example 6-4-3, Green's function for the Dirichlet problem was derived. We use the same method to determine $N$. In this example, the eigenfunctions $w_{mn}$ have to satisfy

$$\frac{\partial w_{mn}}{\partial x}\bigg|_{\substack{y=0 \\ y=b}} = \frac{\partial w_{mn}}{\partial y}\bigg|_{\substack{x=0 \\ x=a}} = 0 \qquad (6.4\text{-}51\text{a,b})$$

The appropriate eigenfunctions are

$$w_{mn} = C \cos \frac{m\pi x}{a} \cos \frac{n\pi y}{b}, \qquad m = 0, 1, \dots, \quad n = 0, 1, \dots \qquad (6.4\text{-}52)$$

Note that, in this case, unlike in Example 6-4-3, $m$ and $n$ can be zero. The normalized eigenfunctions are

$$w_{mn} = \frac{2}{\sqrt{ab}} \cos \frac{m\pi x}{a} \cos \frac{n\pi y}{b}, \qquad m \neq 0, \quad n \neq 0 \qquad (6.4\text{-}53\text{a})$$

$$w_{m0} = \sqrt{\frac{2}{ab}} \cos \frac{m\pi x}{a}, \qquad\qquad m \neq 0, \quad n = 0 \qquad (6.4\text{-}53\text{b})$$

$$w_{0n} = \sqrt{\frac{2}{ab}} \cos \frac{n\pi y}{b}, \qquad\qquad m = 0, \quad n \neq 0 \qquad (6.4\text{-}53\text{c})$$

Proceeding as in Example 6.4-4, we obtain

$$N = -\frac{8\pi}{ab} \sum_{\substack{m=1 \\ n=1}}^{\infty} \frac{\cos(m\pi x/a) \cos(m\pi x_0/a) \cos(n\pi y/b) \cos(n\pi y_0/b)}{(m\pi/a)^2 + (n\pi/b)^2}$$

$$-\frac{4\pi}{ab} \sum_{m=1}^{\infty} \frac{\cos(m\pi x/a) \cos(m\pi x_0/a)}{(m\pi/a)^2} - \frac{4\pi}{ab} \sum_{n=1}^{\infty} \frac{\cos(n\pi y/b) \cos(n\pi y_0/b)}{(n\pi/b)^2} \qquad (6.4\text{-}54)$$

$N$ is defined to the extent of an arbitrary constant.

Once $N$ is known, the solution of Poisson's (or Laplace's) equation is found by substituting $N$ into Equation (6.4-46 or 47).

## Mixed Problems (Robin's Problems)

For mixed problems, (u is given in parts of the boundary and $\partial u / \partial n$ in the remaining parts of the boundary), we combine the techniques explained previously. The extension is straightforward and is illustrated in the next example.

***Example 6.4-6.*** Solve the equation

$$\frac{\partial^2 u}{\partial x^2} + \frac{\partial^2 u}{\partial y^2} = 0, \quad x \geq 0, \ y \geq 0 \tag{6.4-55}$$

subject to

$$u = f(x), \quad \text{on} \ y = 0 \tag{6.4-56a,b}$$

$$\frac{\partial u}{\partial x} = h(y), \quad \text{on} \ x = 0 \tag{6.4-56c,d}$$

by the method of Green's functions.

We first determine the Green's function G for this mixed boundary value problem by the method of images. Let P be the point $(x_0, y_0)$ as shown in Figure 6.4-2. The function G has to satisfy

$$G = 0, \quad \text{on} \ y = 0, \ x \geq 0 \tag{6.4-57a,b}$$

$$\frac{\partial G}{\partial x} = 0, \quad \text{on} \ x = 0, \ y \geq 0 \tag{6.4-57c,d}$$

Relative to $x = 0$, we have a Neumann problem and the image system is another source at Q $(-x_0, y_0)$. The boundary $y = 0$ defines a Dirichlet problem and the image system are charges of opposite sign at R $(-x_0, -y_0)$ and S $(x_0, -y_0)$. The function G $(x, y; x_0, y_0)$ is given by

$$G = \frac{1}{2} \ell n \left[ (x - x_0)^2 + (y - y_0)^2 \right] + \frac{1}{2} \ell n \left[ (x + x_0)^2 + (y - y_0)^2 \right]$$

$$- \frac{1}{2} \ell n \left[ (x + x_0)^2 + (y + y_0)^2 \right] - \frac{1}{2} \ell n \left[ (x - x_0)^2 + (y + y_0)^2 \right] \tag{6.4-58a}$$

$$= \frac{1}{2} \ell n \frac{\left[ (x - x_0)^2 + (y - y_0)^2 \right] \left[ (x + x_0)^2 + (y - y_0)^2 \right]}{\left[ (x + x_0)^2 + (y + y_0)^2 \right] \left[ (x - x_0)^2 + (y + y_0)^2 \right]} \tag{6.4-58b}$$

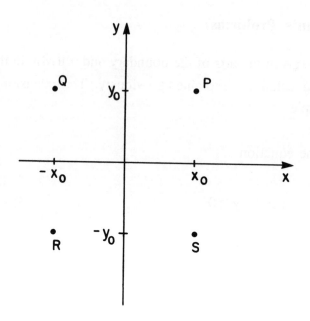

**FIGURE 6.4-2    Image system of a mixed boundary value problem**

From Equations (6.4-4, 47), we deduce that  u (P)  is given by

$$u(P) = \frac{1}{2\pi} \int_C (f\frac{\partial G}{\partial n} - Gh)\, ds \tag{6.4-59}$$

where  C  is the contour enclosing the positive quadrant.  That is to say, along the positive  x  and  y  axes.  From Equations (6.4-57a to d), we deduce that Equation (6.4-59) can be written as

$$u(P) = \frac{1}{2\pi}\left[\int_0^\infty f\frac{\partial G}{\partial n}\, dx - \int_\infty^0 Gh\, dy\right] \tag{6.4-60}$$

Note that the contour integral is taken in the counter clockwise sense, so that the integral along the y-axis is from  ∞  to  0.  On the x-axis

$$\frac{\partial}{\partial n} = -\frac{\partial}{\partial y} \tag{6.4-61}$$

Combining Equations (6.4-60, 61) yields

$$u\,(P) \;=\; \frac{1}{2\pi}\left[-\int_0^\infty f\left.\frac{\partial G}{\partial y}\right|_{y=0} dx \;+\; \int_0^\infty G\big|_{x=0}\,h\;dy\right] \tag{6.4-62}$$

From Equation (6.4-58b), we obtain

$$\left.\frac{\partial G}{\partial y}\right|_{y=0} \;=\; \frac{-2y_0\,[(x+x_0)^2 + (x-x_0)^2 + 2y_0^2]}{[(x+x_0)^2 + y_0^2]\,[(x-x_0)^2 + y_0^2]} \tag{6.4-63a}$$

$$G\big|_{x=0} \;=\; \ell n\,\frac{[x_0^2 + (y-y_0)^2]}{[x_0^2 + (y+y_0)^2]} \tag{6.4-63b}$$

Substituting Equations (6.4-63a, b) into Equation (6.4-62), we obtain

$$u\,(x_0,\,y_0) \;=\; \frac{1}{2\pi}\left\{2y_0\int_0^\infty \frac{[(x+x_0)^2 + (x-x_0)^2 + 2y_0^2]}{[(x+x_0)^2 + y_0^2]\,[(x-x_0)^2 + y_0^2]}\,f\;dx\right.$$

$$\left. +\int_0^\infty \ell n\,\frac{[x_0^2 + (y-y_0)^2]}{[x_0^2 + (y+y_0)^2]}\,h\;dy\right\} \tag{6.4-64}$$

●

So far, we have obtained Green's (or Neumann's) functions by the method of images, by expansion in eigenfunctions, and by separation of variables. The method of transforms discussed in Chapter 5 can also be used to determine  G  (see Problems 12b, 13b).

## Conformal Mapping

In two-dimensional cases, we can use **conformal mapping** to find  G.  If  G,  for the Laplacian operator, is known in a certain domain  R'  and if  R'  can be mapped into  R,  G  is known in  R.  We illustrate this method by an example.

***Example 6.4-7.*** Determine  G  for the Dirichlet problem in the case  $y \geq 0$,  given that  G  is known for a unit circle.

Let  P  be the point  $(x_0, y_0)$  which we denote by  $z_0$.  In Example 3.8-3, we have shown that the mapping

$$w = \frac{z - z_0}{z - \bar{z}_0} \tag{6.4-65}$$

maps the upper half plane into a unit circle. The point $z_0$ is mapped to the centre of the unit circle and the complex conjugate $\bar{z}_0$ is mapped to the region outside the circle. Thus, the singular point $(x_0, y_0)$ is mapped to the origin. The Green's function for a unit circle with the singular point at the origin ($b = 0$) is given by Equation (6.4-25) and can be written as

$$G = \ell n \, r = \ell n \, |w| \tag{6.4-66}$$

Combining Equations (6.4-65, 66) yields

$$G = \ell n \left| \frac{z - z_0}{z - \bar{z}_0} \right|$$

$$= \ell n \left| \frac{x + iy - x_0 - iy_0}{x + iy - x_0 + iy_0} \right| \tag{6.4-67b}$$

$$= \ell n \left| \frac{(x - x_0)^2 + (y - y_0)^2}{(x - x_0)^2 + (y + y_0)^2} \right| \tag{6.4-67c}$$

Equation (6.4-67c) is Equation (6.4-12), as expected.

## 6.5   GREEN'S FUNCTIONS FOR GENERAL LINEAR OPERATORS

The method involving Green's functions can be extended to the wave and the diffusion equations as well. We now discuss the case of the general second order linear differential operator in two variables $(x_1, x_2)$ which is given by Equation (5.3-1). For convenience, we reproduce it here.

$$L(u) = a_{11} \frac{\partial^2 u}{\partial x_1^2} + a_{12} \frac{\partial^2 u}{\partial x_1 \partial x_2} + a_{22} \frac{\partial^2 u}{\partial x_2^2} + b_1 \frac{\partial u}{\partial x_1} + b_2 \frac{\partial u}{\partial x_2} + cu = f(x_1, x_2) \tag{5.3-1}$$

The **adjoint operator** $L^*$ is defined by

$$L^*(v) = \frac{\partial^2}{\partial x_1^2}(a_{11} v) + \frac{\partial^2}{\partial x_1 \partial x_2}(a_{12} v) + \frac{\partial^2}{\partial x_2^2}(a_{22} v) - \frac{\partial}{\partial x_1}(b_1 v) - \frac{\partial}{\partial x_2}(b_2 v) + cv \tag{6.5-1}$$

We can verify that

$$a_{11} v \frac{\partial^2 u}{\partial x_1^2} - u \frac{\partial^2}{\partial x_1^2}(a_{11} v) = \frac{\partial}{\partial x_1} \left[ a_{11} v \frac{\partial u}{\partial x_1} - u \frac{\partial}{\partial x_1}(a_{11} v) \right] \tag{6.5-2}$$

Similarly, we deduce that

$$v\,L\,(u) - u\,L^*(v) \;=\; \frac{\partial r_1}{\partial x_1} + \frac{\partial r_2}{\partial x_2} \tag{6.5-3a}$$

$$r_1 \;=\; a_{11}v\,\frac{\partial u}{\partial x_1} - u\,\frac{\partial}{\partial x_1}\,(a_{11}v) + \frac{a_{12}}{2}\,v\,\frac{\partial u}{\partial x_2} - \frac{u}{2}\,\frac{\partial}{\partial x_2}\,(a_{12}v) + b_1 u\,v \tag{6.5-3b}$$

$$r_2 \;=\; -\frac{u}{2}\,\frac{\partial}{\partial x_1}\,(b_{12}v) + \frac{a_{12}}{2}\,v\,\frac{\partial u}{\partial x_1} + a_{22}v\,\frac{\partial u}{\partial x_2} - u\,\frac{\partial}{\partial x_2}\,(a_{22}v) + b_2 u\,v \tag{6.5-3c}$$

Note that the right side of Equation (6.5-3a) is div $\underline{r}$ and the two components of $\underline{r}$ $(r_1, r_2)$ are given by Equations (6.5-3b, c). Equation (6.5-3a) can be written as

$$v\,L\,(u) - u\,L^*(v) \;=\; \mathrm{div}\ \underline{r} \tag{6.5-3a'}$$

The operator $L$ is **self-adjoint** iff

$$L \equiv L^* \tag{6.5-4}$$

From Equations (6.5-3a, 4.4-23), we obtain

$$\iint\limits_{A} [v\,L\,(u) - u\,L^*(v)]\,dx_1\,dx_2 = \iint\limits_{A}\left(\frac{\partial r_1}{\partial x_1} + \frac{\partial r_2}{\partial x_2}\right)dx_1\,dx_2 = \int\limits_{C}(r_1\,dx_2 - r_2\,dx_1) \tag{6.5-5a,b}$$

The **adjoint Green's function** is defined as

$$L^*\,(G^*) \;=\; 2\pi\,\delta(x_1 - x_{10})\,\delta(x_2 - x_{20}) \tag{6.5-6}$$

The function $G^*$ also satisfies certain homogeneous boundary conditions to be defined in conjunction with the boundary conditions associated with Equation (5.3-1).

Replacing $v$ by $G^*$ and using Equations (5.3-1, 6.5-6), Equations (6.5-5a, b) become

$$\iint\limits_{A} G^*f\,dx_1\,dx_2 - 2\pi\,u\,(x_{10},\,x_{20}) \;=\; \int\limits_{C}(r_1\,dx_2 - r_2\,dx_1) \tag{6.5-7}$$

The functions $r_1$ and $r_2$ are given by Equations (6.5-3b, c) with $G^*$ replacing $v$ and are to be evaluated on the boundary $C$. If $u$ satisfies homogeneous boundary condition $(u = 0$ on $C)$, $r_1$ and $r_2$ simplify to

$$r_1 \;=\; G^*\left(a_{11}\,\frac{\partial u}{\partial x_1} + \frac{a_{12}}{2}\,\frac{\partial u}{\partial x_2}\right) \tag{6.5-8a}$$

$$r_2 = G^* \left( \frac{a_{12}}{2} \frac{\partial u}{\partial x_1} + a_{22} \frac{\partial u}{\partial x_2} \right) \qquad (6.5\text{-}8b)$$

We impose the condition that $G^*$ vanishes on the boundary, resulting in the line integral in Equation (6.5-7) to be zero. The value of $u$ at $(x_{10}, x_{20})$ is then given by

$$u(x_{10}, x_{20}) = \frac{1}{2\pi} \iint\limits_A G^*(x_1, x_2; x_{10}, x_{20}) \, f \, dx_1 \, dx_2 \qquad (6.5\text{-}9)$$

If the boundary conditions on $u$ are non-homogeneous,

$$u(x_{10}, x_{20}) = \frac{1}{2\pi} \left[ \iint\limits_A G^* f \, dx_1 \, dx_2 - \int\limits_C (r_1 \, dx_2 - r_2 \, dx_1) \right] \qquad (6.5\text{-}10)$$

The boundary conditions to be imposed on $G^*$ are of the same type as those imposed on $u$. However, for $G^*$, they are homogeneous. That is to say, if $u$ (or $\partial u / \partial n$) is given on $C$, then $G^*$ (or $\partial G^* / \partial n$) is zero on $C$.

We now show that

$$G^*(x_1, x_2; x_{10}, x_{20}) = G(x_{10}, x_{20}; x_1, x_2) \qquad (6.5\text{-}11)$$

where $G$ is the original Green's function. The function $G(x_1, x_2; \overline{x}_1, \overline{x}_2)$ satisfies the equation

$$L(G) = 2\pi \, \delta(x_1 - \overline{x}_1) \, \delta(x_2 - \overline{x}_2) \qquad (6.5\text{-}12)$$

Multiply Equation (6.5-12) by $G^*$ and Equation (6.5-6) by $G$, subtract the products, and integrate the difference over $A$, to obtain

$$2\pi \iint\limits_A \Big[ G^*(x_1, x_2; x_{10}, x_{20}) \, \delta(x_1 - \overline{x}_1) \, \delta(x_2 - \overline{x}_2) - G(x_1, x_2; \overline{x}_1, \overline{x}_2) \, \delta(x_1 - x_{10})$$

$$\delta(x_2 - x_{20}) \Big] dx_1 \, dx_2 = 2\pi \left[ G^*(\overline{x}_1, \overline{x}_2; x_{10}, x_{20}) - G(x_{10}, x_{20}; \overline{x}_1, \overline{x}_2) \right] \quad (6.5\text{-}13)$$

From Equation (6.5-5a, b) and the homogeneous boundary conditions imposed on $G$ and $G^*$, we deduce that the integral in Equation (6.5-13) is zero. It follows that

$$G^*(\overline{x}_1, \overline{x}_2; x_{10}, x_{20}) = G(x_{10}, x_{20}; \overline{x}_1, \overline{x}_2) \qquad (6.5\text{-}14)$$

Replacing $(\overline{x}_1, \overline{x}_2)$ by $(x_1, x_2)$, we obtain Equation (6.5-11). In the self-adjoint case, we have

$$G(x_1, x_2; x_{10}, x_{20}) = G(x_{10}, x_{20}; x_1, x_2) \qquad (6.5\text{-}15)$$

that is to say, $G$ is symmetric in $(x_1, x_2)$ and $(x_{10}, x_{20})$.

The function $G^*$ in Equations (6.5-9, 10) can be replaced by $G$ via Equation (6.5-11) and, in self-adjoint cases, both $G$ and $G^*$ are equal.

Note that the heat (diffusion) equation is not self-adjoint whereas the Laplace and wave equations are self-adjoint.

The method of Green's functions is applicable to hyperbolic, parabolic, and elliptic equations but the method involving Riemann's functions is restricted to hyperbolic equations. Only hyperbolic equations have two real characteristics.

Further details on Green's functions can be found in Morse and Feshbach (1953), Courant and Hilbert (1966), Greenberg (1971), and Zauderer (1983).

## 6.6  QUANTUM MECHANICS

### Limitations of Newtonian Mechanics

Newton's law of motion are adequate to describe the motion of bodies on a macroscopic scale usually encountered in everyday life. They fail to describe the motion of bodies traveling at high speeds, that is to say, at speeds approaching the velocity of light. In this case, the laws of **relativity**, as developed by Einstein at the turn of the twentieth century, have to be applied. For micro particles, such as atoms and their constituents, Newton's laws have to be replaced by the laws of **quantum mechanics**.

The importance of quantum mechanics needs no stressing. It provides explanations for a variety of physical phenomena ranging from the structure of the atoms to the beginning of the universe. It has led to a number of applications, such as lasers, electron microscopes, silicon chips, and non-linear optics. New devices based on the principles of quantum mechanics are still being developed.

To formulate the laws of quantum mechanics on an acceptable basis, two approaches were adopted in the 1920's. Heisenberg proposed the theory of matrix mechanics, though at that time, he was not aware of the existence of the matrices. It was the mathematician Jordan who saw the connection between the theory proposed by Heisenberg and the theory of matrices. This led to the publication of the famous Born-Heisenberg-Jordan (1926) paper.

At about the same time, Schrödinger, from an apparently different starting point and using the classical continuous formalism, provided an alternative foundation for quantum mechanics. His theory is based on the possibility that matter can have particle and wave properties. In 1924, de Broglie suggested that atomic particles might behave as waves. The relationship between the momentum p of the particle and the wavelength $\lambda$ of the "particle wave" is

$$p = h/\lambda \tag{6.6-1}$$

where h is **Planck's constant**.

From Equation (6.6-1), we can deduce that, for macroscopic particles (p is large), the wavelength $\lambda$ has to be small and the wave properties can be neglected. Thus, the wave properties have to be considered for microscopic particles only.

For a short period, the matrix and the wave mechanics were thought to be unrelated until Schrödinger showed the equivalence of the two formalisms.

We state some of the basic assumptions of quantum mechanics.

(i)     Electromagnetic energy occurs in discrete quantities. The energy E is related to the frequency $\nu$ by

$$E = h \nu = \hbar \omega \tag{6.6-2a,b}$$

where h is the Planck's constant, $\omega = 2\pi\nu$, $\hbar = h/2\pi$.

(ii)    Heisenberg's uncertainty principle

It is not possible to simultaneously measure the momentum p and the position x to arbitrary precision. If $\Delta p$ and $\Delta x$ are the errors in the measurements of p and x respectively,

$$\Delta p \, \Delta x = h \tag{6.6-3}$$

**Schrödinger Equation**

It was stated earlier that Schrödinger based his theory on the possibility of matter having both particle and wave properties. He proposed that the **wave function** $\psi(\underline{x}, t)$ obeys a diffusion equation which can be written as

$$-\frac{\hbar^2}{2m} \nabla^2 \psi + \Phi \psi = i \hbar \frac{\partial \psi}{\partial t} \tag{6.6-4}$$

where m is the mass of the particle, $\Phi$ is the potential energy, and $\nabla^2$ is the Laplacian. The wave function $\psi$ is complex and the observable quantity is not $\psi$ but $|\psi|^2$. The quantity $|\psi(\underline{x}, t)|^2$ is the probability density of the particle being at $\underline{x}$ at time t. The function $\psi$ is normalized such that

$$\int\limits_{V} |\psi(\underline{x}, t)|^2 \, d\underline{x} = 1 \qquad (6.6\text{-}5)$$

where $V$ is the whole space.

Note that it is not claimed that the particle has become a wave but rather that its position is not known with certainty. Its position is given by a probability density and it is this probability density which is associated with a wave. That is to say, $\psi$ can be considered to be a wave of probabilities. These waves are abstract waves in the same sense that crime waves are abstract waves. Equally, a photon, which is an electromagnetic wave, does not have a visible orbit as a particle but its presence can be detected when it hits a target and acts as a particle. Thus, we have a coexistence of wave and particle properties. The property which is manifested depends on the situation under consideration.

Equation (6.6-4) can be solved by the method of separation of variables. We write

$$\psi(\underline{x}, t) = u(\underline{x}) \, f(t) \qquad (6.6\text{-}6)$$

Substituting $\psi$ into Equation (6.6-4) and assuming a static potential $\Phi(\underline{x})$, we have

$$\frac{1}{u} \left[ \frac{-\hbar^2}{2m} \nabla^2 u + \Phi(\underline{x}) u \right] = \frac{i \hbar}{f} \frac{df}{dt} = E \qquad (6.6\text{-}7a,b)$$

where $E$ is a constant representing the total energy (kinetic + potential). Equations (6.6-7a, b) lead to

$$\frac{-\hbar^2}{2m} \nabla^2 u + \Phi u = E u \qquad (6.6\text{-}8a)$$

$$f = C \, e^{-iEt/\hbar} \qquad (6.6\text{-}8b)$$

where $C$ is a constant.

The wave function $\psi(\underline{x}, t)$ can be written as

$$\psi = u(\underline{x}) \, e^{-iEt/\hbar} \qquad (6.6\text{-}9)$$

The probability density $P_r$ is given by

$$P_r = |\psi|^2 = |u(\underline{x})|^2 \qquad (6.6\text{-}10a,b)$$

and is independent of time.

Equation (6.6-8a) is the **time-independent Schrödinger equation**. The solution $\psi(\underline{x}, t)$ depends on the potential $\Phi$ and if $\Phi$ is time independent, the solution is of the form given by

Equation (6.6-9) and stationary states exist. In the next two examples, we consider the case of the hydrogen atom.

***Example 6.6-1.*** Obtain the wave function in the case of an electron moving radially in a potential $\Phi$ given by

$$\Phi = -e^2/r \tag{6.6-11}$$

Angular variables can be neglected.

Since the potential is time independent, $\psi$ is given by Equation (6.6-9). The function $u$ depends on $r$ only. From Equations (6.6-8a, 11, 5.5-32), we deduce that $u$ satisfies

$$-\frac{\hbar^2}{2m}\left(\frac{d^2u}{dr^2} + \frac{2}{r}\frac{du}{dr}\right) - \frac{e^2u}{r} = Eu \tag{6.6-12}$$

To simplify Equation (6.6-12), we introduce the following variables

$$\rho = \alpha r, \qquad \alpha^2 = -8mE/\hbar^2, \qquad \lambda = 2me^2/\alpha\hbar^2 \tag{6.6-13a,b,c}$$

Equation (6.6-12) becomes

$$\frac{d^2u}{d\rho^2} + \frac{2}{\rho}\frac{du}{d\rho} + \left(\frac{\lambda}{\rho} - \frac{1}{4}\right)u = 0 \tag{6.6-14}$$

We require that $u \longrightarrow 0$ as $\rho \longrightarrow \infty$ and this suggests that we look for a solution of the form

$$u = e^{-\beta\rho}v(\rho) \tag{6.6-15}$$

where $\beta$ is a positive constant.

Combining Equations (6.6-14, 15) yields

$$\frac{d^2v}{d\rho^2} + 2\left(\frac{1}{\rho} - \beta\right)\frac{dv}{d\rho} + \left(\beta^2 - \frac{1}{4} + \frac{\lambda}{\rho} - \frac{2\beta}{\rho}\right)v = 0 \tag{6.6-16}$$

The number of terms in Equation (6.6-16) will be reduced, if we choose

$$\beta = 1/2 \tag{6.6-17}$$

Equation (6.6-16) now becomes

$$\frac{d^2 v}{d\rho^2} + \left(\frac{2}{\rho} - 1\right) \frac{dv}{d\rho} + \left(\frac{\lambda - 1}{\rho}\right) v = 0 \tag{6.6-18}$$

We seek a series solution and, since we are interested in the region $\rho \longrightarrow \infty$, we expand $v$ in an inverse power series of $\rho$ and write

$$v = \sum_{m=0}^{\infty} c_m \rho^{k-m} \tag{6.6-19}$$

where $k$ is a constant.

On differentiating term by term and substituting the result into Equation (6.6-18), we obtain

$$\sum_{m=0}^{\infty} c_m [(k-m)(k-m-1) + 2(k-m)] \rho^{k-m-2} - \sum_{m=0}^{\infty} c_m [(k-m) - (\lambda - 1)] \rho^{k-m-1} = 0 \tag{6.6-20}$$

On comparing powers of $\rho$, we deduce that

$$c_0 [k - (\lambda - 1)] = 0 \tag{6.6-21}$$

For a non-trivial solution $c_0 \neq 0$ and this implies that

$$\lambda = 1 + k \tag{6.6-22}$$

Equating terms of $\rho^{k-s-2}$ leads to

$$c_s [(k-s)(k-s-1) + 2(k-s)] = c_{s+1} [(k-s-1) - (\lambda - 1)] \tag{6.6-23a}$$

Substituting Equation (6.6-22) into Equation (6.6-23a), we obtain

$$c_{s+1} = -(k-s)(k+1-s) c_s / (s+1) \tag{6.6-23b}$$

From Equation (6.6-23b), we note that the series solution does not converge. If $k$ is an integer ($k = 0, 1, ...$). the infinite series becomes a polynomial. The valid solution is the polynomial solution. If $k$ is zero (the ground state), $c_1 = c_2 = ... = 0$ and the solution is given by

$$v = \text{constant} = v_0 \tag{6.6-24a,b}$$

For any value of $k$, we can calculate the coefficients $c_s$ from Equation (6.6-23b) and the solution is given by Equation (6.6-19) in the form of a polynomial of degree $k$. We recall a similar situation in Chapter 2 concerning the Legendre equation.

From Equations (6.6-13b, c, 22), we deduce that

$$E = E_k = -me^4/2\hbar^2 (k+1)^2 \qquad\qquad (6.6\text{-}25a)$$

Equation (6.6-25a) gives the energy levels and, in quantum mechanics, the energy levels are discrete and not continuous as in classical mechanics. There are an infinite number of energy levels and they get closer and closer as $k \longrightarrow \infty$. The lowest energy level $E_0$ is given by

$$E_0 = -me^4/2\hbar^2 \qquad\qquad (6.6\text{-}25b)$$

The corresponding wave function $\psi_0$ can be obtained from Equations (6.6-9, 15, 24b) and can be written as

$$\psi_0 = v_0\, e^{-\rho/2}\, e^{-iE_0 t/\hbar} \qquad\qquad (6.6\text{-}26)$$

The probability density $P_r$ is given by

$$P_r = |\psi_0|^2 = v_0^2\, e^{-\rho} \qquad\qquad (6.6\text{-}27a,b)$$

Applying the normalization condition [Equation (6.6-5)] and noting that $V$ is the volume of a sphere of infinite radius, we obtain

$$4\pi \int_0^\infty r^2\, v_0^2\, e^{-\alpha r} dr = 1 \qquad\qquad (6.6\text{-}28)$$

On integrating by parts, we find that $v_0$ is given by

$$v_0^2 = \alpha^3/8\pi = \alpha_0^3/8\pi \qquad\qquad (6.6\text{-}29a,b)$$

$$\alpha_0 = 2me^2/\hbar^2 \qquad\qquad (6.6\text{-}29c)$$

Equation (6.6-29c) is obtained from Equations (6.6-13b, 25b). Combining Equations (6.6-27b, 29) yields

$$P_r = \frac{\alpha_0^3}{8\pi}\, e^{-\alpha_0 r} \qquad\qquad (6.6\text{-}30)$$

The probability $P_\Delta$ of finding an electron in the spherical shell between $r$ and $r + \Delta r$ is given by

$$P_\Delta = 4\pi r^2 P_r\, \Delta r = r^2 \alpha_0^3\, e^{-\alpha_0 r}\, \Delta r/2 \qquad\qquad (6.6\text{-}31a,b)$$

The maximum value of $P_\Delta$ is given by

$$\frac{dP_\Delta}{dr} = 0 \tag{6.6-32}$$

On differentiating $P_\Delta$, we find that the maximum value of $P_\Delta$ is at

$$r = 2/\alpha_0 \tag{6.6-33}$$

The most probable radial distance of the electron from the nucleus is $2/\alpha_0$.

***Example 6.6-2.*** Solve the Schrödinger equation for the potential given in Example 6.6-1 without assuming radial symmetry.

In spherical polar coordinates $(r, \theta, \phi)$, Equation (6.6-8a) can be written as

$$\frac{1}{r^2}\frac{\partial}{\partial r}\left(r^2 \frac{\partial u}{\partial r}\right) + \frac{1}{r^2 \sin\theta}\frac{\partial}{\partial \theta}\left(\sin\theta \frac{\partial u}{\partial \theta}\right) + \frac{1}{r^2 \sin^2\theta}\frac{\partial^2 u}{\partial \phi^2} + \frac{2m}{\hbar^2}(E - \Phi)\,u = 0 \tag{6.6-34}$$

We seek a solution of the form

$$u = R(r)\,F(\theta, \phi) \tag{6.6-35}$$

On differentiating $u$, substituting the results into Equation (6.6-34), and separating the functions of $r$, $\theta$, and $\phi$, we obtain

$$\frac{1}{r^2}\frac{d}{dr}\left(r^2 \frac{dR}{dr}\right) + \frac{2m}{\hbar^2}(E - \Phi)\,R = \frac{\ell(\ell + 1)\,R}{r^2} \tag{6.6-36a}$$

$$\frac{1}{\sin\theta}\frac{\partial}{\partial \theta}\left(\sin\theta \frac{\partial F}{\partial \theta}\right) + \frac{1}{\sin^2\theta}\frac{\partial^2 F}{\partial \phi^2} = -\ell(\ell + 1)\,F \tag{6.6-36b}$$

where $\ell(\ell + 1)$ is the separation constant.

The function $F(\theta, \phi)$ is assumed to be given by

$$F(\theta, \phi) = G(\theta)\,H(\phi) \tag{6.6-37}$$

Proceeding in the usual manner, we find that $G$ and $H$ satisfy

$$\sin\theta \frac{d}{d\theta}\left(\sin\theta \frac{dG}{d\theta}\right) + [\ell(\ell + 1)\sin^2\theta - n^2]\,G = 0 \tag{6.6-38a}$$

$$\frac{d^2H}{d\phi^2} + n^2H = 0 \tag{6.6-38b}$$

where n is another separation constant.

The solution of Equation (6.6-38b) is

$$H = A\,e^{in\phi} + B\,e^{-in\phi} \tag{6.6-39}$$

where A and B are arbitrary constants. For H to be single valued, n has to be an integer.

Equation (6.6-38a) is transformed to the standard form by setting $x = \cos\theta$ and it becomes

$$\frac{d}{dx}\left[(1-x^2)\frac{dG}{dx}\right] + \left[\ell(\ell+1) - n^2/(1-x^2)\right]G = 0 \tag{6.6-40}$$

If n is zero (axial symmetry), Equation (6.6-40) is the standard Legendre equation [Equation (2.7-1)] and the solution is given by Equation (5.5-38). If n is non-zero, Equation (6.6-40) is the associated Legendre equation [Equation (2.7-33)] and the solution can be written as

$$G(x) = C\,P_\ell^n(x) \tag{6.6-41}$$

where C is a constant. The associated Legendre functions $P_\ell^n(x)$ are defined by

$$P_\ell^n(x) = \frac{(-1)^n}{2^\ell\,(\ell!)}(1-x^2)^{n-2}\frac{d^{\ell+n}}{dx^{\ell+n}}\left[(x^2-1)^\ell\right] \tag{6.6-42}$$

We note the following properties of $P_\ell^n$

(i)     n and $\ell$ are integers, $\ell$ is positive [see Equation (2.7-11)].

(ii)    $P_\ell^{-n} = (-1)^n\dfrac{(\ell-n)!}{(\ell+n)!}\,P_\ell^n$ \hfill (6.6-43)

This shows that $P_\ell^{-n}$ is a multiple of $P_\ell^n$ and we need to consider only positive n.

(iii)   If $n > \ell$, we deduce from Equation (6.6-42) that $P_\ell^n$ is zero. [Note that the term inside the square bracket is of order $2\ell$ and its $(\ell+n)$ derivative is zero if $n > \ell$].

The function F can now be written as

$$F = (A e^{in\phi} + B e^{-in\phi}) P_\ell^n (\cos\theta) \tag{6.6-44}$$

Combining Equations (6.6-11, 13a, b, c, 36a) yields

$$\frac{d^2R}{d\rho^2} + \frac{2}{\rho}\frac{dR}{d\rho} + \left(\frac{\lambda}{\rho} - \frac{1}{4}\right) R = \frac{\ell(\ell+1)}{\rho^2} R \tag{6.6-45}$$

Equation (6.6-45) is similar to Equation (6.6-14) and we assume that $R$ is of the form

$$R = e^{-\rho/2} v(\rho) \tag{6.6-46}$$

Substituting Equation (6.6-46) into Equation (6.6-45), we obtain

$$\frac{d^2v}{d\rho^2} + \left(\frac{2}{\rho} - 1\right)\frac{dv}{d\rho} + \left[\frac{(\lambda-1)}{\rho} - \frac{\ell(\ell+1)}{\rho^2}\right] v = 0 \tag{6.6-47}$$

Equation (6.6-47) can be reduced to a standard form by writing

$$v = \rho^\ell w(\rho) \tag{6.6-48}$$

Substituting $v$ into Equation (6.6-47), we obtain

$$\rho \frac{d^2w}{d\rho^2} + (c - \rho)\frac{dw}{d\rho} - a w = 0 \tag{6.6-49a}$$

where $c = 2(\ell+1)$, $a = (\ell+1-\lambda)$ \hfill (6.6-49b,c)

Equation (6.6-49a) is the confluent hypergeometric differential equation and its solution can be written as

$$w = {}_1F_1(a, c, \rho) \tag{6.6-50}$$

The properties of confluent hypergeometric functions are given in Slater (1960).

The solution $\psi$ is obtained by combining Equations (6.6-9, 35, 44, 46, 50) and is

$$\psi = e^{-iEt/\hbar} e^{-(\alpha r/2)} (\alpha r)^\ell {}_1F_1(\ell+1-\lambda, 2\ell+2, \alpha r) [A e^{in\phi} + B e^{-in\phi}] P_\ell^n (\cos\theta) \tag{6.6-51}$$

●

One important result of Schrödinger's theory is the prediction of the energy levels given by Equation (6.5-25a) which are observed in spectroscopic measurements of moderate resolution. The splitting of the energy level observed in high resolution measurements can be accounted for by including the spin effects. Interested readers can consult Eisberg and Resnick (1985). We next consider an example which leads to an application.

***Example 6.6-3***. Solve the one-dimensional Schrödinger equation for a potential given by

$$\Phi(x) = \begin{cases} 0, & x < 0 \\ \Phi_0, & x > 0 \end{cases} \qquad\qquad (6.6\text{-}52a,b)$$

where $\Phi_0$ (constant) is greater than the total energy E.

In this example, the potential has a jump discontinuity at the origin. This potential function can represent the motion of a charged particle between two electrodes kept at different voltages.

Since $\Phi$ is time independent and is discontinuous, we need to solve Equation (6.6-8a) in the regions $x < 0$ and $x > 0$ separately. Equation (6.6-8a) can be written for the given $\Phi$ as

$$-\frac{\hbar^2}{2m}\frac{d^2u}{dx^2} = E\,u\,, \qquad\qquad x < 0 \qquad\qquad (6.6\text{-}53a)$$

$$-\frac{\hbar^2}{2m}\frac{d^2u}{dx^2} + \Phi_0 u = E\,u\,, \qquad x > 0 \qquad\qquad (6.6\text{-}53b)$$

The solution of Equation (6.6-53a) is

$$u = A\,e^{i\lambda_1 x} + B\,e^{-i\lambda_1 x} \qquad\qquad (6.6\text{-}54a)$$

where A and B are constants and $\lambda_1$ is given by

$$\lambda_1 = (\sqrt{2mE}\,)/\hbar \qquad\qquad (6.6\text{-}54b)$$

Since $\Phi_0 > 0$, the solution of Equation (6.6-53b) can be written as

$$u = C\,e^{\lambda_2 x} + D\,e^{-\lambda_2 x} \qquad\qquad (6.6\text{-}55a)$$

where C and D are constants and $\lambda_2$ is given by

$$\lambda_2 = \sqrt{2m(\Phi_0 - E)}\,/\hbar \qquad\qquad (6.6\text{-}55b)$$

The condition that u tends to zero as x tends to infinity implies that C is zero. The conditions requiring u and du/dx to be continuous at the origin yield the following

$$A + B = D \tag{6.6-56a}$$

$$i \lambda_1 (A - B) = - \lambda_2 D \tag{6.6-56b}$$

On solving Equations (6.6-56a, b), we obtain

$$A = \frac{D}{2\lambda_1} (\lambda_1 + i \lambda_2), \qquad B = \frac{D}{2\lambda_1} (\lambda_1 - i \lambda_2) \tag{6.6-57a,b}$$

The constant $D$ can be obtained by applying the normalization condition.

The probability $P_r$ of finding the particle in the region $x > 0$ is

$$P_r = |u|^2 = |D|^2 e^{-2\lambda_2 x} \tag{6.6-58a,b}$$

The probability $P_r$ decreases rapidly with increasing $x$. For small values of $x$, there is a finite probability of finding a particle in this region. This is not predicted in classical mechanics where the total energy $E$ is constant. That is to say, if the potential energy is greater that $E$, the kinetic energy must be negative which is physically impossible. The particle does not enter the region $x > 0$. In quantum mechanics, there is a finite probability that the particle may penetrate the classically excluded region and this phenomenon is called **penetration** (tunneling). This phenomenon can be observed experimentally and has found an application in the tunnel diode which is used in modern electronics. For details, see Eisberg and Resnick (1985).

Note that the statement $\Phi_0$ is greater than $E$ should be subjected to the uncertainty principle. If the depth of penetration is $\Delta x$, we deduce from Equations (6.6-55b, 58a, b) that

$$\Delta x \approx 1/\lambda_2 \approx \hbar / \sqrt{2m (\Phi_0 - E)} \tag{6.6-59a,b}$$

The uncertainty in the momentum $\Delta p$ can be obtained from Equation (6.6-3) and is

$$\Delta p = h / \Delta x \approx \sqrt{2m (\Phi_0 - E)} \tag{6.6-60a,b}$$

The total energy $E$ is proportional to $p^2/2m$ and it follows that

$$\Delta E \approx \Phi_0 - E \tag{6.6-61}$$

In quantum mechanics, it is not possible to state that $\Phi_0$ is definitely greater than $E$ and that the particle cannot enter the region $x > 0$.

## PROBLEMS

1a.    Determine the Riemann function $v$ for the wave equation

$$\frac{\partial^2 u}{\partial t^2} = c^2 \frac{\partial^2 u}{\partial x^2}, \qquad -\infty < x < \infty, \quad t > 0$$

Obtain the solution $u$ that satisfies the following conditions using Riemann's method

$$u(x, 0) = f(x), \qquad \left. \frac{\partial u}{\partial t} \right|_{t=0} = g(x)$$

Answer: $v = 1$

2b.    Show that the characteristics of the equation

$$t^2 \frac{\partial^2 u}{\partial t^2} = x^2 \frac{\partial^2 u}{\partial x^2}, \qquad -\infty < x < \infty, \quad t > 0$$

can be written as

$$\xi = t/x, \qquad \eta = xt$$

Deduce that the Riemann function $v$ satisfies the equation

$$\frac{\partial^2 v}{\partial \xi \partial \eta} + \frac{\partial}{\partial \xi} \left( \frac{v}{2\eta} \right) = \frac{\partial}{\partial \xi} \left( \frac{\partial v}{\partial \eta} + \frac{v}{2\eta} \right) = 0$$

Integrate the partial differential equation and show that $v$ is given by

$$v = \sqrt{(\eta_0 / \eta)}$$

Obtain $u$ by Riemann's method if at time $t = t_1 \ (\neq 0)$, $u$ and $\partial u / \partial t$ are given by

$$u(x, t_1) = f(x), \qquad \left. \frac{\partial u}{\partial t} \right|_{t=t_1} = g(x)$$

3a.    Show that on introducing a new variable $\eta = x^\alpha t^\beta$, the partial differential equation

$$9x \frac{\partial u}{\partial t} - \frac{\partial^2 u}{\partial x^2} = 0, \qquad 0 \leq x < \infty, \quad t > 0$$

can be transformed to

$$\frac{d^2u}{d\eta^2} + 3\eta^2 \frac{du}{d\eta} = 0$$

The boundary and initial conditions are

$$u(0, t) = C_1, \qquad \lim_{x \to \infty} u(x, t) = C_2, \qquad u(x, 0) = C_3$$

where $C_1$, $C_2$, and $C_3$ are constants.

State the conditions that $C_1$, $C_2$, and $C_3$ have to satisfy in order that a similarity solution can exist. Assuming that these conditions are satisfied, find $u$ and express it in the form of an integral.

$$\text{Answer: } C_1 + [3(C_2 - C_1) / \Gamma(1/3)] \int_0^{x/t^{1/3}} \exp(-\xi^3) \, d\xi$$

4b.   Balmer and Kauzlarich (1971) considered a two-dimensional steady flow of a viscoelastic fluid between two non-parallel walls with suction or injection at the walls. Referred to a Cartesian coordinate system $(x, y)$, the walls are given by $y = \pm R(x)$. It is assumed that the velocity component in the y-direction $(v_y)$ is small. The viscoelastic model they chose was a modified form of the White-Metzner model (Carreau et al., 1997). In a viscometric flow $[v_x = v_x(y)$, $v_y = v_z = 0]$, the stress components $\tau_{ij}$ are given by

$$\tau_{xy} = \dot{\gamma}^n / N_{Re}, \qquad \tau_{xx} - \tau_{yy} = \dot{\gamma}^s / N, \qquad \tau_{yy} = \tau_{zz}$$

where $\dot{\gamma} = dv_x/dy$, $N = \rho U^{2-s} L^s / \xi$, $N_{Re}$ (Reynolds number) $= \rho U^{2-n} L^n / \mu$, $\rho$ is the density, $U$ and $L$ are a characteristic velocity and length respectively, $\xi$ and $\mu$ are constants.

From the symmetry of the problem, we need to consider only the region $0 \le y \le R(x)$. The boundary conditions are

$$v_x = 0, \quad v_y = g(x), \qquad \text{at } y = R(x)$$

$$\tau_{xy} = 0, \qquad \text{at } y = 0$$

On introducing a stream function $\psi(x, y)$ defined by

$$v_x = \partial\psi/\partial y, \qquad v_y = -\partial\psi/\partial x$$

and making appropriate assumption, $\psi$ was found to satisfy

$$\frac{\partial \psi}{\partial y}\frac{\partial^2 \psi}{\partial x \partial y} - \frac{\partial \psi}{\partial x}\frac{\partial^2 \psi}{\partial y^2} = (1/N_{Re})\left[ n\left(\frac{\partial^2 \psi}{\partial y^2}\right)^{n-1}\frac{\partial^3 \psi}{\partial y^3}\right] + (1/N)\left[ s\left(\frac{\partial^2 \psi}{\partial y^2}\right)^{s-1}\frac{\partial^3 \psi}{\partial y^2 \partial x}\right] + f(x)$$

where $f(x) = -\dfrac{\partial}{\partial x}[p(x, 0) - \tau_{yy}(x, 0)]$ and $p$ is the pressure.

The following similarity transformation was proposed

$$\psi = g(x)\, F(\eta), \qquad \eta = y/R(x)$$

Determine the conditions that $g$, $R$, $s$, $n$, and $f$ have to satisfy so that the P.D.E. can be transformed to an O.D.E.

Verify that if $R = k_1 x + k_2$, $g = k_3$, $n = s = 1$, where $k_1$, $k_2$, and $k_3$ are constants, the equation for $\psi$ becomes

$$(1 - k_1 \eta N_{ws})\, H''' - 2N_{ws}k_1 H'' + N_{Re}\, k_1 k_3 \alpha\, (H')^2 = 1$$

where $H = F/\alpha$, $\alpha = -N_{Re}R^3 f(x)/g$, $N_{ws} = N_{Re}/N$.

Deduce the boundary conditions in terms of $F$.

The non-linear O.D.E. has to be solved numerically.

Answer: $F(0) = F'(1) = F''(0) = 0$

5a. Use Birkhoff's method to obtain the similarity variables for the boundary layer system defined by

$$\frac{\partial \psi}{\partial y}\frac{\partial^2 \psi}{\partial x \partial y} - \frac{\partial \psi}{\partial x}\frac{\partial^2 \psi}{\partial y^2} - \frac{\partial^3 \psi}{\partial y^3} = 0$$

$$\frac{\partial \psi}{\partial x} = \frac{\partial \psi}{\partial y} = 0 \quad \text{at } y = 0, \qquad \lim_{y \to \infty}\frac{\partial \psi}{\partial y} = 0$$

Transform the P.D.E. to an O.D.E. and obtain the associated boundary conditions.

[Hint: write $\bar{x} = a^{\alpha_1}x$, $\bar{y} = a^{\alpha_2}y$, $\bar{\psi} = a^{\alpha_3}\psi$, $\eta = y/(x^{\alpha_2/\alpha_1})$, $\phi(\eta) = \psi/(x^{\alpha_3/\alpha_1})$]

Answer: $1 - \alpha_2/\alpha_1 = \alpha_3/\alpha_1$

6b. The heat equation in a cylindrical polar coordinate system can be written as

$$\frac{\partial u}{\partial t} = \frac{\partial^2 u}{\partial r^2} + \frac{1}{r}\frac{\partial u}{\partial r}, \qquad 0 \le r < \infty, \quad t \ge 0$$

The initial and boundary conditions are

$$u(0, t) = t, \qquad \lim_{r \to \infty} u(r, t) = 0, \qquad u(r, 0) = 0$$

The time dependent condition at $r = 0$ can be transformed to a constant boundary condition by writing

$$u(r, t) = t\, v(r, t)$$

Determine the partial differential equation, the boundary and initial conditions that $v$ has to satisfy. Show that by introducing the variable $\eta = r/\sqrt{t}$, $v(\eta)$ has to satisfy

$$\frac{d^2 v}{d\eta^2} + \frac{dv}{d\eta}\left(\frac{1}{2}\eta + \frac{1}{\eta}\right) - v = 0$$

Obtain the boundary conditions on $v(\eta)$.

7a. The function $u$ satisfies Laplace's equation inside and on a unit sphere. On the surface of the sphere, $u = f(\theta, \phi)$. Use Equations (6.4-2,14) to obtain $u$ in the form of an integral (Poisson's integral).

$$\text{Answer: } u(b, \theta_0, \phi_0) = \frac{(b^2 - 1)}{4\pi} \int_0^{2\pi}\int_0^{\pi} \frac{f(\theta, \phi)\sin\theta\, d\theta\, d\phi}{(1 + b^2 - 2b\cos\Theta)^{3/2}}$$

8a. Show, using the Green's function given by Equation (6.4-12), that the solution of the boundary value problem

$$\frac{\partial^2 u}{\partial x^2} + \frac{\partial^2 u}{\partial y^2} = \rho(x, y), \qquad -\infty < x < \infty, \quad y > 0$$

$$u = f(x) \quad \text{on } y = 0$$

is

$$u(x_0, y_0) = \frac{y_0}{\pi}\int_{-\infty}^{\infty} \frac{f(x)\, dx}{(x - x_0)^2 + y_0^2} + \frac{1}{4\pi}\int_0^{\infty}\int_{-\infty}^{\infty} \rho\, \ell n\left[\frac{(x - x_0)^2 + (y - y_0)^2}{(x - x_0)^2 + (y + y_0)^2}\right] dx\, dy$$

9a. Show that the solution of the boundary value problem

$$\frac{\partial^2 u}{\partial x^2} + \frac{\partial^2 u}{\partial y^2} = \rho\,(x, y), \quad -\infty < x < \infty, \quad y > 0$$

$$\frac{\partial u}{\partial y} = h\,(x, y) \quad \text{on } y = 0$$

can be written as

$$u\,(x_0, y_0) = \frac{1}{4\pi} \iint \rho\left\{\ell n\,[(x - x_0)^2 + (y - y_0)^2]\,[(x - x_0)^2 + (y + y_0)^2]\right\} dx\,dy$$

$$-\frac{1}{2\pi} \int_{-\infty}^{\infty} h\left\{\ell n\,[(x - x_0)^2 + y_0^2]\right\} dx$$

10b. Use the method of images to show that Neumann's function (N) for the interior of the unit circle is

$$N = \ell n\,(b\,r_1\,r_2)$$

in the notation of Equation (6.4-16).

If u satisfies Laplace's equation inside and on the unit circle and if on the unit circle

$$\left.\frac{\partial u}{\partial r}\right|_{r=1} = h\,(\theta)$$

show that u is given by

$$u\,(b, \theta_0) = -\frac{1}{2\pi} \int_0^{2\pi} h\,(\theta)\left\{\ell n\,[1 + b^2 - 2b\cos\,(\theta - \theta_0)]\right\} d\theta$$

Verify that u satisfies Laplace's equation.

11b. Show that the Neumann function for the three-dimensional upper half plane $z \geq 0$ is

$$N = -\left(\frac{1}{r} + \frac{1}{r'}\right)$$

where r and r' are defined by Equations (6.4-11b, c).

Obtain u (x, y, x) if u satisfies the following conditions

$$\nabla^2 u = 0, \quad z \geq 0, \quad -\infty < x < \infty, \quad -\infty < y < \infty$$

$$\left.\frac{\partial u}{\partial z}\right|_{z=0} = \begin{cases} 1, & \text{if } (x^2 + y^2) < 1 \\ 0, & \text{otherwise} \end{cases}$$

12b.  The Green function $G$ associated with the wave equation satisfies the equation

$$\frac{\partial^2 G}{\partial x^2} - \frac{1}{c^2}\frac{\partial^2 G}{\partial t^2} = \delta(x - x_0)\,\delta(t - t_0)$$

Solve for $G$ by first taking the Laplace transform of $G$ with respect to $t$ and verify that $\overline{G}$ (the Laplace transform of $G$) satisfies the equation

$$\frac{\partial^2 \overline{G}}{\partial x^2} - \frac{s^2\overline{G}}{c^2} = e^{-st_0}\,\delta(x - x_0)$$

Next, take the Laplace transform of $\overline{G}$ with respect to $x$ and verify that $\overline{G}$ is given by

$$\overline{G} = -(c\,e^{-st_0}\,e^{-s|x-x_0|/c})/2s$$

Deduce that $G$ can be written as

$$G(x, t\,; x_0, t_0) = \begin{cases} -c/2, & \text{if } [c\,(t - t_0) > |x - x_0|] \\ 0, & \text{otherwise} \end{cases}$$

Is $G$ finite at $(x_0, t_0)$? Discuss the discontinuity of $G$.

13b.  The one-dimensional Schrödinger equation for a free particle of mass $m$ is [Equation (6.6-4)]

$$i\hbar\frac{\partial \psi}{\partial t} = -\frac{\hbar^2}{2m}\frac{\partial^2 \psi}{\partial x^2}, \quad -\infty < x < \infty, \quad t > 0$$

$$\psi(x, 0) = f(x), \qquad \psi \text{ and } \frac{\partial \psi}{\partial x} \longrightarrow 0 \text{ as } |x| \longrightarrow \infty$$

Take the Fourier transform with respect to $x$ [Equations (5.9-16a, b)] and show that $\Psi(\alpha, t)$, the Fourier transform of $\psi$, is

$$\Psi(\alpha, t) = F(\alpha)\,e^{-i\gamma\alpha^2 t}$$

where $F$ is the Fourier transform of $f$ and $\gamma = \hbar/2m$.

On inverting and using the convolution theorem, $\psi$ can be written as

$$\psi = \frac{1}{2\pi} \int_{-\infty}^{\infty} F(\alpha)\, e^{-i\gamma\alpha^2 t}\, e^{i\alpha x}\, d\alpha = \int_{-\infty}^{\infty} f(\xi)\, G(x - \xi)\, d\xi$$

The function $G$ is Green's function. Show that $G$ is given by

$$G(x) = \left[(1 - i) \exp\left(i x^2 / 4\gamma t\right)\right] / 2\sqrt{2\pi\gamma t}$$

Deduce $\psi$ if $f(x) = \delta(x - x_0)$.

# CHAPTER 7

# NUMERICAL METHODS

## 7.1  INTRODUCTION

Many engineering problems can not be solved exactly by analytical techniques. The knowledge of numerical methodology is essential for determining approximate solutions. Numerical methods, in one form or another, have been studied for several centuries. We will look, for example, at Newton's method of approximating the solution of an algebraic or transcendental equation, at the Gaussian elimination method for solving linear systems of equations and at the Euler and Runge-Kutta methods for solving initial-value problems. Since the arrival of computers, the potential of these methods has been realized and the entire character of numerical methods has changed. We can now use iterative methods with much greater speed. We can also solve large systems of equations numerically. Many non-linear equations which were intractable in the past can now be solved approximately by numerical methods. As a result of these possibilities, mathematical modelling is now more realistic and a new field of numerical simulation has been opened up. Successful simulations of complex processes in science and engineering are now being achieved. In this chapter, we present different methods which will be useful to applied science students.

We first briefly examine the possible sources of error that may occur in the solutions obtained by numerical methods. Most numbers have an infinite decimal representation and for computational purposes they have to be rounded to a finite number of decimal places. This type of error is known as **round off error** and is unavoidable. During the process of calculation, we generate errors. If we subtract two almost equal numbers, the number of significant digits will be reduced. This type of error is known as **loss of significance**. As an example, we consider the function $f(x)$ given by

$$f(x) = x - \sin x \qquad (7.1-1)$$

We now evaluate $f(x)$ when $x$ is equal to 0.150 radians.

$$f(0.150) = 0.150 - 0.149 \qquad (7.1-2a)$$

$$= 0.001 \qquad (7.1-2b)$$

In Equation (7.1-2a), the value of $\sin(0.150)$ is correct to three significant figures but $f(0.150)$ is correct to at most one significant figure. Losing all significant figures during machine computations is not unheard of. It is therefore necessary to check the numbers during the calculations. In some cases, it is possible to rewrite the function to be evaluated in such a way that the loss of significant figures does not occur. For example, we can approximate $\sin x$ by its Taylor series and $f(x)$ is approximated as

$$f(x) \approx x - (x - \frac{1}{6} x^3 + \dots) \tag{7.1-3a}$$

$$\approx \frac{1}{6} x^3 \tag{7.1-3b}$$

Using Equation (7.1-3a), we find that

$$f(0.150) = 0.000562 \tag{7.1-4}$$

The value of $\sin (0.150)$ correct to five significant figures is 0.14944 and it can be seen that Equation (7.1-4) gives a better approximation to $f(0.150)$ than Equation (7.1-2b).

In classical analysis, we consider infinite processes, such as infinite sums. In computational mathematics, we can consider only finite sums. This results in an error called the **truncation error**. It is often possible to provide an estimate of such errors.

As mentioned earlier, iterative methods are used to solve equations numerically. These methods generate a sequence of numbers. We need to establish that the sequence converges to the solution. Ideally we should always perform an error analysis before accepting a particular solution which was generated numerically. Often this is not possible as the problem is too complicated. In such a case, we can rely on intuition and past experience and hope that the results obtained are reliable and useful. In this sense, numerical analysis is an art and not a science. A more detailed discussion on error analysis can be found in Hildebrand (1956) and Elden and Wittmeyer-Koch (1990).

## 7.2   SOLUTIONS OF EQUATIONS IN ONE VARIABLE

We begin our discussion by considering equations of the form

$$f(x) = 0 \tag{7.2-1}$$

where $x$ and $f(x)$ are real or complex. We need to find values of the variable $x$ that satisfy Equation (7.2-1) for a given $f$. This equation could be algebraic or transcendental. If $x$ is a real or complex number satisfying Equation (7.2-1), we say that $x$ is a **root** of the equation. Alternatively we will also say that $x$ is a **zero** of the function $f$. Next we present four different methods of solving Equation (7.2-1).

## Bisection Method (Internal Halving Method)

This method is based upon the use of the **intermediate mean-value theorem** which states that for a continuous function f, defined on the closed interval [a, b], with f(a) and f(b) having opposite signs, there exists at least one number x such that a < x < b for which f(x) = 0. The method consists of a repeated halving of the interval [a, b] with f(a) and f(b) having opposite signs. One starts by letting $x_1$ be the mid-point of [a, b]. If $x_1$ satisfies Equation (7.2-1), $x_1$ is the required root. If not, $f(x_1)$ has the same sign as either f(a) or f(b). If $f(x_1)$ and f(a) have the same sign, the required root lies in the interval $(x_1, b)$, as illustrated in Figure 7.2-1. We then proceed to find the mid-point of $(x_1, b)$. Similarly if $f(x_1)$ and f(b) have the same sign, the required root lies in the interval $(a, x_1)$ and in this case $x_2$ is the mid-point of $(a, x_1)$. We repeat this process of halving until a satisfactory value of the root is obtained. That is to say, until the difference between two consecutive values of $x_i$ is within the required accuracy.

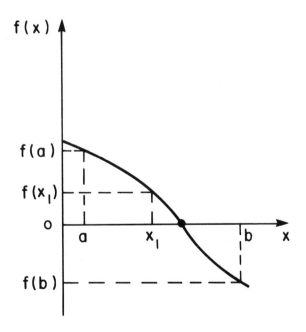

**FIGURE 7.2-1**    **Identification of the root (•) of the equation**
**via the bisection method**

*Example 7.2-1*. Solve the following equation via the bisection method

$$f(x) = x^3 + 4x^2 - 10 = 0 \qquad\qquad (7.2\text{-}2a,b)$$

We note that $f(1)$ is negative $(-5)$ and that $f(2)$ is positive $(14)$. Thus, at least one root lies in the interval $(1, 2)$ and $x_1$, the mid-point of $(1, 2)$, is 1.5. We calculate $f(1.5)$ which is found to be positive $(2.37)$ and the root lies between 1 and 1.5. The second approximation $x_2$ is 1.25. We continue this process. The values of $x_k$ and $f(x_k)$ are given in Table 7.2-1 for various values of k.

### TABLE 7.2-1

### Values of $x_k$ and $f(x_k)$

| k | $x_k$ | $f(x_k)$ |
|---|-------|----------|
| 1 | 1.5 | 2.37 |
| 2 | 1.25 | $-1.797$ |
| 3 | 1.375 | 0.162 |
| 4 | 1.3125 | $-0.848$ |
| 5 | 1.3437 | $-0.351$ |
| 6 | 1.3593 | $-0.102$ |
| 7 | 1.3671 | 0.029 |
| 8 | 1.3632 | $-0.048$ |
| 9 | 1.3651 | $-0.0038$ |
| 10 | 1.3661 | 0.0127 |
| 11 | 1.3655 | 0.0045 |
| 12 | 1.3652 | 0.0003 |
| 13 | 1.3651 | $-0.0017$ |

From Table 7.2-1, we see that the root accurate to three decimal places is 1.365. Note that as k increases the difference $x_k - x_{k-1}$ decreases and $f(x_k)$ approaches zero.

## Secant Method

Consider the graph of the function f as shown in Figure 7.2-2.

In this method, we consider two points $[x_0, f(x_0)]$ and $[x_1, f(x_1)]$. It is not necessary that $x_0 < x_1$ or that $f(x_0)$ and $f(x_1)$ have different signs. That is to say, it is not necessary that the root lies in the interval $[x_0, x_1]$. We draw the **secant line**, that is the line passing through points $[x_0, f(x_0)]$ and $[x_1, f(x_1)]$ which will intersect the x-axis at some point $x_2$. The equation of the secant line is

$$f(x) - f(x_1) = \frac{f(x_0) - f(x_1)}{(x_0 - x_1)}(x - x_1) \tag{7.2-3}$$

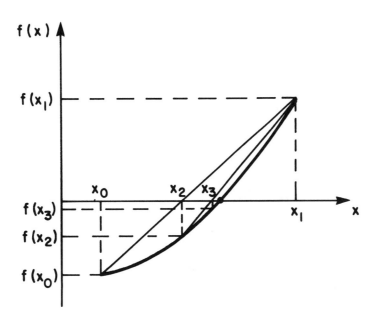

**FIGURE 7.2-2    Identification of the root (•) of the equation via the secant method**

The point $x_2$ for which $f(x_2)$ is zero is given from Equation (7.2-3) by

$$x_2 = x_1 - \frac{f(x_1)(x_1 - x_0)}{f(x_1) - f(x_0)} \tag{7.2-4}$$

If $f(x_2) \neq 0$, we draw the secant line joining the points $[x_1, f(x_1)]$ and $[x_2, f(x_2)]$ and determine $x_3$ as

$$x_3 = x_2 - \frac{f(x_2)(x_2 - x_1)}{f(x_2) - f(x_1)} \tag{7.2-5}$$

Proceeding in this manner, we find in general

$$x_{k+1} = x_k - \frac{f(x_k)(x_k - x_{k-1})}{f(x_k) - f(x_{k-1})}, \qquad (k \geq 1) \tag{7.2-6}$$

The root is determined to the desired degree of accuracy.

***Example 7.2-2***. Use the secant method to solve the equation given in Example 7.2-1.

Taking $x_0$ to be 1 and $x_2$ to be 2, we obtain from Equation (7.2-4)

$$x_2 = 2 - \frac{14\,(2-1)}{14-(-5)} \tag{7.2-7a}$$

$$= 1.26 \tag{7.2-7b}$$

Via Equation (7.2-6), we generate

$$x_3 = 1.26 - \frac{(-1.649)\,(1.26-2)}{-1.649-14} \tag{7.2-8a}$$

$$= 1.338 \tag{7.2-8b}$$

$$x_4 = 1.338 - \frac{(-0.44368)\,(1.338-1.26)}{-0.443678+1.649} \tag{7.2-8c}$$

$$= 1.3667 \tag{7.2-8d}$$

$$x_5 = 1.3667 - \frac{(0.024291)\,(1.3667-1.338)}{0.024291+0.443678} \tag{7.2-8e}$$

$$= 1.3652 \tag{7.2-8f}$$

We can calculate $x_6$ and it is found that $x_6$ is also 1.3652. Thus the root accurate to three decimal places is 1.365. We note that we achieve this accuracy in five iterations whereas in the bisection method we needed twelve iterations.

## Newton's Method

This is a popular and widely used method. In this method, we start with only one point, which we denote by $x_0$. We assume that $f(x)$ is differentiable in the interval containing $x_0$ and the root. Expanding $f(x)$ in a Taylor series about the point $x_0$ yields

$$f(x) = f(x_0) + f'(x_0)(x-x_0) + \ldots \tag{7.2-9}$$

We assume $x_1$ to be an approximate root and set $f(x_1)$ to be zero in Equation (7.2-9) yielding

$$x_1 = x_0 - \frac{f(x_0)}{f'(x_0)} \tag{7.2-10}$$

Generalizing, we obtain

$$x_{k+1} = x_k - \frac{f(x_k)}{f'(x_k)}, \qquad (k \geq 0) \tag{7.2-11}$$

Again, the root can be determined to the desired degree of accuracy.

**Example 7.2-3.** Solve Equation (7.2-2) using Newton's method.

On differentiating Equation (7.2-2), we obtain

$$f'(x) = 3x^2 + 8x \tag{7.2-12}$$

We start the iteration with

$$x_0 = 1 \tag{7.2-13}$$

From Equations (7.2-10, 11), we have

$$x_1 = 1 + \frac{5}{11} \tag{7.2-14a}$$

$$= 1.4545 \tag{7.2-14b}$$

$$x_2 = 1.4545 - 0.0856 \tag{7.2-14c}$$

$$= 1.3689 \tag{7.2-14d}$$

$$x_3 = 1.3689 - 0.0037 \tag{7.2-14e}$$

$$= 1.3652 \tag{7.2-14f}$$

Using Newton's method, we obtain the root 1.365 accurate to three decimal places after three iterations.

●

Newton's method is also used to extract the $n^{th}$ root of real numbers. Suppose that the $n^{th}$ root of a number A is x. That is

$$x^n = A \tag{7.2-15a}$$

or     $$f(x) = x^n - A = 0 \tag{7.2-15b,c}$$

Taking the derivative, we have

$$f'(x) = n\, x^{n-1} \tag{7.2-16}$$

From Equation (7.2-11), we deduce

$$x_{k+1} = x_k - \frac{\left(x_k^n - A\right)}{n\, x_k^{n-1}} \tag{7.2-17}$$

In the case of a square root, we have $n = 2$ and Equation (7.2-17) becomes

$$x_{k+1} = x_k - \frac{\left(x_k^2 - A\right)}{2\, x_k} \tag{7.2-18a}$$

$$= \frac{1}{2}\left(x_k + \frac{A}{x_k}\right) \tag{7.2-18b}$$

*Example 7.2-4.* Find the square root of 3.

Here $A = 3$, and we start with

$$x_0 = 1.5 \tag{7.2-19a}$$

$$x_1 = \frac{1}{2}\left(1.5 + \frac{3}{1.5}\right) = 1.75 \tag{7.2-19b,c}$$

$$x_2 = \frac{1}{2}\left(1.75 + \frac{3}{1.75}\right) = 1.73214 \tag{7.2-19d,e}$$

$$x_3 = \frac{1}{2}\left(1.73214 + \frac{3}{1.73214}\right) = 1.7320508 \tag{7.2-19f,g}$$

which approximates the actual root to six decimal places.

**Fixed Point Iteration Method**

In this method, we rewrite Equation (7.2-1) in the following form

$$x = g(x) \tag{7.2-20}$$

If Equation (7.2-20) holds, $x$ is said to be a **fixed point** of $g(x)$.

As in Newton's method, we choose a starting point $x_0$ and then compute $g(x_0)$. We now label $g(x_0)$ as $x_1$ and compute $g(x_1)$. We continue this process by writing

$$x_{k+1} = g(x_k), \qquad k = 0, 1, 2, \ldots \tag{7.2-21}$$

until the root to the desired degree of accuracy is obtained.

The conditions on $g$ which will ensure converge are

(i)      there is a closed interval $I = [a, b]$ such that for $x \in I$, $g(x)$ is defined and $g(x) \in I$,

(ii)      $g(x)$  is continuous on  I,  and

(iii)     $g(x)$  is differentiable on  I,  $\left| g'(x) \right| < k$  for all  $x \in I$  and  $k$  is a non-negative constant less than one.  If  $g$  satisfies conditions (i to iii), Equation (7.2-20) has exactly one solution.  The mapping  $g$  is said to be a **contraction mapping** in  I.

*Example 7.2-5*.  Using the fixed point iteration method, solve

$$f(x) = x - \cos x = 0 \qquad\qquad\qquad\qquad (7.2\text{-}22a,b)$$

We rewrite Equation (7.2-22b) as

$$x = \cos x \qquad\qquad\qquad\qquad (7.2\text{-}23)$$

In this example,  $g(x)$  is  cos x.  In the interval  $[0, \pi/2]$,  $g(x)$  satisfies all the conditions stated earlier.  Thus, there is exactly one solution in  $[0, \pi/2]$.  Starting with the initial value  $\pi/4$,  we can, via Equation (7.2-21), generate the values of  $x_k$  as given in Table 7.2-2.

## TABLE 7.2-2

## Values of  $x_k$  and  $g(x_k)$

| k | $x_k$ | $g(x_k) = \cos x_k$ |
|---|-------|---------------------|
| 0 | $\pi/4$ | 0.707 |
| 1 | 0.707 | 0.760 |
| 2 | 0.760 | 0.725 |
| 3 | 0.725 | 0.748 |
| 4 | 0.748 | 0.733 |
| 5 | 0.733 | 0.743 |
| 6 | 0.743 | 0.736 |
| 7 | 0.736 | 0.741 |
| 8 | 0.741 | 0.738 |
| 9 | 0.738 | 0.740 |
| 10 | 0.740 | 0.738 |

From Table 7.2-2, we deduce that the root correct to two decimal places is  0.74, which is achieved after six iterations.

●

The transformation from Equation (7.2-1) to Equation (7.2-20) is not unique for any given f. We should choose g so that all the conditions on g stated earlier, are satisfied, in order to ensure that the iterations converge. We note that for the secant method (Equation 7.2-6), $g(x_k)$ is given by

$$g(x_k) = x_k - \frac{f(x_k)(x_k - x_{k-1})}{f(x_k) - f(x_{k-1})}$$

(7.2-24)

In the case of Newton's method (Equation 7.2-11), $g(x_k)$ is

$$g(x_k) = x_k - \frac{f(x_k)}{f'(x_k)}$$

(7.2-25)

**Example 7.2-6.** Solve Equation (7.2-2) by the fixed point iteration method.

We can rewrite Equation (7.2-2) as

$$x = \frac{1}{2}\sqrt{10 - x^3}$$

(7.2-26)

The iteration function $g(x)$ is

$$g(x) = \frac{1}{2}\sqrt{10 - x^3}$$

(7.2-27)

The iteration formula is

$$x_{k+1} = \frac{1}{2}\sqrt{10 - x_k^3}$$

(7.2-28)

Starting with the initial value $x_0 = 1$, we generate, via Equation (7.2-28), the following

$$x_1 = 1.5$$

(7.2-29a)

$$x_2 = 1.287$$

(7.2-29b)

$$x_3 = 1.402$$

(7.2-29c)

$$x_4 = 1.345$$

(7.2-29d)

$$x_5 = 1.375$$

(7.2-29e)

$$x_6 = 1.360$$

(7.2-29f)

$$x_7 = 1.368$$

(7.2-29g)

$$x_8 = 1.364$$

(7.2-29h)

$$x_9 = 1.366$$

(7.2-29i)

$$x_{10} = 1.365$$

(7.2-29j)

We note that the iteration given by Equation (7.2-28) converges to the solution 1.365 in ten iterations. We can verify that $g(x)$ satisfies the conditions for the convergence of the iteration.

An alternative form for Equation (7.2-2) is

$$x = \frac{10}{x(x+4)} \tag{7.2-30}$$

In this case, the iteration formula is

$$x_{k+1} = \frac{10}{x_k(x_k+4)} \tag{7.2-31}$$

Again, starting with an initial value equal to 1, we obtain

$$x_1 = 2 \tag{7.2-32a}$$

$$x_2 = 0.833 \tag{7.2-32b}$$

$$x_3 = 2.484 \tag{7.2-32c}$$

$$x_4 = 0.621 \tag{7.2-32d}$$

From Equations (7.2-32a to d), we note that the $x_k$ are not converging to the solution (1.365) but they are oscillating. We now examine the properties of the iteration function $g(x)$. The function $g(x)$ and $g'(x)$ are given by

$$g(x) = \frac{10}{x(x+4)} \tag{7.2-33}$$

$$g'(x) = \frac{-20(x+2)}{(x^2+4x)^2} \tag{7.2-34}$$

From Equation (7.2-34), we find that $\left| g'(x) \right|$, in the interval $(1, 1.5)$, is greater than one and condition (iii) is violated. The sequences generated by Equation (7.2-31) do not converge to the root.

Other forms of $g(x)$ can be chosen. For Newton's method, $g(x)$, computed via Equation (7.2-25) is given by

$$g(x) = \frac{2x^3 + 4x^2 + 10}{3x^2 + 8x} \tag{7.2-35}$$

It is shown in Example 7.2-3 that the root is obtained after three iterations whereas in Example 7.2-6 ten iterations are needed to achieve the same accuracy. However, it is faster to compute $g(x)$ as given in Equation (7.2-27) compared to that in Equation (7.2-35).

For human computers, it is recommended to start with the bisection or secant method. Once a reasonably good approximation is obtained, Newton's method is used for faster convergence. To avoid dividing by zero, Newton's method should be avoided if $f'(x)$ is close to zero near the root.

## 7.3   POLYNOMIAL EQUATIONS

We now apply the discussion of the previous section to the special case where $f(x)$ is a polynomial of degree n. We want to find the roots (real or complex) of the polynomial equation

$$f(x) = a_n x^n + a_{n-1} x^{n-1} + ... + a_1 x + a_0 = 0 \qquad (7.3\text{-}1a,b)$$

where $a_n \neq 0$, $n > 0$.

We note the following facts regarding Equations (7.3-1a,b) with real or complex coefficients.

1)      There are n (not necessarily distinct) real or complex roots.

2)      If n is odd and all coefficients are real, there is at least one real root.

3)      If all the coefficients are real and complex roots exist, they occur as conjugate pairs.

4)      If $x_0$ is a root of Equation (7.3-1a,b), then necessarily

$$f(x) = (x - x_0) \, g(x) \qquad (7.3\text{-}2)$$

where $g(x)$ is a polynomial of degree $(n - 1)$.

### Newton's Method

To use Newton's method for computing the root of an equation [Equation (7.2-11)], we need to evaluate $f(x_k)$ and $f'(x_k)$. We describe a method, known as **Horner's method**, for calculating $f(x_k)$ and $f'(x_k)$ where f is a polynomial. Recall that for any polynomial as given in Equations (7.3-1a,b) we can write

$$f(x) = (x - x_0) \, q(x) + R \qquad (7.3\text{-}3)$$

where

$$q(x) = b_{n-1} x^{n-1} + b_{n-2} x^{n-2} + ... + b_1 x + b_0 \qquad (7.3\text{-}4)$$

and R is a constant. Equations (7.3-1a, b, 3, 4) are compatible provided

$$b_{n-1} = a_n \qquad (7.3\text{-}5a)$$

$$b_{n-2} = a_{n-1} + x_0\, b_{n-1} \tag{7.3-5b}$$

$$\begin{array}{c} . \\ . \\ . \end{array}$$

$$b_0 = a_1 + x_0\, b_1 \tag{7.3-5c}$$

$$R = a_0 + x_0\, b_0 \tag{7.3-5d}$$

Thus if $(x - x_0)$ is not a factor of $f(x)$, we have from Equation (7.3-3)

$$f(x_0) = R = a_0 + x_0\, b_0 \tag{7.3-6a,b}$$

In Equations (7.3-5a to d), the $b_i$ are given in terms of $a_n$ and $x_0$. Therefore, starting from Equation (7.3-5a), we can successively compute $b_{n-1}, \dots, b_0$. We can then determine $f(x_0)$ from Equation (7.3-6a, b).

To compute $f'(x_0)$, we first differentiate Equation (7.3-3) and obtain

$$f'(x) = q(x) + (x - x_0)\, q'(x) \tag{7.3-7}$$

Thus

$$f'(x_0) = q(x_0) \tag{7.3-8}$$

**Example 7.3-1.** If

$$f(x) = 3x^3 - 4x^2 + x - 3 = 0 \tag{7.3-9a,b}$$

find $f(2)$ and $f'(2)$.

For $x_0 = 2$, we first obtain the constants $b_i$ using Equations (7.3-5a to c)

$$b_2 = 3 \tag{7.3-10a}$$

$$b_1 = -4 + 2\,(3) = 2 \tag{7.3-10b,c}$$

$$b_0 = 1 + 2\,(2) = 5 \tag{7.3-10d,e}$$

Hence

$$f(2) = R = -3 + 2\,(5) = 7 \tag{7.3-10f,g,h}$$

and $\quad f'(2) = b_2\, x_0^2 + b_1\, x_0 + b_0 = 21 \tag{7.3-10i,j}$

**Example 7.3-2.** Solve

$$x^3 - 3x^2 + 3 = 0 \qquad (7.3\text{-}11)$$

given that a root is near $x = 2.5$. Use Newton's method to evaluate it.

We note that

$$a_3 = 1, \quad a_2 = -3, \quad a_1 = 0, \quad a_0 = 3 \qquad (7.3\text{-}12\text{a to d})$$

We first need to calculate $f(2.5)$ and $f'(2.5)$. Equations (7.3-5a to d) yield

$$b_2 = 1 \qquad (7.3\text{-}13\text{a})$$

$$b_1 = -3 + 2.5 = -0.5 \qquad (7.3\text{-}13\text{b,c})$$

$$b_0 = 1 + 2.5(-0.5) = -1.25 \qquad (7.3\text{-}13\text{d,e})$$

$$f(2.5) = R = a_0 + x_0 b_0 = -0.125 \qquad (7.3\text{-}13\text{f,g,h})$$

$$f'(2.5) = b_2 x_0^2 + b_1 x_0 + b_0 = 3.75 \qquad (7.3\text{-}13\text{i,j})$$

According to Equation (7.2-11), we obtain as a first approximation for the root

$$x_1 = 2.5 - \frac{-0.125}{3.75} = 2.533 \qquad (7.3\text{-}14\text{a,b})$$

We now compute $f(2.533)$ and $f'(2.533)$ in a similar manner to obtain

$$f(2.533) = -0.008423 \qquad (7.3\text{-}15\text{a})$$

$$f'(2.533) = 4.0227 \qquad (7.3\text{-}15\text{b})$$

Hence the second approximation for the root is

$$x_2 = 2.533 - \frac{0.008423}{4.0227} = 2.532098 \qquad (7.3\text{-}16\text{a,b})$$

We find that the root, accurate to two decimal places is 2.53.

We can improve the accuracy of the root to any desired number of decimal places by proceeding to compute $x_3$, $x_4$, ...

We now briefly outline a method for locating intervals containing the zeros of Equation (7.3-1) where $f(x)$ is a polynomial of degree greater than 2. We note that if $x \in (0, 1]$, $\frac{1}{x} \in [1, \infty)$; and if $x \in [-1, 0)$, $\frac{1}{x} \in (-\infty, -1)$. Also

$$f\left(\frac{1}{x}\right) = a_n \left(\frac{1}{x}\right)^n + a_{n-1} \left(\frac{1}{x}\right)^{n-1} + ... + a_0 \qquad (7.3\text{-}17a)$$

$$= \frac{1}{x^n} \left[ a_n + a_{n-1} x + ... + a_1 x^{n-1} + a_0 x^n \right] \qquad (7.3\text{-}17b)$$

$$= \frac{1}{x^n} \hat{f}(x) \qquad (7.3\text{-}17c)$$

We note that changing the variable $x$ to $1/x$ allows us to look for solutions on the intervals $[-1, 0)$ and $(0, 1]$ instead of on the original interval $(-\infty, \infty)$. Furthermore if $\hat{x}$ is a zero of $\hat{f}(x)$, $\hat{x} \neq 0$, $1/\hat{x}$ is a zero of $f$. The converse is also true. It thus follows that to approximately determine all the real zeros of Equation (7.3-1), it suffices to search in the intervals $[-1, 0)$ and $(0, 1]$ evaluating both $f$ and $\hat{f}$ to determine sign changes. Recall that if $f(x_i) f(x_j) < 0$, $f$ has a zero between $x_i$ and $x_j$. Similarly if $\hat{f}(x_i) \hat{f}(x_j) < 0$, $\hat{f}$ has a zero between $x_i$ and $x_j$ and consequently Equation (7.3-1) has a zero between $\frac{1}{x_i}$ and $\frac{1}{x_j}$.

***Example 7.3-3.*** Locate the roots of the polynomial equation

$$f(x) = 16x^4 - 40x^3 + 5x^2 + 20x + 6 = 0 \qquad (7.3\text{-}18a,b)$$

As previously noted, we consider the reciprocal variable $\frac{1}{x}$, in order to consider the finite intervals $[-1, 0)$ and $(0, 1]$.

$$f\left(\frac{1}{x}\right) = \hat{f}(x) = 6x^4 + 20x^3 + 5x^2 - 40x + 16 \qquad (7.3\text{-}19a,b)$$

We note that

$$f(-1) = 47, \qquad f(0) = 6 \qquad (7.3\text{-}20a,b)$$

$$\hat{f}(-1) = 47, \qquad \hat{f}(0) = 16 \qquad (7.3\text{-}21a,b)$$

From Equations (7.3-18 to 21b), we deduce that there is no change of sign in the interval $[-1, 0)$. Consequently, the zeros of $\hat{f}$ are in the interval $(0, 1]$. To determine them we evaluate $\hat{f}$ at several points, to identify sign changes. For example, we find

$$\hat{f}(0.5) = 0.125 \qquad (7.3\text{-}22a)$$

$$\hat{f}(0.75) = -0.8516 \tag{7.3-22b}$$

$$\hat{f}(1) = 7 \tag{7.3-22c}$$

Thus $\hat{f}$ has at least one zero in the interval $(0.5, 0.75)$ and at least one other in the interval $(0.75, 1)$. It then follows that $f$ has at least one zero in the interval $\left(\dfrac{1}{0.75}, \dfrac{1}{0.5}\right) = \left(\dfrac{4}{3}, 2\right)$ and at least one other in the interval $\left(1, \dfrac{4}{3}\right)$. We have identified two approximate roots. The quartic can be written as a product of two quadratic expressions. One of them is known and the other is obtained by division.

## 7.4  SIMULTANEOUS LINEAR EQUATIONS

We now describe methods of solving **systems of linear equations** which are written as

$$a_{11} x_1 + a_{12} x_2 + ... + a_{1n} x_n = b_1$$

$$a_{21} x_1 + a_{22} x_2 + ... + a_{2n} x_n = b_2 \tag{7.4-1a,b,c}$$

$$\vdots$$

$$a_{n1} x_1 + a_{n2} x_2 + ... + a_{nn} x_n = b_n$$

A more compact form of writing Equations (7.4-1a,b,c) is

$$\sum_{j=1}^{n} a_{ij} x_j = b_i, \qquad i = 1, 2, 3, ..., n \tag{7.4-2a}$$

or $\quad \underline{\underline{A}}\, \underline{x} = \underline{b}$ $\hspace{6cm}$ (7.4-2b)

where $\underline{\underline{A}} = (a_{ij})$ is the **coefficient matrix**, and $\underline{x}$ and $\underline{b}$ are **column matrices**. Equation (7.4-2b) has a unique solution if the **inverse** of $\underline{\underline{A}}$ exists. The conditions for the existence of $\underline{\underline{A}}^{-1}$ can be stated in the following alternate forms.

1)      The determinant of $\underline{\underline{A}}$ is not zero.

2)      The columns of $\underline{\underline{A}}$ are linearly independent.

3)      The rows of $\underline{\underline{A}}$ are linearly independent.

4)      The homogenous system $(\underline{\underline{A}}\, x = 0)$ has only the trivial solution $(\underline{x} = 0)$.

5)      The rank of $\underline{\underline{A}}$ is $n$.

If $\underline{\underline{A}}^{-1}$ exists, we say that the matrix $\underline{\underline{A}}$ is **non-singular**; otherwise $\underline{\underline{A}}$ is said to be a **singular** matrix. Thus the linear system (7.4-2b) has a unique solution iff $\underline{\underline{A}}$ is non-singular. Recall also that if $\{\underline{v}_1, \underline{v}_2, \dots, \underline{v}_n\}$ is a set of vectors in $R^n$, the following statements are equivalent.

1)      Vectors $\{\underline{v}_1, \underline{v}_2, \dots, \underline{v}_n\}$ are linearly independent.

2)      Vectors $\{\underline{v}_1, \underline{v}_2, \dots, \underline{v}_n\}$ span $R^n$.

3)      Vectors $\{\underline{v}_1, \underline{v}_2, \dots, \underline{v}_n\}$ form a basis for $R^n$.

We now discuss a standard method of solving the set of Equations (7.4-1).

We note that the following operations on a system of linear equations do not change the solution of the system

1)      interchanging two rows,

2)      multiplying a row by a non-zero constant,

3)      adding a multiple of one row to another row.

## Gaussian Elimination Method

The basic idea behind this method is to convert the $n \times n$ coefficient matrix of the given system to an **upper triangular matrix**. We reduce the matrix $\underline{\underline{A}}$ to a triangular form by the following elementary operations. We choose $a_{11}$ as a **pivot** and keep the first row unchanged. We multiply the first row by an appropriate factor and add it to the second row so as to reduce the coefficient of $x_1$ in the second row to zero. Similarly we reduce the coefficients of $x_1$ in all subsequent rows to zero. We then choose the new coefficient of $x_2$ in the second row as pivot and reduce the coefficient of $x_2$ to zero in row three as well as in the subsequent rows. We continue this process until we have only one unknown $x_n$ in the $n^{th}$ row. We can solve for $x_n$. By backward substitution, we can determine the solutions for $x_{n-1}, \dots, x_2, x_1$.

The row operations described earlier are to be applied to column $\underline{b}$ as well. It is economical to consider the matrix $\underline{\underline{A}}$ and the column $\underline{b}$ simultaneously. We then operate on the **augmented matrix** $\underline{\underline{A}}_g$, which is obtained by combining $\underline{\underline{A}}$ and $\underline{b}$, where $\underline{b}$ is the $(n+1)^{th}$ column of $\underline{\underline{A}}_g$. The matrix $\underline{\underline{A}}_g$ is now a rectangular matrix with $n$ rows and $(n+1)$ columns.

***Example 7.4-1***. Solve the following system of equations via the Gaussian elimination method.

$$x_1 + 2x_2 + 3x_3 = 5$$

$$x_1 + 3x_2 + 3x_3 = 4 \qquad\qquad (7.4\text{-}3a)$$

$$2x_1 + 4x_2 + 7x_3 = 12$$

We rewrite Equations (7.4-3a) in matrix form as follows

$$
\begin{bmatrix} 1 & 2 & 3 \\ 1 & 3 & 3 \\ 2 & 4 & 7 \end{bmatrix}
\begin{bmatrix} x_1 \\ x_2 \\ x_3 \end{bmatrix}
=
\begin{bmatrix} 5 \\ 4 \\ 12 \end{bmatrix}
\qquad (7.4\text{-}3b)
$$

To carry out the required operations, we write the augmented matrix

$$
\underline{\underline{A}}_g =
\begin{bmatrix} 1 & 2 & 3 & 5 \\ 1 & 3 & 3 & 4 \\ 2 & 4 & 7 & 12 \end{bmatrix}
\qquad (7.4\text{-}3c)
$$

For the row operations on the augmented matrix, we choose $a_{11} = 1$ as pivot and we do not change the first row. Multiplying row 1 by $-1$ and adding it to row 2, as well as multiplying row 1 by $-2$ and adding it to row 3, yields the modified augmented matrix $\underline{\underline{A}}_g^{(1)}$

$$
\underline{\underline{A}}_g^{(1)} =
\begin{bmatrix} 1 & 2 & 3 & 5 \\ 0 & 1 & 0 & -1 \\ 0 & 0 & 1 & 2 \end{bmatrix}
\qquad (7.4\text{-}3d)
$$

The matrix

$$
\underline{\underline{A}}^{(1)} =
\begin{bmatrix} 1 & 2 & 3 \\ 0 & 1 & 0 \\ 0 & 0 & 1 \end{bmatrix}
\qquad (7.4\text{-}3e)
$$

is the required triangular form. Since $a_{32}$ was already zero, we did not need to carry out further manipulations. From row 3 of Equation (7.4-3d) we can identify

$$x_3 = 2 \qquad\qquad (7.4\text{-}4a)$$

From row 2, we obtain

$$x_2 = -1 \qquad\qquad\qquad\qquad (7.4\text{-}4b)$$

Substituting the values of $x_2$ and $x_3$ in row 1 yields

$$x_1 + 2\,(-1) + 3\,(2) = 5 \qquad\qquad\qquad (7.4\text{-}4c)$$

and the value of $x_1$ is

$$x_1 = 1 \qquad\qquad\qquad\qquad (7.4\text{-}4d)$$

●

The following observations concerning this method are important. At each stage in the Gaussian elimination procedure, a pivot row must be selected. Appropriate multiples of the pivot row are added to other rows to perform the desired elimination. The selection of this pivot row is very important. We illustrate a problem which could arise by considering the following non-singular matrix

$$\underline{\underline{A}} = \begin{bmatrix} 1 & 2 & 4 \\ 1 & 2 & -5 \\ 5 & 1 & 10 \end{bmatrix} \qquad\qquad\qquad (7.4\text{-}5a)$$

Choosing the first row as a pivot row yields

$$\underline{\underline{A}}^{(1)} = \begin{bmatrix} 1 & 2 & 4 \\ 0 & 0 & -9 \\ 0 & -9 & -10 \end{bmatrix} \begin{matrix} \\ (\text{row } 2 - \text{row } 1) \\ (\text{row } 3 - 5 \times \text{row } 1) \end{matrix} \qquad (7.4\text{-}5b)$$

If, in the usual manner, we were to choose the second row as the pivot row, we would not be able to proceed, since the coefficient $a_{22}$ is zero. However, interchanging rows 2 and 3 allows us to eliminate the coefficient of $x_2$ in such a way as to proceed towards the computation of the required triangular matrix. That is, we write Equation (7.4-5b) as

$$\underline{\underline{A}}^{(2)} = \begin{bmatrix} 1 & 2 & 4 \\ 0 & -9 & -10 \\ 0 & 0 & -9 \end{bmatrix} \qquad\qquad\qquad (7.4\text{-}5c)$$

and this allows us to determine $x_3$, $x_2$ and $x_1$ provided the vector $\underline{b}$ is known.

This example illustrates the need for adopting an appropriate pivoting strategy. One such strategy is as follows.

(i)     Scan the first column to find the largest absolute value $|a_{11}|$.

(ii)    Exchange this row with the first row.

(iii)   After the first column of zeros has been determined, scan the second column below the first row to find the largest absolute value $|a_{i2}|$, $i \neq 1$.

(iv)    The row in which this element is found now becomes the second row.

This process is continued until the matrix has been transformed into an upper triangular matrix. If the matrix is non-singular, this strategy will always work.

*Example 7.4-2.* Solve the following system of equations using the Gaussian elimination method.

$$x_1 + 5x_2 + 4x_3 = 23$$

$$2x_1 + \frac{1}{2}x_2 + x_3 = 6 \qquad\qquad (7.4\text{-}6a,b,c)$$

$$5x_1 + 2x_2 - 3x_3 = 0$$

Here the largest absolute value in the first column occurs in the third row. We interchange the third row with the first row and write the augmented matrix as

$$\underline{\underline{A}}_g = \begin{bmatrix} 5 & 2 & -3 & 0 \\ 2 & \frac{1}{2} & 1 & 6 \\ 1 & 5 & 4 & 23 \end{bmatrix} \qquad\qquad (7.4\text{-}6d)$$

Choosing the first row as the pivot row, we proceed to generate zeros in the first column, as described.

$$
\underline{A}_g^{(1)} = \begin{bmatrix} 5 & 2 & -3 & 0 \\[6pt] 0 & -\dfrac{3}{10} & \dfrac{22}{10} & 6 \\[6pt] 0 & \dfrac{46}{10} & \dfrac{46}{10} & 23 \end{bmatrix} \quad \begin{array}{l} \\[6pt] \left(\text{row } 2 - \dfrac{2}{5}\,\text{row } 1\right) \\[6pt] \left(\text{row } 3 - \dfrac{1}{5}\,\text{row } 1\right) \end{array} \tag{7.4-6e}
$$

We now look at the second column and note that $\left|\dfrac{46}{10}\right| > \left|-\dfrac{3}{10}\right|$. We then interchange the second and third rows leaving the first row unchanged. Performing the elimination, as stated earlier, we find the new matrix to be

$$
\underline{A}_g^{(2)} = \begin{bmatrix} 5 & 2 & -3 & 0 \\[4pt] 0 & 4.6 & 4.6 & 23 \\[4pt] 0 & 0 & 2.5 & 7.5 \end{bmatrix} \tag{7.4-6f}
$$

The last row identifies $x_3$ as $7.5/2.5$. Substituting this value of $x_3$ in the second row, we obtain $x_2$ and finally, from the first row, we obtain $x_1$, yielding

$$
x_1 = 1, \qquad x_2 = 2, \qquad x_3 = 3 \tag{7.4-7a,b,c}
$$

The second kind of problem that can arise in the solution of a system of linear equations deals with **singular or ill-conditioned** (nearly singular) matrices. When the matrix is singular, we do not have a unique solution. We examine the case of ill-conditioning by considering the following two systems of equations

(i)
$$
\begin{aligned} x_1 + x_2 &= 1 \\ x_1 + 1.01\,x_2 &= 2 \end{aligned} \tag{7.4-8a,b}
$$

and

(ii)
$$
\begin{aligned} x_1 + x_2 &= 1 \\ 1.01\,x_1 + x_2 &= 2 \end{aligned} \tag{7.4-9a,b}
$$

The exact solution of Equations (7.4-8a,b) is

$$
x_1 = -99, \qquad x_2 = 100 \tag{7.4-10a,b}
$$

The exact solution of Equations (7.4-9a,b) is

$$x_1 = 100, \qquad x_2 = -99 \tag{7.4-11a,b}$$

Note that for a small change in the value of the coefficient in the second equation, the solution $(-99, 100)$ is changed to $(100, -99)$. That is to say, from one quadrant to another quadrant!

Consider the system of equations

$$x_1 + x_2 = 2$$

$$1.001\, x_1 + x_2 = 2.001 \tag{7.4-12a,b}$$

The exact solution here is

$$x_1 = 1, \qquad x_2 = 1 \tag{7.4-13a,b}$$

The equations are also satisfied approximately by

$$x_1 \approx 0, \qquad x_2 \approx 2 \tag{7.4-14a,b}$$

Here, we find that Equations (7.4-12a,b) can have an incorrect approximate solution.

The above examples are cases of ill-conditioning. Ill-conditioning is thus in a sense a measure of the closeness of the coefficient matrix to a singular matrix. The coefficient matrices in the three previous cases were

$$\begin{bmatrix} 1 & 1 \\ 1.01 & 1 \end{bmatrix}, \quad \begin{bmatrix} 1 & 1 \\ 1 & 1.01 \end{bmatrix}, \quad \begin{bmatrix} 1 & 1 \\ 1.001 & 1 \end{bmatrix}$$

which are, in a sense, close to the singular matrix $\begin{bmatrix} 1 & 1 \\ 1 & 1 \end{bmatrix}$. The determinant of each of the coefficient matrices is nearly zero. We define the closeness of two matrices by the **condition number** which in turn is expressed in terms of the **norm** of the matrices. Vector and matrix norms are defined as follows.

To each vector (or matrix) $\underline{x}$, we assign a real number which we call a norm and denote it by $\| \underline{x} \|$.

A norm shares certain properties with absolute values. A norm satisfies the following properties.

1)      $\| \underline{x} \| \geq 0$ for all $\underline{x}$, $\| \underline{x} \| = 0$ iff $\underline{x} = \underline{0}$.

2)      $\| c\,\underline{x} \| = | c | \, \| \underline{x} \|$ for any real constant $c$.

3)      $\| \underline{x} + \underline{y} \| \leq \| \underline{x} \| + \| \underline{y} \|$ (triangle inequality).

The first condition states that all non-zero vectors have a positive norm. The second property implies that the norm of a vector is invariant, irrespective of the direction. The third property is the usual triangle inequality.

The $\mathbf{L_p}$ **norm** is defined as

$$L_p = \| \underline{x} \|_p = \left( \sum |x_i|^p \right)^{1/p} \quad \text{for } p \geq 1 \qquad \text{(7.4-15 a,b)}$$

where $x_i$ are the components of $\underline{x}$.

For $p = 1$, the $L_1$ norm is the sum of the absolute values of the components of $\underline{x}$.

The familiar Euclidean norm, which is the length of the vector $\underline{x}$, is obtained by setting $p = 2$. In this case, Equation (7.4-15b) becomes

$$\| \underline{x} \|_2 = \sqrt{x_1^2 + x_2^2 + \dots + x_n^2} \qquad \text{(7.4-16)}$$

The $\| \underline{x} \|_\infty$ norm is defined as

$$\| \underline{x} \|_\infty = \max_{1 \leq i \leq n} |x_i| \qquad \text{(7.4-17)}$$

The $\| \underline{x} \|_\infty$ chooses the maximum absolute value of $x_i$ as a measure of $\underline{x}$. It is easier to compute $\| \underline{x} \|_\infty$ than $\| \underline{x} \|_2$ and it is not surprising that $\| \underline{x} \|_\infty$ is preferred to $\| \underline{x} \|_2$ in numerical analysis.

The $\mathbf{L_1}$ **and** $\mathbf{L_\infty}$ **norms of a matrix** are defined as

$$L_1 = \| \underline{\underline{A}} \|_1 = \max_{1 \leq j \leq n} \sum_{i=1}^{n} |a_{ij}| \qquad \text{(7.4-18a,b)}$$

$$L_\infty = \| \underline{\underline{A}} \|_\infty = \max_{1 \leq i \leq n} \sum_{j=1}^{n} |a_{ij}| \qquad \text{(7.4-19a,b)}$$

The $\| \underline{\underline{A}} \|_1$ norm, as defined by Equation (7.4-18b), is obtained by adding the absolute values of the elements of each column of $\underline{A}$. The maximum value of the column sum is $\| \underline{\underline{A}} \|_1$. To calculate $\| \underline{\underline{A}} \|_\infty$ we interchange the roles of columns and rows and repeat the process described earlier in the determination of $\| \underline{\underline{A}} \|_1$.

***Example 7.4-3.*** Calculate $\| \underline{\underline{A}} \|_1$ and $\| \underline{\underline{A}} \|_\infty$ if

$$\underset{=}{A} = \begin{bmatrix} 1 & 2 & -2 \\ 0 & -3 & 2 \\ -4 & 3 & 2 \end{bmatrix} \qquad\qquad (7.4\text{-}20)$$

The sum of the absolute values of the columns is 5, 8, and 6 respectively. The maximum is 8 and

$$\| \underset{=}{A} \|_1 = 8 \qquad\qquad (7.4\text{-}21)$$

Similarly, the rows yield a maximum of 9, that is

$$\| \underset{=}{A} \|_\infty = 9 \qquad\qquad (7.4\text{-}22)$$

●

A norm of a vector or a matrix may be considered to be a measure of its size.

We now investigate the effect of slightly changing the vectors $\underline{x}$ and $\underline{b}$ in Equation (7.4-2b). Suppose $\underline{x}$ is perturbed to $\underline{x} + \delta\underline{x}$ and $\underline{b}$ to $\underline{b} + \delta\underline{b}$. Then

$$\underset{=}{A}(\underline{x} + \delta\underline{x}) = \underline{b} + \delta\underline{b} \qquad\qquad (7.4\text{-}23)$$

Combining Equations (7.4-2b) and (7.4-23) yields

$$\underset{=}{A}\,\delta\underline{x} = \delta\underline{b} \qquad\qquad (7.4\text{-}24a)$$

and    $\delta\underline{x} = \underset{=}{A}^{-1}\delta\underline{b}$ $\qquad\qquad (7.4\text{-}24b)$

In terms of norms, we have

$$\| \underset{=}{A}\ \underline{x} \| = \| \underline{b} \| \qquad\qquad (7.4\text{-}25a)$$

and    $\| \underline{b} \| \le \| \underset{=}{A} \|\ \| \underline{x} \|$ $\qquad\qquad (7.4\text{-}25b)$

We can then write

$$\| \delta\underline{x} \| \le \| \underset{=}{A}^{-1} \|\ \| \delta\underline{b} \| \qquad\qquad (7.4\text{-}26a)$$

or    $\dfrac{\| \delta\underline{x} \|}{\| \underset{=}{A} \|\ \| \underline{x} \|} \le \dfrac{\| \underset{=}{A}^{-1} \|\ \| \delta\underline{b} \|}{\| \underline{b} \|}$ $\qquad\qquad (7.4\text{-}26b)$

or $\quad \dfrac{\| \delta \underline{x} \|}{\| \underline{x} \|} \leq \| \underline{\underline{A}} \| \, \| \underline{\underline{A}}^{-1} \| \, \dfrac{\| \delta \underline{b} \|}{\| \underline{b} \|}$          (7.4-26c)

Equation (7.4-26c) shows that a relative change in the solution vector is given by the product of $\| \underline{\underline{A}} \| \, \| \underline{\underline{A}}^{-1} \|$ with the relative change of vector $\underline{b}$. This quantity $\| \underline{\underline{A}} \| \, \| \underline{\underline{A}}^{-1} \|$ is called the **condition number or COND** ($\underline{\underline{A}}$) and is denoted by $K(\underline{\underline{A}})$. If $K(\underline{\underline{A}})$ is large we say that the matrix is ill-conditioned. We can then expect that a small change in $\underline{b}$ will produce a large change in $\underline{x}$ resulting in loss of accuracy. The choice of norm is not important. Note that $K(\underline{\underline{A}})$ is undefined for a singular matrix.

***Example 7.4-4.*** Determine $K(\underline{\underline{A}})$ using the $L_1$ and $L_\infty$ norms where

$$\underline{\underline{A}} = \begin{bmatrix} 1 & 1 \\ 1.01 & 1 \end{bmatrix} \qquad\qquad (7.4\text{-}27)$$

The $L_1$ and $L_\infty$ norms are

$$L_1 = \| \underline{\underline{A}} \|_1 = \max_{1 \leq j \leq 2} \sum_{i=1}^{n} \left| a_{ij} \right| = \max(2.01,\, 2) = 2.01 \qquad (7.4\text{-}28a,b,c)$$

$$L_\infty = \| \underline{\underline{A}} \|_\infty = \max_{1 \leq i \leq 2} \sum_{j=1}^{n} \left| a_{ij} \right| = \max(2,\, 2.01) = 2.01 \qquad (7.4\text{-}29a,b,c)$$

To obtain the inverse of $\underline{\underline{A}}$, we use the Gauss-Jordan method which proceeds as follows. To the matrix $\underline{\underline{A}}$, we adjoin the identity matrix $\underline{\underline{I}}$ and we obtain a $n \times 2n$ matrix which we write as $\left[ \underline{\underline{A}} \;\vdots\; \underline{\underline{I}} \right]$. We separate the matrices $\underline{\underline{A}}$ and $\underline{\underline{I}}$ by a dotted line. We use the Gaussian elimination method to reduce $\underline{\underline{A}}$ to the usual triangular form, performing simultaneously the same operations on $\underline{\underline{I}}$. We divide the resulting last row by an appropriate number so that the element of the last row and last column is one. The other elements in the last row are zeros. We then perform the Gaussian elimination process from the $n^{\text{th}}$ row upwards until $\underline{\underline{A}}$ has been transformed to $\underline{\underline{I}}$ and $\underline{\underline{I}}$ is then transformed to $\underline{\underline{A}}^{-1}$. The resulting $n \times 2n$ matrix is $\left[ \underline{\underline{I}} \;\vdots\; \underline{\underline{A}}^{-1} \right]$. Below we illustrate the method in the case of the matrix $\underline{\underline{A}}$ given by Equation (7.4-27).

$$\left[ \underline{\underline{A}} \;\vdots\; \underline{\underline{I}} \right] = \begin{bmatrix} 1 & 1 & \vdots & 1 & 0 \\ 1.01 & 1 & \vdots & 0 & 1 \end{bmatrix} \qquad\qquad (7.4\text{-}30a)$$

$$\downarrow$$

$$\begin{bmatrix} 1 & 1 & \vdots & 1 & 0 \\ 0 & -0.01 & \vdots & -1.01 & 1 \end{bmatrix} \text{(row 2 – 1.01 row 1)}$$

(7.4-30b)

$$\downarrow$$

$$\begin{bmatrix} 1 & 1 & \vdots & 1 & 0 \\ 0 & 1 & \vdots & 101 & -100 \end{bmatrix} \text{(divide row 2 by -0.01)}$$

(7.4-30c)

$$\downarrow$$

$$\begin{bmatrix} 1 & 0 & \vdots & -100 & 100 \\ 0 & 1 & \vdots & 101 & -100 \end{bmatrix} \begin{matrix} \text{(row 1 – row 2)} \\ \\ \end{matrix}$$

(7.4-30d)

The inverse of $\underline{\underline{A}}$ is

$$\underline{\underline{A}}^{-1} = \begin{bmatrix} -100 & 100 \\ 101 & -100 \end{bmatrix}$$

(7.4-31)

The norm $\| \underline{\underline{A}}^{-1} \|_1$ is given by the maximum of the column sum, (201, 200)

$$\| \underline{\underline{A}}^{-1} \|_1 = 201$$

(7.4-32)

$\| \underline{\underline{A}}^{-1} \|_\infty$ is given by the maximum of the row sum (200, 201)

$$\| \underline{\underline{A}}^{-1} \|_\infty = 201$$

(7.4-33)

Using the $L_1$ norm, we find that $K(\underline{\underline{A}})$ is given by

$$K(\underline{\underline{A}}) = \| \underline{\underline{A}} \|_1 \| \underline{\underline{A}}^{-1} \|_1$$

(7.4-34a)

$$= 2.01 \times 201 = 404.01$$

(7.4-34b,c)

Similarly using the $L_\infty$ norm, $K(\underline{\underline{A}})$ is

$$K(\underline{\underline{A}}) = 2.01 \times 201 = 404.01$$

(7.4-35a,b)

We conclude this section by stating the nearest singular matrix theorem which helps to circumvent the cumbersome problem of finding $\underline{\underline{A}}^{-1}$.

If $\underline{\underline{A}}$ and $\underline{\underline{B}}$ are a non-singular and a singular matrix respectively

$$\frac{1}{\| \underline{\underline{A}} - \underline{\underline{B}} \|} \leq \| \underline{\underline{A}}^{-1} \| \qquad (7.4\text{-}36)$$

Using this result in the above example, we choose $\underline{\underline{B}}$ to be

$$\underline{\underline{B}} = \begin{bmatrix} 1 & 1 \\ 1 & 1 \end{bmatrix} \qquad (7.4\text{-}37)$$

$$\| \underline{\underline{A}} - \underline{\underline{B}} \| = 0.01 \qquad (7.4\text{-}38)$$

So that

$$\| \underline{\underline{A}}^{-1} \| \geq \frac{1}{0.01} = 10^2 \qquad (7.4\text{-}39a,b)$$

With $\| \underline{\underline{A}} \| = 2.01$, we have $K(\underline{\underline{A}}) \geq 2.01 \times 10^2$ which is of the same order of magnitude as that obtained earlier.

## Iterative Method

When the dimension of the system of linear equations to be solved is small, the Gauss elimination method is useful. However, when the number of variables is large and when the coefficient matrix is sparse (has many zero entries), **iterative techniques** are preferable in terms of computer storage and time requirements.

Sparse matrices of high order are of frequent occurrence. An example is when partial differential equations are solved numerically (see Chapter 8).

When using an iterative method, the coefficient matrix $\underline{\underline{A}}$ is written as the difference of two matrices $\underline{\underline{B}}_0$ and $\underline{\underline{C}}_0$. Equation (7.4-2b) is now written as

$$\underline{\underline{B}}_0 \, \underline{x} = \underline{b} + \underline{\underline{C}}_0 \, \underline{x} \qquad (7.4\text{-}40)$$

Premultiplying Equation (7.4-40) by the inverse of $\underline{\underline{B}}_0$ we obtain

$$\underline{x} = \underline{\underline{B}}_0^{-1} \underline{b} + \underline{\underline{B}}_0^{-1} \underline{\underline{C}}_0 \underline{x} \qquad \text{(7.4-41a)}$$

$$= \underline{\underline{B}} \underline{x} + \underline{c} \qquad \text{(7.4-41b)}$$

Starting from an initial guess $\underline{x}_0$, we can generate a sequence of values of $\underline{x}$ from Equation (7.4-41b) as follows

$$\underline{x}_{k+1} = \underline{\underline{B}} \underline{x}_k + \underline{c} \qquad \text{(7.4-42)}$$

A judicious choice of $\underline{\underline{B}}_0$ and $\underline{\underline{C}}_0$ will lead to a sequence of $\underline{x}_k$ which will converge to the required solution $\underline{x}$. We first consider the method of Jacobi and then the Gauss-Seidel method.

### Jacobi's method

In **Jacobi's method**, we solve each equation for only one of the variables, choosing whenever possible, to solve for the variable with the largest coefficient. That is to say, $\underline{\underline{B}}_0$ is chosen to be a diagonal matrix. We begin with some initial approximate values for each variable. Each component of $\underline{x}$ on the right side of Equation (7.4-42) may be taken equal to zero if no information is available. Substituting these values into the right side of the set of equations generates new approximations that are closer to the required value. These new values are then substituted in the right side of each equation to generate successive approximations until the difference of two successive approximations of each variable is within the desired degree of accuracy, as shown in the next example.

***Example 7.4-5.*** Using Jacobi's iteration method, solve the system of equations

$$8x_1 + x_2 - x_3 = 8$$

$$2x_1 + x_2 + 9x_3 = 12 \qquad \text{(7.4-43a,b,c)}$$

$$x_1 - 7x_2 + 2x_3 = -4$$

We first write Equations (7.4-43a, b, c) for one variable in each row, with the largest coefficient

$$x_1 = 1 - \frac{1}{8} x_2 + \frac{1}{8} x_3 \qquad \left(1^{\text{st}} \text{ row}\right) \qquad \text{(7.4-44a)}$$

$$x_2 = \frac{4}{7} + \frac{1}{7} x_1 + \frac{2}{7} x_3 \qquad \left(3^{\text{rd}} \text{ row}\right) \qquad \text{(7.4-44b)}$$

$$x_3 = \frac{4}{3} - \frac{2}{9} x_1 - \frac{1}{9} x_2 \quad \left(2^{\text{nd}} \text{ row}\right) \tag{7.4-44c}$$

We interchange the second and third equations of Equations (7.4-43) so that the diagonal elements of the coefficient matrix $\underset{=}{A}$ are the elements that have the maximum absolute value for each row. The matrix $\underset{=}{A}$ is

$$\underset{=}{A} = \begin{bmatrix} 8 & 1 & -1 \\ 1 & -7 & 2 \\ 2 & 1 & 9 \end{bmatrix} \tag{7.4-45}$$

The matrices $\underset{=}{B}_0$ and $\underset{=}{C}_0$ are chosen to be

$$\underset{=}{B}_0 = \begin{bmatrix} 8 & 0 & 0 \\ 0 & -7 & 0 \\ 0 & 0 & 9 \end{bmatrix} \tag{7.4-46a}$$

$$\underset{=}{C}_0 = \begin{bmatrix} 0 & -1 & 1 \\ -1 & 0 & -2 \\ -2 & -1 & 0 \end{bmatrix} \tag{7.4-46b}$$

and $\quad \underset{=}{A} = \underset{=}{B}_0 - \underset{=}{C}_0 \tag{7.4-46c}$

Equations (7.4-44a to c) are obtained by substituting Equations (7.4-46a, b) into Equations (7.4-41a, b).

We denote the $i^{\text{th}}$ component of $\underline{x}_k$ by $x_i^{(k)}$ and the iterative formulae [Equations (7.4-44a to c) or Equation (7.4-42)] are

$$x_1^{(k+1)} = 1 - \frac{1}{8} x_2^{(k)} + \frac{1}{8} x_3^{(k)} \tag{7.4-47a}$$

$$x_2^{(k+1)} = \frac{4}{7} + \frac{1}{7} x_1^{(k)} + \frac{2}{7} x_3^{(k)} \tag{7.4-47b}$$

$$x_3^{(k+1)} = \frac{4}{3} - \frac{2}{9} x_1^{(k)} - \frac{1}{9} x_2^{(k)} \tag{7.4-47c}$$

As a first approximation, we take

$$x_1^{(0)} = x_2^{(0)} = x_3^{(0)} = 0$$                                                           (7.4-48a,b,c)

Substituting Equations (7.4-48a to c) into Equations (7.4-47a to c) generates

$$x_1^{(1)} = 1, \qquad x_2^{(1)} = \frac{4}{7}, \qquad x_3^{(1)} = \frac{4}{3}$$                (7.4-49a,b,c)

Substituting the values of $x_i^{(1)}$ into Equations (7.4-47a to c), we obtain

$$x_1^{(2)} = 1.095, \quad x_2^{(2)} = 1.095, \quad x_3^{(2)} = 1.048$$                        (7.4-50a,b,c)

Repeating this procedure, we eventually obtain the exact solution which is

$$x_1 = x_2 = x_3 = 1$$                                                                          (7.4-51a,b,c)

●

With the choice of $\underline{\underline{B}}_0$ as described in Example 7.4-5, Equation (7.4-42) can be written in component form as

$$x_i^{(k+1)} = \frac{b_i}{a_{ii}} - \sum_{\substack{j=1 \\ j \neq i}}^{n} \frac{a_{ij}}{a_{ii}} x_j^{(k)}$$                      (7.4-52)

Equation (7.4-52) is the Jacobi's iteration formula for solving the system of Equations (7.4-2a).

### *Gauss-Seidel method*

In Equation (7.4-52) to compute $x_i^{(k+1)}$, we have made use of the values of $x_j^{(k)}$ only, although the values of $x_j^{(k+1)}$ $(j < i)$ are already known. For example, in computing $x_3^{(k+1)}$, we can substitute the values of $x_1^{(k+1)}$ and $x_2^{(k+1)}$ in Equation (7.4-52) instead of $x_1^{(k)}$ and $x_2^{(k)}$. By so doing, we hope to be able to accelerate the convergence.

Equation (7.4-52) is modified to

$$x_i^{(k+1)} = \frac{b_i}{a_{ii}} - \sum_{j=1}^{i-1} \frac{a_{ij}}{a_{ii}} x_j^{(k+1)} - \sum_{j=i+1}^{n} \frac{a_{ij}}{a_{ii}} x_j^{(k)}$$          (7.4-53)

Equation (7.4-53) is the Gauss-Seidel iteration method.

_Example 7.4-6_. Solve the system of Equations (7.4-43) by the Gauss-Seidel method.

The coefficient matrix is rewritten as in Example 7.4-5.

Equation (7.4-47a) remains unchanged. Equations (7.4-47b, c) are modified to

$$x_2^{(k+1)} = \frac{4}{7} + \frac{1}{7} x_1^{(k+1)} + \frac{2}{7} x_3^{(k)} \qquad\qquad (7.4\text{-}54a)$$

$$x_3^{(k+1)} = \frac{4}{3} - \frac{2}{9} x_1^{(k+1)} - \frac{1}{9} x_2^{(k+1)} \qquad\qquad (7.4\text{-}54b)$$

Using the same initial values ($\underline{x}_0 = \underline{0}$), the first iteration gives the same values for $\underline{x}_1$. Using

Equations (7.4-47a, 54a, b), we obtain

$$x_1^{(2)} = 1.095, \quad x_2^{(2)} = 1.109, \quad x_3^{(2)} = 0.967 \qquad\qquad (7.4\text{-}55a,b,c)$$

Continuing the process, we obtain, after five iterations, the exact solution (1, 1, 1) whereas Jacobi's method requires seven iterations.

Table 7.4-1 gives the values of $\underline{x}^{(k)}$ generated by Equations (7.4-47a to c, 47a, 54a, b).

## TABLE 7.4-1

**Values of $x_i^{(k)}$ generated by the methods of Jacobi's and Gauss-Seidel**

| k | Jacobi | | | Gauss-Seidel | | |
|---|---|---|---|---|---|---|
|   | $x_1^{(k)}$ | $x_2^{(k)}$ | $x_3^{(k)}$ | $x_1^{(k)}$ | $x_2^{(k)}$ | $x_3^{(k)}$ |
| 0 | 0 | 0 | 0 | 0 | 0 | 0 |
| 1 | 1.000 | 0.571 | 1.333 | 1.000 | 0.571 | 1.333 |
| 2 | 1.095 | 1.095 | 1.048 | 1.095 | 1.109 | 0.967 |
| 3 | 0.994 | 1.027 | 0.968 | 0.982 | 0.988 | 1.005 |
| 4 | 0.993 | 0.990 | 0.998 | 1.002 | 1.002 | 0.999 |
| 5 | 1.001 | 0.998 | 1.003 | 1.000 | 1.000 | 1.000 |
| 6 | 1.001 | 1.001 | 1.000 | | | |
| 7 | 1.000 | 1.000 | 1.000 | | | |

## 7.5    EIGENVALUE PROBLEMS

Problems involving the determination of eigenvalues and eigenvectors often arise in science and engineering. For example, eigenvalues arise in connection with vibration problems in mechanical engineering, in discussing the stability of an aircraft in aeronautical engineering and in quantum mechanics. In this section, we study the problems of calculating the eigenvalues of a square matrix. The general problem of finding all eigenvalues of a non-symmetric matrix is much more difficult as it easily leads to stability problems with respect to perturbations. Most of the algorithms for estimating eigenvalues of a symmetric matrix $\underline{\underline{A}}$ make use of similarity transformations and are therefore carried out in two stages. In the first stage, the matrix is reduced to a suitable form and, in the second stage, the method of determining the eigenvalues is executed. Before proceeding with the method, we briefly summarize the basic results and definitions.

We say that a matrix $\underline{\underline{A}}$ is **similar** to another matrix $\underline{\underline{B}}$ if there exists a non-singular matrix $\underline{\underline{P}}$ such that

$$\underline{\underline{A}} = \underline{\underline{P}}^{-1} \underline{\underline{B}} \, \underline{\underline{P}} \tag{7.5-1}$$

If $\lambda$ is an **eigenvalue** of $\underline{\underline{A}}$ and $\underline{x}$ is the corresponding **eigenvector**

$$\underline{\underline{A}} \, \underline{x} = \lambda \, \underline{x} \tag{7.5-2}$$

Suppose that matrix $\underline{\underline{A}}$ is similar to $\underline{\underline{B}}$. If $\lambda$ is an eigenvalue of $\underline{\underline{A}}$ with associated eigenvector $\underline{x}$, then $\lambda$ is also an eigenvalue of $\underline{\underline{B}}$ with $\underline{\underline{P}} \, \underline{x}$ as the associated eigenvector of $\underline{\underline{B}}$.

If $\underline{\underline{A}}$ is triangular or diagonal, the diagonal entries of $\underline{\underline{A}}$ are the eigenvalues of $\underline{\underline{A}}$.

A real matrix $\underline{\underline{Q}}$ is **orthogonal** iff

$$\underline{\underline{Q}}^{-1} = \underline{\underline{Q}}^{\dagger} \tag{7.5-3}$$

If $\underline{\underline{A}}$ is a real symmetric matrix, there exists an orthogonal matrix $\underline{\underline{Q}}$ such that

$$\underline{\underline{Q}}^{-1} \underline{\underline{A}} \, \underline{\underline{Q}} = \underline{\underline{C}} \tag{7.5-4}$$

where $\underline{\underline{C}}$ is a diagonal matrix. Thus the diagonal entries of $\underline{\underline{C}}$, in this case, are the eigenvalues of $\underline{\underline{A}}$.

The eigenvalues of a real symmetric matrix are real numbers and their corresponding eigenvectors are mutually orthogonal.

If $\underline{\underline{A}}$ is a matrix with complex entries, the **Hermitian transpose** of $\underline{\underline{A}}$ is the transpose of the complex conjugate of $\underline{\underline{A}}$ and is denoted by $\underline{\underline{A}}^H$.

If $\qquad \underline{\underline{A}} = \underline{\underline{A}}^H$ $\hfill$ (7.5-5)

the eigenvalues of $\underline{\underline{A}}$ are all real and $\underline{\underline{A}}$ is known as a **Hermitian matrix**.

A matrix $\underline{\underline{U}}$ is a **unitary** matrix iff

$$\underline{\underline{U}}^{-1} = \underline{\underline{U}}^H \hfill (7.5\text{-}6)$$

A matrix $\underline{\underline{A}}$ $(= a_{ij})$ is **tridiagonal** if

$$a_{ij} = 0 \quad \text{whenever} \quad |i - j| > 1 \hfill (7.5\text{-}7)$$

The only entries in $\underline{\underline{A}}$ that could be non-zero are the elements along the diagonal, the subdiagonal and the super diagonal. Examples of tridiagonal matrices are

$$\begin{bmatrix} 3 & 1 & 0 \\ 1 & 2 & 4 \\ 0 & 4 & 5 \end{bmatrix}, \qquad \begin{bmatrix} -1 & 2 & 0 & 0 & 0 \\ 2 & 1 & 3 & 0 & 0 \\ 0 & 3 & 4 & 2 & 0 \\ 0 & 0 & 2 & 1 & 5 \\ 0 & 0 & 0 & 5 & -1 \end{bmatrix}$$

We next describe a method of transforming a real symmetric matrix to a tridiagonal matrix.

## Householder Algorithm

It was stated earlier that for any real symmetric matrix $\underline{\underline{A}}$, there exists a real orthogonal matrix $\underline{\underline{Q}}$ such that $\underline{\underline{Q}}^{-1} \underline{\underline{A}} \, \underline{\underline{Q}}$ is diagonal. The diagonal elements of $\underline{\underline{A}}$ are then its eigenvalues, and the columns of $\underline{\underline{Q}}$ are the eigenvectors. Attempts at diagonalizing $\underline{\underline{A}}$ have not always been successful. The methods proposed so far do not always converge. Instead we transform $\underline{\underline{A}}$ to a tridiagonal form via the method of Householder which is economical and reliable.

Let $\underline{v}$ be a unit vector (column matrix). Consider the matrix

$$\underline{\underline{Q}} = \underline{\underline{I}} - 2 \underline{v} \, \underline{v}^\dagger \hfill (7.5\text{-}8)$$

To verify that $\underline{\underline{Q}}$ is orthogonal , we note

$$\underline{\underline{Q}}\,\underline{\underline{Q}}^{\dagger} = (\underline{\underline{I}} - 2\,\underline{v}\,\underline{v}^{\dagger})\,(\underline{\underline{I}} - 2\,\underline{v}\,\underline{v}^{\dagger}) \tag{7.5-9a}$$

$$= \underline{\underline{I}} - 4\,\underline{v}\,\underline{v}^{\dagger} + 4\,\underline{v}\,\underline{v}^{\dagger}\,\underline{v}\,\underline{v}^{\dagger} = \underline{\underline{I}} \tag{7.5-9b,c}$$

Equation (7.5-9c) follows from Equation (7.5-9b), since $\underline{v}$ is a unit vector.

Note that $\underline{v}\,\underline{v}^{\dagger}$ is a symmetric matrix and its transpose is $\underline{v}\,\underline{v}^{\dagger}$. The product $\underline{v}^{\dagger}\underline{v}$ is a scalar.

To transform a matrix $\underline{\underline{A}}$ to a tridiagonal form, we perform the following sequence of transformations

$$\underline{\underline{A}}_0 = \underline{\underline{A}} \tag{7.5-10a}$$

$$\underline{\underline{A}}_k = \underline{\underline{Q}}_k\,\underline{\underline{A}}_{k-1}\,\underline{\underline{Q}}_k^{\dagger} \tag{7.5-10b}$$

$$\underline{\underline{Q}}_k = \underline{\underline{I}} - 2\,\underline{v}_k\,\underline{v}_k^{\dagger} \tag{7.5-10c}$$

$$\underline{v}_k^{\dagger}\underline{v}_k = 1 \tag{7.5-10d}$$

We choose $\underline{v}_k$ such that its first $k$ elements are zero. After $(n-2)$ operations, the matrix $\underline{\underline{A}}_{n-2}$ will be a tridiagonal matrix. We illustrate the method by the following example.

***Example 7.5-1.*** Reduce the matrix $\underline{\underline{A}}$ to a tridiagonal matrix by the method of Householder, if $\underline{\underline{A}}$ is given by

$$\underline{\underline{A}} = \begin{bmatrix} a_{11} & a_{12} & a_{13} \\ a_{12} & a_{22} & a_{23} \\ a_{13} & a_{23} & a_{33} \end{bmatrix} \tag{7.5-11}$$

We choose $\underline{v}_1$ to be

$$\underline{v}_1 = \begin{bmatrix} 0 \\ v_2 \\ v_3 \end{bmatrix} \tag{7.5-12a}$$

$$v_2^2 + v_3^2 = 1 \tag{7.5-12b}$$

From Equation (7.5-10c), we have

$$
\underline{\underline{Q}}_1 = \begin{bmatrix} 1 & 0 & 0 \\ 0 & 1 & 0 \\ 0 & 0 & 1 \end{bmatrix} - \begin{bmatrix} 0 & 0 & 0 \\ 0 & 2v_2^2 & 2v_2v_3 \\ 0 & 2v_2v_3 & 2v_3^2 \end{bmatrix} \tag{7.5-13a}
$$

$$
= \begin{bmatrix} 1 & 0 & 0 \\ 0 & 1-2v_2^2 & -2v_2v_3 \\ 0 & -2v_2v_3 & 1-2v_3^2 \end{bmatrix} \tag{7.5-13b}
$$

Applying Equation (7.5-10b), we obtain

$$
\underline{\underline{A}}_1 = \begin{bmatrix} 1 & 0 & 0 \\ 0 & 1-2v_2^2 & -2v_2v_3 \\ 0 & -2v_2v_3 & 1-2v_3^2 \end{bmatrix} \begin{bmatrix} a_{11} & a_{12} & a_{13} \\ a_{12} & a_{22} & a_{23} \\ a_{13} & a_{23} & a_{33} \end{bmatrix} \begin{bmatrix} 1 & 0 & 0 \\ 0 & 1-2v_2^2 & -2v_2v_3 \\ 0 & -2v_2v_3 & 1-2v_3^2 \end{bmatrix} \tag{7.5-14a}
$$

$$
= \begin{bmatrix} a_{11}^{(1)} & a_{12}^{(1)} & a_{13}^{(1)} \\ a_{12}^{(1)} & a_{22}^{(1)} & a_{23}^{(1)} \\ a_{13}^{(1)} & a_{23}^{(1)} & a_{33}^{(1)} \end{bmatrix} \tag{7.5-14b}
$$

where

$$
a_{11}^{(1)} = a_{11} \tag{7.5-15a}
$$

$$
a_{12}^{(1)} = a_{12} (1-2v_2^2) + a_{13} (-2v_2v_3) \tag{7.5-15b}
$$

$$
a_{13}^{(1)} = a_{12} (-2v_2v_3) + a_{13} (1-2v_3^2) \tag{7.5-15c}
$$

$$
a_{22}^{(1)} = a_{22} (1-2v_2^2)^2 + 2a_{23} (1-2v_2^2)(-2v_2v_3) + a_{33} (-2v_2v_3)^2 \tag{7.5-15d}
$$

$$
a_{23}^{(1)} = a_{23} (1-2v_3^2)^2 + (1-2v_3^2)(-2v_2v_3)(a_{22}+a_{33}) + a_{23} (-2v_2v_3)^2 \tag{7.5-15e}
$$

$$
a_{33}^{(1)} = a_{33} (1-2v_3^2)^2 + 2(1-2v_3^2)(-2v_2v_3) a_{23} + a_{22} (-2v_2v_3)^2 \tag{7.5-15f}
$$

For $\underline{\underline{A}}_{(1)}$ to be tridiagonal, we have from Equation (7.5-15c)

$$a_{13} - 2 v_3 \left( a_{12} v_2 + a_{13} v_3 \right) = 0 \tag{7.5-16}$$

Since $\underset{=}{Q}$ is an orthogonal matrix, it follows that

$$a_{11}^2 + a_{12}^2 + a_{13}^2 = \left( a_{11}^{(1)} \right)^2 + \left( a_{12}^{(1)} \right)^2 \tag{7.5-17}$$

Note that $a_{13}^{(1)}$ is zero

Using Equations (7.5-15a, b), Equation (7.5.17) becomes

$$\left[ a_{12} - 2 v_2 \left( a_{12} v_2 + a_{13} v_3 \right) \right]^2 = a_{12}^2 + a_{13}^2 \tag{7.5-18a}$$

$$= S_1^2 \tag{7.5-18b}$$

Equation (7.5-18b) implies

$$a_{12} - 2 v_2 \left( a_{12} v_2 + a_{13} v_3 \right) = \pm S_1 \tag{7.5-19}$$

Multiplying Equations (7.5-16, 19) by $v_3$ and $v_2$ respectively, adding the resulting expressions and using Equation (7.5-12b), we obtain

$$a_{12} v_2 + a_{13} v_3 = \mp S_1 v_2 \tag{7.5-20}$$

Substituting Equation (7.5-20) into Equation (7.5-19) yields

$$v_2^2 = \frac{1}{2} \left( 1 \mp \frac{a_{12}}{S_1} \right) \tag{7.5-21}$$

From Equations (7.5-16, 20), we deduce

$$v_3 = \mp \frac{a_{13}}{2 S_1 v_2} \tag{7.5-22}$$

Since we have $v_2$ in the denominator, we choose the sign in Equation (7.5-21) such that the absolute value of $v_2$ is the larger of the two possible values to achieve the best accuracy. That is to say, if $a_{12}$ is negative, we choose the negative sign and if $a_{12}$ is positive, we choose the positive sign. From Equation (7.5-21), we obtain

$$v_2 = \pm \sqrt{ \frac{1}{2} \left[ 1 + \frac{a_{12} \left( \text{sign } a_{12} \right)}{S_1} \right] } \tag{7.5-23}$$

We choose the positive sign in Equation (7.5-23) and it follows that we take the negative sign in Equation (7.5-22). Substituting $v_2$ and $v_3$ into Equation (7.5-15a to f), the matrix $\underline{\underline{A}}^{(1)}$ is tridiagonal.

●

In Example 7.5-1, we have considered a $(3 \times 3)$ matrix. If the matrix $\underline{\underline{A}}$ is a $(n \times n)$ matrix $(n > 2)$, we start with $\underline{v}_1$ given by

$$\underline{v}_1^\dagger = (0, v_{12}, v_{13}, \ldots, v_{1n}) \tag{7.5-24}$$

Then from Equation (7.5-10c), $\underline{\underline{Q}}_1$ will be of the form

$$\underline{\underline{Q}}_1 = \begin{bmatrix} 1 & 0 & 0 & \ldots & 0 \\ 0 & x & x & \ldots & x \\ 0 & x & x & \ldots & x \\ \vdots & \vdots & & & \\ 0 & x & x & \ldots & x \end{bmatrix} \tag{7.5-25}$$

where the x denote non-zero entries.

Let $S_1$ be given by

$$S_1^2 = a_{12}^2 + a_{13}^2 + \ldots + a_{1n}^2 \tag{7.5-26}$$

Then

$$v_{12}^2 = \frac{1}{2} \left[ 1 + \frac{a_{12} (\text{sign } a_{12})}{S_1} \right] \tag{7.5-27a}$$

$$v_{1j} = \frac{a_{1j} (\text{sign } a_{12})}{2 S_1 v_{12}}, \qquad j \geq 3 \tag{7.5-27b}$$

Note that Equation (7.5-27b) is valid only for $j \geq 3$. The choice of $v_{12}$ is different from $v_{1j}$ $(j \geq 3)$ because we want to reduce the matrix $\underline{\underline{A}}$ to a tridiagonal form [see also Equations (7.5-22, 23)].

The orthogonal matrix $\underline{\underline{Q}}_1$ is now constructed (Equation 7.5-10c) and the operations defined by Equations (7.5-10a, b) can be performed.

$\underline{\underline{A}}_1$ will then be of the form

$$
\underline{\underline{A}}_1 = \begin{bmatrix} x & x & 0 & \cdots & 0 \\ x & x & x & \cdots & x \\ 0 & x & x & \cdots & x \\ \vdots & \vdots & & & \\ 0 & x & x & \cdots & x \end{bmatrix}
\tag{7.5-28}
$$

In $\underline{\underline{A}}_1$, the first row and column are of the desired form. We now proceed to reduce the second row and column to the form of a tridiagonal matrix. We write $\underline{v}_2$ as a row vector

$$
\underline{v}_2^\dagger = (0, 0, v_{23}, \ldots, v_{2n})
\tag{7.5-29}
$$

We then proceed to construct $\underline{\underline{Q}}_2$ and obtain $\underline{\underline{A}}_2$ as described earlier. The process is continued until we obtain $\underline{\underline{A}}_{n-2}$ which will be of the tridiagonal form, as shown in the next example.

**Example 7.5-2.** Use the Householder method to find a tridiagonal matrix similar to the matrix

$$
\underline{\underline{A}} = \begin{bmatrix} 1 & -1 & 2 & 2 \\ -1 & 2 & 1 & -1 \\ 2 & 1 & 3 & 2 \\ 2 & -1 & 2 & 1 \end{bmatrix}
\tag{7.5-30}
$$

S as defined by Equation (7.5-26) is given by

$$
S_1^2 = (-1)^2 + 2^2 + 2^2 = 9
\tag{7.5-31a,b}
$$

The $v_{ij}$ are given by Equations (7.5-27a, b)

$$
v_{12} = \sqrt{\frac{1}{2}\left[1 + \frac{1}{3}\right]} = \sqrt{\frac{2}{3}}
\tag{7.5-32a,b}
$$

$$
v_{13} = \frac{-2}{6\left(\sqrt{2/3}\,\right)} = -\frac{\sqrt{6}}{6}
\tag{7.5-32c,d}
$$

$$
v_{14} = \frac{-2}{6\left(\sqrt{2/3}\,\right)} = -\frac{\sqrt{6}}{6}
\tag{7.5-32e,f}
$$

From Equation (7.5-10c), $\underline{\underline{Q}}_1$ is

$$Q_1 = \frac{1}{3}\begin{bmatrix} 3 & 0 & 0 & 0 \\ 0 & -1 & 2 & 2 \\ 0 & 2 & 2 & -1 \\ 0 & 2 & -1 & 2 \end{bmatrix} \tag{7.5-33}$$

$$A_1 = \frac{1}{9}\begin{bmatrix} 9 & 27 & 0 & 0 \\ 27 & 34 & 7 & 1 \\ 0 & 7 & 25 & 10 \\ 0 & 1 & 10 & -5 \end{bmatrix} \tag{7.5-34}$$

The first row and column in $A_1$ are in the desired form. We now proceed to reduce to zero the element $a_{24}^{(1)}$. We can partition $A_1$ into a $(3 \times 3)$ matrix $B$, choosing $B$ to be

$$B = \frac{1}{9}\begin{bmatrix} 34 & 7 & 1 \\ 7 & 25 & 10 \\ 1 & 10 & -5 \end{bmatrix} \tag{7.5-35}$$

The matrix $B$ can be reduced to a tridiagonal form as in Example 7.5-1. Alternatively, we can proceed as before by writing $v_2$ as a row vector

$$v_2^\dagger = (0, 0, v_{22}, v_{23}) \tag{7.5-36}$$

$S_2^2$ is now given by

$$S_2^2 = \left(a_{23}^{(1)}\right)^2 + \left(a_{24}^{(1)}\right)^2 = \left(\frac{7}{9}\right)^2 + \left(\frac{1}{9}\right)^2 = \frac{50}{81} \tag{7.5-37a,b,c}$$

The elements $v_{22}$ and $v_{23}$ are

$$v_{22} = \sqrt{\frac{1}{2}\left[1 + \left(\frac{7}{9}\right) \Big/ \left(\frac{5\sqrt{2}}{9}\right)\right]} \tag{7.5-38a}$$

$$= \sqrt{0.99497} \tag{7.5-38b}$$

$$= 0.9975 \tag{7.5-38c}$$

$$v_{23} = \frac{1}{9} \Big/ \left[2\left(\frac{5\sqrt{2}}{9}\right)v_{22}\right] \tag{7.5-38d}$$

$$= 0.0709 \tag{7.5-38e}$$

The orthogonal matrix $\underset{=}{Q}_2$ is given by

$$\underset{=}{Q}_2 = \underset{=}{I} - 2\,\underline{v}_2\,\underline{v}_2^\dagger \tag{7.5-39a}$$

$$= \begin{bmatrix} 1 & 0 & 0 & 0 \\ 0 & 1 & 0 & 0 \\ 0 & 0 & -0.9899 & -0.1414 \\ 0 & 0 & -0.1414 & 0.9899 \end{bmatrix} \tag{7.5-39b}$$

$\underset{=}{A}_2$ is then calculated and is

$$\underset{=}{A}_2 = \begin{bmatrix} 1 & 3 & 0 & 0 \\ 3 & 3.7778 & -0.7857 & 0 \\ 0 & -0.7857 & 3.0222 & -0.6000 \\ 0 & 0 & -0.6000 & -0.8000 \end{bmatrix} \tag{7.5-40}$$

The tridiagonal matrix $\underset{=}{A}_2$ is obtained by the similarity transformation

$$\underset{=}{A}_2 = \underset{=}{Q}_2\,\underset{=}{A}_1\,\underset{=}{Q}_2^\dagger \tag{7.5-41a}$$

$$= \underset{=}{Q}_2\,\underset{=}{Q}_1\,\underset{=}{A}\,\underset{=}{Q}_1^\dagger\,\underset{=}{Q}_2^\dagger \tag{7.5-41b}$$

$$= \underset{=}{P}\,\underset{=}{A}\,\underset{=}{P}^\dagger \tag{7.5-41c}$$

●

We note that in the Householder method, the elements which have been reduced to zero in the first column and row at the first iteration remain zero at the second and subsequent iterations. In general, all the zeros created in previous iterations are not destroyed in subsequent iterations. This property made Householder's method very reliable.

Once the matrix has been reduced to a tridiagonal matrix, the most efficient method of obtaining all the eigenvalues is the QR method.

## The QR Algorithm

We now discuss the powerful QR algorithm for computing eigenvalues of a symmetric matrix. The main idea is based upon the following result of linear algebra.

If $\underline{\underline{A}}$ is an $(n \times n)$ real matrix, there exists an upper (right) triangular matrix $\underline{\underline{R}}$ and an orthogonal matrix $\underline{\underline{Q}}$ such that

$$\underline{\underline{A}} = \underline{\underline{Q}}\,\underline{\underline{R}} \qquad\qquad (7.5\text{-}42)$$

Since rotation matrices are orthogonal matrices, we use rotation matrices to transform $\underline{\underline{A}}$ to the triangular matrix $\underline{\underline{R}}$.

In the case of a $(3 \times 3)$ matrix, a rotation matrix may be one of the following

$$\underline{\underline{P}}_1 = \begin{bmatrix} c & -s & 0 \\ s & c & 0 \\ 0 & 0 & 1 \end{bmatrix}, \quad \underline{\underline{P}}_2 = \begin{bmatrix} 1 & 0 & 0 \\ 0 & c & -s \\ 0 & s & c \end{bmatrix}, \quad \underline{\underline{P}}_3 = \begin{bmatrix} c & 0 & -s \\ 0 & 1 & 0 \\ s & 0 & c \end{bmatrix}$$

$$(7.5\text{-}43\text{a,b,c})$$

where $c = \cos\theta$ and $s = \sin\theta$ $\qquad\qquad (7.5\text{-}43\text{d,e})$

Let $\underline{\underline{A}}$ be a $(3 \times 3)$ tridiagonal matrix and form the product of $\underline{\underline{A}}$ with the rotation matrix $\underline{\underline{P}}_1$

$$\underline{\underline{P}}_1\,\underline{\underline{A}} = \begin{bmatrix} c & -s & 0 \\ s & c & 0 \\ 0 & 0 & 1 \end{bmatrix} \begin{bmatrix} a_{11} & a_{12} & 0 \\ a_{12} & a_{22} & a_{23} \\ 0 & a_{23} & a_{33} \end{bmatrix} \qquad\qquad (7.5\text{-}44\text{a})$$

$$= \begin{bmatrix} c\,a_{11} - s\,a_{12} & c\,a_{12} - s\,a_{22} & -s\,a_{23} \\ s\,a_{11} - c\,a_{12} & s\,a_{12} + c\,a_{22} & c\,a_{23} \\ 0 & a_{23} & a_{33} \end{bmatrix} \qquad\qquad (7.5\text{-}44\text{b})$$

We set the element in the second row and first column of $\underline{\underline{P}}_1\,\underline{\underline{A}}$ to be zero, that is

$$s\,a_{11} - c\,a_{12} = 0 \qquad\qquad (7.5\text{-}45)$$

From Equation (7.5-45), we deduce that

$$c = \frac{a_{11}}{\sqrt{a_{11}^2 + a_{12}^2}} \qquad\qquad (7.5\text{-}46\text{a})$$

$$s = \frac{a_{12}}{\sqrt{a_{11}^2 + a_{12}^2}} \qquad\qquad (7.5\text{-}46\text{b})$$

With this choice of $c$ and $s$ and denoting $\underline{\underline{P}}_1\,\underline{\underline{A}}$ by $\underline{\underline{A}}_1$, we write

$$\underline{\underline{A}}_1 = \begin{bmatrix} a_{11}^{(1)} & a_{12}^{(1)} & a_{13}^{(1)} \\ 0 & a_{22}^{(1)} & a_{23}^{(1)} \\ 0 & a_{32}^{(1)} & a_{33}^{(1)} \end{bmatrix} \tag{7.5-47}$$

We now form the product $\underline{\underline{P}}_2 \, \underline{\underline{A}}_1$ as follows

$$\underline{\underline{P}}_2 \, \underline{\underline{A}}_1 = \begin{bmatrix} 1 & 0 & 0 \\ 0 & c_2 & -s_2 \\ 0 & s_2 & c_2 \end{bmatrix} \begin{bmatrix} a_{11}^{(1)} & a_{12}^{(1)} & a_{13}^{(1)} \\ 0 & a_{22}^{(1)} & a_{23}^{(1)} \\ 0 & a_{32}^{(1)} & a_{33}^{(1)} \end{bmatrix} \tag{7.5-48a}$$

$$= \begin{bmatrix} a_{11}^{(1)} & a_{12}^{(1)} & a_{13}^{(1)} \\ 0 & c_2 a_{22}^{(1)} - s_2 a_{32}^{(1)} & c_2 a_{23}^{(1)} - s_2 a_{33}^{(1)} \\ 0 & s_2 a_{22}^{(1)} + c_2 a_{32}^{(1)} & s_2 a_{23}^{(1)} + c_2 a_{33}^{(1)} \end{bmatrix} \tag{7.5-48b}$$

where $c_2$ and $s_2$ are $\cos \theta_2$ and $\sin \theta_2$.

If we denote the product $\underline{\underline{P}}_2 \, \underline{\underline{A}}_1$ by $\underline{\underline{A}}_2$, we have to set

$$s_2 \, a_{22}^{(1)} + c_2 \, a_{32}^{(1)} = 0 \tag{7.5-49}$$

for the element $a_{32}^{(2)}$ to be zero.

To satisfy Equation (7.5-49), we require

$$c_2 = \frac{a_{22}^{(1)}}{\sqrt{\left(a_{22}^{(1)}\right)^2 + \left(a_{32}^{(1)}\right)^2}} \tag{7.5-50a}$$

$$s_2 = \frac{-a_{32}^{(1)}}{\sqrt{\left(a_{22}^{(1)}\right)^2 + \left(a_{32}^{(1)}\right)^2}} \tag{7.5-50b}$$

Thus $\underline{\underline{A}}_2$ is now of an upper triangular form and we denote it by $\underline{\underline{R}}$. Summarizing we have

$$\underline{\underline{R}} = \underline{\underline{P}}_2 \, \underline{\underline{A}}_1 = \underline{\underline{P}}_2 \, \underline{\underline{P}}_1 \, \underline{\underline{A}} \tag{7.5-51a,b}$$

Generalizing to a $(n \times n)$ matrix, we state that

$$\underline{\underline{R}} = \underline{\underline{P}}_{n-1} \underline{\underline{A}}_{n-2} = \underline{\underline{P}}_{n-1} \underline{\underline{P}}_{n-2} \cdots \underline{\underline{P}}_1 \underline{\underline{A}} \qquad (7.5\text{-}52a,b)$$

The matrices $\underline{\underline{P}}_r$ $(r = 1, \dots, n-1)$ will be of the form given in Equation (7.5-43a, b, c) except that they will be $(n \times n)$ matrices. We note that in Equations (7.5-43a, b, c), the $c$ is along the diagonal elements and the $s$ is off diagonal. Similarly we choose $\underline{\underline{P}}_r$ so as to reduce the element $a_{ij}$ of any matrix $\underline{\underline{A}}_r$ $(r = 0, 1, \dots, n-2)$ to zero. The elements of $\underline{\underline{P}}_r$ are zero everywhere except

(i) along the diagonal where they are one with the exception of $P_{ii}^{(r)}$ where they are $c$;

(ii) $P_{ji}^{(r)}$ is $s$ and $P_{ij}^{(r)}$ is $-s$.

The terms $c$ and $s$ are given by

$$c = \frac{a_{ii}}{\sqrt{a_{ii}^2 + a_{ji}^2}} \qquad (7.5\text{-}53a)$$

$$s = \frac{a_{ji}}{\sqrt{a_{ii}^2 + a_{ji}^2}} \qquad (7.5\text{-}53b)$$

By this process, we can successively reduce the non-diagonal elements in the lower triangular section of the matrix to zero. An upper triangular matrix $\underline{\underline{R}}$ is then obtained and the eigenvalues are the diagonal elements. In a tridiagonal matrix, many of the elements are already zero, and the reduction to an upper triangular matrix is fast.

A standard text on the eigenvalues of a matrix was written by Wilkinson (1965).

## 7.6   INTERPOLATION

In this section, we address the following problem. We wish to estimate the value of $f(x)$ for some $x$ in the interval $(x_0, x_n)$, given a collection of experimental data points $[x_k, f(x_k)]$, $k = 0, 1, \dots, n$. This problem has been studied by well known mathematicians such as Gauss, Lagrange, and Newton. The earliest method requires one to fit a polynomial (an interpolation function) that approximates the function $f$ over $(x_0, x_n)$ or a curve that passes through the data points. The points $x_k$ are sometimes called the **nodes**.

### Lagrange Interpolation

This method is based on choosing a polynomial $p_n(x)$ of degree $n$ for which

$$p_n(x_k) = f(x_k), \qquad k = 0, 1, 2, \dots \qquad (7.6\text{-}1)$$

An existence theorem for polynomial interpolation states that given $(n + 1)$ points $\{x_0, x_1, ..., x_n\}$ in the domain of a function $f$ with $n \geq 1$ and $x_0 < x_1, ..., < x_n$, there exists at least one polynomial $p_n(x)$ of degree less than or equal to $n$ such that Equation (7.6-1) holds. Moreover, if $x_0, x_1, ..., x_n$ are distinct points on the domain of $f$, $p_n(x)$ exists for any function $f$.

A polynomial of **degree one (linear polynomial)** has the form

$$p_1(x) = a_0 + a_1 x \tag{7.6-2}$$

This equation has two coefficients and, therefore, two points $(x_0, f_0)$ and $(x_1, f_1)$ will define $a_0$ and $a_1$ as follows

$$p_1(x_0) = a_0 + a_1 x_0 = f_0 \tag{7.6-3a}$$

$$p_1(x_1) = a_0 + a_1 x_1 = f_1 \tag{7.6-3b}$$

Solving for $a_1$ and $a_0$, we find

$$a_0 = \frac{x_1 f_0 - x_0 f_1}{x_1 - x_0}, \qquad a_1 = \frac{f_0 - f_1}{x_0 - x_1} \tag{7.6-4a,b}$$

Thus, the linear polynomial for the points $(x_0, f_0)$ and $(x_1, f_1)$ is

$$p_1(x) = \frac{x_1 f_0 - x_0 f_1}{x_1 - x_0} + x \frac{f_0 - f_1}{x_0 - x_1} \tag{7.6-5}$$

This equation can also be written as

$$p_1(x) = f_0 \frac{x - x_1}{x_0 - x_1} + f_1 \frac{x - x_0}{x_1 - x_0} \tag{7.6-6a}$$

$$= f_0 \, \ell_0(x) + f_1 \, \ell_1(x) \tag{7.6-6b}$$

where

$$\ell_0(x) = \frac{x - x_1}{x_0 - x_1} \quad \text{and} \quad \ell_1(x) = \frac{x - x_0}{x_1 - x_0} \tag{7.6-7a,b}$$

Observe that

$$\ell_0(x_0) = 1, \qquad \ell_0(x_1) = 0 \tag{7.6-8a,b}$$

$$\ell_1(x_0) = 0, \qquad \ell_1(x_1) = 1 \tag{7.6-8c,d}$$

Generalizing, we can write

$$\ell_j(x_k) = \begin{cases} 0 & \text{if } j \neq k \\ 1 & \text{if } j = k \end{cases} \tag{7.6-9a,b}$$

Similarly generalizing the concepts involved in Equation (7.6-6b), we can construct a polynomial of degree $n$, which passes through $(n + 1)$ points $[(x_0, f_0), (x_1, f_1), \ldots, (x_n, f_n)]$ of the form

$$p_n(x) = f(x_0) \ell_0 + f(x_1) \ell_1(x) + \ldots + f(x_n) \ell_n(x) \tag{7.6-10}$$

The $\ell_j(x)$ are quotients which have to satisfy Equations (7.6-9a, b) and are of degree $n$. A natural choice is

$$\ell_j(x) = \frac{(x - x_0)(x - x_1) \ldots (x - x_{j-1})(x - x_{j+1}) \ldots (x - x_n)}{(x_j - x_0)(x_j - x_1) \ldots (x_j - x_{j-1})(x_j - x_{j+1}) \ldots (x_j - x_n)} \tag{7.6-11a}$$

$$= \prod_{\substack{i=0 \\ i \neq j}}^{n} \frac{(x - x_i)}{(x_j - x_i)} \quad \text{for each } j = 0, 1, 2, \ldots, n \tag{7.6-11b}$$

Equation (7.6-10) is the **Lagrange interpolating polynomial** of degree $n$ with $\ell_j(x)$ given by Equation (7.6-11b), which passes through $(n + 1)$ points.

***Example 7.6-1.*** Construct a Lagrange polynomial that passes through the points $\left(2, \frac{1}{2}\right)$, $\left(\frac{5}{2}, \frac{2}{5}\right)$ and $\left(4, \frac{1}{4}\right)$.

Here the nodes are $2$, $\frac{5}{2}$ and $4$ and $f(x_0)$, $f(x_1)$ and $f(x_2)$ are $\frac{1}{2}$, $\frac{2}{5}$ and $\frac{1}{4}$ respectively. Since we have three points, the polynomial will be of degree 2.

$$p_2(x) = f(x_0) \ell_0(x) + f(x_1) \ell_1(x) + f(x_2) \ell_2(x) \tag{7.6-12}$$

Now

$$\ell_0(x) = \frac{(x - x_1)(x - x_2)}{(x_0 - x_1)(x_0 - x_2)} = \frac{(x - 2.5)(x - 4)}{(2 - 2.5)(2 - 4)} = x^2 - 6.5x + 10 \tag{7.6-13a}$$

$$\ell_1(x) = \frac{(x - 2)(x - 4)}{(2.5 - 2)(2.5 - 4)} = -1.333x^2 + 8x - 10.667 \tag{7.6-13b}$$

$$\ell_2(x) = \frac{(x - 2)(x - 2.5)}{(4 - 2)(4 - 2.5)} = 0.33x^2 - 1.5x + 1.667 \tag{7.6-13c}$$

and

$$p_2(x) = 0.05\, x^2 - 0.425\, x + 1.15 \tag{7.6-14}$$

An approximation of $f(3)\; [= p_2(3)]$ is

$$p_2(3) = 0.05\,(9) - 0.425\,(3) + 1.15 = 0.325 \tag{7.6-15a,b}$$

We note that the three points lie on the curve

$$f(x) = \frac{1}{x} \tag{7.6-16}$$

and hence

$$f(3) = \frac{1}{3} = 0.3333 \tag{7.6-17a,b}$$

A comparison between Equations (7.6-15a, b) and Equations (7.6-17a,b) shows that the value obtained by interpolation is correct to the first two decimal places.

### Newton's Divided Difference Representation

Before we discuss this method we shall state some definitions and introduce the accompanying notation.

The **zeroth divided difference** of the function $f$ with respect to $x_i$ is denoted by $f[x_i]$ and is simply the value of $f$ at $x_i$.

$$f[x_i] = f(x_i) \tag{7.6-18}$$

The **first divided difference** of $f$ with respect to $x_i$ and $x_j$ is denoted by $f[x_i, x_j]$ and is defined as

$$f\left[x_i, x_j\right] = \frac{f\left[x_j\right] - f\left[x_i\right]}{\left(x_j - x_i\right)} = \frac{f\left(x_j\right) - f\left(x_i\right)}{\left(x_j - x_i\right)} \tag{7.6-19a,b}$$

Similarly the $n^{th}$ divided difference is

$$f\left[x_0, x_1, \dots, x_{n-1}\right] = \frac{f\left[x_1, \dots, x_{n-1}\right] - f\left[x_0, x_1, \dots, x_{n-2}\right]}{\left(x_{n-1} - x_0\right)} \tag{7.6-20}$$

For specific values of $x_i$ and $x_j$, we have from Equation (7.6-20)

$$f\left[x_0, x_1\right] = \frac{f\left(x_1\right) - f\left(x_0\right)}{x_1 - x_0} \tag{7.6-21a}$$

$$f[x_0, x_1, x_2] = \frac{f[x_1, x_2] - f[x_0, x_1]}{(x_2 - x_0)} \qquad (7.6\text{-}21b)$$

We recall that the definition of the derivative of $f(x)$ at $x_0$ is

$$f'(x_0) = \lim_{x_1 \to x_0} \frac{f(x_1) - f(x_0)}{x_1 - x_0} \qquad (7.6\text{-}22)$$

In the limit as $x_1$ tends to $x_0$, $f[x_0, x_1]$ is the derivative of $f$ at $x_0$. The second difference (Equation 7.6-21b) is the difference of the first difference and can be associated with the second derivative of $f$. The derivative can be discretized in terms of finite differences. In Section 7.7, we shall approximate the derivative by the finite difference.

Noting that the $n^{th}$ difference is the difference of the $(n-1)^{th}$ difference, the divided difference can be tabulated as in Table 7.6-1.

## TABLE 7.6-1

### Divided difference

| $x_i$ | $f[x_i]$ (zeroth difference) | $f[x_i, x_j]$ (first difference) | $f[x_i, x_j, x_k]$ (second difference) |
|---|---|---|---|
| $x_0$ | $f(x_0)$ | | |
| | | $\dfrac{f(x_1) - f(x_0)}{x_1 - x_0}$ | |
| $x_1$ | $f(x_1)$ | | $\dfrac{f[x_2, x_1] - f[x_1, x_0]}{x_2 - x_0}$ |
| | | $\dfrac{f(x_2) - f(x_1)}{x_2 - x_1}$ | |
| $x_2$ | $f(x_2)$ | | $\dfrac{f[x_3, x_2] - f[x_2, x_1]}{x_3 - x_1}$ |
| | | $\dfrac{f(x_3) - f(x_2)}{x_3 - x_2}$ | |
| $x_3$ | $f(x_3)$ | | $\dfrac{f[x_4, x_3] - f[x_3, x_2]}{x_4 - x_2}$ |
| | | $\dfrac{f(x_4) - f(x_3)}{x_4 - x_3}$ | |
| $x_4$ | $f(x_4)$ | | |

This divided difference scheme lends itself to the construction of a polynomial through the points $\{x_i, f(x_i)\}$. We write

$$p_n(x) = a_0 + a_1(x - x_0) + a_2(x - x_0)(x - x_1) + \ldots + a_{n-1}(x - x_0)(x - x_1) \ldots (x - x_{n-1})$$

$$+ a_n(x - x_0)(x - x_1) \ldots (x - x_{n-1})(x - x_n) \tag{7.6-23}$$

and require this polynomial to pass through the points $\{x_i, f(x_i)\}$.

We note that using Equation (7.6-1)

$$p_0(x_0) = a_0 = f(x_0) = f[x_0] \tag{7.6-24a,b,c}$$

$$p_1(x_1) = a_0 + a_1(x_1 - x_0) = f(x_1) \tag{7.6-25a,b}$$

and via Equation (7.6-24b), we get

$$a_1 = \frac{f(x_1) - f(x_0)}{x_1 - x_0} = f[x_0, x_1] \tag{7.6-26a,b}$$

Similarly

$$p_2(x_2) = a_0 + a_1(x_2 - x_0) + a_2(x_2 - x_0)(x_2 - x_1) = f(x_2) \tag{7.6-27a,b}$$

With the use of Equations (7.6-24 to 27), we find

$$a_2 = \frac{f(x_2) - f(x_0) - f[x_0, x_1](x_2 - x_0)}{(x_2 - x_0)(x_2 - x_1)} = f[x_0, x_1, x_2] \tag{7.6-28a,b}$$

Equations (7.6-24 to 28) can be generalized and

$$a_n = f[x_0, x_1, \ldots, x_n] \tag{7.6-29}$$

Substituting the values of $a_0, a_1, \ldots, a_n$ in Equation (7.6-23), we obtain **Newton's forward divided difference formula** which can be written as

$$p_n(x) = f[x_0] + f[x_0, x_1](x - x_0) + f[x_0, x_1, x_2](x - x_0)(x - x_1)$$

$$+ \ldots + f[x_0, x_1, \ldots, x_n](x - x_0)(x - x_1) \ldots (x - x_{n-1}) \tag{7.6-30}$$

Note that it is easier to construct the polynomial by first constructing a table of divided differences.

***Example 7-6-2.*** Given the four values of $\{x_i, f(x_i)\}$ as $(0,1)$, $(1, 1)$, $(2, 2)$ and $(4, 5)$; construct a divided difference table and then use formula (7.6-30) to find the third degree polynomial through the given points.

## TABLE 7.6-2

### Zeroth to third difference

| $x_i$ | Zeroth difference | First difference | Second difference | Third difference |
|-------|------------------|------------------|-------------------|------------------|
| 0 | 1 | | | |
| | | $\dfrac{1-1}{1} = 0$ | | |
| 1 | 1 | | $\dfrac{1-0}{2-0} = \dfrac{1}{2}$ | |
| | | $\dfrac{2-1}{2-1} = 1$ | | $\dfrac{1/6 - 1/2}{4-0} = -\dfrac{1}{12}$ |
| 2 | 2 | | $\dfrac{3/2 - 1}{4-1} = \dfrac{1}{6}$ | |
| | | $\dfrac{5-2}{4-2} = \dfrac{3}{2}$ | | |
| 4 | 5 | | | |

Note that the points $x_i$ are not equidistant.

From Table 7.6-2, we obtain

$$a_0 = 1, \qquad a_1 = 0, \qquad a_2 = \frac{1}{2}, \qquad a_3 = -\frac{1}{12} \qquad \text{(7.6-31a to d)}$$

Equation (7.6-23) yields

$$p_3(x) = 1 + \frac{1}{2}(x-0)(x-1) - \frac{1}{12}(x-0)(x-1)(x-2) \qquad \text{(7.6-32a)}$$

$$= \frac{1}{12}(-x^3 + 9x^2 - 8x + 12) \qquad \text{(7.6-32b)}$$

We next give an example to show the difference between the two methods discussed so far.

***Example 7.6-3.*** Use Newton's and Lagrange's formulas to evaluate $\sqrt{1.12}$, given the following: $\{x_i, f(x_i)\}$ values where $f(x) = \sqrt{x}$.   (1, 1), (1.05, 1.02470), (1.1, 1.04881), (1.15, 1.07238), (1.20, 1.09544), (1.25, 1.11803) and (1.30, 1.14017).

To apply Newton's divided difference formula, we construct the table of differences.

## TABLE 7.6-3

### Divided differences for $\sqrt{x}$

| x | f(x) | 1st | 2nd | 3rd | 4th | 5th | 6th |
|---|---|---|---|---|---|---|---|
| 1 | 1 | | | | | | |
| | | 2470 | | | | | |
| | | | −59 | | | | |
| 1.05 | 1.02470 | | | 5 | | | |
| | | 2411 | | | −1 | | |
| | | | −54 | | | −1 | |
| 1.10 | 1.04881 | | | 4 | | | 4 |
| | | 2357 | | | −2 | | |
| | | | −50 | | | 3 | |
| 1.15 | 1.07238 | | | 2 | | | |
| | | 2307 | | | 1 | | |
| | | | −48 | | | | |
| 1.20 | 1.09544 | | | 3 | | | |
| | | 2259 | | | | | |
| | | | −45 | | | | |
| 1.25 | 1.11803 | | | | | | |
| | | 2214 | | | | | |
| 1.30 | 1.14017 | | | | | | |

Note: We have omitted the decimal point and the leading zeros. This is customary in writing tables of differences.

Using Equation (7.6-30), we have

$$p_6(x) = 1 + 2470(x-1) - 59(x-1)(x-1.05) + 5(x-1)(x-1.05)(x-1.10)$$
$$- 1(x-1)(x-1.05)(x-1.10)(x-1.15) - 1(x-1)(x-1.05)(x-1.10)$$
$$(x-1.15)(x-1.20) + 4(x-1)(x-1.05)(x-1.10)(x-1.15)(x-1.20)$$
$$(x-1.25) \tag{7.6-33}$$

$$p_6(1.12) \approx \sqrt{1.12} \tag{7.6-34a}$$

$$= 1 + 2470(.12) - 59(.12)(.07) + 5(.12)(.07)(.02) - 1(.12)(.07)(.02)(-.03)$$
$$- 1(.12)(.07)(.02)(-.03)(-.08) + 4(.12)(.07)(.02)(-.03)(-.08)(-.13) \tag{7.6-34b}$$

$$= 1.05735 \tag{7.6-34c}$$

For Lagrange's formula, we approximate $\sqrt{x}$ by a third degree polynomial involving four points. The points closest to 1.12 are: 1.05, 1.10, 1.15 and 1.20. We could consider all the given points but the calculations become lengthy. We write

$$p_3(x) = \frac{(x - x_1)(x - x_2)(x - x_3)}{(x_0 - x_1)(x_0 - x_2)(x_0 - x_3)}f(x_0) + \frac{(x - x_0)(x - x_2)(x - x_3)}{(x_1 - x_0)(x_1 - x_2)(x_1 - x_3)}f(x_1)$$

$$+ \frac{(x - x_0)(x - x_1)(x - x_3)}{(x_2 - x_0)(x_2 - x_1)(x_2 - x_3)}f(x_2) + \frac{(x - x_0)(x - x_1)(x - x_2)}{(x_3 - x_0)(x_3 - x_1)(x_3 - x_2)}f(x_3) \quad \text{(7.6-35)}$$

Thus

$$p_3(1.12) = \frac{(1.12 - 1.10)(1.12 - 1.15)(1.12 - 1.20)}{(1.05 - 1.10)(1.05 - 1.15)(1.05 - 1.20)}(1.02470) + ..... \quad \text{(7.6-36a)}$$

$$= 1.05830 \quad \text{(7.6-36b)}$$

We note that Lagrange's formula produces a more accurate value for $\sqrt{1.12}$ than Newton's formula, although we have used fewer points in Lagrange's method. Using Lagrange's method, we have to decide at the outset of the degree of the polynomial to be considered. If subsequently we wish to improve the approximation by using a higher degree polynomial, we have to perform the calculations again, since we cannot make use of the values already computed. Using Newton's method, we only have to calculate the higher differences.

## Spline Functions

A high degree polynomial interpolation method is not satisfactory when the number of data points becomes too large, or when the data points are associated with a function whose derivatives are large or do not exist. In the recent past, piecewise polynomial approximations have become prominent. Instead of trying to approximate a function over the entire interval by a single polynomial of a high degree, it is more convenient to approximate the function by a piecewise polynomial function in a small interval. Such piecewise low degree polynomials are called **splines**. A spline function is a function consisting of polynomial pieces joined together with certain smoothness conditions. The points at which the low degree polynomials are joined are termed **knots**.

First degree polynomials (straight lines) are continuous but not smooth functions. Their first derivatives are discontinuous at the knots. Second degree polynomials (parabolas) allow us to impose continuity and slope conditions at the knots but the curvature changes abruptly at the knots. **Cubic splines** are the most commonly used splines. The above conditions of continuity, slope and curvature at the knots can be easily satisfied.

Let $(x_0, f_0)$, $(x_1, f_1)$, ... , $(x_n, f_n)$ be the given data points. Instead of using a high degree polynomial to pass through the $(n + 1)$ points, we choose a set of cubic splines $S_i(x)$. Each $S_i(x)$ joins only two points $x_i$ and $x_{i+1}$. Thus over the whole interval $(x_0, x_n)$, the function is approximated by n cubic splines. This is illustrated in Figure 7.6-1.

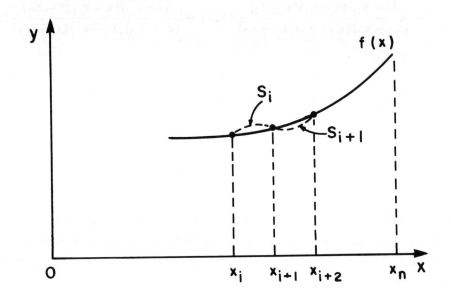

**FIGURE 7.6-1    Approximating f (x) by a set of cubic splines**

We write $S_i(x)$ as

$$S_i(x) = a_i + b_i (x - x_i) + c_i (x - x_i)^2 + d_i (x - x_i)^3, \quad i = 0, 1, ... , (n - 1) \qquad (7.6\text{-}37)$$

where the coefficients $a_i$, $b_i$, $c_i$ and $d_i$ are to be determined from the definition of cubic splines and the following continuity, slope and curvature conditions.

Continuity at the knots requires

$$S_i (x_{i+1}) = S_{i+1} (x_{i+1}) \qquad (7.6\text{-}38a)$$

Combining Equations (7.6-37, 38a) yields

$$a_i + b_i h_i + c_i h_i^2 + d_i h_i^3 = a_{i+1} \qquad (7.6\text{-}38b)$$

where

$$h_i = x_{i+1} - x_i \qquad (7.6\text{-}38c)$$

At the knots, the values of $S_i(x)$ are equal to $f(x)$.

$$S_i(x_i) = f_i \qquad\qquad\qquad (7.6\text{-}39a)$$

Using Equation (7.6-37), Equation (7.6-39a) becomes

$$a_i = f_i \qquad\qquad\qquad (7.6\text{-}39b)$$

The continuity of the slopes at the knots imposes

$$S_i'(x_{i+1}) = S_{i+1}'(x_{i+1}) \qquad\qquad\qquad (7.6\text{-}40a)$$

On differentiating $S_i$ and $S_{i+1}$, we obtain from Equation (7.6-40a)

$$b_i + 2c_i h_i + 3d_i h_i^2 = b_{i+1} \qquad\qquad\qquad (7.6\text{-}40b)$$

Similarly the continuity of the curvature at the knots leads to

$$S_i''(x_{i+1}) = S_{i+1}''(x_{i+1}) \qquad\qquad\qquad (7.6\text{-}41a)$$

$$c_i + 3d_i h_i = c_{i+1} \qquad\qquad\qquad (7.6\text{-}41b)$$

The derivatives of $f$ are not known, the conditions (7.6-40a, b) are applied at the interior points and the two end points $(x_0, x_n)$ have to be excluded. We need to impose two additional conditions which are

$$S_0''(x_0) = 0 \qquad\qquad\qquad (7.6\text{-}42a)$$

$$S_{n-1}''(x_n) = 0 \qquad\qquad\qquad (7.6\text{-}42b)$$

Combining Equations (7.6-37, 42a, b) yields

$$c_0 = 0 \qquad\qquad\qquad (7.6\text{-}43a)$$

$$c_{n-1} + 3d_{n-1} h_{n-1} = 0 \qquad\qquad\qquad (7.6\text{-}43b)$$

Equations (7.6-38b, 39b, 40b, 41b, 43a, b) form a set of $4n$ equations to be solved for the $4n$ unknowns ($a_i$, $b_i$, $c_i$ and $d_i$).

***Example 7.6-4.*** Use cubic splines to determine an interpolation formula for the function $\sqrt{x}$. Consider three points (knots).

We choose the three knots to be 100, 121 and 144. The values of the function are 10, 11 and 12.

$S_0(x)$ and $S_1(x)$ are given by Equation (7.6-37) and are written as

$$S_0(x) = a_0 + b_0(x - 100) + c_0(x - 100)^2 + d_0(x - 100)^3 \qquad \text{(7.6-44a)}$$

$$S_1(x) = a_1 + b_1(x - 121) + c_1(x - 121)^2 + d_1(x - 121)^3 \qquad \text{(7.6-44b)}$$

From Equation (7.6-38c), we have

$$h_0 = 121 - 100 = 21 \qquad \text{(7.6-45a,b)}$$

$$h_1 = 144 - 121 = 23 \qquad \text{(7.6-45c,d)}$$

Equation (7.6-39b) yields

$$a_0 = 10 , \qquad a_1 = 11 , \qquad a_2 = 12 \qquad \text{(7.6-46a,b,c)}$$

From Equation (7.6-38b), we obtain

$$10 + 21b_0 + (21)^2 c_0 + (21)^3 d_0 = 11 \qquad \text{(7.6-47a)}$$

$$11 + 23b_1 + (23)^2 c_1 + (23)^3 d_1 = 12 \qquad \text{(7.6-47b)}$$

Similarly Equations (7.6-40b, 41b) yield respectively

$$b_0 + 42 c_0 + 3 (21)^2 d_0 = b_1 \qquad \text{(7.6-48a)}$$

$$c_0 + 63 d_0 = c_1 \qquad \text{(7.6-48b)}$$

Finally Equations (7.6-43a, b) give respectively

$$c_0 = 0 \qquad \text{(7.6-49a)}$$

$$c_1 + 63 d_1 = 0 \qquad \text{(7.6-49b)}$$

The six unknowns $(b_0, c_0, d_0, b_1, c_1, d_1)$ are to be determined from Equations (7.6-47a, b, 48a, b, 49a, b).

The solution is

$$b_0 = 0.0486 , \qquad b_1 = 0.0456 \qquad \text{(7.6-50a,b)}$$

$$c_0 = 0 , \qquad c_1 = -0.0001 \qquad \text{(7.6-51a,b)}$$

$$d_0 = - 0.0000022 , \qquad d_1 = 0.0000022 \qquad \text{(7.6-52a,b)}$$

Substituting Equations (7.6-46a, b, 50a, b, 51a, b, 52a, b) into Equations (7.6-44a, b), we obtain $S_0$ and $S_1$. For the interval $10 \le x \le 11$, we use $S_0$, and for the interval $11 \le x \le 12$, we use $S_1$.

## Least Squares Approximation

The concept of **least squares approximation** was introduced in Chapter 2. Here, we start with the simplest case of a linear polynomial ($y = ax + b$). We assume that the x's are exact.

Suppose we have a set of $n$ data points $\{x_i, y_i\}$, where $n$ is greater than 2, and a set of weights $\{w_i\}$. The weights express our confidence in the accuracy of the points. If we think that they are all equally accurate, we set $w_i = 1$ for all $i$.

We want to obtain the best values of $a$ and $b$ to fit the data. The deviation at each point is

$$\varepsilon_i = a + bx_i - y_i \tag{7.6-53}$$

The weighted sum of the squares of the deviation is

$$f(a, b) = \sum_{i=1}^{n} \varepsilon_i^2 w_i \tag{7.6-54a}$$

$$= \sum_{i=1}^{n} (a + bx_i - y_i)^2 w_i \tag{7.6-54b}$$

The best values of $a$ and $b$ imply that we choose $a$ and $b$ such that $f(a, b)$ is a minimum. That is

$$\frac{\partial f}{\partial a} = \frac{\partial f}{\partial b} = 0 \tag{7.6-55a,b}$$

Combining Equations (7.6-54a, b, 55a, b) yields

$$2 \sum (a + bx_i - y_i) w_i = 0 \tag{7.6-56a}$$

$$2 \sum (a + bx_i - y_i) x_i w_i = 0 \tag{7.6-56b}$$

Equations (7.6-56a, b) can be rewritten respectively as

$$a \sum_{i=1}^{n} w_i + b \sum_{i=1}^{n} x_i w_i = \sum_{i=1}^{n} y_i w_i \tag{7.6-57a}$$

$$a \sum_{i=1}^{n} x_i w_i + b \sum_{i=1}^{n} x_i^2 w_i = \sum_{i=1}^{n} x_i y_i w_i \tag{7.6-57b}$$

If all the points are equally weighted ($w_i = 1$, for each $i$), we have

$$n a + b \sum_{i=1}^{n} x_i = \sum_{i=1}^{n} y_i \tag{7.6-58a}$$

$$a \sum_{i=1}^{n} x_i + b \sum_{i=1}^{n} x_i^2 = \sum_{i=1}^{n} x_i y_i \tag{7.6-58b}$$

Solving for a and b from Equations (7.6-58a, b), we obtain the desired linear expression, which represents the **best fitting regression line**.

**Example 7.6-5**. Find the best fitting regression line through the set of points (1, 1), (2, 1.5), (3, 1.75) and (4, 2).

We have

$$n = 4, \qquad \sum x_i = 10, \qquad \sum y_i = 6.25 \tag{7.6-59a,b,c}$$

$$\sum x_i^2 = 30, \qquad \sum x_i y_i = 17.25 \tag{7.6-59d,e}$$

Substituting these values in Equations (7.6-58a, b), we obtain

$$4a + 10b = 6.25 \tag{7.6-60a}$$

$$10a + 30b = 17.25 \tag{7.6-60b}$$

The solution of Equations (7.6-60a, b) is

$$a = 0.75, \qquad b = 0.325 \tag{7.6-61a,b}$$

The equation for the regression is

$$y = 0.75 + 0.325x \tag{7.6-62}$$

●

The above method can be extended to multiple regression. In this case, y is a linear function of r variables $x_1, x_2, \dots, x_r$ and is written as

$$y = b_0 + b_1 x_1 + \dots + b_r x_r \tag{7.6-63}$$

We can then form the square of the deviations at each of the n points $\left( x_1^{(i)}, x_2^{(i)}, \dots, x_r^{(i)} \right)$, $i = 1, 2, \dots, n$, and obtain

$$\epsilon^2 (b_0, b_1, \ldots, b_r) = \sum_{i=1}^{n} \left[ b_0 + b_1 x_1^{(i)} + \ldots + b_r x_r^{(i)} - y_i \right]^2 \qquad (7.6\text{-}64)$$

As in the case of one variable, we choose $b_0, b_1, \ldots, b_r$ such that $\epsilon^2$ is a minimum with respect to the constants $b_0, b_1, \ldots, b_r$. This implies that

$$\frac{\partial \epsilon^2}{\partial b_\ell} = 2 \sum_{i=1}^{n} \left[ b_0 + b_1 x_1^{(i)} + \ldots + b_r x_r^{(i)} - y_i \right] \left[ x_\ell^{(i)} \right] = 0 \qquad (7.6\text{-}65\text{a,b})$$

where $\ell = 0, 1, \ldots, r$ and $x_0^{(i)} = 1$ for all i.

Equations (7.6-65a, b) can be written as

$$b_0 \sum_{i=1}^{n} x_\ell^{(i)} + b_1 \sum_{i=1}^{n} x_1^{(i)} x_\ell^{(i)} + \ldots + b_r \sum_{i=1}^{n} x_r^{(i)} x_\ell^{(i)} = \sum_{i=1}^{n} y_i x_\ell^{(i)} \qquad (7.6\text{-}66)$$

where $\ell = 0, 1, \ldots, r$.

Equations (7.6-66) represent $(r + 1)$ equations, from which we can obtain the $(r + 1)$ constants, $b_0, b_1, \ldots, b_r$.

The least squares approximation is not restricted to linear functions only. We can choose polynomials of degree higher than one, orthogonal functions such as Legendre polynomials and trigonometric functions. We refer the readers to Chapter 2 and Hildebrand (1956) for further details.

## 7.7 NUMERICAL DIFFERENTIATION AND INTEGRATION

### Numerical Differentiation

In this section, we describe methods of calculating the derivatives of a function f using only its values at $(n + 1)$ distinct points. We approximate the function by a polynomial and we assume that the derivatives of the function are approximately equal to the derivatives of the polynomial.

Recall that a polynomial of first degree passing through the points $[x_0, f(x_0)]$ and $[x_1, f(x_1)]$ is

$$p_1(x) = \frac{x - x_1}{x_0 - x_1} f(x_0) + \frac{x - x_0}{x_1 - x_0} f(x_1) \qquad (7.6\text{-}7)$$

Differentiating $p_1(x)$ with respect to x, we obtain

$$p_1'(x) = -\frac{f(x_0)}{h} + \frac{f(x_1)}{h} \qquad (7.7\text{-}1a)$$

$$= \frac{f(x_1) - f(x_0)}{h} \qquad (7.7\text{-}1b)$$

where $h \; (= x_1 - x_0)$ is the distance between the points $x_1$ and $x_0$.

A formula for $f'(x_0)$ is then given approximately by

$$f'(x_0) \approx \frac{f(x_0 + h) - f(x_0)}{h} \qquad (7.7\text{-}2)$$

The formula given by Equation (7.7-2) is known as the **two-point formula** since it involves two points $x_0$ and $x_1$. It is also known as the **forward difference formula**, as the derivative at $x_0$ depends on the forward point $x_1$.

From Equations (1.2-11, 12), we have

$$f(x_0 + h) = f(x_0) + hf'(x_0) + \frac{h^2}{2} f''(x_0 + \theta h) \qquad (7.7\text{-}3a)$$

where $0 \le \theta \le 1$.

Equation (7.7-3a) may be written as

$$f'(x_0) = \frac{f(x_0 + h) - f(x_0)}{h} - \frac{h}{2} f''(x + \theta h) \qquad (7.7\text{-}3b)$$

By comparing Equations (7.7-2, 3b), we find that the error is of the order $h$.

The error can be reduced by taking $h$ to be smaller, provided $f''$ behaves well in the interval. Since $f''$ is generally not known, we cannot estimate the error.

Similarly if we approximate $f(x)$ by a polynomial of second degree $p_2(x)$ and differentiate $p_2(x)$, we obtain

$$f'(x_0) \approx \frac{f(x_0 + h) - f(x_0 - h)}{2h} \qquad (7.7\text{-}4)$$

Equation (7.7-4) is known as the **three-point formula** as it involves the three points $(x_0 - h, \; x_0, x_0 + h)$. It is also referred to as the **central difference formula**, since the derivative at $x_0$ depends on the value of $f$ at $(x_0 - h)$ and at $(x_0 + h)$.

A four-point formula is obtained by considering a third degree polynomial. It is found to be (see Problem 16a)

$$f'(x_0) \approx \frac{-f(x_0 + 2h) + 6f(x_0 + h) - 3f(x_0) - 2f(x_0 - h)}{6h} \qquad (7.7-5)$$

Formulas for approximating higher derivatives of $f(x)$ can be obtained in a similar way. Thus by differentiating $p_2(x)$ twice, we obtain

$$f''(x_0) \approx \frac{f(x_0 + h) - 2f(x_0) + f(x_0 - h)}{h^2} \qquad (7.7-6)$$

**Example 7.7-1.** The distance-time (s, t) relation for a moving body is given by (4.807, 0.99), (4.905, 1.00), (5.004, 1.01), (5.103, 1.02) and (5.204, 1.03). Find the velocity at each instant of time.

The velocity $v$ is given by $\frac{ds}{dt}$. We use Equation (7.7-2) to find $v$ at various times $t$. In this case, $h$ is 0.01.

$$v(0.99) = \frac{4.905 - 4.807}{0.01} = 9.8 \qquad (7.7\text{-}7a,b)$$

$$v(1.00) = \frac{5.004 - 4.905}{0.01} = 9.9 \qquad (7.7\text{-}7c,d)$$

$$v(1.01) = \frac{5.103 - 5.004}{0.01} = 9.9 \qquad (7.7\text{-}7e,f)$$

Using the three-point formula (Equation 7.7-4), we obtain

$$v(1.00) = \frac{5.004 - 4.807}{0.02} = 9.85 \qquad (7.7\text{-}7g,h)$$

$$v(1.01) = \frac{5.103 - 4.905}{0.02} = 9.9 \qquad (7.7\text{-}7i,j)$$

Comparing Equations (7.7-7c to j), we find that in this case the difference between using two or three-point formulas is small.

●

In Table 7.7-1, we list the formulas giving the first and second derivatives of f. We use $f_{\pm k}$ to denote $f(x_0 \pm kh)$.

## TABLE 7.7-1

### Formulas for $f'$ and $f''$

| | $f'(x_0)$ | $f''(x_0)$ |
|---|---|---|
| two-point | $\dfrac{f_1 - f_0}{h} + 0(h)$ | $\dfrac{f_2 - 2f_1 - f_0}{h^2} + 0(h)$ |
| three-point (central) | $\dfrac{f_1 - f_{-1}}{2h} + 0(h^2)$ | $\dfrac{f_1 - 2f_0 + f_{-1}}{h^2} + 0(h^2)$ |
| three-point | $\dfrac{-f_2 + 4f_1 - 3f_0}{2h} + 0(h^2)$ | $\dfrac{-f_3 + 4f_2 - 5f_1 + 2f_0}{h^2} + 0(h^2)$ |
| four-point | $\dfrac{-f_2 + 8f_1 - 8f_{-1} + f_{-2}}{12h} + 0(h^4)$ | $\dfrac{-f_2 + 16f_1 - 30f_0 + 16f_{-1} - f_{-2}}{12h^2} + 0(h^4)$ |

*Example 7.7-2.* Given the values of $[x_k, f(x_k)]$ to be (1.3, 3.669), (1.5, 4.489), (1.7, 5.474), (1.9, 6.686) and (2.1, 8.166), find the derivative of $f$ at 1.7.

Using the formulas in the order given in Table 7.7-1, we obtain

$$f'(1.7) \approx \frac{1}{0.2}(6.686 - 5.474) = 6.06 \tag{7.7-8a,b}$$

$$f'(1.7) \approx \frac{1}{0.4}(6.686 - 4.489) = 5.390 \tag{7.7-8c,d}$$

$$f'(1.7) \approx \frac{1}{0.4}(-8.166 + 4 \times 6.686 - 3 \times 5.474) = 5.490 \tag{7.7-8e,f}$$

$$f'(1.7) \approx \frac{1}{2.4}(-8.166 + 8 \times 6.686 - 8 \times 4.489 + 3.669) = 5.475 \tag{7.7-8g,h}$$

We note that $f(x)$ is $e^x$ and its derivative is also $e^x$. Thus $f'(1.7)$ is 5.474 and not surprisingly the four-point formula gives the most accurate result.

### Numerical Integration

We consider numerical methods of evaluating the integral $\int_a^b f(x)\,dx$. These methods are called **numerical quadratures**. We need to resort to numerical integration when the integral cannot be evaluated exactly or $f(x)$ is given only at a finite number of points. Unlike numerical differentiation, numerical integration is a smooth operation and many adequate formulas exist.

A common practice in numerical methods is to approximate the function by a polynomial and then carry out the mathematical operation. If we approximate $f(x)$ by a polynomial of first degree [Equation (7.6-5)], we have

$$\int_{x_0}^{x_1} f(x)\,dx \approx \int_{x_0}^{x_1} \left\{ \frac{x - x_1}{-h}\, f(x_0) + \frac{x - x_0}{h}\, f(x_1) \right\} dx \qquad (7.7\text{-}9a)$$

$$\approx -\frac{f(x_0)}{h} \int_{x_0}^{x_1} (x - x_1)\,dx + \frac{f(x_1)}{h} \int_{x_0}^{x_1} (x - x_0)\,dx \qquad (7.7\text{-}9b)$$

$$\approx -\frac{f(x_0)}{h} \left[ \frac{x^2}{2} - x_1 x \right]_{x_0}^{x_1} + \frac{f(x_1)}{h} \left[ \frac{x^2}{2} - x_0 x \right]_{x_0}^{x_1} \qquad (7.7\text{-}9c)$$

$$\approx \frac{h}{2} \left[ f(x_0) + f(x_1) \right] \qquad (7.7\text{-}9d)$$

Equation (7.7-9d) is known as the **Trapezoidal rule**.

To improve on the Trapezoidal rule, we approximate the function $f(x)$ by a polynomial of second degree. The polynomial $p_2(x)$ may be written as [Equation (7.6-12)]

$$p_2(x) = \frac{(x - x_1)(x - x_2)}{(x_0 - x_1)(x_0 - x_2)}\, f(x_0) + \frac{(x - x_0)(x - x_2)}{(x_1 - x_0)(x_1 - x_2)}\, f(x_1) + \frac{(x - x_0)(x - x_1)}{(x_2 - x_0)(x_2 - x_1)}\, f(x_2)$$
$$(7.6\text{-}12)$$

Let the limits of integration be $a$ and $b$. We denote $a$ by $x_0$, $b$ by $x_2$ and we let $x_1$ be the midpoint of $(x_0, x_2)$. As usual we denote the distance between $(x_1, x_0)$ and $(x_2, x_1)$ by $h$.

Approximating the integral of $f(x)$ by the integral of $p_2(x)$, we have

$$\int_a^b f(x)\,dx \approx \int_a^b p_2(x)\,dx \qquad (7.7\text{-}10a)$$

$$\approx \int_{x_0}^{x_2} \left\{ \frac{(x-x_1)(x-x_2)}{(-h)(-2h)} \, f(x_0) + \frac{(x-x_0)(x-x_2)}{(h)(-h)} \, f(x_1) + \frac{(x-x_0)(x-x_1)}{(2h)(h)} \, f(x_2) \right\} dx \quad (7.7\text{-}10b)$$

$$\approx \frac{f(x_0)}{2h^2} \int_{x_0}^{x_2} (x-x_1)(x-x_2)\, dx - \frac{f(x_1)}{h^2} \int_{x_0}^{x_2} (x-x_0)(x-x_2)\, dx + \frac{f(x_2)}{2h^2} \int_{x_0}^{x_2} (x-x_0)(x-x_2)\, dx$$

$$\quad (7.7\text{-}10c)$$

$$\approx \frac{h}{3} \left[ f(x_0) + 4f(x_1) + f(x_2) \right] \quad (7.7\text{-}10d)$$

Formula (7.7-10d) is the well known **Simpson's rule**.

If we approximate $f(x)$ by a third degree polynomial $p_3(x)$, we obtain the **3/8 rule** which is given by

$$\int_{x_0}^{x_3} f(x)\, dx \approx \frac{3h}{8} \left[ f(x_0) + 3\, f(x_1) + 3\, f(x_2) + f(x_3) \right] \quad (7.7\text{-}11)$$

In Equation (7.7-11), the limits of integration are $(x_0, x_3)$. The points $x_1$ and $x_2$ are in the interval $(x_0, x_3)$ such that the points $x_0, x_1, x_2$ and $x_3$ are equidistant points and the distance between two consecutive points is $h$.

If the interval [a, b] is large, the formulas given so far will not generate good result. We may be tempted to approximate $f(x)$ by a high degree polynomial. This should be resisted. As pointed out in the previous section, it is better to divide the interval into subintervals and approximate the function by a low degree polynomial (cubic splines, for example) over the subintervals. Equally, in the present section, we subdivide the interval [a, b] into $N$ subintervals of equal length $h$. We apply the formulas given earlier to each subinterval and the desired result is the sum of the integrals of each subinterval.

We denote the break-points by $x_i$, that is to say, we write

$$x_i = a + i\, h, \qquad i = 0, 1, \ldots, N \quad (7.7\text{-}12a)$$

$$h = \frac{b-a}{N} \quad (7.7\text{-}12b)$$

The integral $\displaystyle\int_a^b f(x)\, dx$ can be written as

$$\int_a^b f(x) \, dx = \sum_{i=1}^{N} \int_{x_{i-1}}^{x_i} f(x) \, dx \tag{7.7-13}$$

We now apply the Trapezoidal rule [Equation (7.7-9d)] to each subinterval $[x_{i-1}, x_i]$. Equation (7.7-13) becomes

$$\int_a^b f(x) \, dx \approx \sum_{i=1}^{N} \frac{h}{2} \left[ f(x_{i-1}) + f(x_i) \right] \tag{7.7-14a}$$

$$\approx \frac{h}{2} \sum_{i=1}^{N} \left[ f(x_{i-1}) + f(x_i) \right] \tag{7.7-14b}$$

$$\approx \frac{h}{2} \left[ f(x_0) + f(x_n) + 2 \sum_{i=1}^{N-1} f(x_i) \right] \tag{7.7-14c}$$

$$\approx \frac{h}{2} \left[ f(a) + f(b) + 2 \sum_{i=1}^{N-1} f(x_i) \right] \tag{7.7-14d}$$

Equation (7.7-14d) is referred to as the **Trapezoidal composite rule**. Similarly **Simpson's composite rule** can be deduced and is given by

$$\int_a^b f(x) \, dx \approx \frac{h}{6} \left[ f(a) + f(b) + 2 \sum_{i=1}^{N-1} f(x_i) + 4 \sum_{i=1}^{N} f(x_{i-1/2}) \right] \tag{7.7-15a}$$

We note that in Simpson's rule [Equation (7.7-10d)], we need to evaluate the function at the midpoints which we have denoted in Equation (7.7-15a) by $x_{i-1/2}$. To apply Simpson's rule, we need to have an odd number of points and an even number of intervals. Thus we divide the interval $[a, b]$ into $2N$ subintervals and each subinterval is of length $h$. Equation (7.7-15a) can then be written as

$$\int_a^b f(x) \, dx \approx \frac{h}{3} \left[ f(a) + f(b) + 2 \sum_{j=1}^{N-1} f(x_{2j}) + 4 \sum_{j=1}^{N} f(x_{2j-1}) \right] \tag{7.7-15b}$$

***Example 7.7-3***. The values of $f(x_k)$ for various values of $x_k$ is given in Table 7.7-2. Use the Trapezoidal rule and Simpson's rule to evaluate $\int_{1.0}^{1.3} f(x) \, dx$.

## TABLE  7.7-2

### Values of $f(x_k)$ for various $x_k$

| $x_k$ | 1 | 1.05 | 1.10 | 1.15 | 1.20 | 1.25 | 1.30 |
|---|---|---|---|---|---|---|---|
| $f(x_k)$ | 1 | 1.02470 | 1.04881 | 1.07238 | 1.09544 | 1.11803 | 1.14017 |

In this case, h is 0.05. Using Equation (7.7-14d), we have

$$\int_1^{1.3} f(x)\,dx = \frac{0.05}{2}\{f(1) + f(1.30) + 2\,[f(1.05) + f(1.10) + f(1.15) + f(1.20) + f(1.25)]\}$$

(7.7-16a)

$$= 0.33147$$

(7.7-16b)

We note that the number of points is seven, which is an odd number and we can directly apply Simpson's rule. From Equation (7.7-15b), we have

$$\int_1^{1.3} f(x)\,dx = \frac{0.05}{3}\{f(1) + f(1.30) + 2\,[f(1.10) + f(1.20)] + 4\,[f(1.05) + f(1.15) + f(1.25)]\}$$

(7.7-17a)

$$= 0.32149$$

(7.7-17b)

The function given in Table 7.7-2 is $\sqrt{x}$ and by direct integration, we obtain

$$\int_1^{1.3} f(x)\,dx = \int_1^{1.3} \sqrt{x}\,dx$$

(7.7-18a)

$$= \frac{2}{3}\left[x^{3/2}\right]_1^{1.3}$$

(7.7-18b)

$$= 0.32149$$

(7.7-18c)

By comparing Equations (7.7-16b, 17b, 18c), we find that Simpson's rule yields the exact solution whereas the Trapezoidal rule is incorrect in the last decimal place.

●

The error involved in evaluating the integral depends on h, the length of the gap. If we denote the exact value of $\displaystyle\int_a^b f(x)\,dx$ by I and the value obtained using the Trapezoidal formula [Equation (7.7-14d)] by $T_h$, we may write

$$I = T_h + O(h^2) \qquad\qquad (7.7\text{-}19)$$

We now assume that the error which is of $O(h^2)$ is proportional to $h^2$. Equation (7.7-19) can then be written as

$$I = T_h + C h^2 \qquad\qquad (7.7\text{-}20)$$

where C is a constant and is assumed to be independent of h.

We now recalculate the integral using the same Trapezoidal rule but halving h. We denote the value obtained with a gap $h/2$ by $T_{h/2}$. Thus I is given by

$$I = T_{h/2} + C h^2/4 \qquad\qquad (7.7\text{-}21)$$

From Equations (7.7-20, 21) , we obtain

$$C = \frac{4}{3h^2} (T_{h/2} - T_h) \qquad\qquad (7.7\text{-}22)$$

Substituting C into Equation (7.7-21), we obtain an improved value of the integral which we denote by $T^{(2)}$. Thus $T^{(2)}$ $(\approx I)$ is given by

$$T^{(2)} = T_{h/2} + \frac{1}{3} (T_{h/2} - T_h) \qquad\qquad (7.7\text{-}23)$$

We can repeat this process of halving the interval, that is to say, using intervals of h, h/2, h/4, ... , $h/2^n$ successively until two improved values of the integral are within the desired degree of accuracy.

This method of obtaining an improved value of the integral from two approximate values is known as the **Romberg method** (extrapolation method). It is widely used in computer programs.

We recall that the Lagrange interpolation formula for the polynomial $p_n(x)$ can be written as

$$p_n(x) = \sum_{i=0}^{n} f(x_i)\, \ell_i(x) \qquad\qquad (7.6\text{-}10)$$

The integral $\displaystyle\int_a^b f(x)\,dx$ can then be approximated by

$$\int_a^b f(x)\,dx \approx \int_a^b p_n(x)\,dx \tag{7.7-24a}$$

$$\approx \sum_{i=0}^n \int_a^b f(x_i)\,\ell_i(x)\,dx \tag{7.7-24b}$$

$$\approx \sum_{i=0}^n f(x_i) \int_a^b \ell_i(x)\,dx \tag{7.7-24c}$$

$$\approx \sum_{i=0}^n f(x_i)\,A_i \tag{7.7-24d}$$

Note that $f(x_i)$ are the values of $f$ at the points $x_i$ and are constants. They can therefore be taken out of the integral sign. The $A_i \left( = \int_a^b \ell_i(x)\,dx \right)$ are known as the **weights**. So far the points $x_i$ are given and are equally spaced. We now assume that the $x_i$ are not given. We choose $x_i$ such that Equation (7.7-24d) is "exact" or "best" in a sense to be defined later. For simplicity, we transform the interval [a, b] to [−1, 1]. This can be achieved by writing

$$z = \frac{2x - (a + b)}{(b - a)} \tag{7.7-25}$$

We now have to determine the points $z_i$ which lie in the interval [−1, 1] such that

$$\int_{-1}^1 f(z)\,dz \approx \sum_{i=0}^n f(z_i)\,A_i \tag{7.7-26}$$

We consider the case where $f(z)$ is approximated by $p_3(z)$. It can be written as

$$p_3(z) = a_0 + a_1 z + a_2 z^2 + a_3 z^3 \tag{7.7-27}$$

We now choose $z_0$, $z_1$, $A_0$ and $A_1$ such that Equation (7.7-26) is "exact". That is to say

$$\int_{-1}^1 p_3(z)\,dz = p_3(z_0)\,A_0 + p_3(z_1)\,A_1 \tag{7.7-28}$$

Carrying out the required integration, Equation (7.7-28) becomes

$$2a_0 + \frac{2a_2}{3} = A_0 \left[ a_0 + a_1 z_0 + a_2 z_0^2 + a_3 z_0^3 \right] + A_1 \left[ a_0 + a_1 z_1 + a_2 z_1^2 + a_3 z_1^3 \right] \tag{7.7-29}$$

Equation (7.7-29) holds for all $a_0$, $a_1$, $a_2$ and $a_3$ and, by comparing the coefficients of $a_0$, $a_1$, $a_2$ and $a_3$, we obtain respectively

$$2 = A_0 + A_1 \tag{7.7-30a}$$

$$0 = A_0 z_0 + A_1 z_1 \tag{7.7-30b}$$

$$\frac{2}{3} = A_0 z_0^2 + A_1 z_1^2 \tag{7.7-30c}$$

$$0 = A_0 z_0^3 + A_1 z_1^3 \tag{7.7-30d}$$

The set of Equations (7.7-30a to d) involves four unknowns $(A_0, A_1, z_0, z_1)$ and we have exactly four equations to solve. Multiplying Equation (7.7-30b) by $z_1^2$ and subtracting the resulting expression from Equation (7.7-30d) yields

$$A_0 z_0 \left( z_1^2 - z_0^2 \right) = 0 \tag{7.7-31}$$

The possible solutions of Equation (7.7-31) are

$$A_0 = 0, \qquad z_0 = 0, \qquad z_1 = z_0, \qquad z_1 = -z_0 \tag{7.7-32a,b,c,d}$$

The first solution $(A_0 = 0)$ implies, from Equation (7.7-30b), that either $A_1$ or $z_1$ is zero and Equation (7.7-30c) cannot be satisfied. Of the four possible solutions, the only valid solution is the last one.

Form Equation (7.7-30b or d), we deduce

$$A_0 = A_1 \tag{7.7-33}$$

Combining Equations (7.7-30a, 30c, 33), we obtain

$$A_0 = 1, \qquad A_1 = 1, \qquad z_0 = -\frac{1}{\sqrt{3}}, \qquad z_1 = \frac{1}{\sqrt{3}} \tag{7.7-34a,b,c,d}$$

Substituting Equations (7.7-34a - d) into Equation (7.7-26), we obtain

$$\int_{-1}^{1} f(z)\, dz \approx f\left(-\frac{1}{\sqrt{3}}\right) + f\left(\frac{1}{\sqrt{3}}\right) \tag{7.7-35}$$

Equation (7.7-35) is the **Gaussian two point formula** and it involves evaluating the function at two points. Thus the number of evaluations of the integrand is the same as in the Trapezoidal rule, but in this case, we have approximated the function $f(x)$ by a third degree polynomial and we expect better accuracy. This is illustrated in the next example.

***Example 7.7-4***.  Evaluate $\int_1^{1.3} \sqrt{x} \; dx$ using the Romberg method and the Gaussian method.

We use the trapezoidal rule to evaluate $T_h$ using $h = 0.3$.

Using Equation (7.7-9d) and the values given in Table 7.7-2, we have

$$T_{0.3} = \frac{0.3}{2} (1 + 1.14017) = 0.32103 \qquad\qquad (7.7\text{-}36a,b)$$

We now halve the gap and the interval is now 0.15.

From Equation (7.7-14d) and Table 7.7-2, we have

$$T_{0.15} = \frac{0.15}{2} \left(1 + 2 \times 1.07238 + 1.14017\right) = 0.32137 \qquad\qquad (7.7\text{-}37a,b)$$

The improved value $T^{(2)}$ is given from Equation (7.7-23)

$$T^{(2)} = 0.32137 + \frac{1}{3} (0.32137 - 0.32103) = 0.32148 \qquad\qquad (7.7\text{-}38a,b)$$

Comparing $T^{(2)}$ with the values obtained in Example 7.7-3, we find the Romberg method gives an answer nearer to the exact values, though the number of points considered is less.  To use the Gaussian quadrature, we have to change the interval from [1, 1.3] to [−1, 1].  From Equation (7.7-25), we have

$$x = 0.15z + 1.15 \qquad\qquad (7.7\text{-}39)$$

By direct substitution, we have

$$\int_1^{1.3} \sqrt{x} \; dx = 0.15 \int_{-1}^{1} \sqrt{0.15z + 1.15} \; dz \qquad\qquad (7.7\text{-}40)$$

Combining Equations (7.7-35, 40) yields

$$\int_1^{1.3} \sqrt{x} \; dx = 0.15 \left[ \sqrt{0.15\left(-\frac{1}{\sqrt{3}}\right) + 1.15} + \sqrt{0.15\left(\frac{1}{\sqrt{3}}\right) + 1.15} \right] \qquad (7.7\text{-}41a)$$

$$= 0.32149 \qquad\qquad (7.7\text{-}41b)$$

The value we obtain for the integral using only two points is as accurate as Simpson's rule using seven points.

●

We can generalize the two-point formula to an n-point formula. In this case, we have $2n$ unknowns $(z_0, \ldots, z_{n-1}, A_0, \ldots, A_{n-1})$ and we have to choose them such that Equation (7.7-26) is "exact" if $f(x)$ is a polynomial of degree $(2n-1)$. This is possible because a polynomial of degree $(2n-1)$ has $2n$ constants which exactly corresponds to the number of unknowns. The determination of $z_i$ and $A_i$ is simplified by considering Legendre polynomials, which are discussed in Chapter 2. For further details, see Hilderbrand (1956) or Stroud and Secrest (1966).

## 7.8  NUMERICAL SOLUTION OF ORDINARY DIFFERENTIAL EQUATIONS, INITIAL VALUE PROBLEMS

The **order and the degree** of an ordinary differential equation (O.D.E.) have been discussed in Chapter 1.

An O.D.E. (or a system of O.D.E.'s) with all conditions specified at one value of the independent variable is called an **initial value problem**. In the case where time is the independent variable, all conditions may be specified at $t = 0$. However if the conditions are specified at more than one point, the O.D.E. with the conditions specified at different points represents a **boundary value problem**.

For example, the differential equation

$$a_2 y'' + a_1 y' + a_0 y = g(x) \qquad (7.8\text{-}1)$$

subject to

$$y(a) = Y_0, \qquad y(b) = Y_1 \qquad (7.8\text{-}2a,b)$$

is a two point boundary value problem whereas if the conditions to be satisfied were

$$y(a) = Y_0, \qquad y'(a) = Y_2 \qquad (7.8\text{-}2c,d)$$

it would be an initial value problem.

Similarly

$$y'' = (z^2 + 1) y, \qquad 0 \le x \le 1 \qquad (7.8\text{-}3a)$$

$$z'' = z + 1 \qquad (7.8\text{-}3b)$$

with   $y(0) = 1, \ y'(0) = 2, \ z(0) = 0, \ z'(0) = 1$ \qquad (7.8\text{-}3c to f)

is an initial value problem of a second order system of differential equations, whereas

$$y'' = f(x, y, y'), \qquad a \le x \le b \qquad (7.8\text{-}4a)$$

$$a_1 \, y\,(a) + a_2 \, y'\,(a) \; = \; c_1, \qquad\qquad b_1 \, y\,(b) + b_2 \, y'\,(b) \; = \; c_2 \qquad\qquad (7.8\text{-}4b,c)$$

is a boundary value problem (Sturm-Liouville problem).

In this section, only initial value problems will be considered.

### First Order Equations

A **first order initial value problem** can be written as

$$\frac{dy}{dx} \; = \; f(x, y) \qquad\qquad (7.8\text{-}5a)$$

$$y\,(a) \; = \; y_0 \qquad\qquad (7.8\text{-}5b)$$

Formally integrating Equation (7.8-5a) subject to the initial condition given by Equation (7.8-5b), we obtain

$$y(x) \; = \; y_0 + \int_a^x f(t, y) \, dt \qquad\qquad (7.8\text{-}6)$$

Note that the unknown variable $y$ is inside the integral sign in Equation (7.8-6) and we cannot evaluate such an integral. Equation (7.8-6) is known as an **integral equation**. We have transformed the differential equation [Equation (7.8-5a)] into an integral equation.

There is a similarity between Equations (7.8-6, 2-20). As suggested by the iteration given by Equation (7.2-21), we seek a solution of Equation (7.8-6) by considering

$$y_{k+1}(x) \; = \; y_0 + \int_a^x f(t, y_k) \, dt, \qquad k = 0, 1, 2, \dots \qquad\qquad (7.8\text{-}7)$$

Since $y_0$ is given, we can determine $y_1$ and subsequently $y_2, \dots$ . The iteration will converge if

(a)     $f(x, y)$ is continuous in the interval $a \le x \le b$, $-\infty \le y \le \infty$

(b)     there is a constant $L$, such that for any two numbers $y$, $z$ and any $x \in [a, b]$

$$|\, f(x, y) - f(x, z) \,| \; < \; L \,|\, y - z \,| \qquad\qquad (7.8\text{-}8)$$

Condition (7.8-8) is the **Lipschitz condition**.

**Picard's method** involves solving Equations (7.8-5a, b) using Equation (7.8-7). Although it ensures that if f satisfies the conditions stated earlier a solution is generated, it is not suitable for numerical computation because it requires the evaluation of many integrals.

Other numerical methods are available and are considered next.

### Euler's method

We subdivide the interval [a, b] into subintervals $[x_i, x_{i+1}]$ and assume that in each subinterval, which is of length h, f(x, y) is constant. Integrating Equation (7.8-5a), we obtain

$$y_{i+1} = y_i + h f(x_i, y_i) \tag{7.8-9}$$

where $y(x_i)$ is denoted by $y_i$.

Equation (7.8-9) is **Euler's formula**. Its accuracy is not very good. Figure 7.8-1 illustrates this method.

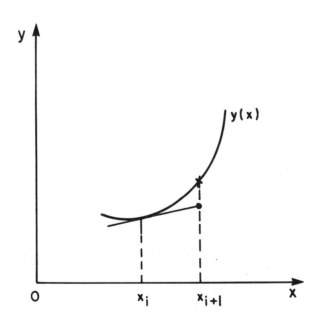

**FIGURE 7.8-1**   **Euler's method.**   $\times$ : **exact value of** $y_{i+1}$ ;
• : **approximate value of** $y_{i+1}$

*Example 7.8-1*. Solve the initial value problem using Euler's method.

$$\frac{dy}{dx} = xy, \qquad y(0) = 1 \tag{7.8-10a,b}$$

We assume $h$ to be $0.1$. Starting with $y(0) (= y_0)$ equals to 1, $x_0 (= a)$ equals to 0, denoting $ih$ by $x_i$ and using Equation (7.8-9), we generate Table 7.8-1.

### TABLE 7.8-1

### Values of $x_i$, $y_i$

| i | $x_i$ | $y_i$ | $f(x_i, y_i)$ | $y_{i+1}$ | Exact $y_i$ |
|---|-------|-------|---------------|-----------|-------------|
| 0 | 0 | 1 | 0 | 1 | 1 |
| 1 | 0.1 | 1 | 0.1 | 1.01 | 1.005 |
| 2 | 0.2 | 1.01 | 0.202 | 1.0302 | 1.020 |
| 3 | 0.3 | 1.0302 | 0.30906 | 1.0611 | 1.046 |
| 4 | 0.4 | 1.0611 | – | – | 1.083 |

The exact solution of Equations (7.8-10a, b) is

$$y = \exp(x^2/2) \tag{7.8-11}$$

The exact solution, accurate to three decimal places, is given in the last column of Table 7.8-1. It can be seen that the numerical solution differs significantly from the exact solution.

### *Taylor's method*

We expand the function $y(x)$ in $f[x, y(x)]$ in a Taylor series about the point $x_0$ and we obtain

$$y(x) = y(x_0) + y'(x_0)(x - x_0) + \frac{y''(x_0)}{2!}(x - x_0)^2 + \dots \tag{7.8-12}$$

From Equation (7.8-5a), we have

$$y'\left(= \frac{dy}{dx}\right) = f(x, y) \quad \text{and} \quad y'(x_0) = f(x_0, y_0) \tag{7.8-13a,b}$$

Differentiating Equation (7.8-13a) with respect to $x$, yields

$$y''(x) = \frac{\partial f}{\partial x} + \frac{\partial f}{\partial y}\frac{dy}{dx} = \frac{\partial f}{\partial x} + \frac{\partial f}{\partial y}f \tag{7.8-14a,b}$$

$$y''(x_0) = f_x(x_0, y_0) + f_y(x_0, y_0) \, f(x_0, y_0) \qquad (7.8\text{-}15)$$

We introduce the following notation

$$y'(x) = f(x, y) = f^{(0)}(x, y) \qquad (7.8\text{-}16a,b)$$

$$y''(x) = \frac{\partial f}{\partial x} + f \frac{\partial f}{\partial y} = f^{(1)}(x, y) \qquad (7.8\text{-}16c,d)$$

$$y'''(x) = f^{(2)}(x, y) \qquad (7.8\text{-}16e)$$

$$\cdot$$
$$\cdot$$
$$\cdot$$

$$y^{(k+1)}(x) = f^{(k)}(x, y) \qquad (7.8\text{-}16f)$$

We define

$$T_p(x, y, h) = f^{(0)}(x, y) + \frac{h}{2!} f^{(1)}(x, y) + \frac{h^2}{3!} f^{(2)}(x, y) + \dots + \frac{h^{p-1}}{p!} f^{(p-1)}(x, y) \qquad (7.8\text{-}17)$$

Since

$$y(x_1) = y(x_0 + h) = y_0 + h \, y'(x_0) + \frac{h^2}{2!} y''(x_0) + \frac{h^3}{3!} y'''(x_0) + \dots \qquad (7.8\text{-}18a,b)$$

we obtain from Taylor's formula

$$y(x_1) = y_0 + h \, T_p(x_0, y_0, h) \qquad (7.8\text{-}19)$$

In general

$$y_{n+1} = y_n + h \, T_p(x_n, y_n, h) \qquad (7.8\text{-}20)$$

***Example 7.8-2.*** Use Taylor's method to solve

$$y' = 1 + y^2, \qquad y(0) = 1 \qquad (7.8\text{-}21a,b)$$

Using the notation given by Equations (7.8-16a to f), we have

$$f(x, y) = 1 + y^2 = f^{(0)}(x, y) \qquad (7.8\text{-}22a,b)$$

$$f^{(1)}(x, y) = y''(x) = 2y(1 + y^2) \qquad (7.8\text{-}22c,d)$$

$$f^{(2)}(x, y) = y'''(x) = 2(1 + y^2)(1 + 3y^2) \tag{7.8-22e,f}$$

$$f^{(3)}(x, y) = y''''(x) = 8y(1 + y^2)(2 + 3y^2) \tag{7.8-22g,h}$$

$$T_4(x, y, h) = (1 + y^2) + \frac{h}{2!} 2y(1 + y^2) + \frac{h^2}{3!} 2(1 + y^2)(1 + 3y^2) + \frac{h^3}{4!} 8y(1 + y^2)(2 + 3y^2) \tag{7.8-22i}$$

With $h = 0.1$, $x_0 = 0$, $y_0 = 1$, we have

$$T_4(x_0, y_0, h) = 2 + .05(2)(2) + \frac{.01}{6} 2(16) + \frac{.001}{24}(80) \approx 2.23042 \tag{7.8-22j,k}$$

and

$$y_1 = y_0 + h T_4(x_0, y_0, h) \tag{7.8-22l}$$

$$= 1 + 0.1(2.23042) \approx 1.22304 \tag{7.8-22m,n}$$

Next we compute

$$y_2 = y_1 + h T_4(x_1, y_1, h) \tag{7.8-22o}$$

$$= 1.22304 + 0.1 T_4(0.1, 1.22304, 0.1) \tag{7.8-22p}$$

Since

$$T_4(0.1, 1.22304, 0.1) = 2.4958 + .05(6.1050218) + \frac{.01}{6}(27.391174)$$

$$+ \frac{.001}{24}(158.42532) \approx 2.8543 \tag{7.8-22q,r}$$

$$y_2 \approx 1.22304 + 0.1(2.8543) \approx 1.50847 \tag{7.8-22s,t}$$

Note that the exact solution of the initial value problem is

$$y(x) = \tan\left(x + \frac{\pi}{4}\right) \tag{7.8-22u}$$

and the values accurate to five decimal places are $y(0.1) = 1.22305$ and $y(0.2) = 1.50848$.

●

Taylor's method has the disadvantage of having to differentiate $f(x, y)$. For computational purposes, Heun's method is more economical.

## Heun's method

In Euler's method [Equation (7.8-9)], we only used the value of $f(x, y)$ $(= y'$, the slope of y) at the point $x_i$ to determine the value of y at $x_{i+1}$. We can improve the accuracy of Euler's method by replacing $f(x_i, y_i)$ by the average value of $f(x_i, y_i)$ and $f(x_{i+1}, y_{i+1})$. Equation (7.8-9) now becomes

$$y_{i+1} = y_i + \frac{h}{2} \left[ f(x_i, y_i) + f(x_{i+1}, y_{i+1}) \right] \qquad (7.8-23)$$

We determine $y_{i+1}$ using Equation (7.8-9). We can then calculate $f(x_{i+1}, y_{i+1})$, and finally using Equation (7.8-23), we obtain an improved value of $y_{i+1}$.

*Example 7.8-3.* Solve the initial value problem of Example 7.8-1 via Heun's method. Use the same value of h.

The values of $y_i$, obtained by using Euler's method, are given in Table 7.8-1. Combining Table 7.8-1 and Equation (7.8-23), we generate Table 7.8-2.

### TABLE 7.8-2

### Values of $x_i$ and improved values of $y_i$

| i | $x_i$ | $f(x_i, y_i)$ | $f(x_{i+1}, y_{i+1})$ | $y_{i+1}$ |
|---|-------|---------------|------------------------|-----------|
| 0 | 0     | 0             | 0.1                    | 1.005     |
| 1 | 0.1   | 0.1           | 0.202                  | 1.020     |
| 2 | 0.2   | 0.202         | 0.3091                 | 1.046     |
| 3 | 0.3   | 0.3091        | 0.4244                 | 1.083     |

Comparing the last column of Tables 7.8-1 and 2, we find that Heun's method gives the exact solution to three decimal places. It represents an improvement over Euler's method.

## Runge-Kutta methods

The Runge-Kutta methods are extensions of Heun's method. They are more accurate than Euler's method and do not involve finding derivatives as in Taylor's method. We only need to evaluate the function $f(x, y)$ at selected points on each subinterval. We derive the formula for the Runge-Kutta method of order two by generalizing Equation (7.8-23). We assume that $y_{i+1}$ is given by

$$y_{i+1} = y_i + a k_1 + b k_2 \qquad (7.8-24)$$

where

$$k_1 = h\, f(x_i, y_i) \tag{7.8-25a}$$

$$k_2 = h\, f(x_i + \alpha\, h,\, y_i + \beta\, k_1) \tag{7.8-25b}$$

with a, b, $\alpha$, $\beta$ as constants.

Expanding $y_{i+1}$ [$= y\,(x_{i+1})$] in a Taylor series about $x_i$, we have

$$y_{i+1} = y_i + h\, y_i' + \frac{h^2}{2}\, y_i'' + \frac{h^3}{6}\, y_i''' + \ldots \tag{7.8-25c}$$

$$= y_i + h\, f_i^{(0)} + \frac{h^2}{2}\, f_i^{(1)} + \frac{h^3}{6}\, f_i^{(2)} + \ldots \tag{7.8-25d}$$

$$= y_i + h\, f_i + \frac{h^2}{2}\left(\frac{\partial f}{\partial x} + f\frac{\partial f}{\partial y}\right)_i + \frac{h^3}{6}\left[\frac{\partial^2 f}{\partial x^2} + 2f\frac{\partial^2 f}{\partial x \partial y} + f^2\frac{\partial^2 f}{\partial y^2} + \frac{\partial f}{\partial x}\frac{\partial f}{\partial y} + f\left(\frac{\partial f}{\partial y}\right)^2\right]_i + \ldots \tag{7.8-25e}$$

We have used the notation introduced in Equations (7.8-16a to d). All the functions on the right side of Equations (7.8-25c to e) are to be evaluated at $(x_i, y_i)$.

Expanding $f(x_i + \alpha\, h,\, y_i + \beta\, k_1)$ about $(x_i, y_i)$ yields

$$f\left(x_i + \alpha h,\, y_i + \beta k_1\right) = f_i + \alpha h\left(\frac{\partial f}{\partial x}\right)_i + \beta k_1\left(\frac{\partial f}{\partial y}\right)_i + \frac{\alpha^2 h^2}{2}\left(\frac{\partial^2 f}{\partial x^2}\right)_i + \alpha\beta h k_1\left(\frac{\partial^2 f}{\partial x \partial y}\right)_i$$

$$+ \frac{\beta^2 k_1^2}{2}\left(\frac{\partial^2 f}{\partial y^2}\right)_i + \ldots \tag{7.8-26a}$$

$$= f_i + \alpha h\left(\frac{\partial f}{\partial x}\right)_i + \beta h\left(f\frac{\partial f}{\partial y}\right)_i + \frac{\alpha^2 h^2}{2}\left(\frac{\partial^2 f}{\partial x^2}\right)_i + \alpha\beta h^2\left(f\frac{\partial^2 f}{\partial x \partial y}\right)_i$$

$$+ \frac{\beta^2 h^2}{2}\left(f\frac{\partial^2 f}{\partial y^2}\right)_i + \ldots \tag{7.8-26b}$$

We have substituted $k_1$ by $hf_i$.

Substituting Equation (7.8-26b) into Equation (7.8-24) yields

$$y_{i+1} = y_i + (a+b)\,h\,f_i + bh^2\left(\alpha\frac{\partial f}{\partial x} + \beta f\frac{\partial f}{\partial y}\right)_i + bh^3\left[\frac{\alpha^2}{2}\frac{\partial^2 f}{\partial x^2} + \alpha\beta f\frac{\partial^2 f}{\partial x\partial y} + \frac{\beta^2}{2}f^2\frac{\partial^2 f}{\partial y^2}\right]_i + \dots$$

$$(7.8\text{-}27)$$

We choose the constants a, b, $\alpha$ and $\beta$ such that Equation (7.8-25e) matches exactly Equation (7.8-27) up to the terms of order $h^2$. By comparing powers of h, we obtain

$$(a+b) = 1 \tag{7.8-28a}$$

$$b\,\alpha = \frac{1}{2} \tag{7.8-28b}$$

$$b\,\beta = \frac{1}{2} \tag{7.8-28c}$$

We have three equations [Equations (7.8-28a to c)] to solve for four unknowns. It is clear that it is impossible to match Equations (7.8-25e, 27) for all f. Thus there is an infinite number of solutions to Equations (7.8-28a to c). We may choose, for example,

$$a = b = \frac{1}{2} \tag{7.8-29a,b}$$

$$\alpha = \beta = 1 \tag{7.8-29c,d}$$

Substituting Equations (7.8-29a to d) into Equation (7.8-24) results in Equation (7.8-23) which is the formula for Heun's method. Note also that when a and b are set to one and zero respectively, we obtain Euler's formula.

Other sets of values of a, b, $\alpha$ and $\beta$ can be chosen. The most widely used formula is not of order two, but of order four, and is written as

$$y_{i+1} = y_i + \frac{1}{6}(k_1 + 2k_2 + 2k_3 + k_4) \tag{7.8-30a}$$

where

$$k_1 = h\,f(x_i, y_i) \tag{7.8.30b}$$

$$k_2 = h\,f(x_i + h/2, \ y_i + k_1/2) \tag{7.8.30c}$$

$$k_3 = h\,f(x_i + h/2, \ y_i + k_2/2) \tag{7.8.30d}$$

$$k_4 = h\,f(x_i + h, \ y_i + k_3) \tag{7.8.30e}$$

Knowing the initial values $(x_0, y_0)$, we can calculate $k_1$, then $k_2$ and successively $k_3$, $k_4$ and $y_1$. We can then determine $y_2$, $y_3$, ... , $y_n$.

***Example 7.8-4***. Solve the initial value problem

$$y' = x^2 + y^2 \qquad\qquad (7.8\text{-}31\text{a})$$

$$y(0) = 0 \qquad\qquad (7.8\text{-}31\text{b})$$

using the fourth order Runge-Kutta method. Take $h$ to be 0.1.

The values of $k_1 - k_4$ are given by

$$k_1 = x_0^2 + y_0^2 = 0 \qquad\qquad (7.8\text{-}31\text{c,d})$$

$$k_2 = 0.1 \left(\frac{0.1}{2}\right)^2 = 0.00025 \qquad\qquad (7.8\text{-}31\text{e,f})$$

$$k_3 = 0.1 \left[\left(\frac{0.1}{2}\right)^2 + \left(\frac{0.0025}{2}\right)^2\right] = 0.00025 \qquad\qquad (7.8\text{-}31\text{g,h})$$

$$k_4 = 0.1 \left[(0.1)^2 + (0.0025)^2\right] = 0.0010 \qquad\qquad (7.8\text{-}31\text{i,j})$$

From Equation (7.8-30a), we have

$$y_1 = y(0.1) = \frac{1}{6}(0.0005 + 0.0005 + 0.001) = 0.00033 \qquad\qquad (7.8\text{-}31\text{k})$$

We now use $(x_1, y_1)$ to determine $k_1, \ldots, k_4$ and then $y_2 \ [= y(0.2)]$ and so on until we have values of $y$ for the required interval.

### Adams-Bashforth method

We divide the interval $[a, b]$ into subintervals $[x_i, x_{i+1}]$. Equation (7.8-6) becomes

$$y_{i+1} = y_i + \int_{x_i}^{x_{i+1}} f(x, y)\, dx \qquad\qquad (7.8\text{-}32)$$

We now approximate $f(x, y)$ by an interpolation formula as discussed in Section 7.6, assuming that somehow we can determine the values of $f$ at some points in the interval. We can then integrate the interpolation polynomial and obtain the **Adams-Bashforth formula**, which can be written as (see Problem 17b)

$$y_{i+1} = y_i + \frac{h}{24}\left[55 f_i - 59 f_{i-1} + 37 f_{i-2} - 9 f_{i-3}\right] \qquad\qquad (7.8\text{-}33)$$

Note that in Equation (7.8-33), we not only need to know $(x_i, y_i)$ to calculate $y_{i+1}$ but also $(x_{i-1}, y_{i-1})$, $(x_{i-2}, y_{i-2})$ and $(x_{i-3}, y_{i-3})$. This type of formula is known as a **multistep formula**. The methods considered previously are **self-starting methods** as they require only the values of $(x_i, y_i)$ to determine $y_{i+1}$. To use the Adams-Bashforth method, we need to use other methods, such as Runge-Kutta methods, to determine $(y_{i-1}, y_{i-2}, y_{i-3})$, prior to applying Equation (7.8-33). This is the disadvantage of this method, but it involves less computation than the Runge-Kutta methods of order four. The Adams-Bashforth method requires only one computation of $f$ per step, once the computation can be started, whereas the Runge-Kutta method of order four requires four evaluations of $f$ at each step.

Several multistep formulas have been proposed [see, for example, Gerald and Wheatley (1994)]. In Problem 17b, we indicate the derivation of the **Adams-Moulton formula**.

A common practice is to start with a Runge-Kutta method and once the values of $f$ at a sufficient number of points have been generated, a multistep methods is used. This is illustrated in Examples 7.8-5 and 7.

**Example 7.8-5.** Solve the initial value problem

$$y' = 2x + y \tag{7.8-34a}$$

$$y(0) = 1 \tag{7.8-34b}$$

using the Adams-Bashforth method. Take $h$ to be $0.1$.

To get the method started, we need to know $f_0, f_1, f_2$ and $f_3$ so as to be able to calculate $y_4$, as can be seen from Equation (7.8-33). We use the Runge-Kutta method of order four to calculate $y_1 - y_3$. The calculations are given in Table 7.8-3.

## TABLE 7.8-3

### Values of $x_i$, $y_i$ using the Runge-Kutta method

| i | $x_i$ | $k_1$ | $k_2$ | $k_3$ | $k_4$ | $y_{i+1}$ |
|---|-------|---------|----------|---------|----------|-----------|
| 0 | 0.0 | 0.10000 | 0.115000 | 0.11575 | 0.131575 | 1.1155 |
| 1 | 0.1 | 0.13155 | 0.148125 | 0.14896 | 0.166450 | 1.2642 |
| 2 | 0.2 | 0.16642 | 0.184740 | 0.18566 | 0.204990 | 1.4496 |
| 3 | 0.3 | 0.20496 | – | – | – | – |

From Equation (7.8-33), we have

$$y_4 = y_3 + \frac{0.1}{24} \left[ 55f_3 - 59f_2 + 37f_1 - 9f_0 \right]$$  (7.8-35)

From Table 7.8-3, we obtain $f_0 - f_3$. From Equation (7.8-30b), we have

$$f(x_i, y_i) = k_1/h$$  (7.8-36)

Equation (7.8-35) becomes

$$y_4 = 1.4496 + \frac{0.1}{24} \left[ 112.728 - 98.1878 + 48.6735 - 9 \right]$$  (7.8-37a)

$$= 1.6754$$  (7.8-37b)

Continuing, we have

$$y_5 = y_4 + \frac{0.1}{24} \left[ 55f_4 - 59f_3 + 37f_2 - 9f_1 \right]$$  (7.8-38a)

$$= 1.6754 + \frac{0.1}{24} \left[ 136.147 - 120.891 + 61.5754 - 11.8395 \right]$$  (7.8-38b)

$$= 1.9462$$  (7.8-38c)

We can similarly proceed to calculate $y_6$, .... Note that to determine $y_5$, we need to evaluate only $f_4$, since $f_1 \ldots f_3$ have previously been evaluated. As stated earlier, at each step, we need to perform only one calculation.

● 

We next extend the methods employed to solve first order equations to higher order equations.

**Higher Order or Systems of First Order Equations**

Any **r**th **order differential equation** can be written as a **system of r first order equations**. For example, the third order differential equation

$$\frac{d^3y}{dx^3} + y \frac{d^2y}{dx^2} + \left( \frac{dy}{dx} \right)^2 + y = g(x)$$  (7.8-39)

can be written as

$$y_1 = y$$

$$y_1' = y_2 \ (= y')$$

$$y_2' = y_3 \ (= y'')$$  (7.8-40a,b,c,d)

$$y_3' = -y_1 y_3 - y_2^2 - y_1 + g(x)$$

Equations (7.8-40b to d) form a system of three first order equations. Instead of considering higher order differential equations, we consider systems of first order equations which are written as

$$y_j' = f_j (x, y_1, \ldots , y_j \ldots), \qquad j = 1, 2, \ldots \qquad (7.8-41)$$

Equation (7.8-41) can be written in vector form as

$$\underline{y}' = \underline{f} (x, \underline{y}) \qquad (7.8-42)$$

Since we are dealing with initial value problems only, all conditions are given at $x = x_0$ and they can be written as

$$\underline{y} (x_0) = \underline{y}_0 \qquad (7.8-43)$$

All the formulas considered earlier can be extended to the present case. The Runge-Kutta method of order four [Equations (7.8-30a to e)] becomes

$$\underline{y}_{i+1} = \underline{y}_i + \frac{1}{6} (\underline{k}_1 + 2\underline{k}_2 + 2\underline{k}_3 + \underline{k}_4) \qquad (7.8-44a)$$

where

$$\underline{k}_1 = \underline{f} (x_i, \underline{y}_i) \qquad (7.8-44b)$$

$$\underline{k}_2 = h \, \underline{f} (x_i + h/2, \underline{y}_i + \underline{k}_1/2) \qquad (7.8-44c)$$

$$\underline{k}_3 = h \, \underline{f} (x_i + h/2, \underline{y}_i + \underline{k}_2/2) \qquad (7.8-44d)$$

$$\underline{k}_4 = h \, \underline{f} (x_i + h, \underline{y}_i + \underline{k}_3) \qquad (7.8-44e)$$

To clarify the notation in Equations (7.8-44a to e), we consider a second order differential equation or equivalently a system of two first order equations.

The system of equations [Equation (7.8-42)] is

$$y_1' = f_1(x, y_1, y_2) \tag{7.8-45a}$$

$$y_2' = f_2(x, y_1, y_2) \tag{7.8-45b}$$

We denote $y_1(x_i)$ and $y_2(x_i)$ by $y_{1,i}$ and $y_{2,i}$ respectively.

Equations (7.8-44a to e) in component form become

$$\begin{bmatrix} y_{1,i+1} \\ y_{2,i+1} \end{bmatrix} = \begin{bmatrix} y_{1,i} \\ y_{2,i} \end{bmatrix} + \frac{1}{6} \begin{bmatrix} k_{11} + 2k_{12} + 2k_{13} + k_{14} \\ k_{21} + 2k_{22} + 2k_{23} + k_{24} \end{bmatrix} \tag{7.8-46a}$$

where

$$k_{11} = h\,f_1\left(x_i, y_{1,i}, y_{2,i}\right); \qquad\qquad k_{21} = h\,f_2\left(x_i, y_{1,i}, y_{2,i}\right) \tag{7.8-46b,c}$$

$$k_{12} = h\,f_1\left(x_i + \frac{h}{2},\, y_{1,i} + \frac{k_{11}}{2},\, y_{2,i} + \frac{k_{21}}{2}\right); \qquad k_{22} = h\,f_2\left(x_i + \frac{h}{2},\, y_{1,i} + \frac{k_{11}}{2},\, y_{2,i} + \frac{k_{21}}{2}\right)$$

$$\tag{7.8-46d,e}$$

$$k_{13} = h\,f_1\left(x_i + \frac{h}{2},\, y_{1,i} + \frac{k_{12}}{2},\, y_{2,i} + \frac{k_{22}}{2}\right); \qquad k_{23} = h\,f_2\left(x_i + \frac{h}{2},\, y_{1,i} + \frac{k_{12}}{2},\, y_{2,i} + \frac{k_{22}}{2}\right)$$

$$\tag{7.8-46f,g}$$

$$k_{14} = h\,f_1\left(x_i + h,\, y_{1,i} + k_{13},\, y_{2,i} + k_{23}\right); \qquad k_{24} = h\,f_2\left(x_i + h,\, y_{1,i} + k_{13},\, y_{2,i} + k_{23}\right)$$

$$\tag{7.8-46h,i}$$

The order of computation is to calculate $k_{j1}$ $(j = 1, 2)$, then $k_{j2}$, $k_{j3}$, $k_{j4}$ and finally $y_{j, i+1}$. The index $j$ in $k_{jm}$ is associated with the $j$ in $f_j$ and the $m$ is associated with the $k_m$ in Equations (7.8-30a to e).

**Example 7.8-6.** Solve the initial value problem

$$y'' - (1 - y^2)\,y' + y = 0 \tag{7.8-47a}$$

$$y(0) = 1, \qquad y'(0) = 0 \tag{7.8-47b,c}$$

We rewrite Equations (7.8-47a to c) as a system of first order equations

$$y_1 = y \tag{7.8-48a}$$

$$y_1' = y_2 \tag{7.8-48b}$$

$$y_2' = \left(1 - y_1^2\right) y_2 - y_1 \tag{7.8-48c}$$

$$y_1(0) \, (= y_{1,0}) = 1 \tag{7.8-48d}$$

$$y_2(0) \, (= y_{2,0}) = 0 \tag{7.8-48e}$$

We choose h to be 0.2 and we compute $k_{j\,m}$, identifying

$$f_1 = y_2, \qquad\qquad f_2 = \left(1 - y_1^2\right) y_2 - y_1 \tag{7.8-49a,b}$$

$$k_{11} = 0.2\,(0) = 0\,; \qquad\qquad k_{21} = 0.2\,[(-1)] = -0.2 \tag{7.8-50a,b}$$

$$k_{12} = 0.2\,(-0.1) = -0.02\,; \qquad k_{22} = 0.2\left[(1-1)\left(-\frac{0.2}{2}\right) - 1\right] = -0.2 \tag{7.8-50c,d}$$

$$k_{13} = 0.2\left(-\frac{0.2}{2}\right) = -0.02\,; \qquad k_{23} = 0.2\left[(1-0.99^2)\,(-0.1) - 0.99\right] = -0.1984 \tag{7.8-50e,f}$$

$$k_{14} = 0.2\,(-0.1984) = -0.03968\,; \qquad k_{24} = 0.2\left[(1-0.98^2)\,(-0.1984) - 0.98\right] = -0.1976 \tag{7.8-50g,h}$$

Substituting Equations (7.8-50a to h) into Equation (7.8-46a) yields

$$\begin{bmatrix} y_{1,1} \\ y_{2,1} \end{bmatrix} = \begin{bmatrix} 1 \\ 0 \end{bmatrix} + \frac{1}{6} \begin{bmatrix} -0.11968 \\ -1.1944 \end{bmatrix} \tag{7.8-51a}$$

$$= \begin{bmatrix} 0.98 \\ -0.199 \end{bmatrix} \tag{7.8-51b}$$

The solution is

$$y(0.2) = 0.98\,, \qquad y'(0.2) = -0.199 \tag{7.8-52a,b}$$

**Example 7.8-7.** Solve the set of equations proposed by Hodgkin and Huxley (1952).

They studied the phenomenon of electric signaling by individual nerve cells and were awarded a Nobel prize for their work. In a state of rest, a nerve cell has a resting potential difference and this potential difference is attributed to the unequal distribution of ions across its membrane,. When activated by a current, the permeability properties of the membrane change resulting in an alteration of the potential difference across the membrane. This change in potential can be measured. Based on experimental observations, Hodgkin and Huxley (1952) assumed that the mechanism that determines ionic fluxes across the membrane can be described by a set of first order equations where the rate constants depend on the voltage. The equations they proposed are

$$\frac{dY_1}{dt} = a_1 (1 - Y_1) - b_1 Y_1 \qquad (7.8\text{-}53a)$$

$$\frac{dY_2}{dt} = a_2 (1 - Y_2) - b_2 Y_2 \qquad (7.5\text{-}53b)$$

$$\frac{dY_3}{dt} = a_3 (1 - Y_3) - b_3 Y_3 \qquad (7.8\text{-}53c)$$

$$C \frac{dY_4}{dt} = -P_{Na} (Y_4 - V_{Na}) - P_K (Y_4 - V_K) - P_L (Y_4 - V_L) \qquad (7.8\text{-}53d)$$

where $C$ is the capacitance, $P_{Na}$, $P_K$, $P_L$, $V_{Na}$, $V_K$, $V_L$ are the permeability coefficients and resting potentials for $Na^+$, $K^+$, and another unspecified ion respectively. $Y_4$ is the potential difference across the membrane, $Y_1$, $Y_2$, $Y_3$ represent the mole fractions of unspecified chemicals that regulate ionic permeability. The coefficients $a_i$ and $b_i$ ($i = 1, 2, 3$) are empirical constants and are functions of $Y_4$. The permeabilities $P_{Na}$ and $P_K$ are assumed to be given by

$$P_{Na} = g_{Na} Y_2^3 Y_3 , \qquad P_K = g_K Y_1^4 \qquad (7.8\text{-}54a,b)$$

where $g_{Na}$ and $g_K$ are the conductance for $Na^+$ and $K^+$ respectively. For computational purposes, we have set

$$g_{Na} = 120 , \qquad g_K = 36 , \qquad P_L = 0.6 , \qquad V_{Na} = -115 , \qquad V_K = 12 , \qquad V_L = -10.6 ,$$

$$a_1 = 0.01 \, (Y_4 + 10) / [\exp (0.1 Y_4 + 1) - 1] , \qquad a_2 = 0.1 \, (Y_4 + 25) / [\exp (0.1 Y_4 + 2.5) - 1] ,$$

$$a_3 = 0.07 \exp (Y_4 / 20) , \qquad b_1 = 0.125 \exp (Y_4 / 80) , \qquad b_2 = 4 \exp (Y_4 / 18) ,$$

$$b_3 = 1 / [\exp (0.1 Y_4 + 8) + 1] .$$

The initial conditions are assumed to be

$$Y_1 = 0.318 , \qquad Y_2 = 0.053 \qquad (7.8\text{-}55a,b)$$

$$Y_3 = 0.596 , \qquad Y_4 = -6.507 \qquad (7.8\text{-}55c,d)$$

Equations (7.8-53a to d) are solved numerically. A Runge-Kutta formula is used to generate values of $Y_i$ at a sufficient number of points. A multistep formula (Adams-Moulton) is then applied to obtain values of $Y_i$ up to $t = 20$. The following FORTRAN program is used to compute $Y_i$. The results are shown in Figure 7.8-2.

```
C       PROGRAM TO SOLVE EXAMPLE 7.8-7

        REAL Y1,Y2,Y3,Y4,Y1K,Y2K,Y3K,Y4K,Y1P,Y2P,Y3P,Y4P
        REAL YIPZ,Y2PZ,Y3PZ,Y4PZ
        REAL D11,D12,D13,D14,D21,D22,D23,D24,D31,D32,D33,D34,D41,D42,D43,D44
        REAL T,R,TAU
        INTEGER I,J,K
        OPEN(UNIT=1,FILE='A:/RESULT.DAT',STATUS='NEW')

C       INITIALIZATION OF THE VARIABLES

        Y1=0.318
        Y2=0.053
        Y3=0.596
        Y4=-6.507
        TAU=0.0005
        T=0.0
        WRITE(6,*)T,Y4
        I=1

C       STARTUP WITH RUNGE-KUTTA 44

        D11=TAU*F1(Y1,Y4)
        D12=TAU*F2(Y2,Y4)
        D13=TAU*F3(Y3,Y4)
        D14=TAU*F4(Y1,Y2,Y3,Y4)

        D21=TAU*F1(Y1+D11/2,Y4+D14/2)
        D22=TAU*F2(Y2+D12/2,Y4+D14/2)
        D23=TAU*F3(Y3+D13/2,Y4+D14/2)
        D24=TAU*F4(Y1+D11/2,Y2+D12/2,Y3+D13/2,Y4+D14/2)

        D31=TAU*F1(Y1+D21/2,Y4+D24/2)
        D32=TAU*F2(Y2+D22/2,Y4+D24/2)
        D33=TAU*F3(Y3+D23/2,Y4+D24/2)
        D34=TAU*F4(Y1+D21/2,Y2+D22/2,Y3+D23/2,Y4+D24/2)

        D41=TAU*F1(Y1+D31,Y34+D34)
        D42=TAU*F2(Y2+D32,Y4+D34)
        D43=TAU*F3(Y3+D33,Y4+D34)
        D44=TAU*F4(Y1+D31,Y2+D32,Y3+D33,Y4+D34)
```

```
        Y1K=Y1+(D11+2*D21+2*D31+D41)/6
        Y2K=Y2+(D12+2*D22+2*D32+D42)/6
        Y3K=Y3+(D13+2*D23+2*D33+D43)/6
        Y4K=Y4+(D14+2*D24+2*D34+D44)/6

C       ADAMS-MOULTON

        U=2*TAU

        DO 10 T=U,20,TAU

        Y1P=Y1K
        Y2P=Y2K
        Y3P=Y3K
        Y4P=Y4K

        R=1.0
        WHILE(R.GT.1E–5)

        Y1PZ=Y1K+TAU/12*(5*F1(Y1P,Y4P)+8*F1(Y1K,Y4K)–F1(Y1,Y4))
        Y2PZ=Y2K+TAU/12*(5*F2(Y2P,Y4P)+8*F2(Y2K,Y4K)–F2(Y2,Y4))
        Y3PZ=Y3K+TAU/12*(5*F3(Y3P,Y4P)+8*F3(Y3K,Y4K)–F3(Y3,Y4))
        Y4PZ=Y4K+TAU/12*(5*F4(Y1P,Y2P,Y3P,Y4P)+8*F4(Y1K,Y2K,Y3K,Y4K)–
      EF4(Y1,Y2,Y3,Y4))

        R=(Y1PZ–Y1P)**2+(Y2PZ–Y2P)**2+(Y3PZ–Y3P)**2+(Y4PZ–Y4P)**2
        R=SQRT(R)

C       WRITE(6,1000)Y1PZ,Y2PZ,Y3PZ,Y4PZ
C       WRITE(6,1000)Y1P,Y2P,Y3P,Y4P

1000    FORMAT(F9.6,3X,F9.6,3X,F9.6,3X,F9.6)

        Y1P=Y1PZ
        Y2P=Y2PZ
        Y3P=Y3PZ
        Y4P=Y4PZ

        ENDWHILE

        J=I/10
        IF(J.EQ.1)THEN
        WRITE(1,*)T,Y4P
        I=0
        ENDIF
        I=I+1
```

```
        Y1=Y1K
        Y2=Y2K
        Y3=Y3K
        Y4=Y4K

        Y1K=Y1P
        Y2K=Y2P
        Y3K=Y3P
        Y4K=Y4P

10      CONTINUE
        STOP
        END

        FUNCTION F1(X,Y)
        REAL A1,B1,X,Y
        A1=0.01*(Y+10)/(EXP(0.1*Y+1)–1)
        B1=0.125*EXP(Y/80)
        F1=A1*(1–X)–B1*X
        RETURN
        END

        FUNCTION F2(X,Y)
        REAL A2,B2,X,Y
        A2=0.1*(Y+25)/(EXP(0.1*Y+2.5)–1)
        B2=4*EXP(Y/18)
        F2=A2*(1–X)–B2*X
        RETURN
        END

        FUNCTION F3(X,Y)
        REAL A3,B3,X,Y
        A3=0.07*EXP(Y/20)
        B3=1/(EXP(0.1*Y+3)+1)
        F3=A3*(1–X)–B3*X
        RETURN
        END

        FUNCTION F4(W,X,Y,Z)
        REAL W, X, Y,Z,
        F4=–36*W**4*(Z–12)–0.3*(Z+10.6)–120*X**3*Y*(Z+115)
        RETURN
        END
```

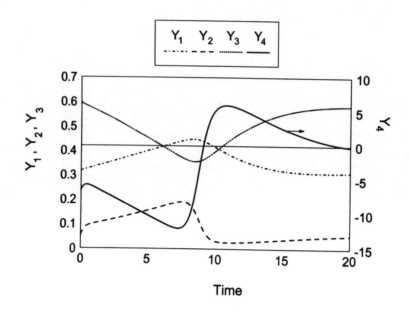

**FIGURE 7.8-2    Plot of $Y_1$, $Y_2$, $Y_3$ and $Y_4$ versus time**

***Example 7.8-8.*** Compute the shape of a drop in an extensional flow of a viscoelastic fluid.

Gonzalez-Nunez et al. (1996) studied the deformation of nylon drops in an extensional flow of a polyethylene melt. The drop is assumed to be spherical of radius $R$ prior to entering the extensional flow. Choosing the centre of the drop as the origin of a spherical polar coordinate system $(r, \theta, \phi)$, the radial distance $r$ of any point on the surface of the deformed drop in the extensional flow can be written as

$$r = R + \xi \tag{7.8-56}$$

We assume the velocity field to be continuous across the boundary and a force balance yields

$$[\underline{\underline{\pi}}_m - (\eta_d / \eta_m)\, \underline{\underline{\pi}}_d] \cdot \underline{n} = k\, Ca\, \underline{n} \tag{7.8-57}$$

where $\underline{\underline{\pi}}_m$, $\underline{\underline{\pi}}_d$ are the stress fields associated with the matrix (polyethylene) and the dispersed phase (nylon) respectively, $\eta_m$, $\eta_d$ are the viscosities of the matrix and dispersed phase respectively, $\underline{n}$ is the unit outward normal to the surface, $k$ is the surface curvature and $Ca$ is the capillary number.

For small deformations, $k$ is approximately given by

$$k = \frac{2}{R} - \frac{2\xi}{R^2} - \frac{1}{R^2}\left[\frac{1}{\sin\theta}\frac{d}{d\theta}\left(\sin\theta\,\frac{d\xi}{d\theta}\right)\right] \tag{7.8-58}$$

The stress fields are calculated by assuming that the polymer melts can be modeled as upper convected Maxwell fluids (Carreau et al., 1997). Substituting the values of $\underline{\underline{\pi}}_m$, $\underline{\underline{\pi}}_d$, and $k$, we obtain the following equation for $\xi$

$$\frac{d^2\xi}{d\theta^2} + \cot\theta \, \frac{d\xi}{d\theta} + 2\xi - 2R - R^2 \, Ca \, (1 - \eta_d/\eta_m) \, T_0 \exp\left[(T_0/L)\,(R+\xi)\,\cos\theta\right] = 0 \quad (7.8\text{-}59)$$

Assuming the deformed drop is symmetrical, the two appropriate boundary conditions are

$$\xi = R_0, \qquad \frac{d\xi}{d\theta} = 0, \qquad \text{at } \theta = \pi/2 \qquad\qquad (7.8\text{-}60a,b)$$

Equations (7.8-59, 60a, b) are solved numerically using a Runge-Kutta formula. Having found $\xi$, the major ($L'$) and the minor ($B'$) axes of the drop can be calculated. The deformation $D'$ is given by

$$D' = (L' - B') / (L' + B') \qquad\qquad\qquad (7.8\text{-}61)$$

Figure 7.8-3 illustrates $D'$ as a function of $Ca$ for various values of $R$.

```
C       PROGRAM TO SOLVE EXAMPLE 7.8-8 USING RK22

        IMPLICIT NONE
        INTEGER N,I,J
        REAL*8 PHI1(5000),PHI2(5000),Z1(5000),Z2(5000),RR(5000),TE(5000)
        REAL*8 TAU,TAU2,TEST1,TEST2,FUNC2,TEMP,X,XK,TO,R,CA,D,R0
        COMMON R,XK,TO

C       DATA

        R=3.35D0
        DO 1 J=1,10
        CA=0.1D0*J
        TO=2.967D0
        XK=R*R*0.01D0*CA*TO
        R0=1.D0/(4.D0*CA)

C       INITIAL VALUES

        PHI1(1)=R0-R
        PHI2(1)=0.D0
        TAU=0.00157
        TAU2=TAU/2.D0
        N=1000
        X=1.57D0
```

```
C        START INTEGRATION

         DO 99 I=1,N+1
         TEST1=PHI1(I)+TAU*PHI2(I)
         TEMP=FUNC2(PHI1(I),PHI2(I),X)
         TEST2=PHI2(I)+TAU*TEMP
         PHI1(I+1)=PHI1(I)+TAU2*(PHI2(I)+TEST2)
         PHI2(I+1)=PHI2(I)+TAU2*(TEMP+FUNC2(TEST1,TEST2,X))
         IF(ABS(PHI1(I+1)).1E.1.0D-4)GO TO 100
         RR(I)=R+CA*PHI1(I)
         TE(I)=X
         Z1(I)=RR(I)*COS(X)
         Z2(I)=RR(I)*SIN(X)
         X=X-TAU
99       CONTINUE

C        RESULTS

100      WRITE(6,*)R,X
         DO 110 I=1,N+1,100
         D=(Z1(I)-R0)/(Z1(I)+R0)
110      CONTINUE
         WRITE(6,*) CA,D
1        CONTINUE
         STOP
         END

C        FUNCTION

         REAL*8 FUNCTION FUNC2(Y1,Y2,X)
         IMPLICIT NONE
         RERAL*8 Y1,Y2,X,C1,C2,C3
         COMMON C1,C2,C3
         FUNC2=0.D0
         FUNC2=2*C1-(COS(X)/SIN(X))*Y2-2*Y1-C2*EXP(C3*(C1+Y1)*
       1 COS(X)/24.D0)
         RETURN
         END
```

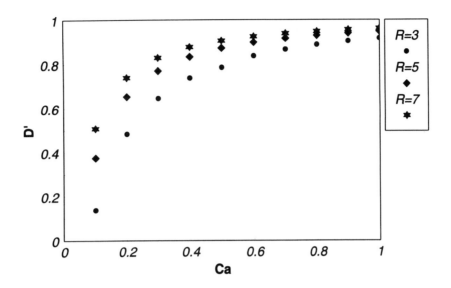

**FIGURE 7.8-3**   **Deformation of D' versus capillary number Ca**
                   **for various values of radius R**

## 7.9   BOUNDARY VALUE PROBLEMS

**Boundary value problems** have been discussed in Chapter 2. Unlike **initial value problems**, the conditions here are given at more than one point and the formulas given in Section 7.8 cannot be applied. We described two methods which can be used to solve boundary value problems.

**Shooting Method**

This method is widely used. It consists of transforming the boundary value problem to an initial value problem.

We consider a second order linear equation

$$\frac{d^2 y}{dx^2} + p(x) \frac{dy}{dx} + q(x) y = r(x) , \qquad a \leq x \leq b \qquad (7.9\text{-}1a)$$

$$y(a) = A , \qquad y(b) = B \qquad (7.9\text{-}1b,c)$$

We convert the boundary value problem to an initial value problem by dropping the boundary condition given by Equation (7.9-1c) and replacing it by imposing a condition at $x = a$. We assume that

$$y'(a) = \alpha_1 \tag{7.9-2}$$

Equations (7.9-1a, b, 2) form an initial value problem and we can integrate the differential equation using one of the formulas given in the previous section. Since $\alpha_1$ is a guessed value of $y'(a)$, it is unlikely that the boundary condition given by Equation (7.9-1c) will be satisfied. We now guess another value of $y'(a)$ (say $\alpha_2$) and we proceed with the integration. Unless we are lucky, Equation (7.9-1c) will still not be satisfied.

In the linear case, from these two values of $\alpha$ ($\alpha_1$ and $\alpha_2$) it is possible to obtain the exact value of $y'(a)$. Let $y_1$ be the solution obtained by using $\alpha_1$ and $y_2$ be the solution with $\alpha_2$, as illustrated in Figure 7.9-1. Since the system is linear, a linear combination of $y_1$ and $y_2$ is also a solution of Equation (7.9-1a). That is to say

$$y = c_1 y_1 + c_2 y_2 \tag{7.9-3}$$

where $c_1$ and $c_2$ are constants, is a solution of Equation (7.9-1a). Both $y_1$ and $y_2$ satisfy Equation (7.9-1b) and $y$ must also satisfy this condition. Combining Equations (7.9-1b, 3) yields

$$1 = c_1 + c_2 \tag{7.9-4}$$

Imposing condition (7.9-1c), we have

$$B = c_1 y_1(b) + c_2 y_2(b) \tag{7.9-5}$$

From Equations (7.9-4, 5), we obtain

$$c_1 = \frac{B - y_2(b)}{y_1(b) - y_2(b)}, \qquad c_2 = \frac{y_1(b) - B}{y_1(b) - y_2(b)} \tag{7.9-6a,b}$$

Differentiating $y$ in Equation (7.9-3) and using Equations (7.9-6a, b), we obtain

$$y'(a) = y_1'(a) + \frac{[B - y_1(b)]}{y_2(b) - y_1(b)} \left[ y_2'(a) - y_1'(a) \right] \tag{7.9-7}$$

In Equation (7.9-7), all the terms on the right side are known, and the value of $y'(a)$ that will satisfy Equation (7.9-1c) is known. With $y'(a)$ given by Equation (7.9-7), we compute the solution $y$ that satisfies Equations (7.9-1a to c).

The above method is applicable only for linear equations. For non-linear equations, we cannot apply the principle of superposition and Equation (7.9-3) is not valid. However, we can compute $y_1$ and $y_2$ as in the linear case and we use Equation (7.9-7) to obtain an improved value of $y'(a)$ which we denote as $y_3'(a)$. Note that $y_3'(a)$ is not the exact value of $y'(a)$. Using $y_3'(a)$, we can compute $y_3$

in the same way as $y_1$ and $y_2$ were computed. We repeat this process until successive solutions $y_i$ and $y_{i+1}$ are within the required degree of accuracy.

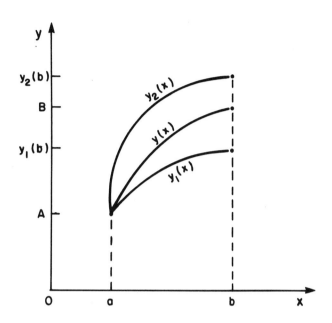

**FIGURE 7.9-1      Shooting  method**

On generalizing Equation (7.9-7), we obtain

$$y'_{i+2}(a) = y'_i(a) + \frac{[B - y_i(b)]}{y_{i+1}(b) - y_i(b)} \left[y'_{i+1}(a) - y'_i(a)\right], \qquad i = 1, 2, \dots \qquad (7.9\text{-}8)$$

Equation (7.9-8) can also be obtained as follows. The solution $y$ is a function of $y'(a) (= \alpha)$ and the solution can be written as $y(\alpha, x)$. We need to determine $\alpha$ such that Equation (7.9-1c) is satisfied. Equation (7.9-1c) can be written as

$$g(\alpha) = y(\alpha, 1) - B = 0 \qquad\qquad (7.9\text{-}9a,b)$$

Equations (7.9-9a, b) can be solved by the secant method [Equation (7.2-6)] and the result is Equation (7.9-8).

***Example 7.9-1.*** Solve the boundary value problem

$$y'' = x + (1 - x/5)\, y \qquad\qquad (7.9\text{-}10a)$$

$$y(1) = 2, \qquad y(3) = -1 \qquad\qquad (7.9\text{-}10b,c)$$

We start the process of integration with the guessed value of $y'(1)$ $(= \alpha_1)$ to be $-1.5$. The resulting $y_1(-1.5, 3)$ is 4.811. Next, we take $y'(1)$ $(= \alpha_2)$ to be $-3.0$ and we obtain $y_1(-3, 3)$ which is 0.453. Then using Equation (7.9-7), we obtain the value of $y'(1)$ which is $-3.5$, and $y(-3.5, 3)$ satisfies Equation (7.9-10c). Values of $y(\alpha, x)$ for various values of $\alpha$ and $x$ are given in Table 7.9-1. Equation (7.9-10a) is written as a set of first order equations and the integration formula used is that of Heun [Equation (7.8-23)].

### TABLE 7.9-1

### Values of $y(\alpha, x)$

| x | $y(-1.5, x)$ | $y(-3.0, x)$ | $y(-3.5, x)$ |
|---|---|---|---|
| 1.0 | 2.000 | 2.000 | 2.000 |
| 1.2 | 1.751 | 1.499 | 1.348 |
| 1.4 | 1.605 | 0.991 | 0.787 |
| 1.6 | 1.561 | 0.619 | 0.305 |
| 1.8 | 1.625 | 0.328 | $-0.104$ |
| 2.0 | 1.803 | 0.118 | $-0.443$ |
| 2.2 | 2.105 | $-0.007$ | $-0.712$ |
| 2.4 | 2.542 | $-0.045$ | $-0.908$ |
| 2.6 | 3.128 | 0.013 | $-1.026$ |
| 2.8 | 3.880 | 0.175 | $-1.000$ |
| 3.0 | 4.811 | 0.453 | $-1.000$ |

●

Equations (7.9-9a, b) can also be solved by Newton's method. Combining Equations (7.2-11, 9-9a, b), we obtain

$$\alpha_{k+1} = \alpha_k - \frac{g(\alpha_k)}{g'(\alpha_k)} \tag{7.9-11}$$

where the prime denotes differentiation with respect to $\alpha$.

To apply Equation (7.9-11), we need to determine $g'(\alpha_k)$ and since B is constant, $g'(\alpha_k)$ is given, from Equation (7.9-9a), by

$$g'(\alpha_k) = y'(\alpha_k, 1) \tag{7.9-12}$$

To obtain $g'(\alpha_k)$, we need to solve an initial value problem. We demonstrate this by considering the boundary value problem given by

$$\frac{d^2y}{dx^2} = f\left(x, y, \frac{dy}{dx}\right) \tag{7.9-13a}$$

$$y(a) = A, \qquad y(b) = B \tag{7.9-13b,c}$$

We replace Equation (7.9-13c) by the initial condition

$$\frac{dy}{dx}\bigg|_a = \alpha \tag{7.9-13d}$$

Equations (7.9-13a, b, d) define an initial value problem. We need to obtain a good approximation to $\alpha$ via Equation (7.9-11), which in turn implies finding $g'(\alpha_k)$. We denote $\frac{\partial y}{\partial \alpha}$ by $\xi(\alpha, x)$. Then on differentiating $\xi$ with respect to $x$ and interchanging the order of differentiation, we obtain

$$\xi'(\alpha, x) = \frac{\partial}{\partial \alpha}\left[y'(\alpha, x)\right] \tag{7.9-14}$$

The prime denotes differentiation with respect to $x$ and this notation will be kept from now on.

Differentiating Equation (7.9-13a) with respect to $\alpha$ and using Equation (7.9-14) yields

$$\xi''(\alpha, x) = \frac{\partial f}{\partial y}\frac{\partial y}{\partial \alpha} + \frac{\partial f}{\partial y'}\frac{\partial y'}{\partial \alpha} \tag{7.9-15a}$$

$$= \frac{\partial f}{\partial y}\xi + \frac{\partial f}{\partial y'}\xi' \tag{7.9-15b}$$

Similarly differentiating Equations (7.9-13b, d) with respect to $\alpha$ yields

$$\xi(\alpha, a) = 0 \tag{7.9-16a}$$

$$\xi'(\alpha, a) = 1 \tag{7.9-16b}$$

The initial value system defined by Equations (7.9-15b, 16a, b) can be solved and $\xi(\alpha, b)$ is obtained. From Equation (7.9-12), we note that $g'(\alpha)$ is $\xi(\alpha, b)$ and then $g'(\alpha)$ is obtained. This can be substituted in Equation (7.9-11) to obtain an improved value of $\alpha$. The iteration is repeated until two iterates yield values within the desired degree of accuracy.

Equations (7.9-15b, 16a, b) are the **associated variational equations** which need to be solved simultaneously with Equations (7.9-13a, b, d).

***Example 7.9-2***. Use the shooting method to solve the non-linear boundary value problem

$$y" = \frac{2y'^2}{y} - y, \quad -1 \le x \le 1 \tag{7.9-17a}$$

$$y(-1) = \left(e + e^{-1}\right)^{-1} \tag{7.9-17b}$$

$$y(1) = \left(e + e^{-1}\right)^{-1} \tag{7.9-17c}$$

The associated variational equations are

$$\xi" = \left[\frac{\partial}{\partial y}\left(\frac{2y'^2}{y} - y\right)\right]\xi + \left[\frac{\partial}{\partial y'}\left(\frac{2y'^2}{y} - y\right)\right]\xi' \tag{7.9-18a}$$

$$= -\left[2\left(\frac{y'}{y}\right)^2 + 1\right]\xi + \frac{4y'}{y}\xi' \tag{7.9-18b}$$

$$\xi(-1) = 0 \tag{7.9-18c}$$

$$\xi'(-1) = 1 \tag{7.9-18d}$$

We denote $y$ by $y_1$ and $\xi$ by $y_3$. Equations (7.9-17a, 18a) can now be written as a system of four first order equations as follows

$$y_1' = y_2 \tag{7.9-19a}$$

$$y_2' = \frac{2y_2^2}{y_1} - y_1 \tag{7.9-19b}$$

$$y_3' = y_4 \tag{7.9-19c}$$

$$y_4' = -\left[2\left(\frac{y_2}{y_1}\right)^2 + 1\right]y_3 + 4\left(\frac{y_2}{y_1}\right)y_4 \tag{7.8-19d}$$

Equation (7.9-17c) has to be replaced by the initial condition

$$y'(-1) = \alpha \tag{7.9-20}$$

Equations (7.9-17c, 18c, 18d, 20) are written as

$$y_1(-1) = \left(e + e^{-1}\right)^{-1} \tag{7.9-21a}$$

$$y_2(-1) = \alpha \qquad\qquad\qquad (7.9\text{-}21b)$$

$$y_3(-1) = 0 \qquad\qquad\qquad (7.9\text{-}21c)$$

$$y_4(-1) = 1 \qquad\qquad\qquad (7.9\text{-}21d)$$

The initial value problem given by Equations (7.9-19a to d, 21 a to d) can be solved using the Runge-Kutta method with an initial value of $\alpha\ (=\alpha_0)$. The value of $\alpha_0$ chosen is 0.2 and after six iterations the value of $y(1)$ was found to converge to 0.324, which is close enough to the value given by Equation (7.9-17c).

●

A method widely used to solve boundary value problems is the method of finite differences which is considered next.

## Finite Difference Method

In transforming a two-point boundary value problem to an initial value problem, we have replaced the boundary condition(s) at one end by guessed initial condition(s) at the other end. If the order of the differential equation is high, the number of initial conditions to be guessed is also high. The shooting method can then be very laborious. Further, if we did not make a good guess of the initial conditions, the convergence might be slow.

An alternative numerical method of solving boundary value problems is the **method of finite differences**. In this method, we replace the **derivatives** by **finite differences**.

We again consider the boundary value problem given by Equations (7.9-1a to c). We divide the interval [a, b] into (n + 1) equal intervals, each of length h. We denote a by $x_0$ and b by $x_{n+1}$. We use the central difference scheme and the derivatives at $x_j\ (= x_0 + jh)$ are given by

$$\left.\frac{dy}{dx}\right|_{x_j} = \frac{y(x_{j+1}) - y(x_{j-1})}{2h} = \frac{y_{j+1} - y_{j-1}}{2h} \qquad\qquad (7.9\text{-}22a,b)$$

$$\left.\frac{d^2y}{dx^2}\right|_{x_j} = \frac{y_{j+1} - 2y_j + y_{j-1}}{h^2} \qquad\qquad (7.9\text{-}23)$$

Substituting Equations (7.9-22b, 23) into Equation (7.9-1a), we obtain

$$\frac{1}{h^2}\left[y_{j+1} - 2y_j + y_{j-1}\right] + \frac{1}{2h}\left[p(x_j)\,(y_{j+1} - y_{j-1})\right] + q(x_j)\,y_j = r(x_j) \qquad (7.9\text{-}24)$$

On multiplying throughout by $h^2$ and collecting like terms together, Equation (7.9-24) becomes

$$a_j y_{j-1} + b_j y_j + c_j y_{j+1} = h^2 r(x_j) \qquad (7.9\text{-}25a)$$

$$a_j = 1 - \frac{h}{2} p(x_j) \qquad (7.9\text{-}25b)$$

$$b_j = -2 + h^2 q(x_j) \qquad (7.9\text{-}25c)$$

$$c_j = 1 + \frac{h}{2} p(x_j) \qquad (7.9\text{-}25d)$$

The boundary conditions (Equations 7.9-1b, c) are now written as

$$y_0 = A \qquad (7.9\text{-}26a)$$

$$y_{n+1} = B \qquad (7.9\text{-}26b)$$

Substituting Equations (7.9-26a, b) into Equation (7.9-25a), and writing out the first two and the last two lines of the latter, we have

$$a_1 A + b_1 y_1 + c_1 y_2 = h^2 r(x_1) \qquad (7.9\text{-}27a)$$

$$a_2 y_1 + b_2 y_2 + c_2 y_3 = h^2 r(x_2) \qquad (7.9\text{-}27b)$$

$$\vdots$$

$$a_{n-1} y_{n-2} + b_{n-1} y_{n-1} + c_{n-1} y_n = h^2 r(x_{n-1}) \qquad (7.9\text{-}27c)$$

$$a_n y_{n-1} + b_n y_n + c_n B = h^2 r(x_n) \qquad (7.9\text{-}27d)$$

Equation (7.9-27a to d) can be written in matrix form as

$$\underline{\underline{A}} \, \underline{y} = \underline{s} \qquad (7.9\text{-}28a)$$

$$\underline{y} = \begin{bmatrix} y_1 \\ \vdots \\ y_n \end{bmatrix}, \qquad \underline{s} = \begin{bmatrix} s_1 \\ \vdots \\ s_n \end{bmatrix} = h^2 \begin{bmatrix} r(x_1) \\ \vdots \\ r(x_n) \end{bmatrix} - \begin{bmatrix} a_1 A \\ 0 \\ \vdots \\ 0 \\ c_n B \end{bmatrix} \qquad (7.9\text{-}28b,c,d)$$

$$\underline{\underline{A}} = \begin{bmatrix} b_1 & c_1 & 0 & 0 & \cdots & 0 & 0 & 0 \\ a_2 & b_2 & c_2 & 0 & \cdots & 0 & 0 & 0 \\ 0 & a_3 & b_3 & c_3 & \cdots & 0 & 0 & 0 \\ \vdots & \vdots & \vdots & \vdots & \cdots & \vdots & \vdots & \vdots \\ 0 & 0 & 0 & 0 & \cdots & a_{n-1} & b_{n-1} & c_{n-1} \\ 0 & 0 & 0 & 0 & \cdots & 0 & a_n & b_n \end{bmatrix}$$

(7.9-28e)

The matrix $\underline{\underline{A}}$ is a tridiagonal matrix whose elements are known.

Equation (7.9-28a) can be solved by the method of Gaussian elimination (discussed in Section 7.4). The unknown values of y at intermediate points $(y_1, \ldots, y_n)$ can thus be determined.

**Example 7.9-3.** Solve the boundary value problem

$$\frac{d^2y}{dx^2} - y = 0$$

(7.9-29a)

$$y(0) = 0, \qquad y(1) = \sinh 1$$

(7.9-29b,c)

by the method of finite difference. Choose h to be 0.25.

The exact solution of the boundary value problem is

$$y = \sinh x$$

(7.9-30)

The finite difference equation [Equation (7.9-25a)] is

$$y_{j-1} - (2 + h^2) y_j + y_{j+1} = 0$$

(7.9-31)

Since h is 0.25, we divide the interval into four equal intervals and label the points $x_0 (= 0)$, $x_1$, $x_2$, $x_3$ and $x_4 (= 1)$.

In matrix form [Equation (7.9-28a)], Equation (7.9-31) can be written as

$$\begin{bmatrix} -(2+h^2) & 1 & 0 \\ 1 & -(2+h^2) & 1 \\ 0 & 1 & -(2+h^2) \end{bmatrix} \begin{bmatrix} y_1 \\ y_2 \\ y_3 \end{bmatrix} = \begin{bmatrix} 0 \\ 0 \\ -\sinh 1 \end{bmatrix} \tag{7.9-32}$$

Solving Equation (7.9-32), we obtain $y_1$, $y_2$ and $y_3$ and their values are given in Table 7.9-2 together with the values of the exact solution.

### TABLE 7.9-2

### Values of $y_i$

| $x_i$ | Numerical solution | Exact solution |
|---|---|---|
| 0.25 | 0.2528 | 0.2526 |
| 0.50 | 0.5214 | 0.5211 |
| 0.75 | 0.8226 | 0.8223 |

For non-linear equations, Equation (7.9-28a) will be non-linear. Thus by discretizing Equation (7.9-13a), we obtain, using Equations (7.9-22b, 23),

$$y_{j+1} - 2y_j + y_{j-1} = h^2 f\left(x_j, y_j, \frac{y_{j+1} - y_{j-1}}{2h}\right) \tag{7.9-33}$$

The boundary conditions are given by Equations (7.9-26a, b). For a given $f$, Equation (7.9-33) can be written out for each value of $j$ ($j = 1, 2, \dots, n$) resulting in a set of non-linear algebraic (or transcendental) equations. This set of equations are solved by iteration techniques, some of which are described in Sections 7.2 and 7.4.

Nowadays numerical software is available which implements the formulas given in this chapter. However, we still have to use our judgment in choosing the appropriate method. This is illustrated in the next example.

***Example 7.9-4***. Spence et al. (1993) modeled a catalytic combustion in a monolith reactor. Catalytic monoliths are widely used in the automobile industry to control the emissions from the vehicle exhaust systems. The monolith usually consists of a number of cells through which the exhaust gases flow. We consider only a single cell and a schematic diagram of such a cell is shown in Figure 7.9-2.

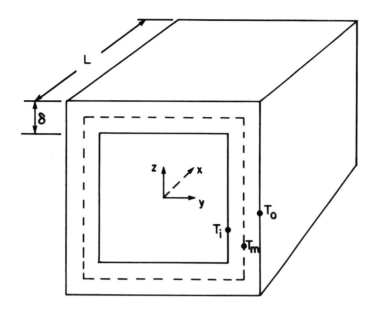

**FIGURE 7.9-2    Schematic diagram of a cell**

The reaction considered is the combustion of propane and is given by

$$C_3H_8 + 5\,O_2 \;\longrightarrow\; 3\,CO_2 + 4\,H_2O \tag{7.9-34}$$

The heterogeneous reaction rate $r_c$ at the catalytic surface is assumed to be

$$r_c = Z_c \exp\left[-E_c/R\,T_i\,(x)\right] C_w\,(x) \tag{7.9-35}$$

where $Z_c$ is a pre-exponential factor, $E_c$ is the activation energy, $R$ is the ideal gas constant, $C_w(x)$ is the fuel concentration at the wall, and $T_i$ is the reacting wall temperature as shown in Figure 7.9-2.

For simplicity, a one-dimensional model is considered. This implies that all variables are functions of x only and only a cross-section of a single cell is considered for deriving the basic equations. Using a series of energy and mass balances, Spence et al. (1993) deduced that the equations to be solved are

$$\frac{1}{P_e J_k} \frac{d^2 y}{d\xi^2} = 2y\,(\xi) - w\,(\xi) - \theta_0\,(\xi)\,, \quad 0 < \xi < 1 \tag{7.9-36a}$$

$$\frac{dx_2}{d\xi} = J_H\big[w\,(\xi) - x_2\,(\xi)\big] \tag{7.9-36b}$$

$$\frac{dx_1}{d\xi} = -J_D\left[x_1(\xi) - \alpha_w(\xi)\right] \tag{7.9-36c}$$

$$0 = J_H\left[x_2(\xi) - w(\xi)\right] + J_k\left[y(\xi) - w(\xi)\right] - \Gamma D_c \exp\left[\gamma_c(w-1)/w\right]\alpha_w \tag{7.9-36d}$$

$$\alpha_w = x_1\left\{1 + (D_c/J_D)\exp\left[\gamma_c(w-1)/w\right]\right\}^{-1} \tag{7.9-36e}$$

where $\xi = x/L$, $L$ is the length of the cell, $y = T_m(x)/T_{B0}$, $x_2 = T_B(x)/T_{B0}$, $w = T_i(x)/T_{B0}$, $\theta_0 = T_0/T_{B0}$, $x_1 = C_B(x)/C_0$, $T_0$, $T_m$, $T_i$ are the temperatures at the outer wall, the mean solid wall temperature, and the temperature at the inner wall, as shown in Figure 7.9-2. $T_B(x)$ is the bulk gas temperature and $T_{B0}$ is $T_B(0)$. $C_B(x)$ is the fuel concentration in the bulk gas and $C_0$ is $C_B(0)$. $P_e$ is the Péclet number, $J_k$, $J_H$, $J_D$, $\Gamma$, $D_c$, and $\gamma_c$ are dimensionless parameters.

The associated boundary conditions are

$$\left.\frac{dy}{d\xi}\right|_{\xi=0} = \left.\frac{dy}{d\xi}\right|_{\xi=1} = 0 \tag{7.9-37a,b}$$

$$x_1(0) = x_2(0) = 1 \tag{7.9-37c,d}$$

Note that Equation (7.9-36d) is an algebraic equation and allows us to determine $w(0)$. Equation (7.9-36a) is written as a system of first order equation as follows

$$\frac{dy_1}{d\xi} = y_2 \tag{7.9-38a}$$

$$\frac{dy_2}{d\xi} = P_e J_k\left\{2y_1 - w - \theta_0\right\} \tag{7.9-38b}$$

The boundary value problem defined by Equations (7.9-36b to 38b) can be solved by the shooting method. As an initial guess, $y_1(0)$ is assumed to be

$$y_1(0) = [w(0) + \theta_0(0)]/2 \tag{7.9-39}$$

The shooting method is found to work only for $P_e < 1$ and $P_e > 500$. The reasons for the failure are as follows. The outside wall temperature $(\theta_0)$ is given and Equations (7.9-36a, 37a, b) form a standard boundary value problem for a known $w(x)$. The complementary function of Equation (7.9-36a) is a linear combination of $\exp\left[\pm\xi\sqrt{2P_e J_k}\right]$ and, in the shooting method, round off errors propagate as multiples of these exponentials. For small values of $P_e$, the errors remain small and the method works. For large values of $P_e$, the left side of Equation (7.9-36a) is almost zero and the starting value of $y_1(0)$ given by Equation (7.9-39) is almost exact and the method works, though unreliable. Spence et al. (1993) have used alternative methods to overcome this instability problem. In the next section, we discuss the problem of instability in more details.

## 7.10 STABILITY

In Example 7.9-4, we have seen that due to round-off errors, the shooting method does not work when the Péclet number is in the range of 1 and 500. This is an example of instability. The round-off errors (or truncation errors) amplify as the integration proceeds and the magnitude of the errors exceeds the solution. We illustrate this situation further by considering the following example.

***Example 7.10-1***. Write the finite difference equation for the system

$$\frac{dy}{dx} = -2y \tag{7.10-1a}$$

$$y(0) = 1 \tag{7.10-1b}$$

Examine the stability of the system.

The exact solution is

$$y = e^{-2x} \tag{7.10-2}$$

We note that $y$ is a decreasing function of $x$.

Using Equation (7.9-22b), the finite difference equation is

$$y_{j+1} - y_{j-1} = -2h\,y_j \tag{7.10-3}$$

Instead of solving Equation (7.10-3) numerically, we seek an analytic solution. The solution is of the form

$$y_j = C_1 \alpha_1^j + C_2 \alpha_2^j \tag{7.10-4}$$

Substituting Equation (7.10-4) into Equation (7.10-3), we obtain

$$C_1 \alpha_1^{j-1} \left[\alpha_1^2 + 2h\alpha_1 - 1\right] + C_2 \alpha_2^{j-1} \left[\alpha_2^2 + 2h\alpha_2 - 1\right] = 0 \tag{7.10-5}$$

From Equation (7.10-5), we deduce that $\alpha_{1,2}$ are the roots of

$$\alpha^2 + 2h\alpha - 1 = 0 \tag{7.10-6}$$

The values of $\alpha_{1,2}$ are

$$\alpha_{1,2} = -2h \pm \sqrt{1+4h^2} \tag{7.10-7}$$

Since  h  is assumed to be small,  we expand  $\sqrt{1+4h^2}$  in powers of  h.

The roots are then approximately given by

$$\alpha_1 = 1 - 2h \tag{7.10-8a}$$

$$\alpha_2 = -(1 + 2h) \tag{7.10-8b}$$

The solution is given approximately by

$$y_j = C_1 (1-2h)^j + C_2 (-1)^j (1+2h)^j \tag{7.10-9}$$

Replacing  j  by  $x_j$  ($x_j = jh$),  and noting that

$$\lim_{h \to 0} (1-2h)^{2x_j/2h} = e^{-2x_j} \tag{7.10-10a}$$

$$\lim_{h \to 0} (1+2h)^{2x_j/2h} = e^{2x_j} \tag{7.10-10b}$$

The solution  $y_j$  can be written as

$$y_j = C_1 e^{-2x_j} + C_2(-1)^j e^{2x_j} \tag{7.10-11}$$

Applying boundary condition (7.10-1b), we find  $C_2$  to be zero and  $y_j$  is a decreasing function of  $x_j$.

In an actual computation, there is round-off error and the value of  $C_2$  will not be exactly zero, but will be small.  It is multiplied by an exponentially increasing function of  $x_j$.  When  $x_j$  exceeds a certain critical value, the second term (which is the error term) on the right side of Equation (7.10-11) becomes the dominant term and the solution  $y_j$  oscillates.  This shows that the difference equation is unstable.  This arises because of the error in the initial condition and of replacing the first order differential equation by a two-point difference equation.

If Equation (7.10-3) were solved numerically, the solution would be oscillatory for some values of  $x_j$  greater than a certain value.  Instability depends on the differential equation and the method used.  If one method is unstable, choose another one.

## PROBLEMS

1a.    One root $x_1$ of a quadratic equation

$$a x^2 + b x + c = 0$$

is $x_1 = [-b + \sqrt{b^2 - 4ac}\,]/2a$.

Calculate $x_1$ if $a = 1.11$, $b = 111$ and $c = 0.111$.

Is the answer reliable? Note that $x_1$ is given by the difference of two almost equal numbers and this could result in loss of significant figures.

By expanding $\sqrt{b^2 - 4ac}$, show that $x_1$ can be approximated by

$$x_{1a} = -c/b$$

By multiplying the numerator and denominator by $b + \sqrt{b^2 - 4ac}$, show that $x_1$ can be written as

$$x_{1b} = [-2c/(b + \sqrt{b^2 - 4ac}\,)]$$

For the given values of a, b, c, calculate $x_{1a}$ and $x_{1b}$. Evaluate $a x^2 + b x + c$ for $x = x_1$, $x = x_{1a}$, and $x = x_{1b}$. Comment on the results.

2a.    The equation $e^x - 4x^2 = 0$ has a root in the interval $[0, 1]$. Obtain the root (i) by the method of bisection and (ii) by Newton's method.

Answer: 0.7148

3a.    The cubic $x^3 - 2x - 1$ has a root in the interval $[1, 2]$. Use (i) the secant method and (ii) Newton's method to find the root.

Answer: 1.618

4a     Deduce that an iterative formula for finding the cube root of a real number A is

$$x_{k+1} = x_k - (x_k^3 - A)/3x_k^2$$

Use the formula to calculate the cube root of 30.                    Answer: 3.107

5a.    Obtain the three roots of $x^3 - 4x + 1$ by the fixed point iteration method. Test the convergence in each case.

Answer: $-2.1149$, $0.2541$, $1.8608$

6b.     The Redlich-Kwong equation [Equation (1.6-2)] can be written as

$$P = \frac{RT}{V-b} - \frac{A}{V(V+b)}$$

Calculate $b$ if $P = 87$, $T = 486$, $V = 12$, $A = 0.08$, and $R = 1.98$.

7b.     Try to solve the equation $x^3 - x - 1 = 0$, using Newton's method with starting values of $x_0 = 0$ and $x_0 = 1$. Plot the function $f(x) = x^3 - x - 1$ and explain why Newton's method does not work with $x_0 = 0$. Are the other two roots real or complex? Compute them.

                                                                            Answer:  1.3247

8a.     Write the following set of equations in matrix form and solve by the method of elimination

$$x_1 + x_2 + x_3 = 0$$

$$x_1 + x_2 + 3x_3 = 0$$

$$3x_1 + 5x_2 + 7x_3 = 1$$                          Answer: $-\frac{1}{2}$, $\frac{1}{2}$, $0$

9a.     Show that the following set of equations has no solution

$$x_1 + x_2 + x_3 = 0$$

$$x_1 + 2x_2 + 3x_3 = 0$$

$$3x_1 + 5x_2 + 7x_3 = 1$$

Is the coefficient matrix $\underline{\underline{A}}$ singular? What is the rank of $\underline{\underline{A}}$?

10b.    Solve the following two sets of equations

(i)      $28x_1 + 25x_2 = 30$

          $19x_1 + 17x_2 = 20$

(ii)     $28x_1 + 25x_2 = 30$

          $19x_1 + 17x_2 = 19$

Note that the two sets are almost identical. Are their solutions almost identical? Compute the determinant of the coefficient matrix $\underline{\underline{A}}$. Did you expect this value of $|\underline{\underline{A}}|$? Compute the condition number $K(\underline{\underline{A}})$.

                                                                            Answer:  2491

11b.   Compute the inverse of

$$\underline{\underline{A}} = \begin{bmatrix} 1 & 1/2 & 1/3 \\ 1/2 & 1/3 & 1/4 \\ 1/3 & 1/4 & 1/5 \end{bmatrix}$$

by the method of elimination in (a) exact computation, (b) rounding off each number to three figures. The matrix $\underline{\underline{A}}$ is a Hilbert matrix and its elements $a_{ij} = 1/(i+j-1)$. Calculate its condition number $K(\underline{\underline{A}})$.

12a.   In structural mechanics, the flexibility matrix $\underline{\underline{F}}$ of a cantilever is given by

$$\underline{\underline{F}} = \begin{bmatrix} L/(EA) & 0 & 0 \\ 0 & L^3/(3EI) & L^2/(2EI) \\ 0 & L^2/(2EI) & L/(EI) \end{bmatrix}$$

where $L$ is the length of the cantilever, $E$ is Young's modulus, $A$ is the cross-sectional area, and $I$ is the second moment. Compute the stiffness matrix $\underline{\underline{K}}$, which is the inverse of $\underline{\underline{F}}$.

13a.   Solve the system of equations

$$-x_1 + 4x_2 - x_3 \qquad = 1$$

$$-x_2 + 4x_3 - x_4 = 1$$

$$4x_1 - x_2 \qquad\qquad = 1$$

$$-x_3 + 4x_4 = 1$$

by (i) Jacobi and (ii) Gauss-Seidel iteration methods. Compare their rates of convergence.

Answer: 0.3636, 0.4545, 0.4545, 0.3636

14a.   Use the QR method to find the eigenvalues of the tridiagonal matrix

$$\begin{bmatrix} 2 & 1 & 0 \\ 1 & 2 & 1 \\ 0 & 1 & 2 \end{bmatrix}$$

Answer: $2, \ 2 \pm \sqrt{2}$

15b.    Show that the matrices $\underline{\underline{A}}$ and $\underline{\underline{B}}$ defined by Equation (7.5-1) have the same eigenvalues. The Gaussian elimination method transforms a matrix to an upper diagonal matrix. Can we replace the QR method by the Gaussian elimination in the calculation of eigenvalues?

16a.    From Equation (7.6-10), deduce the expressions $p_2(x)$ and $p_3(x)$ for equidistant points $x_0$, $x_1$, $x_2$, and $x_3$. Choose $x_1 = x_0 - h$, $x_2 = x_0 + h$, and $x_3 = x_0 + 2h$. Calculate $\ell_i'(x_0)$ and $\ell_i''(x_0)$ and verify Equations (7.7-5, 6).

17b.    The solution of the differential equation

$$\frac{dy}{dx} = f(x, y)$$

subject to the condition

$$y(x_i) = y_i$$

is      $$y(x_{i+1}) = y_{i+1} = y_i + \int_{x_i}^{x_{i+1}} f(x, y)\, dx$$

Approximate $f(x, y)$ by a polynomial $p_n(x)$, in the form given by Equation (7.6-10), taking the interpolation points to be $x_i$, $x_{i+1}$, ... with a constant stepsize $h$. Denote $f(x_i, y_i)$ by $f_i$.

By approximating $f$ by a polynomial of degree 2 (through 3 points) and of degree 3 (through 4 points), show that $y_{i+1}$ is given respectively by

$$y_{i+1} = y_i + \frac{h}{12}(23f_i - 16f_{i-1} + 5f_{i-2})$$

$$y_{i+1} = y_i + \frac{h}{24}(55f_i - 59f_{i-1} + 37f_{i-2} - 9f_{i-3})$$

These formulae are the Adams-Bashforth formulae [Equation (7.8-33)]. In the derivation of the Adams-Bashforth formulae, we have interpolated $f$ through the points $x_i$, $x_{i-1}$, ... and not through the point $x_{i+1}$. If the point $x_{i+1}$ is included, we obtain the Adams-Moulton formulae. Show that if $f$ is approximated by a quadratic expression passing through $x_{i+1}$, $x_i$ and $x_{i-1}$, $y_{i+1}$ is given by

$$y_{i+1} = y_i + \frac{h}{12}(5f_{i+1} + 8f_i - f_{i-1})$$

18a.    Show that the following functions are not cubic splines

(i) $\quad f(x) = \begin{cases} 11 - 24x + 18x^2 - 4x^3, & 1 \leq x \leq 2 \\ \\ -54 + 72x - 30x^2 + 4x^3, & 2 < x \leq 3 \end{cases}$

(ii) $\quad f(x) = \begin{cases} 13 - 31x + 23x^2 - 5x^3, & 1 \leq x \leq 2 \\ \\ -35 + 51x - 22x^2 + 3x^3, & 2 \leq x \leq 3 \end{cases}$

19a. The viscosity $\eta$ of water at various temperatures $T$ is given in the following table

| T (°C) | 10 | 20 | 30 |
|---|---|---|---|
| $\eta$ (cp) | 1.3 | 1.0 | 0.8 |

Compute the cubic splines that pass through the three given points. Evaluate the viscosity at 25°C. The measured value has been reported to be 0.8904 cp.

20b. The function $f(x) = e^x$, $0 \leq x \leq 1$ is approximated by $y(x) = a_0 + a_1 x + a_2 x^2$. Obtain $a_0$, $a_1$, and $a_2$ such that $\int_0^1 [f(x) - y(x)]^2 \, dx$ is a minimum (least squares approximation). Expand $f(x)$ in a Taylor series about $x = 0.5$ up to the term $x^2$. Compare this series with $y(x)$.

21a. The demand for a certain product is a linear function of the price. The sales of the product for three different prices are given in the following table

| Price ($) | 1.00 | 1.25 | 1.50 |
|---|---|---|---|
| Demand | 450 | 375 | 330 |

Find the least squares regression line and estimate the demand when the price is $1.40.

Answer: 349

22b. Show that the recurrence formula for evaluating the integral $I_n$ defined by

$$I_n = \int_0^1 \frac{x^n}{10 + x} \, dx$$

is $\quad I_n = \dfrac{1}{n} - 10 I_{n-1}$

Use this formula to calculate $I_4$ retaining (i) three significant and (ii) five significant figures.

Use Simpson's rule to evaluate $I_4$.

Compare the two methods and proposed a modified recurrence formula so as to avoid loss of significant figures.

[Hint: $\int_0^1 [x^n/(10+x)]\,dx = \int_0^1 [1-10/(10+x)]\,x^{n-1}\,dx$]     Answer: 0.0185

23a.   Evaluate $\int_0^1 e^{-x^2}\,dx$ (a) by the trapezoidal rule, (b) by the Simpson's rule, (c) by the Romberg's method, and (d) by the Gaussian method.

                                                                                       Answer: 0.7468

24a.   Show, using Euler's method, that the initial value problem

$$\frac{dy}{dx} = -ay + b, \quad a>0, \quad 0<x<\infty$$

$$y(0) = 1 + b/a$$

leads to the difference equation

$$y_{i+1} - (1-ah)\,y_i = bh$$

where $y_i \approx y(x_i)$, $x_i = ih$, $h$ is the constant interval $(x_{i+1} - x_i)$. Verify that $y_i = (1-ah)^i + b/a$ is a solution of the difference equation. Find conditions on $h$ such that $y_i$ tends to the proper limit as $i$ tends to infinity.

25a.   Use Euler's and Heun's methods to solve the equation

$$\frac{dy}{dx} = x+y, \quad y(0) = 1$$

with $h = 0.01$.

Solve the equation analytically and compare the values of $y$ at $x = 0.1$.

                                                                       Answer: $2e^x - x - 1$

26a.   Solve the equation

$$\frac{dy}{dx} = x+y+xy, \quad y(0) = 1$$

using (i) Euler's method with $h = 0.025$ and (ii) the fourth Runge-Kutta method with $h = 0.1$ to find $y(0.1)$.

Work to five decimal places, compare the accuracy and amount of computations required associated with the two methods, given that $y(0.1)$, accurate to four decimal places, is 1.11587.

Use the Adams-Bashforth method to obtain $y$ at $x = 1$ with $h = 0.1$.

27a.   A vibrating system with a periodic forcing term is given by

$$\frac{d^2y}{dt^2} + 64y = 16 \cos 8t$$

$$y(0) = y'(0) = 0$$

Find $y(0.5)$, analytically and numerically, using the Runge-Kutta method.

                                                                        Answer: $-0.3784$

28b.   The steady one-dimensional heat equation can be written as

$$\frac{d^2T}{dx^2} - 0.01\,(T - 20) = 0$$

The boundary conditions are

$$T(0) = 40, \qquad T(10) = 200$$

Find the analytical solution and solve the problem numerically using (a) the shooting method, and (b) the finite difference method. Compare the results obtained by the three methods.

29a.   Solve the boundary value problem

$$\frac{d^2y}{dx^2} + 2\,(2 - x)\,\frac{dy}{dx} = 2\,(2 - x), \qquad 0 \le x \le 4$$

$$y(0) = -1, \qquad y(4) = 3$$

by the method of finite differences.

Choose an appropriate value of $h$. Is $h = 1$ appropriate?

30b.   In Problem 17b, we can integrate from $x_{i-1}$ to $x_{i+1}$ instead of from $x_i$ to $x_{i+1}$. Show that, under certain conditions which should be stated, we obtain a multistep formula which can be written as

$$y_{i+1} = y_{i-1} + 2hf_i$$

where $h = x_{i+1} - x_i$.

Solve the initial value problem

$$\frac{dy}{dx} = -2y + 1, \qquad y(0) = 1$$

by the formula obtained in this problem and by Euler's formula. Which of the two methods gives the correct solution? Discuss the stability of the two methods.

# CHAPTER 8

# NUMERICAL SOLUTION
# OF PARTIAL DIFFERENTIAL EQUATIONS

## 8.1  INTRODUCTION

In Chapters 5 and 6, we have discussed several analytical methods for solving P.D.E.'s. These methods cannot be used to solve all P.D.E.'s. For example, the method of separation of variables can generally be applied only to linear homogeneous equations with homogeneous boundary conditions. Under favorable conditions, non-homogeneity can be transformed to homogeneity via auxiliary functions as shown in Section 5.7.

Many equations cannot be solved exactly and we have to be satisfied with approximate solutions. Even some of the exact solutions given in Chapter 5 are in reality approximate solutions. For instance, the infinite Fourier series solution obtained in Chapter 5 has the appearance of an exact solution, but, in many cases, we can sum only a finite number of terms. The solution is then an approximate solution. Also, the Fourier coefficients are expressed as integrals and some of them cannot be evaluated analytically, and we have to integrate them numerically.

Using computers, numerical methods are the most appropriate methods of solving some P.D.E.'s. In this chapter, we shall extend the method of **finite differences** described in Chapter 7 to P.D.E.'s. We shall also consider the method of **finite elements**.

## 8.2  FINITE DIFFERENCES

For simplicity, we consider the unknown function $u$ to be a function of two variables $x$ and $y$. We divide the xy-plane into a grid consisting of $(n \times m)$ rectangles with sides $\Delta x = h$ and $\Delta y = k$ as shown in Figure 8.2-1. We denote the value of $u(x_i, y_j)$ by $u_{i,j}$, the value of $u(x_i+h, y_j)$ by $u_{i+1,j}$, the value of $u(x_i, y_j+k)$ by $u_{i,j+1}$, and that of $u(x_i+h, y_j+k)$ by $u_{i+1,j+1}$. Similarly, $u_{i+s,j+t}$ denotes the value of $u(x_i+sh, y_j+tk)$, where $s$ and $t$ are integers.

Using Taylor's theorem, we have

$$u(x_i+h, y_j) = u_{i+1,j} = u_{i,j} + h \left. \frac{\partial u}{\partial x} \right|_{i,j} + \frac{h^2}{2} \left. \frac{\partial^2 u}{\partial x^2} \right|_{i,j} + R_n \qquad (8.2\text{-}1a,b)$$

where $R_n$ is the remainder term.

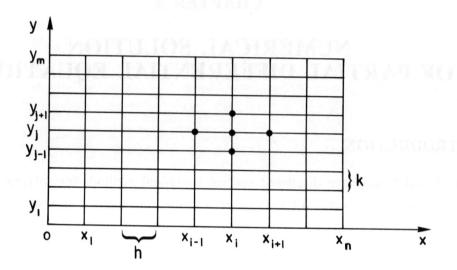

**FIGURE 8.2-1    Grid system**

Similarly,

$$u(x_i-h, y_j) = u_{i-1,j} = u_{i,j} - h \left.\frac{\partial u}{\partial x}\right|_{i,j} + \frac{h^2}{2} \left.\frac{\partial^2 u}{\partial x^2}\right|_{i,j} + R_n \qquad (8.2\text{-}2a,b)$$

From Equations (8.2-1a,b), we deduce that $\left.\dfrac{\partial u}{\partial x}\right|_{i,j}$ can be approximated as

$$\left.\frac{\partial u}{\partial x}\right|_{i,j} \approx (u_{i+1,j} - u_{i,j})/h \qquad (8.2\text{-}3a)$$

The partial derivative $\partial u / \partial x$ can also be approximated from Equations (8.2-2a,b) by

$$\left.\frac{\partial u}{\partial x}\right|_{i,j} \approx (u_{i,j} - u_{i-1,j})/h \qquad (8.2\text{-}3b)$$

In deriving Equations (8.2-3a, b), we have neglected terms of $O(h^2)$. To include term of $O(h^2)$, we find that the difference of Equations (8.2-1a, b, 2a, b) yields

$$\left.\frac{\partial u}{\partial x}\right|_{i,j} = (u_{i+1,j} - u_{i-1,j})/2h + O(h^2) \qquad (8.2\text{-}3c)$$

Formulae (8.2-3a, b, c) are the forward, backward, and central difference forms of $\partial u / \partial x$ respectively.

Adding Equations (8.2-1a, b, 2a, b) yields

$$\left.\frac{\partial^2 u}{\partial x^2}\right|_{i,j} \approx (u_{i+1,j} - 2u_{i,j} + u_{i-1,j})/h^2 \tag{8.2-4}$$

Similarly, $\dfrac{\partial u}{\partial y}$ and $\dfrac{\partial^2 u}{\partial y^2}$ can be approximated as

$$\left.\frac{\partial u}{\partial y}\right|_{i,j} \approx \begin{cases} (u_{i,j+1} - u_{i,j})/k \\[2mm] (u_{i,j} - u_{i,j-1})/k \\[2mm] (u_{i,j+1} - u_{i,j-1})/2k \end{cases} \tag{8.2-5a,b,c}$$

$$\left.\frac{\partial^2 u}{\partial y^2}\right|_{i,j} \approx (u_{i,j+1} - 2u_{i,j} + u_{i,j-1})/k^2 \tag{8.2-6}$$

Using the central difference form, the mixed second order derivative is given by

$$\left.\frac{\partial^2 u}{\partial x \partial y}\right|_{i,j} \approx (u_{i+1,j+1} - u_{i-1,j+1} - u_{i+1,j-1} + u_{i-1,j-1})/4hk \tag{8.2-7}$$

The partial derivatives are replaced by the finite differences and a partial differential equation becomes a finite difference equation. We now consider the **parabolic, elliptic,** and **hyperbolic** types of equations.

## 8.3   PARABOLIC EQUATIONS

The **canonical form of the parabolic equation** is

$$\frac{\partial u}{\partial t} = \alpha^2 \frac{\partial^2 u}{\partial x^2} \tag{8.3-1}$$

where $\alpha$ is a constant.

We suppose the region of interest is

$$0 \le x \le a, \qquad t > 0 \tag{8.3-2a,b}$$

The boundary and initial conditions are assumed to be given by

$$u(0, t) = g_1(t), \qquad u(a, t) = g_2(t), \qquad u(x, 0) = f(x) \tag{8.3-3a,b,c}$$

We subdivide the region of interest into rectangles as shown in Figure 8.2-1 with the y-axis being replaced by the t-axis. In this case, we have a semi-finite region with $t_m$ tending to infinity and $x_n (= n\,h) = a$.

## Explicit Method

Using Equations (8.2-4, 5a), Equation (8.3-1) becomes

$$u_{i,j+1} - u_{i,j} = \frac{\alpha^2 k}{h^2} (u_{i+1,j} - 2u_{i,j} + u_{i-1,j}) \tag{8.3-4a}$$

$$= r (u_{i+1,j} - 2u_{i,j} + u_{i-1,j}) \tag{8.3-4b}$$

where $r = \alpha^2 k / h^2$. $\tag{8.3-4c}$

Equation (8.3-4b) can be written as

$$u_{i,j+1} = (1 - 2r) u_{i,j} + r (u_{i+1,j} + u_{i-1,j}) \tag{8.3-5}$$

Equation (8.3-5) gives the value of $u_{i,j+1}$ in terms of $u_{i,j}$, $u_{i+1,j}$, and $u_{i-1,j}$; that is to say, the values of u along the $(j + 1)^{th}$ row can be determined if the values of u along the $j^{th}$ row are known. The initial condition gives the values of u along the zeroth row, so the values of u along the first row can be computed. Once the values of u along the first row are known, we can compute u along the second row. This process can be repeated for the subsequent rows.

Formula (8.3-5) which gives the value of $u_{i,j+1}$ in terms of known values of $u_{i,j}$, $u_{i+1,j}$, and $u_{i-1,j}$ is an **explicit formula**.

Note that Equations (8.3-3a, b, c) can be written respectively as

$$u_{0,j} = g_1(kj), \qquad u_{n,j} = g_2(kj), \qquad u_{i,0} = f(ih) \tag{8.3-6a,b,c}$$

***Example 8.3-1.*** Solve Equation (8.3-1) with $\alpha = 1$ by the method of finite differences. The boundary and initial conditions are

$$u(0, t) = u(1, t) = 0, \qquad 0 \le x \le 1 \tag{8.3-7a,b}$$

$$u(x, 0) = \begin{cases} 2x, & 0 \le x \le \dfrac{1}{2} \\[2mm] 2(1-x), & \dfrac{1}{2} \le x \le 1 \end{cases} \tag{8.3-8a,b}$$

Choose

(a) $h = 0.1$, $k = 0.005$;     (b) $h = 0.1$, $k = 0.01$           (8.3-9a,b,c,d)

In case (a)

$$r = k/h^2 = 0.5 \tag{8.3-10a,b}$$

Equation (8.3-5) can be written as

$$u_{i,j+1} = 0.5 (u_{i+1,j} + u_{i-1,j}) \tag{8.3-11}$$

From Equations (8.3-8a, b), we can calculate the values of $u_{i,0}$ and substituting these values into Equation (8.3-11) yields the values of $u_{i,1}$. The values of $u_{i,1}$ then generate the values of $u_{i,2}$, and this process is repeated to obtain the values of $u_{i,m}$, that is to say, the values of u at time $t_m$.

We note that, in this example, u is symmetrical about the midpoint ($x = 0.5$) and so we need to compute the values of u in the interval $0 \le x \le 0.5$. By symmetry, u (0.6, t) is equal to u (0.4, t).

The computed values of u are given in Table 8.3-1.

### TABLE 8.3-1

### Values of u for two values of r, (a) r = 0.5, (b) r = 1.0

| $t_j$ | $x_0 = 0.0$ (a) | (b) | $x_1 = 0.1$ (a) | (b) | $x_2 = 0.2$ (a) | (b) | $x_3 = 0.3$ (a) | (b) | $x_4 = 0.4$ (a) | (b) | $x_5 = 0.5$ (a) | (b) |
|---|---|---|---|---|---|---|---|---|---|---|---|---|
| 0.0 | 0.0 | 0.0 | 0.2 | 0.2 | 0.4 | 0.4 | 0.6 | 0.6 | 0.8 | 0.8 | 1.0 | 1.0 |
| 0.005 | 0.0 | 0.0 | 0.200 | — | 0.400 | — | 0.600 | — | 0.800 | — | 0.800 | — |
| 0.010 | 0.0 | 0.0 | 0.200 | 0.200 | 0.400 | 0.400 | 0.600 | 0.600 | 0.700 | 0.800 | 0.800 | 0.600 |
| 0.015 | 0.0 | 0.0 | 0.200 | — | 0.400 | — | 0.550 | — | 0.700 | — | 0.700 | — |
| 0.020 | 0.0 | 0.0 | 0.200 | 0.200 | 0.375 | 0.400 | 0.550 | 0.600 | 0.625 | 0.400 | 0.700 | 1.000 |
| 0.025 | 0.0 | 0.0 | 0.187 | — | 0.375 | — | 0.500 | — | 0.625 | — | 0.625 | — |
| 0.030 | 0.0 | 0.0 | 0.187 | 0.200 | 0.344 | 0.400 | 0.500 | 0.200 | 0.563 | 1.200 | 0.531 | –0.200 |
| 0.035 | 0.0 | 0.0 | 0.172 | — | 0.344 | — | 0.453 | — | 0.516 | — | 0.531 | — |
| 0.040 | 0.0 | 0.0 | 0.172 | 0.200 | 0.312 | 0.000 | 0.430 | 1.400 | 0.492 | –1.200 | 0.484 | 2.600 |

In case (b)

$$r = 1 \tag{8.3-12}$$

Equation (8.3-5) becomes

$$u_{i,j+1} = -u_{i,j} + (u_{i+1,j} + u_{i-1,j}) \tag{8.3-13}$$

Proceeding as in case (a), we can compute the values of $u_{i,1}$, $u_{i,2}$, ... from Equation (8.3-13) and these values are also tabulated in Table 8.3-1.

From Table 8.3-1, it can be seen that the values of $u$ obtained in case (b) cannot be the solution of Equations (8.3-1, 7a, b, 8a, b). The analytical solution [see Problem 11a, Chapter 5] is

$$u(x, t) = \frac{8}{\pi^2} \sum_{n=0}^{\infty} \frac{(-1)^n}{(2n+1)^2} \sin\left[(2n+1)\pi x\right] \exp\left[-(2n+1)^2 \pi^2 t\right] \tag{8.3-14}$$

Equation (8.3-14) predicts that $u$ is an exponentially decreasing function of $t$. The values of $u$ obtained in case (a) are in qualitative agreement with the analytical solution whereas the values of $u$ obtained in case (b) are not. This example can be associated with the heat transfer in a rod of unit length with its two ends kept at zero degrees. Initially, the temperature distribution is given by Equations (8.3-8a, b) and the temperature is positive. From the physics of the problem, the temperature can never be negative and we conclude that the values of $u$ in column (b) are unacceptable.

This example shows the importance of the choice of the value of $r$ and the explicit method is not valid for all values of $r$. A **convergence**, **stability**, and **compatibility** analysis is necessary. Such an analysis can be found in a more advanced text, such as the one by Richtmyer and Morton (1967).

In the present example, the explicit method is valid only when $0 < r \le 1/2$. We next consider an implicit method which is valid for a wider range of values of $r$.

## Crank-Nicolson Implicit Method

**Crank and Nicolson** (1947) proposed that $\left.\dfrac{\partial^2 u}{\partial x^2}\right|_{i,j}$ be approximated by the mean value of $\left.\dfrac{\partial^2 u}{\partial x^2}\right|_{i,j}$ and $\left.\dfrac{\partial^2 u}{\partial x^2}\right|_{i,j+1}$. Equation (8.3-4b) is now replaced by

$$u_{i,j+1} - u_{i,j} = \frac{r}{2}\left(u_{i+1,j} - 2u_{i,j} + u_{i-1,j} + u_{i+1,j+1} - 2u_{i,j+1} + u_{i-1,j+1}\right) \tag{8.3-15}$$

On simplifying Equation (8.3-15), we obtain

$$-r u_{i-1,j+1} + 2(1+r) u_{i,j+1} - r u_{i+1,j+1} = r u_{i-1,j} + 2(1-r) u_{i,j} + r u_{i+1,j} \qquad (8.3\text{-}16)$$

In the explicit method, the right side of Equation (8.3-5) involves three known terms, while the left side contains one unknown term. In the implicit method [Equation (8.3-16)], the right side also involves three known terms, but the left side contains three unknown terms. To simplify the writing of Equation (8.3-16), we introduce the following notation

$$u_{i,j+1} = v_i \quad \text{and} \quad u_{i,j} = w_i \qquad (8.3\text{-}17a,b)$$

Equation (8.3-16) can be written as

$$-r v_{i-1} + 2(1+r) v_i - r v_{i+1} = r w_{i-1} + 2(1-r) w_i + r w_{i+1} = c_i \qquad (8.3\text{-}18a,b)$$

In Equations (8.3-18a, b), the left side is in terms of $v_{i-1}$, $v_i$, and $v_{i+1}$ only, which are values of u at time $(j+1)$ at points $x_{i-1}$, $x_i$, and $x_{i+1}$. On the right side, we have only the values of u at an earlier time $t_j$. From the initial conditions, we can calculate the initial values of $c_i$.

We write the first few as well as the last few rows of Equation (8.3-18b) and we obtain

$$-r v_0 + 2(1+r) v_1 - r v_2 = c_1 \qquad (8.3\text{-}19a)$$

$$-r v_1 + 2(1+r) v_2 - r v_3 = c_2 \qquad (8.3\text{-}19b)$$

$$-r v_2 + 2(1+r) v_3 - r v_4 = c_3 \qquad (8.3\text{-}19c)$$

$$\vdots$$

$$-r v_{n-3} + 2(1+r) v_{n-2} - r v_{n-1} = c_{n-2} \qquad (8.3\text{-}19d)$$

$$-r v_{n-2} + 2(1+r) v_{n-1} - r v_n = c_{n-1} \qquad (8.3\text{-}19e)$$

If the boundary conditions are given in terms of u, then $v_0 \, (= u_{0,j+1})$ and $v_n \, (= u_{n,j+1})$ are known and the unknowns are $v_1, v_2, \ldots, v_{n-1}$. Writing all the known quantities on the right side, Equations (8.3-19a to e) become

$$2(1+r) v_1 - r v_2 = c_1 + r v_0 = d_1 \qquad (8.3\text{-}20a,b)$$

$$-r v_1 + 2(1+r) v_2 - r v_3 = d_2 \qquad (8.3\text{-}20c)$$

$$-r v_2 + 2(1+r) v_3 - r v_4 = d_3 \qquad (8.3\text{-}20d)$$

$$\vdots$$

$$-r\,v_{n-3} + 2\,(1+r)\,v_{n-2} - r\,v_{n-1} = d_{n-2} \tag{8.3-20e}$$

$$-r\,v_{n-2} + 2\,(1+r)\,v_{n-1} = c_{n-1} + r\,v_n = d_{n-1} \tag{8.3-20f,g}$$

Equations (8.3-20 a to g) can be written in matrix form as

$$\underline{\underline{A}}\,\underline{v} = \underline{d} \tag{8.3-21a}$$

where

$$
\underline{\underline{A}} =
\begin{bmatrix}
2\,(1+r) & -r & 0 & 0 & 0 & - & - & - & 0 \\
-r & 2\,(1+r) & -r & 0 & 0 & - & - & - & 0 \\
0 & -r & 2\,(1+r) & -r & 0 & - & - & - & 0 \\
 & & & \vdots & & & & & \\
0 & 0 & 0 & 0 & 0 & 0 & -r & 2\,(1+r) & -r \\
0 & 0 & 0 & 0 & 0 & 0 & 0 & 2\,(1+r) & -r
\end{bmatrix}
$$

$$\tag{8.3-21b}$$

$$
\underline{v} =
\begin{bmatrix}
v_1 \\
v_2 \\
\vdots \\
v_{n-2} \\
v_{n-1}
\end{bmatrix},
\qquad
\underline{d} =
\begin{bmatrix}
d_1 \\
d_2 \\
\vdots \\
d_{n-2} \\
d_{n-1}
\end{bmatrix}
\tag{8.3-21c,d}
$$

The matrix $\underline{\underline{A}}$ is a $(n-1) \times (n-1)$ tridiagonal matrix, $\underline{v}$ and $\underline{d}$ are column vectors, each having $(n-1)$ elements. Equation (8.3-21a) can be solved by one of the methods discussed in Chapter 7. Among the most widely used methods are the Gaussian elimination method and the iterative Gauss-Seidel method. If the Gauss-Seidel method is chosen [Equation (7.4-53)], the $(k+1)$ iterate of $v_i$ is given by

$$v_i^{(k+1)} = \frac{d_i}{2\,(1+r)} + \frac{r}{2\,(1+r)}\,\left[v_{i-1}^{(k+1)} + v_{i+1}^{(k)}\right] \tag{8.3-22}$$

From the initial and boundary conditions, we can calculate $d_i$ at time $t_0 \, (= 0)$ and solving Equation (8.3-21a), we obtain $v_i$ which are the values of $u_i$ at time $t_1$. These values can be used to compute $d_i$ at time $t_1$ and from Equation (8.3-21a), we can obtain the values of $u_i$ at time $t_2$. This process can be repeated to obtain the values of $u_i$ at time $t_3, \dots, t_m$.

***Example 8.3-2.*** Solve Equation (8.3-1) subject to Equations (8.3-7a, b, 8a, b), using the Crank-Nicolson formula. Assume $\alpha$ to be unity and choose $h = 0.1$ and $k = 0.01$.

With the given choice of $h$ and $k$, the value of $r$ is unity. From the symmetry of the problem, we need to calculate up to $x = 0.5$ only.

In this case, using the symmetry property $(v_6 = v_4)$, $\underset{=}{A}$ can be written as

$$\underset{=}{A} = \begin{bmatrix} 4 & -1 & 0 & 0 & 0 \\ -1 & 4 & -1 & 0 & 0 \\ 0 & -1 & 4 & -1 & 0 \\ 0 & 0 & -1 & 4 & -1 \\ 0 & 0 & 0 & -2 & 4 \end{bmatrix} \qquad (8.3\text{-}23)$$

From Equations (8.3-18b, 20a to g), we find that initially the $d_i$ are

$$\underline{d} = \begin{bmatrix} 0.4 \\ 0.8 \\ 1.2 \\ 1.6 \\ 1.6 \end{bmatrix} \qquad (8.3\text{-}24)$$

For the given $\underset{=}{A}$ and $\underline{d}$ [Equations (8.3-23, 24)], we can solve, by the elimination method, Equation (8.3-21a) to obtain $v_i$ which are the values of $u$ at points $x_i$ at time $t = 0.01$. We can now calculate $d_i$ at time $t = 0.01$ and solving Equation (8.3-21a), we obtain the values of $u$ at time $t = 0.02$. We can repeat this process and obtain the values of $u$ at subsequent times.

Table 8.3-2 gives the values of $u$ at various times $t_j$ and the analytical solution at the point $x = 0.5$.

## TABLE 8.3-2

### Numerical (N) and analytical (A) values of u

| $t_j$ | N $x_0 = 0$ | N $x_1 = 0.1$ | N $x_2 = 0.2$ | N $x_3 = 0.3$ | N $x_4 = 0.4$ | $x_5 = 0.5$ N | A |
|---|---|---|---|---|---|---|---|
| 0.0 | 0.0 | 0.2 | 0.4 | 0.6 | 0.8 | 1.0 | 1.0 |
| 0.01 | 0 | 0.1989 | 0.3956 | 0.5834 | 0.7381 | 0.7691 | 0.7743 |
| 0.02 | 0 | 0.1936 | 0.3789 | 0.5400 | 0.6461 | 0.6921 | 0.6809 |

We noted that for $r = 1$, the explicit method did not generate a meaningful solution, but the implicit method does. From Tables 8.3-1 and 2, it is seen that the Crank-Nicolson formula provides more accurate values of $u$ than the explicit method for $r = 0.5$. The Crank-Nicolson formula is convergent for all values of $r$, though the smaller the value of $r$ the better the accuracy.

●

Compared to the explicit method [Equation (8.3-5)], the Crank-Nicolson method [Equation (8.3-18a)] involves more computation; however, in the Crank-Nicolson method, there is no restriction on $r$, the time interval can be larger, and this generally compensates for the increase in computational effort at each grid point.

## Derivative Boundary Conditions

So far we have assumed that $u$ is given on the boundary. In some cases, it could be that the derivative of $u$ is given on the boundary. For example, in the heat conduction problem, if the material is insulated, there is no heat flow across the surface and the boundary condition is

$$\frac{\partial u}{\partial n} = 0 \qquad\qquad\qquad (8.3\text{-}25)$$

where $n$ is the unit outward normal to the surface.

We now replace the boundary condition (8.3-3a) by

$$\frac{\partial u}{\partial x}\bigg|_{x=0} = \beta u, \qquad t \geq 0 \qquad\qquad\qquad (8.3\text{-}26)$$

where $\beta$ is a constant.

Equation (8.3-25) corresponds to the case $\beta = 0$.

To obtain better accuracy, we use the central difference form of $\partial u / \partial x$ [Equation (8.2-3c)] and Equation (8.3-26) becomes

$$(u_{1,j} - u_{-1,j}) = 2h\beta u_{0,j}, \qquad j = 0, 1, 2, \dots \qquad\qquad (8.3\text{-}27)$$

We note that $u_{-1,j}$ is outside our region and can be eliminated via Equation (8.3-1) as shown later.

If we employ the explicit formula, the difference equation at the origin ($x_0 = 0$) is given by [Equation (8.3-5)]

$$u_{0,j+1} = (1 - 2r) u_{0,j} + r (u_{1,j} + u_{-1,j}) \qquad\qquad (8.3\text{-}28)$$

Combining Equations (8.3-27, 28), we eliminate $u_{-1,j}$ to obtain

$$u_{0,j+1} = (1 - 2r) u_{0,j} + r (u_{1,j} + u_{1,j} - 2h\beta u_{0,j}) \qquad (8.3-29a)$$

$$= u_{0,j} + 2r [u_{1,j} - (1 + h\beta) u_{0,j}] \qquad (8.3-29b)$$

From the initial condition $u_{0,0}$, $u_{1,0}$ are known and from Equation (8.3-29b), the values of u at later times at the origin can be computed. Suppose that at the other end $(x_n = 1)$ the boundary condition is given by

$$\left. \frac{\partial u}{\partial x} \right|_{x=1} = \gamma u, \qquad t \geq 0 \qquad (8.3-30)$$

where $\gamma$ is a constant.

Using the same technique as at the origin, we deduce that

$$u_{n,j+1} = u_{n,j} + 2r [u_{n-1,j} - (1 + h\gamma) u_{n,j}] \qquad (8.3-31)$$

From Equation (8.3-31) and using the initial condition, the values of u at subsequent times can be computed.

If the Crank-Nicolson formula is used, we obtain from Equation (8.3-18a)

$$-r v_{-1} + 2 (1 + r) v_0 - r v_1 = r w_{-1} + 2 (1 - r) w_0 + r w_1 \qquad (8.3-32)$$

Equation (8.3-27) is true for all times and we have

$$v_1 - v_{-1} = 2h\beta v_0 \qquad (8.3-33a)$$

$$w_1 - w_{-1} = 2h\beta w_0 \qquad (8.3-33b)$$

Combining Equations (8.3-32, 33a, b) yields

$$(1 + r + rh\beta) v_0 - r v_1 = (1 - r - h\beta r) w_0 + r w_1 = c_0 \qquad (8.3-34a, b)$$

The boundary condition at the other end $(x_n = 1)$ can be treated similarly.

We have two equations to determine $v_0 (= u_{0,j+1})$ and $v_n (= u_{n,j+1})$.

The matrix $\underline{\underline{A}}$ has been augmented by two columns and two rows, the vector $\underline{v}$ has two new elements $(v_0$ and $v_n)$, and similarly vector $\underline{c}$ is increased by two elements $(c_0$ and $c_n)$. Note that $v_0$ and $v_n$ are not given and we work with Equations (8.3-19a to e) instead of with Equations (8.3-20a to g).

***Example 8.3-3.*** Solve the equation

$$\frac{\partial u}{\partial t} = \frac{\partial^2 u}{\partial x^2} , \qquad 0 \le x \le 1 \tag{8.3-35}$$

subject to the conditions

$$u(x, 0) = \begin{cases} 2x, & 0 \le x \le 1/2 \\ 2(1-x), & 1/2 \le x \le 1 \end{cases} \tag{8.3-36a,b}$$

$$\left.\frac{\partial u}{\partial x}\right|_{x=0} = \left.\frac{\partial u}{\partial x}\right|_{x=1} = 0 , \qquad \text{for all } t \tag{8.3-36c,d}$$

Use the Crank-Nicolson formula and the central difference scheme for the boundary conditions. Choose $h = 0.1$ and $k = 0.01$. As discussed before, the problem is symmetrical about $x = 0.5$ and we need to compute the values of $u$ between $0 \le x \le 0.5$. The value of $r$ is one and the value of $\beta$ is zero. Equations (8.3-34a, b) become

$$2 v_0 - v_1 = w_1 = c_0 \tag{8.3-37a,b}$$

Equations (8.3-37a, b, 19a to e) can be written as

$$\begin{bmatrix} 2 & -1 & 0 & 0 & 0 & 0 \\ -1 & 4 & -1 & 0 & 0 & 0 \\ 0 & -1 & 4 & -1 & 0 & 0 \\ 0 & 0 & -1 & 4 & -1 & 0 \\ 0 & 0 & 0 & -1 & 4 & -1 \\ 0 & 0 & 0 & 0 & -2 & 4 \end{bmatrix} \begin{bmatrix} v_0 \\ v_1 \\ v_2 \\ v_3 \\ v_4 \\ v_5 \end{bmatrix} = \begin{bmatrix} c_0 \\ c_1 \\ c_2 \\ c_3 \\ c_4 \\ c_5 \end{bmatrix} \tag{8.3-38}$$

The initial values of $c_i$ can be computed from the initial conditions [Equations (8.3-36a, b)] and Equation (8.3-37b). Writing $c_i$ as a row vector, we have

$$c_i = [0.2, \ 0.4, \ 0.6, \ 0.8, \ 1.0, \ 0.8] , \qquad (i = 0, \dots, 5) \tag{8.3-39}$$

Solving Equation (8.3-38) with the given values of $c_i$ [Equation (8.3-39)], we obtain $v_i$ which are the values of $u_i$ at time $t_1 (= 0.01)$. With these values of $v_i$, we can compute $c_i$ at time $t_1$ and, on solving Equation (8.3-38), we obtain the values of $u_i$ at time $t_2 (= 0.02)$. We repeat the process and we obtain the values of $u$ at times $t_3, t_4, \dots, t_m$. Table 8.3-3 gives the values of $u_i$ at various times.

## TABLE 8.3-3

### Values of u for r = 1

| $t_j$ | $x_0 = 0$ | $x_1 = 0.1$ | $x_2 = 0.2$ | $x_3 = 0.3$ | $x_4 = 0.4$ | $x_5 = 0.5$ |
|-------|-----------|-------------|-------------|-------------|-------------|-------------|
| 0.0   | 0.0       | 0.2         | 0.4         | 0.6         | 0.8         | 1.0         |
| 0.01  | 0.212     | 0.234       | 0.274       | 0.390       | 0.454       | 0.427       |
| 0.02  | 0.176     | 0.140       | 0.151       | 0.189       | 0.214       | 0.214       |

*Example 8.3-4*. Solve the heat equation

$$\frac{\partial u}{\partial t} = \frac{\partial^2 u}{\partial x^2}, \qquad t > 0, \qquad 0 \le x \le 1 \tag{8.3-40}$$

subject to

$$u(x, 0) = 0, \quad u(0, t) = u(1, t) = 1 \tag{8.3-41a,b,c}$$

Using the Crank-Nicolson method, Equation (8.3-40) is replaced by the finite difference Equation (8.3-16) which on dividing by r can be written as

$$-u_{i-1,j+1} + 2(1 + 1/r)u_{i,j+1} - u_{i+1,j+1} = u_{i-1,j} - 2(1 - 1/r)u_{i,j} + u_{i+1,j} \tag{8.3-42}$$

It is Equation (8.3-42) which is adopted in the following program. The results are illustrated in Figure 8.3-1.

The analytical solution is

$$u = 1 - \frac{4}{\pi} \sum_{n=0}^{\infty} \frac{\exp\left[-\pi^2 (2n+1)^2 t\right]}{(2n+1)} \sin(2n+1)\pi x \tag{8.3-43}$$

From Equation (8.3-43), we deduce that $u \longrightarrow 1$ as $t \longrightarrow \infty$. This is confirmed in Figure 8.3-1.

```
C       MAIN PROGRAM TO SOLVE EXAMPLE 8.3-4
C       SOLUTION METHOD IS CRANK-NICOLSON

C       Y          – THE SOLUTION VECTOR
C       A,B,C      – THE COEFFICIENTS OF THE TRIDIAGONAL SYSTEM OF EQUATIONS
C       D          – THE RIGHT SIDE OF THE EQUATIONS
C       YNEW       – THIS VECTOR WILL CONTAIN THE NEW SOLUTION VECTOR
C       BETA, GAMM – WORK SPACE FOR THE EQUATION SOLVER
```

```
C       FOR ILLUSTRATIVE PURPOSES ALL VECTORS HAVE THE SAME DIMENSIONS
C       ALTHOUGH SOME ELEMENTS ARE NOT USED

        REAL*8 Y(21),YNEW(21),BETA(21),GAMM(21),A(21),B(21),C(21),D(21)
        REAL*8 DT,DX,ALPHA,CF1,CF2

C       INITIALIZE THE SOLUTION VECTORS

        DO 1   I=1,21
        Y(I)=0.D0
        YNEW(I)=0.D0
1       CONTINUE

C       BOUNDARY CONDITIONS AND INTEGRATION PARAMETERS

        YNEW(1)=1.D0
        YNEW(21)=1.D0
        DT=.002D0
        DX=.05D0
        T=0.
        NITER=25
        ALPHA=DT/(2.D0*(DX**2))
        CF1=2.D0+(1.D0/ALPHA)
        CF2=2.D0-(1.D0/ALPHA)

C       INITIALIZE THE VECTORS FOR THE TRIDIAGONAL SYSTEM

        DO 2   I=1,21
        A(I)=-1.D0
        B(I)=CF1
        C(I)=-1.D0
        D(I)=0.D0
        BETA(I)=0.D0
        GAMM(I)=0.D0
2       CONTINUE

C       INITIAL OUTPUT

        WRITE(6,998)
        WRITE(6,999) T,Y(1),Y(6),Y(11),Y(16),Y(21)

C       START INTEGRATION

        DO 3   L=1,10
        DO 31  K=1,NITER
```

```
C       THE RIGHT SIDES FOR THE FIRST AND LAST EQUATIONS

        D(2)=Y(1)+Y(3)–CF2*Y(2)+YNEW(1)
        D(20)=Y(19)+Y(21)–CF2*Y(20)+YNEW(21)

C       THE RIGHT SIDES FOR THE OTHER EQUATIONS

        DO 311  J=3,19
        D(J)=Y(J–1)+Y(J+1)–CF2*Y(J)
311     CONTINUE

C       SOLVE THE EQUATIONS AND UPDATE TIME

        CALL TRDIAG(19,A(2),B(2),C(2),D(2),YNEW(2),BETA,GAMM)
        T=T+DT

C       UPDATE THE SOLUTION VECTOR

        DO 312  J=1,21
        Y(J)=YNEW(J)
312     CONTINUE

31      CONTINUE

C       OUTPUT RESULTS

        WRITE(6,999) T,Y(1),Y(6),Y(11),Y(16),Y(21)

3       CONTINUE

        STOP

998     FORMAT(7X,'T',7X,'X=0',7X,'X=0.25',4X,'X=0.5',5X,'X=0.75',4X,
     *  'X=1.0'/)
999     FORMAT(1X,6F10.5)

        END

C       SUBPROGRAM TRDIAG SOLVES A TRIDIAGONAL SYSTEM OF EQUATIONS

C       N               – THE NUMBER OF EQUATIONS
C       A,B,C           – THE VECTORS OF COEFFICIENTS OF THE LEFT SIDE
C       D               – THE RIGHT SIDE VECTOR
C       BETA,GAMM       – WORK SPACE VECTORS
C       YNEW            – THE SOLUTION VECTOR

        SUBROUTINE TRDIAG(N,A,B,C,D,YNEW,BETA,GAMM)
        REAL*8 A(N),B(N),C(N),D(N),YNEW(N),BETA(N),GAMM(N)
```

```
        BETA(1)=B(1)
        GAMM(1)=D((1)/BETA(1)
        DO 1   I=2,N
        BETA(I)=B(I)-(A(I)*C(I-1)/BETA(I-1))
        GAMM(I)=(D(I)-(A(I)*GAMM(I-1)))/BETA(I)
1       CONTINUE
        YNEW(N)=GAMM(N)
        DO 2   J=1,N-1
        I=N-J
        YNEW(I)=GAMM(I)-(C(I)*YNEW(I+1)/BETA(I))
2       CONTINUE
        RETURN
        END
```

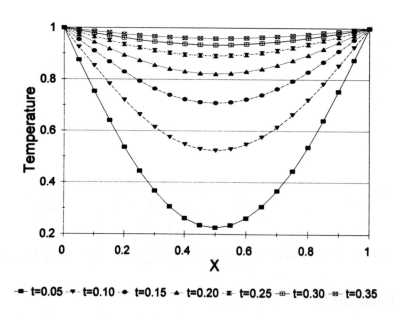

**FIGURE 8.3-1    Temperature distribution at various times**

## 8.4    ELLIPTIC EQUATIONS

An example of an elliptic equation is Laplace's equation which can be written in two-dimensions as

$$\frac{\partial^2 u}{\partial x^2} + \frac{\partial^2 u}{\partial y^2} = 0 \tag{8.4-1}$$

Substituting Equations (8.2-4, 6) into Equation (8.4-1), we obtain the difference form of Laplace's equation and it is

$$\frac{u_{i+1,j} - 2u_{i,j} + u_{i-1,j}}{h^2} + \frac{u_{i,j+1} - 2u_{i,j} + u_{i,j-1}}{k^2} = 0 \qquad (8.4\text{-}2)$$

It is usual to take $h = k$ and Equation (8.4-2) simplifies to

$$u_{i+1,j} + u_{i-1,j} + u_{i,j+1} + u_{i,j-1} - 4u_{i,j} = 0 \qquad (8.4\text{-}3a)$$

or

$$u_{i,j} = \frac{1}{4}(u_{i+1,j} + u_{i-1,j} + u_{i,j+1} + u_{i,j-1}) \qquad (8.4\text{-}3b)$$

Equation (8.4-3a) is known as a **five-point formula** as it involves five points. Equation (8.4-3b) shows that the value of u at point (i, j) is the average value of u computed from its four nearest surrounding points, as shown in Figure 8.2-1. This implies that the value of u at an interior point cannot be a maximum or a minimum (see Chapter 3, maximum modulus principle).

The appropriate boundary conditions associated with elliptic equations are (see Chapter 5, Section 6):

(i)      values of u given on the boundary (**Dirichlet problem**);

(ii)     normal derivative of u given on the boundary (**Neumann problem**);

(iii)    a combination of u and its normal derivative is given on the boundary (**Robin problem**).

## Dirichlet Problem

We assume the region to be rectangular and given by $0 \le x \le a, \ 0 \le y \le b$.

We subdivide the region into small squares each of length $h \ (= k)$, such that

$$x_n = a, \qquad y_m = b, \qquad x_0 = y_0 = 0 \qquad (8.4\text{-}4a,b,c,d)$$

The boundary conditions can be written as

$$u_{0,j} = \alpha_{0,j}, \qquad u_{n,j} = \alpha_{n,j}, \qquad u_{i,0} = \alpha_{i,0}, \qquad u_{i,m} = \alpha_{i,m} \qquad (8.4\text{-}5a,b,c,d)$$

where $\alpha_{i,j}$ are given.

For simplicity, we consider a coarse mesh and assume $m = n = 3$.

Equation (8.4-3a) now becomes, using the row-by-row ordering,

$$-4u_{1,1} + u_{2,1} + u_{0,1} + u_{1,2} + u_{1,0} = 0 \tag{8.4-6a}$$

$$-4u_{2,1} + u_{3,1} + u_{1,1} + u_{2,2} + u_{2,0} = 0 \tag{8.4-6b}$$

$$-4u_{1,2} + u_{2,2} + u_{0,2} + u_{1,3} + u_{1,1} = 0 \tag{8.4-6c}$$

$$-4u_{2,2} + u_{3,2} + u_{1,2} + u_{2,3} + u_{2,1} = 0 \tag{8.4-6d}$$

We label the four unknowns ($u_{1,1}$, $u_{2,1}$, $u_{1,2}$, and $u_{2,2}$) as ($v_1$, $v_2$, $v_3$, and $v_4$). Combining Equations (8.4-5a to d, 6a to d), we obtain

$$-4v_1 + v_2 + v_3 \quad = -\alpha_{0,1} - \alpha_{1,0} \quad = r_1 \tag{8.4-7a,b}$$

$$v_1 - 4v_2 + v_4 \quad = -\alpha_{2,0} - \alpha_{3,1} \quad = r_2 \tag{8.4-7c,d}$$

$$v_1 - 4v_3 + v_4 \quad = -\alpha_{1,3} - \alpha_{0,2} \quad = r_3 \tag{8.4-7e,f}$$

$$v_2 + v_3 - 4v_4 \quad = -\alpha_{3,2} - \alpha_{2,3} \quad = r_4 \tag{8.4-7g,h}$$

Equations (8.4-7a to h) can be written in matrix form as

$$\begin{bmatrix} -4 & 1 & . & 1 & 0 \\ 1 & -4 & . & 0 & 1 \\ . & . & . & . & . \\ 1 & 0 & . & -4 & 1 \\ 0 & 1 & . & 1 & -4 \end{bmatrix} \begin{bmatrix} v_1 \\ v_2 \\ v_3 \\ v_4 \end{bmatrix} = \begin{bmatrix} r_1 \\ r_2 \\ r_3 \\ r_4 \end{bmatrix} \tag{8.4-8a}$$

or $\quad \underline{\underline{A}}\,\underline{v} = \underline{r}$ \hfill (8.4-8b)

Note that the matrix $\underline{\underline{A}}$ can be partitioned into four matrices as shown in Equation (8.4-8a). An alternative way of writing Equation (8.4-8b) is

$$\begin{bmatrix} \underline{\underline{B}} & \underline{\underline{I}} \\ \underline{\underline{I}} & \underline{\underline{B}} \end{bmatrix} \underline{v} = \underline{r} \tag{8.4-9}$$

where $\underline{\underline{I}}$ is the (2 × 2) identity matrix and $\underline{\underline{B}}$ is another (2 × 2) matrix.

For any value of n and m, $\underline{\underline{A}}$ is a (n – 1) (m – 1) rows × (n – 1) (m – 1) columns matrix. It can be partitioned into (m – 1) identity matrices each of size (n – 1) × (n – 1) and (m – 1) square matrices each of size (n – 1) × (n – 1) denoted by $\underline{\underline{B}}$. $\underline{\underline{A}}$ and $\underline{\underline{B}}$ are of the form

$$
\underline{\underline{A}} = \begin{bmatrix} \underline{\underline{B}} & \underline{\underline{I}} & & & 0 \\ \underline{\underline{I}} & \underline{\underline{B}} & \underline{\underline{I}} & & \\ & & \cdot & & \\ & & & \cdot & \\ & & & \cdot & \\ 0 & & \underline{\underline{I}} & \underline{\underline{B}} & \underline{\underline{I}} \\ & & & \underline{\underline{I}} & \underline{\underline{B}} \end{bmatrix} \tag{8.4-10a}
$$

$$
\underline{\underline{B}} = \begin{bmatrix} -4 & 1 & & & 0 \\ 1 & -4 & 1 & & \\ & & \cdot & & \\ & & & \cdot & \\ & & & \cdot & \\ 0 & & 1 & -4 & 1 \\ & & & 1 & -4 \end{bmatrix} \tag{8.4-10b}
$$

If the size of $\underline{\underline{A}}$ is small, Equation (8.4-8b) can be solved by the method of Gaussian elimination. But if $\underline{\underline{A}}$ is large, and given that it is sparse, it is more economical to adopt the method of iteration. This method is discussed in Chapter 7, Section 4. In this chapter, we describe the method of **successive over relaxation** (S.O.R.).

The **S.O.R.** method is essentially a modified version of the Gauss-Seidel method (see, Section 7.4). We use the same notation as in Chapter 7 and the $k^{th}$ iterate is written as $u_{i,j}^{(k)}$. From Equation (8.4-3b), we see that a possible iteration formula is

$$
u_{i,j}^{(k+1)} = \frac{1}{4} \left[ u_{i+1,j}^{(k)} + u_{i-1,j}^{(k+1)} + u_{i,j+1}^{(k)} + u_{i,j-1}^{(k+1)} \right] \tag{8.4-11}
$$

Note that, as in the Gauss-Seidel method, we have used the new values (k + 1 iterate) of $u_{i-1,j}$ and $u_{i,j-1}$.

Equation (8.4-11) can also be written as

$$
u_{i,j}^{(k+1)} = u_{i,j}^{(k)} + \frac{1}{4} \left[ u_{i+1,j}^{(k)} + u_{i-1,j}^{(k+1)} + u_{i,j+1}^{(k)} + u_{i,j-1}^{(k+1)} - 4u_{i,j}^{(k)} \right] \tag{8.4-12}
$$

To improve convergence, Equation (8.4-12) is modified to

$$
u_{i,j}^{(k+1)} = u_{i,j}^{(k)} + \frac{\omega}{4} \left[ u_{i+1,j}^{(k)} + u_{i-1,j}^{(k+1)} + u_{i,j+1}^{(k)} + u_{i,j-1}^{(k+1)} - 4u_{i,j}^{(k)} \right] \tag{8.4-13}
$$

The optimum value of $\omega$ depends on the mesh size and lies between 1 and 2.  For the Dirichlet problem in a rectangular region, the optimum value of $\omega$ ($\omega_{op}$) is given by

$$\omega_{op} = \frac{4}{2 + \sqrt{4 - (\cos\frac{\pi}{n} + \cos\frac{\pi}{m})^2}} \qquad (8.4\text{-}14)$$

For details on $\omega_{op}$, see Hageman and Young (1981).

To start the iterative process, we need to assume the values of the zeroth iterative of the interior points. One possibility is to assume that the value of all the interior points is the same and equal to the average value of the boundary points.

***Example 8.4-1.***  Solve the equation

$$\frac{\partial^2 u}{\partial x^2} + \frac{\partial^2 u}{\partial y^2} = 0, \qquad 0 < x < 3, \quad 0 < y < 3 \qquad (8.4\text{-}15)$$

subject to

$$u(x, 0) = 10, \qquad u(x, 3) = 90, \qquad u(0, y) = 70, \qquad u(3, y) = 0 \qquad (8.4\text{-}16a,b,c,d)$$

We divide the region into 9 unit squares.  The values of $u$ at the four interior points are given by Equations (8.4-7a to h).  The transpose of the column vector $\underline{r}$ is given by

$$\underline{r}^T = [-80, -10, -160, -90] \qquad (8.4\text{-}17)$$

Solving Equation (8.4-8a) directly (Gaussian elimination), we obtain

$$v_1 = u_{1,1} = 41.25, \qquad v_2 = u_{2,1} = 23.75 \qquad (8.4\text{-}18a,b)$$

$$v_3 = u_{1,2} = 61.25, \qquad v_4 = u_{2,2} = 43.75 \qquad (8.4\text{-}18c,d)$$

We now solve the same set of equations by the S.O.R. method.  From Equation (8.4-14), we deduce that $\omega_{op}$ is given approximately by

$$\omega_{op} = 1.156 \qquad (8.4\text{-}19)$$

Taking into account the boundary conditions, Equation (8.4-13) for the four unknowns ($u_{1,1}$, $u_{2,1}$, $u_{1,2}$, and $u_{2,2}$) can be written as

$$u_{1,1}^{(k+1)} = u_{1,1}^{(k)} + 0.289\,[u_{2,1}^{(k)} + 70 + u_{1,2}^{(k)} + 10 - 4u_{1,1}^{(k)}] \qquad (8.4\text{-}20a)$$

$$u_{2,1}^{(k+1)} = u_{2,1}^{(k)} + 0.289 \ [u_{1,1}^{(k+1)} + u_{2,2}^{(k)} + 10 - 4u_{2,1}^{(k)}] \tag{8.4-20b}$$

$$u_{1,2}^{(k+1)} = u_{1,2}^{(k)} + 0.289 \ [u_{2,2}^{(k)} + 70 + 90 + u_{1,1}^{(k+1)} - 4u_{1,2}^{(k)}] \tag{8.4-20c}$$

$$u_{2,2}^{(k+1)} = u_{2,2}^{(k)} + 0.289 \ [u_{1,2}^{(k+1)} + 90 + u_{2,1}^{(k+1)} - 4u_{2,2}^{(k)}] \tag{8.4-20d}$$

To start the iteration, we assume that the value of u at the interior points is equal to the average value at the boundary points. The average value in this case is 42.5. The values of $u_{i,j}$ at subsequent iterations are given in Table 8.4-1. It can be seen that the convergence is rapid.

### TABLE 8.4-1

### Values of $u_{i,j}$ using the S.O.R. method

| $u_{i,j}$ | k = 0 | k = 1 | k = 2 | k = 3 | k = 4 |
|-----------|-------|-------|-------|-------|-------|
| $u_{1,1}$ | 42.5 | 41.055 | 41.039 | 41.277 | 41.252 |
| $u_{2,1}$ | 42.5 | 20.407 | 24.197 | 23.689 | 23.761 |
| $u_{1,2}$ | 42.5 | 63.757 | 60.784 | 61.331 | 61.240 |
| $u_{2,2}$ | 42.5 | 43.703 | 43.752 | 43.755 | 43.750 |

## Neumann Problem

In this case, the values of u at the boundary points are not given and have to be determined at the same time as the interior points. We can proceed in the same way as in the case of parabolic equations and write the normal derivative in a finite difference form. This will introduce points which are outside the region of interest and these fictitious outside points can be eliminated as previously shown in the parabolic case. The next example shows how this is done.

***Example 8.4-2.*** Solve the equation

$$\frac{\partial^2 u}{\partial x^2} + \frac{\partial^2 u}{\partial y^2} = 0, \quad 0 < x < 1, \ 0 < y < 1 \tag{8.4-21}$$

subject to the conditions

$$\left. \frac{\partial u}{\partial x} \right|_{x=0} = \left. \frac{\partial u}{\partial y} \right|_{y=0} = 0 \tag{8.4-22a,b}$$

$$\frac{\partial u}{\partial x}\bigg|_{x=1} = 1, \qquad \frac{\partial u}{\partial y}\bigg|_{y=1} = -1 \qquad\qquad\qquad (8.4\text{-}22c,d)$$

We subdivide the region of interest, which is a unit square, into four squares, each of size (1/2 × 1/2). That is to say

$$h = k = 1/2 \qquad\qquad\qquad\qquad\qquad\qquad\qquad (8.4\text{-}23a,b)$$

Equation (8.4-21) in difference form is given by Equation (8.4-3a). Note that only the single interior point (1, 1) has all its surrounding points in the region of interest, as shown in Figure 8.4-1. On writing the difference equation for points on the boundary, we will introduce fictitious outside points which lie on the dotted lines as shown in Figure 8.4-1. The first and last few difference equations [Equation (8.4-3a)] are

$$u_{1,0} + u_{-1,0} + u_{0,1} + u_{0,-1} - 4u_{0,0} = 0 \qquad\qquad\qquad (8.4\text{-}24a)$$

$$u_{2,0} + u_{0,0} + u_{1,1} + u_{1,-1} - 4u_{1,0} = 0 \qquad\qquad\qquad (8.4\text{-}24b)$$

$$\begin{matrix} \cdot \\ \cdot \\ \cdot \end{matrix}$$

$$u_{2,2} + u_{0,2} + u_{1,3} + u_{1,1} - 4u_{1,2} = 0 \qquad\qquad\qquad (8.4\text{-}24c)$$

$$u_{3,2} + u_{1,2} + u_{2,3} + u_{2,1} - 4u_{2,2} = 0 \qquad\qquad\qquad (8.4\text{-}24d)$$

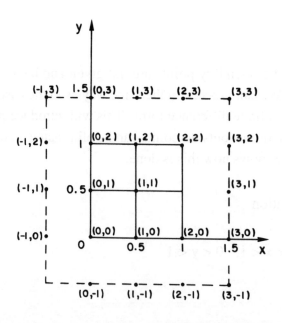

FIGURE 8.4-1    Grid for solving Laplace's equation in the Neumann case

We eliminate $u_{-1,0}$, $u_{0,-1}$, $u_{3,2}$, and $u_{2,3}$ in Equations (8.4-24a to d) via the boundary conditions. Using the central difference form for the derivatives [Equations (8.2-3c, 5c)], the boundary conditions [Equations (8.4-22a to d)] can be written respectively as

$$u_{1,j} - u_{-1,j} = 0, \qquad u_{i,1} - u_{i,-1} = 0 \tag{8.4-25a,b}$$

$$u_{3,j} - u_{1,j} = 1, \qquad u_{i,3} - u_{i,1} = -1 \tag{8.4-25c,d}$$

The values of $u$ at the outside points ($u_{-1,j}$, $u_{i,-1}$, $u_{3,j}$, and $u_{i,3}$) can be replaced by ($u_{1,j}$, $u_{i,1}$, $1 + u_{1,j}$, and $-1 + u_{i,1}$) respectively. We now have a set of 9 equations to solve for the 9 unknowns ($u_{0,0}$, $u_{1,0}$, ... , $u_{2,2}$). This set of 9 equations is

$$\begin{bmatrix} -4 & 2 & 0 & 2 & 0 & 0 & 0 & 0 & 0 \\ 1 & -4 & 1 & 0 & 2 & 0 & 0 & 0 & 0 \\ 0 & 2 & -4 & 0 & 0 & 2 & 0 & 0 & 0 \\ 1 & 0 & 0 & -4 & 2 & 0 & 1 & 0 & 0 \\ 0 & 1 & 0 & 1 & -4 & 1 & 0 & 1 & 0 \\ 0 & 0 & 1 & 0 & 2 & -4 & 0 & 0 & 1 \\ 0 & 0 & 0 & 2 & 0 & 0 & -4 & 2 & 0 \\ 0 & 0 & 0 & 0 & 2 & 0 & 1 & -4 & 1 \\ 0 & 0 & 0 & 0 & 0 & 2 & 0 & 2 & -4 \end{bmatrix} \begin{bmatrix} u_{0,0} \\ u_{1,0} \\ u_{2,0} \\ u_{0,1} \\ u_{1,1} \\ u_{2,1} \\ u_{0,2} \\ u_{1,2} \\ u_{2,2} \end{bmatrix} = \begin{bmatrix} 0 \\ 0 \\ -1 \\ 0 \\ 0 \\ -1 \\ 1 \\ 1 \\ 0 \end{bmatrix} \tag{8.4-26a}$$

or $\qquad \underline{\underline{A}}\, \underline{u} = \underline{b}$          (8.4-26b)

Equation (8.4-26a) can be solved either by a direct method or an iterative method. Since the size of $\underline{\underline{A}}$ is manageable, we have obtained the solution by a direct method and it is

$$\underline{u}^T = \left[ 0, \frac{1}{8}, \frac{1}{2}, -\frac{1}{8}, 0, \frac{3}{8}, -\frac{1}{2}, -\frac{3}{8}, 0 \right] \tag{8.4-27}$$

**Poisson's and Helmholtz's Equations**

**Poisson's equation** can be written as

$$\nabla^2 u - g(x, y) = 0 \tag{8.4-28}$$

Proceeding as in the case of Laplace's equation, the finite difference form for Equation (8.4-28) can be written as

$$u_{i+1,j} - 4u_{i,j} + u_{i-1,j} + u_{i,j+1} + u_{i,j-1} - h^2 g_{ij} = 0 \tag{8.4-29a}$$

where $g_{ij} = g(x_i, y_i)$ (8.4-29b)

**Helmholtz's equation** is

$$\nabla^2 u + f(x, y) u - g(x, y) = 0 \tag{8.4-30}$$

Its finite difference form is

$$u_{i+1,j} - (4 - h^2 f_{ij}) u_{i,j} + u_{i-1,j} + u_{i,j+1} + u_{i,j-1} - h^2 g_{ij} = 0 \tag{8.4-31a}$$

where $f_{ij} = f(x_i, y_i)$ (8.4-31b)

The difference equations [Equations (8.2-29a, 31a)] can be solved by the methods discussed earlier.

***Example 8.4-3.*** Solve the equation

$$\frac{\partial^2 u}{\partial x^2} + \frac{\partial^2 u}{\partial y^2} = 0, \quad 0 \le x \le 1, \quad 0 \le y \le 1 \tag{8.4-32}$$

subject to the following boundary conditions

$$u(x, 0) = x^2, \qquad u(x, 1) = x^2 - 1 \tag{8.4-33a,b}$$

$$u(0, y) = -y^2, \qquad u(1, y) = 1 - y^2 \tag{8.4-33c,d}$$

The finite difference form of Equation (8.4-32) can be written in the form of Equation (8.4-13), that is

$$u_{i,j}^{(k+1)} = \omega \left( u_{i-1,j}^{(k)} + u_{i+1,j}^{(k)} + u_{i,j-1}^{(k)} + u_{i,j+1}^{(k)} \right) / 4 + (1 - \omega) u_{i,j}^{(k)} \tag{8.4-34}$$

The value of $\omega$ is chosen to be 1.5. The program used to solve the present problem is next listed. The results are shown in Figure 8.4-2.

The analytical solution is

$$u = x^2 - y^2 \tag{8.4-35}$$

```
C     MAIN PROGRAM FOR SOLUTION OF EXAMPLE 8.4-3 USING SIMULTANEOUS
C     OVER-RELAXATION TO SOLVE THE FINITE DIFFERENCE EQUATION

      REAL*8U(11,11),OMEGA,X,Y,TEMP
      OPEN(UNIT=3,FILE='MATRIX3.OUT')

C     SET BOUNDARY CONDITIONS
```

```
        DO 1   I=2,10
        DO 1   J=2,10
1       U(J,I)=0.D0
        DO 2   I=1,11
        TEMP=(I-1)**2/1.D2
        U(1,I)=-TEMP
        U(I,1)=TEMP
        U(11,I)=1.D0-TEMP
        U(I,11)=TEMP-1.D0
2       CONTINUE

C       SET OVER-RELAXATION PARAMETER AND ITERATE THE SOLUTION

        OMEGA=1.5D0
        DO 3   NITER=1,40
        CALL SOR(U,OMEGA,2,11,2,11)
3       CONTINUE

C       OUTPUT THE FINAL RESULTS

        WRITE(3,999)U
        STOP
999     FORMAT(11E12.3)
        END

C       SUBROUTINE SOR FOR SIMULTANEOUS OVER-RELAXATION

C       THE CALCULATIONS ARE DONE IN PLACE RATHER THAN COPYING TO
C       ANOTHER MATRIX

        SUBROUTINE SOR(U,OMEGA,NSTART,N,MSTART,M)
        REAL*8 U(N,M),OMEGA

C       LOOP THROUGH THE MATRIX

        DO 1   I=NSTART,N-1
        DO 1   J=MSTART,M-1
        U(I,J)=OMEGA*(U(I-1,J)+U(I+1,J)+U(I,J-1)+U(I,J+1))/4.D0
     *  +(1.D0-OMEGA)*U(I,J)
1       CONTINUE
        RETURN
        END
```

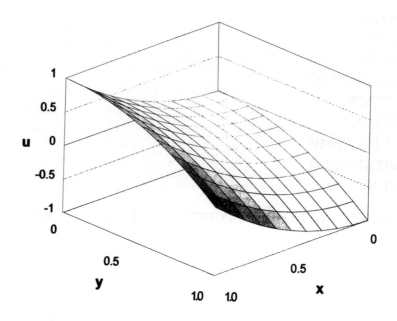

**FIGURE 8.4-2    Values of u as a function of x and y**

## 8.5    HYPERBOLIC EQUATIONS

### Difference Equations

The wave equation [Equation (5.1-1)] is an example of a hyperbolic equation. Associated with this P.D.E. is a set of boundary and initial conditions. We write the whole system as

$$\frac{\partial^2 u}{\partial t^2} = c^2 \frac{\partial^2 u}{\partial x^2}, \quad t > 0, \quad 0 \le x \le \ell \tag{8.5-1a}$$

$$u\,(0, t) \,=\, u\,(\ell, t) \,=\, 0 \tag{8.5-1b,c}$$

$$u\,(x, 0), \,=\, f\,(x), \qquad \left.\frac{\partial u}{\partial t}\right|_{t=0} = g\,(x) \tag{8.5-1d,e}$$

where c and $\ell$ are constants.

The finite difference form of Equation (8.5-1a) is

$$\frac{u_{i,j+1} - 2\,u_{i,j} + u_{i,j-1}}{k^2} = c^2 \left(\frac{u_{i+1,j} - 2\,u_{i,j} + u_{i-1,j}}{h^2}\right) \tag{8.5-2}$$

where the first subscript denotes position (x) and the second subscript denotes time (t). As usual, the space of interest (x, t) is subdivided into rectangles of size (h × k). Recall that

$$x_0 = 0, \qquad x_n = \ell, \qquad t_0 = 0 \tag{8.5-3a,b,c}$$

Equations (8.5-1b to e) are now written as

$$u_{0,j} = u_{n,j} = 0, \quad \forall\, j \geq 0 \tag{8.5-4a,b}$$

$$u_{i,0} = f(x_i) = f_i \tag{8.5-4c,d}$$

$$\frac{u_{i,1} - u_{i,-1}}{2k} = g(x_i) = g_i \tag{8.5-4e,f}$$

Note that Equations (8.5-4e, f) involve the fictitious point of $t_{-1}$. Equation (8.5-2) can be written as

$$u_{i,j+1} = 2u_{i,j} - u_{i,j-1} + r^2(u_{i+1,j} - 2u_{i,j} + u_{i-1,j}) \tag{8.5-5a}$$

where $r^2 = c^2 k^2 / h^2$ $\tag{8.5-5b}$

We choose r to be unity and combining Equations (8.5-4c to f, 5a) yields

$$u_{i,1} = 2u_{i,0} - u_{i,-1} + u_{i+1,0} - 2u_{i,0} + u_{i-1,0} \tag{8.5-6a}$$

$$= f_{i+1} + f_{i-1} - u_{i,1} + 2kg_i \tag{8.5-6b}$$

Equation (8.5-6b) simplifies to

$$u_{i,1} = \frac{1}{2}(f_{i+1} + f_{i-1} + 2kg_i) \tag{8.5-7}$$

Note that $u_{0,1}$ and $u_{n,1}$ are given by Equations (8.5-4a, b) and need not be computed.

Having determined the value of u at time $t_1$, we can similarly determine the values of u at subsequent times $t_2$, $t_3$, ...

***Example 8.5-1.*** Solve the equation

$$\frac{\partial^2 u}{\partial t^2} = \frac{\partial^2 u}{\partial x^2} \tag{8.5-8a}$$

subject to

$$u(0, t) = u(1, t) = 0 \tag{8.5-8b,c}$$

$$u(x, 0) = \frac{1}{2} x(1 - x), \qquad \frac{\partial u}{\partial t}\bigg|_{t=0} = 0 \qquad\qquad\qquad (8.5\text{-}8d,e)$$

We choose

$$h = k = 0.1 \qquad\qquad\qquad (8.5\text{-}9a,b)$$

The value of $r$ is then unity. In this example, $g$ is zero and Equation (8.5-7) simplifies to

$$u_{i,1} = \frac{1}{4}(x_{i+1} + x_{i-1} - x_{i+1}^2 - x_{i-1}^2) \qquad\qquad\qquad (8.5\text{-}10)$$

Equation (8.5-5a) simplifies to

$$u_{i,j+1} = u_{i+1,j} + u_{i-1,j} - u_{i,j-1} \qquad\qquad\qquad (8.5\text{-}11)$$

The initial condition [Equation (8.5-8d)] provides $u_{i,0}$, Equation (8.5-10) determines $u_{i,1}$, then from Equation (8.5-11), the values of $u$ at all positions and subsequent times can be computed. The values of $u$ from time $t = 0$ to $t = 0.5$ are tabulated in Table 8.5-1. From the symmetry of the problem, we need to compute for the interval $0 < x \le 0.5$ only.

### TABLE 8.5-1

### Values of $u$ at various positions and times

| t \ x | 0 | 0.1 | 0.2 | 0.3 | 0.4 | 0.5 |
|---|---|---|---|---|---|---|
| 0 | 0 | 0.045 | 0.080 | 0.105 | 0.120 | 0.125 |
| 0.1 | 0 | 0.040 | 0.075 | 0.100 | 0.115 | 0.120 |
| 0.2 | 0 | 0.030 | 0.060 | 0.085 | 0.100 | 0.105 |
| 0.3 | 0 | 0.020 | 0.040 | 0.060 | 0.075 | 0.080 |
| 0.4 | 0 | 0.01 | 0.02 | 0.03 | 0.04 | 0.045 |
| 0.5 | 0 | 0 | 0 | 0 | 0 | 0 |

Equation (8.5-8a), subject to Equations (8.5-8b to e), can be solved analytically (see also Chapter 5, Problem 9a) and the solution is

$$u = \frac{4}{\pi^3} \sum_{s=0}^{\infty} \frac{\cos(2s+1)\pi t \, \sin(2s+1)\pi x}{(2s+1)^3} \qquad\qquad\qquad (8.5\text{-}12)$$

From Equation (8.5-12), we can deduce that at

$$t = 0.5 , \qquad u = 0 \qquad\qquad\qquad\qquad (8.5\text{-}13a,b)$$

From Table 8.5-1, we observe that there is complete agreement between the analytical and numerical results.

●

We now show that Equation (8.5-11) is an exact difference equation of the problem. D'Alembert's solution (Chapter 5, Section 4) of Equations (8.5-1a to e) with $g(x) = 0$ can be written as

$$u = f(x - t) + f(x + t) \qquad\qquad\qquad\qquad (8.5\text{-}14)$$

Using Equation (8.5-14), the right side of Equation (8.5-11) is

$$u_{i+1,j} + u_{i-1,j} - u_{i,j-1} = f(x_{i+1} - t_j) + f(x_{i+1} + t_j) + f(x_{i-1} - t_j) + f(x_{i-1} + t_j)$$

$$- f(x_i - t_{j-1}) - f(x_i + t_{j-1}) \qquad\qquad (8.5\text{-}15a)$$

$$= f(x_i + h - t_j) + f(x_i + h + t_j) + f(x_i - h - t_j) + f(x_i - h + t_j)$$

$$- f(x_i - t_j + h) - f(x_i + t_j - h) \qquad\qquad (8.5\text{-}15b)$$

$$= f(x_i + h + t_j) + f(x_i - h - t_j) \qquad\qquad\qquad (8.5\text{-}15c)$$

$$= u_{i,j+1} \qquad\qquad\qquad\qquad (8.5\text{-}15d)$$

We note that the right side of Equation (8.5-15d) is exactly the left side of Equation (8.5-11). This result is due to $r$ being unity ($h = k$).

Another way of showing that Equation (8.5-11) provides an exact solution of the problem is to consider the domain of dependence of the solution. This concept is discussed in Chapter 5, Section 4.

In Figure 8.5-1, we have marked the pivotal triangular domains which determine the value of $u_{i,j+1}$. Noting that the characteristics in this case have gradients $\pm 1$, we observe by comparing Figures 8.5-1 and 5.4-3 that the domain of dependence of Equation (8.5-11) is exactly that of Equation (8.5-8a).

The parameter $r$ is the **Courant parameter** and it has been shown by Courant et al. (1928) that if $r$ is less than or equal to one, Equation (8.5-5a) is a stable algorithm for solving Equation (8.5-1a). In Chapters 5 and 6, we have used the method of characteristics to solve hyperbolic equations. This method has the advantage that discontinuities in the initial data are propagated along the characteristics. Thus, this method is generally more suitable for cases of discontinuities in the given data.

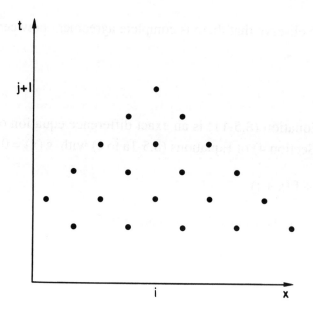

**FIGURE 8.5-1  Pivotal points determining $u_{i,j+1}$**

## Method of Characteristics

We consider a quasi-linear P.D.E. which can be written as

$$a_{11} \frac{\partial^2 u}{\partial x^2} + a_{12} \frac{\partial^2 u}{\partial x \, \partial y} + a_{22} \frac{\partial^2 u}{\partial y^2} = f \qquad (8.5\text{-}16)$$

where $a_{ij}$ and $f$ are functions of $x$, $y$, $u$, $\dfrac{\partial u}{\partial x}$ and $\dfrac{\partial u}{\partial y}$.

For simplicity, we denote

$$p = \frac{\partial u}{\partial x}, \qquad q = \frac{\partial u}{\partial y} \qquad (8.5\text{-}17a,b)$$

Writing out the differentials of $p$ and $q$ yields

$$dp = \frac{\partial p}{\partial x} dx + \frac{\partial p}{\partial y} dy \qquad (8.5\text{-}18a)$$

$$= \frac{\partial^2 u}{\partial x^2} dx + \frac{\partial^2 u}{\partial x \, \partial y} dy \qquad (8.5\text{-}18b)$$

$$dq = \frac{\partial q}{\partial x} dx + \frac{\partial q}{\partial y} dy \tag{8.5-18c}$$

$$= \frac{\partial^2 u}{\partial x \, \partial y} dx + \frac{\partial^2 u}{\partial y^2} dy \tag{8.5-18d}$$

From Equations (8.5-18b, d), we deduce

$$\frac{\partial^2 u}{\partial x^2} = \frac{dp}{dx} - \frac{\partial^2 u}{\partial x \, \partial y} \frac{dy}{dx} \tag{8.5-19a}$$

$$\frac{\partial^2 u}{\partial y^2} = \frac{dq}{dy} - \frac{\partial^2 u}{\partial x \, \partial y} \frac{dx}{dy} \tag{8.5-19b}$$

Substituting Equations (8.5-19a, b) into Equation (8.5-16) yields

$$a_{11} \frac{dp}{dx} + a_{22} \frac{dq}{dy} - \frac{\partial^2 u}{\partial x \, \partial y} \left( a_{11} \frac{dy}{dx} - a_{12} + a_{22} \frac{dx}{dy} \right) = f \tag{8.5-20}$$

Equation (8.5-20) can be written as

$$a_{11} \frac{dp}{dx} \frac{dy}{dx} + a_{22} \frac{dq}{dy} \frac{dy}{dx} - f \frac{dy}{dx} - \frac{\partial^2 u}{\partial x \, \partial y} \left[ a_{11} \left( \frac{dy}{dx} \right)^2 - a_{12} \left( \frac{dy}{dx} \right) + a_{22} \right] = 0 \tag{8.5-21}$$

We now define a curve $\Gamma$ in the xy-plane, as shown in Figure 8.5-2, such that the slope of the tangent at every point of $\Gamma$ is a root of the equation

$$a_{11} \left( \frac{dy}{dx} \right)^2 - a_{12} \left( \frac{dy}{dx} \right) + a_{22} = 0 \tag{8.5-22}$$

Equation (8.5-22) has two roots

$$\frac{dy}{dx} = \frac{a_{12} \pm \sqrt{a_{12}^2 - 4a_{11}a_{22}}}{2a_{11}} = m_{\pm} \tag{8.5-23a,b,c,d}$$

Since the equation is hyperbolic, the two roots are real. We denote the root given by the positive sign by $m_+$ and the curve corresponding to $m_+$ by $\Gamma_+$. Similarly, $m_-$ and $\Gamma_-$ denote the root and the curve corresponding to the negative sign. $\Gamma_{\pm}$ are the characteristics of Equation (8.5-16) and are discussed in Chapter 5, Section 3.

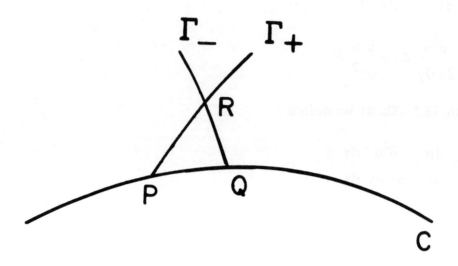

**FIGURE 8.5-2    Method of characteristics**

It follows from Equation (8.5-21) that along a characteristic

$$a_{11} \frac{dp}{dx} \frac{dy}{dx} + a_{22} \frac{dq}{dx} - f \frac{dy}{dx} = 0 \qquad (8.5\text{-}24a)$$

or $\quad a_{11} \dfrac{dy}{dx} dp + a_{22} dq - f\, dy = 0 \qquad (8.5\text{-}24b)$

Let C be a curve along which initial values for u, p, and q are known. C is not a characteristic. Let P and Q be two neighboring points on C. The characteristic $\Gamma_+$ through P intersects $\Gamma_-$ through Q at R, as shown in Figure 8.5-2. As a first approximation, we assume PR and QR as straight lines with slopes $m_+$ and $m_-$. The equation of these lines are

$$y_R - y_P = m_{+P} (x_R - x_P) \qquad (8.5\text{-}25a)$$

$$y_R - y_Q = m_{-Q} (x_R - x_Q) \qquad (8.5\text{-}25b)$$

where the suffixes P, Q, and R denote that the quantities are evaluated at P, Q, and R respectively.

From Equations (8.5-25a, b), we can determine $(x_R, y_R)$ and they are

$$x_R = \frac{(y_Q - y_P) + (m_+ x_P - m_- x_Q)}{(m_+ - m_-)} \qquad (8.5\text{-}26a)$$

$$y_R = \frac{m_+ m_- (x_P - x_Q) + (m_+ y_Q - m_- y_P)}{(m_+ - m_-)} \qquad (8.5\text{-}26b)$$

where $m_+$ and $m_-$ denote $m_{+P}$ and $m_{-P}$.

Equation (8.5-24b) along $\Gamma_+$ and $\Gamma_-$ are respectively given by

$$a_{11} m_+ dp + a_{22} dq - f\, dy = 0 \qquad (8.5\text{-}27a)$$

$$a_{11} m_- dp + a_{22} dq - f\, dy = 0 \qquad (8.5\text{-}27b)$$

Equations (8.5-27a, b) can be approximated by

$$a_{11} m_+ (p_R - p_P) + a_{22} (q_R - q_P) - f\, (y_R - y_P) = 0 \qquad (8.5\text{-}28a)$$

$$a_{11} m_- (p_R - p_Q) + a_{22} (q_R - q_Q) - f\, (y_R - y_Q) = 0 \qquad (8.5\text{-}28b)$$

Combining Equations (8.5-26a, b, 28 a, b) yields values of $p_R$ and $q_R$. The values of $p_P$, $p_Q$, $q_P$, and $q_Q$ are known. To determine $u_R$ we use the relation

$$du = \frac{\partial u}{\partial x} dx + \frac{\partial u}{\partial y} dy \qquad (8.5\text{-}29a)$$

$$= p\, dx + q\, dy \qquad (8.5\text{-}29b)$$

The values of $p$ and $q$ along PR are assumed to be given by the average values evaluated at P and R. Equation (8.5-29b) is then approximated by

$$u_R - u_P = \frac{1}{2} (p_R + p_P)(x_R - x_P) + \frac{1}{2} (q_R + q_P)(y_R - y_P) \qquad (8.5\text{-}30a)$$

or $\quad u_R = u_P + \frac{1}{2} (p_R + p_P)(x_R - x_P) + \frac{1}{2} (q_R + q_P)(y_R - y_P) \qquad (8.5\text{-}30b)$

To improve the accuracy, we replace the values of $m_+$ and $m_-$ evaluated at P in Equations (8.5-25a, b) by the average values of $m_+$ and $m_-$ evaluated at P and R. Equations (8.5-25a, b) now read

$$y_R - y_P = \frac{1}{2} (m_{+P} + m_{+R})(x_R - x_P) \qquad (8.5\text{-}31a)$$

$$y_R - y_Q = \frac{1}{2} (m_{-Q} + m_{-R})(x_R - x_Q) \qquad (8.5\text{-}31b)$$

The value of $u_R$ is obtained as described earlier. This process can be repeated until the required accuracy is achieved. Similarly, by considering other points along C, the solution u in the xy-plane can be determined.

***Example 8.5-2.*** Solve the problem stated in Example 8.5-1 by the method of characteristics.

In this problem, y is replaced by t. The characteristics are given by

$$\Gamma_+: \quad \frac{dt}{dx} = m_+ = 1 \qquad\qquad\qquad (8.5\text{-}32a,b)$$

$$\Gamma_-: \quad \frac{dt}{dx} = m_- = -1 \qquad\qquad\qquad (8.5\text{-}32c,d)$$

The initial values of u are given along the x-axis which is not a characteristic. We choose the points P and Q to be (0.1, 0) and (0.3, 0) respectively as shown in Figure 8.5-3. The values u, p, and q at P and Q can be computed from Equations (8.5-8d, e) and are

$$u_P = 0.045, \quad u_Q = 0.105, \quad p_P = 0.4 \qquad\qquad (8.5\text{-}33a,b,c)$$

$$p_Q = 0.2, \quad q_P = q_Q = 0 \qquad\qquad\qquad (8.5\text{-}33d,e,f)$$

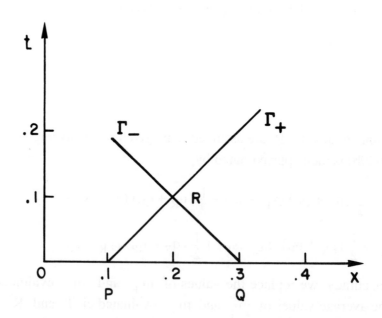

**FIGURE 8.5-3     Numerical solution of the wave equation
by the method of characteristics**

The point $(x_R, y_R)$ is determined from Equations (8.5-26a, b) and is found to be

$$x_R = 0.2, \qquad y_R = 0.1 \qquad\qquad\qquad (8.5\text{-}34a,b)$$

This can also be seen from Figure 8.5-3.

Equations (8.5-28a, b) reduce to

$$p_R - 0.4 - q_R = 0 \qquad\qquad (8.5\text{-}35a)$$

$$0.2 - p_R - q_R = 0 \qquad\qquad (8.5\text{-}35b)$$

On solving Equations (8.5-35a, b), we obtain

$$p_R = 0.3, \qquad q_R = -0.1 \qquad\qquad (8.5\text{-}36a,b)$$

The approximate value of $u_R$ is given approximately by Equation (8.5-30b) and is found to be

$$u_R = 0.075 \qquad\qquad (8.5\text{-}37)$$

From Table 8.5-1, we note that this is exactly the value obtained by the finite difference method. The values of $u$ at other points in the $xt$-plane can similarly be obtained.

## 8.6   IRREGULAR BOUNDARIES AND HIGHER DIMENSIONS

So far, we have considered only the case where the grid points lie on the boundaries. This is usually the case where the geometry is simple. For curved boundaries, it is not always possible to arrange for the grid points to be on the boundaries. Figure 8.6-1 shows an example of such a situation.

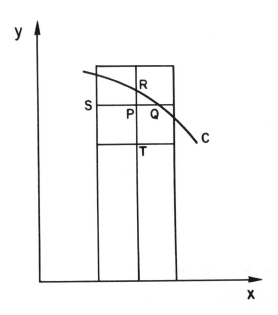

**FIGURE 8.6-1   Irregular boundary**

We consider Laplace's equation with Dirichlet boundary conditions. The value of $u$ is given on the boundary and let $Q$ and $R$ be two points on $C$. The points $P$, $S$ and $T$ are grid points as shown in Figure 8.6-1. Let the length of $PQ$ and $PR$ be $\theta_1 h$ and $\theta_2 k$ respectively, where, as usual, $h$ and $k$ refer to the grid size and where $\theta_1$ and $\theta_2$ satisfy the inequality

$$0 < \theta_1 < 1, \qquad 0 < \theta_2 < 1 \tag{8.6-1a,b}$$

For points next to the boundary, the grid is not rectangular and we cannot use Equation (8.4-2). Instead we need to deduce the difference form of Laplace's equation at the point $P$ involving the given values of $u$ at $R$ and $Q$. For compactness, we use the suffixes, $P$, $Q$, ... to denote that the quantities are evaluated at $P$, $Q$, ... Expanding $u_R$ in a Taylor series about $P$, we have

$$u_R = u\,(x_P, y_P + \theta_2 k) = u_P + \theta_2 k \frac{\partial u}{\partial y} + \frac{\theta_2^2 k^2}{2} \frac{\partial^2 u}{\partial y^2} + \ldots \tag{8.6-2a,b}$$

Similarly, $u_T$, $u_Q$, and $u_S$ are given by

$$u_T = u\,(x_P, y_P - k) = u_P - k \frac{\partial u}{\partial y} + \frac{k^2}{2} \frac{\partial^2 u}{\partial y^2} \tag{8.6-3a,b}$$

$$u_Q = u\,(x_P + \theta_1 h, y_P) = u_P + \theta_1 h \frac{\partial u}{\partial x} + \frac{\theta_1^2 h^2}{2} \frac{\partial^2 u}{\partial x^2} \tag{8.6-4a,b}$$

$$u_S = u\,(x_P - h, y_P) = u_P - h \frac{\partial u}{\partial x} + \frac{h^2}{2} \frac{\partial^2 u}{\partial x^2} \tag{8.6-5a,b}$$

In Equations (8.6-2b, 5b), the partial derivatives are evaluated at $P$. From Equations (8.6-2b, 3b), we deduce that

$$\frac{\partial^2 u}{\partial y^2} = \frac{2\,[u_R + \theta_2 u_T - (1 + \theta_2)\,u_P]}{k^2\,\theta_2\,(1 + \theta_2)} \tag{8.6-6}$$

Similarly, we have

$$\frac{\partial^2 u}{\partial x^2} = \frac{2\,[u_Q + \theta_1 u_S - (1 + \theta_1)\,u_P]}{h^2\,\theta_1\,(1 + \theta_1)} \tag{8.6-7}$$

The difference form of Laplace's equation at $P$ is

$$\frac{[u_Q + \theta_1 u_S - (1 + \theta_1) u_P]}{h^2 \theta_1 (1 + \theta_1)} + \frac{[u_R + \theta_2 u_T - (1 + \theta_2) u_P]}{k^2 \theta_2 (1 + \theta_2)} = 0 \tag{8.6-8}$$

The difference equation for the other points closest to the boundary are deduced in the same way.

For Neumann conditions, the difference equation is complicated and it is preferable to use the finite elements method which will be described later.

The derivation of the difference equation for three and higher dimensional problems is straightforward. Laplace's equation in three dimensions for a cubical grid of size $h^3$ is

$$\frac{u_{i+1,j,k} - 2u_{i,j,k} + u_{i-1,j,k}}{h^2} + \frac{u_{i,j+1,k} - 2u_{i,j,k} + u_{i,j-1,k}}{h^2} + \frac{u_{i,j,k+1} - 2u_{i,j,k} + u_{i,j,k-1}}{h^2} = 0 \tag{8.6-9}$$

From Equation (8.6-9), we deduce that

$$u_{i,j,k} = \frac{1}{2} [u_{i+1,j,k} + u_{i-1,j,k} + u_{i,j+1,k} + u_{i,j-1,k} + u_{i,j,k+1} + u_{i,j,k-1}] \tag{8.6-10}$$

As in the two-dimensional case, the value of $u$ at the point $(i, j, k)$ is the average of the values of $u$ at the six neighboring points that surround it.

Equation (8.6-10) can be solved by the methods described earlier, however it involves a larger number of points and the direct method is not recommended. Other methods are available and are described in more advanced books.

## 8.7  NON-LINEAR EQUATIONS

One of the advantages of numerical methods is that many of the methods based on linear equations with constant coefficients can be carried over directly to non-linear equations. Some of the methods described in this chapter can be used for non-linear equations.

The difference form of the quasi-linear equation

$$\frac{\partial u}{\partial t} = \frac{\partial}{\partial x} \left[ D(x, t, u) \frac{\partial u}{\partial x} \right] \tag{8.7-1}$$

can be written as

$$\frac{u_{i,j+1} - u_{i,j}}{k} = D(x_i, t_j, u_{i,j}) \left[ \frac{u_{i+1,j} - 2u_{i,j} + u_{i-1,j}}{h^2} \right] \tag{8.7-2a}$$

or $\quad u_{i,j+1} = u_{i,j} + \dfrac{k}{h^2} D(x_i, t_j, u_{i,j})(u_{i+1,j} - 2u_{i,j} + u_{i-1,j})$ $\qquad$ (8.7-2b)

Equations (8.7-2a, b) are the simplest form of the explicit difference for Equation (8.7-1). In fact, we have approximated Equation (8.7-1) by

$$\frac{\partial u}{\partial t} = D(x, t, u)\frac{\partial^2 u}{\partial x^2}$$ $\qquad$ (8.7-3)

We have evaluated $D(x, t, u)$ at the grid point $(i, j)$. Note that Equation (8.7-2b) is linear. Using the initial and boundary conditions, Equation (8.7-2b) generates values of $u$ at all points for subsequent times. Comparing Equations (8.7-2b, 3-4b), we expect the numerical scheme to be stable if

$$\frac{k}{h^2} D(x_i, t_j, u_{i,j}) < \frac{1}{2}$$ $\qquad$ (8.7-4)

Experience has shown that this expectation is justified.

***Example 8.7-1.*** Ames (1969) has considered the flow past a solid plate in the region $-\ell < x < 0$ and over a porous plate in the region $x > 0$, as shown in Figure 8.7-1. The velocity components along the $x$ and $y$ axes are respectively $(v_x, v_y)$. The equation of continuity and the equation of motion, assuming no pressure gradient and making the usual boundary layer approximations, are obtained form Equations (A.I-1, II-1) and are

$$\frac{\partial v_x}{\partial x} + \frac{\partial v_y}{\partial y} = 0$$ $\qquad$ (8.7-5a)

$$v_x \frac{\partial v_x}{\partial x} + v_y \frac{\partial v_x}{\partial y} = v\frac{\partial^2 v_x}{\partial y^2}$$ $\qquad$ (8.7-5b)

where $v \; (= \mu/\rho)$ is the kinematic viscosity.

The appropriate boundary conditions are

$\quad y = 0, \; v_x = 0, \qquad v_y = V \text{ (a constant)}, \quad \text{for } x > 0$ $\qquad$ (8.7-6a,b)

$\quad y \longrightarrow \infty, \qquad\qquad v_x \longrightarrow v_\infty \text{ (free stream velocity)}, \quad \text{for } x > 0$ $\qquad$ (8.7-6c)

$\quad x = 0, \qquad\qquad\quad v_x = \text{Blasius solution (Rosenhead, 1963)}$ $\qquad$ (8.7-6d)

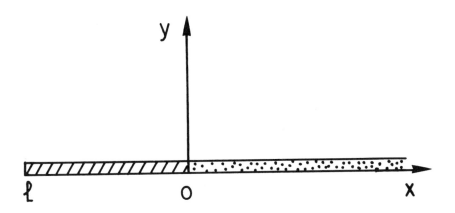

**FIGURE 8.7-1**    **Flow over a solid plate and a porous plate**

We introduce a stream function $\psi\,(x, y)$, defined by

$$v_x = \frac{\partial \psi}{\partial y}, \qquad v_y = -\frac{\partial \psi}{\partial x} \tag{8.7-7a,b}$$

Combining Equations (8.7-5a, b, 7a, b) yields

$$\frac{\partial \psi}{\partial y} \frac{\partial^2 \psi}{\partial x\, \partial y} - \frac{\partial \psi}{\partial x} \frac{\partial^2 \psi}{\partial y^2} = v \frac{\partial^3 \psi}{\partial y^3} \tag{8.7-8}$$

Equation (8.7-8) is a third order equation and it can be reduced to a second order equation by the **von Mises transformation** which makes a change of variables from $(x, y)$ to $(x, \psi)$. The components $v_x$ and $v_y$ can be functions of $x$ and $y$ or functions of $x$ and $\psi$.

Using the chain rule, we can write

$$\frac{\partial v_x\,(x, y)}{\partial x} = \frac{\partial v_x\,(x, \psi)}{\partial x} + \frac{\partial v_x\,(x, \psi)}{\partial \psi} \frac{\partial \psi}{\partial x} \tag{8.7-9}$$

Using Equation (8.7-7b), Equation (8.7-9) can be written as

$$\frac{\partial v_x}{\partial x} = \frac{\partial v_x}{\partial x} - v_y \frac{\partial v_x}{\partial \psi} \tag{8.7-10}$$

Similarly

$$\frac{\partial v_x}{\partial y} = \frac{\partial v_x}{\partial \psi} \frac{\partial \psi}{\partial y} = v_x \frac{\partial v_x}{\partial \psi} \qquad\qquad (8.7\text{-}11a,b)$$

$$\frac{\partial^2 v_x}{\partial y^2} = v_x \frac{\partial}{\partial \psi}\left( v_x \frac{\partial v_x}{\partial \psi}\right) \qquad\qquad (8.7\text{-}11c)$$

Note that in Equations (8.7-9 to 11c), $v_x$ is a function of $x$ and $y$ on the left side and a function of $x$ and $\psi$ on the right side.

Substituting Equations (8.7-9 to 11c) into Equation (8.7-5b), we obtain

$$\frac{\partial v_x}{\partial x} = v \frac{\partial}{\partial \psi}\left( v_x \frac{\partial v_x}{\partial \psi}\right) \qquad\qquad (8.7\text{-}12)$$

The boundary conditions also have to be transformed to $(x, \psi)$ coordinates. Integrating Equations (8.7-7a, b), we obtain respectively

$$\psi = \int_0^y v_x(x, \xi)\, d\xi + f(x) \qquad\qquad (8.7\text{-}13a)$$

$$\psi = -\int_0^x v_y(\xi, y)\, d\xi + g(y) \qquad\qquad (8.7\text{-}13b)$$

where $f(x)$ and $g(y)$ are arbitrary functions.

Imposing the conditions given by Equations (8.7-6a, b), we obtain

$$f(x) = -\int_0^x V\, d\xi + g(0) \qquad\qquad (8.7\text{-}14a)$$

$$= -Vx + g(0) \qquad\qquad (8.7\text{-}14b)$$

Since $\psi$ is defined to the extent of an arbitrary constant, we can choose $g(0)$ to be zero and $\psi(x, y)$ is given by

$$\psi = \int_0^y v_x(x, \xi)\, d\xi - Vx \qquad\qquad (8.7\text{-}15)$$

The conditions at $y = 0$ and $y \longrightarrow \infty$ correspond to $\psi = -Vx$ and $\psi \longrightarrow \infty$ respectively and the boundary conditions [Equations (8.7-6a, b, c)] are transformed to

$$\psi = -Vx, \qquad\qquad v_x = 0 \tag{8.7-16a}$$

$$\psi \longrightarrow \infty, \qquad\qquad v_x \longrightarrow v_\infty \tag{8.7-16b}$$

Equation (8.7-6d) remains unchanged.

We now introduce the following dimensionless variables

$$\eta = \frac{x}{\ell}, \qquad \mathscr{S} = \frac{\psi}{v_\infty \ell}\left(\frac{v_\infty \ell}{v}\right)^{1/2}, \qquad u = 1 - \left(\frac{v_x}{v_\infty}\right)^2 \tag{8.7-17a,b,c}$$

Equation (8.7-12) can now be written as

$$\frac{\partial u}{\partial \eta} = \sqrt{1-u}\;\frac{\partial^2 u}{\partial \mathscr{S}^2} \tag{8.7-18}$$

The boundary conditions become

$$\mathscr{S} = -\beta\eta, \qquad u = 1, \qquad \beta = \frac{V}{v_\infty}\left(\frac{v_\infty \ell}{v}\right)^{1/2} \tag{8.7-19a,b,c}$$

$$\mathscr{S} \longrightarrow \infty, \qquad u = 0 \tag{8.7-19d}$$

$$\eta = 0, \qquad u \text{ is given by the Blasius solution} \tag{8.7-19e}$$

Equation (8.7-18) can be approximated as

$$u_{i,j+1} = u_{i,j} + \frac{k}{h^2}\sqrt{1-u_{i,j}}\;[u_{i+1,j} - 2u_{i,j} + u_{i-1,j}] \tag{8.7-20}$$

In discretizing Equation (8.7-18), we have divided the region of interest into rectangles each of size $(h \times k)$. The quantity $u_{i,j}$ refers to the value of $u(\mathscr{S}_i, \eta_j)$.

From Equations (8.7-19a to 20), we can compute the values of $u_{i,j}$. As expected, it is found that if

$$\frac{k\sqrt{1-u_{i,j}}}{h^2} < \frac{1}{2} \tag{8.7-21}$$

we have computational stability.

Note that Equations (8.7-19a to c) imply that we have irregular (triangular) boundaries and we need to make the modifications proposed in Section 6. Further details can be found in Ames (1969).

The accuracy of Equation (8.7-2b) can be improved by replacing $D(x_i, t_j, u_{i,j})$ by an average value of $D$ over the points considered. Several formulae have been proposed which are more complicated than Equation (8.7-2b) but which give results closer to the solution of Equation (8.7-1). Implicit formulae which are unconditionally stable have also been derived. These formulae can be found in more advanced books on numerical solutions of P.D.E.

## 8.8   FINITE ELEMENTS

The **finite element method** was developed to study the stresses in complex discrete structures. More recently, it has been widely used to obtain approximate solutions to continuum problems. The battle between **finite differences** and **finite elements** is over. For regular boundaries and simple equations, the method of finite differences is preferable because finite difference equations are easier to set up. It was pointed out earlier that if the boundary is irregular, the finite element method is the obvious choice. The reason for this choice will become apparent after we have introduced the method.

Zienkiewicz and Morgan (1983) have shown that a **generalized finite element method** can be defined which includes all the numerical methods and it is left to the user to choose the optimum method.

We introduce the basic concepts of the finite element method by considering one-dimensional problems. It must be pointed out that one does not usually solve one-dimensional problems by this method. The one-dimensional case is used to ease the way to the understanding of two and higher dimensional problems.

### One-Dimensional Problems

In a **one-dimensional problem**, we have only one independent variable and the equation we need to solve is an ordinary differential equation. We consider the differential equation

$$\frac{d^2u}{dx^2} = -g(x), \qquad 0 \leq x \leq 1 \tag{8.8-1}$$

subject to

$$u(0) = A, \qquad u(1) = B \tag{8.8-2a,b}$$

By an appropriate scaling, we can always transform any arbitrary interval into an interval of [0, 1].

## *Variational method*

In the finite difference method, we replace the derivative by a finite difference. In the finite element method, we reformulate the problem into a **variational problem**. In Chapters 9 and 10, we give examples of such variational formulations. Instead of looking for a function that satisfies the differential equation, we look for a function that extremizes an integral. In Chapter 9, it is shown that the function $y(x)$ that yields the extreme values of the integral I, defined by

$$I = \int_a^b f(x, y, y') \, dx \tag{8.8-3}$$

is

$$\frac{\partial f}{\partial y} - \frac{d}{dx}\left(\frac{\partial f}{\partial y'}\right) = 0 \tag{8.8-4}$$

Consider the integral

$$I = \int_0^1 \left[\frac{1}{2}\left(\frac{du}{dx}\right)^2 - gu\right] dx \tag{8.8-5}$$

Applying the condition given by Equation (8.8-4) yields

$$\frac{\partial}{\partial u}\left[\frac{1}{2}\left(\frac{du}{dx}\right)^2 - gu\right] - \frac{d}{dx}\left[\frac{\partial}{\partial u'}\left(\frac{1}{2}u'^2 - gu\right)\right] = 0 \tag{8.8-6}$$

Simplifying Equation (8.8-6), we obtain Equation (8.8-1). This means that the solution of Equation (8.8-1) is the function that extremizes the integral I defined by Equation (8.8-5).

We approximate u by

$$u \approx \sum_{i=1}^m a_i N_i(x) \tag{8.8-7}$$

where $a_i$ are constants and the $N_i(x)$ are the **shape (trial) functions**. They can be polynomials, trigonometric functions, or other functions whose properties are known. They are usually chosen such that the boundary conditions are satisfied.

On substituting Equation (8.8-7) into Equation (8.8-5), we obtain an expression of the form

$$I = F(a_1, a_2, \ldots, a_m) \tag{8.8-8}$$

It can now be considered as a function of $a_i$ and the extreme values of I are given by

$$\frac{\partial F}{\partial a_i} = 0, \qquad i = 1, 2, \dots, m \tag{8.8-9}$$

From Equation (8.8-9), we obtain the coefficients $a_i$ and on substituting them into Equation (8.8-7), we obtain an approximate solution for Equation (8.8-1). This method is known as the **Rayleigh-Ritz method** (see Example 9.11-2).

In the finite element method, we subdivide the interval [0, 1] into a number of subintervals and each subinterval is called an **element**. In each element, we approximate u by Equation (8.8-7). Since each element is shorter than the whole interval, we can expect that a better accuracy can be obtained for the same number of terms in Equation (8.8-7). We illustrate this method in Example 8.8-1.

## *Galerkin method*

Not all problems can be formulated as extremum problems. We now describe a method which provides a **weak solution** to Equation (8.8-1). We approximate the solution u in the interval $0 \le x \le 1$ by $\hat{u}(x)$. The function $\hat{u}$ generally does not satisfy Equation (8.8-1) and we write

$$\frac{d^2\hat{u}}{dx^2} + g(x) = R \tag{8.8-10}$$

where R is the residual. If R is identically zero, then $\hat{u}$ is an exact solution of Equation (8.8-1). The function $\hat{u}$ is a **weak solution** if the integral of R with respect to a weight function $w(x)$ is zero. That is to say

$$\int_0^1 w(x) \left[ \frac{d^2\hat{u}}{dx^2} + g(x) \right] dx = \int_0^1 w(x) R(x) dx = 0 \tag{8.8-11a,b}$$

In the **Galerkin method**, we choose the weight function to be the approximate solution $(w = \hat{u})$. In the method of finite elements, we subdivide the interval [0, 1] into subintervals (elements) and on each element, $\hat{u}$ is approximated by Equation (8.8-7).

Suppose we subdivide the interval [0, 1] into $[0, x_0)$ and $(x_0, 1]$ and assume that two terms in Equation (8.8-7) provide sufficient accuracy. On the first interval, Equation (8.8-11a) becomes

$$\int_0^{x_0} N_1 [a_1 N_1'' + a_2 N_2'' + g] dx = 0 \tag{8.8-12a}$$

$$\int_0^{x_0} N_2 \left[ a_1 N_1'' + a_2 N_2'' + g \right] dx = 0 \qquad (8.8\text{-}12b)$$

Integrating Equations (8.8-12a, b) by parts yields

$$a_1 \left[ N_1 N_1' \right]_0^{x_0} + a_2 \left[ N_1 N_2' \right]_0^{x_0} - a_1 K_{11} - a_2 K_{12} + G_1 = 0 \qquad (8.8\text{-}13a)$$

$$a_1 \left[ N_2 N_1' \right]_0^{x_0} + a_2 \left[ N_2 N_2' \right]_0^{x_0} - a_1 K_{12} - a_2 K_{22} + G_2 = 0 \qquad (8.8\text{-}13b)$$

where

$$K_{11} = \int_0^{x_0} N_1'^2 \, dx \qquad K_{12} = \int_0^{x_0} N_1' N_2' \, dx \qquad K_{22} = \int_0^{x_0} N_2'^2 \, dx \qquad (8.8\text{-}14a,b,c)$$

$$G_1 = \int_0^{x_0} g N_1 \, dx \qquad G_2 = \int_0^{x_0} g N_2 \, dx \qquad (8.8\text{-}14d,e)$$

Similarly, for the interval $(x_0, 1]$, we obtain another set of equations. We can generalize Equations (8.8-13a, b) for any interval $x_k \le x \le x_{k+1}$ and for any number of shape functions $N_i$ ($i = 1, \dots, m$) and we obtain

$$\left\{ K_{ij} - \left[ N_i N_j' \right]_{x_k}^{x_{k+1}} a_j \right\} = G_i \qquad (8.8\text{-}15a)$$

where

$$K_{ij} = \int_{x_k}^{x_{k+1}} N_i' N_j' \, dx, \qquad G_j = \int_{x_k}^{x_{k+1}} g N_i \, dx \qquad (8.8\text{-}15b,c)$$

The quantities $K_{ij}$ are the elements of the **stiffness matrix** $\underline{\underline{K}}$ and $G_i$ are the components of the **force vector** $\underline{G}$. The shape functions $N_i$ are usually chosen such that $\underline{\underline{K}}$ can be evaluated easily (analytically or numerically), is sparse, and well conditioned. The coefficients $a_j$ can be determined by assembling Equation (8.8-15a) for all the elements, subject to the boundary and continuity conditions. Once $a_j$ have been obtained, the approximate solution $\hat{u}$ is given by Equation (8.8-7).

Note that, on integrating Equations (8.8-12a, b) by parts, we have avoided having to evaluate the second derivative of the shape functions. This implies that we can choose linear functions as shape functions.

***Example 8.8-1.*** Solve the equation

$$\frac{d^2u}{dx^2} + x = 0 \tag{8.8-16a}$$

subject to the conditions

$$u(0) = 0, \qquad u(1) = 1 \tag{8.8-16b,c}$$

by the method of finite elements.

We subdivide the interval $[0, 1]$ into three elements of equal length $[0, 1/3)$, $(1/3, 2/3)$, and $(2/3, 1]$. We denote the elements by $E_1$, $E_2$, and $E_3$ and the values of $u$ at the nodal points $x_i$ by $u_i$.

We approximate the function $u$ by a piecewise linear function. In $E_1$, $E_2$, and $E_3$, $u$ is approximated respectively by

$$u \approx 3u_1 x \tag{8.8-17a}$$

$$u \approx u_1(2 - 3x) + u_2(3x - 1) \tag{8.8-17b}$$

$$u \approx 3u_2(1 - x) + (3x - 2) \tag{8.8-17c}$$

Note that in Equations (8.8-17a to c), $u$ satisfies the boundary and the continuity conditions at the nodal points (Problem 14a shows how these functions are derived).

We solve the problem by the variational method. Equation (8.8-16a) is equivalent to finding the extremum of

$$I = \int_0^1 \left[ \frac{1}{2}\left(\frac{du}{dx}\right)^2 - xu \right] dx \tag{8.8-18}$$

Substituting Equations (8.8-17a to c) into Equation (8.8-18) yields

$$I = I_1 + I_2 + I_3 \tag{8.8-19a}$$

$$= \left( \frac{3u_1^2}{2} - \frac{u_1}{27} \right) + \left[ \frac{3}{2}(u_2 - u_1)^2 - \frac{2u_1}{27} - \frac{5u_2}{54} \right] + \left[ \frac{3}{2}(1 - u_2)^2 - \frac{7u_2}{54} - \frac{4}{27} \right] \tag{8.8-19b}$$

The extremum of $I$ is given by

$$\frac{\partial I}{\partial u_1} = \frac{\partial I}{\partial u_2} = 0 \tag{8.8-20a,b}$$

On carrying out the differentiation, we obtain

$$6u_1 - 3u_2 = 1/9 \tag{8.8-21a}$$

$$-3u_1 + 6u_2 = 29/9 \tag{8.8-21b}$$

The solution is

$$u_1 = 62/162 = 0.383 \tag{8.8-22a}$$

$$u_2 = 59/81 = 0.728 \tag{8.8-22b}$$

The exact solution is

$$u = \frac{x}{6}(7 - x^2) \tag{8.8-23}$$

By comparing Equations (8.8-22a, b, 23), we note that the solution we have obtained by the finite element method is exact! The exact solution is cubic, the approximate solution is linear, and at the nodal points, the two coincide. This is not generally true. Strang and Fix (1973) have observed that in general, if the shape functions satisfy exactly the homogeneous differential equation, the nodal values are the exact values.

## Two-Dimensional Problems

In two-dimensional problems, we subdivide the domain into elements and we consider triangular elements. We can also consider rectangular or other geometrical shapes. Triangular elements allow for an easier fit to non-rectangular domains. This is shown in Figure 8.8-1.

We propose linear shape functions. Consider a typical element E with nodes i, j, and k as shown in Figure 8.8-1. We choose the shape function $N_i$ such that at node i, it is one and at nodes j and k, it is zero. That is, $N_i(x, y)$ has the following properties

$$N_i(x, y) = a_i + b_i x + c_i y \tag{8.8-24a}$$

$$N_i(x_i, y_i) = 1, \qquad N_i(x_j, y_j) = 0, \qquad N_i(x_k, y_k) = 0 \tag{8.8-24b,c,d}$$

Combining Equations (8.8-24a to d) yields

$$\begin{bmatrix} 1 & x_i & y_i \\ 1 & x_j & y_j \\ 1 & x_k & y_k \end{bmatrix} \begin{bmatrix} a_i \\ b_i \\ c_i \end{bmatrix} = \begin{bmatrix} 1 \\ 0 \\ 0 \end{bmatrix} \tag{8.8-25}$$

Solving Equation (8.8-25), we obtain

$$a_i = (x_j y_k - x_k y_j)/D, \quad b_i = (y_j - y_k)/D, \quad c_i = (x_k - x_j)/D \qquad (8.8\text{-}26a,b,c)$$

$$D = \begin{vmatrix} 1 & x_i & y_i \\ 1 & x_j & y_j \\ 1 & x_k & y_k \end{vmatrix} = 2 \,(\text{area of element E}) = 2\,\Delta \qquad (8.8\text{-}26d,e,f)$$

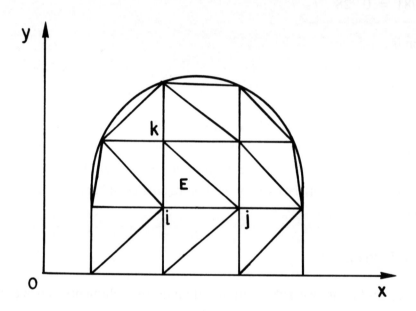

**FIGURE 8 8-1    Domain subdivided into triangular elements.**
**E is a typical element with nodes i, j, and k**

For each element E, we define a shape function $N_i$ given by Equations (8.8-24a to e) and if i does not belong to E, then we set $N_i$ to be zero. With this definition of $N_i$, we can approximate u as

$$u = \sum_i u_i N_i \qquad (8.8\text{-}27)$$

where $u_i$ is the value of u at node i.

Next, we generalize the variational and the Galerkin methods to two-dimensional problems.

We consider Poisson's equation which can be written as

$$\frac{\partial^2 u}{\partial x^2} + \frac{\partial^2 u}{\partial y^2} = -g(x, y) \qquad (8.8\text{-}28a)$$

$$u = 0 \quad \text{on the boundary} \qquad (8.8\text{-}28b)$$

The function $u$ that satisfies Equation (8.8-28a) is the function that extremizes the integral [see Equation (9.11-13)]

$$I = \iint \left[ \frac{1}{2}\left(\frac{\partial u}{\partial x}\right)^2 + \frac{1}{2}\left(\frac{\partial u}{\partial y}\right)^2 - gu \right] dx \, dy \tag{8.8-29}$$

Substituting Equation (8.8-27) into Equation (8.8-29) yields

$$I = \iint \left[ \frac{1}{2}\left(u_1\frac{\partial N_1}{\partial x} + \dots + u_m\frac{\partial N_m}{\partial x}\right)^2 + \frac{1}{2}\left(u_1\frac{\partial N_1}{\partial y} + \dots + u_m\frac{\partial N_m}{\partial y}\right)^2 \right.$$

$$\left. - g\left(u_1 N_1 + \dots + u_m N_m\right) \right] dx \, dy \tag{8.8-30a}$$

$$= \iint \left[ \frac{1}{2}u_1^2\left(\frac{\partial N_1}{\partial x}\frac{\partial N_1}{\partial x} + \frac{\partial N_1}{\partial y}\frac{\partial N_1}{\partial y}\right) + u_1 u_2\left(\frac{\partial N_1}{\partial x}\frac{\partial N_2}{\partial x} + \frac{\partial N_1}{\partial y}\frac{\partial N_2}{\partial y}\right) + \dots \right.$$

$$\left. + \frac{1}{2}u_m^2\left(\frac{\partial N_m}{\partial x}\frac{\partial N_m}{\partial x} + \frac{\partial N_m}{\partial y}\frac{\partial N_m}{\partial y}\right) - (u_1 g N_1 + \dots + u_m g N_m) \right] dx \, dy \tag{8.8-30b}$$

Equation (8.8-30b) can be written in a compact form as

$$I = \sum_{i,j} K_{ij} u_i u_j - \sum_i u_i G_i \tag{8.8-31a}$$

or     $I = \underline{u}^T \underline{\underline{K}} \, \underline{u} - \underline{u}^T \underline{G}$            (8.8-31b)

where

$$K_{ij} = \iint \left(\frac{\partial N_i}{\partial x}\frac{\partial N_j}{\partial x} + \frac{\partial N_i}{\partial y}\frac{\partial N_j}{\partial y}\right) dx \, dy \tag{8.8-31c}$$

$$G_i = \iint g N_i \, dx \, dy \tag{8.8-31d}$$

The extreme values of $I$ are given by

$$\frac{\partial I}{\partial u_i} = 0 \tag{8.8-32}$$

Combining Equations (8.8-31a, 32) yields

$$\sum_j K_{ij} u_j - G_i = 0 \tag{8.8-33a}$$

or      $\underline{\underline{K}}\,\underline{u} - \underline{G} = 0$                                                                                                (8.8-33b)

To obtain a weak solution $\hat{u}\,(x, y)$, we define a weight function $w\,(x, y)$, such that

$$\iint \left[ w\,(x, y) \left( \frac{\partial^2 \hat{u}}{\partial x^2} + \frac{\partial^2 \hat{u}}{\partial y^2} + g\,(x, y) \right) \right] dx\, dy = 0 \tag{8.8-34}$$

On integrating by parts and assuming that $w$ vanishes on the boundary, Equation (8.8-34) becomes

$$\iint \left( \frac{\partial w}{\partial x} \frac{\partial \hat{u}}{\partial x} + \frac{\partial w}{\partial y} \frac{\partial \hat{u}}{\partial y} - wg \right) dx\, dy = 0 \tag{8.8-35}$$

The function $\hat{u}$ is chosen to be given by Equation (8.8-27) and using Galerkin's method (setting $w = N_i$ in turn), Equation (8.8-35) can be written as

$$\iint \left[ \left( u_1 \frac{\partial N_1}{\partial x} + \ldots + u_m \frac{\partial N_m}{\partial x} \right) \frac{\partial N_i}{\partial x} + \left( u_1 \frac{\partial N_1}{\partial y} + \ldots + u_m \frac{\partial N_m}{\partial y} \right) \frac{\partial N_i}{\partial y} - gN_i \right] dx\, dy = 0 \tag{8.8-36a}$$

or      $$\iint \left[ u_1 \left( \frac{\partial N_1}{\partial x} \frac{\partial N_i}{\partial x} + \frac{\partial N_1}{\partial y} \frac{\partial N_i}{\partial y} \right) + \ldots + u_m \left( \frac{\partial N_m}{\partial x} \frac{\partial N_i}{\partial x} + \frac{\partial N_m}{\partial y} \frac{\partial N_i}{\partial y} \right) - gN_i \right] dx\, dy = 0 \tag{8.8-36b}$$

In the usual compact form, Equation (8.8-36b) can be written as

$$\sum_j K_{ij} u_j - G_i = 0 \tag{8.8-37}$$

where $K_{ij}$ and $G_i$ are defined by Equations (8.8-31c, d).

Note that in this case both the variational and the Galerkin methods yield the same discretized equations [Equations (8.8-33b, 37)].

This is true for all symmetric problems. For non-symmetric problems, the variational method may not be applicable but the Galerkin method can still be used. In this case, the matrix $\underline{\underline{K}}$ is no longer symmetric.

By combining Equations (8.8-31c, 24a), we find that for each element E, $K_{ij}$ is given by

$$K_{ij} = \iint \left[ b_i b_j + c_i c_j \right] dx\, dy \tag{8.8-38a}$$

$$= \left[b_i b_j + c_i c_j\right] \text{ (area of the element)} \tag{8.8-38b}$$

$$= \Delta \left[b_i b_j + c_i c_j\right] \tag{8.8-38c}$$

Using Equation (8.8-24a), Equation (8.8-31d) can be written as

$$G_i = \iint g\left[a_i + b_i x + c_i y\right] dx\, dy \tag{8.8-39}$$

which simplifies the computation of $G_i$.

If the triangular element is small enough, then $g$ can be approximated by its values at the centre of mass of the triangular element and we denote it by $g(c)$. Similarly, $x$ and $y$ can be approximated as the average values at the nodes. That is to say, $x$ and $y$ are approximated as

$$x \approx \bar{x} = \frac{1}{3}(x_i + x_j + x_k) \tag{8.8-40a,b}$$

$$y \approx \bar{y} = \frac{1}{3}(y_i + y_j + y_k) \tag{8.8-41a,b}$$

The approximate value of $G_i$ is thus

$$G_i \approx g(\bar{c})\left[a_i + b_i \bar{x} + c_i \bar{y}\right] \iint dx\, dy \tag{8.8-42a}$$

$$\approx g(\bar{c})\left[a_i + b_i \bar{x} + c_i \bar{y}\right] \Delta \tag{8.8-42b}$$

$$\approx \frac{1}{3} g(\bar{c}) \Delta \tag{8.8-42c}$$

Equation (8.8-42c) is obtained by substituting Equations (8.8-26a to c) into Equation (8.8-42b).

Thus using Equations (8.8-38c, 42c), we can compute $K_{ij}$ and $g_i$ for each element E. We then have to assemble them to obtain the global $\underline{\underline{K}}_{(g)}$ and $\underline{G}_{(g)}$. We note that since each triangular element has three nodes, $\underline{\underline{K}}$ is a $(3 \times 3)$ matrix for each element, and $\underline{G}$ is a vector with three components. If the domain has $m$ nodes, then the global matrix $\underline{\underline{K}}_{(g)}$ is a $(m \times m)$ matrix and the global $\underline{G}_{(g)}$ is a vector with $m$ components. By numbering the nodes appropriately, we can obtain a sparse matrix $\underline{\underline{K}}_{(g)}$.

**Example 8.8-2.** Solve the equation

$$\frac{\partial^2 u}{\partial x^2} + \frac{\partial^2 u}{\partial y^2} = -1, \quad -2 \le x \le 2, \quad -2 \le y \le 2 \tag{8.8-43}$$

subject to the condition

$$u = 0 \quad \text{on the boundary} \tag{8.8-44}$$

From the symmetry of the problem, we may limit our calculations to one octant only. We subdivide this reduced domain in four triangles and label the elements and nodes as shown in Figure 8.8-2. Node 1 is the origin of the coordinate system. The coordinates of the nodes are

$$1 \ (0,0), \quad 2 \ (1, 0), \quad 3 \ (1, 1), \quad 4 \ (2, 0), \quad 5 \ (2, 1), \quad \text{and} \quad 6 \ (2, 2)$$

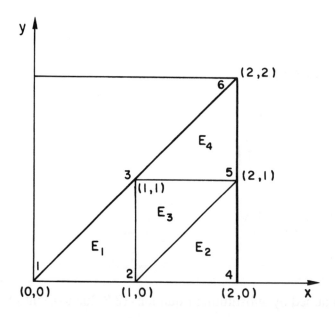

**FIGURE 8.8-2   Division of a domain in triangular elements**

We now use Equations (8.8-26a to c, 38c, 42c) to calculate $\underline{K}$ and $\underline{G}$ for each element starting with element $E_1$. The area of each triangular element is 0.5.

$E_1$:     $a_1 = 1$,           $b_1 = -1$,           $c_1 = 0$                                        (8.8-45a,b,c)

       $a_2 = 0$,           $b_2 = 1$,            $c_2 = -1$                                        (8.8-45d,e,f)

       $a_3 = 0$,           $b_3 = 0$,            $c_3 = 1$                                         (8.8-45g,h,i)

$$K_{11} = \frac{1}{2}, \quad K_{12} = -\frac{1}{2}, \quad K_{13} = 0, \quad K_{22} = 1, \quad K_{23} = -\frac{1}{2}, \quad K_{33} = \frac{1}{2} \qquad \text{(8.8-46a to f)}$$

$$G_1 = G_2 = G_3 = \frac{1}{6} \qquad \text{(8.8-47a,b,c)}$$

$E_2$:     $a_2 = 2$,           $b_2 = -1$,           $c_2 = 0$               (8.8-48a,b,c)

$\quad\quad\quad a_4 = -1$,           $b_4 = 1$,            $c_4 = -1$             (8.8-48d,e,f)

$\quad\quad\quad a_5 = 0$,           $b_5 = -1$,           $c_5 = 0$              (8.8-48g,h,i)

$$K_{22} = \frac{1}{2}, \quad K_{24} = -\frac{1}{2}, \quad K_{25} = \frac{1}{2}, \quad K_{44} = 1, \quad K_{45} = -\frac{1}{2}, \quad K_{55} = \frac{1}{2} \qquad \text{(8.8-49a to f)}$$

$$G_2 = G_4 = G_5 = \frac{1}{6} \qquad \text{(8.8-50a,b,c)}$$

Similarly, for $E_3$ and $E_4$, we have

$E_3$:     $K_{22} = \frac{1}{2}, \quad K_{23} = -\frac{1}{2}, \quad K_{25} = -\frac{1}{2}, \quad K_{33} = 1, \quad K_{35} = -\frac{1}{2}, \quad K_{55} = \frac{1}{2}$    (8.8-51a to f)

$$G_2 = G_3 = G_5 = \frac{1}{6} \qquad \text{(8.8-52a,b,c)}$$

$E_4$:     $K_{33} = \frac{1}{2}, \quad K_{35} = -\frac{1}{2}, \quad K_{36} = 0, \quad K_{55} = 1, \quad K_{56} = -\frac{1}{2}, \quad K_{66} = \frac{1}{2}$    (8.8-53a to f)

$$G_3 = G_5 = G_6 = \frac{1}{6} \qquad \text{(8.8-54a,b,c)}$$

We assemble the $\underline{\underline{K}}$ and $\underline{G}$ from each element so as to obtain the global $\underline{\underline{K}}_{(g)}$ and $\underline{G}_{(g)}$. The result is

$$
\begin{bmatrix}
\frac{1}{2} & -\frac{1}{2} & 0 & \vert & 0 & 0 & 0 \\
-\frac{1}{2} & 2 & -1 & \vert & -\frac{1}{2} & 0 & 0 \\
0 & -1 & 2 & \vert & 0 & -1 & 0 \\
- & - & - & - & - & - & - \\
0 & -\frac{1}{2} & 0 & \vert & 1 & -\frac{1}{2} & 0 \\
0 & 0 & -1 & \vert & -\frac{1}{2} & 2 & -\frac{1}{2} \\
0 & 0 & 0 & \vert & 0 & -\frac{1}{2} & -\frac{1}{2}
\end{bmatrix}
\begin{bmatrix}
u_1 \\ u_2 \\ u_3 \\ - \\ u_4 \\ u_5 \\ u_6
\end{bmatrix}
= \frac{1}{6}
\begin{bmatrix}
1 \\ 3 \\ 3 \\ - \\ 1 \\ 3 \\ 1
\end{bmatrix}
\qquad \text{(8.8-55)}
$$

Note that the matrix $\underline{\underline{K}}_{(g)}$ in Equation (8.8-55) is singular and this implies that the $u_i$ cannot be determined. The components $u_4$, $u_5$, and $u_6$ are the values of $u$ on the boundary and are known. The only unknowns are $u_1$, $u_2$, and $u_3$ which can be determined by partitioning the matrix $\underline{\underline{K}}_{(g)}$ into four $(3 \times 3)$ matrices. The vectors $\underline{u}$ and $\underline{G}_{(g)}$ are each divided into two vectors each with three components as shown by the dotted lines in Equation (8.8-55). We choose to consider the upper half of $\underline{\underline{K}}_{(g)}\underline{u}$ and $\underline{G}_{(g)}$ since we require only three equations to solve for $u_1$, $u_2$, and $u_3$. Equation (8.8-55) now reduces to

$$
\begin{bmatrix} \frac{1}{2} & -\frac{1}{2} & 0 \\ -\frac{1}{2} & 2 & -1 \\ 0 & -1 & 2 \end{bmatrix} \begin{bmatrix} u_1 \\ u_2 \\ u_3 \end{bmatrix} + \begin{bmatrix} 0 & 0 & 0 \\ -\frac{1}{2} & 0 & 0 \\ 0 & -1 & 0 \end{bmatrix} \begin{bmatrix} u_4 \\ u_5 \\ u_6 \end{bmatrix} = \frac{1}{6} \begin{bmatrix} 1 \\ 3 \\ 3 \end{bmatrix} \tag{8.8-56}
$$

Using the boundary conditions $(u_4 = u_5 = u_6 = 0)$, we deduce from Equation (8.8-56) that

$$
\begin{bmatrix} u_1 \\ u_2 \\ u_3 \end{bmatrix} = \begin{bmatrix} \frac{1}{2} & -\frac{1}{2} & 0 \\ -\frac{1}{2} & 2 & -1 \\ 0 & -1 & 2 \end{bmatrix}^{-1} \begin{bmatrix} \frac{1}{6} \\ \frac{1}{2} \\ \frac{1}{2} \end{bmatrix} = \begin{bmatrix} \frac{5}{4} \\ \frac{11}{12} \\ \frac{17}{24} \end{bmatrix} \tag{8.8-57a,b}
$$

Note that the matrix in Equation (8.8-57a) is non-singular (its inverse exists) and $u_1$, $u_2$, and $u_3$ can be determined.

$\underline{\underline{K}}$ and $\underline{G}$ are calculated for each element, they are transferred to $\underline{\underline{K}}_{(g)}$ and $\underline{G}_{(g)}$: that is to say, $\underline{\underline{K}}_{(g)}$ and $\underline{G}_{(g)}$ are being assembled as the calculation for each element proceeds. Thus $\underline{\underline{K}}$ and $\underline{G}$ for each element need not be stored. Strang and Fix (1973) have discussed the computational aspects of finite element methods.

## PROBLEMS

1a.     Use the explicit method to solve the initial-boundary value problem defined by

$$
\frac{\partial u}{\partial t} = \frac{\partial^2 u}{\partial x^2} , \quad t > 0 , \quad 0 \le x \le 1
$$

$$
u(x, 0) = x^2 , \quad u(0, t) = 0 , \quad u(1, t) = 1
$$

Choose $h = 0.1$ and $k = 0.05$ so as to ensure the validity of the method.

2b.    Solve the non-linear parabolic partial differential equation

$$\frac{\partial u}{\partial t} = \frac{\partial^2 u}{\partial x^2} - (u + \cos u), \quad 0 < x < 5, \quad 0 < t < 1$$

subject to the following conditions

$$u(x, 0) = 1, \quad 0 \le x \le 5; \quad u(0, t) = u(5, t) = 1, \quad 0 < t \le 1$$

3a.    Solve the following initial-boundary problem

$$\frac{\partial u}{\partial t} = \frac{\partial^2 u}{\partial x^2}, \quad 0 \le x \le 1, \quad t > 0$$

$$u(x, 0) = 0, \quad u(1, t) = 1, \quad \left. \frac{\partial u}{\partial x} \right|_{x=0} = 0$$

by the method of Crank-Nicolson.

4b.    Use the Crank-Nicolson method to solve the problem examined in Example 5.7-2. By introducing the following dimensionless variables

$$\bar{c} = (c - c_f)/(c_i - c_f), \quad \bar{x} = x/L, \quad \bar{z} = z/L$$

show that Equations (5.7-21, 22 a to c) become

$$\frac{\partial \bar{c}}{\partial \bar{z}} = D \frac{\partial^2 \bar{c}}{\partial \bar{x}^2}, \quad D = \mathcal{D}_{OB}/(Lv_z)$$

$$\bar{c}(0, \bar{z}) = 0, \quad \bar{c}(\bar{x}, 0) = 1, \quad \left. \frac{\partial \bar{c}}{\partial \bar{x}} \right|_{\bar{x}=1} = 0$$

Obtain $\bar{c}$ for $D = 10^{-4}$ and calculate the average concentration $\langle c \rangle$ defined by Equation (5.7-36a) by numerical integration.

5a.    Use the finite difference method to solve the following elliptic equation

$$\frac{\partial^2 u}{\partial x^2} + \frac{\partial^2 u}{\partial y^2} - y \frac{\partial u}{\partial x} - x \frac{\partial u}{\partial y} - u = 0, \quad 0 \le x \le 1, \quad 0 \le y \le 1$$

$u = x - y$ on the boundary of the square.

Choose $h = k = 0.5$.

6a.     Solve Poisson's equation

$$\frac{\partial^2 u}{\partial x^2} + \frac{\partial^2 u}{\partial y^2} = -2, \quad 0 \le x \le 6, \quad 0 \le y \le 8$$

u = 0 on the boundary

(i) by direct elimination and (ii) by the S.O.R. methods.

Choose h = k = 2, and ω = 1.383.

7b.     Obtain a central difference approximation to the biharmonic equation

$$\frac{\partial^4 u}{\partial x^4} + 2\frac{\partial^4 u}{\partial x^2 \partial y^2} + \frac{\partial^4 u}{\partial y^4} = 0$$

8b.     Derive a finite difference approximation to Laplace's equation in the polar coordinate system. The Laplacian is

$$\frac{\partial^2 u}{\partial r^2} + \frac{1}{r}\frac{\partial u}{\partial r} + \frac{1}{r^2}\frac{\partial^2 u}{\partial \theta^2}$$

9a.     Use the finite difference method to solve the hyperbolic equation

$$\frac{\partial^2 u}{\partial t^2} = \frac{\partial^2 u}{\partial x^2}, \quad 0 \le x \le 1, \quad t > 0$$

subject to

$$u(0, t) = 0, \quad u(1, t) = 1, \quad u(x, 0) = x, \quad \left.\frac{\partial u}{\partial t}\right|_{t=0} = 0$$

Choose h = k = 0.2. Is it possible to choose k > h?

10b.    Show that by approximating $(\partial^2 u / \partial x^2)$ at $(i, j)$ by its mean value at $(i, j)$ and $(i, j + 1)$, an implicit formula for Equation (8.5-1a) is

$$u_{i-1,j+1} - 2[1 + (h/ck)^2] u_{i,j+1} + u_{i+1,j+1}$$

$$= -u_{i-1,j-1} - u_{i+1,j-1} - 4(h/ck)^2 u_{i,j} + 2[1 + (h/ck)^2] u_{i,j-1}$$

Use the implicit formula to solve

$$\frac{\partial^2 u}{\partial t^2} = \frac{\partial^2 u}{\partial x^2}, \qquad 0 \le x \le 1, \qquad t > 0$$

$$u\,(0,\, t) \;=\; u\,(1,\, t) \;=\; t\,(1 - t)\,, \qquad u\,(x,\, 0) \;=\; x\,(1 - x)\,, \qquad \left.\frac{\partial u}{\partial t}\right|_{t=0} = 1$$

Choose $h = 1/4$, $k = 1/8$.

11b.   The function $u\,(x,\, t)$ satisfies the following conditions

$$\frac{\partial^2 u}{\partial x^2} + \frac{\partial^2 u}{\partial x\,\partial t} - \frac{\partial^2 u}{\partial t^2} = 1$$

$$u\,(x,\, 0) = 0\,, \qquad \left.\frac{\partial u}{\partial t}\right|_{t=0} = x\,(1 - x)$$

Use the method of characteristics to find $u$ at the intersection of the characteristics through (0.4, 0), (0.5, 0) and (0.6, 0).

Answer: 0.01, 0.01, 0.02

12a.   Show that the solution to the following boundary value problem

$$\frac{d^2 u}{dx^2} + a_0(x)\,u \;=\; r\,(x)$$

$$u\,(a) \;=\; \alpha\,, \qquad u\,(b) \;=\; \beta$$

is given by the extremum of the integral $I$ which is defined by

$$I = \int_a^b \left[ \left(\frac{du}{dx}\right)^2 - a_0 u^2 + 2ru \right] dx$$

13a.   Solve the boundary value problem

$$\frac{d^2 u}{dx^2} = 3x + 1$$

$$u\,(0) \;=\; u\,(1) \;=\; 0$$

(i) by the Rayleigh-Ritz method and (ii) by the Galerkin method.

Use a quadratic in  x  as the approximating function.

Compare the approximate solutions with the exact solution.

Answer: $5x\,(x-1)/4$ ; $x\,(x^2+x-2)/2$

14a.   Deduce Equations (8.8-17a, b, c).  Note that  u  is approximated by a linear function  $p_1\,(x)$  and, from Section 7.6, one deduces that  $p_1\,(x)$  can be written as

$$p_1\,(x) \;=\; u_i + (u_{i+1} - u_i)\,(x - x_i)/h$$

in the interval $[x_i,\,x_{i+1}]$.

Impose the boundary conditions and obtain the required results.

15a.   Solve the boundary value problem defined in Problem 13a by the method of finite elements.  Consider three elements and approximate  u  by a piecewise linear function.  Compare the values of  $u\,(1/3))$  and $u\,(2/3)$  with those obtained in Problem 13a.

16b.   Solve the non-linear system

$$\frac{d^2u}{dx^2} + 0.5\left[\left(\frac{du}{dx}\right)^2 + u^2\right] \;=\; 1\,, \qquad 0 \le x \le \pi/2$$

$$u\,(0) \;=\; 1\,, \qquad u\,(\pi/2) \;=\; 0$$

by the method of finite elements.

Divide the domain into five elements and choose linear shape functions.

Verify that  $1 - \sin x$  is a solution of the boundary value problem.  Compare the analytical solution with the numerical one.

17a.   The temperature  u (x, y)  on a square plate of size  2m $\times$ 2m  has reached a steady state and satisfies the equation

$$\frac{\partial^2u}{\partial x^2} + \frac{\partial^2u}{\partial y^2} \;=\; 0\,, \qquad 0 \le x \le 2\,, \qquad 0 \le y \le 2$$

The sides  x = 0  and  x = 2  are kept at 0°C and 100°C respectively.  The sides  y = 0  and y = 2  are insulated  $(\partial u/\partial y = 0)$.  Use the method of finite elements to find  u  at the points (1,0),  (1, 1)  and  (1, 2).

Answer: 50°C

# CHAPTER 9

# CALCULUS OF VARIATIONS

## 9.1  INTRODUCTION

Calculus of variations was introduced by Johann Bernouilli two centuries ago and is now widely used in optimization, control theory, and for computing approximate solutions of differential equations. In differential calculus we learned to determine the extreme values (maxima or minima) of a function. For example, the function $f(x) = x^2$ has a minimum at $x = 0$, where its value is 0, $f(x)$ is greater than zero everywhere else.

Calculus of variations is an extension of the above concept. Suppose we wish to determine the shortest plane curve joining two points A and B in the xy-plane as shown in Figure 9.1-1. If the coordinates of A and B are $(a, y(a))$ and $(b, y(b))$ respectively, then the length s of the curve $y = y(x)$ joining A and B is given by

$$s = \int_a^b \sqrt{1 + \left(\frac{dy}{dx}\right)^2} \; dx \tag{9.1-1}$$

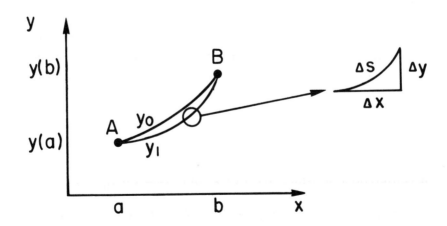

FIGURE 9.1-1    Curves joining two points A and B

For each curve $y(x)$, we can obtain the corresponding value of $s$ and we have to determine the curve $y(x)$ which yields the minimum value of $s$. Figure 9.1-1 shows two such curves, $y_0$ and $y_1$. In variational calculus we need to determine a function, if it exists, such that an integral is an extremum. That is to say, we wish to find the extreme values of a **functional**, where a functional can loosely be defined as a rule which assigns a real number to a function. In this example, we assign a number $s$ to each function $y(x)$ and the rule refers to an integration with respect to $x$ from $a$ to $b$. The integral on the right side of Equation (9.1-1) is referred to as a functional.

In engineering we encounter similar variational problems. In kinetics we can consider the reactions $A \rightarrow B \rightarrow C$. The reaction rates $r_A$, $r_B$ and $r_C$ may be functions of temperature and concentration. We may wish to determine the reaction temperature that will produce the maximum amount of $B$ in a fixed time interval from $0$ to $t_1$. The production $I$ of $B$ in the time interval $(0, t_1)$ is given by

$$I = \int_0^{t_1} r_B \, dt \tag{9.1-2}$$

Equation (9.1-2) is similar to Equation (9.1-1). We need to determine the function $r_B$ that will maximize the integral.

Prior to considering variational calculus, we review the theory of extreme values in differential calculus.

## 9.2   FUNCTION OF ONE VARIABLE

In Chapter 1, it is stated that a function of one variable $f(x)$ has an **extreme value** in the neighborhood of $x_0$ if $\Delta f$ given by

$$\Delta f = f(x_0 + \Delta x) - f(x_0) \tag{9.2-1}$$

retains the same sign for all small values of $\Delta x$, irrespective of the sign of $\Delta x$. Figure 9.2-1 illustrates this for the situations where $x_0$ is at $P$ (both $\Delta f$ and $\Delta x$ change sign) or at $Q$ (only $\Delta x$ changes sign). Hence $Q$ represents an extremum. Expanding the right side of Equation (9.2-1) about $x_0$ yields

$$f(x) - f(x_0) = \Delta f = f'(x_0)\,\Delta x + \frac{1}{2}\,f''(x_0)\,(\Delta x)^2 + \dots \tag{9.2-2a,b}$$

where $(\,')$ denotes differentiation with respect to $x$ and $\Delta x = x - x_0$.

It can be seen from Equation (9.2-2) that the condition for $\Delta f$ to preserve its sign, irrespective of the sign of $\Delta x$ is

$$f'(x_0) = 0 \tag{9.2-3}$$

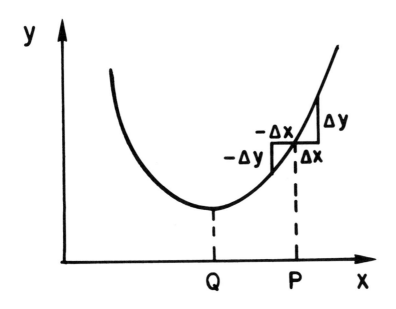

**FIGURE 9.2-1    Extremum for a function of one variable**

Thus the condition for point $x_0$ to give an extremum of $f$ is given by Equation (9.2-3). It follows from Equation (9.2-2) that $f(x_0)$ is a **maximum** if $f''(x_0)$ is negative and $f(x_0)$ is a **minimum** if $f''(x_0)$ is positive. In Figure 9.2-1, Q represents a minimum since $\Delta f$ is always positive.

***Example 9.2-1.*** Find the extreme values of $f(x) = \sin x$.

On differentiating, we have

$$f'(x) = \cos x \qquad \text{and} \qquad f''(x) = -\sin x \tag{9.2-4, 5}$$

Thus the extreme values of $\sin x$ are at $f'(x) = 0$; that is, $\cos x = 0$, so that

$$x_n = (2n + 1)\,\pi/2\,, \qquad n = 0, 1, 2, \dots \tag{9.2-6}$$

At these points

$$f''(x_n) = -\sin x_n = -\sin(2n+1)\,\pi/2 = -\sin\pi/2\,\cos n\pi = (-1)^{n+1} \tag{9.2-7a,b,c,d}$$

These extreme values are maxima when $n$ is even and minima when $n$ is odd. Thus $x_0 = \pi/2$ is maximum $(\sin\pi/2 = 1)$ and $x_1 = 3\pi/2$ is a minimum $(\sin 3\pi/2 = -1)$.

## 9.3    FUNCTION OF SEVERAL VARIABLES

If $f$ is a function of several variables $x_1, x_2, \dots, x_n$ then $f$ is said to have an **extreme value** in the neighborhood of $\overline{x}_1, \overline{x}_2, \dots, \overline{x}_n$ if $\Delta f$ defined as

$$\Delta f = f\,(\overline{x}_1 + \Delta x_1, \overline{x}_2 + \Delta \overline{x}_2, \dots \overline{x}_n + \Delta x_n) - f\,(\overline{x}_1, \overline{x}_2, \dots \overline{x}_n) \qquad (9.3\text{-}1)$$

preserves its sign for all $\Delta x_i$, $i = 1, \dots n$.

Expanding the right side of Equation (9.3-1), we obtain

$$\Delta f = \sum_{i=1}^{n} \frac{\partial f}{\partial x_i}\, \Delta x_i + \frac{1}{2} \sum_{i=1}^{n} \sum_{j=1}^{n} \frac{\partial^2 f}{\partial x_i \partial x_j}\, \Delta x_i\, \Delta x_j + \dots \qquad (9.3\text{-}2)$$

In Equation (9.3-2), all the derivatives are to be evaluated at $x_i = \overline{x}_i$, $i = 1, \dots n$.

From Equation (9.3-2), we deduce the condition for $f$ to have an extremum at $x_i = \overline{x}_i$ is

$$\left. \frac{\partial f}{\partial x_i} \right|_{x_i = \overline{x}_i} = 0, \qquad \text{for all } i = 1, 2, \dots n \qquad (9.3\text{-}3)$$

The function $f$ will be a **minimum** if the **quadratic expression** $\displaystyle \sum_{i=1}^{n} \sum_{j=1}^{n} \frac{\partial^2 f}{\partial x_i \partial x_j}\, \Delta x_i\, \Delta x_j$ is positive definite.  The conditions for this are

$$D_1 = \frac{\partial^2 f}{\partial x_1^2} > 0 \qquad (9.3\text{-}4)$$

$$D_2 = \begin{vmatrix} \dfrac{\partial^2 f}{\partial x_1^2} & \dfrac{\partial^2 f}{\partial x_1 \partial x_2} \\[4mm] \dfrac{\partial^2 f}{\partial x_1 \partial x_2} & \dfrac{\partial^2 f}{\partial x_2^2} \end{vmatrix} > 0 \qquad (9.3\text{-}5)$$

$$\begin{array}{c} \cdot \\ \cdot \\ \cdot \end{array}$$

$$D_n = \begin{vmatrix} \dfrac{\partial^2 f}{\partial x_1^2} & \dfrac{\partial^2 f}{\partial x_1 \partial x_2} & \cdots & \dfrac{\partial^2 f}{\partial x_1 \partial x_n} \\[2ex] \dfrac{\partial^2 f}{\partial x_1 \partial x_2} & \dfrac{\partial^2 f}{\partial x_2^2} & \cdots & \dfrac{\partial^2 f}{\partial x_2 \partial x_n} \\[2ex] \vdots & & & \\[1ex] \dfrac{\partial^2 f}{\partial x_1 \partial x_n} & \cdots & & \dfrac{\partial^2 f}{\partial x_n^2} \end{vmatrix} > 0 \qquad (9.3\text{-}6)$$

All the derivatives are to be evaluated at $x_i = \bar{x}_i$.

The conditions for $f$ to have a **maximum** at $\bar{x}_i$ is that the quadratic expression must be negative definite, that is

$$D_1 < 0, \qquad D_2 > 0, \qquad D_3 < 0 \qquad \cdots \qquad (9.3\text{-}7a,b,c)$$

and so on with the sign of the determinant alternating.

When the quadratic expression is not definite we have to use ad hoc methods or consider higher order terms to determine the nature of the extrema.

***Example 9.3-1.*** Find the extreme values of

$$f(x_1, x_2) = x_1^2 + x_2^2 + x_1 x_2 \qquad (9.3\text{-}8)$$

On differentiating, we obtain

$$\frac{\partial f}{\partial x_1} = 2x_1 + x_2, \quad \frac{\partial f}{\partial x_2} = 2x_2 + x_1, \quad \frac{\partial^2 f}{\partial x_1^2} = 2, \quad \frac{\partial^2 f}{\partial x_2^2} = 2, \quad \frac{\partial^2 f}{\partial x_1 \partial x_2} = 1 \qquad (9.3\text{-}9a \text{ to } e)$$

The extreme values of $f$ are at

$$2x_1 + x_2 = 0, \qquad 2x_2 + x_1 = 0 \qquad (9.3\text{-}10a,b)$$

The solutions are

$$x_1 = \bar{x}_1 = 0, \qquad x_2 = \bar{x}_2 = 0 \qquad (9.3\text{-}11a,b,c,d)$$

So the extreme value of $f$ is at the origin. At the origin, both $D_1$ and $D_2$ are positive and $f$ has a minimum at the origin.

***Example 9.3-2.*** Find the extreme values of

$$f(x_1, x_2) = x_2^2 + x_1^2 x_2 + x_1^4 \tag{9.3-12}$$

The partial derivatives of $f$ are

$$\frac{\partial f}{\partial x_1} = 2x_1 x_2 + 4x_1^3, \quad \frac{\partial f}{\partial x_2} = 2x_2 + x_1^2, \quad \frac{\partial^2 f}{\partial x_1^2} = 2x_2 + 12x_1^2 \tag{9.3-13a,b,c}$$

$$\frac{\partial^2 f}{\partial x_2^2} = 2, \quad \frac{\partial^2 f}{\partial x_1 \partial x_2} = 2x_1 \tag{9.3-13d,e}$$

The extreme values of $f$ are given by

$$\frac{\partial f}{\partial x_1} = 0 \implies 2x_1 x_2 + 4x_1^3 = 0 \tag{9.3-14a,b}$$

$$\frac{\partial f}{\partial x_2} = 0 \implies 2x_2 + x_1^2 = 0 \tag{9.3-14c,d}$$

The solutions are

$$x_1 = 0, \qquad x_2 = 0 \tag{9.3-15a,b}$$

Thus $f$ has its extreme value at the origin. Both $D_1 = D_2 = 0$ at the origin, and so we need to investigate further to determine if the origin is a maximum, a minimum or neither.

$\Delta f$ is given approximately by

$$\Delta f = (\Delta x_2)^2 + (\Delta x_1)^2 (\Delta x_2) + (\Delta x_1)^4 = \left[ \Delta x_2 + \frac{1}{2}(\Delta x_1)^2 \right]^2 + \frac{3}{4}(\Delta x_1)^4 \tag{9.3-16a,b}$$

and $\Delta f$ will always be positive and the origin is a minimum.

***Example 9.3-3.*** Find the extreme values of

$$f(x_1, x_2) = x_1 x_2 \tag{9.3-17}$$

On differentiating, we obtain

$$\frac{\partial f}{\partial x_1} = x_2, \quad \frac{\partial f}{\partial x_2} = x_1, \quad \frac{\partial^2 f}{\partial x_1^2} = 0, \quad \frac{\partial^2 f}{\partial x_2^2} = 0, \quad \frac{\partial^2 f}{\partial x_1 \partial x_2} = 1 \tag{9.3-18a to e}$$

The extreme value of f is at the origin where $D_1$ and f are zero and the quadratic expression is not definite. f is greater than zero if $x_1$ and $x_2$ are of the same sign and less than zero if $x_1$ and $x_2$ are of opposite sign. The origin is a **saddle point**, it is a maximum along one line and a minimum along another line, in the neighborhood of the origin.

## 9.4   CONSTRAINED EXTREMA AND LAGRANGE MULTIPLIERS

We have assumed earlier that $x_1, x_2, \ldots, x_n$ are all independent and can take any arbitrary value. This might not necessarily be the case. For example, if $x_1$ is the cost of a product, then $x_1$ cannot be negative. There might be a relationship between the $x_i$. Thus if $x_1$ represents the length of a rod and $x_2$ is the temperature, then $x_1$ is a function of $x_2$, $x_1$ and $x_2$ are not completely independent. In theory it is possible to eliminate all dependent variables and retain only the independent variables. In practice, this can be very cumbersome and the process of elimination can be very time consuming. It is preferable to use the **method of Lagrange multipliers**.

Suppose we wish to find the extremum of $f(x_1, x_2, \ldots, x_n)$ subject to the condition

$$g(x_1, x_2, \ldots, x_n) = 0 \qquad (9.4\text{-}1)$$

We now consider the function L, sometimes called a Lagrangian, and is given by

$$L(x_1, x_2, \ldots, x_n, \lambda) = f(x_1, x_2, \ldots, x_n) + \lambda g(x_1, x_2, \ldots, x_n) \qquad (9.4\text{-}2)$$

where $\lambda$ is the **Lagrange multiplier**.

The extremum of L is then given by

$$\frac{\partial L}{\partial x_i} = 0, \quad i = 1, 2, \ldots, n \qquad (9.4\text{-}3a)$$

$$\frac{\partial L}{\partial \lambda} = 0 \qquad (9.4\text{-}3b)$$

From the $(n + 1)$ equations [Equations (9.4-3a,b)], we can solve for $x_i = \bar{x}_i$ and $\lambda$.

If we have m constraints, instead of one, and the constraints are given by

$$g_j(x_1, x_2, \ldots, x_n) = 0, \qquad j = 1, 2, \ldots, m \qquad (9.4\text{-}4)$$

The Lagrangian becomes

$$L = f(x_1, x_2, \ldots, x_n) + \sum_{j=1}^{m} \lambda_j g_j(x_1, x_2, \ldots, x_n) \qquad (9.4\text{-}5)$$

The extremum of L is then given by

$$\frac{\partial L}{\partial x_i} = 0 , \qquad i = 1, 2, \ldots , n \tag{9.4-6a}$$

$$\frac{\partial L}{\partial \lambda_j} = 0 , \qquad j = 1, 2, \ldots , m \tag{9.4-6b}$$

From the $(n + m)$ equations, we can solve for $\bar{x}_i$ and $\lambda_j$.

Since the $g_j (x_1, x_2, \ldots , x_n)$ are identically equal to zero, the extremum of L is the extremum of f and the Lagrangian has been introduced to ensure that the conditions of constraints are automatically satisfied.

***Example 9.4-1.*** Find the extremum of

$$f (x_1, x_2) = x_1 x_2 \tag{9.4-7}$$

subject to

$$x_1^2 + x_2^2 = 1 \tag{9.4-8}$$

In this example, we are required to find the maximum or minimum of the rectangle inscribed in a circle of unit radius. The equation of constraint is

$$g (x_1, x_2) = x_1^2 + x_2^2 - 1 = 0 \tag{9.4-9}$$

The Lagrangian and its derivative are

$$L = x_1 x_2 + \lambda (x_1^2 + x_2^2 - 1) \tag{9.4-10}$$

$$\frac{\partial L}{\partial x_1} = x_2 + 2\lambda x_1 , \qquad \frac{\partial L}{\partial x_2} = x_1 + 2\lambda x_2 , \qquad \frac{\partial L}{\partial \lambda} = x_1^2 + x_2^2 - 1 = 0 \tag{9.4-11a,b,c}$$

From Equations (9.4-6a, 11a, b), we deduce that

$$x_1 = x_2 \tag{9.4-12}$$

Substituting Equation (9.4-12) in Equation (9.4-11c), we obtain

$$x_1 = \frac{1}{\sqrt{2}} , \qquad f = \frac{1}{2} \tag{9.4-13a,b}$$

Note that, in this example, the point $(x_1, x_2)$ has to be on the circle of unit radius with the centre located at the origin. In Example 9.3-3, $(x_1, x_2)$ can be any point and thus $f = x_1 x_2$ has a saddle point at the origin. In the present problem, $(x_1, x_2)$ cannot be the origin.

## 9.5    EULER-LAGRANGE EQUATIONS

We now deduce the conditions for the functional (integral) I given by

$$I = \int_a^b f(x, y, y')dx \tag{9.5-1}$$

to have an extreme value. The points a and b are fixed and we also assume $y(a)$ and $y(b)$ to be fixed. Let the function $y_0(x)$ represent an extreme value for I. We can consider a neighboring curve $y_1 = y_0 + \varepsilon \eta(x)$ as shown in Figure 9.1-1. The increment $\Delta I$ is given by

$$\Delta I = \int_a^b f\left[x, y_0 + \varepsilon\eta(x), y_0' + \varepsilon\eta'(x)\right] dx - \int_a^b f\left[x, y_0(x), y_0'(x)\right] dx \tag{9.5-2}$$

where $\varepsilon$ is a small quantity and $\eta$ is an arbitrary continuous function of x.

On expanding the function $f[x, y_0 + \varepsilon\eta(x), y_0' + \varepsilon\eta'(x)]$ in powers of $\varepsilon$, according to the Taylor expansion [Equation (1.7-7)] and substituting into Equation (9.5-2), we obtain

$$\Delta I = \int_a^b \left[\varepsilon\eta \frac{\partial f}{\partial y} + \varepsilon\eta' \frac{\partial f}{\partial y'} + O(\varepsilon^2)\right] dx \tag{9.5-3}$$

Integrating the second term by parts yields

$$\int_a^b \varepsilon\eta' \frac{\partial f}{\partial y'} dx = \left[\varepsilon\eta \frac{\partial f}{\partial y'}\right]_a^b - \int_a^b \varepsilon\eta \frac{d}{dx}\left(\frac{\partial f}{\partial y'}\right) dx \tag{9.5-4}$$

Since $y(a)$ and $y(b)$ are fixed, $\eta(a) = \eta(b) = 0$. Substituting Equation (9.5-4) into Equation (9.5-3), we obtain

$$\Delta I = \int_a^b \left[\varepsilon\eta\left\{\frac{\partial f}{\partial y} - \frac{d}{dx}\left(\frac{\partial f}{\partial y'}\right)\right\}\right] dx + O(\varepsilon^2) \tag{9.5-5}$$

From Equation (9.5-5), we deduce that $\Delta I$ preserves its sign for all arbitrary functions $\varepsilon\eta(x)$ if

$$\frac{\partial f}{\partial y} - \frac{d}{dx}\left(\frac{\partial f}{\partial y'}\right) = 0 \qquad\qquad\qquad (9.5\text{-}6)$$

Equation (9.5-6) is the **Euler-Lagrange equation** and gives the condition for $I$ to have an extremum. We can also deduce Equation (9.5-6) by considering $I$ to be a function of $\varepsilon$. The extremum of the integral

$$I = \int_a^b f\,(x,\, y_0 + \varepsilon\eta,\, y_0' + \varepsilon\eta')\,dx \qquad\qquad (9.5\text{-}7a)$$

is given by

$$\left.\frac{dI}{d\varepsilon}\right|_{\varepsilon=0} = 0 \qquad\qquad\qquad (9.5\text{-}7b)$$

$$\frac{dI}{d\varepsilon} = \int_a^b \frac{d}{d\varepsilon}\,[f\,(x,\, y_0 + \varepsilon\eta,\, y_0' + \varepsilon\eta')]\,dx \qquad\qquad (9.5\text{-}8a)$$

$$= \int_a^b \left[\frac{\partial f}{\partial y_1}\frac{\partial y_1}{\partial \varepsilon} + \frac{\partial f}{\partial y_1'}\frac{\partial y_1'}{\partial \varepsilon}\right] dx \qquad\qquad (9.5\text{-}8b)$$

$$= \int_a^b \left[\eta\frac{\partial f}{\partial y_1} + \eta'\frac{\partial f}{\partial y_1'}\right] dx \qquad\qquad (9.5\text{-}8c)$$

We recall that $y_1 = y_0 + \varepsilon\eta$ and, on setting $\varepsilon = 0$, we obtain $y_1 = y_0$. We now denote $y_0$ by $y$. Equation (9.5-8c) becomes

$$\left.\frac{dI}{d\varepsilon}\right|_{\varepsilon=0} = \int_a^b \left[\eta\frac{\partial f}{\partial y} + \eta'\frac{\partial f}{\partial y'}\right] dx \qquad\qquad (9.5\text{-}9)$$

On again integrating $\displaystyle\int_a^b \eta'\,\frac{\partial f}{\partial y'}\,dx$ by parts, we obtain

$$\left.\frac{dI}{d\varepsilon}\right|_{\varepsilon=0} = \int_a^b \eta\left[\frac{\partial f}{\partial y} - \frac{d}{dx}\left(\frac{\partial f}{\partial y'}\right)\right] dx = 0 \qquad\qquad (9.5\text{-}10a,b)$$

Thus the condition for $\left.\dfrac{dI}{d\varepsilon}\right|_{\varepsilon=0} = 0$ for all $\eta(x)$ is also given by Equation (9.5-6). The variation $\dfrac{dI}{d\varepsilon}$ is also known as the **Gâteaux variation**.

Expanding Equation (9.5-6), we have

$$\frac{\partial f}{\partial y} - \frac{\partial^2 f}{\partial x \partial y'} - y' \frac{\partial^2 f}{\partial y \partial y'} - y'' \frac{\partial^2 f}{(\partial y')^2} = 0 \tag{9.5-11}$$

Hint: Note that $\dfrac{\partial f}{\partial y'}$ can be written as $g[x, y(x), y'(x)]$ and compute $\dfrac{dg}{dx}$.

Thus to determine the extremum of the functional I, we need to solve a second-order differential equation [Equation (9.5-11)], subject to the boundary conditions

$$\text{at } x = a, \quad y = y(a); \qquad \text{at } x = b, \quad y = y(b) \tag{9.5-12a, b}$$

In practice, not all the arguments $x, y, y'$ of $f$ are present and Equation (9.5-11) is then simplified. We now consider these special cases.

## 9.6 SPECIAL CASES

### Function f Does Not Depend on y' Explicitly

In this case, $f$ is a function of $x$ and $y$ only and it follows

$$\frac{\partial f}{\partial y'} = 0 \tag{9.6-1}$$

Equation (9.5-11) reduces to

$$\frac{\partial f}{\partial y} = 0 \tag{9.6-2}$$

From Equation (9.6-2), we obtain the relation $y = y(x)$. The solution $y = y(x)$ has to satisfy the boundary conditions [Equations (9.5-12a, b)].

If $y$ does not satisfy Equations (9.5-12a, b), $y$ is not an admissible solution and the functional I does not have an extreme value.

***Example 9.6-1.*** Find the extreme values of

$$I = \int_1^2 (xy^2 - y)\, dx \qquad\qquad (9.6\text{-}3)$$

The boundary conditions are

(a)     $y(1) = 1,$          $y(2) = 1$                                                                          (9.6-4a)

(b)     $y(1) = 1/2,$          $y(2) = 1/4$                                                                    (9.6-4b)

In this problem

$$f = x y^2 - y \qquad\qquad (9.6\text{-}5)$$

which is not an explicit function of $y'$.

The extremum of $I$ is then given by Equation (9.6-2), which yields

$$2xy - 1 = 0 \qquad\qquad (9.6\text{-}6)$$

The solution of Equation (9.6-6) [$y = (1/2\,x)$] does not satisfy Equation (9.6-4a) and, in this case, $I$ has no extremum. However, it satisfies Equation (9.6-4b) and $y = (1/2\,x)$ gives the extreme value of $I$. In this case

$$I = I_1 = \int_1^2 \left( x\,\frac{1}{4x^2} - \frac{1}{2x} \right) dx \;=\; -\int_1^2 \frac{1}{4x}\, dx \qquad\qquad (9.6\text{-}7a,b,c)$$

$$= -\frac{1}{4}\, \ell n\, 2 = -0.1733 \qquad\qquad (9.6\text{-}7d,e)$$

For the purpose of comparison, suppose we have a straight line joining the points $(1, 1/2)$ to $(2, 1/4)$. The equation of such a line is

$$y = \frac{1}{4}\,(3 - x) \qquad\qquad (9.6\text{-}8)$$

Substituting this $y$ into $I$, we obtain

$$I = I_2 = \int_1^2 \left\{ \frac{x}{16}\,(3 - x)^2 - \frac{1}{4}\,(3 - x) \right\} dx \;=\; -\frac{11}{64} = -0.1719 \qquad\qquad (9.6\text{-}9a,b,c,d)$$

Comparing Equations (9.6-7e, 9d), it is seen that $I_1$ is a minimum.

This means that there is no function (curve) $y(x)$ that extremizes $I$ and satisfies Equation (9.6-4a). Of all the functions that satisfy Equation (9.6-4b), the function $y = (1/2\,x)$ minimizes $I$.

## Function f Does Not Depend on y Explicitly

In this case, Equation (9.5-6) simplifies to

$$\frac{d}{dx}\left(\frac{\partial f}{\partial y'}\right) = 0 \qquad (9.6\text{-}10)$$

which on integrating yields

$$\frac{\partial f}{\partial y'} = \text{constant} = c_1 \qquad (9.6\text{-}11)$$

From Equation (9.6-11), we can solve for $y'$ and $y'$ can be further integrated to obtain y. Since we have integrated twice, we introduce two arbitrary constants which are determined by the end conditions [Equations (9.5-12a, b)]. Unlike the previous case, here we always have a solution.

*Example 9.6-2.* Find the extremum of

$$I = \int_1^2 \frac{\sqrt{1 + (y')^2}}{x}\, dx \qquad (9.6\text{-}12)$$

In this example, $f\left[= \dfrac{\sqrt{1 + (y')^2}}{x}\right]$ is not an explicit function of y. Equation (9.6-11) yields

$$\frac{\partial}{\partial y'}\left(\frac{\sqrt{1 + (y')^2}}{x}\right) = \frac{y'}{x\sqrt{1 + (y')^2}} = c_1 \qquad (9.6\text{-}13a,b)$$

Solving for $y'$, we obtain

$$y' = \frac{c_1\, x}{\sqrt{1 - c_1^2\, x^2}} \qquad (9.6\text{-}14)$$

Integrating yields

$$y = \int \frac{c_1\, x}{\sqrt{1 - c_1^2\, x^2}}\, dx = -\frac{1}{c_1}\sqrt{1 - c_1^2\, x^2} + c_2 \qquad (9.6\text{-}15a,b)$$

where $c_2$ is a constant.

We may rewrite Equation (9.6-15b) as

$$c_1^2 (y - c_2)^2 = (1 - c_1^2 x^2) \qquad\qquad (9.6\text{-}16)$$

and this represents a family of circles with centres along the y-axis. The constants $c_1$ and $c_2$ are determined by the boundary conditions. If the curve passes through the origin $(0, 0)$ and the point $(1, 1)$, we obtain from Equation (9.6-16)

$$c_1 = c_2 = 1 \qquad\qquad (9.6\text{-}17a,b)$$

Substituting the values of $c_1$ and $c_2$ into Equation (9.6-16), we obtain

$$(y - 1)^2 + x^2 = 1 \qquad\qquad (9.6\text{-}18)$$

which is the equation of unit circle with centre at $(0, 1)$.

## Function  f  Does Not Depend on  x  Explicitly

Since  f  is not an explicit function of  x

$$\frac{\partial f}{\partial x} = 0 \qquad\qquad (9.6\text{-}19)$$

Equation (9.5-11) simplifies to

$$\frac{\partial f}{\partial y} - y' \frac{\partial^2 f}{\partial y \partial y'} - y'' \frac{\partial^2 f}{(\partial y')^2} = 0 \qquad\qquad (9.6\text{-}20)$$

Multiplying Equation (9.6-20) by $y'$, we find that the resulting expression is an exact differential which may be written as

$$\frac{d}{dx} \left[ f - y' \frac{\partial f}{\partial y'} \right] = 0 \qquad\qquad (9.6\text{-}21)$$

The solution of the above equation is

$$f - y' \frac{\partial f}{\partial y'} = \text{constant} = c_1 \qquad\qquad (9.6\text{-}22)$$

From Equation (9.6-22), we can solve for $y'$ and this can be integrated so as to obtain the required  y  as a function of  x.  As usual, the two integration constants are determined from the boundary conditions.

**Example 9.6-3.** Find the extremum of

$$I = \frac{1}{\sqrt{2g}} \int_0^a \frac{\sqrt{1+(y')^2}}{\sqrt{y}}\, dx \tag{9.6-23}$$

This example is known as the **brachistochrone problem**. It was first proposed by Johann Bernouilli in 1696. It is usually accepted that this problem marks the beginning of calculus of variations. In this example, we have to determine the path that a particle will take, while falling under gravity, so that the time taken is a minimum. If we take one of the points to be the origin of the Cartesian coordinate system, x the horizontal axis, and y the vertical axis, then the velocity v at height y is computed through the conservation of energy (potential + kinetic = constant) to be

$$v = \sqrt{2gy} \tag{9.6-24}$$

where g is the gravity. Then the time t needed for the particle to fall from the origin (0, 0) to a point (a, b) is, using Equations (9.1-1, 6-24)

$$t = \int_0^a \frac{ds}{v} = \frac{1}{\sqrt{2g}} \int_0^a \frac{\sqrt{1+(y')^2}}{\sqrt{y}}\, dx \tag{9.6-25a,b}$$

Since

$$f = \frac{1}{\sqrt{2g}} (1 + (y')^2)^{1/2}\, y^{-1/2} \tag{9.6-26}$$

$$\frac{\partial f}{\partial y'} = \frac{y'\, y^{-1/2}(1 + (y')^2)^{-1/2}}{\sqrt{2g}} \tag{9.6-27}$$

Substituting f and $\dfrac{\partial f}{\partial y'}$ and absorbing $\sqrt{2g}$ into the constant $c_1$, Equation (9.6-22) becomes

$$(1 + (y')^2)^{1/2}\, y^{-1/2} - (y')^2\, (1 + (y')^2)^{-1/2}\, y^{-1/2} = c_1 \tag{9.6-28}$$

On simplifying the above expression, we obtain

$$1 = c_1\, y^{1/2}\, (1 + (y')^2)^{1/2} \tag{9.6-29}$$

Squaring both sides of the above expression, we can then solve for y'. Integrating y' yields

$$\int \sqrt{\frac{c_1^2\, y}{1 - c_1^2\, y}}\, dy = \int dx \tag{9.6-30}$$

To evaluate the integral on the left side of the above equation, we use the substitution

$$y = \frac{1}{c_1^2} \sin^2\theta \qquad\qquad (9.6\text{-}31)$$

$$\int \sqrt{\frac{c_1^2 y}{1 - c_1^2 y}} \, dy = \frac{2}{c_1^2} \int \frac{\sin\theta}{\cos\theta} \sin\theta \cos\theta \, d\theta = \frac{1}{c_1^2} \left[ \theta - \frac{1}{2} \sin 2\theta \right] \qquad (9.6\text{-}32a,b)$$

Equation (9.6-30) can now be written as

$$\frac{1}{c_1^2} \left[ \theta - \frac{1}{2} \sin 2\theta \right] = x + c_2 \qquad\qquad (9.6\text{-}33)$$

where $c_2$ is a constant. The curve has to pass through the origin, $(0, 0)$. Therefore, $\theta = 0$, and $c_2 = 0$. The parametric equation of the curve is

$$x = \frac{1}{c_1^2} \left[ \theta - \frac{1}{2} \sin 2\theta \right] \qquad\qquad (9.6\text{-}34)$$

$$y = \frac{1}{c_1^2} \sin^2\theta = \frac{1}{2c_1^2} \left[ 1 - \cos 2\theta \right] \qquad\qquad (9.6\text{-}35a,b)$$

The constant $c_1$ is to be determined from the condition $x = a$, $y = b$. The parametric equations of the curve as given above is that of a **cycloid**. Thus the curve of quickest descent is a cycloid.

## Function  f  Is a Linear Function of  y'

Let us write  f  as

$$f = p (x, y) + y' \, q (x, y) \qquad\qquad (9.6\text{-}36)$$

Then

$$\frac{\partial f}{\partial y} = \frac{\partial p}{\partial y} + y' \frac{\partial q}{\partial y} \qquad\qquad (9.6\text{-}37)$$

$$\frac{\partial f}{\partial y'} = q (x, y) \qquad\qquad (9.6\text{-}38)$$

Equation (9.5-6) becomes

$$\frac{\partial p}{\partial y} + y' \frac{\partial q}{\partial y} = \frac{d}{dx} (q(x, y)) = \frac{\partial q}{\partial x} + \frac{\partial q}{\partial y} y' \qquad\qquad (9.6\text{-}39a,b)$$

The condition for I to be an extremum is

$$\frac{\partial p}{\partial y} = \frac{\partial q}{\partial x} \tag{9.6-40}$$

Equation (9.6-40) gives y as a function of x and there is no integration constant. Thus an extremum of I will exist only if y satisfies the end conditions [Equations (9.5-12a,b)]. This is similar to the case where f is not an explicit function of y'.

If Equation (9.6-40) is identically satisfied, then the expression given by Equation (9.6-36) is an exact differential, and there exists a function $\phi$ such that

$$\frac{d\phi}{dx} = f \tag{9.6-41}$$

The integral becomes

$$I = \int_a^b f(x, y, y') \, dx = \int_a^b \frac{d\phi}{dx} \, dx = \phi_b - \phi_a \tag{9.6-42a,b,c}$$

Thus I is independent of the path taken and all curves yield the same value of I.

The above result is equivalent to the statement in mechanics, that the work done by a conservative force is independent of the path taken. If we take (p, q) to be the components of a vector $\underline{a}$ in a two-dimensional space, then

$$\text{curl } \underline{a} = \left( \frac{\partial q}{\partial x} - \frac{\partial p}{\partial y} \right) \tag{9.6-43}$$

If Equation (9.6-40) is identically satisfied, then

$$\text{curl } \underline{a} = \underline{0} \tag{9.6-44}$$

So $\underline{a}$ is a conservative force and there exists a potential $\phi$, such that

$$\underline{a} = \pm \text{ grad } \phi \tag{9.6-45}$$

Then the work done by $\underline{a}$ from one point to another is independent of the path and is only a function of the difference of potential between the two points.

***Example 9.6-4.*** Find the extremum of

$$I = \int_0^1 (y^2 + x^2 y') \, dx \qquad\qquad (9.6\text{-}46)$$

The requirement for an extremum of $I$ is given by Equation (9.6-40) with

$$p = y^2 \quad \text{and} \quad q = x^2 \qquad\qquad (9.6\text{-}47a,b)$$

Thus the condition is

$$2y = 2x \quad \text{or} \quad x - y = 0 \qquad\qquad (9.6\text{-}48a,b)$$

If the end conditions are represented by the points $(0, 0)$ and $(1, 2)$, then the first end condition is satisfied, but the second one is not. So in this case there is no extremum. If the second end condition is modified to $(1, 1)$, then the curve $y = x$ will give the extremum of $I$.

***Example 9.6-5***.  Find the extremum of

$$I = \int_0^1 (y + xy') \, dx \qquad\qquad (9.6\text{-}49)$$

In this example, $p = y$, $q = x$, so Equation (9.6-40) is identically satisfied. In this case, there is no extremum.

$$I = \int_0^1 (y + xy') \, dx = \int_0^1 [y \, dx + x \, dy] = \int_0^1 d(xy) = [xy]_0^1 = y(1) \qquad (9.6\text{-}50a \text{ to } e)$$

Therefore, $I$ does not depend on which curve $y(x)$ is taken.

●

In general, if $f$ is an explicit function of all three variables $x$, $y$, and $y'$, then we have to solve the second order differential Equation (9.5-11) and the two end conditions will determine the two arbitrary constants.

## 9.7   EXTENSION TO HIGHER DERIVATIVES

We extend the theory to $f$ being an explicit function of higher derivatives of $y$, such as $y''$, $y'''$, ... , $y^{(n)}$. Thus we find the extremum of

$$I = \int_a^b f(x, y, y', y'', y''', \ldots y^{(n)}) \, dx \qquad (9.7\text{-}1)$$

If $y = y_0(x)$ is the function that will make $I$ an extremum, then by considering another function

$$y = y_0 + \varepsilon \eta(x) \qquad (9.7\text{-}2)$$

where $\varepsilon$ is small and $\eta(x)$ is an arbitrary function of $x$ and proceeding as in Section 9.5, we can show that the condition for $I$ to be an extremum is

$$\frac{\partial f}{\partial y} - \frac{d}{dx}\left(\frac{\partial f}{\partial y'}\right) + \frac{d^2}{dx^2}\left(\frac{\partial f}{\partial y''}\right) - \frac{d^3}{dx^3}\left(\frac{\partial f}{\partial y'''}\right) + \ldots (-1)^n \frac{d^n}{dx^n}\left(\frac{\partial f}{\partial y^{(n)}}\right) = 0 \qquad (9.7\text{-}3)$$

This equation is essentially a generalization of Equation (9.5-6). The alternating signs are a result of several integrations by parts of the derivatives of $\eta$.

In deducing Equation (9.7-3), we have assumed that $y, y', y'', \ldots$ are given at $x = a$ and $x = b$. The existence and continuity of $y$ and its derivatives, as required, are assumed to be satisfied.

## 9.8    TRANSVERSALITY (MOVING BOUNDARY) CONDITIONS

We reconsider the problem of finding the extremum of $I$ [see Equation (9.5-1)] with the end points $x = a$ and $x = b$ fixed, but the values of $y$ and its derivatives at the end points not specified. The y-coordinate of the curve will have its ends at any point along the ordinates $x = a$ and $x = b$ as shown in Figure 9.8-1. The extremum curve has a greater degree of freedom and we cannot determine the two arbitrary constants in the solution of the Euler-Lagrange equation from the end conditions. Instead we need to deduce two new conditions.

We proceed as before by considering two curves $y_0$ and $y_0 + \varepsilon \eta$. We obtain Equations (9.5-3, 4), but we can now no longer impose the boundary conditions $\eta(a) = \eta(b) = 0$, since $y(a)$ and $y(b)$ are not fixed. We impose a new **transversality condition**

$$\left[\eta \frac{\partial f}{\partial y'}\right]_a^b = 0 \qquad (9.8\text{-}1)$$

Using Equation (9.8-1), we obtain Equations (9.5-5, 6). Thus the conditions for $I$ to be an extremum are now given by Equations (9.5-6, 8-1). If the value of $y$ is fixed at one of the end points (say at $x = a$) and is free at the other end, then the condition given by Equation (9.8-1) simplifies to

$$\left.\frac{\partial f}{\partial y'}\right|_{x=b} = 0 \qquad (9.8\text{-}2)$$

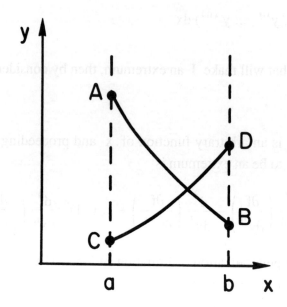

**FIGURE 9.8-1     Variable values of  y  and its derivatives at constant  x**

*Example 9.8-1*.  Find the extremum of

$$I = \int_0^1 \left[ (y')^2 + 1 \right]^2 dx \tag{9.8-3}$$

given  $y(0) = 1$  but  $y(1)$  is not prescribed.

Since  $f = \left[ (y')^2 + 1 \right]^2$  is not an explicit function of  y, Equation (9.6-11) implies

$$4y' \left[ (y')^2 + 1 \right] = c_1 \tag{9.8-4}$$

At  $x = 1$,  the transversality condition imposes

$$\frac{\partial f}{\partial y'} = 0 \quad \Rightarrow \quad c_1 = 0 \tag{9.8-5a,b}$$

It follows that

$$y' = 0 , \qquad y = \text{constant} = c_2 \tag{9.8-6a,b,c}$$

The solution that satisfies the prescribed condition  $[y(0) = 1]$  is

$$y = 1 \tag{9.8-7}$$

Thus the extremum of $I$ is given by

$$I = \int_0^1 dx = 1 \tag{9.8-8}$$

Let us determine $I$ when the values of $y$ are specified at both ends. We suppose $y(1) = a$. The condition for $I$ to be an extremum is still given by Equation (9.8-4), which is a cubic equation in $y'$. At least one of the roots must be real and since $c_1$ is an arbitrary constant, we can assume that $y'$ is a constant. Thus

$$y = k_1 x + k_0 \tag{9.8-9}$$

where $k_1$ and $k_0$ are arbitrary.

Imposing the end conditions

$$y(0) = 1, \qquad y(1) = a \tag{9.8-10a,b}$$

we find that

$$k_0 = 1, \qquad k_1 = (a-1) \tag{9.8-11a,b}$$

The function $y$ and the integral $I$ are given by

$$y = (a-1)x + 1 \tag{9.8-12}$$

$$I = \left[a^2 - 2a + 2\right]^2 \int_0^1 dx = \left[a^2 - 2a + 2\right]^2 \tag{9.8-13a,b}$$

We regard $a$ as a variable and determine the values of $a$ that will make $I$ a minimum. That is

$$\frac{dI}{da} = 0 \tag{9.8-14a}$$

or $\quad 2 [a^2 - 2a + 2] [2a - 2] = 0 \tag{9.8-14b}$

The solution is $a = 1$ and $I = 1$ which is the solution given by Equation (9.8-6). Using the transversality condition, we obtain automatically the value of $y(1)$ which renders $I$ an extremum.

## 9.9  CONSTRAINTS

So far we have been concerned with finding the extremum of

$$I = \int_a^b f(x, y, y') \, dx \tag{9.5-1}$$

with $y$ satisfying certain end conditions.

However, in many problems $y$ has to satisfy additional conditions and these conditions may take the form of integral relations, algebraic equations or differential equations. If the supplementary conditions are of an integral type, then they are known as **isoperimetric constraints** which we shall consider next.

Suppose we have to find the extremum of $I$ as given by Equation (9.5-1), subject to the constraint

$$J = \int_a^b g(x, y, y') \, dx = \text{constant} = C \tag{9.9-1a,b}$$

We assume that $y_0(x)$ is the function that extremizes $I$. Consider a neighboring curve to $y_0$ given by

$$y = y_0(x) + \varepsilon_1 \eta_1(x) + \varepsilon_2 \eta_2(x) \tag{9.9-2}$$

In this case, we have introduced two parameters $\varepsilon_1$ and $\varepsilon_2$. The functions $\eta_1(x)$ and $\eta_2(x)$ are arbitrary functions. In extremizing $I$ we can allow $\varepsilon_1$ to be completely arbitrary as before and then adjust $\varepsilon_2$ so that the condition given by Equation (9.9-1b) is satisfied.

We define $I^*$ and $J^*$ as

$$I^* = \int_a^b f(x, y_0 + \varepsilon_1 \eta_1 + \varepsilon_2 \eta_2, \ y_0' + \varepsilon_1 \eta_1' + \varepsilon_2 \eta_2') \, dx \tag{9.9-3}$$

$$J^* = \int_a^b g(x, y_0 + \varepsilon_1 \eta_1 + \varepsilon_2 \eta_2, \ y_0' + \varepsilon_1 \eta_1' + \varepsilon_2 \eta_2') \, dx = C \tag{9.9-4a,b}$$

$I^*$ and $J^*$ can be considered as functions of $\varepsilon_1$ and $\varepsilon_2$ and the extremum of $I^*$ is given by $\varepsilon_1 = \varepsilon_2 = 0$. We need to find the extremum of $I^*$ subject to the constraint given by Equation (9.9-4b). This problem is equivalent to the case of finding the extremum of a function of two variables subject to a constraint as discussed in Section 9.4. The method of Lagrange multipliers can thus be used. We therefore define a function $H^*$ as

$$H^* = I^* + \lambda J^* = \int_a^b (f + \lambda g) \, dx \qquad (9.9\text{-}5a,b)$$

$$= \int_a^b L(x, y_0 + \varepsilon_1 \eta_1 + \varepsilon_2 \eta_2, \, y_0' + \varepsilon_1 \eta_1' + \varepsilon_2 \eta_2') \, dx \qquad (9.9\text{-}5c)$$

where $\lambda$ is a constant.

The extremum of $I^*$ is given

$$\left. \frac{\partial H^*}{\partial \varepsilon_i} \right|_{\varepsilon_i = 0} = 0 \qquad (i = 1, 2) \qquad (9.9\text{-}6)$$

$$\frac{\partial H^*}{\partial \lambda} = C \qquad (9.9\text{-}7)$$

Following the development leading to Equation (9.5-8), we obtain

$$\frac{\partial H^*}{\partial \varepsilon_i} = \int_a^b \left[ \eta_i \frac{\partial L}{\partial y_0} + \eta_i' \frac{\partial L}{\partial y_0'} \right] dx \qquad (9.9\text{-}8a)$$

$$= \int_a^b \eta_i \left[ \frac{\partial L}{\partial y_0} - \frac{d}{dx} \left( \frac{\partial L}{\partial y_0'} \right) \right] dx \qquad (9.9\text{-}8b)$$

as a result of integrating $\int_a^b \eta_i' \dfrac{\partial L}{\partial y_0'} \, dx$ by parts and using the boundary conditions

$$\eta_i(a) = \eta_i(b) = 0 \qquad (9.9\text{-}9a,b)$$

Thus the extremum of $I$ is given by

$$\frac{\partial L}{\partial y} - \frac{d}{dx} \left( \frac{\partial L}{\partial y'} \right) = 0 = \frac{\partial}{\partial y} (f + \lambda g) - \frac{d}{dx} \left[ \frac{\partial}{\partial y'} (f + \lambda g) \right] \qquad (9.9\text{-}10a,b)$$

and $\quad \displaystyle\int_a^b g\ dx = C$ $\hspace{10cm}$ (9.9-11)

***Example 9.9-1.*** Given a curve of fixed length  s  joining the points (0, 0) and (1, 0) as shown in Figure (9.9-1), find the extremum of the shaded area enclosed by the curve and the x-axis.

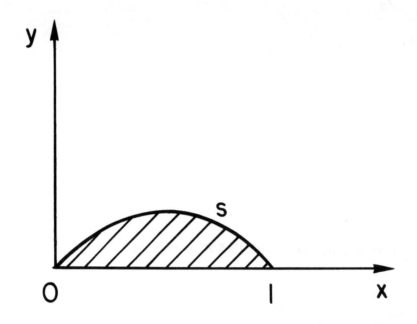

**FIGURE 9.9-1     Area (shaded) enclosed by a curve of length s**

The area

$$A = \int_0^1 y\ dx \hspace{8cm} (9.9\text{-}12)$$

and the length  s  is given by Equation (9.1-1). Using Equation (9.9-10), the extremum is given by

$$\frac{\partial}{\partial y}\left(y + \lambda \sqrt{1 + (y')^2}\right) - \frac{d}{dx}\left[\frac{\partial}{\partial y'}\left\{y + \lambda \sqrt{1 + (y')^2}\right\}\right] = 0 \hspace{2cm} (9.9\text{-}13)$$

Taking the partial derivatives, remembering that  y  and  y'  are to be treated as independent variables, we obtain

$$1 - \lambda \frac{d}{dx}\left[ \frac{y'}{\sqrt{1+(y')^2}} \right] = 0 \tag{9.9-14}$$

This can be integrated to yield

$$\frac{y'}{\sqrt{1+(y')^2}} = \frac{(x-c_1)}{\lambda} \tag{9.9-15}$$

where $c_1$ is an arbitrary constant.

On squaring both sides of Equation (9.9-15), we obtain

$$\left( \frac{dy}{dx} \right)^2 = \frac{(x-c_1)^2}{\lambda^2 - (x-c_1)^2} \tag{9.9-16}$$

To integrate the above expression, we use the substitution

$$z = (x-c_1) \tag{9.9-17}$$

Then the integral becomes

$$y = \pm \int \frac{z\,dz}{\sqrt{\lambda^2 - z^2}} \tag{9.9-18}$$

On further substituting

$$z = \lambda \cos\theta \tag{9.9-19}$$

we obtain

$$y = \mp \int \lambda \cos\theta \, d\theta = \mp \lambda \sin\theta + c_2 \tag{9.9-20a,b}$$

On squaring both sides of Equations (9.9-20a, b) and returning to the original variables, we finally obtain

$$(y-c_2)^2 + (x-c_1)^2 = \lambda^2 \tag{9.9-21}$$

Thus the curve is an arc of a circle. The end conditions implies

$$c_2^2 + c_1^2 = \lambda^2 \tag{9.9-22}$$

$$c_2^2 + (1 - c_1)^2 = \lambda^2 \tag{9.9-23}$$

with solutions $c_1 = \dfrac{1}{2}$ and $c_2 = \pm \sqrt{\lambda^2 - \dfrac{1}{4}}$ .

To obtain $\lambda$, we impose the isoperimetric constraint given by Equation (9.1-1).  Differentiating Equation (9.9-21) with respect to $x$ and substituting $y'$ into Equation (9.1-1) yields

$$s = \lambda \int_0^1 \frac{dx}{\sqrt{\lambda^2 - (x - 1/2)^2}} \tag{9.9-24}$$

This integral is of a standard form and its value is

$$s = \lambda \left[ \sin^{-1} \frac{(x - 1/2)}{\lambda} \right]_0^1 \tag{9.9-25a}$$

$$= 2\lambda \sin^{-1} \left( \frac{1}{2\lambda} \right) \tag{9.9-25b}$$

[since $\sin^{-1}(-x) = -\sin^{-1}(x)$].

We can now express $\lambda$ in terms of $s$ and the required curve is a circle with centre $\left( 1/2, \pm\sqrt{\lambda^2 - 1/4} \right)$ and radius $\lambda$.

If $s = \pi/2$, $\lambda = 1/2$ and the equation of the curve is given by

$$y^2 + (x - 1/2)^2 = \frac{1}{4} \tag{9.9-26}$$

Thus the required curve is the semi-circle with centre $(1/2, 0)$ and radius $1/2$.  This result can also be obtained from geometrical considerations.

●

If the constraint is of algebraic type or in the form of a differential equation, then we have to choose the Lagrange multiplier as a function of $x$, instead of a constant.  The constraints can be of the form

$$f(x, y, y') = \text{constant} \tag{9.9-27a}$$

or     $$y' = g(x, y) \tag{9.9-27b}$$

***Example 9.9-2***.  Find the extremum of

$$I = \int_0^1 [1 + (y'')^2] \, dx \qquad\qquad (9.9\text{-}28)$$

We reduce the integrand to the usual form in terms of $y'$. We write

$$y_1 = y' \qquad\qquad (9.9\text{-}29)$$

Equation (9.9-28) becomes

$$I = \int_0^1 [1 + (y_1')^2] \, dx \qquad\qquad (9.9\text{-}30)$$

Subject to the constraint

$$y_1 - y' = 0 \qquad\qquad (9.9\text{-}31)$$

The Lagrangian $L$ now is given by

$$L = (1 + (y_1')^2) + \lambda (y_1 - y') = 0 \qquad\qquad (9.9\text{-}32)$$

Thus the problem of having a higher derivative in the integrand has been transformed to one involving two dependent variables $y$ and $y_1$ subject to a constraint. We first extend the theory of calculus of variations to the case of more than one dependent variables, and then we shall be able to complete the above example.

## 9.10 SEVERAL DEPENDENT VARIABLES

We generalize to the case of several dependent variables. The function $f$ depends on $n$ dependent variables $y_1, y_2, \dots, y_n$. We need to determine the $n$ functions $y_1, y_2, \dots, y_n$ that extremize the functional $I$ given by

$$I = \int_a^b f(x, y_1, y_2, \dots, y_n, y_1', y_2', \dots, y_n') \, dx \qquad\qquad (9.10\text{-}1)$$

We assume the end points are fixed and the conditions are

$$y_1(a) = a_1, \quad y_2(a) = a_2, \quad \dots \quad, \quad y_n(a) = a_n \qquad\qquad (9.10\text{-}2a,b,c)$$

$$y_1(b) = b_1, \quad y_2(b) = b_2, \quad \dots \quad, \quad y_n(b) = b_n \qquad\qquad (9.10\text{-}3a,b,c)$$

We proceed as in the case of one variable. Suppose $y_{01}, y_{02}, \ldots, y_{0n}$ are the functions that extremize $I$ and consider the functions

$$y_i = y_{0i} + \varepsilon_i \, \eta_i(x), \qquad i = 1, \ldots, n \tag{9.10-4}$$

where the $\eta_i$ are arbitrary functions, satisfying the conditions

$$\eta_i(a) = \eta_i(b) = 0 \tag{9.10-5a,b}$$

The functional $I^*(\varepsilon_i)$ is given by

$$I^*(\varepsilon_i) = \int_a^b f(x, y_{01} + \varepsilon_1\eta_1, \ldots, y_{0n} + \varepsilon_n\eta_n, y'_{01} + \varepsilon_1\eta'_1, \ldots, y'_{0n} + \varepsilon_n\eta'_n) \, dx \tag{9.10-6}$$

The extremum of $I$ is then given by

$$\left. \frac{\partial I^*}{\partial \varepsilon_i} \right|_{\varepsilon_i = 0} = 0 \tag{9.10-7}$$

Equation (9.10-7) can be written, again following the development leading to Equation (9.5-9), as

$$\int_a^b \left[ \eta_i \frac{\partial f}{\partial y_i} + \eta'_i \frac{\partial f}{\partial y'_i} \right] dx = 0 \tag{9.10-8}$$

The conditions for $I$ to be an extremum are obtained by integrating the second term in Equation (9.10-8) by parts and the result is

$$\frac{\partial f}{\partial y_i} - \frac{d}{dx} \left( \frac{\partial f}{\partial y'_i} \right) = 0, \qquad i = 1, 2, \ldots, n \tag{9.10-9}$$

***Example 9.10-1.*** Complete Example 9.9-2. The extremum of $I$ is now given by

$$\frac{\partial L}{\partial y} - \frac{d}{dx} \left( \frac{\partial L}{\partial y'} \right) = 0 \tag{9.10-10}$$

$$\frac{\partial L}{\partial y_1} - \frac{d}{dx} \left( \frac{\partial L}{\partial y'_1} \right) = 0 \tag{9.10-11}$$

Using Equation (9.9-32), Equations (9.10-10, 11) become

$$\frac{d}{dx}(\lambda) = 0 \tag{9.10-12}$$

$$\lambda - \frac{d}{dx}(2y_1') = 0 \tag{9.10-13}$$

Equation (9.10-12) implies that $\lambda$ is a constant and we can integrate Equation (9.10-13) to obtain

$$y = a_3 x^3 + a_2 x^2 + a_1 x + a_0 \tag{9.10-14}$$

where $a_0$, $a_1$, $a_2$, $a_3$ are constants. Remember that $y_1 = y'$.

If the end conditions were

$$y(0) = 0, \quad y'(0) = 1, \quad y(1) = 1, \quad y'(1) = 1 \tag{9.10-15a,b,c,d}$$

then the constants $a_0$ to $a_3$ will be given by

$$a_0 = a_2 = a_3 = 0 \quad \text{and} \quad a_1 = 1 \tag{9.10-16a,b,c,d}$$

Thus, the function that will extremize $I$ [Equation (9.9-30)], subject to the end conditions given by Equation (9.10-15a to d), is

$$y = x \tag{9.10-17}$$

***Example 9.10-2.*** Find the extremum of

$$I = \int_0^{\pi/2} [(y_1')^2 + (y_2')^2 + 2y_1 y_2] \, dx \tag{9.10-18}$$

subject to the following conditions

$$y_1(0) = y_2(0) = 0, \qquad y_1(\pi/2) = y_2(\pi/2) = 1 \tag{9.10-19a,b,c,d}$$

Via Equation (9.10-9), we compute

$$\frac{\partial f}{\partial y_1} = 2y_2, \qquad\qquad \frac{\partial f}{\partial y_1'} = 2y_1' \tag{9.10-20, 21}$$

$$\frac{\partial f}{\partial y_2} = 2y_1, \qquad\qquad \frac{\partial f}{\partial y_2'} = 2y_2' \tag{9.10-22, 23}$$

Thus the extremum of $I$ are given by

$$2y_2 - \frac{d}{dx}(2y_1') = 0 \quad \Rightarrow \quad y_2 - 2y_1'' = 0 \tag{9.10-24a,b}$$

$$2y_1 - \frac{d}{dx}(2y_2') = 0 \quad \Rightarrow \quad y_1 - y_2'' = 0 \tag{9.10-25a,b}$$

Eliminating $y_1$ from this system, we obtain

$$y_2'''' - y_2 = 0 \tag{9.10-26}$$

The solution of Equation (9.10-26) is

$$y_2 = a_1 \cos x + a_2 \sin x + a_3 \cosh x + a_4 \sinh x \tag{9.10-27}$$

where $a_1$ to $a_4$ are arbitrary constants.

From Equation (9.10-25), we deduce that

$$y_1 = -a_1 \cos x - a_2 \sin x + a_3 \cosh x + a_4 \sinh x \tag{9.10-28}$$

Imposing the boundary conditions given by Equations (9.10-19a to d) yields

$$a_1 = a_2 = a_3 = 0, \qquad a_4 = \frac{1}{\sinh \pi/2} \tag{9.10-29a,b,c,d}$$

Thus $y_1$ and $y_2$ are given by

$$y_1 = y_2 = \frac{\sinh x}{\sinh \pi/2} \tag{9.10-30a,b}$$

I is given by

$$I = 2 \int_0^{\pi/2} \frac{\cosh^2 x + \sinh^2 x}{\sinh^2 \pi/2} \, dx = \frac{2 \cosh \pi/2}{\sinh \pi/2} \tag{9.10-31a,b}$$

***Example 9.10-3***. Derive Lagrange's equations.

In elementary mechanics, one usually starts with Newton's laws of motion. It is also possible to develop a theory of mechanics starting from a **variational principle** and this type of mechanics is known as **analytical mechanics**. In advanced mechanics, the variational approach has certain advantages [Lanczos (1966)]. One of the possible variational principles we may start with is **Hamilton's principle**, which may be stated as follows.

The path that a dynamical system takes in moving from one point at time $t_1$ to another point at time $t_2$ is such that the line integral

$$I = \int_{t_1}^{t_2} L \, dt \tag{9.10-32}$$

in an extremum.  L  is the **Lagrangian** and is given by

$$L = K - \varphi \tag{9.10-33}$$

where  K  is the **kinetic energy** and  $\varphi$  is the **generalized potential energy**.

The **generalized coordinates** of the dynamical system are  $q^1, q^2, \ldots, q^n$  and the **generalized velocities** are  $\dot{q}^1, \dot{q}^2, \ldots, \dot{q}^n$.  The dot denotes differentiation with respect to time.  L  is then a function of  $q^i$  and  $\dot{q}^i$.  Thus the condition for  I  to be an extremum is

$$\frac{\partial L}{\partial q^i} - \frac{d}{dt}\left(\frac{\partial L}{\partial \dot{q}^i}\right) = 0, \qquad i = 1, 2, \ldots, n \tag{9.10-34}$$

The  n  equations given by Equation (9.10-34) are known as **Lagrange's equations**.

***Example 9.10-4.*** A particle  P  of mass  m  lies on a smooth horizontal table and is connected to another particle  Q  of mass  3m  by an inextensible string which passes through a smooth hole in the table as shown in Figure 9.10-1. If the particle  P  is projected from rest with a speed  $\sqrt{8ag}$  along the table at right angle to the string when it is at a distance  a  from the hole, determine the ensuing motion of  P  and  Q.

Let  O, the position of the hole, be the origin. The polar coordinates of  P  are  $(r, \theta)$  as shown. The components of the velocity of  P  are  $(\dot{r}, r\dot{\theta})$. Taking the level of the table to be the zero-level of potential energy, the potential energy of particle  Q  is  $-3\,mg(\ell - r)$, where  $\ell$  is the length of string. Since the string is inextensible,  Q  will move up or down with speed  $\dot{r}$. The kinetic energy of the system is

$$K = \frac{1}{2} m (\dot{r}^2 + r^2 \dot{\theta}^2) + \frac{3m}{2} (\dot{r}^2) \tag{9.10-35}$$

The potential energy  $\varphi$  is

$$\varphi = -3 \, mg \, (\ell - r) \tag{9.10-36}$$

The Lagangian  L  becomes

$$L = \frac{1}{2} m (\dot{r}^2 + r^2 \dot{\theta}^2) + \frac{3m}{2} \dot{r}^2 + 3 \, mg \, (\ell - r) \tag{9.10-37}$$

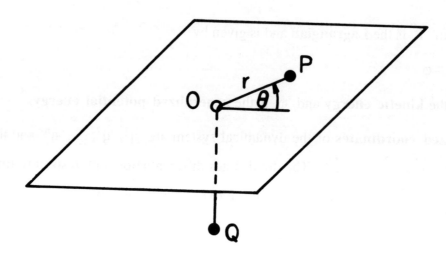

**FIGURE 9.10-1    Particles P and Q connected by a string of constant length, passing through a hole in the table**

The Lagrange equations are therefore given by

$$\frac{\partial L}{\partial r} - \frac{d}{dt}\left(\frac{\partial L}{\partial \dot{r}}\right) = 0 \quad \Rightarrow \quad mr\dot{\theta}^2 - 3mg - \frac{d}{dt}(4m\dot{r}) = 0 \qquad (9.10\text{-}38a,b)$$

$$\frac{\partial L}{\partial \theta} - \frac{d}{dt}\left(\frac{\partial L}{\partial \dot{\theta}}\right) = 0 \quad \Rightarrow \quad \frac{d}{dt}(mr^2\dot{\theta}) = 0 \qquad (9.10\text{-}39a,b)$$

Integrating Equation (9.10-39b) yields

$$r^2\dot{\theta} = \text{constant} = C_1 \qquad (9.10\text{-}40)$$

Initially $r = a$, $a\dot{\theta} = \sqrt{8ag}$, so

$$C_1 = a\sqrt{8ag} \qquad (9.10\text{-}41)$$

Combining Equations (9.10-40, 41), we find that $\dot{\theta}$ is given by

$$\dot{\theta} = \frac{a\sqrt{8ag}}{r^2} \tag{9.10-42}$$

Substituting the value of $\dot{\theta}$ into Equation (9.10-38b), we obtain

$$4\ddot{r} = \frac{8a^3g}{r^3} - 3g \tag{9.10-43}$$

The term $\ddot{r}$ can be written as

$$\ddot{r} = \frac{d}{dt}(\dot{r}) = \frac{d}{dr}(\dot{r})\frac{dr}{dt} = \dot{r}\frac{d\dot{r}}{dr} = \frac{1}{2}\frac{d}{dr}(\dot{r}^2) \tag{9.10-44a,b,c,d}$$

Equation (9.10-43) may now be written as

$$2\frac{d}{dr}(\dot{r}^2) = \frac{8a^3g}{r^3} - 3g \tag{9.10-45}$$

which can be integrated to yield

$$2\dot{r}^2 = -\frac{4a^3g}{r^2} - 3gr + C_2 \tag{9.10-46}$$

where $C_2$ is a constant.

From the initial conditions $r = a$, $\dot{r} = 0$, $C_2$ is calculated to be

$$C_2 = 7ag \tag{9.10-47}$$

From Equations (9.10-46, 47), we find that $\dot{r}^2$ is given by

$$\dot{r}^2 = \frac{7ag}{2} - \frac{2a^3g}{r^2} - \frac{3}{2}gr \tag{9.10-48}$$

We can deduce from Equation (9.10-43), that initially since $r = a$, $\ddot{r}$ is positive. That is to say, although the radial velocity is zero initially (at time $t = 0$), this does not imply that the radial acceleration at $t = 0$ is zero. Consequently, the particle Q will initially more up and r will increase. $\ddot{r}$ will be zero when $r^3 = 8a^3/3$. From Equation (9.10-48), we deduce that $\dot{r} = 0$, when $r = 2a$ and Q will come momentarily to rest. At $r = 2a$, $\ddot{r}$ is negative ($-2g$) [see Equation (9.10-43)] and so Q will fall. Thus Q will oscillate between the heights $(\ell - a)$ and $(\ell - 2a)$. That is to say, because of the lack of friction, we allow the cycle to repeat when r is again equal to a. The particle P has a radial velocity given by Equation (9.10-48) and a tangential velocity $\dfrac{a\sqrt{8ag}}{r}$. The radial position of P lies therefore between a and 2a.

## 9.11 SEVERAL INDEPENDENT VARIABLES

The integrand might be a function of two or more independent variables. Geometrically it means that we are considering a surface instead of a curve. For simplicity, let us consider a function $u(x_1, x_2)$ which is a function of two independent variables $x_1$ and $x_2$. We now have to find the extremum of

$$I = \iint\limits_{R} f\left[x_1, x_2, u(x_1, x_2), u_{x_1}(x_1, x_2), u_{x_2}(x_1, x_2)\right] dx_1 \, dx_2 \qquad (9.11-1)$$

where $u_{x_i} = \dfrac{\partial u}{\partial x_i}$ $(i = 1, 2)$, and R is a region in the $x_1 x_2$-plane bounded by a curve C as shown in Figure 9.11-1. The value of $u(x_1, x_2)$ is given on C.

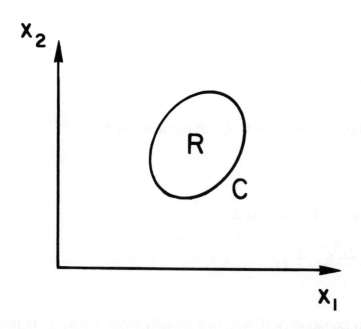

**FIGURE 9.11-1    Region R bounded by curve C**

We need to find the function u which extremizes I over the region R and satisfies the boundary condition

$$u(x_1, x_2) = g(x_1, x_2) \text{ on } C \tag{9.11-2}$$

where $g$ is a given function.

Suppose $u_0(x_1, x_2)$ is the function that renders $I$ an extremum. We consider another function $u_1(x_1, x_2)$ (a neighboring surface) given by

$$u_1(x_1, x_2) = u_0(x_1, x_2) + \varepsilon\eta(x_1, x_2) \tag{9.11-3}$$

where $\varepsilon$ is a constant and $\eta$ is an arbitrary function satisfying the condition

$$\eta(x_1, x_2) = 0 \text{ on } C \tag{9.11-4}$$

Then the extremum of $I$ is given by

$$\left.\frac{d}{d\varepsilon} I(u_0 + \varepsilon\eta)\right|_{\varepsilon=0} = 0 \tag{9.11-5}$$

Substituting Equation (9.11-3) into Equation (9.11-1), we obtain

$$\left.\frac{d}{d\varepsilon} I(u_0 + \varepsilon\eta)\right|_{\varepsilon=0} = \left.\frac{d}{d\varepsilon} \iint\limits_R f(x_1, x_2, u_0 + \varepsilon\eta, u_{0x_1} + \varepsilon\eta_{x_1}, u_{0x_2} + \varepsilon\eta_{x_2}) \, dx_1 \, dx_2\right|_{\varepsilon=0} \tag{9.11-6}$$

Since $\varepsilon$ is independent of $x_1$ and $x_2$, we can introduce the operator $\frac{d}{d\varepsilon}$ inside the integration. Note that

$$\frac{df}{d\varepsilon} = \frac{\partial f}{\partial u_1} \frac{\partial u_1}{\partial \varepsilon} + \sum_{i=1}^{2} \frac{\partial f}{\partial u_{1x_i}} \frac{\partial u_{1x_i}}{\partial \varepsilon} \quad \text{with} \quad \frac{\partial u_1}{\partial \varepsilon} = \eta \tag{9.11-7}$$

Equation (9.11-6) now becomes

$$\left.\frac{d}{d\varepsilon} I(u_1)\right|_{\varepsilon=0} = \iint\limits_R \left[ \eta \frac{\partial f}{\partial u} + \eta_{x_1} \frac{\partial f}{\partial u_{x_1}} + \eta_{x_2} \frac{\partial f}{\partial u_{x_2}} \right] dx_1 \, dx_2 \tag{9.11-8}$$

Making use of the fact that

$$\eta_{x_i} \frac{\partial f}{\partial u_{x_i}} = \frac{\partial}{\partial x_i}\left( \eta \frac{\partial f}{\partial u_{x_i}} \right) - \eta \frac{\partial}{\partial x_i}\left( \frac{\partial f}{\partial u_{x_i}} \right), \quad i = 1, 2 \tag{9.11-9}$$

we finally obtain

$$\frac{dI(u_1)}{d\epsilon}\bigg|_{\epsilon=0} = \iint_R \left[ \eta \left\{ \frac{\partial f}{\partial u} - \frac{\partial}{\partial x_1}\left(\frac{\partial f}{\partial u_{x_1}}\right) - \frac{\partial}{\partial x_2}\left(\frac{\partial f}{\partial u_{x_2}}\right) \right\} \right.$$

$$\left. + \left\{ \frac{\partial}{\partial x_1}\left(\eta \frac{\partial f}{\partial u_{x_1}}\right) + \frac{\partial}{\partial x_2}\left(\eta \frac{\partial f}{\partial u_{x_2}}\right) \right\} \right] dx_1 \, dx_2 \qquad (9.11\text{-}10)$$

We can write the terms inside the second brace of the double integral as a line integral using Green's theorem. We are interested in converting the double integral to a line integral, because the condition on $\eta$ is given on the curve $C$, by Equation (9.11-4). Green's theorem may be stated as follows

$$\iint_R \left( \frac{\partial Q}{\partial x_1} - \frac{\partial P}{\partial x_2} \right) dx_1 \, dx_2 = \int_C (P \, dx_1 + Q \, dx_2) \qquad (9.11\text{-}11)$$

The double integral becomes

$$\iint_R \left[ \frac{\partial}{\partial x_1}\left(\eta \frac{\partial f}{\partial u_{x_1}}\right) + \frac{\partial}{\partial x_2}\left(\eta \frac{\partial f}{\partial u_{x_2}}\right) \right] dx_1 \, dx_2 = \int_C \left[ \eta \frac{\partial f}{\partial u_{x_1}} dx_2 - \eta \frac{\partial f}{\partial u_{x_2}} dx_1 \right] \qquad (9.11\text{-}12)$$

The boundary condition given by Equation (9.11-4) implies that the integral in Equation (9.11-12) is zero. From Equation (9.11-10), we deduce that the condition for $I$ to be an extremum is

$$\frac{\partial f}{\partial u} - \frac{\partial}{\partial x_1}\left(\frac{\partial f}{\partial u_{x_1}}\right) - \frac{\partial}{\partial x_2}\left(\frac{\partial f}{\partial u_{x_2}}\right) = 0 \qquad (9.11\text{-}13)$$

We can extend to $m$ independent variables $x_1, x_2, \dots, x_m$. Then we consider the functional

$$I = \iint_R \cdots \int f(x_1, x_2, \dots, x_m, u, u_{x_1}, u_{x_2}, \dots, u_{x_m}) \, dx_1, \dots, dx_m \qquad (9.11\text{-}14)$$

The condition for $I$ to be an extremum is

$$\frac{\partial f}{\partial u} - \frac{\partial}{\partial x_1}\left(\frac{\partial f}{\partial u_{x_1}}\right) - \frac{\partial}{\partial x_2}\left(\frac{\partial f}{\partial u_{x_2}}\right) \cdots - \frac{\partial}{\partial x_m}\left(\frac{\partial f}{\partial u_{x_m}}\right) = 0 \qquad (9.11\text{-}15)$$

where, as before, $u_{x_i} = \dfrac{\partial u}{\partial x_i}$.

***Example 9.11-1.*** Find the minimum surface area of the surface $x_3 = u(x_1, x_2)$ bounded by a curve $C$ as shown in Figure 9.11-2. This problem is known as **Plateau's problem**. Plateau, a Belgian physicist (1801-1883), did many experiments with soap films to determine minimal surfaces.

We shall first derive the formula for the surface element $\Delta S$, enclosed by the points $P_0, P_1, P_2$ and $P_3$. That is to say, we consider four neighboring points $P_0, P_1, P_2$ and $P_3$ on the surface as shown in Figure 9.11-2. The projections of these points on the $x_1 x_2$-plane are denoted by primes. Let the vector positions of $OP_0, OP_1, OP_2, OP_3$ be $\underline{r}_0, \underline{r}_1, \underline{r}_2$ and $\underline{r}_3$ respectively.

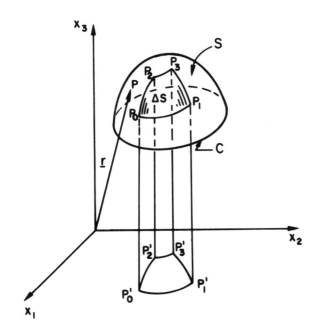

**FIGURE 9.11-2    Surface S bounded by curve C**

Further let

$$\underline{r}_0 = \underline{r}\,(x_1, x_2)\,, \qquad\qquad \underline{r}_1 = \underline{r}\,(x_1 + \Delta x_1, x_2) \qquad\qquad (9.11\text{-}16a,b)$$

$$\underline{r}_2 = \underline{r}\,(x_1, x_2 + \Delta x_2)\,, \qquad \underline{r}_3 = \underline{r}\,(x_1 + \Delta x_1, x_2 + \Delta x_2) \qquad (9.11\text{-}16c,d)$$

The vectors $\overrightarrow{P_0 P_1}$ and $\overrightarrow{P_0 P_2}$ are given by

$$\overrightarrow{P_0 P_1} = (\underline{r}_1 - \underline{r}_0) = \underline{r}\,(x_1 + \Delta x_1, x_2) - \underline{r}\,(x_1, x_2) \approx \frac{\partial \underline{r}}{\partial x_1}\,\Delta x_1 \qquad (9.11\text{-}17a,b,c)$$

$$\overrightarrow{P_0 P_2} = (\underline{r}_2 - \underline{r}_0) = \underline{r}\,(x_1, x_2 + \Delta x_2) - \underline{r}\,(x_1, x_2) \approx \frac{\partial \underline{r}}{\partial x_2}\,\Delta x_2 \qquad (9.11\text{-}18a,b,c)$$

according to Taylor's expansion.

The surface element $\Delta S$, neglecting curvature, is

$$\Delta S = \left| \overrightarrow{P_0 P_1} \wedge \overrightarrow{P_0 P_2} \right| = \left| \frac{\partial r}{\partial x_1} \wedge \frac{\partial r}{\partial x_2} \right| \Delta x_1 \Delta x_2 \qquad (9.11\text{-}19a,b)$$

On the surface, the vector $\underline{r}$ can be written as

$$\underline{r} = x_1 \underline{i} + x_2 \underline{j} + u(x_1, x_2) \underline{k} \qquad (9.11\text{-}20)$$

where $\underline{i}$, $\underline{j}$, and $\underline{k}$ are the unit vectors along the coordinate curves.

From Equation (9.11-20), we obtain

$$\frac{\partial \underline{r}}{\partial x_1} = \underline{i} + \frac{\partial u}{\partial x_1} \underline{k} \qquad (9.11\text{-}21)$$

$$\frac{\partial \underline{r}}{\partial x_2} = \underline{j} + \frac{\partial u}{\partial x_2} \underline{k} \qquad (9.11\text{-}22)$$

Substituting Equations (9.11-21, 22) into Equation (9.11-19) and carrying out the cross product, we find

$$\Delta S = \sqrt{1 + \left(\frac{\partial u}{\partial x_1}\right)^2 + \left(\frac{\partial u}{\partial x_2}\right)^2} \; \Delta x_1 \Delta x_2 \qquad (9.11\text{-}23)$$

The surface area $S$ is given by

$$S = \iint \sqrt{1 + u_{x_1}^2 + u_{x_2}^2} \; dx_1 \, dx_2 \qquad (9.11\text{-}24)$$

From Equation (9.11-13), we deduce the condition for the extremum of $S$ to be

$$\frac{\partial}{\partial u} \sqrt{1 + u_{x_1}^2 + u_{x_2}^2} - \frac{\partial}{\partial x_1} \left( \frac{u_{x_1}}{\sqrt{1 + u_{x_1}^2 + u_{x_2}^2}} \right) - \frac{\partial}{\partial x_2} \left( \frac{u_{x_2}}{\sqrt{1 + u_{x_1}^2 + u_{x_2}^2}} \right) = 0 \quad (9.11\text{-}25)$$

Noting that $\sqrt{1 + u_{x_1}^2 + u_{x_2}^2}$ is a function of $u_{x_i}$ only and differentiating the last two terms of Equation (9.11-25), we find that Equation (9.11-25) simplifies to

$$\left[ 1 + \left(\frac{\partial u}{\partial x_2}\right)^2 \right] \frac{\partial^2 u}{\partial x_1^2} + \left[ 1 + \left(\frac{\partial u}{\partial x_1}\right)^2 \right] \frac{\partial^2 u}{\partial x_2^2} - 2 \frac{\partial u}{\partial x_1} \frac{\partial u}{\partial x_2} \frac{\partial^2 u}{\partial x_1 \partial x_2} = 0 \qquad (9.11\text{-}26)$$

In general the non-linear Equation (9.11-26) is difficult to solve. We shall consider a special case when the surface is a surface of revolution with the $x_3$-axis as the axis of symmetry. In this case the equation of the surface is given by

$$x_3 = u(r) \tag{9.11-27}$$

where $r = \sqrt{x_1^2 + x_2^2}$ .

We have

$$\frac{\partial u}{\partial x_1} = \frac{du}{dr} \frac{\partial r}{\partial x_1} = \frac{x_1}{r} \frac{du}{dr} \tag{9.11-28a,b}$$

$$\frac{\partial u}{\partial x_2} = \frac{x_2}{r} \frac{du}{dr} \tag{9.11-29}$$

$$\frac{\partial^2 u}{\partial x_1^2} = \frac{x_1^2}{r^2} \frac{d^2 u}{dr^2} + \frac{1}{r} \frac{du}{dr} - \frac{x_1^2}{r^3} \frac{du}{dr} = \frac{x_1^2}{r^2} \frac{d^2 u}{dr^2} + \frac{x_2^2}{r^3} \frac{du}{dr} \tag{9.11-30a,b}$$

$$\frac{\partial^2 u}{\partial x_2^2} = \frac{x_2^2}{r^2} \frac{d^2 u}{dr^2} + \frac{x_1^2}{r^3} \frac{du}{dr} \tag{9.11-31}$$

$$\frac{\partial^2 u}{\partial x_1 \partial x_2} = \frac{x_1 x_2}{r^2} \frac{d^2 u}{dr^2} - \frac{x_1 x_2}{r^3} \frac{du}{dr} \tag{9.11-32}$$

Substituting Equations (9.11-28 to 32) into Equation (9.11-26) and simplifying, we obtain

$$r \frac{d^2 u}{dr^2} + \frac{du}{dr} + \left( \frac{du}{dr} \right)^3 = 0 \tag{9.11-33}$$

Equation (9.11-33) can be transformed to a first order equation, by substituting

$$\frac{du}{dr} = v \tag{9.11-34}$$

Equation (9.11-33) becomes

$$r \frac{dv}{dr} + v + v^3 = 0 \tag{9.11-35}$$

Equation (9.11-35) is a separable equation and we have

$$\int \frac{dv}{v\,(1+v^2)} = -\int \frac{dr}{r}$$

(9.11-36)

which on integrating yields

$$\ln v - \frac{1}{2}\ln(1+v^2) = -\ln r + \ln c$$

(9.11-37)

where c is an arbitrary constant.

This can also be written as

$$\frac{v}{\sqrt{1+v^2}} = \frac{c}{r}$$

(9.11-38)

We can solve for v and $\frac{du}{dr}$ is then given by

$$\frac{du}{dr} = \frac{c}{\sqrt{r^2-c^2}}$$

(9.11-39)

Integration gives u as

$$u = c\cosh^{-1}(r/c) + k$$

(9.11-40)

where k is a constant.

Thus the generating curve is

$$r = c\cosh\left(\frac{u-k}{c}\right)$$

(9.11-41)

This is the equation of a **catenary**.  Thus the minimum surface is a catenoid.

The extension to several functions $(u_1, u_2, \dots, u_n)$ of several independent variables $(x_1, x_2, \dots, x_m)$ is straight forward.  In this case, the functional is given by

$$\int_m \cdots \int f\left(x_1 \dots x_m, u_1 \dots u_m, \frac{\partial u_1}{\partial x_1} \dots \frac{\partial u_1}{\partial x_m}, \dots \frac{\partial u_n}{\partial x_1} \dots \frac{\partial u_n}{\partial x_m}\right) dx_1 \dots dx_m$$

(9.11-42)

The conditions for I to be an extremum are then

$$\frac{\partial f}{\partial u_i} - \sum_{j=1}^{m} \frac{\partial}{\partial x_j} \left( \frac{\partial f}{\partial V_{ij}} \right) = 0, \qquad i = 1, \dots, n, \quad j = 1, \dots, m \tag{9.11-43}$$

where $V_{ij} = \dfrac{\partial u_i}{\partial x_j}$.

***Example 9.11-2.*** Show that the equations governing the slow flow of a generalized Newtonian fluid are equivalent to extremizing a functional.

The equations of motion, neglecting inertia, can be written, when referred to a rectangular Cartesian coordinate system $Ox_1x_2x_3$, as

$$\frac{\partial p}{\partial x_i} + \frac{\partial \tau_{ij}}{\partial x_j} - \rho f_i = 0, \qquad i = 1, 2, 3 \tag{9.11-44}$$

where $p$ is the pressure, $\tau_{ij}$ is the deviatoric stress tensor, $\rho$ is the density and $f_i$ are the external applied forces (usually only gravity). Remember the convention, requiring summation over repeated indices.

The equation of continuity is given by

$$\frac{\partial v_i}{\partial x_i} = v_{i,i} = 0 \tag{9.11-45a,b}$$

The constitutive equation of a generalized Newtonian fluid may be written a

$$\tau_{ij} = -\eta \, (II_{\dot{\gamma}}) \, \dot{\gamma}_{ij} \tag{9.11-46}$$

where $\dot{\gamma}_{ij} \left( = \dfrac{\partial v_i}{\partial x_j} + \dfrac{\partial v_j}{\partial x_i} \right)$ is the rate of deformation tensor, $v_i$ are the velocity components, and

$II_{\dot{\gamma}} \left( = \sqrt{\dfrac{1}{2} \dot{\gamma}_{ij} \dot{\gamma}_{ji}} \right)$ is the second invariant of $\dot{\gamma}_{ij}$.

Let $V$ be the flow region (volume) bounded by two non-overlapping surfaces $S_v$ and $S_t$. The velocity is given on $S_v$ and the traction (surface) force $t_i$ $[= -(p\,n_i + \tau_{ij}\,n_j)]$ is given on $S_t$, $n_i$ is the component of the unit normal to the surface $S_t$. Note that in this example, we consider different boundary conditions on different parts of the enclosing surface.

The system given by Equations (9.11-44, 46) subject to the boundary conditions stated above will be shown to be equivalent to finding the extremum of

$$\int_V \left[ g(II_{\dot\gamma}) - \rho f_i v_i - p v_{i,i} \right] dV - \int_{S_t} t_i v_i \, dS \qquad (9.11\text{-}47)$$

where $g(II_{\dot\gamma}) = \displaystyle\int_0^{II_{\dot\gamma}} u \, \eta(u) \, du$ .

The functions inside the integral are functions of the velocity components $v_i$ and the pressure $p$. The quantities $f_i$ and $t_i$ are given. To find the extremum of $I$, we consider neighboring functions $v_i^*$ and $p^*$ which are given by

$$v_i^* = v_i + \varepsilon w_i , \qquad p^* = p + \varepsilon q \qquad (9.11\text{-}48\text{a,b})$$

where $v_i$ and $p$ extremize $I$.

Substituting Equations (9.11-48a, b) into Equation (9.11-47) yield

$$I^* = \int_V \left[ g(II_{\dot\gamma}^*) - \rho f_i (v_i + \varepsilon w_i) - (p + \varepsilon q)(v_i + \varepsilon w_i)_{,i} \right] dV - \int_{S_t} t_i (v_i + \varepsilon w_i) \, dS \quad (9.11\text{-}49)$$

The extremum of $I$ is given by

$$\left. \frac{dI^*}{d\varepsilon} \right|_{\varepsilon = 0} = 0 \qquad (9.11\text{-}50)$$

In the computation of this extremum, we will have to determine $\dfrac{dg(II_{\dot\gamma}^*)}{d\varepsilon}$. Since $II_{\dot\gamma}^*$ is defined in terms of $v_i^*$, we will be required to calculate $\dfrac{\partial g}{\partial II_{\dot\gamma}^*}$ and hence $\dfrac{dg}{d\varepsilon}$. Next we illustrate these manipulations.

From the definition of $II_{\dot\gamma}$, we have

$$\frac{\partial II_{\dot\gamma}}{\partial \dot\gamma_{ij}} = \frac{\partial}{\partial \dot\gamma_{ij}} \left[ \tfrac{1}{2} \dot\gamma_{ij} \dot\gamma_{ji} \right]^{1/2} = \frac{1}{2} \left[ \tfrac{1}{2} \dot\gamma_{ij} \dot\gamma_{ji} \right]^{-1/2} \dot\gamma_{ij} = \frac{\dot\gamma_{ij}}{2 II_{\dot\gamma}} \qquad (9.11\text{-}51\text{a,b,c})$$

Since $g(II_{\dot\gamma}) = \displaystyle\int_0^{II_{\dot\gamma}} u \, \eta(u) \, du$, so

$$\frac{\partial g}{\partial \dot{\gamma}_{ij}} = \frac{\partial g}{\partial II_{\dot{\gamma}}} \frac{\partial II_{\dot{\gamma}}}{\partial \dot{\gamma}_{ij}} = \frac{II_{\dot{\gamma}} \eta (II_{\dot{\gamma}}) \dot{\gamma}_{ji}}{2 II_{\dot{\gamma}}} = \frac{1}{2} \eta (II_{\dot{\gamma}}) \dot{\gamma}_{ji} = -\frac{1}{2} \tau_{ij} \qquad (9.11\text{-}52a,b,c,d)$$

$$\frac{\partial g}{\partial \varepsilon} = \frac{\partial g}{\partial \dot{\gamma}_{ij}^*} \frac{\partial \dot{\gamma}_{ij}^*}{\partial \varepsilon} = -\frac{1}{2} \tau_{ij} (w_{i,j} + w_{j,i}) = -\tau_{ij} w_{i,j} \qquad (9.11\text{-}53a,b,c)$$

since $\tau_{ij}$ is symmetric.

Combining Equations (9.11-50, 53), we obtain

$$\int_V \left[ -\tau_{ij} w_{i,j} - \rho f_i w_i - p w_{i,i} - q v_{i,i} \right] dV - \int_{S_t} t_i w_i \, dS = 0 \qquad (9.11\text{-}54)$$

Recognizing that

$$(\tau_{ij} w_i)_{,j} = \tau_{ij,j} w_i + \tau_{ij} w_{i,j} \qquad (9.11\text{-}55a)$$

or $\qquad \tau_{ij} w_{i,j} = (\tau_{ij} w_i)_{,j} - \tau_{ij,j} w_i \qquad (9.11\text{-}55b)$

It follows that

$$\int_V \tau_{ij} w_{i,j} \, dV = \int_V \left[ (w_i \tau_{ij})_{,j} - \tau_{ij,j} w_i \right] dV \qquad (9.11\text{-}56a)$$

$$= \int_{S_t + S_v} (w_i \tau_{ij}) n_j \, dS - \int_V \tau_{ij,j} w_i \, dV \qquad (9.1156b)$$

where we used the divergence theorem on the first integral on the right side.

On the surface $S_v$, the velocity is prescribed, and so $w_i = 0$. Equation (9.11-56b) can then be written as

$$\int_V \tau_{ij} w_{i,j} \, dV = \int_{S_t} (w_i \tau_{ij}) n_j \, dS - \int_V (\tau_{ij,j}) w_i \, dV \qquad (9.11\text{-}57)$$

Similarly the integral

$$\int_V p \, w_{i,i} \, dV = \int_V (p w_i)_{,i} \, dV - \int_V p_{,i} \, w_i \, dV \qquad (9.11\text{-}58a)$$

$$= \int_{S_t} p \, w_i \, n_i \, dS - \int_V p_{,i} \, w_i \, dV \qquad (9.11\text{-}58b)$$

Substituting Equations (9.11-57, 58b) into Equation (9.11-54), we obtain

$$\int_V \left[ \left\{ -\tau_{ij,j} + \rho f_i - p_{,i} \right\} w_i + q v_{i,i} \right] dV + \int_{S_t} \left[ \left\{ (\tau_{ij} + p \delta_{ij}) \, n_j + t_i \right\} \right] w_i \, dS = 0 \qquad (9.11\text{-}59)$$

Since $w_i$ and $q$ are arbitrary it follows that

$$\tau_{ij,j} - \rho f_i + p_{,i} = 0 \qquad (9.11\text{-}44)$$

$$v_{i,i} = 0 \qquad (9.11\text{-}45)$$

$$t_i = -(\tau_{ij} + p \delta_{ij}) \, n_j \qquad (9.11\text{-}60)$$

Thus instead of solving Equations (9.11-44, 46), we can instead extremize the integral $I$ given by Equation (9.11-47). Note that the boundary conditions are satisfied automatically, since we used them in formulating the integral given by Equation (9.11-47).

**Example 9.11-3.** Use the variational method to solve the axial annular flow of a power law fluid between two co-axial cylinders, as illustrated by Figure 9.11-3.

We assume the flow to be in the z-direction. The velocity distribution is

$$v_r = 0 , \qquad v_\theta = 0 , \qquad v_z = v_z(r) \qquad (9.11\text{-}61a,b,c)$$

The radii of the inner and outer cylinders are $kR$ and $R$. The flow region to be considered is defined by

$$kR \le r \le R , \qquad 0 \le \theta \le 2\pi , \qquad 0 \le z \le L \qquad (9.11\text{-}62)$$

The boundary conditions are

$$r = kR , \qquad v_z = 0 \qquad (9.11\text{-}63a,b)$$

$$r = R , \qquad v_z = 0 , \qquad \text{for all } \theta \text{ and } z \qquad (9.11\text{-}64a,b)$$

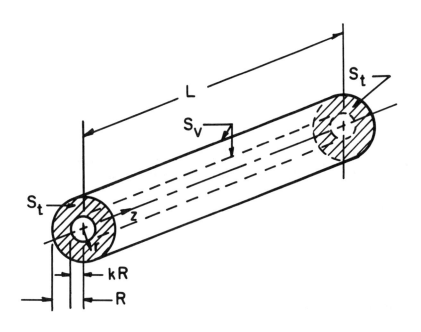

**FIGURE 9.11-3**    **Annular flow bounded by cylindrical surfaces $S_v$ and end surfaces $S_t$**

On the shaded surfaces at $z = 0$ and $z = L$, the pressure is given. Suppose the potential drop from $z = 0$ to $z = L$ to be $\Delta\mathcal{P}$, where the potential $\mathcal{P}$ combines the effects of pressure and gravity according to

$$\mathcal{P} = p + \rho gh \qquad\qquad (9.11\text{-}65)$$

The viscosity function $\eta$ for a power-law fluid may be written as (Carreau et al., 1997)

$$\eta(\dot{\gamma}) = m\left|\dot{\gamma}\right|^{n-1} \qquad\qquad (9.11\text{-}66)$$

We non-dimensionalize $r$ by writing

$$\bar{r} = \frac{r}{R} \qquad\qquad (9.11\text{-}67)$$

We select a trial solution which satisfies automatically the boundary conditions [Equations (9.11-64, 65)]. This is the usual procedure in variational calculus. Thus an appropriate $v_z$ can be chosen to be

$$v_z = a\left(1 - \left|\sigma\right|^{(1+n)/n}\right) \qquad\qquad (9.11\text{-}68a)$$

$$\sigma = \frac{2\bar{r} - (1 + k)}{(1 - k)} \tag{9.11-68b}$$

where a is a constant to be determined.

Then $\dot{\gamma}$ is given by

$$\dot{\gamma} = \pm \frac{dv_z}{dr} = \pm \frac{1}{R} \frac{dv_z}{d\bar{r}} = \pm \frac{2}{R(1-k)} \frac{dv_z}{d\sigma} = \mp \frac{2a(1+n)}{nR(1-k)} \left| \sigma \right|^{1/n} \tag{9.11-69a,b,c,d}$$

where the upper sign is for $\sigma < 0$ and the lower one for $\sigma > 0$.

From Equation (9.11-66), we have

$$g(\dot{\gamma}) = \int_0^{\dot{\gamma}} u \, \eta(u) \, du = \frac{m}{n+1} \dot{\gamma}^{n+1} \tag{9.11-70a,b}$$

Thus the integral I [Equation (9.11-47)] we need to extremize is now given by

$$I = \int_V \frac{m}{n+1} \dot{\gamma}^{n+1} \, dV - \int_{S_t} \Delta \mathcal{P} \, v_z \, dS \tag{9.11-71}$$

where the mass balance for an incompressible fluid $(v_{i,i} = 0)$ eliminates the pressure term in the first integral of Equation (9.11-47) and where the boundary conditions on the surface $S_t$ result in a potential drop. We now rewrite I as follows

$$I = \int_0^L \int_0^{2\pi} \int_{kR}^R \frac{m}{n+1} \dot{\gamma}^{n+1} r \, dr \, d\theta \, dz - \Delta \mathcal{P} \int_0^{2\pi} \int_{kR}^R v_z \, r \, dr \, d\theta \tag{9.11-72a}$$

$$= \frac{2\pi L m R^2}{(n+1)} \int_k^1 \dot{\gamma}^{n+1} \bar{r} \, d\bar{r} - \Delta \mathcal{P} \, 2\pi \, R^2 \int_k^1 \bar{r} \, v_z \, d\bar{r} \tag{9.11-72b}$$

$$= \frac{2\pi L m R^2}{(n+1)} \int_{-1}^1 \dot{\gamma}^{n+1} \left[ \frac{(1-k)}{2} \sigma + \frac{(1+k)}{2} \right] \left( \frac{1-k}{2} \right) d\sigma$$

$$- 2\pi \, \Delta \mathcal{P} \, R^2 \int_{-1}^1 v_z \left[ \frac{(1-k)}{2} \sigma + \frac{(1+k)}{2} \right] \left( \frac{1-k}{2} \right) d\sigma \tag{9.11-72c}$$

Since we have defined $\dot{\gamma}$ and $v_z$ as functions of $|\sigma|$, the first term in both integrals of Equation (9.11-72) will be zero when integrated from $\sigma = -1$ to $\sigma = 1$, since they are odd functions of $\sigma$. Integration yields

$$I = C_0 \left[ 2a(n+1) \right]^{n+1} - C_1 a \tag{9.11-73}$$

where $C_0 = \dfrac{\pi L R^2 m (1-k^2) n}{(n+1)(1+2n) \left[ nR(1-k) \right]^{n+1}}$, $\quad C_1 = \dfrac{(n+1)}{(2n+1)} \pi \Delta \mathcal{P} R^2 (1-k^2)$.

The integral $I$ is now a function of $a$. This integral has to be minimized and we need to compute $\dfrac{dI}{da}$. Differentiating yields

$$\frac{dI}{da} = C_0 (n+1) \left[ 2a(n+1) \right]^n \bullet 2(n+1) - C_1 = 0 \tag{9.11-74a,b}$$

Hence

$$a = \left( \frac{R \Delta \mathcal{P}}{2mL} \right)^{1/n} \frac{R(1-k)^{(1+n)/n} \, n}{2(n+1)} \tag{9.11-75}$$

We can now calculate the volumetric flow rate $Q$, which is given by

$$Q = \int_{kR}^{R} 2\pi r \, v_z \, dr = \pi R^3 \left( \frac{R \Delta \mathcal{P}}{2mL} \right)^{1/n} \frac{(1-k)^{(1+2n)/n} \, n(1+k)}{2(1+2n)} \tag{9.11-76a,b}$$

using $v_z$, given by Equations (9.11-68a, b, 75).

In Bird et al. (1987), the value of $Q$ for various values of $n$ and $k$ are compared with the exact solution. The above approximate solution is within 2% of the exact solution for $k > 0.5$ and $n > 0.25$. For smaller values of $k$ and $n$, the approximate solution gets worse and could probably be improved by considering more terms in the trial solution.

●

The method we have used to solve the present boundary-value problem consists of rewriting the governing differential equations and boundary conditions as a variational problem. This is followed by expressing the unknown functions as a series of simple functions that satisfy the boundary conditions. The unknown coefficients are determined by satisfying the variational principle. This technique is known as the **Rayleigh-Ritz method**.

Although simple problems, such as this flow problem, can be solved exactly via the equations of continuity and motion, more complex problems require the use of an approximate solution. Variational methods are widely used to obtain approximate solutions to boundary value problems in continuum mechanics.

## 9.12 TRANSVERSALITY CONDITIONS WHERE THE FUNCTIONAL DEPENDS ON SEVERAL FUNCTIONS

We extend the theory considered in Section 9.8 to the case where the functional depends on n functions, $y_1, y_2, \ldots, y_n$. We further relax the condition that the end points can move only along the ordinates (that is, only along the straight lines $x = a$, $x = b$). We allow the end points to move along given curves or surfaces and we label the end points as P and Q.

For simplicity we derive the transversality conditions for the case of two functions $y_1$ and $y_2$. We aim to find the extremum of

$$I = \int_P^Q f(x, y_1, y_2, y_1', y_2') \, dx \tag{9.12-1}$$

The end points P and Q are not fixed nor are the values of $y_1$, $y_2$ at these points. We assume that P and Q lie on two surfaces respectively given by

$$x = \phi(y_1, y_2), \qquad x = \psi(y_1, y_2) \tag{9.12-2a,b}$$

We proceed as in Section 9.10, but now the end points are free to move on the surfaces given by Equation (9.12-2a, b). Using Leibnitz's rule, Equation (9.10-7) becomes

$$\frac{\partial I^*}{\partial \varepsilon_i} = \left[ \int_P^Q \frac{\partial f}{\partial \varepsilon_i} \, dx + f \frac{\partial Q}{\partial \varepsilon_i} \bigg|_Q - f \frac{\partial P}{\partial \varepsilon_i} \bigg|_P \right] = 0, \qquad i = 1, 2 \tag{9.12-3a,b}$$

where P and Q are functions of $\varepsilon_i$.

The quantity $\dfrac{\partial f}{\partial \varepsilon_i}$ is given by

$$\frac{\partial f}{\partial \varepsilon_i} = \sum_{j=1}^2 \left[ \frac{\partial f}{\partial y_j} \frac{\partial y_j}{\partial \varepsilon_i} + \frac{\partial f}{\partial y_j'} \frac{\partial y_j'}{\partial \varepsilon_i} \right] \tag{9.12-4a}$$

$$= \sum_{j=1}^2 \left[ \frac{\partial f}{\partial y_j} \frac{\partial y_j}{\partial \varepsilon_i} + \frac{\partial f}{\partial y_j'} \frac{\partial}{\partial \varepsilon_i} \left( \frac{dy_j}{dx} \right) \right] \tag{9.12-4b}$$

$$= \sum_{j=1}^{2} \left[ \frac{\partial f}{\partial y_j} \frac{\partial y_j}{\partial \epsilon_i} + \frac{d}{dx} \left( \frac{\partial f}{\partial y_j'} \frac{\partial y_j}{\partial \epsilon_i} \right) - \frac{\partial y_j}{\partial \epsilon_i} \frac{d}{dx} \left( \frac{\partial f}{\partial y_j'} \right) \right] \qquad (9.12\text{-}4c)$$

Substituting Equation (9.12-4c) into Equation (9.12-3) and simplifying, we obtain

$$0 = \left[ \int_P^Q \frac{\partial y_j}{\partial \epsilon_i} \left\{ \frac{\partial f}{\partial y_j} - \frac{d}{dx} \left( \frac{\partial f}{\partial y_j'} \right) \right\} dx \right] + \left[ \frac{\partial f}{\partial y_j'} \frac{\partial y_j}{\partial \epsilon_i} \right]_P^Q + f \frac{\partial Q}{\partial \epsilon_i} \bigg|_Q - f \frac{\partial P}{\partial \epsilon_i} \bigg|_P \qquad (9.12\text{-}5)$$

using the convention of summation over repeated indices. At any point in space, except at $P$ and $Q$, $y_1$ and $y_2$ are independent. It can be seen from Equation (9.10-4) that

$$\frac{\partial y_j}{\partial \epsilon_i} = \eta_j \, \delta_{ij} \qquad (9.12\text{-}6)$$

where $\delta_{ij}$ is the Kronecker delta.

Substituting Equation (9.12-6) into Equation (9.12-5), we deduce the extremum conditions to be

$$\frac{\partial f}{\partial y_i} - \frac{d}{dx} \left( \frac{\partial f}{\partial y_i'} \right) = 0, \qquad i = 1, 2 \qquad (9.12\text{-}7)$$

Next, we deduce the transversality conditions.

The transversality conditions are to be imposed at the points $P$ and $Q$ which are on the surfaces given by Equation (9.12-2a, b). Thus $y_1$, $y_2$ and $x$ are related through Equation (9.12-2a, b) and we can regard $y_1$ and $y_2$ to be functions of $x$ and $\epsilon$ instead of considering them individually as functions of $\epsilon_1$ and $\epsilon_2$. That is to say, if equations $y_1$ and $y_2$ [see Equation (9.10-4)] are related for example by an expression such as $y_1^2 + y_2^2 = c$, then a change in $\epsilon_1$ automatically induces a change in $\epsilon_2$ and we can indeed consider the $y_i$ to be functions of $x$ and $\epsilon$. Therefore, derivatives with respect to $\epsilon_i$ are now replaced by derivatives with respect to $\epsilon$. The transversality conditions may now be written as

$$\left[ \frac{\partial f}{\partial y_1'} \frac{\partial y_1}{\partial \epsilon} + \frac{\partial f}{\partial y_2'} \frac{\partial y_2}{\partial \epsilon} + f \frac{\partial Q}{\partial \epsilon} \right]_Q = 0 \qquad (9.12\text{-}8)$$

$$\left[ \frac{\partial f}{\partial y_1'} \frac{\partial y_1}{\partial \epsilon} + \frac{\partial f}{\partial y_2'} \frac{\partial y_2}{\partial \epsilon} + f \frac{\partial P}{\partial \epsilon} \right]_P = 0 \qquad (9.12\text{-}9)$$

Since $y_1$ is a function of $x$ and $\epsilon$, we differentiate to obtain

$$\frac{dy_1}{d\varepsilon} = \frac{\partial y_1}{\partial \varepsilon} + y_1' \frac{dx}{d\varepsilon} \tag{9.12-10}$$

We can also do this for $y_2$.

The point $Q$ is given by Equation (9.12-2b) and $\dfrac{\partial Q}{\partial \varepsilon}$ is given by

$$\frac{\partial Q}{\partial \varepsilon} = \frac{dx}{d\varepsilon} = \frac{\partial \psi}{\partial y_1} \frac{dy_1}{d\varepsilon} + \frac{\partial \psi}{\partial y_2} \frac{dy_2}{d\varepsilon} \tag{9.12-11a,b}$$

Substituting Equations (9.12-10, 11a, b) into Equation (9.12-8), we obtain

$$\left[ \frac{\partial f}{\partial y_1'} \left\{ \frac{dy_1}{d\varepsilon} - y_1' \left( \frac{\partial \psi}{\partial y_1} \frac{dy_1}{d\varepsilon} + \frac{\partial \psi}{\partial y_2} \frac{dy_2}{d\varepsilon} \right) \right\} + \frac{\partial f}{\partial y_2'} \left\{ \frac{dy_2}{d\varepsilon} - y_2' \left( \frac{\partial \psi}{\partial y_1} \frac{dy_1}{d\varepsilon} + \frac{\partial \psi}{\partial y_2} \frac{dy_2}{d\varepsilon} \right) \right\} \right.$$
$$\left. + f \left\{ \frac{\partial \psi}{\partial y_1} \frac{dy_1}{d\varepsilon} + \frac{\partial \psi}{\partial y_2} \frac{dy_2}{d\varepsilon} \right\} \right]_Q = 0 \tag{9.12-12}$$

Collecting similar terms, Equation (9.12-12) becomes

$$\left[ \frac{dy_1}{d\varepsilon} \left\{ \frac{\partial f}{\partial y_1'} + \frac{\partial \psi}{\partial y_1} \left( -\frac{\partial f}{\partial y_1'} y_1' - \frac{\partial f}{\partial y_2'} y_2' + f \right) \right\} \right.$$
$$\left. + \frac{dy_2}{d\varepsilon} \left\{ \frac{\partial f}{\partial y_2'} + \frac{\partial \psi}{\partial y_2} \left( -y_1' \frac{\partial f}{\partial y_1'} - y_2' \frac{\partial f}{\partial y_2'} + f \right) \right\} \right]_Q = 0 \tag{9.12-13}$$

Thus, the transversality conditions at $Q$ are given by

$$\left[ \frac{\partial f}{\partial y_1'} + \frac{\partial \psi}{\partial y_1} \left( f - y_1' \frac{\partial f}{\partial y_1'} - y_2' \frac{\partial f}{\partial y_2'} \right) \right]_Q = 0 \tag{9.12-14}$$

$$\left[ \frac{\partial f}{\partial y_2'} + \frac{\partial \psi}{\partial y_2} \left( f - y_1' \frac{\partial f}{\partial y_1'} - y_2' \frac{\partial f}{\partial y_2'} \right) \right]_Q = 0 \tag{9.12-15}$$

Similarly, the transversality conditions at $P$ are given by

$$\left[ \frac{\partial f}{\partial y_1'} + \frac{\partial \phi}{\partial y_1} \left( f - y_1' \frac{\partial f}{\partial y_1'} - y_2' \frac{\partial f}{\partial y_2'} \right) \right]_P = 0 \tag{9.12-16}$$

$$\left[ \frac{\partial f}{\partial y_2'} + \frac{\partial \phi}{\partial y_2} \left( f - y_1' \frac{\partial f}{\partial y_1'} - y_2' \frac{\partial f}{\partial y_2'} \right) \right]_P = 0 \tag{9.12-17}$$

The above derivations can be extended to more than two functions. Thus if $f$ is a function of $y_1$, $y_2$, ... , $y_n$, the transversality conditions at $Q$ are

$$\left[ \frac{\partial f}{\partial y_i'} + \frac{\partial \psi}{\partial y_i} \left( f - \sum_{j=1}^{n} y_j' \frac{\partial f}{\partial y_j'} \right) \right]_Q = 0 , \qquad i = 1, ..., n \tag{9.12-18}$$

Similar expression for the condition at $P$ can be established.

If $f$ is a function of only one dependent variable $y$, then the end points $P$ and $Q$ will be free to move on the curves, respectively given by

$$x = \overline{\varphi}(y) , \qquad x = \overline{\psi}(y) \tag{9.12-19a,b}$$

The extremum condition will be given by Equation (9.12-7) with $y_i = y$. The transversality conditions then reduce to

$$\left[ \frac{\partial f}{\partial y'} + \frac{\partial \overline{\psi}}{dy} \left( f - y' \frac{\partial f}{\partial y'} \right) \right]_Q = 0 \tag{9.12-20}$$

$$\left[ \frac{\partial f}{\partial y'} + \frac{\partial \overline{\varphi}}{dy} \left( f - y' \frac{\partial f}{\partial y'} \right) \right]_P = 0 \tag{9.12-21}$$

***Example 9.12-1***. Find the shortest distance from the origin to a point on the circle of radius 3 and centre $(4, 0)$. The equation of the circle is

$$(x - 4)^2 + y^2 = 9 \tag{9.12-22}$$

The distance from the origin to any point $Q$ on the circle is

$$S = \int_0^Q \sqrt{1 + (y')^2} \, dx \tag{9.12-23}$$

Since $y$ does not appear explicitly in the integrand, Equation (9.12-7) simplifies to

$$\frac{d}{dx} \left( \frac{\partial f}{\partial y'} \right) = 0 \quad \Rightarrow \quad \frac{\partial f}{\partial y'} = \text{constant} \tag{9.12-24a,b}$$

With $f$ given by Equation (9.12-23), we have

$$\frac{y'}{\sqrt{1 + y'^2}} = C_0 \quad \Rightarrow \quad y' = k_1 \text{ (constant)} \tag{9.12-25a,b}$$

Integrating Equation (9.12-25b) and using the end condition at the origin (x = 0, y = 0) yields

$$y = k_1 x \tag{9.12-26}$$

To determine the slope $k_1$, we make use of Equation (9.12-20). Q lies on the circle given by Equation (9.12-22) as well as on the line given by Equation (9.12-26). Thus

$$\frac{d\overline{\psi}}{dy} = \frac{dx}{dy} = \frac{-y}{(x-4)} = \frac{-k_1 x}{(x-4)} \tag{9.12-27a,b,c}$$

Substituting the value of y into Equation (9.12-20), we obtain

$$\frac{k_1}{\sqrt{1 + k_1^2}} - \frac{k_1 x}{(x-4)} \left( \sqrt{1 + k_1^2} - \frac{k_1^2}{\sqrt{1 + k_1^2}} \right) = 0 \tag{9.12-28}$$

The solution is $k_1 = 0$, $y = 0$ and the point Q is on the x-axis. From Equation (9.12-22), we deduce that Q is the point $(1, 0)$. Then the minimum distance is 1.

## 9.13   SUBSIDIARY CONDITIONS WHERE THE FUNCTIONAL DEPENDS ON SEVERAL FUNCTIONS

In Section 9.9, we established the conditions for the extremum of a functional which depends on only one function. The function was subjected to certain supplementary conditions, in addition to the end points conditions. In this section, we consider the case of a functional which depends on several functions. We shall, as in Section 9.9, consider two types of constraints: an integral type and an algebraic type.

We now look for an extremum of the functional

$$I = \int_a^b f(x, y_1, y_2, ..., y_n, y_1', ..., y_n') \, dx \tag{9.13-1}$$

subject to

$$J_j = \int_a^b g_j(x, y_1, ..., y_n, y_1', ..., y_n') \, dx = \text{constant} = c_j, \qquad j = i, ..., k \tag{9.13-2a,b}$$

The end conditions are prescribed.

We use the method of Lagrange multipliers and introduce $\lambda_1, \lambda_2, ..., \lambda_k$. The condition for the extremum of I subject to the isoperimetric constraint is

$$\frac{\partial}{\partial y_i}\left(f + \sum_{j=1}^{k}\lambda_j\, g_j\right) - \frac{d}{dx}\left\{\frac{\partial}{\partial y_i'}\left(f + \sum_{j=1}^{k}\lambda_j\, g_j\right)\right\} = 0\,, \qquad i = 1, ..., n \qquad (9.13\text{-}3)$$

We have to solve Equations (9.13-2a, b, 3). Equation (9.13-3) is in general a second order differential equation. The 2n arbitrary constants resulting from the solution of the n differential equations are to be determined from the end point conditions. The k Lagrange multipliers can be determined from Equations (9.13-2a,b).

***Example 9.13-1.*** Find the extremum of

$$I = \int_0^1 \left[y_1'^2 + y_2'^2 - 4x y_2' - 4y_2\right] dx \qquad (9.13\text{-}4)$$

subject to

$$J = \int_0^1 \left[y_1'^2 - x y_1' - y_2'^2\right] dx = 2 \qquad (9.13\text{-}5a,b)$$

The end point conditions are

$$y_1(0) = y_2(0) = 0\,, \qquad y_1(1) = y_2(1) = 1 \qquad (9.13\text{-}6a,b,c,d)$$

We have one constraint and so we need to introduce only one Lagrange multiplier $\lambda$. Substituting f and g from Equations (9.13-4, 5) into Equation (9.13-3) and noting that f and g are not explicit functions of $y_1$, we have for $i = 1$ and 2

$$\frac{d}{dx}\left[\frac{\partial}{\partial y_1'}\left\{(y_1'^2 + y_2'^2 - 4x y_2' - 4y_2) + \lambda\,(y_1'^2 - x y_1' - y_2'^2)\right\}\right] = 0 \qquad (9.13\text{-}7)$$

$$-4 - \frac{d}{dx}\left[\frac{\partial}{\partial y_2'}\left\{(y_1'^2 + y_2'^2 - 4x y_2' - 4y_2) + \lambda\,(y_1'^2 - x y_1' - y_2'^2)\right\}\right] = 0 \qquad (9.13\text{-}8)$$

Carrying out the differentiations in Equations (9.13-7, 8) and simplifying, we write

$$\frac{d}{dx}\left[2y_1' + 2\lambda y_1' - \lambda x\right] = 0 \qquad (9.13\text{-}9)$$

$$4 + \frac{d}{dx}\left[2y_2' - 4x - 2\lambda y_2'\right] = 0 \qquad (9.13\text{-}10)$$

Integrating Equations (9.13-9, 10) yields

$$2y_1' + 2\lambda y_1' - \lambda x = c_1 \tag{9.13-11}$$

$$2y_2' - 2\lambda y_2' = c_2 \tag{9.13-12}$$

where $c_1$ and $c_2$ are arbitrary constants.

On further integrating Equations (9.13-11, 12), we obtain

$$2y_1 (1 + \lambda) = \frac{\lambda x^2}{2} + c_1 x + c_3 \tag{9.13-13}$$

$$2y_2 (1 - \lambda) = c_2 x + c_4 \tag{9.13-14}$$

where $c_3$ and $c_4$ are constants.

The constants $c_1$ to $c_4$ can be determined from the end point conditions given by Equations (9.13-6a to d) and are

$$c_1 = \frac{(4 + 3\lambda)}{2} , \qquad c_2 = 2 (1 - \lambda) , \qquad c_3 = c_4 = 0 \tag{9.13-15a,b,c,d}$$

Substituting the values of $c_1$ and $c_2$ into Equations (9.13-11, 12) respectively, we find that $y_1'$ and $y_2'$ are given by

$$y_1' = \frac{(3\lambda + 4 + 2\lambda x)}{4 (1 + \lambda)} , \qquad y_2' = 1 \tag{9.13-16a,b}$$

The Lagrange multiplier can now be determined from Equation (9.13-5) since $y_1'$ and $y_2'$ are known in terms of $\lambda$. The value of $\lambda$ is

$$\lambda = -10/11 \tag{9.13-17}$$

Thus the required functions $y_1$ and $y_2$ are

$$y_1 = -\frac{5}{2} x^2 + \frac{7}{2} x , \qquad y_2 = x \tag{9.13-18a,b}$$

●

If the constraint is not of an integral form, we introduce Lagrange multipliers which are functions of $x$ rather than being constants as in the case of a functional with only one function. Suppose we need to find the extremum of

$$I = \int_a^b f(x, y_1, y_2, ..., y_n, y_1', ..., y_n') \, dx \qquad (9.13\text{-}19)$$

Subject to the constraints

$$g_j(x, y_1, ..., y_n, y_1', ..., y_n') = 0, \qquad j = 1, 2, ..., m \qquad (9.13\text{-}20)$$

The end point conditions are given as follows

$$y_1(a) = a_1, \; ..., \; y_n(a) = a_n \qquad\qquad (9.13\text{-}21a,b)$$

$$y_1(b) = b_1, \; ..., \; y_n(b) = b_n \qquad\qquad (9.13\text{-}22a,b)$$

We now introduce a Lagragian function L given by

$$L = f + \sum_{j=1}^m \lambda_j(x) \, g_j \qquad (9.13\text{-}23)$$

The conditions for the extremum can be shown to be

$$\frac{\partial L}{\partial y_i} - \frac{d}{dx}\left(\frac{\partial L}{\partial y'}\right) = 0, \qquad i = 1, 2, ..., n \qquad (9.13\text{-}24)$$

Finding of the extremum of I now consists of solving Equation (9.13-24) subject to Equations (9.13-20, 22).

*Example 9.13-2.* Determine the shortest curve between two fixed points on the surface of a sphere of unit radius and centre at the origin.

In a rectangular Cartesian coordinate system, the equation of the sphere is

$$x^2 + y^2 + z^2 = 1 \qquad (9.13\text{-}25)$$

If the curve joins the points $(x_0, y_0, z_0)$ and $(x_1, y_1, z_1)$, the distance s is given by

$$s = \int_{x_0}^{x_1} \sqrt{1 + (y')^2 + (z')^2} \; dx \qquad (9.13\text{-}26)$$

The Lagrangian function L is then

$$L = \sqrt{1 + (y')^2 + (z')^2} + \lambda(x)(x^2 + y^2 + z^2 - 1) \qquad (9.13\text{-}27)$$

It is useful to introduce the parametric representation of the curves as

$$x = x(t), \qquad y = y(t), \qquad z = z(t) \tag{9.13-28a,b,c}$$

where $t$ is now the independent variable and $x$, $y$, $z$ are the dependent variables. Noting that $\dfrac{dx_i}{dx_j} = \dfrac{dx_i}{dt}\dfrac{dt}{dx_j}$, Equation (9.13-26) becomes

$$s = \int_{t_0}^{t_1} \sqrt{\dot{x}^2 + \dot{y}^2 + \dot{z}^2} \; dt = \int_{t_0}^{t_1} \sigma \, dt \tag{9.13-29a,b}$$

where the dot denotes differentiation with respect to $t$. The fixed points are

$$x(t_0) = t_0, \qquad x(t_1) = t_1 \tag{9.13-30a,b}$$

Equation (9.13-27) can be written as

$$L = \sqrt{\dot{x}^2 + \dot{y}^2 + \dot{z}^2} + \mu(t)\,(x^2 + y^2 + z^2) \tag{9.13-31}$$

where $\mu(t)$ is the t-dependent Lagrange multiplier.

Equation (9.13-24) identifies the extremum conditions as

$$\frac{\partial L}{\partial x} - \frac{d}{dt}\left(\frac{\partial L}{\partial \dot{x}}\right) = 0 \qquad \text{i.e.} \qquad 2\mu x - \frac{d}{dt}\left[\frac{\dot{x}}{\sigma}\right] = 0 \tag{9.13-32a,b}$$

$$\frac{\partial L}{\partial y} - \frac{d}{dt}\left(\frac{\partial L}{\partial \dot{y}}\right) = 0 \qquad \text{i.e.} \qquad 2\mu y - \frac{d}{dt}\left[\frac{\dot{y}}{\sigma}\right] = 0 \tag{9.13-33a,b}$$

and $\quad \dfrac{\partial L}{\partial z} - \dfrac{d}{dt}\left(\dfrac{\partial L}{\partial \dot{z}}\right) = 0 \qquad \text{i.e.} \qquad 2\mu z - \dfrac{d}{dt}\left[\dfrac{\dot{z}}{\sigma}\right] = 0 \tag{9.13-34a,b}$

Eliminating $\mu(t)$ yields

$$\frac{\dfrac{d}{dt}\left[\dfrac{\dot{x}}{\sigma}\right]}{2x} = \frac{\dfrac{d}{dt}\left[\dfrac{\dot{y}}{\sigma}\right]}{2y} = \frac{\dfrac{d}{dt}\left[\dfrac{\dot{z}}{\sigma}\right]}{2z} \tag{9.13-35a,b}$$

On differentiating $\left[\dfrac{\dot{x}}{\sigma}\right]$, we obtain

$$\frac{d}{dt} \left[ \frac{\dot{x}}{\sigma} \right] = \frac{\sigma \ddot{x} - \dot{\sigma} \dot{x}}{\sigma^2} \tag{9.13-36}$$

We have similar terms for $\frac{d}{dt} \left[ \frac{\dot{y}}{\sigma} \right]$ and $\frac{d}{dt} \left[ \frac{\dot{z}}{\sigma} \right]$.

Equation (9.13-35a,b) becomes

$$\frac{\sigma \ddot{x} - \dot{\sigma} \dot{x}}{2x\sigma^2} = \frac{\sigma \ddot{y} - \dot{\sigma} \dot{y}}{2y\sigma^2} = \frac{\sigma \ddot{z} - \dot{\sigma} \dot{z}}{2z\sigma^2} \tag{9.13-37a,b}$$

From the first two sets of terms in Equation (9.13-37), we have

$$y (\sigma \ddot{x} - \dot{\sigma} \dot{x}) = x (\sigma \ddot{y} - \dot{\sigma} \dot{y}) \tag{9.13-38a}$$

or $\quad \sigma (\ddot{x} y - x \ddot{y}) = \dot{\sigma} (\dot{x} y - x \dot{y}) \tag{9.13-38b}$

or $\quad \dfrac{\dot{\sigma}}{\sigma} = \dfrac{(\ddot{x} y - x \ddot{y})}{(\dot{x} y - x \dot{y})} \tag{9.13-38c}$

Similarly, from the last 2 sets of terms in Equation (9.13-37), we find

$$\frac{\dot{\sigma}}{\sigma} = \frac{(\ddot{z} y - z \ddot{y})}{(\dot{z} y - z \dot{y})} \tag{9.13-39}$$

Also, from Equation (9.13-38, 39), we deduce

$$\frac{\frac{d}{dt} (\dot{x} y - x \dot{y})}{(\dot{x} y - x \dot{y})} = \frac{\frac{d}{dt} (\dot{z} y - z \dot{y})}{(\dot{z} y - z \dot{y})} \tag{9.13-40}$$

Note that in deriving Equation (9.13-40), we have made use of the following expressions

$$\frac{d}{dt} (\dot{x} y - x \dot{y}) = \ddot{x} y + \dot{x} \dot{y} - \dot{x} \dot{y} - x \ddot{y} = \ddot{x} y - x \ddot{y} \tag{9.13-41a,b}$$

Integrating Equation (9.13-40), we obtain

$$\ell n (\dot{x} y - x \dot{y}) = \ell n (\dot{z} y - z \dot{y}) + \ell n \, c_1 \tag{9.13-42a}$$

or      $\dot{x}y - x\dot{y} = c_1(\dot{z}y - z\dot{y})$                                                 (9.13-42b)

where  $c_1$  is a constant.

Equation (9.13-42b) may be written as

$$y(\dot{x} - c_1\dot{z}) = \dot{y}(x - c_1z)$$                                                       (9.13-43a)

or      $$\frac{\dot{x} - c_1\dot{z}}{x - c_1z} = \frac{\dot{y}}{y}$$                                  (9.13-43b)

Equation (9.13-43b) can be integrated to yield

$$\ell n(x - c_1z) = \ell n\, y + \ell n\, c_2$$                                                    (9.11-44a)

or      $x - c_1z = c_2\, y$                                                                          (9.11-44b)

or      $x - c_2y - c_1z = 0$                                                                         (9.13-44c)

Equation (9.13-44c) is the equation of a plane passing through the origin.  The arbitrary constants $c_1$ and $c_2$ can be determined if the fixed points $(x_0, y_0, z_0)$ and $(x_1, y_1, z_1)$ are given.  The intersection of the plane given by Equation (9.13-44c) and the sphere given by Equation (9.13-25) is the arc of the great circle.

***Example 9.13-3***.  Formulate the problem of silastic implant and determine the optimum design of the implant.

Silastic implants are widely used in plastic surgery (Atkinson, 1988).  The technique consists of inserting a deflated silastic balloon underneath the skin.  The balloon is then filled with saline over a period of several weeks until the skin has been sufficiently stretched so as to cover an adjacent defect area.  The plastic surgeon needs to know the optimum design of the implant so as to cover the damaged area for a given volume of saline.  This problem is similar to the Plateau problem considered in Section 9.11.

Let  A  be the area of the base of the implant which is in the xy-plane, as shown in Figure 9.13-1.

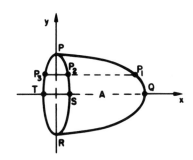

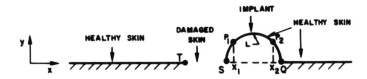

**FIGURE 9.13-1    Geometry of an implant**

The base of the implant is PQRP and the defect area to be covered is PTRP. Once the skin has been sufficiently expanded, the surgeon will perform the operation and advance the skin to cover the defect area. Let $h(x, y)$ be the height of the expanded skin above the base, then the surface area $S$ according to Equation (9.11-24) is given by

$$S = \iint_A \left[1 + h_x^2 + h_y^2\right]^{1/2} dx\, dy \qquad (9.13\text{-}45)$$

where $h_x = \dfrac{\partial h}{\partial x}, \quad h_y = \dfrac{\partial h}{\partial y}$ .

The volume $V$ of saline is

$$V = \iint_A h(x, y)\, dx\, dy = V_0 \qquad (9.13\text{-}46a,b)$$

To ensure that the defect area will be covered we need to know the minimum surface $S$ for a given volume $V$. This isoperimetric problem can be solved by introducing the Lagrange multiplier and is equivalent to finding the extremum of

$$L(h, \lambda) = S - \lambda V \qquad (9.13\text{-}47a)$$

$$= \iint\limits_{A} \left[ \left( 1 + h_x^2 + h_y^2 \right)^{1/2} - \lambda h \right] dx \, dy \qquad (9.13\text{-}47b)$$

$$= \iint\limits_{A} F \, dx \, dy \qquad (9.13\text{-}47c)$$

The condition for L to be an extremum is given by Equation (9.11-13), that is

$$\frac{\partial F}{\partial h} - \frac{\partial}{\partial x} \left( \frac{\partial F}{\partial h_x} \right) - \frac{\partial}{\partial y} \left( \frac{\partial F}{\partial h_y} \right) = 0 \qquad (9.13\text{-}48)$$

Substituting F from Equation (9.13-47b) into Equation (9.13-48) yields

$$\frac{\partial}{\partial x} \left[ \left( 1 + h_x^2 + h_y^2 \right)^{-1/2} h_x \right] + \frac{\partial}{\partial y} \left[ \left( 1 + h_x^2 + h_y^2 \right)^{-1/2} h_y \right] = \lambda \qquad (9.13\text{-}49)$$

We assume that the implant is made up of sections of ellipses and the curves PQR and PSR are given respectively by

PQR:  $y^2 = 1 - \dfrac{(x-d)^2}{\alpha_1^2}$ \qquad (9.13-50)

PSR:  $y^2 = \beta^2 - \dfrac{\beta^2 x^2}{\alpha^2}$ \qquad (9.13-51)

where $\alpha$, $\beta$, $\alpha_1$ and d are constants and $\alpha - \alpha_1 \le d \le \alpha + \alpha_1$.

Further we assume that the y-axis is the axis of symmetry for PSRTP and thus the equation of the curve PTR is given by Equation (9.13-51) with $-x$ replacing x. We now use the Rayleigh-Ritz method to solve for h. We choose the trial function for h to be

$$h = \frac{(A_1 + B_1 x + C_1 y) \, (\beta^2 x^2 + \alpha^2 y^2 - \alpha^2 \beta^2) \, (\alpha_1^2 - \alpha_1^2 y^2 - (x-d)^2)}{2 \, (\alpha^2 y^2 + \beta^2 x^2)} \qquad (9.13\text{-}52)$$

The boundary conditions $h = 0$ on the perimeters PQR and PSR are automatically satisfied by our choice of h. The constants $A_1$, $B_1$ and $C_1$ are to be determined by satisfying the variational condition of Equation (9.13-49). This non-linear minimization problem is solved numerically.

The parameter $\lambda$ can be determined in terms of $V_0$, since once h has been obtained in term of $\lambda$, it can be substituted into Equation (9.13-46) and the relation between $\lambda$ and $V_0$ can be obtained. Thus varying the values of $\lambda$ is equivalent to varying the values of $V_0$. The shape of the implant is determined by the values of $\alpha$, $\beta$, d and $\alpha_1$.

The values of $A_1$, $B_1$ and $C_1$ are obtained numerically for various values of $\alpha$, $\beta$, d, $\alpha_1$ and $\lambda$. The length of the curve L on the inflated balloon corresponding to a horizontal (y = constant) section $P_1P_2$ (Figure 9.13-1) is given by

$$L = \int_{x_2}^{x_1} \left(1 + h_x^2\right)^{1/2} dx \tag{9.13-53}$$

where the coordinates of $P_1$ and $P_2$ are $(x_1, y)$ and $(x_2, y)$ respectively.

The extension E produced by the inflation of the balloon in the x-direction for a given y is

$$E = L - (x_1 - x_2) \tag{9.13-54}$$

The balloon must be inflated to a volume such that E is sufficient to cover the defect.

For the values of $\alpha = \alpha_1 = d = 1.5$, $\beta = 1$, $\lambda = 2.85$, Atkinson has found the following values for $A_1$, $B_1$ and $C_1$: $A_1 = 1.8186$, $B_1 = 0.4632$, $C_1 = 0.2192$. For other values of the geometric parameters and $\lambda$, the reader is referred to the original paper. Also the values of E and the required values of E for various conditions are tabulated in Atkinson's paper.

*Example 9.13-4.* Solve the problem of optimization of thermal conductivities of isotropic solids.

We consider an isotropic body of volume V with non-homogeneous thermal conductivity $k(x_i)$ which is unknown. We need to determine the optimal distribution of $k(x_i)$ such that it satisfies a desired distribution $k_d(x_i)$ and a desired temperature distribution $T_d(x_i)$ in the body. This is a problem in the manufacture of composite insulating materials. This problem has been considered by Meric (1985). Mathematically we are required to minimize the so-called **performance index** J defined by

$$J = \int_V \left[ (T - T_d)^2 + \beta (k - k_d)^2 \right] dV \tag{9.13-55}$$

where T is the temperature and $\beta$ is a weighting factor.

Small values of $\beta$ imply that more weight is given to attaining the desired temperature distribution $T_d$. At the other extreme, large values of $\beta$ imply that more importance is given to achieving the desired thermal conductivity distribution $k_d$. At steady state, the equation governing the temperature throughout the body of volume $V$ is

$$\underline{\nabla} \cdot (k \underline{\nabla} T) + Q = 0 \tag{9.13-56}$$

where $Q$ is the energy, associated with the heat source.

We assume that the body $(V)$ is enclosed by two surfaces $S_1$ and $S_2$ and the boundary conditions on $S_1$ and $S_2$ are given by

on $S_1$:   $T = T_s$ \hfill (9.13-57)

on $S_2$:   $k \dfrac{\partial T}{\partial n} + q + \alpha (T - T_\infty) = 0$ \hfill (9.13-58)

where $T_s$ is the prescribed temperature on $S_1$, $\dfrac{\partial}{\partial n}$ is the directional derivative along the outward normal to $S_2$, $q$ is the boundary heat flux, $\alpha$ is the heat transfer coefficient and $T_\infty$ is the ambient temperature.

Finding the minimum of $J$ subject to the constraint given by Equation (9.13-56) is equivalent to finding the extremum of

$$I = J + \int_V \lambda [\underline{\nabla} \cdot (k \underline{\nabla} T) + Q] \, dV \tag{9.13-59}$$

where $\lambda$ is the Lagrange multiplier.

To show this equivalence we proceed as in Example 9.11-2 and consider the neighboring functions

$$T^* = T + \varepsilon T', \qquad k^* = k + \varepsilon k', \qquad \lambda^* = \lambda + \varepsilon \lambda' \tag{9.13-60a,b,c}$$

$$I^* = \int_V \left[ \frac{1}{2} \left\{ (T^* - T_d)^2 + \beta (k^* - k_d)^2 \right\} + \lambda^* \left\{ \underline{\nabla} \cdot (k^* \underline{\nabla} T^*) + Q \right\} \right] dV \tag{9.13-61}$$

Then the extremum condition, $\dfrac{dI^*}{d\varepsilon}\bigg|_{\varepsilon=0} = 0$, leads to

$$\int_V \left\{ T'(T - T_d) + \beta k'(k - k_d) + \lambda'(\underline{\nabla} \cdot k \underline{\nabla} T + Q) + \lambda [\underline{\nabla} \cdot (k' \underline{\nabla} T + k \underline{\nabla} T')] \right\} dV = 0 \tag{9.13-62}$$

We now apply Green's generalized first and second identities to transform part of the volume integral to a surface integral because the conditions are specified on the surface. The Green's identities may be written as follows

$$\int_V \left[ w \underline{\nabla} u \cdot \underline{\nabla} v + u \underline{\nabla} \cdot (w \underline{\nabla} v) \right] dV = \int_S w u \frac{\partial v}{\partial n} \, dS \tag{9.13-63}$$

$$\int_V \left[ u \underline{\nabla} \cdot (w \underline{\nabla} v) - v \underline{\nabla} \cdot (w \underline{\nabla} u) \right] dV = \int_S w \left( u \frac{\partial v}{\partial n} - v \frac{\partial u}{\partial n} \right) dS \tag{9.13-64}$$

Using Equations (9.13-63, 64), we have

$$\int_V \lambda \underline{\nabla} \cdot (k' \underline{\nabla} T) \, dV = \int_{S_1+S_2} k' \lambda \frac{\partial T}{\partial n} \, dS - \int_V k' \underline{\nabla} \lambda \cdot \underline{\nabla} T \, dV \tag{9.13-65}$$

$$\int_V \lambda \underline{\nabla} \cdot (k \underline{\nabla} T') \, dV = \int_{S_1+S_2} k \left( \lambda \frac{\partial T'}{\partial n} - T' \frac{\partial \lambda}{\partial n} \right) dS + \int_V T' \underline{\nabla} \cdot (k \underline{\nabla} \lambda) \, dV \tag{9.13-66}$$

Substituting Equations (9.13-65, 66) into Equation (9.13-62), we have

$$\int_V \left\{ T'(T - T_d + \underline{\nabla} \cdot (k \underline{\nabla} \lambda)) + k' [\beta (k - k_d) - \underline{\nabla} \lambda \cdot \underline{\nabla} T] + \lambda' [\underline{\nabla} \cdot (k \underline{\nabla} T) + Q] \right\} dV$$

$$+ \int_{S_1} \left\{ k' \lambda \frac{\partial T}{\partial n} + k \left( \lambda \frac{\partial T'}{\partial n} - T' \frac{\partial \lambda}{\partial n} \right) \right\} dS + \int_{S_2} \left\{ k' \lambda \frac{\partial T}{\partial n} + k \left( \lambda \frac{\partial T'}{\partial n} - T' \frac{\partial \lambda}{\partial n} \right) \right\} dS = 0$$

$$\tag{9.13-67}$$

On $S_1$, $T$ is prescribed ($= T_s$) and $T' = 0$. On $S_2$, the boundary condition given by Equation (9.13-58) must also be satisfied by $k^*$ and $T^*$. This implies that

$$(k + \varepsilon k') \frac{\partial}{\partial n} (T + \varepsilon T') + q + \alpha (T + \varepsilon T' - T_\infty) = 0 \tag{9.13-68}$$

Expanding and comparing powers of $\varepsilon$, we deduce from Equation (9.13-68) that on $S_2$

$$k' \frac{\partial T}{\partial n} + k \frac{\partial T'}{\partial n} + \alpha T' = 0 \tag{9.13-69}$$

Equation (9.13-67) reduces to

$$\int_V \left\{ T'(T - T_0 + \underline{\nabla} \cdot (k \underline{\nabla} \lambda)) + k' [\beta (k - k_d) - \underline{\nabla} \lambda \cdot \underline{\nabla} T] + \lambda'[\underline{\nabla} \cdot (k \underline{\nabla} T) + Q] \right\} dV$$

$$+ \int_{S_1} k'\lambda \frac{\partial T}{\partial n} \, dS - \int_{S_2} T'\left( \lambda\alpha + k \frac{\partial \lambda}{\partial n} \right) dS = 0 \tag{9.13-70}$$

Since $T'$, $k'$ and $\lambda'$ are arbitrary, Equation (9.13-70) implies that Equations (9.13-56, 58) have to be satisfied, in addition to an extra set of equations which may be written as

in V:        $\underline{\nabla} \cdot (k \underline{\nabla} \lambda) + (T - T_0) = 0$ \hfill (9.13-71)

$\beta (k - k_d) - \underline{\nabla} \lambda \cdot \underline{\nabla} T = 0$ \hfill (9.13-72)

on $S_1$:     $\lambda = 0$ \hfill (9.13-73)

on $S_2$:     $\lambda\alpha + k \dfrac{\partial \lambda}{\partial n} = 0$ \hfill (9.13-74)

We note that Equations (9.13-71, 74) are the equations governing $\lambda$. They are similar to Equations (9.13-56, 58). Equation (9.13-72) is the so-called **gradient condition**.

This non-linear system has been solved numerically and details are given in the original paper by Meric (1985). Numerical results for two geometries, an infinite plate and an infinite cylinder with $T_d = T_s = k_d = 1$, $Q = 0.01$, $q = T_\infty$ for various values of $\alpha$, $\beta$ and for several layers of material are discussed.

In the case of the infinite plate, the average optimal thermal conductivity decreases slowly with increasing number of layers and increases with increasing heat transfer coefficient.

## PROBLEMS

1a.     A processing operation produces twenty cups per minute at 420 K. An extra two cups per minute can be produced for each degree above 420 K. At a given temperature T, it costs $\left[15 + \dfrac{(T-420)^2}{30}\right]$ dollars per minute to operate the machine. If the cups are sold at one dollar each, determine the profit as a function of T. What is the most profitable temperature to run the machine?

Answer: T = 450 K

2a.     A cylindrical can holds $V\,m^3$ of soup. Find the radius r and the height h so that the surface area of the can is a minimum. Use the following methods

   (i)     determine the surface area of the can in terms of r and h and substitute r or h in terms of V and hence write the surface area as a function of one variable only. Then determine the minimum surface area;

   (ii)    use the method of Lagrange multipliers. The constraint is the fixed volume V.

Answer: h = 2r

3a.     A volume temperature profile is represented by n experimental points $(V_i, T_i)$, $i = 1, 2, ..., n$. We are required to find the straight line $V = mT + c$ which will "best" approximate the data. That is, if we denote the error at the point $V_i$ by $e_i$, then $e_i = (V(T_i) - V_i) = (mT_i + c - V_i)$. The "best" line can be defined such that $E^2 = \sum\limits_{i=1}^{n} e_i^2$ is a minimum. By considering $E^2$ as a function of m and c, show that for $E^2$ to be a minimum, m and c must satisfy the equations

$$m \sum_{i=1}^{n} T_i + nc = \sum_{i=1}^{n} V_i$$

$$m \sum_{i=1}^{n} T_i^2 + c \sum_{i=1}^{n} T_i = \sum_{i=1}^{n} T_i V_i$$

The above equations are known as normal equations (see Section 7.6).

4a.     Evaluate the integral

$$I = \int_0^1 (y^2 + x^2 y')\, dx$$

along the following curves joining the points (0, 0) to (1, 1):

(i)     $y = x$                                                          Answer: 0.667

(ii)    $y = x^2$                                                        Answer: 0.700

(iii)   $y = 1/2 \, (x + x^2)$                                           Answer: 0.675

Find the curve that will extremize $I$. From the values of $I$ determined using (i to iii), deduce if $I$ is a maximum or a minimum.

Answer: $y = x$

5a.  The curve $y = y(x)$ passes through the points $(a, c)$, $(b, d)$ and is rotated about the x-axis. The surface area $S$ thus generated is given by

$$S = 2\pi \int_a^b y \sqrt{1 + (y')^2} \; dx$$

Find the curve $y(x)$ which will extremize $S$.          Answer: $y = c_1 \cosh \dfrac{(x - c_2)}{c_1}$

6b.  Any point $(x, y, z)$ on the surface of a sphere of radius $a$ can be expressed in a spherical polar coordinate system $(r, \theta, \phi)$ as

$$x = a \sin \phi \cos \theta, \qquad y = a \sin \phi \sin \theta, \qquad z = a \cos \phi$$

Let $\theta = F(\phi)$ be a curve lying on the surface of the sphere joining two points $A$ and $B$. The length of the curve is given by

$$\int_A^B ds = \int_A^B \sqrt{dx^2 + dy^2 + dz^2}$$

Show that in the spherical polar coordinate system

$$\int_A^B ds = a \int_A^B \sqrt{1 + (F'(\phi))^2 \sin^2 \phi} \; d\phi$$

By solving the Euler-Lagrange equation, determine the function $F(\phi)$ that minimizes the length of the curve.

Answer: $\cot \phi = \alpha \cos (\theta + c_2)$

7a.  Find the curve $y(x)$ which extremizes the functional

$$I = \int_0^1 [y' + (y'')^2] \, dx$$

by the following methods

(i)   using Equation (9.7-3),

(ii)   using the method of constraints explained in Sections 9.9 and 9.10.

The boundary conditions are

$$y(0) = y(1) = 0 \; ; \qquad y'(0) = y'(1) = 1. \qquad\qquad \text{Answer: } y = 2x^3 - 3x^2 + x$$

8a.   Find the extremum of

$$I = \int_0^1 \left( \tfrac{1}{2} \, y'^2 + y'y + y' + y \right) dx$$

The boundary conditions are

(i)   $y(0) = y(1) = 1$                                                         Answer: $y = \dfrac{x^2}{2} - \dfrac{x}{2} + 1$

(ii)   $y(0) = 1$ , $y(1)$ is unspecified.                        Answer: $y = \dfrac{x^2}{2} - \dfrac{7x}{4} + 1$

9a.   Show that the boundary-value problem

$$y'' - y = x, \qquad y(0) = y(1) = 0$$

is equivalent to finding the extremum of

$$I = \int_0^1 [y'^2 + y^2 + 2xy] \, dx$$

An approximate solution $y_a$ that satisfies the boundary conditions is given by

$$y_a = c_0 \, x \, (1 - x)$$

Substitute $y_a$ into I. I can now be considered as a function of $c_0$. Determine $c_0$ by extremizing I. Compare the approximate solution $y_a$ with the exact solution.

Answer: $y_a = -\dfrac{5}{22} x (1 - x)$

10a.   Find the shortest distance from the origin $(0, 0)$ to any point $P$ on the line $2y = x - 4$. Hint: the distance $s = \int_0^P \sqrt{1 + (y')^2} \, dx$, $y(0) = 0$ and the point $P$ is on the line $x = 4 + 2y = \bar{\phi}(y)$.

Answer: $\frac{4}{5}\sqrt{5}$

11b.   Show that extremizing the functional

$$I = \iiint_R \left[ \frac{k^2}{2m} \left( \psi_{x_1}^2 + \psi_{x_2}^2 + \psi_{x_3}^2 \right) + \varphi \psi^2 \right] dx_1 \, dx_2 \, dx_3$$

subject to the condition

$$\iiint_R \psi^2 \, dx_1 \, dx_2 \, dx_3 = 1$$

leads to

$$-\frac{k^2}{2m} \nabla^2 \psi + \varphi \psi = -\lambda \psi$$

The operator $\nabla^2$ is the Laplacian $\left( = \sum_{i=1}^{3} \frac{\partial^2}{\partial x_i^2} \right)$. The differential equation is known as the **Schrödinger equation** (see Section 6.6). The constraint is known as the normalization condition. $\varphi$ is the potential, $k$ and $m$ are constants.

12b.   Consider the irreversible reaction $A \rightarrow B$ in a tubular reactor of length $\ell$. We wish to determine the best operating temperature $T$ as a function of the axial position $x$ along the tube so as to maximize the production of $B$. If $C_A$ is the concentration of $A$, then the relation between $C_A$ and $T$ is given by a kinetic equation of the form

$$\frac{dC_A}{dx} = g(C_A, T)$$

The production of $A$ is given by

$$I = \int_0^\ell f(C_A, T) \, dx$$

Thus to maximize the production of B, we have to minimize the production of A, that is to say, we have to extremize I subject to the kinetic equation.

In this example, the dependent variables are $C_A$ and T. The equation of constraint is in the form of a differential equation and has to be satisfied at each point along the tube. The Lagrange multiplier is thus a function of x and is not a constant. Write down the Lagrange function L. Show that the Euler-Lagrange equations are

$$\frac{\partial f}{\partial T} + \lambda \frac{\partial g}{\partial T} = 0$$

$$\frac{\partial f}{\partial C_A} + \lambda \frac{\partial g}{\partial C_A} + \frac{d\lambda}{dx} = 0$$

Determine T if $f = C_A^2 + T^2$, $g = T - \sqrt{3}\ C_A$ and the boundary conditions are: $C_A = 1$ at $x = 0$ and $T = T_0$ at $x = \ell$.

Answer: $T = 2 \sinh 2x + \sqrt{3}\ \cosh 2x + C\,(2 \cosh 2x + \sqrt{3}\ \sinh 2x)$

$C = [T_0 - (2 \sinh 2\ell + \sqrt{3}\ \cosh 2\ell)] / (2 \cosh 2\ell + \sqrt{3}\ \sinh 2\ell)$

13b.   In Example 9.11-3, we used a variational method to solve the axial flow of a power-law fluid between two co-axial circular cylinders. In the present problem we consider the axial flow of two immiscible fluids in a circular pipe of radius a and length $\ell$. This flow situation was investigated by Bentwich (1976). Bentwich has shown that the power required to transport a required volume of liquid, such as crude oil, is reduced if a small volume of less viscous liquid is added to the crude.

The flow is still an axial flow but the interface between the two liquids is not circular and the axial velocity $v_z$ is a function of r and $\theta$. If $v_z$ is non-dimensionalized and $\Delta p$ is the pressure drop, then $\bar{v}_z = v_z \ell \eta_1 / (a^2 \Delta p)$ has to satisfy the following equations of motion

$$\nabla^2 \bar{v}_z = -1 \qquad \text{in phase (liquid) 1}$$

$$\eta_2 / \eta_1 \nabla^2 \bar{v}_z = -1 \qquad \text{in phase (liquid) 2}$$

where $\eta_i$ is the viscosity of liquid i (i = 1, 2), $\nabla^2 \left( = \dfrac{\partial^2}{\partial r^2} + \dfrac{1}{r}\dfrac{\partial}{\partial r} + \dfrac{1}{r^2}\dfrac{\partial^2}{\partial \theta^2} \right)$ is the Laplacian.

The boundary conditions are

$\bar{v}_z = 0$ on the surface of the pipe which is taken to be at $\bar{r} = \frac{r}{a} = 1$.

At the interface

$$\bar{v}_z\big|_1 = \bar{v}_z\big|_2$$

$$\eta_1 \frac{\partial \bar{v}_z}{\partial n}\bigg|_1 = \eta_2 \frac{\partial \bar{v}_z}{\partial n}\bigg|_2$$

where $n$ is the outward normal to the surface, 1 and 2 indicate the evaluation is to be made at the interface in liquid 1 and 2 respectively.

Show that the above boundary value problem corresponds to extremizing the functional

$$I = \iint\limits_{R_1} \frac{1}{2} \left[ \left( \frac{\partial \bar{v}_z}{\partial \bar{r}} \right)^2 + \frac{1}{\bar{r}^2} \left( \frac{\partial \bar{v}_z}{\partial \theta} \right)^2 \right] \bar{r}\, d\bar{r}\, d\theta - \iint\limits_{R_1+R_2} \bar{r}\, \bar{v}_z\, d\bar{r}\, d\theta$$

$$+ \frac{\eta_2}{\eta_1} \iint\limits_{R_2} \frac{1}{2} \left[ \left( \frac{\partial \bar{v}_z}{\partial \bar{r}} \right)^2 + \frac{1}{\bar{r}^2} \left( \frac{\partial \bar{v}_z}{\partial \theta} \right)^2 \right] \bar{r}\, d\bar{r}\, d\theta$$

where $R_i$ denote the surface occupied by liquid $i$. An approximate solution can be expressed as

$$\bar{v}_z = A_0\, \phi_0 + \sum_{n=0}^{N} \sum_{m=1}^{M} A_{nm}\, \phi_{nm}$$

where $\phi_0 = 1 - \bar{r}^2$, $\phi_{nm} = J_n(\lambda_{nm}\bar{r}) \cos n\theta$, $J_n$ is the Bessel function of the first kind of order $n$, $\lambda_{nm}$ are the roots of the equations $J_n(\lambda_{nm}) = 0$ and the coefficients $A_0$, $A_{nm}$ are constants which are to be determined. On substituting the expression for $\bar{v}_z$ in the double integral, we obtain $I$ as a function of $A_0$ and $A_{nm}$. We now require to extremize $I$ with respect to $A_0$ and $A_{nm}$ (properties of $J_n$ are given in Section 2.7).

Find the equations that extremize $I$ and write down the algebraic equations that $A_0$ and $A_{nm}$ have to satisfy.

Note that for one phase flow $\bar{v}_z$ is given by: $\bar{v}_z = 1 - \bar{r}^2$

For a fixed $\bar{r}$, the above expansion of $\bar{v}_z$ corresponds to a Fourier series expansion.

# CHAPTER 10

# SPECIAL TOPICS

## 10.1 INTRODUCTION

It is not uncommon nowadays to require students in engineering to take a course in statistical mechanics. The present chapter does not intend to provide such a course. Our aim is more modest. We plan to provide the necessary background for the students to be able to follow with ease a more demanding course in statistical mechanics.

In the first few sections, we shall extend the analytical mechanics introduced in Chapter 9. Hamiltonian mechanics will be introduced. This can also be used to pave the way for a course in chaos, a topic which is gaining widespread popularity. Next, we shall consider probability. It will be at an elementary level, but we shall emphasize the interpretation of probability. Students used to deterministic concepts find it hard to think in terms of probability. The idea of an ensemble average will be discussed. Finally, some topics in thermodynamics and Brownian motion will be examined.

## 10.2 PHASE SPACE

Consider a system consisting of $N$ particles. Let the vector position of particle $i$ be denoted by $\underline{r}_i$ relative to an origin $O$. In a three-dimensional space, each particle is defined by three coordinates. If there is no constraint between the particles, we need $3N$ quantities to describe the configuration of the system. We denote these $3N$ quantities by $q^1$, $q^2$, ... , $q^{3N}$. The $q^i$ are independent variables and are known as **generalized coordinates**. We may regard the system to be in a $3N$ dimensional space and to have $3N$ degrees of freedom. If there are $m$ constraints, the number of generalized coordinates will be reduced to $3N - m$. The generalized coordinates, unlike the Cartesian coordinates need not all have the same physical dimension. The choice of the generalized coordinates depends on the geometry of the problem under consideration. Thus if a particle is forced to move on the curved surface of a cylinder of radius $a$, as shown in Figure 10.2-1, the only two generalized coordinates are

$$q^1 = z , \qquad q^2 = \theta \qquad\qquad (10.2\text{-}1a,b)$$

If the particle is constrained to move on a flat surface of the cylinder, say the surface $z = h$, then the two generalized coordinates are, as shown in Figure 10.2-1,

$$q^1 = r , \qquad q^2 = \theta \qquad\qquad (10.2\text{-}2a,b)$$

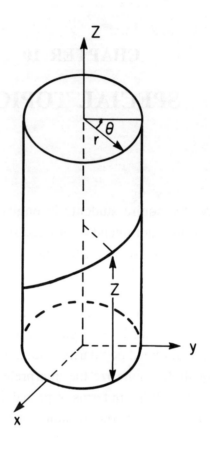

**FIGURE 10.2-1    Motion of a particle on the curved and flat surface
of a circular cylinder**

We now consider the system to have  n  (= 3N – m)  degrees of freedom and we have  n  generalized

coordinates which are related to the  N  vector positions  $\underline{r}_1$,  $\underline{r}_2$, ... ,  $\underline{r}_N$.  The relationship can be

expressed as (see Section 4.6)

$$\underline{r}_1 = \underline{r}_1 (q^1, q^2, ... , q^n, t) \tag{10.2-3a}$$

$$\vdots$$

$$\underline{r}_N = \underline{r}_N (q^1, q^2, ... , q^n, t) \tag{10.2-3b}$$

where  t  is the time.

The velocity of particle  i  can be obtained by differentiating  $\underline{r}_i$  and we obtain

$$\underline{v}_i = \dot{\underline{r}}_i = \frac{\partial \underline{r}_i}{\partial q^j} \dot{q}^j + \frac{\partial \underline{r}_i}{\partial t} \qquad (10.2\text{-}4a,b)$$

In Equation (10.2-4b), we have adopted the summation convention.

Note that the indices on the vectors $\underline{v}_i$ and $\underline{r}_i$ appear as subscripts, as they refer to a particle $i$.

Following the notation introduced in Chapter 4, the indices denoting generalized coordinates are written as superscripts. Some authors dealing with analytical mechanics, adopt a different notation. They employ subscripts throughout. As they consider mostly scalar quantities such as kinetic and potential energies, there is less of a compelling need to distinguish between subscripts and superscripts.

The kinetic energy of the system is given by

$$K = \frac{1}{2} \sum_i m_i \, \underline{v}_i \cdot \underline{v}_i \qquad (10.2\text{-}5a)$$

$$= \frac{1}{2} \sum_i m_i \left\{ \sum_{j,\,k} \frac{\partial \underline{r}_i}{\partial q^j} \cdot \frac{\partial \underline{r}_i}{\partial q^k} \, \dot{q}^j \dot{q}^k + \frac{\partial \underline{r}_i}{\partial t} \cdot \frac{\partial \underline{r}_i}{\partial t} \right\} \qquad (10.2\text{-}5b)$$

In Example 9.10-3, we have deduced **Lagrange's equation** which is

$$\frac{\partial L}{\partial q^i} - \frac{d}{dt} \left( \frac{\partial L}{\partial \dot{q}^i} \right) = 0 \qquad (9.10\text{-}34)$$

where $L = K - \varphi$ is the Lagrangian and $\varphi$ is the **generalized potential**.

Consider a single particle of mass $m$ moving in a potential field which depends on the position only, then in Cartesian coordinates $(x, y, z)$, $L$ is given by

$$L = \frac{1}{2} m (\dot{x}^2 + \dot{y}^2 + \dot{z}^2) - \varphi \qquad (10.2\text{-}6)$$

It follows that

$$\frac{\partial L}{\partial \dot{x}} = m \dot{x} = p_x \qquad (10.2\text{-}7a,b)$$

In Equation (10.2-7b), $p_x$ is the x-component of the momentum.

This suggests that a **generalized momentum** associated with the $q^i$ coordinate can be defined as

$$p^i = \frac{\partial L}{\partial \dot{q}^i} \tag{10.2-8}$$

If the coordinate $q^j$ does not appear explicitly in $L$ ($\dot{q}^j$ may be present in $L$), $q^j$ is known as a **cyclic (or ignorable) coordinate**. Equation (9.10-34) simplifies to

$$\frac{\partial L}{\partial \dot{q}^j} = p^j = \text{constant} \tag{10.2-9}$$

Thus the generalized momentum of a cyclic coordinate is a constant.

The configuration of the system is specified by the $n$ generalized coordinates $q^1, \dots, q^n$ and the momenta (mass $\times$ velocity) by the $n$ generalized momenta $p^1, \dots, p^n$. The **dynamical state** of the system is thus defined by the $2n$ quantities, $q^1, \dots, q^n, p^1, \dots, p^n$, which defines a space of $2n$ dimensions. The motion of the system is thus described by a path (trajectory or orbit) in this $2n$-dimensional space, known as the **phase space**.

***Example 10.2-1.*** A particle of mass $m$ is attached to the end of a linear spring of natural length $\ell_0$ and of negligible mass. The spring is fixed at the other end and lies on a smooth horizontal table. The mass is displaced from its equilibrium by a distance $q_0$ along the length of the spring and is then released from rest. Determine the subsequent motion.

Let $q$ be the displacement of the particle from its equilibrium position at any time $t$. The kinetic energy

$$K = \frac{1}{2} m (\dot{q})^2 \tag{10.2-10}$$

The potential energy is

$$\varphi = \frac{1}{2} G (q)^2 \tag{10.2-11}$$

where $G$ is the modulus of the spring.

The Lagrangian is given by

$$L = \frac{1}{2} [m (\dot{q})^2 - G(q)^2] \tag{10.2-12}$$

The equation of motion is

$$\frac{1}{2} \frac{\partial}{\partial q} [m(\dot{q})^2 - G(q)^2] - \frac{1}{2} \frac{d}{dt} \left\{ \frac{\partial}{\partial \dot{q}} [m(\dot{q})^2 - G(q)^2] \right\} = 0 \qquad (10.2\text{-}13a)$$

or $\quad -Gq - \dfrac{d}{dt}(m\dot{q}) = 0$ $\qquad\qquad\qquad\qquad\qquad\qquad\qquad$ (10.2-13b)

or $\quad \ddot{q} + \omega^2 q = 0$ $\qquad\qquad\qquad\qquad\qquad\qquad\qquad\qquad\qquad$ (10.2-13c)

where $\omega^2 = \dfrac{G}{m}$.

Equation (10.2-13c) is the simple harmonic equation and is the equation governing the motion of all linear oscillators.

The solution of Equation (10.2-13c) is

$$q = A \cos \omega t + B \sin \omega t \qquad\qquad\qquad\qquad\qquad\qquad (10.2\text{-}14a)$$

$$\dot{q} = -A \omega \sin \omega t + B \omega \cos \omega t \qquad\qquad\qquad\qquad\qquad (10.2\text{-}14b)$$

where A and B are constants.

Using the initial conditions

$$q = q_0 \qquad\qquad\qquad\qquad\qquad\qquad\qquad\qquad\qquad\qquad (10.2\text{-}15a)$$

$$\dot{q} = 0 \qquad\qquad\qquad\qquad\qquad\qquad\qquad\qquad\qquad\qquad\quad (10.2\text{-}15b)$$

we obtain

$$q = q_0 \cos \omega t \qquad\qquad\qquad\qquad\qquad\qquad\qquad\qquad (10.2\text{-}16a)$$

$$p = \frac{\partial L}{\partial \dot{q}} = m\dot{q} = -m \omega q_0 \sin \omega t \qquad\qquad\qquad\qquad (10.2\text{-}16b,c,d)$$

Eliminating the time t between Equations (10.2-16a, d) yields

$$\left(\frac{q}{q_0}\right)^2 + \left(\frac{p}{m q_0 \omega}\right)^2 = 1 \qquad\qquad\qquad\qquad\qquad (10.2\text{-}17)$$

Thus the trajectory in the phase space is an ellipse (closed curve), as shown in Figure 10.2-2. At any time t, the position and momentum of the particle is given by a point on the ellipse. Thus the ellipse describes the dynamical state of the particle (or oscillator). All oscillators with the same total energy describe the same ellipse. The starting point on the ellipse depends on the initial conditions. If we have  n  oscillators not necessarily all with the same energy, we have  2n  variables  $(q^1, p^1), \dots ,$ $(q^n, p^n)$. Each set of  $(q^i, p^i)$  corresponds to an oscillator.

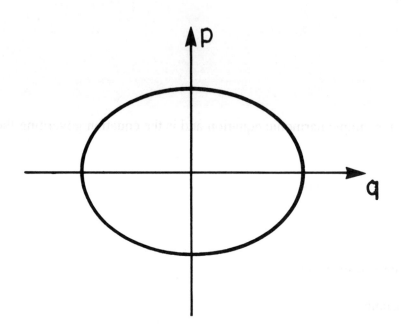

**FIGURE 10.2-2   The  trajectory  of  an  oscillator**

## 10.3  HAMILTON'S EQUATIONS OF MOTION

In Lagrangian dynamics, the variables are  $(q^i, \dot{q}^i, t)$. In the phase space, the variables are $(q^i, p^i)$. It is desirable to transform from the  $(q^i, \dot{q}^i, t)$  system to the  $(q^i, p^i, t)$  system. This can be accomplished by the use of the **Legendre transformation**. A new function  H  known as the **Hamiltonian** is defined as

$$H\,(p^j, q^j, t) \;=\; p^m\,\dot{q}^m - L\,(q^j, \dot{q}^j, t)\,, \qquad j = 1, 2, \dots, n \qquad\qquad (10.3\text{-}1)$$

and we sum over  m.

The differential of  H  is given by

$$dH \;=\; \frac{\partial H}{\partial t} + \frac{\partial H}{\partial q^j}\,dq^j + \frac{\partial H}{\partial p^j}\,dp^j \qquad\qquad (10.3\text{-}2)$$

Differentiating Equation (10.3-1), we have

$$dH = p^m \, d\dot{q}^m + \dot{q}^m \, dp^m - \frac{\partial L}{\partial q^j} \, dq^j - \frac{\partial L}{\partial \dot{q}^j} \, d\dot{q}^j - \frac{\partial L}{\partial t} \qquad (10.3\text{-}3a)$$

$$= \left( p^j - \frac{\partial L}{\partial \dot{q}^j} \right) d\dot{q}^j + \dot{q}^m \, dp^m - \frac{\partial L}{\partial q^j} \, dq^j - \frac{\partial L}{\partial t} \qquad (10.3\text{-}3b)$$

$$= \dot{q}^m \, dp^m - \frac{\partial L}{\partial q^j} \, dq^j - \frac{\partial L}{\partial t} \qquad (10.3\text{-}3c)$$

Equation (10.3-3c) is obtained using Equation (10.2-8).

From Equations (9.10-34, 10.2-8), we have

$$\frac{\partial L}{\partial q^j} = \frac{d}{dt} \left( \frac{\partial L}{\partial \dot{q}^j} \right) \qquad (10.3\text{-}4a)$$

$$= \frac{d}{dt} (p^j) = \dot{p}^j \qquad (10.3\text{-}4b,c)$$

Substituting Equation (10.3-4c) into Equation (10.3-3c) yields

$$dH = \dot{q}^j \, dp^j - \dot{p}^j \, dq^j - \frac{\partial L}{\partial t} \qquad (10.3\text{-}5)$$

Comparing Equations (10.3-2, 5), we deduce

$$\dot{q}^j = \frac{\partial H}{\partial p^j} \qquad (10.3\text{-}6a)$$

$$\dot{p}^j = -\frac{\partial H}{\partial q^j} \qquad (10.3\text{-}6b)$$

$$-\frac{\partial L}{\partial t} = \frac{\partial H}{\partial t} \qquad (10.3\text{-}6c)$$

Equations (10.3-6a to c) are known as **Hamilton's equations of motion**. They are a set of first order equations, whereas Lagrange's equations are second order equations. Note also the symmetry in $p^j$ and $q^j$. We now consider the case where $t$ does not occur explicitly in Equation (10.2-3). Such a situation may arise if the constraints are time independent. Then from Equation (10.2-5b), we deduce that $K$ is a quadratic function in $\dot{q}^j$. We also assume that the system is conservative, then $\varphi$ is a function of $q^j$ but not of $\dot{q}^j$. It then follows that Equation (10.2-8) may be written as

$$p^m = \frac{\partial}{\partial \dot{q}^m}(K - \varphi) = \frac{\partial K}{\partial \dot{q}^m} \qquad (10.3\text{-}7a,b)$$

$K$ is a quadratic in $\dot{q}^m$. Using Euler's theorem (Section 1.7), we have

$$\dot{q}^m p^m = \dot{q}^m \frac{\partial K}{\partial \dot{q}^m} = 2K \qquad (10.3\text{-}8a,b)$$

Substituting Equation (10.3-8b) into Equation (10.3-1) yields

$$H = 2K - (K - \varphi) \qquad (10.3\text{-}9a)$$

$$= K + \varphi \qquad (10.3\text{-}9b)$$

Thus the Hamiltonian $H$ represents the total energy of the system and is a constant for the system under consideration.

***Example 10.3-1***. Obtain the Hamiltonian $H$ from Equation (10.3-1) for the oscillator described in Example 10.2-1. Write down Hamilton's equations of motion and show that they reduced to Equation (10.2-13c).

The Lagrangian $L$ is given by Equation (10.2-12). The generalized momentum $p$ is given by

$$p = \frac{\partial L}{\partial \dot{q}} = m\dot{q} \qquad (10.3\text{-}10a,b)$$

Thus $H$ is given by

$$H = p\dot{q} - L = m(\dot{q})^2 - \frac{1}{2}[m(\dot{q})^2 - G(q)^2] \qquad (10.3\text{-}11a,b)$$

$$= \frac{1}{2}[m(\dot{q})^2 + G(q)^2] \qquad (10.3\text{-}11c)$$

$$= \frac{1}{2}\left[\frac{(p)^2}{m} + G(q)^2\right] \qquad (10.3\text{-}11d)$$

From Equations (10.3-6a, b), we have

$$\dot{q} = \frac{p}{m} \qquad (10.3\text{-}12a)$$

$$\dot{p} = -Gq \qquad (10.3\text{-}12b)$$

Eliminating p between Equations (10.3-12a, b), we obtain the result. Equation (10.2-17) can also be deduced from Equation (10.3-11d). Since H is a constant, Equation (10.3-11d) may be written as

$$Gq^2 + \frac{p^2}{m} = C_0 \tag{10.3-13}$$

where $C_0$ is a constant depending on the initial conditions.

At time $t = 0$, we have

$$p = 0, \qquad q = q_0 \tag{10.3-14a,b}$$

and $\quad C_0 = Gq_0^2 \tag{10.3-14c}$

Substituting Equation (10.3-14c) into Equation (10.3-13) and dividing both sides of the resulting equation by $Gq_0^2$, we obtain Equation (10.2-17). From Equations (10.3-13, 14c), we observe that the level of energy determines the ellipse.

## 10.4  POISSON BRACKETS

Let f be a function of $p^1, p^2, \dots, p^n, q^1, q^2, \dots, q^n$ and t. Then

$$\frac{df}{dt} = \frac{\partial f}{\partial t} + \frac{\partial f}{\partial p^i} \dot{p}^i + \frac{\partial f}{\partial q^i} \dot{q}^i \tag{10.4-1}$$

Combining Equations (10.3-6a, b, 4-1) yields

$$\frac{df}{dt} = \frac{\partial f}{\partial t} + \frac{\partial f}{\partial q^i} \frac{\partial H}{\partial p^i} - \frac{\partial f}{\partial p^i} \frac{\partial H}{\partial q^i} \tag{10.4-2a}$$

$$= \frac{\partial f}{\partial t} + [H, f] \tag{10.4-2b}$$

where

$$[H, f] = \frac{\partial f}{\partial q^i} \frac{\partial H}{\partial p^i} - \frac{\partial f}{\partial p^i} \frac{\partial H}{\partial q^i} \tag{10.4-2c}$$

is the **Poisson bracket** of f with H.

We note that the Poisson bracket is the Jacobian of the transformation (H, f) to $(p^i, q^i)$. If f does not depend explicitly on t and the Poisson bracket [H, f] is zero, we deduce from Equation (10.4-2b) that f is a **constant of motion**. The Hamiltonian and the total energy are examples of constants of motion. Setting f = H into Equation (10.4-2c), we observe as expected that H is a constant of motion. The converse is also true. The Poisson brackets of all constants of motion with H vanish. This provides a test for constants of motion.

Hamilton's equations of motion, Equations (10.3-6a, b), may be written in terms of Poisson brackets. From Equation (10.4-2c), we obtain

$$\left[H, p^j\right] = -\delta_{ij} \frac{\partial H}{\partial q^i} = -\frac{\partial H}{\partial q^j} \qquad (10.4\text{-}3a,b)$$

$$\left[H, q^j\right] = \delta_{ij} \frac{\partial H}{\partial p^i} = \frac{\partial H}{\partial p^j} \qquad (10.4\text{-}3c,d)$$

Combining Equations (10.3-6a, b, 4-3a to d) yields

$$\dot{q}^j = \left[H, q^j\right] \qquad (10.4\text{-}4a)$$

$$\dot{p}^j = \left[H, p^j\right] \qquad (10.4\text{-}4b)$$

The Poisson bracket, for any pair of dynamical quantities f and g, is defined as

$$[g, f] = \frac{\partial g}{\partial p^i} \frac{\partial f}{\partial q^i} - \frac{\partial g}{\partial q^i} \frac{\partial f}{\partial p^i} \qquad (10.4\text{-}5)$$

From Equation (10.4-5), we deduce that, for any three quantities f, g and h, we have

$$[g, f] = -[f, g] \qquad (10.4\text{-}6a)$$

$$[f + g, h] = [f, h] + [g, h] \qquad (10.4\text{-}6b)$$

$$[fg, h] = f[g, h] + g[f, h] \qquad (10.4\text{-}6c)$$

$$[f, [g, h]] + [g, [h, f]] + [h, [f, g]] = 0 \qquad (10.4\text{-}6d)$$

$$[p^j, p^k] = 0 \qquad (10.4\text{-}6e)$$

$$[q^j, q^k] = 0 \qquad (10.4\text{-}6f)$$

$$[p^j, q^k] = \delta_{jk} \qquad (10.4\text{-}6g)$$

Equation (10.4-6d) is known as **Jacobi's identity**. In theory it is possible to construct from this identity a constant of motion, if two constants of motion are known. Thus if in Equation (10.4-6d), we put h = H (Hamiltonian), f and g are two constants of motion and their Poisson brackets with H vanish. Equation (10.4-6d) reduces to

$$[H, [f, g]] = 0 \qquad (10.4\text{-}7)$$

Equation (10.4-7) shows that $[f, g]$, the Poisson bracket of f with g, is a constant of motion. Thus the Poisson bracket of a constant of motion with another constant of motion yields another constant of motion. However, often the constants of motion so obtained are trivial functions of the original constants of motion. The properties of Poisson brackets as given by Equations (10.4-6a to g) are the properties of **Lie algebra** [see Lipkin (1985)]. The Poisson bracket corresponds to the commutator in quantum mechanics.

## 10.5  CANONICAL TRANSFORMATIONS

In Section 10.2, we have shown that if a generalized coordinate is cyclic, its corresponding momentum is a constant. Thus it will be desirable to transform all the generalized coordinates to cyclic coordinates. Normally the most obvious generalized coordinates will not be cyclic. We should then look for a set of generalized coordinates which contains the largest subset of cyclic coordinates. In the Hamiltonian formulation, both the generalized coordinates and the generalized momenta have the same status. Thus we shall consider the transformation from the set $(p^i, q^i)$ to a new set $(P^i, Q^i)$ given by

$$P^i = P^i (q^1, \dots, q^n, p^1, \dots, p^n, t) \qquad (10.5\text{-}1a)$$

$$Q^i = Q^i (q^1, \dots, q^n, p^1, \dots, p^n, t) \qquad (10.5\text{-}1b)$$

We also require that the new $P^i$ and $Q^i$ satisfy Hamilton's equations of motion. That is to say, there exists a function $M (P^i, Q^i)$ such that

$$\dot{Q}^i = \frac{\partial M}{\partial P^i} \qquad (10.5\text{-}2a)$$

$$\dot{P}^i = -\frac{\partial M}{\partial Q^i} \qquad (10.5\text{-}2b)$$

Transformations given by Equations (10.5-2a, b) are known as **canonical transformations**.

If now all the $Q^i$ are cyclic coordinates, then M is a function of the $P^i$ only and the $P^i$ are constants. Equation (10.5-2a) can then be integrated to yield

$$Q^i = f_i (P^1, \dots, P^n) \, t + c_i \qquad (10.5\text{-}3)$$

where $f_i = \dfrac{\partial M}{\partial P^i}$ and $c_i$ is a set of arbitrary constants to be determined from the initial conditions.

Thus if we can find the momenta $P^i$, the coordinates $Q^i$ as well as $M$, we can integrate the equations of motion. Those quantities can be obtained via the so-called **generating function** $F$.

Since the sets $\{p^i, q^i\}$ and $\{P^i, Q^i\}$ both satisfy Hamilton's equations, we expect them to satisfy Hamilton's principle which is given in Example 9.10-3. Using Equation (10.3-1), Hamilton's principle can be expressed as

$$\delta I = \delta \int_{t_1}^{t_2} \left\{ p^m \, \dot{q}^m - H \, (p^j, q^j, t) \right\} dt = 0 \qquad (10.5\text{-}4a)$$

$$\delta I = \delta \int_{t_1}^{t_2} \left\{ P^m \, \dot{Q}^m - M \, (P^j, Q^j, t) \right\} dt = 0 \qquad (10.5\text{-}4b)$$

The two integrals given in Equations (10.5-4a, b) can differ at most by an arbitrary function of the form $\dfrac{dF}{dt}$. This can be seen by observing that on integrating $\dfrac{dF}{dt}$ with respect to $t$ between the limits we obtain $F(t_2) - F(t_1)$ and since the variation at the end points is zero, the variation of the integral $\dfrac{dF}{dt}$ is zero. Thus we have

$$p^m \, \dot{q}^m - H = P^m \, \dot{Q}^m - M + \frac{dF}{dt} \qquad (10.5\text{-}5)$$

where $F$ is the **generating function**.

The generating function $F$ can be a function of $(4n + 1)$ variables, the old set $\{p^i, q^i\}$, the new set $\{P^i, Q^i\}$ and the time $t$. Equations (10.5-1a, b) impose $2n$ constraints, so that we only have $(2n + 1)$ independent variables. The generating function can be one of the following forms

$$F = F_1 \, (q^j, Q^j, t), \qquad F = F_2 \, (q^j, P^j, t), \qquad (10.5\text{-}6a,b)$$

$$F = F_3 \, (p^j, Q^j, t), \qquad F = F_4 \, (p^j, P^j, t) \qquad (10.5\text{-}6c,d)$$

The form of $F$ to be chosen depends on the problem.

If the form $F_1$ is chosen, then

$$\frac{dF_1}{dt} = \frac{\partial F_1}{\partial t} + \frac{\partial F_1}{\partial q^j} \, \dot{q}^j + \frac{\partial F_1}{\partial Q^j} \, \dot{Q}^j \qquad (10.5\text{-}7)$$

Substituting Equation (10.5-7) into Equation (10.5-5) yields

$$p^m \, \dot{q}^m - H = P^m \, \dot{Q}^m - M + \frac{\partial F_1}{\partial t} + \frac{\partial F_1}{\partial q^m} \dot{q}^m + \frac{\partial F_1}{\partial Q^m} \dot{Q}^m \qquad (10.5\text{-}8)$$

On the right side of the Equation (10.5-8), we have changed the dummy index from $j$ to $m$. Since we consider the $q^m$ and $Q^m$ to be independent, the coefficients of $\dot{q}^m$ and $\dot{Q}^m$ in Equation (10.5-8) must be identically zero. We thus have

$$p^m = \frac{\partial F_1}{\partial q^m} \qquad (10.5\text{-}9a)$$

$$P^m = -\frac{\partial F_1}{\partial Q^m} \qquad (10.5\text{-}9b)$$

$$H = M - \frac{\partial F_1}{\partial t} \qquad (10.5\text{-}9c)$$

The other generating functions can be obtained by using the Legendre transformation and is described in Goldstein (1972). We note that if $F$ is not an explicit function of time, then

$$H = M \qquad (10.5\text{-}10)$$

**Example 10.5-1.** Obtain a cyclic coordinate for the problem considered in Example 10.2-1. The Hamiltonian $H$ is given by

$$H = \frac{1}{2}\left[ \frac{(p)^2}{m} + G(q)^2 \right] \qquad (10.5\text{-}11)$$

Note that $q$ is not a cyclic coordinate.

We choose a generating function of type $F_1$ and write

$$F_1 = \frac{m\omega}{2}(q)^2 \cot Q \qquad (10.5\text{-}12)$$

where $\omega$ is defined in Equation (10.2-13c).

From Equations (10.5-9a to c), we obtain

$$p = \frac{\partial F_1}{\partial q} = m\omega q \cot Q \qquad (10.5\text{-}13a,b)$$

$$P = -\frac{\partial F_1}{\partial Q} = \frac{m\omega(q)^2}{2\sin^2 Q} \qquad (10.5\text{-}13c,d)$$

$$H = M \tag{10.5-13e}$$

Equations (10.5-13a to d) may be rewritten as

$$(q)^2 = \frac{2P \sin^2 Q}{m\omega} \tag{10.5-14a}$$

$$(p)^2 = 2m\omega P \cos^2 Q \tag{10.5-14b}$$

Combining Equations (10.5-11, 13e, 14 a, b) yields

$$M = \frac{1}{2}\left(2\omega P \cos^2 Q + 2\omega P \sin^2 Q\right) \tag{10.5-15a}$$

$$= \omega P \tag{10.5-15b}$$

Thus $Q$ does not appear explicitly in $M$ and is a cyclic coordinate.

From Equation (10.5-2a), we obtain

$$\dot{Q} = \frac{\partial M}{\partial P} = \omega \tag{10.5-16a,b}$$

Integrating Equations (10.5-16a,b) yields

$$Q = \omega t + c \tag{10.5-17}$$

where $c$ is an arbitrary constant to be determined from initial condition.

From Equation (10.5-2b), we deduce that $P$ is a constant and we denote it by $\delta_0$. From Equations (10.5-13c, d), we then deduce

$$(q)^2 = \frac{2\delta_0 \sin^2 Q}{m\omega} \tag{10.5-18}$$

Substituting Equation (10.5-17) into Equation (10.5-18) yields

$$q = \sqrt{\frac{2\delta_0}{m\omega}} \; \sin(\omega t + c) \tag{10.5-19a}$$

$$= \sqrt{\frac{2\delta_0}{m\omega}} \; [\sin\omega t \, \cos c + \cos\omega t \, \sin c] \tag{10.5-19b}$$

Equation (10.5-19b) is identical to Equation (10.2-14a) with

$$A = \sqrt{\frac{2\delta_0}{m\omega}} \sin c \qquad\qquad\qquad (10.5\text{-}20\text{a})$$

$$B = \sqrt{\frac{2\delta_0}{m\omega}} \cos c \qquad\qquad\qquad (10.5\text{-}20\text{b})$$

Note that in Example (10.2-1), we had to solve a differential equation, as opposed to using a canonical transformation in the present example. Such a transformation obviates solving a differential equation.

## 10.6  LIOUVILLE'S THEOREM

The dynamical evolution of a particle is given by a trajectory in the phase space and any point on that trajectory corresponds to the state of that particle at a particular time. Thus if the position and momentum of a particle are known at a given instant of time, which can be taken to be the initial time, the motion of that particle is determined. Similarly if the system consists of $N$ particles and has $n$ degrees of freedom, we need $2n$ initial conditions, namely: $p^i(0)$ and $q^i(0)$ ($i = 1, 2, \ldots, n$). We also need to solve $2n$ equations of motion so as to obtain the trajectory in the phase space, which is a $2n$-dimensional space. If $n$ is very large, it is very unlikely that we can accurately determine the $2n$ initial conditions. Further it will be time consuming to solve the $2n$ first order equations of motion. Thus we do not have a precise knowledge of the system. This leads us to the concept of an **ensemble** introduced by Gibbs.

We consider the system we are investigating to be made up of a large number of subsystems; that is to say, we consider an ensemble of subsystems such that each subsystem of the ensemble has the same structure as the original system. We can associate a trajectory in the phase space to the motion of a particular subsystem of the ensemble. No trajectories may intersect since this would violate the uniqueness of the system. That is to say, reversing the motion from a point of intersection would lead to the unacceptable possibility of more than one motion for the subsystems. We may also define a density $\rho$ in the phase space which describes the condition of the ensemble. If $N$ is the number of particles in the actual system, we write

$$N = \int \rho \, dp^1 \ldots dp^n \, dq^1 \ldots dq^n \qquad\qquad\qquad (10.6\text{-}1)$$

The **density** $\rho$ is a function of the generalized coordinates $q^1, \ldots, q^n$, momenta $p^1, \ldots, p^n$ and time $t$. We have assumed that the systems in the ensemble are numerous enough so that $\rho$ can be considered to be a continuous function. The integral in Equation (10.6-1) is a volume integral in the $2n$-space.

Let $\Omega$ be a volume fixed in the phase space and S be the surface enclosing $\Omega$. Let $\underline{r}$ be a vector whose components are $q^1, \ldots, q^n, p^1, \ldots, p^n$, that is to say, $\underline{r}$ is a vector in the 2n-space. The law of conservation of subsystems in the ensemble leads to

$$\frac{\partial}{\partial t} \int_{\Omega} \rho \, dr_1 \ldots dr_{2n} = -\int_{S} \rho \, \underline{\dot{r}} \cdot \underline{n} \, dS \tag{10.6-2}$$

where $\underline{n}$ is the unit outward normal to the surface S.

Applying the divergence theorem to Equation (10.6-2), we obtain

$$\frac{\partial}{\partial t} \int_{\Omega} \rho \, dr_1 \ldots dr_{2n} + \int_{\Omega} \text{div} \, (\rho \, \underline{\dot{r}}) \, dr_1 \ldots dr_{2n} = 0 \tag{10.6-3}$$

Since $\Omega$ is fixed in space and is arbitrary, Equation (10.6-3) implies

$$\frac{\partial \rho}{\partial t} + \frac{\partial}{\partial r_i} (\rho \, \underline{\dot{r}}_i) = 0 \tag{10.6-4}$$

In Equation (10.6-4), we have used the summation convention and we have sum from $i = 1$ to $i = 2n$. Replacing $r_1, \ldots, r_{2n}$ by $q^1, \ldots, q^n, p^1, \ldots, p^n$, Equation (10.6-4) becomes

$$\frac{\partial \rho}{\partial t} + \frac{\partial}{\partial q^i} (\rho \, \dot{q}^i) + \frac{\partial}{\partial p^i} (\rho \, \dot{p}^i) = 0 \tag{10.6-5a}$$

or 
$$\frac{\partial \rho}{\partial t} + \dot{q}^i \frac{\partial \rho}{\partial q^i} + \dot{p}^i \frac{\partial \rho}{\partial p^i} + \rho \left( \frac{\partial \dot{q}^i}{\partial q^i} + \frac{\partial \dot{p}^i}{\partial p^i} \right) = 0 \tag{10.6-5b}$$

From Equation (10.3-6a,b) we have

$$\frac{\partial \dot{q}^i}{\partial q^i} = \frac{\partial^2 H}{\partial p^i \, \partial q^i} = -\frac{\partial \dot{p}^i}{\partial p^i} \tag{10.6-6a,b}$$

It follows that

$$\frac{\partial \dot{q}^i}{\partial q^i} + \frac{\partial \dot{p}^i}{\partial p^i} = 0 \tag{10.6-7}$$

Note the similarity of Equation (10.6-7) to the equation of continuity in fluid dynamics. Combining Equations (10.6-5b, 7) yields

$$\frac{\partial \rho}{\partial t} + \dot{q}^i \frac{\partial \rho}{\partial q^i} + \dot{p}^i \frac{\partial \rho}{\partial p^i} = 0 \qquad (10.6\text{-}8)$$

Since $\rho$ is a function of $q^i$, $p^i$ and t, Equation (10.6-8) can be written as

$$\frac{D\rho}{Dt} = 0 \qquad (10.6\text{-}9)$$

where $\frac{D}{Dt}$ is the **material derivative** (total or substantial derivative).

Thus along a trajectory, $\rho$ is a constant.

Using Equations (10.3-6a, b, 4-2c), Equation (10.6-8) can be written as

$$\frac{\partial \rho}{\partial t} + \frac{\partial H}{\partial p^i} \frac{\partial \rho}{\partial q^i} - \frac{\partial H}{\partial q^i} \frac{\partial \rho}{\partial p^i} = 0 \qquad (10.6\text{-}10a)$$

$$\frac{\partial \rho}{\partial t} + [H, \rho] = 0 \qquad (10.6\text{-}10b)$$

If the system is in equilibrium, $\rho$ is not an explicit function of t and Equation (10.6-10b) reduces to

$$[H, \rho] = 0 \qquad (10.6\text{-}11)$$

Thus in the state of equilibrium, $\rho$ is a constant of motion. The total energy or the Hamiltonian H are constants of motion. Functions of these quantities are possible forms for $\rho$.

Equation (10.6-8) expresses **Liouville's theorem**. It is the analogue of the conservation of mass in continuum mechanics and is a fundamental theorem in statistical mechanics.

It should be pointed out that Liouville's theorem is valid only in the phase space. There is no equivalent theorem in the configuration or momentum space. This suggests that the Hamiltonian formalism is the most appropriate one in statistical mechanics.

The ideas and methods used to solve problems in Hamiltonian mechanics can be extended to other fields of science and technology. Chemical reactions, for example, are modeled by kinetic rate equations, which form a set of first order equations, with time as the independent variable. They are not unlike Hamilton's equations of motion. If we consider n chemical species which we denote by $A_1, \ldots, A_n$ and the mass of $A_i$ by $m_{A_i}$, we have n equations involving n, $m_{A_i}$, and time. Since the total mass is conserved, we have one constraint and the number of degrees of freedom is $n - 1$. The reaction can be described in the $(n - 1)$-dimensional space, known as the reaction space, with coordinates $m_{A_1}, \ldots, m_{A_{n-1}}$ and the time has been eliminated. The reaction space corresponds to the phase space. Further, the equations are coupled and are difficult to solve. We therefore look for equivalent generalized coordinates. In this case we choose a transformation that will transform $m_{A_i}$ to $b_i$ so that the equations will be decoupled. If the equations are linear, the required transformation is the transformation that will diagonalise the rate-constant matrix. Wei and Prater (1962) have eloquently shown how this can be done.

In recent years, there is a growing acceptance of oscillating chemical reactions and the occurrence of chaos in non-equilibrium chemical reactions [Epstein (1983), Roux (1983)]. A knowledge of Hamiltonian mechanics can be helpful in the understanding of these processes.

## 10.7 DISCRETE PROBABILITY THEORY

The theory of probability was started in the 17th century by Fermat, Pascal, Leibnitz and Bernouilli and was associated with gambling. It is now applied in many fields of science, technology and commerce.

We shall start from the simplest experiment which is tossing a coin. The outcome of this experiment is either head or tail. If the coin is unbiased, then it is as likely to fall on its head as on its tail. The two events are equally likely and we assign to each the probability of 1/2. Two events are said to be independent if they do not interact. If we toss a coin twice, the outcome of the first toss does not influence the outcome of the second toss. Thus the first toss and the second toss are independent. If we denote two events by $A$ and $B$ and if the probability that event $A$ occurs is $p(A)$ and that event $B$ occurs is $p(B)$ then the probability that both events occur, which we denote by $AB$, is the product $p(A) p(B)$. The probability of having two successive heads on tossing a coin twice is 1/4.

Two events $A$, $B$ are said to be **mutually exclusive** if when event $A$ occurs, event $B$ cannot occur. In the experiment of tossing a coin, the events of the coin landing on its head and landing on its tail are two mutually exclusive events. The two events cannot occur simultaneously. If $A$ and $B$ are two mutually exclusive events then the probability of $AB$ is zero, that is to say, event $AB$ cannot occur. We assign the value zero to the probability of an event which cannot happen.

We denote by $A \cup B$ the event that either $A$ or $B$ or both occur. If $A$ and $B$ are mutually exclusive, then the probability of $A \cup B$ is the sum of the probabilities of $A$ and $B$. In our coin experiment, if $A$ is the event of landing on its head and $B$ on its tail, the probability of $A \cup B$ is one, the sum of two halves. The probability of certainty is one. We have excluded the possibility that the coin may roll away. It must fall either on its head or on its tail. Thus the event $A \cup B$ is certain to occur.

The basic rules can be summarized as follows

(a)    $p(A)$ denotes the probability that event $A$ occurs;

(b)    $p(A) \geq 0$;

(c)    $p(A) = 0$, event $A$ does not occur;

(d)    $p(A) = 1$, event $A$ is certain to occur;

(e)    $p(AB) = p(A) \, p(B)$, if $A$ and $B$ are **independent**;

(f)    $p(A \cup B) = p(A) + p(B)$, if $A$ and $B$ are mutually exclusive;

(g)    $p(A \cup B) = p(A) + p(B) - p(AB)$, if $A$ and $B$ are not mutually exclusive;

The **conditional probability** of $A$ given $B$ is defined as the probability of $A$ occurring given that $B$ has happened and is denoted by $p(A|B)$.

(h)    $p(A|B) = \dfrac{p(AB)}{p(B)}$

If $A$ and $B$ are independent, then $p(A|B) = p(A)$.

To compute the probability of an event $A$ we have to define the sample space. Since the sample space is discrete, we can determine all the elements of the space. The probability of $A$ is the ratio of the number of elements of $A$ to the total number of elements of the space. We have thus used the **relative frequency** of event $A$ as its probability. This is the definition in practice. In theory we need the number of elements in the sample space to tend to infinity. To facilitate this computation we state briefly the rules of **combination** and **permutation**.

Suppose we have $n$ elements $a_1, \ldots, a_n$ and out of that set of $n$ elements we want to draw a sample of $r$ elements. There are two procedures of selecting those elements. There is the sampling with replacement. In this case, once an element has been selected, it is returned and is available to be drawn again. If the sampling is random, that is to say, each element is equally likely to be selected, then the first element can be chosen in $n$ ways. Since there is replacement and once again the sampling is random, there are $n$ ways of choosing the second element. The total number of ways for choosing the first two elements is thus $n^2$. Thus the total number of possible samples of size $r$ is $n^r$.

In the second procedure, replacement is not allowed. In this case, once an element has been chosen, it is no longer available to be selected. Thus the first element can be chosen in $n$ ways and the second element in $n - 1$ ways, since the number of elements available is now $n - 1$. The third element can be chosen in $n - 2$ ways and continuing along the same reasoning we find that the number of choices is $n(n-1) \ldots (n-r+1)$ and this product is denoted by $P_{n,r}$, that is to say

$$P_{n,r} = n(n-1)....(n-r+1) \qquad\qquad (10.7\text{-}1a)$$

$$= \frac{n!}{(n-r)!} \qquad\qquad (10.7\text{-}1b)$$

where $n! = n(n-1)...1$.

$P_{n,r}$ denotes the **number of permutations** of n elements taken r at a time.

If $n = r$, Equation (10.7-1b) becomes

$$P_{n,n} = n! \qquad\qquad (10.7\text{-}2)$$

Thus there are n! ways of permuting n elements, that is to say, n! ways of ordering n elements.

In the determination of the number of permutations, the order of selection is important. A change in the order of the r elements will result in a new selection. If a change of order in the r elements does not make a difference to the final result, we need to find the number of **combinations** of n elements taken r at a time.. This is denoted by $C_{n,r}$. Such a situation would occur if all r elements are alike. Since the number of permutations of r elements is r! and the number of permutations of n elements taken r at a time is $P_{n,r}$, $C_{n,r}$ is given by

$$C_{n,r} = \frac{P_{n,r}}{r!} \qquad\qquad (10.7\text{-}3a)$$

$$= \frac{n!}{r!\,(n-r)!} \qquad\qquad (10.7\text{-}3b)$$

***Example 10.7-1.*** Find (a) the number of permutations, (b) the number of combinations of the letters A, B, C, D in sets of two.

(a)      The number of permutations is

$$P_{4,2} = 12 \qquad\qquad (10.7\text{-}4)$$

(b)      The number of combinations is

$$C_{4,2} = \frac{12}{2} = 6 \qquad\qquad (10.7\text{-}5)$$

As the sample is small, we can list the elements in each case.

In case (a), we have: AB, BA, AC, CA, AD, DA, BC, CB, BD, DB, CD, DC.

In case (b), we make no distinction between AB and BA, the order is not important (similarly for the other pairs AC, CA, ...) and the number of elements in this case is half that of (a).

*Example 10.7-2.* We have n undistinguishable balls to be distributed randomly in n cells. What is the probability of having one ball in each cell?

The first ball can placed in any of the n cells. That is to say, there are n possibilities. Similarly the second ball can be placed in any of the n cells. Reasoning along the same line, we find that the total number of ways of randomly distributing the n balls in n cells is $n^n$. To have only one ball in each cell is equivalent to sampling without replacement: once a cell is filled it cannot be used again. Thus the number of permutations is $P_{n,n}$ which is given by Equation (10.7-2). The total number of possibilities is $n^n$ of which n! are favorable. Thus the probability p of having one ball in each cell is

$$p = \frac{n!}{n^n} \qquad (10.7\text{-}6)$$

If $n = 6$, p turns out to be 0.015, which shows that if we have six balls and six cells, the probability that each cell will contain a ball is 0.0154, a small probability!

*Example 10.7-3.* A sample of n elements is partitioned into k subgroups. The first subgroup contains $r_1$ elements, the second $r_2$ elements, ... , the $k^{th}$ subgroup contains $r_k$ elements. Determine the number of ways the partition can be done.

To form the first subgroup, we have to select $r_1$ elements from a sample of n elements. Within the subgroup, the order of the $r_1$ elements is not important. The number of ways of forming such a subgroup is $C_{n,r_1}$. The number of elements in the sample is now reduced to $(n - r_1)$ out of which we have to choose $r_2$ elements to form the second subgroup. The number of ways of accomplishing this is $C_{n-r_1, r_2}$. Similarly the $k^{th}$ subgroup can be obtained in $C_{n-r_1-r_2-...-r_{k-1}, r_k}$ ways. Thus the partition can be achieved in $C_{n,r_1} C_{n-r_1, r_2} \cdots C_{n-r_1-r_2-...-r_{k-1}, r_k}$ ways. Using Equation (10.7-3b), we have

$$C_{n,r_1} \cdots C_{n-r_1-r_2-...-r_{k-1}, r_k} = \frac{n!}{r_1! \, (n - r_1)!} \frac{(n - r_1)!}{r_2! \, (n - r_1 - r_2)!} \cdots \frac{(n - r_1 - r_2 - ... - r_{k-1})!}{r_k! \, (n - r_1 - ... - r_k)!} \qquad (10.7\text{-}7a)$$

$$= \frac{n!}{r_1! \, r_2! \, ... \, r_k!} \qquad (10.7\text{-}7b)$$

Note that

$$r_1 + r_2 + ... + r_k = n \qquad (10.7\text{-}8)$$

***Example 10.7-4***. In Example 10.7-2, we deduced the probability of having one ball in each cell. We now consider the case of having six balls distributed in six cells with three of the cells containing two balls. Determine the probability of such a partition.

The cells are divided into two groups, those that contain balls and those that are empty, and each group has three elements. The number of ways of doing this is given by Equation (10.7-7b) and is $6!/(3!\,3!)$. The partition of the six balls into three subsets of two elements each and three subsets that are empty can be achieved in $6!/(2!\,2!\,2!)$ ways. Thus the total number of ways (N) of obtaining a distribution of three cells containing two balls each is

$$N = \frac{6!\;6!}{3!\;3!\;2!\;2!\;2!} \qquad\qquad (10.7\text{-}9)$$

The six balls can be put randomly in the six cells in $6^6$ ways. Thus the probability $p$ of obtaining the desired distribution is

$$p = \frac{N}{6^6} = 0.0386 \qquad\qquad (10.7\text{-}10a,b)$$

Comparing with Example 10.7-2, we find that the distribution considered in the present example is more than twice as likely to happen as the distribution considered in the previous example, though intuition might suggest otherwise.

## 10.8   BINOMIAL, POISSON AND NORMAL DISTRIBUTION

### Binomial   Distribution

We now consider the case where the trials can have only two outcomes, such as tossing a coin. Also, the placing of balls in cells give rise to either empty cells or cells containing one or more balls. If we throw a dice we can have a six or not. Of the two outcomes, one is a success with probability $p$ and the other a failure with probability $q$. Since there are only two possible outcomes

$$p + q = 1 \qquad\qquad (10.8\text{-}1)$$

We assume that $p$ remains constant in $n$ trials and that the trials are independent. We can now calculate the probability $b\,(k; n, p)$ of $k$ successes in $n$ trials.

In the $n$ trials, we have $k$ successes with probability $p$ and $(n-k)$ failures with probability $q$. The events are independent and the probability of $k$ successes and $(n-k)$ failures is $p^k q^{n-k}$. The order of the successes is not important, so that the $k$ successes can be arranged in $C_{n,k}$ ways. Thus $b\,(k; n, p)$ is given by

$$b\,(k; n, p) = C_{n,k}\,p^k\,q^{n-k} \qquad\qquad (10.8\text{-}2a)$$

$$= \frac{n!}{k! \, (n-k)!} \, p^k \, q^{n-k} \tag{10.8-2b}$$

If we expand $(p+q)^n$ as a binomial expansion, we obtain

$$(q+p)^n = q^n + nq^{n-1}p + \frac{n\,(n-1)}{2} q^{n-2}p^2 + \dots + \frac{n\,(n-1)\,\dots\,(n-k+1)}{k!} q^{n-k}p^k + \dots + p^n \tag{10.8-3a}$$

$$= q^n + C_{n,1}\, q^{n-1}p + C_{n,2}\, q^{n-2}p^2 + \dots + C_{n,k}\, q^{n-k}p^k + \dots + p^n \tag{10.8-3b}$$

Comparing Equations (10.8-3a, 2a), we find that $b\,(k;\,n,\,p)$ is the $k^{th}$ term of the expansion of $(q+p)^n$. Thus this distribution is known as the **binomial distribution**.

Using Equations (10.8-1, 2a, 3b), we obtain

$$\sum_{k=0}^{n} b(k;\, n,\, p) = 1 \tag{10.8-4}$$

*Example 10.8-1.* Compute the probability of obtaining one six in twelve throws of a dice.

The probability of obtaining a six in one throw, if the dice is perfect, is 1/6. The trials are independent and the probability of obtaining a six in each throw remains the same. Then from Equation (10.8-2b), we have

$$b(1;\, 12,\, 1/6) = \frac{12!}{1! \, 11!} \left(\frac{1}{6}\right) \left(\frac{5}{6}\right)^{11} = 0.269 \tag{10.8-5a,b}$$

To compare the probability of having $k$ successes with $k-1$ successes, we have from Equation (10.8-2b)

$$\frac{b(k;\, n,\, p)}{b(k-1;\, n,\, p)} = \frac{n!}{k! \, (n-k)!} \, p^k \, q^{n-k} \, \frac{(k-1)! \, (n-k+1)!}{n! \, p^{k-1} \, q^{n-k+1}} \tag{10.8-6a}$$

$$= \frac{p\,(n-k+1)}{k\,q} \tag{10.8-6b}$$

$$= 1 + \frac{(n+1)p - k}{k\,q} \tag{10.8-6c}$$

From Equation (10.8-6c), we note that as $k$ increases from $0$ to $n$, $b\,(k;\,n,\,p)$ at first increases until $k$ is equal to $m$ and then decreases. The value of $m$ is given by

$$(n+1)p - 1 < m < (n+1)p \tag{10.8-7}$$

The number  m  is the **most probable number of successes**. If  $(n + 1)p$  is an integer and is equal to  m,  we deduce from Equation (10.8-6c)

$$b\,(m;\,n,\,p) \; = \; b\,(m-1;\,n,\,p) \tag{10.8-8}$$

***Example 10.8-2***.  A coin is tossed 40 times.  Calculate the probabilities of obtaining (a) 25 heads, (b) 20 heads.

(a)       We assume that the coin is perfect and that each toss is independent.

The probability of obtaining 25 heads is

$$b(25;\,40,\,1/2) \; = \; \frac{40!}{25!\;15!}\left(\frac{1}{2}\right)^{25}\left(\frac{1}{2}\right)^{15} \; = \; 0.0366 \tag{10.8-9a,b}$$

(b)       The probability of obtaining 20 heads is

$$b(20;\,40,\,1/2) \; = \; \frac{40!}{20!\;20!}\left(\frac{1}{2}\right)^{20}\left(\frac{1}{2}\right)^{20} \; = \; 0.1254 \tag{10.8-10a,b}$$

We note that 20 is the most probable number of heads but its probability is still small, though greater than for 25 heads.  Only in one out of eight experiments can we expect to have twenty heads.

●

To calculate  n!  can be tedious, especially when  n  is greater than ten.  We can then use **Stirling's formula** which is

$$n! \; \approx \; (2\pi n)^{1/2}\,n^n\,e^{-n} \tag{10.8-11}$$

The percentage error decreases dramatically as  n  increases.  The exact value of 10! is 3 628 800. Stirling's formula yields 3 598 600 resulting in a percentage error of 0.8%.  For 100! the percentage error reduces to 0.08%.

## Poisson  Distribution

In many cases,  n  is large and successes are rare; that is to say,  p  is small.  In such a case, it is desirable to find an approximation to the binomial distribution.  The product  n p  is the average or expected value and is denoted by  $\lambda$.  We assume that  $\lambda$  is not small  The probability of failure is, using Equations (10.8-2b, 1)

$$b(0;\,n,\,p) \; = \; \frac{n!}{n!}\,p^0\,q^n \tag{10.8-12a}$$

$$= \; (1-p)^n \tag{10.8-12b}$$

$$= (1 - \lambda /n)^n \tag{10.8-12c}$$

Taking the logarithms of both sides of Equation (10.8-12c) and expanding the resulting right side, we obtain

$$\ell n \, b \, (0; n, p) = n \, \ell n \, (1 - \lambda /n) \tag{10.8-13a}$$

$$= n \left( -\frac{\lambda}{n} - \frac{\lambda^2}{2n^2} - \ldots \right) \tag{10.8-13b}$$

as $n \longrightarrow \infty$, Equation (10.8-13b) implies

$$b \, (0; n, p) \approx e^{-\lambda} \tag{10.8-14}$$

Equation (10.8-6b) may be written as

$$\frac{b \, (k; n, p)}{b \, (k-1; n, p)} = \frac{\lambda + p \, (1 - k)}{k \, (1 - p)} \tag{10.8-15a}$$

$$\approx \frac{\lambda}{k} \tag{10.8-15b}$$

since $p$ is small.

From Equations (10.8-14, 15b), we obtain

$$b \, (1; n, p) \approx e^{-\lambda} \lambda \tag{10.8-16}$$

Similarly

$$b(2; n, p) \approx \frac{\lambda}{2} \, b(1; n, p) \tag{10.8-17a}$$

$$\approx \frac{\lambda^2}{2} \, e^{-\lambda} \tag{10.8-17b}$$

Generalizing, we obtain

$$b(k; n, p) \approx \frac{\lambda^k}{k!} \, e^{-\lambda} \tag{10.8-18}$$

This can be shown by induction as follows.

Assume that Equation (10.8-18) is true for $k = s$, from Equations (10.8-15b, 18), we then have

$$b(s+1; n, p) \approx \frac{\lambda}{(s + 1)} \, b(s; n, p) \tag{10.8-19a}$$

$$\approx \frac{\lambda}{(s + 1)} \, \frac{\lambda^s}{s!} \, e^{-\lambda} \tag{10.8-19b}$$

$$\approx \frac{\lambda^{s+1}}{(s + 1)!} \, e^{-\lambda} \tag{10.8-19c}$$

Thus Equation (10.8-18) is true for $(s + 1)$ and from Equation (10.8-16), we know that it is true for $s = 1$ and thus, by induction, it is true for all $s$. We denote the approximation of $b(k; n, p)$ under the conditions $p \longrightarrow 0$, $n \longrightarrow \infty$ and $np \longrightarrow$ finite $\lambda$ by $p(k; \lambda)$. Equation (10.8-18) is now written as

$$p(k; \lambda) = \frac{\lambda^k}{k!} \, e^{-\lambda} \tag{10.8-20}$$

The distribution $p(k; \lambda)$ given by Equation (10.8-20) is known as the **Poisson distribution**.

The expansion of $e^\lambda$ is

$$e^\lambda = 1 + \lambda + \dots + \frac{\lambda^k}{k!} + \dots \tag{10.8-21}$$

Combining Equations (10.8-20, 21) yields

$$\sum_{k=0}^{\infty} p(k; \lambda) = 1 \tag{10.8-22}$$

for any fixed $\lambda$.

Since the sum of $p(k; \lambda)$ is one, the Poisson distribution is also a distribution of random rare events and is not merely an approximation of the binomial distribution. The number of atoms disintegrating per second in a quantity of radioactive substance, the number of misprints in a book, the number of hits on a target, the number of wrong telephonic connections, the number of reported cases of a rare disease , and the number of automobile accidents in a fixed time interval at a particular spot, are all examples of the Poisson distribution.

We now derive an approximation of $b(k; n, p)$ relaxing the condition that $p$ has to be small.

**Normal Distribution**

Let $u$ denote the deviation from the mean value $np$.

$$u = k - np \tag{10.8-23}$$

If $\ell$ denotes the number of failures, then $\ell$ is given by

$$\ell = n - k \tag{10.8-24a}$$

$$= n - u - np \tag{10.8-24b}$$

$$= nq - u \tag{10.8-24c}$$

Substituting Equations (10.8-23, 24c) into Equation (10.8-2b) yields

$$b(k; n, p) = \frac{n!}{(u + np)! \, (nq - u)!} \, p^{np+u} \, q^{nq-u} \tag{10.8-25}$$

Taking the logarithms on both sides gives

$$\ell n \, b = \ell n \, n! + (np + u) \, \ell n \, p + (nq - u) \, \ell n \, q - \ell n \, (np + u)! - \ell n \, (nq - u)! \tag{10.8-26}$$

Using Stirling's formula [Equation (10.8-11)], we have

$$\ell n \, n! \approx \frac{1}{2} \, \ell n \, (2\pi n) + n \, \ell n \, n - n \tag{10.8-27}$$

Similarly

$$\ell n \, (np + u)! \approx \frac{1}{2} \, \ell n \, [2\pi \, (np + u)] + (np + u) \, \ell n \, (np + u) - (np + u) \tag{10.8-28a}$$

$$\approx \frac{1}{2} \, \ell n \, [2\pi \, (np + u)] + np \left(1 + \frac{u}{np}\right) \ell n \left[np \left(1 + \frac{u}{np}\right)\right] - (np + u) \tag{10.8-28b}$$

$$\approx \frac{1}{2} \, \ell n \, [2\pi \, (np + u)] + np \left(1 + \frac{u}{np}\right) \left[\ell n \, np + \ell n \left(1 + \frac{u}{np}\right)\right] - (np + u) \tag{10.8-28c}$$

$$\approx \frac{1}{2} \, \ell n \, [2\pi \, (np + u)] + np \left(1 + \frac{u}{np}\right) \left[\ell n \, np + \frac{u}{np} - \frac{u^2}{2n^2 p^2}\right] - (np + u) \tag{10.8-28d}$$

$$\ell n \, (nq - u)! \approx \frac{1}{2} \, \ell n \, [2\pi \, (nq - u)] + nq \left(1 - \frac{u}{nq}\right) \left[\ell n \, nq - \frac{u}{nq} - \frac{u^2}{2n^2 q^2}\right] - (nq - u) \tag{10.8-28e}$$

Substitution of Equations (10.8-27, 28d, 28e) in Equation (10.8-26) yields

$$\ell n \, b \approx \frac{1}{2} \left[\ell n \, (2\pi n) - \ell n \, 2\pi \, (np + u) - \ell n \, 2\pi \, (nq - u)\right] + n \, \ell n \, n$$

$$- np \left(1 + \frac{u}{np}\right) \left[\ell n \, np + \frac{u}{np} - \frac{u^2}{2n^2 p^2}\right] - nq \left(1 - \frac{u}{nq}\right) \left[\ell n \, nq - \frac{u}{nq} - \frac{u^2}{2n^2 q^2}\right]$$

$$- n + (np + u) + (nq - u) + np \left(1 + \frac{u}{np}\right) \ell n \, p + nq \left(1 - \frac{u}{nq}\right) \ell n \, q \tag{10.8-29a}$$

$$\approx -\frac{1}{2}\left[\ell n \frac{4\pi^2\,(np + u)\,(nq - u)}{2\pi n}\right] + n\,\ell n\,n - np\left(1 + \frac{u}{np}\right)\left[\ell n\,np - \ell n\,p\right]$$

$$-\left(1 + \frac{u}{np}\right)\left(u - \frac{u^2}{2np}\right) - nq\left(1 - \frac{u}{nq}\right)\left[\ell n\,nq - \ell n\,q\right] + \left(1 - \frac{u}{nq}\right)\left(u + \frac{u^2}{2nq}\right) \qquad (10.8\text{-}29b)$$

$$\approx -\frac{1}{2}\,\ell n\,2\pi npq - \frac{u^2}{2np} - \frac{u^2}{2nq} \qquad (10.8\text{-}29c)$$

$$\approx -\frac{1}{2}\,\ell n\,2\pi npq - \frac{u^2}{2npq} \qquad (10.8\text{-}29d)$$

Taking the anti-logarithm of Equation (10.8-29d), we obtain

$$b\,(k;\,n,\,p)\ \approx\ \frac{1}{\sqrt{2\pi npq}}\ \exp\,(-u^2\,/\,2npq) \qquad (10.8\text{-}30)$$

To simplify the expression on the right side of Equation (10.8-30), we introduce

$$\sigma\ =\ \sqrt{npq} \qquad (10.8\text{-}31a)$$

$$x\ =\ \frac{u}{\sigma} \qquad (10.8\text{-}31b)$$

Equation (10.8-30) now becomes

$$b\,(k;\,n,\,p)\ \approx\ \frac{1}{\sigma\,\sqrt{2\pi}}\ e^{-x^2/2} \qquad (10.8\text{-}32a)$$

$$\approx\ \frac{1}{\sigma}\ \varphi\,(x) \qquad (10.8\text{-}32b)$$

In Equation (10.8-32b), $\varphi\,(x)$ is given by

$$\varphi\,(x)\ =\ \frac{1}{\sqrt{2\pi}}\ e^{-x^2/2} \qquad (10.8\text{-}33)$$

The function $\varphi\,(x)$ is known as the **standard normal density function**. The quantity $\sigma$ is the **standard deviation of the binomial distribution**.

In the binomial distribution, we calculated the probability of $k$ successes in a discrete sample space. The function $\varphi\,(x)$ is now a continuous function of a continuous variable $x$. The variable $x$ is a measure of the departure from the mean value $np$ [see Equation (10.8-23)]. $x$ can be on either side of the mean value $np$, so that its value can range from $-\infty$ to $\infty$. Using the result given in Chapter 4, Problem 9b, we obtain

$$\int_{-\infty}^{\infty} \varphi(x)\, dx = 1 \tag{10.8-34}$$

Thus $\varphi(x)$ satisfies the condition that the sum (integral) of all probabilities is a certainty and is equal to one.

The **cumulative distribution function** of a standard normal random variable is denoted by $\Phi(x)$ and is defined as

$$\Phi(x) = \frac{1}{\sqrt{2\pi}} \int_{-\infty}^{x} e^{-y^2/2}\, dy \tag{10.8-35}$$

$\Phi$ is also known as the **normal distribution function**. $\Phi(x)$ increases steadily from 0 to 1. The values of $\Phi(x)$ for positive values of $x$ are tabulated.

We list some of the properties of $\varphi(x)$ and $\Phi(x)$

(a)     $\varphi(x)$ is symmetrical about $x = 0$;

(b)     $\varphi(x)$ has its maximum at $x = 0$, that is, when $k = np$, the most probable number of successes;

(c)     $\Phi(-x) = 1 - \Phi(x)$; \hfill (10.8-36)

(d)     The probability that $x$ lies between $a$ and $b$, $a < b$, is

$$p(a \le x \le b) = \Phi(b) - \Phi(a) \tag{10.8-37}$$

Equation (10.8-37) is known as **Laplace's limit theorem**.

In terms of the original variables, Equation (10.8-32b, 33) can be written as

$$b(k; n, p) \approx \frac{1}{\sqrt{2\pi npq}} \exp\left(-\frac{(k-np)^2}{2npq}\right) \tag{10.8-38a}$$

$$\varphi\left(\frac{k-np}{\sqrt{npq}}\right) = \frac{1}{\sqrt{2\pi}} \exp\left(-\frac{(k-np)^2}{2npq}\right) \tag{10.8-38b}$$

Equations (10.8-38a, b) can be used in the case of discrete variables whereas Equations (10.8-32b, 33) are for continuous variables.

***Example 10.8-3***. The probability of being dealt a full house in a hand of poker is 0.0014. Compute the probability of being dealt at least two full houses in one thousand hands.

Getting a full house is a rare event which can be assumed to follow the Poisson distribution. The parameter $\lambda$ is

$$\lambda = 0.0014 \times 1000 = 1.4 \qquad\qquad (10.8\text{-}39a,b)$$

The probability of having at least two houses is

$$p = 1 - p\,(0;\lambda) - p\,(1;\lambda) \qquad\qquad (10.8\text{-}40a)$$

$$= 1 - e^{-1.4} - 1.4\,e^{-1.4} \qquad\qquad (10.8\text{-}40b)$$

$$= 1 - 2.4\,e^{-1.4} = 0.408 \qquad\qquad (10.8\text{-}40c,d)$$

***Example 10.8-4***. A machine is known to produce on an average one faulty item out of twenty. Calculate the probability that out of a sample of twenty items, there will be at most one defective item.

We assume that the distribution is Poisson and the parameter $\lambda$ is given by

$$\lambda = 20 \times \frac{1}{20} = 1 \qquad\qquad (10.8\text{-}41a,b)$$

The probability of having at most one faulty item is

$$p = p\,(0;1) + p\,(1;1) \qquad\qquad (10.8\text{-}42a)$$

$$= e^{-1} + e^{-1} = 0.736 \qquad\qquad (10.8\text{-}42b,c)$$

The probability of having at least one defective item is

$$q = 1 - p = 0.264 \qquad\qquad (10.8\text{-}43a,b)$$

Thus it is almost twice more likely to find at most one faulty item than to find more than one faulty item.

***Example 10.8-5***. A perfect coin is tossed 40 times. Determine the probability of having twenty heads.

The probability of getting a head is 1/2. In this case, the distribution is normal and not Poisson since the value of $p$ associated with each toss is 0.5, which is not small.

From Equation (10.8-38a), we obtain

$$p = \frac{1}{\sqrt{20\pi}} \exp\left(-\frac{(20-20)^2}{20}\right) \approx 0.13 \tag{10.8-44a,b}$$

We could also use Equation (10.8-37) to determine the probability of twenty heads. In this case, x has to be transformed into a continuous variable. The number 19.5 and 20.5 span the range closest to the integer 20. The number of heads should then be between 19.5 and 20.5 and this leads us to calculate a and b via Equations (10.8-23, 31a, b)

$$b = \frac{(20.5-20)}{\sqrt{10}} \approx 0.16 \tag{10.8-45a,b}$$

$$a = \frac{(19.5-20)}{\sqrt{10}} \approx -0.16 \tag{10.8-45c,d}$$

The probability of having twenty heads is thus

$$p = \Phi(0.16) - \Phi(-0.16) \tag{10.8-46a}$$

$$= 2\Phi(0.16) - 1 \approx 0.13 \tag{10.8-46b,c}$$

We have used Equation (10.8-36) and a Table of $\Phi$.

Both methods, as expected, give the same answer.

***Example 10.8-6.*** The heights of 10 000 young men follow closely a normal distribution with a mean of 69 inches and a standard deviation of 2.5 inches. Find

(a)     the expected number of men $N_2$ to be at least 6 feet tall;
(b)     the expected number of men to be exactly 6 feet tall;
(c)     the expected number of men $N$ to be between 70 inches and 74 inches tall.

(a)     We first calculate the probability of the number of men whose heights are 6 feet (= 72 inches) and below. From Equation (10.8-37), we have

$$p\left(-\frac{69}{2.5} \le x \le \frac{72-69}{2.5}\right) = \Phi(1.2) - \Phi(-28) \approx 0.885 \tag{10.8-47a,b}$$

The expected number of men who are at most 6 feet tall is

$$N_1 = 10\,000 \times 0.885 = 8\,850 \tag{10.8-48a,b}$$

The number of men that are at least 6 feet tall is

$$N_2 = 10\,000 - 8\,850 = 1\,150 \tag{10.8-49a,b}$$

(b)      In this case,  $a = b = 1.2$  and the probability (area of a line) is zero.  We do not expect
         anybody to be exactly 6 feet tall, since we cannot measure to that level of accuracy.

(c)                    $$p\left(\frac{70-69}{2.5} \leq x \leq \frac{74-69}{2.5}\right) = \Phi(2) - \Phi(0.4)$$                    (10.8-50a)

$$\approx 0.977 - 0.655 \approx 0.322$$                    (10.8-50b,c)

The number of men  N  whose heights are between 70 and 74 inches is

$$N = 0.322 \times 10\,000 = 3\,200$$                    (10.8-51a,b)

## 10.9  SCOPE OF STATISTICAL MECHANICS

In Section 10.8, we have seen that statistical methods can be employed to solve real problems.  These
same methods are used to solve problems in mechanics.  In statistical mechanics, we predict the
macroscopic properties of the system from its microscopic properties.  For example, the pressure on
the wall of a gas container is due to the impacts of the gas molecules on the wall.  A pressure gauge is
not able to measure the pressure exerted by each individual molecule but it measures only an average
force due to a large number of impacts.  Since the number of impacts per second per square centimeter
is of the order of a few billion, statistical methods are most appropriate.

In Section 10.2, we established that the state of a dynamical system is defined if the generalized
coordinates and momenta are known.  Since we are dealing with a system of a large number of
particles, it is impossible to know the initial conditions of each particle and it is prohibitively
impossible to solve the equations of motion.  Consequently the idea of an ensemble was introduced in
Section 10.6.

We illustrate the idea of an ensemble by considering a simple example.  Suppose the oscillator in
Example 10.2-1 is isolated and is in equilibrium.  Since it is isolated, it does not exchange energy with
any other system.  Its energy is therefore constant.  Let its energy be in the range  E  and  $E + \Delta E$.  In
quantum mechanics,  E  cannot be determined exactly.  In classical mechanics,  E  can theoretically be
determined exactly but in practice there is bound to be an error  $\Delta E$  in its determination.  In the phase
plane, as shown in Example 10.2-1, the trajectory for a constant  E  is an ellipse.  In our example, we
have two ellipses, one corresponding to  E  and the other to  $E + \Delta E$  as shown in Figure 10.9-1.  If
we know the initial conditions, we can solve the equations of motion and predict the coordinate  q  and
momentum  p  of the oscillator at any time.  But suppose as stated earlier, that we do not know the
initial conditions with sufficient precision and we are unable to solve the equations of motion.  All we
know is that the oscillator is isolated, in equilibrium and its energy lies between  E  and  $E + \Delta E$.
Though we cannot predict its coordinate  q  and its momentum  p,  we know they must be in the area
between the two ellipses in the phase plane.  There are many oscillators which could satisfy the
conditions of being isolated and having energy in the specified range and they form an ensemble.

They could have distinct initial conditions but they all will be confined to the area between the two ellipses in the phase plane. We can now introduce statistical methods. We subdivide the area between the ellipses into small cells, each of size $\Delta q \, \Delta p$, as shown in Figure 10.9-1. We introduce a probability distribution $f(q, p)$, which is a continuous function of q and p. The probability of an oscillator occupying the cell of size $\Delta q \, \Delta p$ at $(q, p)$ is

$$\text{prob.} = f(q, p) \, \Delta q \, \Delta p \qquad\qquad (10.9\text{-}1)$$

If there is one oscillator then the integral $\iint f(q, p) \, dq \, dp$ over the relevant area is one.

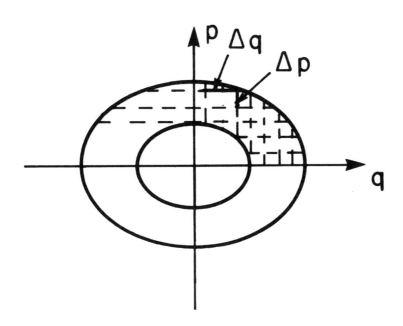

**FIGURE 10.9-1    Phase diagram of an oscillator at two different energy levels**

We can extend to the case of N particles with n degrees of freedom. We have n generalized coordinates $q_1, \dots, q_n$ and n generalized momenta $p_1, \dots, p_n$, and the dimension of the phase space is 2n. The phase space is divided into cells of size $\Delta p_1, \dots, \Delta p_n$ and $\Delta q_1, \dots, \Delta q_n$ and the probability distribution function is $f(q_1, \dots, q_n, p_1, \dots, p_n)$. For brevity, we write $\underline{q}$ and $\underline{p}$ for $q_1, \dots, q_n$ and $p_1, \dots, p_n$ respectively. Similarly $\Delta \underline{q}$ and $\Delta \underline{p}$ stand for $\Delta q_1, \dots, \Delta q_n$ and $\Delta p_1, \dots, \Delta p_n$ respectively. Then $f(\underline{q}, \underline{p}) \, \Delta \underline{q} \, \Delta \underline{p}$ is the probable number of particles in the cell of size $\Delta \underline{q} \, \Delta \underline{p}$ at $(\underline{q}, \underline{p})$. The function $f(\underline{q}, \underline{p})$ is known as the **phase space distribution function**. The integral of $f(\underline{q}, \underline{p})$ over the whole phase space is N, the total number of particles. In many cases, f is normalized such that its integral over the complete phase space is one, and in this case, f is the **probability density**.

If $B(\underline{q}, \underline{p})$ is any quantity associated with the system, then the **average value** of $B$, denoted by $<B>$, is

$$<B> = \frac{\iint B(\underline{q}, \underline{p})\, f(\underline{q}, \underline{p})\, d\underline{q}\, d\underline{p}}{\iint f(\underline{q}, \underline{p})\, d\underline{q}\, d\underline{p}} \tag{10.9-2}$$

Based on some plausible physical assumptions, it is possible to determine f. The average value of the physical quantities can be calculated.

## 10.10   BASIC ASSUMPTIONS

One of the basic assumptions is that all microstates having the same energy have the same probability of existence. Let us consider an isolated system consisting of $N$ particles in a volume $V$. Since the system is isolated, its energy is constant. Suppose it lies in the range $E$ and $E + \Delta E$. All systems having $N$ particles occupying volume $V$ with energy in the range $E$ and $E + \Delta E$ form an ensemble. The probability assumption for such an ensemble in equilibrium is that each of its accessible states is equally likely. Thus the probability $P_r$ of finding a system with energy $E_r$ is given by

$$P_r = \begin{cases} C & \text{if } E < E_r < E + \Delta E \\ \\ 0 & \text{otherwise} \end{cases} \tag{10.10-1a,b}$$

where $C$ is a constant. Such an ensemble is called a **microcanonical ensemble**.

If the system is in thermal contact with other systems or heat reservoirs then there will be an exchange of energy and its energy will not be constant. An ensemble of such systems which are in thermal equilibrium is known as a **canonical ensemble**.

Consider two systems $A$ and $B$ which are in thermal equilibrium with the same heat reservoir. We denote the probability of finding the system $A$ with energy $E_A$ by $P_A(E_A)$. Similarly the probability of finding $B$ with energy $E_B$ is $P_B(E_B)$. The systems $A$ and $B$ can be regarded as one system with energy $E_{A+B}$ and the probability of the composite system with energy $E_{A+B}$ is denoted by $P_{A+B}(E_{A+B})$. Since $A$ and $B$ are macroscopic system, we can assume that the interaction energy is negligible compared to $E_A$ and $E_B$, so that

$$E_{A+B} = E_A + E_B \tag{10.10-2}$$

Using Equation (10.10-2), we have

$$P_{A+B}(E_{A+B}) = P_{A+B}(E_A + E_B) \tag{10.10-3}$$

The right side of Equation (10.10-3) can be interpreted as expressing the probability of system A having energy $E_A$ while, at the same time, system B has energy $E_B$. We assume systems A and B to be independent . Using relation (e) of Section 10.7, we have

$$P_{A+B}(E_A + E_B) = P_A(E_A)\, P_B(E_B) \tag{10.10-4}$$

Since we have assumed systems A and B to be independent, $E_A$ and $E_B$ are independent. Differentiating the left side of Equation (10.10-4) with respect to $E_A$, we have

$$\frac{\partial}{\partial E_A}\left[P_{A+B}(E_A + E_B)\right] = \frac{\partial}{\partial(E_A + E_B)}\left[P_{A+B}(E_A + E_B)\right]\frac{\partial(E_A + E_B)}{\partial E_A} \tag{10.10-5a}$$

$$= P'_{A+B}(E_A + E_B) \tag{10.10-5b}$$

where the prime refers to differentiation with respect to $(E_A + E_B)$.

Similarly

$$\frac{\partial}{\partial E_B}\left[P_{A+B}(E_A + E_B)\right] = P'_{A+B}(E_A + E_B) \tag{10.10-6}$$

Differentiating both sides of Equation (10.10-4) and using Equations (10.10-5b, 6), we have

$$P'_A(E_A)\, P_B(E_B) = P_A(E_A)\, P'_B(E_B) \tag{10.10-7}$$

where the prime refers to differentiation with respect to the argument.

Equation (10.10-7) implies

$$\frac{P'_A(E_A)}{P_A(E_A)} = \frac{P'_B(E_B)}{P_B(E_B)} = -\beta \tag{10.10-8a,b}$$

Since systems A and B are independent, $\beta$ is a constant, that is to say, it is a quantity independent of systems A and B.

Integrating Equations (10.10-8a,b), we have

$$P_A(E_A) = C_A\, e^{-\beta E_A} \tag{10.10-9a}$$

$$P_B(E_B) = C_B\, e^{-\beta E_B} \tag{10.10-9b}$$

where $C_A$ and $C_B$ are constants and depend on the systems. The quantity $\beta$ is independent of A and B but may depend on the heat reservoir. The temperature T is the quantity that characterizes a reservoir, thus $\beta$ can depend on T.

Generalizing Equations (10.10-9a, b), we state that the probability $P_r$ of a system in thermal equilibrium in a particular state with energy $E_r$ is

$$P_r = C e^{-\beta E_r} \tag{10.10-10}$$

To determine $C$ we sum over all possible microstates, and we obtain

$$\sum_r C e^{-\beta E_r} = 1 \tag{10.10-11}$$

Thus $C$ is given by

$$C = \frac{1}{\sum_r e^{-\beta E_r}} \tag{10.10-12}$$

Equation (10.10-10) may be written as

$$P_r = Z^{-1} e^{-\beta E_r} \tag{10.10-13}$$

where $Z = \sum_r e^{-\beta E_r}$.

If the energy can vary continuously, the distribution function $f(E)$ is given by

$$f(E) = D e^{-\beta E} \tag{10.10-14}$$

where $E$ is a continuous function of the coordinates $q_i$ and momenta $p_i$. The constant $D$ can be obtained by integrating over the whole phase space. Thus, using the notation of Equation (10.9-2), we have

$$D \iint e^{-\beta E} d\underline{q} \, d\underline{p} = 1 \tag{10.10-15}$$

Using Equation (10.10-15), Equation (10.10-14) can be written as

$$f(E) = Z^{-1} D_N e^{-\beta E} \tag{10.10-16}$$

where $Z = D_N \iint e^{-\beta E} d\underline{q} \, d\underline{p}$.

The quantity $D_N$ has been introduced so that $Z$ becomes dimensionless. Note that in Equation (10.10-13), $Z$ is dimensionless. In classical mechanics, it is usual to assume $D_N$ to be unity. The quantity $Z$ (which stands for Zustandssumme, state sum) is called the **partition function**. The exponential factor $e^{-\beta E}$ is known as the **Boltzmann factor**. The probability distribution given by

Equations (10.10-13, 16) is known as the **canonical distribution**, and $D_N$ is assumed to be unity. The factor $\beta$ is given by

$$\beta = 1/kT \tag{10.10-17}$$

where $k$ is **Boltzmann's constant** and $T$ is the temperature.

## 10.11  STATISTICAL THERMODYNAMICS

The internal energy $U$ of a system is identified as the statistical average energy $<E>$. Thus combining Equations (10.9-2, 10-13), we have

$$U = <E> = Z^{-1} \sum_r E_r e^{-\beta E_r} \tag{10.11-1a,b}$$

On differentiating the partition function $Z$ with respect to $\beta$, we obtain

$$\frac{\partial Z}{\partial \beta} = -\sum_r E_r e^{-\beta E_r} \tag{10.11-2}$$

Comparing Equations (10.11-1a, b, 2), we find that

$$U = -Z^{-1} \frac{\partial Z}{\partial \beta} \tag{10.11-3a}$$

$$= \frac{\partial}{\partial \beta} (-\ln Z) \tag{10.11-3b}$$

The partition function $Z$ is a function of the temperature $T$ and the volume $V$. Thus

$$d(\ln Z) = \frac{\partial}{\partial T} (\ln Z) dT + \frac{\partial}{\partial V} (\ln Z) dV \tag{10.11-4a}$$

$$= \frac{\partial}{\partial \beta} (\ln Z) dT \frac{d\beta}{dT} + \frac{\partial}{\partial V} (\ln Z) dV \tag{10.11-4b}$$

$$= \frac{\partial}{\partial \beta} (\ln Z) d\beta + \frac{\partial}{\partial V} (\ln Z) dV \tag{10.11-4c}$$

$$= -U d\beta + \frac{\partial}{\partial V} (\ln Z) dV \tag{10.11-4d}$$

$$= -d(U\beta) + \beta \, dU + \frac{\partial}{\partial V} (\ln Z) dV \tag{10.11-4e}$$

Using Equation (10.10-17), Equation (10.11-4e) can be written as

$$dU = kT\,d\left(\ell n\,Z + U\beta\right) - kT\frac{\partial}{\partial V}\left(\ell n\,Z\right)dV \qquad (10.11\text{-}5)$$

From the laws of thermodynamics, we can relate the quantities in Equation (10.11-5) to other thermodynamic variables. We recall that the first law of thermodynamics may be written as

$$dU = -dW + dQ \qquad (10.11\text{-}6)$$

where $dW$ is the infinitesimal work done by the system and $dQ$ is the infinitesimal heat absorbed by the system.

The second law of thermodynamics introduces the **entropy** $S$ and $dS$ is related to $dQ$ by

$$dS = \frac{dQ}{T} \qquad (10.11\text{-}7)$$

For an ideal gas, the work done $dW$ is given by

$$dW = p\,dV \qquad (10.11\text{-}8)$$

where $p$ is the pressure.

Substituting Equations (10.11-7, 8) into Equation (10.11-6) yields

$$dU = -p\,dV + T\,dS \qquad (10.11\text{-}9)$$

Comparing Equations (10.11-5, 9), we deduce

$$p = kT\frac{\partial}{\partial V}\left(\ell n\,Z\right) \qquad (10.11\text{-}10a)$$

$$dS = k\,d\left(\ell n\,Z + U\beta\right) \qquad (10.11\text{-}10b)$$

On integrating Equation (10.11-10b), we obtain

$$S = k\,\ell n\,Z + kU\beta + \text{constant} \qquad (10.11\text{-}11)$$

The constant term in Equation (10.11-11) has no physical significance since only a change in $S$ can be measured.

We now introduce the **Helmholtz free energy** $A$ which is defined as

$$A = U - TS \qquad (10.11\text{-}12)$$

Differentiating A yields

$$dA = dU - T\,dS - S\,dT \qquad\qquad (10.11\text{-}13)$$

Combining Equations (10.11-9, 13) results in

$$dA = -p\,dV - S\,dT \qquad\qquad (10.11\text{-}14)$$

We consider A to be a function of two independent variables T and V, that is to say

$$A = A\,(T, V) \qquad\qquad (10.11\text{-}15)$$

It then follows that

$$dA = \left(\frac{\partial A}{\partial T}\right)_V dT + \left(\frac{\partial A}{\partial V}\right)_T dV \qquad\qquad (10\text{-}11\text{-}16)$$

Comparing Equations (10.11-14, 16) yields

$$\left(\frac{\partial A}{\partial V}\right)_T = -p \qquad\qquad (10.11\text{-}17a)$$

$$\left(\frac{\partial A}{\partial T}\right)_V = -S \qquad\qquad (10.11\text{-}17b)$$

Since A is a continuous function of T and V with continuous partial derivatives, we deduce from Equations (10.11-17a, b)

$$\left(\frac{\partial p}{\partial T}\right)_V = \left(\frac{\partial S}{\partial V}\right)_T \qquad\qquad (10.11\text{-}18)$$

The four thermodynamic variables p, V, S and T are not all independent. They are related via Equation (10.11-9). We can consider any two of them as independent variables. In the definition of A, we have assumed T and V to be the independent variables.

A function H (S, p) is now introduced. H is known as the **enthalpy** and is defined as

$$H = U + pV \qquad\qquad (10.11\text{-}19)$$

On differentiating, we have

$$dH = dU + p\,dV + V\,dp \qquad\qquad (10.11\text{-}20)$$

Since H is a function of S and p, we can write

$$dH = \left(\frac{\partial H}{\partial S}\right)_p dS + \left(\frac{\partial H}{\partial p}\right)_S dp \qquad (10.11\text{-}21)$$

Combining Equations (10.11-9, 20, 21), we deduce

$$T = \left(\frac{\partial H}{\partial S}\right)_p \qquad (10.11\text{-}22a)$$

$$V = \left(\frac{\partial H}{\partial p}\right)_S \qquad (10.11\text{-}22b)$$

Cross differentiation of Equations (10.11-22a, b) yields

$$\left(\frac{\partial T}{\partial p}\right)_S = \left(\frac{\partial V}{\partial S}\right)_p \qquad (10.11\text{-}23)$$

The **Gibbs free energy**  G  is usually assumed to be a function of  T  and  p  and is defined as

$$G = H - TS \qquad (10.11\text{-}24a)$$

or     $$G = A + pV \qquad (10.11\text{-}24b)$$

Proceeding as in the case of  A  and  H,  we deduce that

$$S = -\left(\frac{\partial G}{\partial T}\right)_p \qquad (10.11\text{-}25a)$$

$$V = \left(\frac{\partial G}{\partial p}\right)_T \qquad (10.11\text{-}25b)$$

$$\left(\frac{\partial S}{\partial p}\right)_T = -\left(\frac{\partial V}{\partial T}\right)_p \qquad (10.11\text{-}25c)$$

In an isothermal process,  $dT = 0$  and Equation (10.11-13) becomes

$$dU = dA + T\,dS \qquad (10\text{-}11\text{-}26)$$

From Equations (10.11-6, 7, 26), we deduce

$$dA = -dW \qquad (10.11\text{-}27)$$

Substituting Equation (10.11-11) into Equation (10.11-12), choosing the constant term to be zero and $\beta = 1/kT$, we obtain

$$A = -kT \,\ell n\, Z \qquad (10.11\text{-}28)$$

From Equation (10.11-28), $Z$ can be written as

$$Z = e^{-A/kT} \qquad (10.11\text{-}29a)$$

$$= e^{-\beta A} \qquad (10.11\text{-}29b)$$

The probability $P_r$ as given by Equation (10.10-13) can now be written as

$$P_r = e^{\beta(A-E_r)} \qquad (10.11\text{-}30)$$

This form of $P_r$ is quite common in statistical mechanics.

***Example 10.11-1.*** Calculate the partition function $Z$ of a perfect monoatomic gas containing $N$ atoms, each of mass $m$.

Let $\underline{v}_i$ be the velocity of atom i, $E_i$ be the kinetic energy. The potential energy is zero since, for a perfect gas, the interatomic forces are negligible.

$$E_i = \frac{1}{2} m \underline{v}_i \bullet \underline{v}_i \qquad (10.11\text{-}31a)$$

$$= \frac{1}{2} m \left(v_x^2 + v_y^2 + v_z^2\right)_i \qquad (10.11\text{-}31b)$$

The partition function $Z$ is given by Equation (10.10-16) with $D_N = 1$

$$Z = \iint e^{-\beta E_1} e^{-\beta E_2} ... e^{-\beta E_n} \, d\underline{r} \, d\underline{v} \qquad (10.11\text{-}32)$$

where $d\underline{r}$ and $d\underline{v}$ are $d\underline{r}_1 ... d\underline{r}_n$ and $d\underline{v}_1 ... d\underline{v}_n$ respectively.

Since the energy does not depend on the position of the atoms, the $E_i$ are independent of the $\underline{r}_i$, and we can carry out the integral with respect to $d\underline{r}$. There are $N$ atoms and since each integral with respect to $d\underline{r}_i$ yields the volume $V$,

$$Z = V^N \int \underset{N}{...} \int e^{-\beta E_1} ... e^{-\beta E_N} \, d\underline{v}_1 ... d\underline{v}_n \qquad (10.11\text{-}33)$$

Each of the integrals in Equation (10.11-33) is given by

$$I = \int\limits_{-\infty}^{\infty} \int\limits_{-\infty}^{\infty} \int\limits_{-\infty}^{\infty} e^{-\beta(1/2)\,m\left(v_x^2 + v_y^2 + v_z^2\right)} \, dv_x \, dv_y \, dv_z \tag{10.11-34}$$

The limits of the integral are the same for each of the velocity components $v_x$, $v_y$, $v_z$, so that Equation (10.11-34) can be written as

$$I = \left(\int_{-\infty}^{\infty} e^{-(1/2)\,\beta m v^2} \, dv\right)^3 \tag{10.11-35}$$

Using the results given in Chapter 4, Problem 9b, we find

$$I = \left(\frac{2\pi}{\beta m}\right)^{3/2} \tag{10.11-36}$$

Substituting $I$ into Equation (10.11-33) yields

$$Z = V^N \left(\frac{2\pi}{\beta m}\right)^{3N/2} \tag{10.11-37}$$

$$\ln Z = N \ln V + \frac{3N}{2} \ln\left(\frac{2\pi}{\beta m}\right) \tag{10.11-38}$$

The internal energy $U$, Equation (10.11-3b), is given by

$$U = -\frac{\partial}{\partial \beta}\left[N \ln V + \frac{3N}{2} \ln\left(\frac{2\pi}{\beta m}\right)\right] \tag{10.11-39a}$$

$$= -\frac{3N}{2} \frac{\beta m}{2\pi}\left(-\frac{2\pi}{m\beta^2}\right) \tag{10.11-39b}$$

$$= \frac{3N}{2\beta} \tag{10.11-39c}$$

The pressure $p$ is given by Equation (10.11-10a). Using Equation (10.11-38), we obtain

$$p = kT \frac{\partial}{\partial V}\left[N \ln V + \frac{3N}{2} \ln\left(\frac{2\pi}{\beta m}\right)\right] \tag{10.11-40a}$$

$$= \frac{kTN}{V} \tag{10.11-40b}$$

Equation (10.11-40b) is the equation of state for a perfect gas. Thus if the partition function $Z$ can be calculated, all the other thermodynamic variables can be deduced. In Example 10.11-1, it was easy to calculate $Z$ since the atoms did not interact. In the liquid state, for example, where there are strong interactions, the calculation of $Z$ is difficult and the integral in Equation (10.10-16) cannot be evaluated.

## 10.12   THE EQUIPARTITION THEOREM

The energy $E$ of a system is a function of the $n$ generalized coordinates $q_i$ and the $n$ generalized momenta $p_i$. Suppose that $E$ can be written as the sum of two parts

$$E = cp_1^2 + E'(q_1, \dots, q_n, p_2, \dots, p_n) \tag{10.12-1}$$

where $c$ is a positive constant.

In many actual situations, $E$ is a quadratic function of the momenta.

The partition function $Z$, Equation (10.10-16), can be written as

$$Z = \int \dots \int e^{-\beta(cp_1^2 + E')} \, dp_1 \, d\underline{q} \, dp_2 \dots dp_n \tag{10.12-2a}$$

$$= \int_{-\infty}^{\infty} e^{-\beta cp_1^2} \, dp_1 \int \dots \int e^{-\beta E'} \, d\underline{q} \, dp_2 \dots dp_n \tag{10.12-2b}$$

$$= \left(\frac{\pi}{\beta c}\right)^{1/2} Z' \tag{10.10-2c}$$

where $Z'$ stands for the remaining multiple integrals.

The internal energy $U$ is given by

$$U = -\frac{\partial}{\partial \beta} \left[\ell n \, Z\right] \tag{10.12-3a}$$

$$= -\frac{\partial}{\partial \beta} \left[\frac{1}{2} \ell n \left(\frac{\pi}{\beta c}\right) + \ell n \, Z'\right] \tag{10.12-3b}$$

$$= \frac{1}{2\beta} - \frac{\partial}{\partial \beta} \ell n \, Z' \tag{10.12-3c}$$

$$= \frac{1}{2} kT + U' \tag{10.12-3d}$$

Thus the internal energy (average energy) is the sum of the two terms $\frac{1}{2}kT$ and $U'$. The contribution to $U$ from $cp_1^2$ is $\frac{1}{2}kT$ and is independent of $c$. Similarly if other variables in $Z$ appear as quadratic forms in $E$, they too will contribute $\frac{1}{2}kT$ each to $U$. Thus the contribution to $U$ of each quadratic term in $E$ is $\frac{1}{2}kT$. This result is known as the **equipartition theorem**.

*Example 10.12-1*. Calculate the internal energy $U$ of a harmonic oscillator. The problem of an oscillator was introduced in Example 10.2-1.

It is seen that, from Equations (10.2-10, 11), the energy $E$ is

$$E = \frac{1}{2} m \dot{q}^2 + \frac{1}{2} Gq^2 \qquad\qquad (10.12\text{-}4)$$

From the partition theorem each of the quadratic terms will contribute $\frac{1}{2}kT$ to $U$, so

$$U = kT \qquad\qquad (10.12\text{-}5)$$

## 10.13    MAXWELL VELOCITY DISTRIBUTION

Consider a monoatomic gas in a container of volume $V$, kept a temperature $T$. Interactions between molecules can be neglected. The total energy of the system is the sum of the energies of the individual molecules. If the mass of a molecule is $m$ and its velocity is $\underline{v}$, then its energy $E$, which is purely kinetic, is

$$E = \frac{1}{2} m \underline{v}^2 \qquad\qquad (10.13\text{-}1a)$$

$$= \frac{1}{2m} \underline{p}^2 \qquad\qquad (10.13\text{-}1b)$$

where $\underline{p}$ is the momentum.

We assume that there is no preferred location in the container, that is to say, we do not consider a density gradient. The probability of a molecule lying in a cell of size $\Delta\underline{q}\,\Delta\underline{p}$ at $(\underline{p}, \underline{q})$ is independent of the generalized coordinate $\underline{q}$.

The probability of finding a molecule in a region $(d\underline{q}, d\underline{p})$ about $(\underline{q}, \underline{p})$ is given by

$$\text{Prob.} = C \exp\left(-\underline{p}^2/2m\,kT\right) d\underline{q}\,d\underline{p} \qquad\qquad (10.13\text{-}2)$$

where $C$ is a constant.

To obtain C, we use the normalization condition. It is certain that there is a molecule in the whole phase space. The integration of Equation (10.13-2) yields

$$1 = \iint C \exp\left(-\underline{p}^2/2m\,kT\right) d\underline{q}\ d\underline{p} \tag{10.13-3a}$$

$$= CV \int \exp\left(-\underline{p}^2/2m\,kT\right) d\underline{p} \tag{10.13-3b}$$

where V is the generalized volume.

From Chapter 4, Problem 9b, we have

$$\int \exp\left(-\underline{p}^2/2mkT\right) d\underline{p} = (2\pi mkT)^{3/2} \tag{10.13-4}$$

Substituting Equation (10.13-4) into Equation (10.13-3b), we deduce that C is given by

$$C = \frac{1}{V\,(2\pi mkT)^{3/2}} \tag{10.13-5}$$

Equation (10.13-2) can now be written as

$$\text{Prob.} = \frac{1}{V}\left(\frac{1}{2\pi mkT}\right)^{3/2} \exp\left(-\underline{p}^2/mkT\right) d\underline{q}\ d\underline{p} \tag{10.13-6}$$

The probable number of molecules $N_p$ in a region $(d\underline{q}, d\underline{p})$ around $(\underline{q}, \underline{p})$ is given by

$$N_p = \chi_M\, d\underline{q}\ d\underline{p} \tag{10.13-7a}$$

$$\chi_M = \left(\frac{N}{V}\right)\left(\frac{1}{2\pi mkT}\right)^{3/2} \exp\left(-\underline{p}^2/2mkT\right) \tag{10.13-7b}$$

The distribution function $\chi_M$ is known as the **Maxwell velocity distribution** and it is normalized to N, the total number of molecules in V. It is also known as a **Gaussian distribution**. We note that the Equation (10.13-7b) is similar to Equation (10.8-33) and $\chi_M$ is also referred to as a **normal distribution**.

## 10.14   BROWNIAN MOTION

In 1828, Robert Brown, a biologist, drew attention to the movements of pollen suspended in a fluid. Some of his contemporaries attributed the motion of the pollen to the fact that they were alive. It was later found that all sufficiently small particles, be they organic or inorganic, will move when suspended in a fluid. Such a motion became known as **Brownian motion**.

It is now known that the motion of these minute particles is due to the collision of the fluid molecules with these particles. The fluid molecules are in random motion and thus will occasionally hit the particles. One molecule by itself may not have the momentum to displace the particle, but the combined contribution of the collision of several fluid molecules on the same particle at the same time could displace the particle. The probability of such an event is not zero. Brownian motion can be explained in probabilistic terms.

The mathematics developed to provide an understanding of Brownian motion has led to the opening of a new field in mathematics known as **stochastic processes**. As a result, the application of Brownian motion is no longer confined to pollen dispersal. Its present range of application extends from diffusion processes to the modelling of stock prices. Exploring all the avenues that it has opened up is beyond the scope of this book. We shall consider only the Langevin approach to Brownian diffusion.

Suppose that the mass of the suspended particle is m. The particle is acted on by a rapid fluctuating force $\underline{K}(t)$ caused by the bombardment of the particle by the fluid molecules. The mean value of $\underline{K}(t)$ over a suitable time interval is zero. In addition, a resistive force $\underline{F}$ acts on the particle. The equation of motion is

$$m \frac{d\underline{v}}{dt} = \underline{K}(t) - \underline{F} \tag{10.14-1}$$

where $\underline{v}$ is the velocity.

It is usual to assume that the particle is spherical with radius a, and the resistive force $\underline{F}$ is given by Stoke's formula

$$\underline{F} = 6\pi\eta a \underline{v} \tag{10.14-2}$$

where $\eta$ is the viscosity of the fluid.

Substituting Equation (10.14-2) into Equation (10.14-1), we obtain

$$m \frac{d\underline{v}}{dt} = \underline{K}(t) - 6\pi\eta a \underline{v} \tag{10.14-3}$$

Forming the scalar product of Equation (10.14-3) with $\underline{r}$, the vector position of the particle, yields

$$m\underline{r} \cdot \frac{d\underline{v}}{dt} = \underline{r} \cdot \underline{K} - 6\pi\eta a \underline{r} \cdot \underline{v} \tag{10.14-4}$$

The following results will be needed

$$\frac{d}{dt}(\underline{r} \cdot \underline{r}) = 2\underline{r} \cdot \frac{d\underline{r}}{dt} = 2\underline{r} \cdot \underline{v} \qquad (10.14\text{-}5a,b)$$

$$\frac{d^2}{dt^2}(\underline{r} \cdot \underline{r}) = 2\underline{r} \cdot \frac{d\underline{v}}{dt} + 2\underline{v} \cdot \underline{v} \qquad (10.14\text{-}5c)$$

Substituting Equations (10.14-5a to c) into Equation (10.14-4) yields

$$\frac{m}{2}\left[\frac{d^2}{dt^2}(\underline{r} \cdot \underline{r}) - 2\underline{v} \cdot \underline{v}\right] = \underline{r} \cdot \underline{K} - 3\pi\eta a \frac{d}{dt}(\underline{r} \cdot \underline{r}) \qquad (10.14\text{-}6)$$

The time average of $\underline{r} \cdot \underline{K}$ is zero since $\underline{r}$ and $\underline{K}$ are uncorrelated and the average value of $\underline{K}$ is zero. From the equipartition theorem (Section 10.12), we deduce in the present case, where $\underline{v}$ has three components, that the average $\langle \underline{v} \cdot \underline{v} \rangle$ is

$$m \langle \underline{v} \cdot \underline{v} \rangle = 3kT \qquad (10.14\text{-}7)$$

Taking the average of all the terms in Equation (10.14-6) leads to

$$\frac{d^2}{dt^2}\langle \underline{r} \cdot \underline{r} \rangle + \frac{6\pi\eta a}{m}\frac{d}{dt}\langle \underline{r} \cdot \underline{r} \rangle = \frac{6kT}{m} \qquad (10.14\text{-}8)$$

The solution of Equation (10.14-8) is

$$\langle \underline{r} \cdot \underline{r} \rangle = \frac{6kT}{m\alpha^2}(\alpha t - 1) + \frac{C_1}{\alpha} + C_2 e^{-\alpha t} \qquad (10.14\text{-}9)$$

where $\alpha = \frac{6\pi\eta a}{m}$, $C_1$ and $C_2$ are arbitrary constants to be determined from initial conditions.

In the limit as $t \longrightarrow \infty$, $\langle \underline{r} \cdot \underline{r} \rangle$ tends to

$$\lim_{t \to \infty} \langle \underline{r} \cdot \underline{r} \rangle = \frac{6kTt}{m\alpha} \qquad (10.14\text{-}10a)$$

$$= \frac{kTt}{\pi a \eta} \qquad (10.14\text{-}10b)$$

We note from Equation (10.14-10b) that $\langle \underline{r} \cdot \underline{r} \rangle$ is proportional to the temperature $T$ and time $t$ and inversely proportional to the radius $a$ of the particle and the viscosity $\eta$. Thus the larger the particle the smaller the contribution $\langle \underline{r} \cdot \underline{r} \rangle$, that is to say, the distance the particle travels is reduced. That $\langle \underline{r} \cdot \underline{r} \rangle$ is proportional to t has been observed experimentally and is also a result of the diffusion process. We also note the absence of the density in Equation (10.14-10b).

A modern survey on the development of Brownian motion in mechanics is given by Russel (1983). Lavenda (1985) has written an article popularizing the importance of Brownian motion.

## PROBLEMS

1a.     In Example 10.2-1, it was shown that the trajectory of the system

$$\ddot{q} + \omega^2 q = 0$$

in the phase plane is a closed curve.

Sketch the trajectory of the system

$$\ddot{q} - \omega^2 q = 0$$

in the phase plane.

2b.     For a conservative system, the total energy is constant.  This is expressed as

$$\frac{1}{2} m (\dot{q})^2 + \phi(q) = C$$

By differentiating with respect to time, show that the energy equation can be written as

$$m \ddot{q} = f(q)$$

Express $f(q)$ in terms of $\phi(q)$.

The points of equilibrium of the system are the points at which the acceleration is zero ($\ddot{q} = 0$). What can you deduce about the potential $\phi$ at the points of equilibrium? If $m = 1$ and $C = 1$, show that the equation of the trajectories is

$$p = \pm \sqrt{2 - 2\phi}$$

What happens when $\phi > 1$?

Sketch the function $\phi$ and the trajectories for the following cases

(i) $\phi = (q)^2$,     (ii) $\phi = 1 - \frac{1}{2}(q)^2$,     (iii) $\phi = (q)^3$,     (iv) $\phi = \frac{1}{q}$

3a.     A particle of unit mass moves under a force $\underline{F}$ given by

$$\underline{F} = -\frac{K}{r^3}\,\underline{r}$$

where $\underline{r}$ is the vector position of the particle relative to an origin and $K$ is a constant.

If the motion is two-dimensional and we choose a polar coordinate system $(r, \theta)$, show that the Hamiltonian $H$ is given by

$$H = \frac{1}{2}\left[(p^r)^2 + \left(\frac{p^\theta}{r}\right)^2\right] - \frac{K}{r}$$

Obtain the $r$ and $\theta$ components of the equation of motion.

4a.     Show that the transformation

$$Q = q \tan p, \qquad P = \ell n \sin p$$

is a canonical transformation.

5a.     Using the definition of the Poisson bracket given by Equation (10.4-5), verify Equations (10.4-6e to g).

6b.     Liouville's theorem is given, in Bird et al. (1987), by Equation (17.2.2) and is written as

$$\frac{\partial f}{\partial t} = -\sum_i \frac{\partial}{\partial x_i}(\dot{x}^i\, f)$$

where the $x_i$ are the generalized coordinates and momenta.

Show that in the case of a single particle, the equation given by Bird et al. (1987) is the same as Equation (10.6-10b) with $\rho = f$.

Verify that a steady solution of Equation (10.6-10b) for a single harmonic oscillator is given by

$$\rho = \rho_0 e^{-\beta H}$$

where $\rho_0$ and $\beta$ are constants and $H$ is the Hamiltonian.

7a.    Twelve identical balls are randomly distributed in a box.  Calculate the probability of

   (i)      having three balls in the first quarter of the box;

   (ii)     having six balls in the first quarter of the box;

   (iii)    having all twelve balls in the first quarter of the box.

8a.    A dilute suspension contains ten spheres per liter.  A sample (one tenth of that solution) is removed.  What is the probability that there is (a) at least one sphere, (b) at most one sphere in the sample.

   You may assume that the particles are distributed according to a Poisson distribution.

9b.    In Section 10.9, we have defined the mean value for a continuous distribution.  The mean value $\langle x \rangle$ of a random variable which assumes values $x_1$, $x_2$, ... with corresponding finite probabilities $f(x_1)$, $f(x_2)$, ... is given by

$$\langle x \rangle = \sum_i x_i \, f(x_i)$$

   Show that, for a Poisson distribution, the mean value $\langle x \rangle$ is $\lambda$.

10a.   There are $3 \times 10^3$ spheres in a box.  Calculate the number of different combinations for the spheres in order to have 1100 spheres in the front third of the box.

   Assuming that the distribution is normal, calculate the probability of having exactly 1100 spheres in the front third of the box.

   Answer: $10^{858}$ ; $8.8 \times 10^{-6}$

11a.   A polymer chain may be idealized by $n + 1$ beads, joined by $n$ links, each of length a. $\underline{r}$ is the end-to-end vector of the chain and we assume spherical symmetry.  The probability density of the chain distribution $f(n, \underline{r})$ can be shown to be

$$f(n, r) = \left( \frac{3}{2} \pi a^2 n \right)^{-3/2} \exp \left( \frac{-3 r^2}{2 a^2 n} \right)$$

   where $r = |\underline{r}|$.

   Calculate $\langle r^2 \rangle$, the mean value of $r^2$.

   Answer: $a^2 n$

12b.   Assuming a Maxwell distribution, calculate the average values of (a) the x-component of the velocity $v_x$; (b) the speed $v$ $(= \sqrt{v_x^2 + v_y^2 + v_z^2})$; (c) the square of the velocity $v^2$ of a molecule in a gas at temperature T.

Why is $\langle v_x \rangle$ not equal to $\frac{1}{3} \langle v \rangle$ ?                    Answer: $0$ ; $\sqrt{\dfrac{8kT}{m\pi}}$ ; $\dfrac{3kT}{m}$

13b.   In Bird et al. (1977), the partition function Z for a macromolecular solid is given by

$$Z = \frac{1}{J_{eq}} \exp\left[ -\sum_{\substack{v,\,\mu \\ v<\mu}} \frac{H_{v\mu}}{2kT} (\underline{r}_v - \underline{r}_\mu) \bullet (\underline{r}_v - \underline{r}_\mu) \right]$$

where $J_{eq}$, $H_{v\mu}$ are constants, $\underline{r}_v$ and $\underline{r}_\mu$ are positions vectors of the junctions.

Combining Equations (10.11-11, 12), show that the Helmholtz free energy A is given by

$$A = -kT \ln\left(\frac{1}{J_{eq}}\right) + \frac{1}{2} \sum_{\substack{v,\,\mu \\ v<\mu}} H_{v\mu} (\underline{r}_v - \underline{r}_\mu) \bullet (\underline{r}_v - \underline{r}_\mu)$$

Under what condition(s) is the change in A equal to the work done?

# REFERENCES

ABBOT, M.B., An Introduction to the Method of Characteristics (Thames and Hudson, London, 1966).

ABRAMOWITZ, M. and STEGUN, I.A., Handbook of Mathematical Functions (Dover, New York, 1964).

AMES, W.F., Non-Linear Partial Differential Equations in Engineering, Vol. I and II (Academic Press, New York, 1972).

AMES, W.F., Numerical Methods for Partial Differential Equations (Nelson, London, 1969).

ATKINSON, C., Quart. J. Mech. Appl. Math., **41**, 301 (1988).

BALMER, R.T. and KAUZLARICH, J.J., AIChE J., **17**, 1181 (1971).

BATCHELOR, G.K., An Introduction to Fluid Mechanics (Cambridge University Press, New York, 1967).

BENTWICH, M., Chem. Eng. Sci., **31**, 71 (1976).

BIRD, R.B., ARMSTRONG, R.C., and HASSAGER, O., Dynamics of Polymeric Liquids, Vol. 1 - Fluid Mechanics, 2nd Ed. (John Wiley, New York, 1987).

BIRD, R.B., HASSAGER, O., ARMSTRONG, R.C., and CURTISS, C.F., Dynamics of Polymeric Liquids, Vol. 2 - Kinetic Theory (John Wiley, New York, 1977).

BIRD, R.B., HASSAGER, O., ARMSTRONG, R.C., and CURTISS, C.F., Dynamics of Polymeric Liquids, Vol. 2 - Kinetic Theory, 2nd Ed. (John Wiley, New York, 1987).

BIRD, R.B., STEWART, W.E., and LIGHTFOOT, E.N., Transport Phenomena (John Wiley, New York, 1960).

BLASIUS, H., Z. Math. Phys., **56**, 1 (1908).

BLUMAN, G.W. and COLE, J.D., Similarity Methods for Differential Equation (Springer Verlag, New York, 1974).

BORN, M., HEISENBERG, W., and JORDAN, P., Z·Physik, **35**, 557 (1926).

BOWEN, J.R., ACRIVOS, A., and OPPENHEIM, A.K., Chem. Eng. Sci., **18**, 177 (1963).

BURGERS, J.M., Adv. Appl. Mech., **1**, 171 (1948).

CARREAU, P.J., DE KEE, D., and CHHABRA, R.P., Polymer Rheology: Principles and Applications (Hanser, New York, 1997).

CARSLAW, H.S. and JAEGER, J.C., Conduction of Heat in Solids, 2nd Ed. (Clarendon Press, Oxford, 1973).

CESARI, L., Asymptotic Behavior and Stability Problems in Ordinary Differential Equations, 3rd Ed. (Springer Verlag, Berlin, 1971).

CHAN MAN FONG, C.F., DE KEE, D., and GRYTE, C., J. Non-Newtonian Fluid Mech., **46**, 111 (1993).

CHAN MAN FONG, C.F., DE KEE, D., and MARCOS, B., J. Applied Phys., **74**, 40 (1993).

CHIAPPETTA, L.M. and SOBEL, D.R., "The Temperature Distribution Within a Hemisphere Exposed to a Hot Gas Stream", in Mathematical Modelling, edited by M.S. Klamkin (Siam, Philadelphia, 1987), 60.

CHOUDHURY, S.R. and JALURIA, Y., Int. J. Heat and Mass Transf., **37**, 1193 (1994).

COURANT, R. and HILBERT, D., Methods of Mathematical Physics, Vol. II (Interscience Pub., New York, 1966).

COURANT, R., FRIEDRICHS, K., and LEWY, H., Mat. Ann., **100**, 32 (1928).

CRANK, J. and NICOLSON, P., Proc. Camb. Phil. Soc., **43**, 50 (1947).

CRANK, J., Free and Moving Boundaries (Clarendon Press, Oxford, 1984).

CROWE, M.J., A History of Vector Analysis (University of Notre Dame Press, Indiana, 1967).

EISBERG, R. and RESNICK, R., Quantum Physics, 2nd Ed. (John Wiley, New York, 1985).

ELDEN, L. and WITTMEYER-KOCH, L., Numerical Analysis – An Introduction (Academic Press, New York, 1990).

EPSTEIN, I.R., Physica, **7D**, 47 (1983).

ERDÉLYI, A. (ed.), Tables of Integral Transforms, Vol. I and II (McGraw Hill, New York, 1954).

ERDÉLYI, A., MAGNUS, W., OBERHETTINGER, F., and TRICOMI, F.G., Higher Transcendental Functions, 3 Vols. (McGraw Hill, New York, 1953, 1955).

ERICKSEN, J.L., "Tensor Fields", in Handbuch der Physik, edited by S. Flügge (Springer Verlag, Berlin, 1960), Vol. III/1.

FERRARO, V.C.A., Electromagnetic Theory (Athlone Press, University of London, London, 1956).

GERALD, C.F. and WHEATLEY, P.O., Applied Numerical Analysis, 5th Ed. (Addison-Wesley, Reading, Massachusetts, 1994).

GOLDSTEIN, H., Classical Mechanics (Addison-Wesley, Reading, Masschusetts, 1972).

GONZALES-NUNEZ, R., CHAN MAN FONG, C.F., FAVIS, B.D., and DE KEE, D., J. Applied Polym. Sci., **62**, 1627 (1996).

GREENBERG, M.D., Applications of Green's Functions in Science and Engineering (Prentice Hall, New Jersey, 1971).

GUPTA, V.P. and DOUGLAS, W.J.M., AIChE J., **13**, 883 (1967).

HAGEMAN, L.A. and YOUNG, D.M., Applied Iterative Methods (Academic Press, New York, 1981).

HERSHEY, D., MILLER, C.J., MENKE, R.C., and HESSELBERTH, J.F., "Oxygen Diffusion Coefficients for Blood Flowing Down a Wetted-Wall Column", in Chemical Engineering in Medicine and Biology, edited by J. Hershey (Plenum Press, New York, 1967), 117.

HILDEBRAND, F., Introduction to Numerical Analysis (McGraw Hill, New York, 1956).

HODGKIN, A.L. and HUXLEY, A.F., J. Physiol., **117**, 500 (1952).

JENSON, V.G. and JEFFREYS, G.V., Mathematical Methods in Chemical Engineering (Academic Press, New York, 1963).

KAMKE, E., Differentialgleichungen. Lösungmethoden und Lösungen (Chelsea Pub. Co., New York, 1959).

KOBER, H., Dictionary of Conformal Representations (Dover, New York, 1952).

LANCZOS, C., The Variational Principles of Mechanics (University of Toronto Press, Toronto, 1966).

LAVENDA, B.H., Scientific American, **252/2**, 70 (1985).

LEVENSPIEL, O., Chemical Reaction Engineering (John Wiley, New York, 1972).

LIPKIN, H.J., Lie Groups for Pedestrians (John Wiley, New York, 1965).

MERIC, R., J. Heat Transfer, **107**, 508 (1985).

MILNE-THOMSON, L.M., Theoretical Hydrodynamics, 5th Printing (Macmillan, New York, 1965).

MORSE, P.M. and FESHBACH, H., Methods of Theoretical Physics (McGraw Hill, New York 1953).

MURPHY, G.M., Ordinary Differential Equations and their Solutions (van Nostrand, New York, 1960).

NAYFEH, A.H., Perturbation Methods (John Wiley, New York, 1973).

PEARSON, C.E., Handbook of Applied Mathematics (van Nostrand, New York, 1974).

REDLICH, O. and KWONG, J.N.S., Chem. Rev., **44**, 233 (1949).

RHEE, H.K., ARIS, R., and AMUNDSON, N.R., First Order Partial Differential Equations, Vol. I and II (Prentice Hall, New Jersey, 1986).

RICHTMYER, R.D. and MORTON, K.W., Difference Methods for Initial Value Problems, 2nd Ed. (Interscience Pub., New York, 1967).

ROSENHEAD, L. (ed.), Laminar Boundary Layers (Oxford University Press, Oxford, 1963).

ROUX, J.C., Physica, **7D**, 57 (1983).

RUSSEL, W.R., Annual Rev. Fluid Mech., **13**, 425 (1981).

SCHWARTZ, L., Théorie des distributions (Hermann, Paris, 1957).

SEDAHMED, G.H., Can. J. Chem. Eng., **64**, 75 (1986).

SHEPPARD, A.J. and EISENKLAM, P., Chem. Eng. Sci., **38**, 169 (1983).

SLATER, L.J., Confluent Hypergeometric Functions (Cambridge University Press, Cambridge, 1960).

SPENCE, A., WORTH, D.J., KOLACZKOWSKI, S.T., and CRUMPTON, P.I., Computers Chem. Eng., **17**, 1057 (1993).

STASTNA, J., DE KEE, D., and HARRISON, B., Chem. Eng. Commun., **110**, 111 (1991).

STRANG, G. and FIX, G., An Analysis of the Finite Element Method (Prentice Hall, New Jersey, 1973).

STROUD, A.H. and SECREST, D., Gaussian Quadrature Formulas (Prentice-Hall, New Jersey, 1966).

TAYLOR, G.I., Proc. Roy. Soc., **A146**, 501 (1934).

TRUESDELL, C. and NOLL, W., "The Non-Linear Field Theories of Mechanics", in Handbuch der Physik, edited by S. Flügge (Springer Verlag, Berlin, 1965), Vol. III/3.

VAN DYKE, M., Perturbation Methods in Fluid Mechanics (Parabolic Press, Stanford, 1975).

VON KARMAN, T. and BIOT, M., Mathematical Methods in Engineering (McGraw Hill, New York, 1940).

WASOW, W., Asymptotic Expansions for Ordinary Differential Equations (Interscience Pub., New York, 1965).

WATSON, G.N., A Treatise on the Theory of Bessel Functions, 2nd Ed. (Cambridge University Press, Cambridge, 1966).

WEI, J. and PRATER, C.D., Advances in Catalysis, **13**, 203 (1962).

WILKINSON, J., The Algebraic Eigenvalue Problem (Oxford University Press, Oxford, 1965).

WILSON, S.D.R., Int. J. Heat and Mass Transf., **22**, 207 (1979).

WIMMERS, O.J., PAULUSSEN, R., VERMEULEN, D.P., and FORTUIN, J.M.H., Chem. Eng. Sci., **39**, 1415 (1984).

ZAUDERER, E., Partial Differential Equations of Applied Mathematics (John Wiley, New York, 1983).

ZIENKIEWICZ, O.C. and MORGAN, K., Finite Elements and Approximation (John Wiley, New York, 1983).

# APPENDIX I

# THE EQUATION OF CONTINUITY
# IN SEVERAL COORDINATE SYSTEMS

## RECTANGULAR COORDINATES

$$\frac{\partial \rho}{\partial t} + \frac{\partial}{\partial x}(\rho v_x) + \frac{\partial}{\partial y}(\rho v_y) + \frac{\partial}{\partial z}(\rho v_z) = 0 \qquad \text{(A.I-1)}$$

## CYLINDRICAL COORDINATES

$$\frac{\partial \rho}{\partial t} + \frac{1}{r}\frac{\partial}{\partial r}(\rho r v_r) + \frac{1}{r}\frac{\partial}{\partial \theta}(\rho v_\theta) + \frac{\partial}{\partial z}(\rho v_z) = 0 \qquad \text{(A.I-2)}$$

## SPHERICAL COORDINATES

$$\frac{\partial \rho}{\partial t} + \frac{1}{r^2}\frac{\partial}{\partial r}(\rho r^2 v_r) + \frac{1}{r \sin\theta}\frac{\partial}{\partial \theta}(\rho v_\theta \sin\theta) + \frac{1}{r \sin\theta}\frac{\partial}{\partial \phi}(\rho v_\phi) = 0 \qquad \text{(A.I-3)}$$

# APPENDIX II

# THE EQUATION OF MOTION
# IN SEVERAL COORDINATE SYSTEMS

## RECTANGULAR COORDINATES

*In terms of velocity gradients for a Newtonian fluid with constant $\rho$ (density) and $\mu$ (viscosity)*

### x-component

$$\rho\left(\frac{\partial v_x}{\partial t} + v_x\frac{\partial v_x}{\partial x} + v_y\frac{\partial v_x}{\partial y} + v_z\frac{\partial v_x}{\partial z}\right) = -\frac{\partial p}{\partial x} + \mu\left(\frac{\partial^2 v_x}{\partial x^2} + \frac{\partial^2 v_x}{\partial y^2} + \frac{\partial^2 v_x}{\partial z^2}\right) + \rho g_x \qquad \text{(A.II-1)}$$

### y-component

$$\rho\left(\frac{\partial v_y}{\partial t} + v_x\frac{\partial v_y}{\partial x} + v_y\frac{\partial v_y}{\partial y} + v_z\frac{\partial v_y}{\partial z}\right) = -\frac{\partial p}{\partial y} + \mu\left(\frac{\partial^2 v_y}{\partial x^2} + \frac{\partial^2 v_y}{\partial y^2} + \frac{\partial^2 v_y}{\partial z^2}\right) + \rho g_y \qquad \text{(A.II-2)}$$

### z-component

$$\rho\left(\frac{\partial v_z}{\partial t} + v_x\frac{\partial v_z}{\partial x} + v_y\frac{\partial v_z}{\partial y} + v_z\frac{\partial v_z}{\partial z}\right) = -\frac{\partial p}{\partial z} + \mu\left(\frac{\partial^2 v_z}{\partial x^2} + \frac{\partial^2 v_z}{\partial y^2} + \frac{\partial^2 v_z}{\partial z^2}\right) + \rho g_z \qquad \text{(A.II-3)}$$

## CYLINDRICAL COORDINATES

*In terms of velocity gradients for a Newtonian fluid with constant $\rho$ and $\mu$*

**r-component**

$$\rho\left(\frac{\partial v_r}{\partial t} + v_r\frac{\partial v_r}{\partial r} + \frac{v_\theta}{r}\frac{\partial v_r}{\partial \theta} - \frac{v_\theta^2}{r} + v_z\frac{\partial v_r}{\partial z}\right)$$

$$= -\frac{\partial p}{\partial r} + \mu\left[\frac{\partial}{\partial r}\left(\frac{1}{r}\frac{\partial}{\partial r}(rv_r)\right) + \frac{1}{r^2}\frac{\partial^2 v_r}{\partial \theta^2} - \frac{2}{r^2}\frac{\partial v_\theta}{\partial \theta} + \frac{\partial^2 v_r}{\partial z^2}\right] + \rho g_r \qquad \text{(A.II-4)}$$

**$\theta$-component**

$$\rho\left(\frac{\partial v_\theta}{\partial t} + v_r\frac{\partial v_\theta}{\partial r} + \frac{v_\theta}{r}\frac{\partial v_\theta}{\partial \theta} + \frac{v_r v_\theta}{r} + v_z\frac{\partial v_\theta}{\partial z}\right)$$

$$= -\frac{1}{r}\frac{\partial p}{\partial \theta} + \mu\left[\frac{\partial}{\partial r}\left(\frac{1}{r}\frac{\partial}{\partial r}(rv_\theta)\right) + \frac{1}{r^2}\frac{\partial^2 v_\theta}{\partial \theta^2} + \frac{2}{r^2}\frac{\partial v_r}{\partial \theta} + \frac{\partial^2 v_\theta}{\partial z^2}\right] + \rho g_\theta \qquad \text{(A.II-5)}$$

**z-component**

$$\rho\left(\frac{\partial v_z}{\partial t} + v_r\frac{\partial v_z}{\partial r} + \frac{v_\theta}{r}\frac{\partial v_z}{\partial \theta} + v_z\frac{\partial v_z}{\partial z}\right)$$

$$= -\frac{\partial p}{\partial z} + \mu\left[\frac{1}{r}\frac{\partial}{\partial r}\left(r\frac{\partial v_z}{\partial r}\right) + \frac{1}{r^2}\frac{\partial^2 v_z}{\partial \theta^2} + \frac{\partial^2 v_z}{\partial z^2}\right] + \rho g_z \qquad \text{(A.II-6)}$$

## SPHERICAL COORDINATES

*In terms of velocity gradients for a Newtonian fluid with constant $\rho$ and $\mu$*

**r-component**

$$\rho\left(\frac{\partial v_r}{\partial t} + v_r\frac{\partial v_r}{\partial r} + \frac{v_\theta}{r}\frac{\partial v_r}{\partial \theta} + \frac{v_\phi}{r\sin\theta}\frac{\partial v_r}{\partial \phi} - \frac{v_\theta^2 + v_\phi^2}{r}\right)$$

$$= -\frac{\partial p}{\partial r} + \mu\left(\nabla^2 v_r - \frac{2}{r^2}v_r - \frac{2}{r^2}\frac{\partial v_\theta}{\partial \theta} - \frac{2}{r^2}v_\theta\cot\theta - \frac{2}{r^2\sin\theta}\frac{\partial v_\phi}{\partial \phi}\right) + \rho g_r \qquad \text{(A.II-7)}$$

**θ-component**

$$\rho\left(\frac{\partial v_\theta}{\partial t} + v_r\frac{\partial v_\theta}{\partial r} + \frac{v_\theta}{r}\frac{\partial v_\theta}{\partial \theta} + \frac{v_\phi}{r\sin\theta}\frac{\partial v_\theta}{\partial \phi} + \frac{v_r v_\theta}{r} - \frac{v_\phi^2\cot\theta}{r}\right)$$

$$= -\frac{1}{r}\frac{\partial p}{\partial \theta} + \mu\left(\nabla^2 v_\theta + \frac{2}{r^2}\frac{\partial v_r}{\partial \theta} - \frac{v_\theta}{r^2\sin^2\theta} - \frac{2\cos\theta}{r^2\sin^2\theta}\frac{\partial v_\phi}{\partial \phi}\right) + \rho g_\theta \qquad \text{(A.II-8)}$$

**φ-component**

$$\rho\left(\frac{\partial v_\phi}{\partial t} + v_r\frac{\partial v_\phi}{\partial r} + \frac{v_\theta}{r}\frac{\partial v_\phi}{\partial \theta} + \frac{v_\phi}{r\sin\theta}\frac{\partial v_\phi}{\partial \phi} + \frac{v_\phi v_r}{r} + \frac{v_\theta v_\phi}{r}\cot\theta\right)$$

$$= -\frac{1}{r\sin\theta}\frac{\partial p}{\partial \phi} + \mu\left(\nabla^2 v_\phi - \frac{v_\phi}{r^2\sin^2\theta} + \frac{2}{r^2\sin\theta}\frac{\partial v_r}{\partial \phi} + \frac{2\cos\theta}{r^2\sin^2\theta}\frac{\partial v_\theta}{\partial \phi}\right) + \rho g_\phi \qquad \text{(A.II-9)}$$

$$\nabla^2 s = \frac{1}{r^2}\frac{\partial}{\partial r}\left(r^2\frac{\partial s}{\partial r}\right) + \frac{1}{r^2\sin\theta}\frac{\partial}{\partial \theta}\left(\sin\theta\frac{\partial s}{\partial \theta}\right) + \frac{1}{r^2\sin^2\theta}\frac{\partial^2 s}{\partial \phi^2} \qquad \text{(A.II-10)}$$

# APPENDIX III

## THE EQUATION OF ENERGY
## IN TERMS OF THE TRANSPORT PROPERTIES
## IN SEVERAL COORDINATE SYSTEMS

*For Newtonian fluids of constant ρ (density) and k (thermal conductivity)*

## RECTANGULAR COORDINATES

$$\rho \hat{C}_p \left( \frac{\partial T}{\partial t} + v_x \frac{\partial T}{\partial x} + v_y \frac{\partial T}{\partial y} + v_z \frac{\partial T}{\partial z} \right) = k \left( \frac{\partial^2 T}{\partial x^2} + \frac{\partial^2 T}{\partial y^2} + \frac{\partial^2 T}{\partial z^2} \right)$$

$$+ 2\mu \left[ \left( \frac{\partial v_x}{\partial x} \right)^2 + \left( \frac{\partial v_y}{\partial y} \right)^2 + \left( \frac{\partial v_z}{\partial z} \right)^2 \right] + \mu \left[ \left( \frac{\partial v_x}{\partial y} + \frac{\partial v_y}{\partial x} \right)^2 + \left( \frac{\partial v_x}{\partial z} + \frac{\partial v_z}{\partial x} \right)^2 + \left( \frac{\partial v_y}{\partial z} + \frac{\partial v_z}{\partial y} \right)^2 \right]$$

$$\text{(A.III-1)}$$

## CYLINDRICAL COORDINATES

$$\rho \hat{C}_p \left( \frac{\partial T}{\partial t} + v_r \frac{\partial T}{\partial r} + \frac{v_\theta}{r} \frac{\partial T}{\partial \theta} + v_z \frac{\partial T}{\partial z} \right) = k \left[ \frac{1}{r} \frac{\partial}{\partial r} \left( r \frac{\partial T}{\partial r} \right) + \frac{1}{r^2} \frac{\partial^2 T}{\partial \theta^2} + \frac{\partial^2 T}{\partial z^2} \right]$$

$$+ 2\mu \left\{ \left( \frac{\partial v_r}{\partial r} \right)^2 + \left[ \frac{1}{r} \left( \frac{\partial v_\theta}{\partial \theta} + v_r \right) \right]^2 + \left( \frac{\partial v_z}{\partial z} \right)^2 \right\} + \mu \left\{ \left( \frac{\partial v_\theta}{\partial z} + \frac{1}{r} \frac{\partial v_z}{\partial \theta} \right)^2 + \left( \frac{\partial v_z}{\partial r} + \frac{\partial v_r}{\partial z} \right)^2 \right.$$

$$\left. + \left[ \frac{1}{r} \frac{\partial v_r}{\partial \theta} + r \frac{\partial}{\partial r} \left( \frac{v_\theta}{r} \right) \right]^2 \right\}$$

$$\text{(A.III-2)}$$

## SPHERICAL COORDINATES

$$\rho \widehat{C}_p \left( \frac{\partial T}{\partial t} + v_r \frac{\partial T}{\partial r} + \frac{v_\theta}{r} \frac{\partial T}{\partial \theta} + \frac{v_\phi}{r \sin\theta} \frac{\partial T}{\partial \phi} \right) = k \left[ \frac{1}{r^2} \frac{\partial}{\partial r} \left( r^2 \frac{\partial T}{\partial r} \right) + \frac{1}{r^2 \sin\theta} \frac{\partial}{\partial \theta} \left( \sin\theta \frac{\partial T}{\partial \theta} \right) \right.$$

$$\left. + \frac{1}{r^2 \sin^2\theta} \frac{\partial^2 T}{\partial \phi^2} \right] + 2\mu \left[ \left( \frac{\partial v_r}{\partial r} \right)^2 + \left( \frac{1}{r} \frac{\partial v_\theta}{\partial \theta} + \frac{v_r}{r} \right)^2 + \left( \frac{1}{r \sin\theta} \frac{\partial v_\phi}{\partial \phi} + \frac{v_r}{r} + \frac{v_\theta \cot\theta}{r} \right)^2 \right]$$

$$+ \mu \left\{ \left[ r \frac{\partial}{\partial r} \left( \frac{v_\theta}{r} \right) + \frac{1}{r} \frac{\partial v_r}{\partial \theta} \right]^2 + \left[ \frac{1}{r \sin\theta} \frac{\partial v_r}{\partial \phi} + r \frac{\partial}{\partial r} \left( \frac{v_\phi}{r} \right) \right]^2 + \left[ \frac{\sin\theta}{r} \frac{\partial}{\partial \theta} \left( \frac{v_\phi}{\sin\theta} \right) + \frac{1}{r \sin\theta} \frac{\partial v_\theta}{\partial \phi} \right]^2 \right\}$$

$$(A.III-3)$$

# APPENDIX IV

## THE EQUATION OF CONTINUITY OF SPECIES A IN SEVERAL COORDINATE SYSTEMS

*For constant $\rho$ (density) and $\mathcal{D}_{AB}$ (binary diffusivity)*

## RECTANGULAR COORDINATES

$$\frac{\partial c_A}{\partial t} + \left( v_x \frac{\partial c_A}{\partial x} + v_y \frac{\partial c_A}{\partial y} + v_z \frac{\partial c_A}{\partial z} \right) = \mathcal{D}_{AB} \left( \frac{\partial^2 c_A}{\partial x^2} + \frac{\partial^2 c_A}{\partial y^2} + \frac{\partial^2 c_A}{\partial z^2} \right) + R_A \qquad \text{(A.IV-1)}$$

## CYLINDRICAL COORDINATES

$$\frac{\partial c_A}{\partial t} + \left( v_r \frac{\partial c_A}{\partial r} + v_\theta \frac{1}{r} \frac{\partial c_A}{\partial \theta} + v_z \frac{\partial c_A}{\partial z} \right) = \mathcal{D}_{AB} \left[ \frac{1}{r} \frac{\partial}{\partial r} \left( r \frac{\partial c_A}{\partial r} \right) + \frac{1}{r^2} \frac{\partial^2 c_A}{\partial \theta^2} + \frac{\partial^2 c_A}{\partial z^2} \right] + R_A$$

$$\text{(A.IV-2)}$$

## SPHERICAL COORDINATES

$$\frac{\partial c_A}{\partial t} + \left( v_r \frac{\partial c_A}{\partial r} + v_\theta \frac{1}{r} \frac{\partial c_A}{\partial \theta} + v_\phi \frac{1}{r \sin\theta} \frac{\partial c_A}{\partial \phi} \right)$$

$$= \mathcal{D}_{AB} \left[ \frac{1}{r^2} \frac{\partial}{\partial r} \left( r^2 \frac{\partial c_A}{\partial r} \right) + \frac{1}{r^2 \sin\theta} \frac{\partial}{\partial \theta} \left( \sin\theta \frac{\partial c_A}{\partial \theta} \right) + \frac{1}{r^2 \sin^2\theta} \frac{\partial^2 c_A}{\partial \phi^2} \right] + R_A$$

$$\text{(A.IV-3)}$$

# AUTHOR INDEX

# SUBJECT INDEX

# STANDARD DERIVATIVES AND INTEGRALS

$$\frac{d}{dx}\,(\ell n\; x) = \frac{1}{x}$$

$$\frac{d}{dx}\,(\sin x) = \cos x$$

$$\frac{d}{dx}\,(\cos x) = -\sin x$$

$$\frac{d}{dx}\,(\tan x) = \sec^2 x$$

$$\frac{d}{dx}\,(\sinh x) = \cosh x$$

$$\frac{d}{dx}\,(\cosh x) = \sinh x$$

$$\frac{d}{dx}\,(\tanh x) = \text{sech}^2 x$$

$$\frac{d}{dx}\,[\text{arc}\sin (x/a)] = 1/\sqrt{a^2 - x^2}$$

$$\frac{d}{dx}\,[\text{arc}\cos (x/a)] = -1/\sqrt{a^2 - x^2}$$

$$\frac{d}{dx}\,[\text{arc}\tan (x/a)] = a/\sqrt{a^2 + x^2}$$

$$\frac{d}{dx}\,[\text{arc}\sinh (x/a)] = 1/\sqrt{a^2 + x^2}$$

$$\frac{d}{dx}\,[\text{arc}\cosh (x/a)] = \pm 1/\sqrt{x^2 - a^2}$$

$$\frac{d}{dx}\,[\text{arc}\tanh (x/a)] = a/(a^2 - x^2)$$

---

$$\int \frac{1}{x}\, dx = \ell n\; x$$

$$\int \cos x\; dx = \sin x$$

$$\int \sin x\; dx = -\cos x$$

$$\int \sec^2 x\; dx = \tan x$$

$$\int \cosh x\; dx = \sinh x$$

$$\int \sinh x\; dx = \cosh x$$

$$\int \text{sech}^2 x\; dx = \tanh x$$

$$\int \frac{dx}{\sqrt{a^2 - x^2}} = \text{arc}\sin (x/a)$$

$$\int \frac{dx}{\sqrt{a^2 - x^2}} = -\text{arc}\cos (x/a)$$

$$\int \frac{dx}{\sqrt{a^2 + x^2}} = \frac{1}{a}\,\text{arc}\tan (x/a)$$

$$\int \frac{dx}{\sqrt{a^2 + x^2}} = \text{arc}\sinh (x/a)$$

$$\int \frac{dx}{\sqrt{x^2 - a^2}} = \pm\, \text{arc}\cosh (x/a)$$

$$\int \frac{dx}{a^2 - x^2} = \frac{1}{a}\,\text{arc}\tanh (x/a)$$

---

$$\int \sin^n x\; dx = -\frac{\cos x\, \sin^{n-1} x}{n} + \frac{(n-1)}{n}\int \sin^{n-2} x\; dx$$

$$\int \cos^n x\; dx = \frac{\sin x\, \cos^{n-1} x}{n} + \frac{(n-1)}{n}\int \cos^{n-2} x\; dx$$

$$\int \sin px\, \cos qx\; dx = -\frac{1}{2}\left[\frac{\cos (p+q)\,x}{p+q} + \frac{\cos (p-q)\,x}{p-q}\right]$$

$$\int \sin px\, \sin qx\; dx = \frac{1}{2}\left[\frac{\sin (p-q)\,x}{p-q} - \frac{\sin (p+q)\,x}{p+q}\right]$$

$$\int \cos px\, \cos qx\; dx = \frac{1}{2}\left[\frac{\sin (p+q)\,x}{p+q} + \frac{\sin (p-q)\,x}{p-q}\right]$$

$$\int e^{ax} \cos (px+q)\; dx = \frac{e^{ax}}{a^2 + p^2}\left[a \cos (px+q) + p \sin (px+q)\right]$$

$$\int e^{ax} \sin (px+q)\; dx = \frac{e^{ax}}{a^2 + p^2}\left[a \sin (px+q) - p \cos (px+q)\right]$$

$$\int_0^\infty e^{-\alpha x^2}\; dx = \frac{1}{2}\sqrt{\frac{\pi}{\alpha}}$$

$$\int_0^\infty x^{2n+1}\, e^{-\alpha x^2}\; dx = \frac{n!}{2\alpha^{n+1}}$$

$$\int_0^\infty e^{-(\alpha x^2 + \beta/x^2)}\; dx = \frac{1}{2}\sqrt{\frac{\pi}{\alpha}}\; e^{-2\sqrt{\alpha\beta}}$$

| TYPE OF EQUATION | METHOD OF SOLUTION |
|---|---|
| **Separable First-Order**<br>$M(x)\,dx + N(y)\,dy = 0$ | Straightforward term by term integration<br>$\int M(x)\,dx + \int N(y)\,dy = 0$ |
| **Homogeneous First-Order**<br>$M(x, y)\,dx + N(x, y)\,dy = 0$<br><br>$M(x, y)$ and $N(x, y)$ are homogeneous polynomials of the same degree. | The substitution $y = ux$ and consequently<br>$\dfrac{dy}{dx} = u + x\dfrac{du}{dx}$<br><br>generates a separable first-order equation. |
| **Reducible First-Order**<br>$M(x, y) = a_1x + b_1y + c_1$<br>$N(x,y) = a_2x + b_2y + c_2$ | a) $\begin{vmatrix} a_1 & b_1 \\ a_2 & b_2 \end{vmatrix} \neq 0$  i) solve the homogeneous first-order equation<br>$\dfrac{dY}{dX} = \dfrac{a_1X + b_1Y}{a_2X + b_2Y}$<br><br>ii) determine the solution $(\alpha, \beta)$ of the system<br>$\begin{cases} a_1x + b_1y + c_1 = 0 \\ a_2x + b_2y + c_2 = 0 \end{cases}$<br><br>iii) replace X and Y in the solution of (i) by $x-\alpha$ and $y-\beta$.<br><br>b) $\begin{vmatrix} a_1 & b_1 \\ a_2 & b_2 \end{vmatrix} = 0$<br><br>Introduce a variable $z = a_1x + b_1y$.<br>This generates a first-order separable equation of the form<br>$\dfrac{dz}{dx} = f(z)$ |
| **Total or Exact**<br>$M(x,y)\,dx + N(x,y)\,dy = 0$<br><br>$\dfrac{\partial M}{\partial y} = \dfrac{\partial N}{\partial x}$ | a) The solution $F(x,y) = c$ can be determined as follows:<br>$\dfrac{\partial F}{\partial x} = M$<br>$F = \int M\,dx + f(y)$<br>substitute into $\dfrac{\partial F}{\partial y} = N$ to compute $f(y)$.<br><br>b) If $\dfrac{\partial M}{\partial y} \neq \dfrac{\partial N}{\partial x}$ but $\left(\dfrac{\partial M}{\partial y} - \dfrac{\partial N}{\partial x}\right)/N = g(x)$<br>(a function of x only) then, the equation can be made to be exact after multiplication by an integrating factor<br>$I(x) = e^{\int g(x)\,dx}$. That is:<br>$\dfrac{\partial}{\partial y}(IM) = \dfrac{\partial}{\partial x}(IN)$<br>Similarly, if $\left(\dfrac{\partial M}{\partial y} - \dfrac{\partial N}{\partial x}\right)/M = -h(y)$ (a function of y only), we take<br>$I(y) = e^{\int h(y)\,dy}$ |

| TYPE OF EQUATION | METHOD OF SOLUTION |
|---|---|
| **Linear First-Order**<br>$\dfrac{dy}{dx} + P(x)y = Q(x)$ | The integrating factor $I(x) = e^{\int P(x)\,dx}$<br><br>Multiply the differential equation by $I(x)$ and integrate. The left side should be $\dfrac{d(Iy)}{dx}$. |
| **Bernoulli's Equation**<br>$\dfrac{dy}{dx} + P(x)y = Q(x)\,y^n$ | First divide both sides of the equation by $y^n$.<br>Let $v = y^{1-n}$. The equation will reduce to a linear first-order equation. |
| **Second-Order Linear with Constant Coefficient**<br>$y'' + Ay' + By = Q(x)$ | First, solve $y'' + Ay' + By = 0$ to obtain $y_h$.<br>The characteristic equation $\alpha^2 + A\alpha + B = 0$ has two solutions: $\alpha_1$ and $\alpha_2$.<br>i) $\alpha_1$ and $\alpha_2$ are real and distinct:<br>$y_h = c_1 e^{\alpha_1 x} + c_2 e^{\alpha_2 x}$<br>ii) $\alpha_1 = \alpha_2$  $y_h = c_1 e^{\alpha_1 x} + c_2 x e^{\alpha_1 x}$<br>iii) $\alpha_1$ and $\alpha_2$ are complex conjugate roots, $(a + ib)$ and $(a - ib)$:<br>$y_h = e^{ax}(c_3 \cos bx + c_4 \sin bx)$<br><br>Secondly, one proposes a solution $y_P$, based on the form $Q(x)$. $y_P$ can also be obtained by the method of variation of parameters. That is:<br>$y_P = c_1(x)e^{\alpha_1 x} + c_2(x)e^{\alpha_2 x}$<br>Substitution of the proposed $y_P$ and its derivatives into the equation to be solved, allows for the determination of $c_1$ and $c_2$, introduced by the proposed $y_P$. The solution of the problem is $y = y_h + y_P$.<br><br>Other methods of solution are:<br>i) the Laplace transform method, which involves:<br>  - taking the Laplace transform of all terms in the equation<br>  - solving the resulting algebraic equation<br>  - inverting the transform.<br>ii) the Green's function method for non-homogeneous equations involves finding the Green's function G, associated with the differential equation $L(y) = f(x)$. The solution is then given by:<br>$$y(x) = \int_{x_0}^{x_1} G(x, t)\, f(t)\, dt$$<br>where $x_0$ and $x_1$ are the boundaries at which $y$ is given. G has to satisfy a number of conditions outlined in §1.18. |